Calcium Signaling in Human Health and Diseases

Calcium Signaling in Human Health and Diseases

Special Issue Editor

Francesco Moccia

MDPI • Basel • Beijing • Wuhan • Barcelona • Belgrade

MDPI

Special Issue Editor
Francesco Moccia
University of Pavia
Italy

Editorial Office
MDPI
St. Alban-Anlage 66
4052 Basel, Switzerland

This is a reprint of articles from the Special Issue published online in the open access journal *International Journal of Molecular Sciences* (ISSN 1422-0067) from 2017 to 2018 (available at: https://www.mdpi.com/journal/ijms/special_issues/calcium_signaling)

For citation purposes, cite each article independently as indicated on the article page online and as indicated below:

LastName, A.A.; LastName, B.B.; LastName, C.C. Article Title. *Journal Name* **Year**, *Article Number*, Page Range.

ISBN 978-3-03897-537-3 (Pbk)
ISBN 978-3-03897-538-0 (PDF)

Contents

About the Special Issue Editor

Francesco Moccia (PhD) was awarded a BSc in 1995 from the University of Pavia and a PhD in Physiology in 2000 from the University of Turin, followed by Postdoctoral training at the Katholieke Universiteit of Leuven (2000–2001) and at the Zoological Station "Anton Dohrn", Naples, where he was appointed Junior Staff Scientist in 2003. He came back to academia in 2006 as Appointed Professor of Biology and Physiology at the Faculty of Medicine and Surgery, University of Molise, while keeping a research fellowship at the University of Naples "Federico II" until late 2008. He then became Assistant Professor of Physiology at the University of Pavia, where he is the Scientific Director of the Laboratory of Ca^{2+} Signaling. Dr. Francesco Moccia has been a member and Secretary of the Directive Board of the Italian Society of Cardiovascular Research and is regularly invited to give lectures and talks at national and international level. In collaboration with an extensive international network, his work has recently been awarded a Future Emerging Technologies (FET) Open grant from the European Commission. His work is devoted to understanding the role of Ca^{2+} signaling in physiological and oncological vascularization, in neurovascular coupling and in neoplastic transformation.

Preface to "Calcium Signaling in Human Health and Diseases"

Intracellular Ca^{2+} signals regulate a myriad of cellular functions, ranging from short-term responses, such as excitation-contraction coupling and stimulus-secretion coupling, to long-term processes, such as proliferation, gene expression, differentiation, motility, synaptic plasticity, programmed cell death (or apoptosis), and metabolism. It is, therefore, not surprising that any derangement of the multifaceted Ca^{2+} toolkit that shapes the elevation in intracellular Ca^{2+} concentration ($[Ca^{2+}]_i$) may lead to severe pathological disorders, including cancer, chronic heart failure, epileptic and neurodegenerative disorders, immunodeficiency, and developmental defects. An increase in $[Ca^{2+}]i$ is shaped by the concerted interaction among the components of an extremely versatile network of channels, transporters, pumps, and buffers that can be uniquely assembled by each cell type to generate intracellular Ca^{2+} signals with spatio-temporal properties precisely tailored to regulate specific functions. We are witnessing a fascinating period of ground-breaking discoveries in the field of Ca^{2+} signaling which enormously improved our knowledge of the molecular components of the Ca^{2+} toolkit, e.g., store-operated Ca^{2+} entry (SOCE) and mitochondrial Ca^{2+} uniporter, of novel Ca^{2+}-dependent pathways, such as mammalian target of rapamycin (mTOR) and endothelium-dependent vasodilation in the brain, and of the pathologies associated with mutations in Ca^{2+}-permeable channels and/or Ca^{2+}-regulated pathways.

The time is therefore right for this book to showcase the "state of the art" research that is being undertaken by leading researchers in the field of Ca^{2+} signaling from all around the world. Our participant authors are at the forefront of their studies and are furthering the understanding of the molecular processes that undergo the formation of intracellular Ca^{2+} signals and translate them into a biological response in both health and disease. This book presents the latest developments in several exciting themes of research in the field of Ca^{2+} signaling, which are subdivided into original research articles and review articles.

Among the original research articles, **Layhadi and Fountain** reported the role of P2X4 receptors in mediating extracellular Ca^{2+} entry in human macrophages. By exploring P2X2 receptors, **Rokic and coworkers** identified the opposing role of cytosolic Ca^{2+} (facilitates) and ATP (inhibits) in the desensitization process observed under naïve conditions. **Meraviglia and coworkers** illustrated how SERCA activity is improved by histone deacetylase (HDAC) inhibition. Similarly, **Sorriento and coworkers** demonstrated that the NH2-terminal domain of G protein-coupled receptor kinase (GRK) type 5, (GRK5-NT), inhibits cardiac hypertrophy by modulating NFAT and GATA-4 activity. These studies could, therefore, exert a remarkable pharmacological impact on the treatment of chronic heart failure. **Izquierdo-Serra and coworkers** found that stroke-like episodes and ataxia in phosphomannomutase deficiency (PMM2-CDG) are associated with hypoglycosylation of both CaV2.1 subunits (α1A and α2α), which induce gain-of-function effects on channel gating that mirror those observed in familial hemiplegic migraine (FHM). **Gerbino and coworkers** demonstrated for the first time that dandelion (*Taraxacum officinale* Weber), a traditional ethnomedical remedy for several diseases, may act through an increase in $[Ca^{2+}]_i$ that is shaped by endogenous Ca^{2+} release and extracellular Ca^{2+} entry, which in turn stimulates phospholipase C (PLC) activation. **Banciu and coworkers** discovered that estradiol down-regulates voltage-gated Ca^{2+} channels (Cav3.1, Ca3.2, Cav3.3, Cav2.1), estrogen receptors (ESR1, ESR2, GPR30, and nuclear receptor coactivator 3 (NCOA3) in telocytes from human pregnant uterine myometrium. This finding might be helpful in understanding labor and pregnancy mechanisms and to develop more effective strategies to reduce the risk of premature birth. **Huang and coworkers** unexpectedly found platelet-derived growth factor (PDGF)-induced Ca^{2+} oscillations and migration in mouse embryonic fibroblasts deficient of STIM1, but not STIM2. This finding indicates that STIM2 may replace STIM2 in transgenic animals and suggests that extreme caution is warranted when interpreting results obtained from animals engineered to block the expression of SOCE components. **Hajdú and coworkers** demonstrated that NR1 and NR3B proteins are selectively expressed in nuclear fractions of human melanoma cells, but not healthy melanocytes. This finding hints at an unexpected role for NMDA receptors in human melanoma and, possibly, other malignancies. **Maqoud and coworkers** revealed that large-conductance Ca^{2+}-activated K^+ channels (BKCa) stimulate proliferation in SH-5YSY neuroblastoma cells by preventing Ca^{2+} entry and AKT1pser473 dephosphorylation by a Ca^{2+}-dependent phosphatase. This finding further endorses the view that BKCa channels could represent a promising anticancer target. **Vazquez-de-Lara and coworkers** provided the evidence that phosphatidylethanolamine induces an antifibrotic phenotype in normal human lung fibroblasts and ameliorates lung fibrosis in vivo through an increase in $[Ca^{2+}]i$. **Ito and coworkers** found that ATP promotes skeletal muscle hypertrophy by stimulating P2Y receptor-dependent endogenous Ca^{2+} release, followed by mTOR activation. This signaling pathway might help explain how physical exercise exploits intracellular Ca^{2+} signaling to promote skeletal muscle hypertrophy. **Lim and coworkers** provided the proof of concept that neural activity can stimulate calcineurin activation in adjacent astrocytes through an increase in $[Ca^{2+}]_I$ that is due to NMDA and mGluR5 receptor activation. This finding could shed novel light on the mechanisms of information processing in the brain. **Sanchez-Tecuatl and coworkers** described a novel technology to automatically draw regions of interests (ROIs) around endothelial cells exposed to mechanical artifacts. This procedure could be helpful to reduce time consumption during analysis of hundreds of cells, such as those covering the intimal layer of blood vessels.

Among the review articles, **Moccia** illustrated how remodeling of the endothelial Ca^{2+} toolkit contributes to the resistance to anticancer treatments. **Guerra and coworkers** presented the novel and unexpected role of endothelial Ca^{2+} signaling in neurovascular coupling in the brain. **Gerbino and Colella** provided a comprehensive analysis of the signaling role of the Ca^{2+}-Sensing Receptor (CaR) by focusing on CaR expression in the heart, CaR trafficking, and pharmacology. **Samuel and coworkers** detailed the progression of SERCA2a gene therapy from its inception in plasmid and animal models, to its clinical trials in CHF patients, highlighting potential avenues for future work. **Maly and Hoffmann** surveyed the role of nuclear calcium signaling in prostate cancer by updating the molecular mechanisms behind malignant prostate cancer progression and the unfavorable response to anti-androgen treatment. **Tellez Freitas and coworkers** reviewed current knowledge on the role of CD5 as a negative regulator of T cell activation in terms of altered Ca^{2+} signal. They describe how improved understanding of CD5 Ca^{2+} signaling regulation could be useful for basic and clinical research. Similarly, **D'Elia and Weinrauch** described how Ca^{2+} channel blockers could be used to reduce the risk of sepsis in immunocompromised patients at risk for infection. **Catacuzzeno and Franciolini** described how intermediate conductance Ca^{2+}-activated K^+ (KCa3.1) channels control the electrochemical gradient sustaining Ca^{2+} entry through Orai1 channels, thereby sustaining proliferation and migration in glioblastoma cells. **Paudel and coworkers** examine different patterns of Ca^{2+} activity during early development, as well as potential medical conditions associated with its dysregulation to understand the involvement of altered Ca^{2+} activity in infertility, abortive pregnancy, and developmental defects.

All the authors are gratefully acknowledged for their enthusiasm and remarkable contribution to this book that we hope will be useful for all colleagues dealing with these important themes of research related to Ca^{2+} signaling. There is much that remains to be done in this field and, if this book can help us move even one step forward, our mission will be accomplished.

Francesco Moccia
Special Issue Editor

International Journal of
Molecular Sciences

MDPI

Article

Neuronal Activity-Dependent Activation of Astroglial Calcineurin in Mouse Primary Hippocampal Cultures

Dmitry Lim [1,*], Lisa Mapelli [2], Pier Luigi Canonico [1], Francesco Moccia [3] and Armando A. Genazzani [1]

[1] Department of Pharmaceutical Sciences, Università del Piemonte Orientale "Amedeo Avogadro", Via Bovio 6, 28100 Novara, Italy; pierluigi.canonico@uniupo.it (P.L.C.); armando.genazzani@uniupo.it (A.A.G.)
[2] Department of Brain and Behavioral Sciences, University of Pavia, Via Forlanini 6, 27100 Pavia, Italy; lisa.mapelli@unipv.it
[3] Department of Biology and Biotechnology "Lazzaro Spallanzani", University of Pavia, Via Forlanini 6, 27100 Pavia, Italy; francesco.moccia@unipv.it
*   Correspondence: dmitry.lim@uniupo.it; Tel.: +39-0321-375822

Received: 31 August 2018; Accepted: 29 September 2018; Published: 30 September 2018

Abstract: Astrocytes respond to neuronal activity by generating calcium signals which are implicated in the regulation of astroglial housekeeping functions and/or in modulation of synaptic transmission. We hypothesized that activity-induced calcium signals in astrocytes may activate calcineurin (CaN), a calcium/calmodulin-regulated protein phosphatase, implicated in neuropathology, but whose role in astroglial physiology remains unclear. We used a lentiviral vector expressing NFAT-EYFP (NY) fluorescent calcineurin sensor and a chemical protocol of LTP induction (cLTP) to show that, in mixed neuron-astrocytic hippocampal cultures, cLTP induced robust NY translocation into astrocyte nuclei and, hence, CaN activation. NY translocation was abolished by the CaN inhibitor FK506, and was not observed in pure astroglial cultures. Using Fura-2 single cell calcium imaging, we found sustained Ca^{2+} elevations in juxtaneuronal, but not distal, astrocytes. Pharmacological analysis revealed that both the Ca^{2+} signals and the nuclear NY translocation in astrocytes required NMDA and mGluR5 receptors and depended on extracellular Ca^{2+} entry via a store-operated mechanism. Our results provide a proof of principle that calcineurin in astrocytes may be activated in response to neuronal activity, thereby delineating a framework for investigating the role of astroglial CaN in the physiology of central nervous system.

Keywords: astrocytes; calcineurin; neuronal activity; NMDA; mGluR5; store-operated calcium entry

1. Introduction

Astrocytes are an abundant non-neuronal cellular type in the brain [1]. They exert fundamental housekeeping and homeostatic functions in the central nervous system (CNS) and are also involved in the pathogenesis of many neurological diseases. Astrocytes are non-excitable cells as they are largely unable to generate action potentials in response to electrical or chemical stimulation. Conversely, astrocytes respond to extracellular stimuli by generating intracellular calcium signals by exploiting two main mechanisms: (i) activation of metabotropic receptors on the plasma membrane, leading to liberation of calcium ions from internal calcium stores; and (ii) a receptor/store-operated mechanism of calcium entry from the extracellular milieu through the plasma membrane [1]. Astroglial calcium signals are thought to have a number of implications for CNS pathophysiology, including modulation of synaptic release [2], synchronization of neuronal activity [3], regulation of frequency of spontaneous α-amino-3-hydroxy-5-methyl-4-isoxazolepropionic acid AMPA receptor currents [4] and participation in vesicular glutamate release [5]. Although the physiological role of calcium signals in astrocytes is still a matter of debate [6–8], there is broad consensus about their role and their alterations in brain pathology [9–15].

Calcineurin (CaN) is a calcium/calmodulin-activated serine-threonine phosphatase, which is highly expressed in the brain [16]. In neurons, CaN regulates neuronal excitability and synaptic transmission [17]. Moreover, CaN activation is associated with long-term depression [18], while CaN inactivation is required for establishment of aversive memory [19]. In astrocytes, calcineurin is principally involved in setting up reactive gliosis and neuroinflammation in a number of neuropathological conditions [20–22]. Conversely, activation of CaN in astrocytes has so far not been documented under physiological conditions and its role in housekeeping and homeostatic functions of astrocytes is currently unknown.

The present work is designed as an in vitro proof of principle of astroglial CaN activation in response to neuronal activity. We show that long-term potentiation (LTP)-like neuronal activity robustly activates CaN in adjacent astrocytes and that store-operated Ca^{2+} entry through the astroglial plasma membrane is required for this process to occur.

2. Results

2.1. Neuronal Activity Leads to Activation of CaN in Astrocytes

Fluorescent CaN probe based on a transcription-deficient truncated variant of nuclear factor of activated T-cells c2 (NFATc2) fused with EYFP (NY) (Figure 1A), has been used in immune cells to monitor CaN activation in vivo [23]. To evaluate whether NY works in astrocytes, we have transduced an astroglial culture with NY-expressing lentiviral particles and monitored EYFP fluorescence in response to stimulation with ionomycin (1 μM), a Ca^{2+} ionophore widely used to activate CaN in different cellular types [24–27]. Figure 1B and Video S1 show that, after about 10 min of treatment, NY robustly translocates into the nucleus. We also checked the tropism of lentiviral particles pseudotyped with vesicular stomatitis virus protein G (VSVG) to neuronal and astroglial cells in culture. We found that astrocytes are efficiently transduced by the vector with occasional transduction of some neurons (<1%). Taking advantage of the preferential astrocytic transduction of the NY probe, we investigated whether chemical induction of neuronal activity was able to activate CaN in astrocytes. To achieve this goal, we have chosen a widely used protocol for chemical induction of long term potentiation (cLTP) in brain slices and in cultured neurons which consisted in short (4 min) application of a cocktail containing bicuculline (20 μM), strychnine (1 μM) and glycine (20 μM) in Mg^{2+}-free Krebs-Ringer buffer (KRB) buffer supplemented with 2 mM Ca^{2-} (cLTP cocktail, the induction phase), followed by washout of the cLTP cocktail and addition of KRB containing 1 mM Mg^{2+} and 2 mM Ca^{2+} (KRB + Mg + CaMg, the expression phase, Figure 2A). We applied the cLTP protocol to mixed hippocampal astrocyte-neuronal cultures, expressing NY probe, at DIV12–14. At different time-points after stimulation, the cells were fixed and NY nuclear translocation was quantified by calculating nuclear-to-cytosol fluorescence ratio (Nuc/Cyt ratio, Figure 1C, see also Methods section for details). Co-expression of histone 2B-fused mCherry allowed reliable localization of nuclear NY even in astrocytes expressing low NY levels or superimposed with neurons. Already 10 min after cLTP induction, nuclear localization of NY was detected as judged by the increased Nuc/Cyt NY fluorescence ratio (Figure 3A). 15 min after cLTP induction, a robust nuclear NY translocation was observed (Figure 2B). The histogram in Figure 3A shows that elevated Nuc/Cyt NY ratio can be observed at least until 1 h after cLTP induction indicating that the protocol induced robust and long lasting CaN activation. This activation was completely abolished if either FK506 (200 nM; Figure 2B) or cyclosporine A (500 μM), two well-known CaN inhibitors, were added to KRB + Mg + Ca during cLTP development.

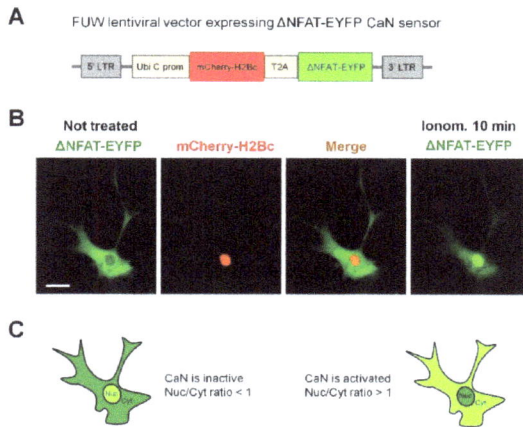

Figure 1. Validation of lentiviral vector expressing NY CaN sensor in astrocytes. (**A**) A scheme of lentiviral vector expressing ΔNFAT-EYFP (NY) CaN sensor. (**B**) Hippocampal astrocytes were transduced with NY CaN sensor and then stimulated with 2 μM ionomycin. Nuclear translocation of NY in astrocyte was observed after 10 min of incubation. A representative astrocyte is shown from three independent preparations. Bar, 10 μm. (**C**) Quantification of nuclear-to-cytosol ratio (Nuc/Cyt ratio) of NY in the green channel. CaN was considered inactive with Nuc/Cyt ratio <1, while Nuc/Cyt ratio >1 indicates CaN activation.

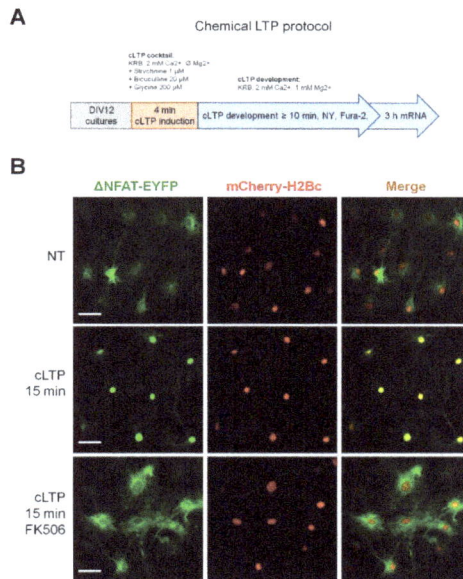

Figure 2. cLTP-induced NY nuclear translocation in astrocytes is CaN-dependent. (**A**) Scheme depicting cLTP induction protocol with consequent readouts. (**B**) Application of cLTP protocol to mixed neuron-astroglial hippocampal primary cultures induces robust NY translocation into the astrocyte nucleus at 15 min after cLTP induction phase. Note the complete inhibition of NY translocation when CaN inhibitor, FK506, was added during the cLTP development phase. Bar, 30 μm.

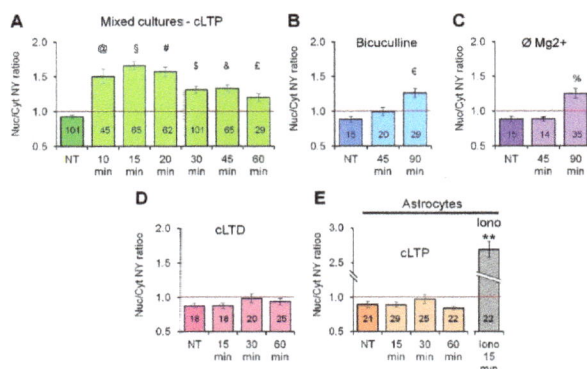

Figure 3. Quantification of NY Cyt/Nuc ratio after cLTP induction. (**A**) Time-course quantification of NY Cyt/Nuc ratio after cLTP induction in hippocampal mixed neuron-astroglial cultures. Note, nuclear localization of NY and hence, CaN activation, is evident already 10 min after cLTP induction, reaches maximum at 15–20 min and is still evident after 1 h of cLTP induction. Continuous bicuculline (40 μM) treatment (**B**) and Mg^{2+} withdrawal (**C**) are sufficient to induce NY translocation into astrocyte nucleus, although more than 1 h is required for the translocation to occur. Nuclear NY translocation in astrocytes was not observed upon application of chemical long-term depression protocol (cLTD) (**D**) or when purified astroglial cultures were stimulated with cLTP protocol (**E**). Ionomycin-induced (2 μM) NY translocation ratio is included as a positive control (**E**). Data are expressed as mean ± SEM of indicated number of cells measured, as indicated in materials and methods, from 2–3 coverslips (technical replicates) for each of three independent culture preparation (biological replicates). @, $p = 1.7 \times 10^{-8}$; §, $p = 2.1 \times 10^{-18}$; #, $p = 2.15 \times 10^{-14}$; $, $p = 1.5 \times 10^{-8}$; &, $p = 4.01 \times 10^{-8}$; £, $p = 0.0058$; €, $p = 0.00015$; %, $p = 0.00109$; **, 1.3×10^{-10}, all vs. NT.

Alongside the cLTP protocol, we used other strategies to induce neuronal activity, namely bicuculline alone at 40 μM in Mg^{2+}-free KRB (Figure 3B) and KRB with Ø Mg^{2+} without bicuculline (Figure 3C). Both protocols produced delayed (90 min of continuous treatment) translocation of NY, indicating that the increase in neuronal activity alongside the de-inactivation of *N*-methyl-D-aspartate receptors is sufficient to produce activation of CaN in astrocytes. Interestingly, NY nuclear translocation was not observed after application of a widely used protocol of chemical induction of long-term depression (cLTD), which consisted in application of *N*-methyl-D-aspartate (NMDA, 50 μM) for 20 min in Mg^{2+}-containing KRB [28], indicating that a spontaneous or an LTP-like neuronal activity is required to induce CaN activation in astrocytes (Figure 3D). Most importantly, application of the cLTP protocol to purified astroglial cultures produced no effect on the localization of NY probe which remained concentrated to the cytosolic compartment with no significant increase in Nuc/Cyt NY ratio, indicating that the effect of cLTP on astroglial CaN activation required the presence of neurons to occur (Figure 3E). Stimuli which were used to induce nuclear NY translocation in astrocytes are summarized in Table 1.

Table 1. Stimuli which induced nuclear NY translocation in astrocytes.

Drug/Condition	Astroglial Ca^{2+} Elevation	NY Nuclear Translocation
cLTP	Yes Ca^{2+} signals	Translocation after 10 min
LTD	NA	No translocation
Bicuculline alone 40 μM	NA	Translocation after 1 h
Ø Mg^{2+}	NA	Translocation after 2 h
cLTP on pure astrocytes	No Ca^{2+} signals	No translocation

Unless otherwise stated, the condition is applied to mixed neuron-astroglial cultures. NA, not assayed.

Given that the cLTP protocol produced the most robust and fast translocation of NY in astrocytes, we further used it for detailed characterization of cLTP-like activity-induced astroglial CaN activation.

2.2. Neuronal Activity Induces Elevation of Cytosolic Calcium in Astrocytes

For activation, CaN requires elevations of calcium concentrations in the cytosol. We investigated if the cLTP protocol induces calcium signals in astrocytes. Fura-2-loaded mixed hippocampal cultures were placed on the stage of epifluorescence setup and, after recording was started, KRB + Ca + Mg was changed first to Mg^{2+}-free KRB + Ca solution to wash out Mg^{2+} ions. Then, cLTP cocktail was applied and, after 4 min, it was changed to KRB + Ca + Mg solution. After registration, Ca^{2+} signals were analyzed separately in neurons and astrocytes. As shown in Figure 4A, Mg^{2+}-free KRB application induced a single Ca^{2+} spike, while cLTP cocktail induced a burst of Ca^{2+} spikes which lasted for the whole duration of the cLTP induction phase in neurons. After cLTP cocktail was washed-out, neurons either did not exhibit any Ca^{2+} activity (36.18%, $n = 72$), or generated single spikes (63.82%, $n = 127$) with a frequency of 0.32 ± 0.14 spikes/min. Example of an experiment without neuronal Ca^{2+} spikes in the cLTP development phase is shown in Figure 4, while in Figure S1 experiments are demonstrated in which neurons generated Ca^{2+} spikes. In astrocytes, robust Ca^{2+} signals were generated during the induction phase, which followed the neuronal burst of Ca^{2+} spikes and slightly decayed before removal of the cLTP cocktail. After the re-addition of KRB + Mg + Ca, the cultures were registered for longer time (25 min) and delayed Ca^{2+} transients were observed in a fraction of astrocytes. Closer examination of Fura-2 images revealed that only astrocytes which were juxtaposed to neuronal bodies or neuronal processed generated delayed Ca^{2+} signals (Figure 4B). These delayed astrocyte Ca^{2+} signals were not synchronous and could consist of more than one Ca^{2+} transient of different duration.

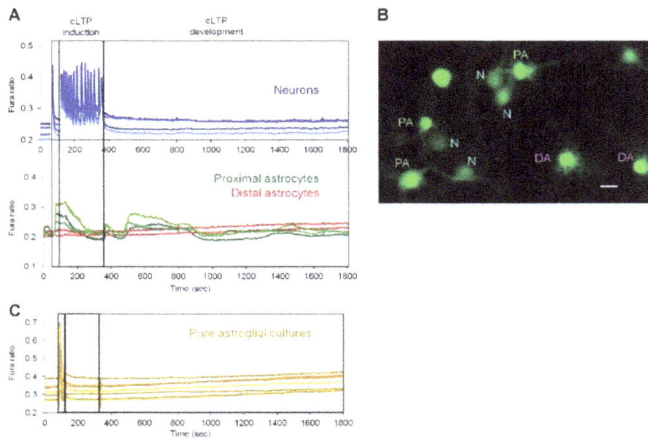

Figure 4. Delayed Ca^{2+} signals in proximal-to-neurons astrocytes induced by cLTP. (**A**) Ca^{2+} signals were detected by Fura-2 single cells imaging and analyzed separately in neurons (blue traces) and astrocytes (green and red traces). Note appearance of delayed Ca^{2+} signals in astrocytes which are in close apposition with neurons (green traces in panel **A** and "PA" (Proximal Astrocytes) label in **B**) but not in distal astrocytes, which are not in contact with neurons ("N") (red traces in panel **A** and "DA" label in panel **B**). Representative traces and image are shown of cLTP induction experiments from 2–3 coverslips (technical replicates) for each of three independent culture preparations (biological replicates). Bar, 10 μm. **C**, Ca^{2+} signals were not detected in pure astroglial cultures stimulated with cLTP protocol.

2.3. Activity-Induced Ca^{2+} Transients and CaN Activation in Astrocytes Depend on NMDA, mGluR5 and Store-Operated Ca^{2+} Entry

Next, we used pharmacological inhibition to investigate the molecular mechanisms of neuronal activity-induced CaN activation in astrocytes. Ca^{2+} traces were registered on Fura-2 loaded cells in a separate set of experiments and the correlation of Ca^{2+} signals with CaN activation was achieved analyzing the responses of astrocytes to withdrawal of Ca^{2+} during the cLTP development phase or to pharmacological treatments. First of all, we investigated if application of FK506 would interfere with astroglial Ca^{2+} signals, and this was not the case since the Ca^{2+} signals in astrocytes during the development phase of cLTP could still be observed in presence of FK506 (Figure 5B). Further, we investigated the requirement of neuronal NMDA receptors. For this, MK801, a specific NMDA receptors inhibitor (50 µM), was applied either for only the phase of cLTP induction (Figure 5C) or only during the cLTP development phase after cLTP cocktail washout (Figure 5D) or during both the induction and the development phases (Figure 5E). When MK801 was applied for the entire duration of the experiment, Ca^{2+} activity was abrogated in both neurons and astrocytes, and no nuclear NY translocation was observed (Figure 5E). When MK801, instead, was applied solely during the phases of cLTP induction (Figure 5C) or cLTP development (Figure 5D and Table 2), neither astrocyte Ca^{2+} transients nor nuclear NY translocation were inhibited indicating that (1) astrocyte Ca^{2+} signals during the induction phase were secondary to neuronal Ca^{2+} signals; and (2) neuronal Ca^{2+} signals during the induction phase are necessary for the astroglial Ca^{2+} signals to occur during the phase of cLTP development. Next, involvement of the metabotropic glutamate receptor, mGluR5, was investigated. Figure 5F and Table 2 show that application of MTEP, a specific mGluR5 antagonist (100 µM), during the cLTP development phase abolished both astroglial Ca^{2+} signals and NY nuclear translocation. This result suggests the requirement of mGluR5 for the activity-induced CaN activation, although, in the current setting, it does not discriminate between neuronal and astroglial localization of mGluR5. Previously, we published that in rat hippocampal mixed neuron-astroglial culture neurons preferentially responded to NMDA while astrocytes responded to 3,5-Dihydroxyphenylglycine (DHPG) [29], a powerful mGluR agonist. We, therefore, stimulated our mixed cultures either with NMDA (50 µM) or DHPG (20 µM). Expectedly, only neurons responded to NMDA stimulation (Figure 6A) while only astrocytes responded to DHPG (Figure 6B), providing thus an indirect evidence that NMDA inhibitor MK801 acted on neuronal NMDA receptors, while MTEP inhibited astroglial mGluR5.

Table 2. Pharmacological characterization of cLTP-induced astroglial Ca^{2+} signals and NY translocation.

Pharmacological Treatment	Astroglial Ca^{2+} Elevation	NY Nuclear Translocation
cLTP → KRB + Mg + Ca	Yes	Yes
cLTP → Ø Ca^{2+}	No	No
cLTP → KRB + Mg + Ca + FK506	Yes	No
cLTP → KRB + Mg + Ca + cyclosporine A	Yes	No
cLTP (MK801) → KRB + Mg + Ca	Yes	Yes
cLTP → KRB + Mg + Ca + MK801	Yes	Yes
cLTP (MK801) → KRB + Mg + Ca + MK801	No	No
cLTP → KRB + Mg + Ca + MTEP	No	No
cLTP → KRB + Mg + Ca + 2APB	No	No
cLTP → KRB + Mg + Ca + Pyr3	No	No
cLTP → KRB + Mg + Ca + Pyr6	No	No
cLTP → KRB + Mg + Ca + Pyr10	Yes	Yes
cLTP → KRB + Mg + Ca + Xestospongin	Yes	Yes
cLTP → KRB + Mg + Ca + U73122	Yes	Yes

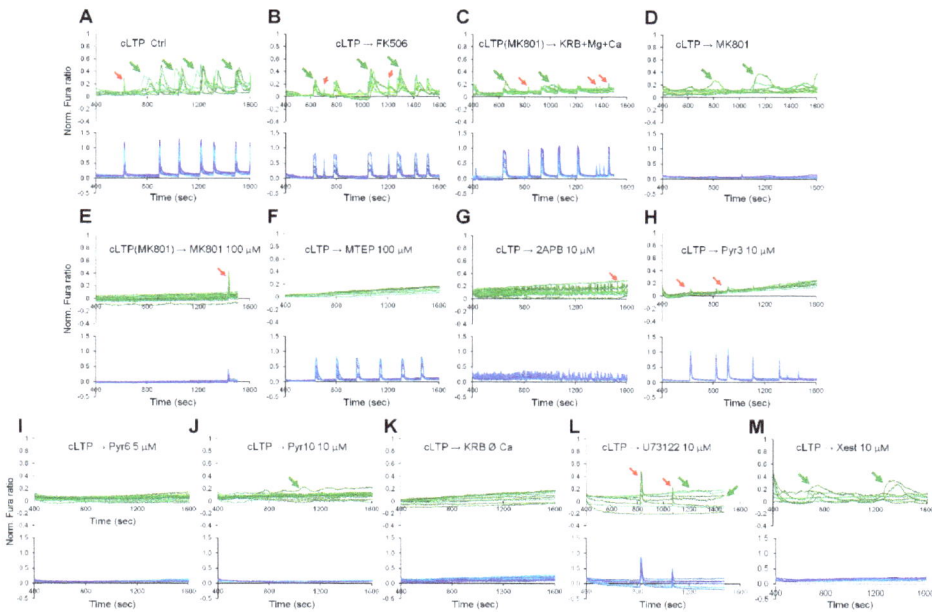

Figure 5. Pharmacological characterization of cLTP-induced astroglial Ca^{2+} signals. Fura-2 loaded hippocampal mixed neuron-astroglial cultures were stimulated with cLTP protocol and either not treated (**A**) or treated as indicated (unless otherwise stated cells were treated during cLTP development phase): FK506 (200 nM) (**B**); MK801 (100 μM, only in the phase of cLTP induction) (**C**); MK801 (100 μM, only in the phase of cLTP development) (**D**); MK801 (100 μM, during both phases, cLTP induction and development) (**E**); MTEP (100 μM) (**F**); 2APB (10 μM) (**G**); Pyr3 (10 μM) (**H**); Pyr6 (5 μM) (**I**); Pyr10 (10 μM) (**J**); Ø Ca^{2+} (**K**); U73122 (10 μM) (**L**); Xestospongin C (10 μM) (**M**). Ca^{2+} signals were analyzed separately in astrocytes (green traces) and in neurons (blue traces). Only parts of traces related to cLTP development phase are shown. cLTP induced Ca^{2+} elevation in astrocytes are indicated by green arrows. Artifacts related to neuronal Ca^{2+} spikes are indicated with red arrows. For corresponding full-length traces, see Figure S2.

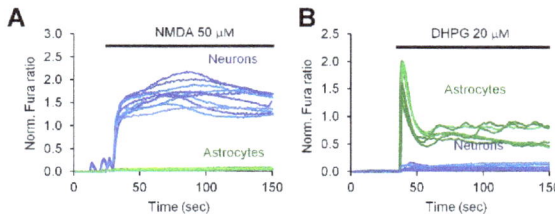

Figure 6. Cell type determination of NMDA and 3,5-Dihydroxyphenylglycine (DHPG) induced Ca^{2+} responses. Hippocampal mixed neuron-astroglial cultures were stimulated with 50 μM NMDA (**A**) or 20 μM DHPG (**B**). Ca^{2+} signals were detected using Fura-2 single cells imaging in astrocytes (green traces) and in neurons (blue traces).

Store-operated Ca^{2+} entry has been shown to mediate long lasting Ca^{2+} elevations in cultured astrocytes [30]. We, therefore, used a panel of drugs known to inhibit SOCE (Table 2) albeit by different mechanisms. Three of them, namely, 2APB (50 μM, a non-specific transient receptor potential (TRP) receptor and Orai1 inhibitor, that inhibits also InsP3 receptors [31] (Figure 5G), Pyr3 (10 μM, inhibitor of Orai1 and TRPC3) (Figure 5H) and Pyr6 (5 μM, specific Orai1-mediated SOCE inhibitor) [32] (Figure 5I), when applied after washout of the cLTP cocktail, efficiently inhibited both Ca^{2+} signals and CaN activation in astrocytes. However, Pyr10 (10 μM, specific TRPC3-mediated SOCE inhibitor [32]) (Figure 5J), failed to inhibit Ca^{2+} elevations, thereby ruling out TRPC3 involvement in astrocyte activation by neuronal activity. Note that 2APB somewhat strongly augmented frequency of neuronal Ca^{2+} spikes during the cLTP development phase (Figure S2G) resulting in the appearance of numerous artefacts in neighboring astrocytes (Figure 5E). The requirement of extracellular Ca^{2+} was also confirmed by withdrawal of Ca^{2+} from the extracellular buffer (Figure 5K). Of note, we could not detect any discernible ER-dependent Ca^{2+} release in the absence of extracellular Ca^{2+}, as recently shown and discussed in [33]. We also attempted to investigate the involvement of astroglial phospholipase C (PLC), which synthetize InsP3, and InsP3 receptors (InsP3Rs) by exploiting specific cell permeant inhibitors. However, U73122 (10 μM; Figure 5L) and xestospongin C (10 μM; Figure 5M), which, respectively, target PLC and InsP3Rs, failed to completely inhibit either astroglial Ca^{2+} signals or CaN activation. These last experiments, however, may not be conclusive as both U73122 and xestospongin C may require up to 10 min to efficiently inhibit InsP3 production and InsP3Rs, respectively. Unfortunately, their addition before the phase of cLTP induction would have compromised InsP3R-mediated signaling in neurons. In Figure 5, the traces are shown separately for astrocytes and neurons and are limited to the cLTP development phase. Full length Ca^{2+} traces are provided in Supplementary Figure S2.

3. Discussion

In this report, we provide an in vitro proof of principle of activation of astroglial CaN by neuronal activation. The main findings are: (1) in mixed neuron-astroglial hippocampal primary cultures, cLTP induction protocol, which specifically stimulates neuronal activity, induced intracellular Ca^{2+} signals and robust CaN activation in astrocytes, and (2) astroglial Ca^{2+} signals and CaN activation required extracellular Ca^{2+} entry via the SOCE mechanism.

Although, to our knowledge, there is no data published to date that neuronal activity may result in CaN activation in astrocytes, it has been reported that the increase in neuronal activity is able to induce CaN activation and nuclear translocation of NFAT in pericytes in cortical slices [34]. This landmark contribution suggests that neuronal activity may, in fact, activate CaN in non-neuronal cells. Now, we demonstrate that, in an in vitro setting, CaN may be activated also in astrocytes.

Calcineurin is activated by a specific pattern of calcium signaling which is characterized by low and sustained (minutes) elevations of baseline cytosolic calcium levels [35], while the specificity of downstream CaN targets activation can be further achieved by a specific temporal pattern of Ca^{2+} elevations [36]. Stimulation of neuronal activity is known to produce calcium signals in astrocytes in vivo [37–45], including awake animals [46,47], in brain slice preparations [34,48–51] and in mixed neuron-astrocyte primary cultures [52,53]. Most of these signals have been registered as short single or oscillatory transients with duration from several milliseconds to seconds [38,39,44]. Some of them, however, lasted long enough (tens of seconds to minutes) [41,54] to speculate that they would be sufficient for CaN activation. Recent experiments employing fast 3D calcium imaging suggest that the spatio-temporal pattern of Ca^{2+} signals in astrocytes is extremely complex, and depends on the nature of Ca^{2+}-related receptors/channels expressed in a particular subdomain of astroglial plasma membrane [55]. Accordingly, it can be speculated that localized CaN activation may be necessary to achieve spatial specificity of processes controlled by the astrocytes. Further experiments are needed to demonstrate and characterize in vivo activity-dependent CaN activation in astrocytes.

Our findings suggest that neuronal activity induces Ca^{2+} entry in astrocytes via SOCE mechanism. SOCE is one of the fundamental mechanisms of astroglial Ca^{2+} signalling [56] and is involved in the generation of Ca^{2+} oscillations, refilling of the endoplasmic reticulum with Ca^{2+} [57], astroglial cytokine production [58] and astroglial metabolism [59]. Spontaneous Ca^{2+} oscillations in vivo in fine astroglial processes were shown to involve Ca^{2+} entry through the plasma membrane, probably through store-operated channels [60,61]. Concerning pathological conditions, SOCE is involved in the invasion of human glioblastoma [62], and is augmented in primary [30] astrocytes from an AD mouse model. Our results provide the physiological rationale for SOCE activation in astrocytes by neuronal activity. We also show that ionotropic NMDA and metabotropic mGluR5 receptors are involved in SOCE generation by LTP-like neuronal activity. Although it was impossible to discriminate the cell type on which NMDA or mGluR5 reside in the current setting, our previous observations [29] and direct stimulation of mixed cultures either with NMDA or DHPG (Figure 6), suggest that NMDAR are expressed in neurons while mGluR5 is located in astrocytes. Furthermore, cultured astrocytes are somewhat more sensitive than neurons to DHPG, as 20 μM DHPG is enough to induce Ca^{2+} increase in astrocytes, but not in neurons [29] (Figure 6), while 200 μM DHPG were used to elicit mGluR5-dependent Ca^{2+} transients in neurons [63]. Last, astroglial mGluR5 receptors are required to sustain long lasting Ca^{2+} entry in cultures astrocytes that was efficiently blocked by SOCE inhibitors [30].

SOCE is known to activate CaN/NFAT axis and modulate gene transcription in a number of cell types, including T-lymphocytes and mast cells [64], cardiomyocytes [65], skeletal muscle cells [66] and in neural progenitor cells [67]. We show now that SOCE is also required for neuronal activity-induced CaN activation in astrocytes in vitro while future experiments will show if SOCE activates CaN in astrocytes also in intact brain. Regarding the nature of SOCE channels, our pharmacological survey suggests that Orai1-containing channels are operative in astrocytes (efficiently inhibited by Orai1-blocking Pyr3 and Pyr6, [32]), while participation of TRPC3 is questionable, since Pyr10, a specific TRPC3 inhibitor [32]), was not as efficient as other SOCE blockers. In line with this, in our previous report, Pyr10 failed to inhibit the DHPG-induced after-peak Ca^{2+} entry, while it was efficiently inhibited by Pyr6 [30]).

In neurons, CaN is involved in long-term changes during neuronal plasticity, e.g., forebrain neuronal deletion of CaN specifically affects bidirectional synaptic plasticity and episodic-like working memory [18], and inactivation of CaN is essential for the onset of transcriptional remodeling during long-term plasticity and memory formation [17,19]. Our results suggest that also in astrocytes activity-induced CaN activation may be involved in long-term transcriptional remodeling leading to structural, biochemical and functional astroglial plasticity [68–70]. Further studies are necessary to confirm this hypothesis.

The present report is a proof of principle in vitro study, and is not devoid of limitations, the two principal of which are: (1) the in vitro setup, which, obviously, only in part replicates the complex LTP phenomenon occurring in intact brain or even in brain slices; and (2) chemical instead of electrical LTP induction. We have consciously accepted these limitations for the following reasons. The in vitro setting proved to be simple and highly reproducible both in terms of lentiviral NY infection and in terms of cLTP and NY nuclear translocation. Thereby, it allowed us to rule out astrocytes as a primary target of cLTP protocol. Yet, we used hippocampal primary cultures and the effect may be different in cultures prepared from the other brain regions. Regarding cLTP induction, there are several protocols which are basically the modifications of two principal variants: the first consists in the application of a cocktail containing protein kinase A (PKA) activators, like forskolin and rolipram [28,71,72], which recruits downstream targets of PKA-dependent phosphorylation involved in LTP induction. For obvious reasons, this protocol is not specific to neurons and would induce PKA-dependent phosphorylation also in astrocytes. The second variant consists in the relieve of GABA-dependent inhibition by blocking ionotropic GABA(A) receptors with bicuculline and strychnine and in facilitation of NMDA receptors with glycine and Mg^{2+}-free buffer [73–76]. We have chosen this second protocol because it minimally interferes with astrocyte biochemistry and does not result in NY translocation in pure astroglial cultures. Relieve of GABA-dependent inhibition alone is known to increase neuronal firing rate [77] and, notably, it was sufficient to induce nuclear translocation of NY and, hence, CaN activation, although significantly later than it was achieved by cLTP. Similar effect was achieved by

Mg^{2+} withdrawal, which indicates that different strategies to facilitate neuronal activity may lead to CaN activation in astrocytes, although with different temporal pattern.

An important question which remains unanswered in this work is the nature of the mediator which is released by neurons to induce the astroglial response. Assuming that astrocytic mGluR5 receptors are involved in astrocyte activation, it is plausible to speculate that the messenger is glutamate released during neuronal activity. This has been demonstrated in vitro [78] and in situ [79]. However, in vivo it has been shown that mGluR5 is downregulated during postnatal development and no longer active in adult astrocytes [80]. Therefore, the mechanisms of the activity-induced generation of Ca^{2+} signals in astrocytes, as well as of CaN activation, may differ between ex vivo and in vivo preparations, and may also depend on age at which the preparation is made.

In summary, we propose a model (Figure 7) in which LTP-like neuronal activity, possibly through activation of NMDA receptors and glutamate release form neuronal terminals, induces Ca^{2+} entry through the plasma membrane, possibly implicating astrocytic mGluR5 receptors. The Ca^{2+} entry occurs through Orai1-containing SOCE channels and results in activation of CaN inside the astrocytes, which, in turn, leads to activation and nuclear translocation of CaN sensor NY. CaN activation in astrocytes may lead to transcriptional remodeling and long-term changes analogous to what occur during neuronal plasticity. In this model, we provide a framework for future investigation of astroglial CaN activation during neuronal activity and plasticity in physiology and pathology of CNS.

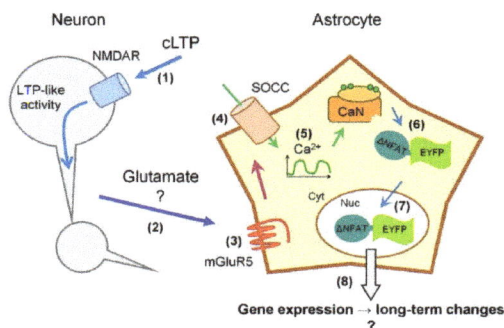

Figure 7. Model of LTP-like neuronal activity-induced CaN activation in astrocytes. LTP-like neuronal activity, possibly through activation of NMDA receptors (**1**) and glutamate release form neuronal terminals (**2**), induces Ca^{2+} entry through the plasma membrane of adjacent astrocytes, possibly implicating astrocytic mGluR5 receptors (**3**). The Ca^{2+} entry occurs through Orai1-containing SOCE channels (SOCC) (**4**) resulting in delayed long-lasting Ca^{2+} transients (**5**) and activation of CaN inside the astrocytes (**6**), which, in turn, leads to nuclear translocation of CaN sensor NY (ΔNFAT-EYFP) (**7**). The latter event is quantified as an increase in ratio between NY fluorescence in the nucleus (Nuc) and the cytoplasm (Cyt). Increase in CaN activity in astrocytes may lead to transcriptional remodeling and long-term changes (**8**). "?" in (**2**) and (**8**) indicates questionable points.

4. Materials and Methods

4.1. Animals

C57Bl/6 mice have been purchased from Charles River Laboratories (Calco (Lecco), Italy). The animals have been given food and water ad libitum; light/dark cycle has been automatically controlled in respect with the natural circadian rhythm and the temperature has been thermostatically regulated. Animals were managed in accordance with European directive 2010/63/UE and with Italian low D.l. 26/2014. The procedures were approved by the local animal-health and ethical committee

(Università del Piemonte Orientale) and were authorized by the national authority (Istituto Superiore di Sanità; authorization number N. 22/2013, 3 February 2013).

4.2. Primary Hippocampal Mixed and Astroglial Cultures

To prepare mixed neuron-astroglial hippocampal primary cultures, new-born mice (less than one day old) were used. Pups were sacrificed by decapitation and hippocampi were dissected in cold HBSS (Sigma, Darmstadt, Germany, Cat. H6648). After dissection, hippocampi were digested in 0.1% Trypsin (Sigma, Cat. T4049) for 20 min in 37 °C water bath. Then, trypsin was neutralized by addition of Dulbecco's Modified Eagle's Medium (DMEM, Sigma, Cat. D5671) supplemented with 10% heat-inactivated fetal bovine serum (FBS, Life Technologies, Monza, Italy, Cat. 10270-106) and tissue was spun at $300 \times g$ for 5 min. Tissue pellet was resuspended in HBSS supplemented with 10% FBS and dissociated by 30 strokes of a 1000 μL automatic pipette. After pipetting, the tissues were left for 5 min to allow sedimentation of un-dissociated tissue, and cell suspension was transferred to a new tube and centrifuged at $250 \times g$ for 5 min. Hippocampal cells were resuspended in Neurobasal-A medium supplemented with 2% B-27 and with 2 mg/mL glutamine, 10 U/mL penicillin and 100 μg/mL streptomycin, and plated onto Poly-L-lysine-coated coverslips (0.1 mg/mL). For NY translocation time-course, the cells were plated onto 13 mm round coverslips in 24 well plates, 2×10^4 cells per well. For Fura-2 imaging, the cells were spotted (2×10^4 cells per spot) on 24 mm round coverslips in 6 well plates. Half of medium was changed every 5 days. Cultures were used for treatments and experiments after 12 days in vitro (DIV12). At DIV12-DIV14 the neuron/astrocyte ratio was 1.29 ± 0.49 as was counted in 34 coverslips from at least 8 independent cultures.

To prepare purified astroglial cultures, one to three-day old pups were sacrificed by decapitation, hippocampi were rapidly dissected and placed in cold HBSS. Tissue was digested in 0.25% trypsin (37 °C, 20 min), washed in complete culture medium ((DMEM), supplemented with 10% FBS, and with 2 mg/mL glutamine, 10 U/mL penicillin and 100 μg/mL streptomycin (all from Sigma) and resuspended in cold HBSS supplemented with 10% FBS. After 30 strokes of dissociation with an automatic pipette, cell suspension was centrifuged ($250 \times g$, 5 min), pellet was resuspended in complete medium and plated in 100 mm Falcon culture dishes, pretreated with 0.1 mg/mL Poly-L-lysine (Sigma). At sub-confluence (DIV5-10), cells were detached with trypsin and microglial cells were removed by magnetic-activated cell sorting (MACS) negative selection using anti-CD11b conjugated beads and MS magnetic columns (Miltenyi Biotech, Bologna, Italy, Cat. 130-093-634). After MACS, astrocytes were counted and plated for experiments as described above. Virtually no microglial cells have been detected in purified astroglial cultures after MACS procedure as was assessed by immunostaining with anti-Iba1 antibody (1:500, D.B.A., Segrate, Italy, Cat. 019-19741) [81].

4.3. Lentivirus Production and Transduction

mCherry-H2Bc-ΔNFAT-EYFP-expressing pFUW (pFUW-NY) third generation lentiviral vector was a kind gift from Prof. Alexander Flügel, Göttingen [23]. The production of infectious lentiviral particles was done using polyethylene glycol (PEG) method as describe elsewhere [82]. Briefly, HEK293T cells were transfected with four plasmids (2.75 μg pMDLg.pRRE, 1 μg pRSV.Rev, 1.5 μg pMD2.VSVG, and 6 μg pFUW-NY per 100 mm dish). After 48–72 h medium was collected and viral particles were precipitated by adding 1/5 of 5× PEG solution (200 g PEG (Sigma, Cat. n. 89510), 12 g NaCl, 1 mL Tris 1 M, pH 7.5, in ddH$_2$O to a final volume of 500 mL) overnight. On the next day the precipitate was concentrated by centrifugation ($2800 \times g$ at 4 °C for 30 min) and resuspended in HBSS (1/100 of the original medium volume, aliquoted and stored at −80 °C. All manipulations with lentiviral vector were conducted in biosafety level 2 environment in accordance with Italian ministry of health-approved protocol.

DIV10-12 mixed neuron-astroglial cultures or 50% confluent purified astrocytes were transduced by adding (50 μL/mL) concentrated NY-expressing viral particles. Experiments were conducted

2–4 days after infection. For each condition and time-point, at least 2 coverslips were used (technical replicates) in three independent culture preparations (biological replicates).

4.4. Induction of cLTP and NY Translocation Quantification

cLTP was induced by a cLTP cocktail containing 20 µM bicuculline, 1 µM strychnine, 200 µM glycine in Mg^{2+}-free Krebs-Ringer buffer (KRB, 135 mM NaCl, 5 mM KCl, 0.4 mM KH_2PO_4, 5.5 mM glucose, 20 mM HEPES, pH 7.4). First the cells were rinsed by Mg^{2+}-free KRB along to wash out Mg^{2+} ions. Then, cLTP cocktail was applied for 4 min to induce LTP (cLTP induction phase). After this, the cLTP cocktail was changed to Mg^{2+}- and Ca^{2+}-containing KRB (KRB + Mg + Ca) and kept until the cells were fixed or imaged (cLTP development phase). At indicated time-points after cLTP induction, KRB + Mg + Ca was quickly removed and cells were fixed with 4% formalin in PBS (20 min, room temperature (RT)). Fixed cells were washed 3 times with PBS and mounted on microscope slides using SlowFade® Gold Antifade mountant (Life Technologies, Monza, Italy).

Fixed cells were imaged on a Leica DMI6000 epifluorescent microscope equipped with Polychrome V monochromator (Till Photonics, Graefelfing, Germany) and S Fluor ×40/1.3 objective (Leica, Buccinasco, Italy). The cells were alternatively excited by 488 and 546 nm and emission light was filtered through 520/20 and 600/40 nm bandpass filters, respectively, and collected by a cooled CCD camera (Leica, Hamamatsu, Japan). The fluorescence signals were acquired and processed using MetaMorph software (Molecular Device, Sunnyvale, CA, USA).

To quantify NY translocation, 5 random fields were photographed from each coverslip, and the astrocytes with clear expression of NY sensor were used for analysis (3–15 astrocytes per field). For each cell two regions of interest (ROIs) were placed inside the nucleus (which was evidenced by mCherry-H2Bc expression) and two in different sides of the cytosol close to the nucleus. For each cell, the fluorescence intensity of two nuclear (Nuc) and two cytosolic (Cyt) ROIs, respectively, measured in green channel, was averaged and Nuc/Cyt ratio was obtained by dividing the resulting Nuc fluorescence to Cyt fluorescence. The data are expressed as mean ± SEM for each cell analyzed form three independent culture preparations (e.g., Figure 3).

4.5. Induction of cLTP and NY Translocation Quantification

For Fura-2 Ca^{2+} imaging experiments, the cells were first loaded with 2 µM Fura-2-AM in presence of 0.02% of Pluronic-127 (both from Life Technologies) and 10 µM sulfinpyrazone (Sigma) for 20 min at RT. Fura-2 loaded cells were washed in KRB + Mg + Ca and allowed to de-esterify for 20 min before cLTP induction.

After that, the coverslips were mounted into acquisition chamber and placed on the stage of a Leica DMI6000 epifluorescence microscope equipped with S Fluor ×40/1.3 objective. The probe was excited by alternate 340 and 380 nm using a Polychrome IV monochromator and the Fura-2 emission light was filtered through 520/20 bandpass filter and collected by a cooled CCD camera (Hamamatsu, Japan). The fluorescence signals were acquired and processed using MetaFluor software (Molecular Device, Sunnyvale, CA, USA). To quantify the differences in the amplitudes of Ca^{2+} transients the ratio values were normalized using the formula $\Delta F/F0$ (referred to as normalized Fura-2 ratio, "Norm. Fura ratio"). At least two coverslips for each of three independent culture preparation were imaged for each condition. In mixed neuron-astroglial cultures, astrocytes were recognized as flat polygonal or star-like cells, while neurons were recognized by round bodies with few processes located on upper focal plane above the astrocytes. The cells with uncertain morphology were not taken in consideration.

4.6. Pharmacological Reagents

MK801 (Cat. 0924, stock 100 mM in DMSO), MTEP (Cat. 2921, stock 100 mM in DMSO), 2APB (Cat. 1224, stock 50 mM in DMSO), Xestospongin C (Cat. 1280, stock 2 mM in DMSO), U73122 (Cat. 1268, stock 10 mM in DMSO), NMDA (Cat. 0114, stock 100 mM in H_2O) and DHPG (Cat. 0342, stock

10 mM in PBS) were from Tocris (Bristol, UK). Pyr3 (Cat. P0032, stock 10 mM in DMSO), Pyr6 (Cat. SML1241, stock 10 mM in DMSO) and Pyr10 (Cat. SML1243, stock 10 mM in DMSO) were from Sigma.

4.7. Statistical Analysis

Statistical analysis was performed using GraphPad Prism software v.7. For analysis of Nuc/Cyt NY ratio (Figure 3) each dataset at indicated time-points was compared with respective control using a two-tailed unpaired Students's *t*-test. Differences were considered significant at $p < 0.05$. Data are expressed as mean ± SEM.

Supplementary Materials: Supplementary materials can be found at http://www.mdpi.com/1422-0067/19/10/2997/s1.

Author Contributions: Conceptualization, D.L., F.M., L.M. and A.A.G.; Methodology, D.L., F.M. and L.M.; Investigation, D.L.; Data Curation, F.M., L.M.; Writing—Original Draft Preparation, D.L. and F.M.; Writing—Review & Editing, D.L., F.M., and A.A.G.; Supervision, P.L.C. and A.A.G.; Funding Acquisition, D.L., A.A.G. and P.L.C.

Funding: This work had the following financial supports: grants 2013-0795 to AAG and 2014-1094 to DL from the Fondazione Cariplo; grant 2016 to DL from The Università del Piemonte Orientale; and grant PRIN-2015N4FKJ4 to PLC from the Italian Ministry of Education.

Conflicts of Interest: The authors declare no conflict of interest.

Abbreviations

CaN	Calcineurin
cLTP	Chemical Long-Term Potentiation
cLTD	Chemical Long-Term Depression
KRB	Modified krebs-Ringer Buffer
MK801	(5*S*,10*R*)-(+)-5-Methyl-10,11-dihydro-5*H*-dibenzo[a,d]cyclohepten-5,10-imine Maleate
MTEP	3-((2-Methyl-1,3-thiazol-4-yl) ethynyl) Pyridine Hydrochloride
DHPG	(*RS*)-3,5-Dihydroxyphenylglycine
U73122	1-[6-[[(17β)-3-Methoxyestra-1,3,5(10)-trien-17-yl]amino]hexyl]-1*H*-pyrrole-2,5-dione
NMDA	*N*-Methyl-D-aspartic Acid
TRPC3	Transient Receptor Potential C3
2APB	2-Aminoethoxydiphenylborane
Pyr3	1-[4-[(2,3,3-Trichloro-1-oxo-2-propen-1-yl)amino]phenyl]-5-(trifluoromethyl)-1*H*-pyrazole-4-carboxylic Acid
Pyr6	*N*-[4-[3,5-Bis(trifluoromethyl)-1*H*-pyrazol-1-yl]phenyl]-3-fluoro-4-pyridinecarboxamide
Pyr10	*N*-[4-[3,5-Bis(trifluoromethyl)-1*H*-pyrazol-1-yl]phenyl]-4-methyl-benzenesulfonamide
DMSO	Dimethyl Sulfoxide
SOCE	Store-Operated Calcium Entry

References

1. Verkhratsky, A.; Nedergaard, M. Physiology of Astroglia. *Physiol. Rev.* **2018**, *98*, 239–389. [CrossRef] [PubMed]
2. Di Castro, M.A.; Chuquet, J.; Liaudet, N.; Bhaukaurally, K.; Santello, M.; Bouvier, D.; Tiret, P.; Volterra, A. Local Ca²⁺ detection and modulation of synaptic release by astrocytes. *Nat. Neurosci.* **2011**, *14*, 1276–1284. [CrossRef] [PubMed]
3. Fellin, T.; Pascual, O.; Gobbo, S.; Pozzan, T.; Haydon, P.G.; Carmignoto, G. Neuronal synchrony mediated by astrocytic glutamate through activation of extrasynaptic NMDA receptors. *Neuron* **2004**, *43*, 729–743. [CrossRef] [PubMed]
4. Fiacco, T.A.; McCarthy, K.D. Intracellular astrocyte calcium waves in situ increase the frequency of spontaneous AMPA receptor currents in CA1 pyramidal neurons. *J. Neurosci. Off. J. Soc. Neurosci.* **2004**, *24*, 722–732. [CrossRef] [PubMed]

5. Parpura, V.; Grubišić, V.; Verkhratsky, A. Ca($2+$) sources for the exocytotic release of glutamate from astrocytes. *Biochim. Biophys. Acta* **2011**, *1813*, 984–991. [CrossRef] [PubMed]

6. Agulhon, C.; Petravicz, J.; McMullen, A.B.; Sweger, E.J.; Minton, S.K.; Taves, S.R.; Casper, K.B.; Fiacco, T.A.; McCarthy, K.D. What is the role of astrocyte calcium in neurophysiology? *Neuron* **2008**, *59*, 932–946. [CrossRef] [PubMed]

7. Fiacco, T.A.; Agulhon, C.; McCarthy, K.D. Sorting out astrocyte physiology from pharmacology. *Annu. Rev. Pharmacol. Toxicol.* **2009**, *49*, 151–174. [CrossRef] [PubMed]

8. Agulhon, C.; Fiacco, T.A.; McCarthy, K.D. Hippocampal short- and long-term plasticity are not modulated by astrocyte Ca^{2+} signaling. *Science* **2010**, *327*, 1250–1254. [CrossRef] [PubMed]

9. Takano, T.; Han, X.; Deane, R.; Zlokovic, B.; Nedergaard, M. Two-photon imaging of astrocytic Ca^{2+} signaling and the microvasculature in experimental mice models of Alzheimer's disease. *Ann. N. Y. Acad. Sci.* **2007**, *1097*, 40–50. [CrossRef] [PubMed]

10. Rossi, D.; Volterra, A. Astrocytic dysfunction: Insights on the role in neurodegeneration. *Brain Res. Bull.* **2009**, *80*, 224–232. [CrossRef] [PubMed]

11. Kuchibhotla, K.V.; Lattarulo, C.R.; Hyman, B.T.; Bacskai, B.J. Synchronous hyperactivity and intercellular calcium waves in astrocytes in Alzheimer mice. *Science* **2009**, *323*, 1211–1215. [CrossRef] [PubMed]

12. Vincent, A.J.; Gasperini, R.; Foa, L.; Small, D.H. Astrocytes in Alzheimer's disease: Emerging roles in calcium dysregulation and synaptic plasticity. *J. Alzheimers Dis.* **2010**, *22*, 699–714. [CrossRef] [PubMed]

13. Lim, D.; Ronco, V.; Grolla, A.A.; Verkhratsky, A.; Genazzani, A.A. Glial calcium signalling in Alzheimer's disease. *Rev. Physiol. Biochem. Pharmacol.* **2014**, *167*, 45–65. [CrossRef] [PubMed]

14. Lim, D.; Rodríguez-Arellano, J.J.; Parpura, V.; Zorec, R.; Zeidán-Chuliá, F.; Genazzani, A.A.; Verkhratsky, A. Calcium signalling toolkits in astrocytes and spatio-temporal progression of Alzheimer's disease. *Curr. Alzheimer Res.* **2016**, *13*, 359–369. [CrossRef] [PubMed]

15. Vardjan, N.; Verkhratsky, A.; Zorec, R. Astrocytic Pathological Calcium Homeostasis and Impaired Vesicle Trafficking in Neurodegeneration. *Int. J. Mol. Sci.* **2017**, *18*, 358. [CrossRef] [PubMed]

16. Klee, C.B.; Crouch, T.H.; Krinks, M.H. Calcineurin: A calcium- and calmodulin-binding protein of the nervous system. *Proc. Natl. Acad. Sci. USA* **1979**, *76*, 6270–6273. [CrossRef] [PubMed]

17. Baumgärtel, K.; Mansuy, I.M. Neural functions of calcineurin in synaptic plasticity and memory. *Learn. Mem.* **2012**, *19*, 375–384. [CrossRef] [PubMed]

18. Zeng, H.; Chattarji, S.; Barbarosie, M.; Rondi-Reig, L.; Philpot, B.D.; Miyakawa, T.; Bear, M.F.; Tonegawa, S. Forebrain-specific calcineurin knockout selectively impairs bidirectional synaptic plasticity and working/episodic-like memory. *Cell* **2001**, *107*, 617–629. [CrossRef]

19. Baumgärtel, K.; Genoux, D.; Welzl, H.; Tweedie-Cullen, R.Y.; Koshibu, K.; Livingstone-Zatchej, M.; Mamie, C.; Mansuy, I.M. Control of the establishment of aversive memory by calcineurin and Zif268. *Nat. Neurosci.* **2008**, *11*, 572–578. [CrossRef] [PubMed]

20. Kipanyula, M.J.; Kimaro, W.H.; Seke Etet, P.F. The Emerging Roles of the Calcineurin-Nuclear Factor of Activated T-Lymphocytes Pathway in Nervous System Functions and Diseases. *J. Aging Res.* **2016**, *2016*, 5081021. [CrossRef] [PubMed]

21. Furman, J.L.; Norris, C.M. Calcineurin and glial signaling: Neuroinflammation and beyond. *J. Neuroinflamm.* **2014**, *11*, 158. [CrossRef] [PubMed]

22. Lim, D.; Rocchio, F.; Lisa, M.; Fcancesco, M. From Pathology to Physiology of Calcineurin Signalling in Astrocytes | Opera Medica et Physiologica. Available online: http://operamedphys.org/OMP_2016_02_0029 (accessed on 24 August 2018).

23. Lodygin, D.; Odoardi, F.; Schläger, C.; Körner, H.; Kitz, A.; Nosov, M.; van den Brandt, J.; Reichardt, H.M.; Haberl, M.; Flügel, A. A combination of fluorescent NFAT and H2B sensors uncovers dynamics of T cell activation in real time during CNS autoimmunity. *Nat. Med.* **2013**, *19*, 784–790. [CrossRef] [PubMed]

24. Palkowitsch, L.; Marienfeld, U.; Brunner, C.; Eitelhuber, A.; Krappmann, D.; Marienfeld, R.B. The Ca^{2+}-dependent phosphatase calcineurin controls the formation of the Carma1-Bcl10-Malt1 complex during T cell receptor-induced NF-kappaB activation. *J. Biol. Chem.* **2011**, *286*, 7522–7534. [CrossRef] [PubMed]

25. Friday, B.B.; Pavlath, G.K. A calcineurin- and NFAT-dependent pathway regulates Myf5 gene expression in skeletal muscle reserve cells. *J. Cell Sci.* **2001**, *114*, 303–310. [PubMed]

26. Hojayev, B.; Rothermel, B.A.; Gillette, T.G.; Hill, J.A. FHL2 binds calcineurin and represses pathological cardiac growth. *Mol. Cell. Biol.* **2012**, *32*, 4025–4034. [CrossRef] [PubMed]

27. Halpain, S.; Hipolito, A.; Saffer, L. Regulation of F-actin stability in dendritic spines by glutamate receptors and calcineurin. *J. Neurosci. Off. J. Soc. Neurosci.* **1998**, *18*, 9835–9844. [CrossRef]

28. Dinamarca, M.C.; Guzzetti, F.; Karpova, A.; Lim, D.; Mitro, N.; Musardo, S.; Mellone, M.; Marcello, E.; Stanic, J.; Samaddar, T.; et al. Ring finger protein 10 is a novel synaptonuclear messenger encoding activation of NMDA receptors in hippocampus. *eLife* **2016**, *5*, e12430. [CrossRef] [PubMed]

29. Grolla, A.A.; Fakhfouri, G.; Balzaretti, G.; Marcello, E.; Gardoni, F.; Canonico, P.L.; DiLuca, M.; Genazzani, A.A.; Lim, D. Aβ leads to Ca^{2+} signaling alterations and transcriptional changes in glial cells. *Neurobiol. Aging* **2013**, *34*, 511–522. [CrossRef] [PubMed]

30. Ronco, V.; Grolla, A.A.; Glasnov, T.N.; Canonico, P.L.; Verkhratsky, A.; Genazzani, A.A.; Lim, D. Differential deregulation of astrocytic calcium signalling by amyloid-β, TNFα, IL-1β and LPS. *Cell Calcium* **2014**, *55*, 219–229. [CrossRef] [PubMed]

31. Dragoni, S.; Laforenza, U.; Bonetti, E.; Lodola, F.; Bottino, C.; Berra-Romani, R.; Carlo Bongio, G.; Cinelli, M.P.; Guerra, G.; Pedrazzoli, P.; et al. Vascular endothelial growth factor stimulates endothelial colony forming cells proliferation and tubulogenesis by inducing oscillations in intracellular Ca^{2+} concentration. *Stem Cells* **2011**, *29*, 1898–1907. [CrossRef] [PubMed]

32. Schleifer, H.; Doleschal, B.; Lichtenegger, M.; Oppenrieder, R.; Derler, I.; Frischauf, I.; Glasnov, T.N.; Kappe, C.O.; Romanin, C.; Groschner, K. Novel pyrazole compounds for pharmacological discrimination between receptor-operated and store-operated $Ca^{(2+)}$ entry pathways. *Br. J. Pharmacol.* **2012**, *167*, 1712–1722. [CrossRef] [PubMed]

33. Zuccolo, E.; Lim, D.; Kheder, D.A.; Perna, A.; Catarsi, P.; Botta, L.; Rosti, V.; Riboni, L.; Sancini, G.; Tanzi, F.; et al. Acetylcholine induces intracellular Ca^{2+} oscillations and nitric oxide release in mouse brain endothelial cells. *Cell Calcium* **2017**, *66*, 33–47. [CrossRef] [PubMed]

34. Filosa, J.A.; Bonev, A.D.; Nelson, M.T. Calcium dynamics in cortical astrocytes and arterioles during neurovascular coupling. *Circ. Res.* **2004**, *95*, e73–e81. [CrossRef] [PubMed]

35. Negulescu, P.A.; Shastri, N.; Cahalan, M.D. Intracellular calcium dependence of gene expression in single T lymphocytes. *Proc. Natl. Acad. Sci. USA* **1994**, *91*, 2873–2877. [CrossRef] [PubMed]

36. Dolmetsch, R.E.; Lewis, R.S.; Goodnow, C.C.; Healy, J.I. Differential activation of transcription factors induced by Ca^{2+} response amplitude and duration. *Nature* **1997**, *386*, 855–858. [CrossRef] [PubMed]

37. Fam, S.R.; Gallagher, C.J.; Kalia, L.V.; Salter, M.W. Differential frequency dependence of P2Y1- and P2Y2-mediated Ca^{2+} signaling in astrocytes. *J. Neurosci. Off. J. Soc. Neurosci.* **2003**, *23*, 4437–4444. [CrossRef]

38. Hirase, H.; Qian, L.; Barthó, P.; Buzsáki, G. Calcium dynamics of cortical astrocytic networks in vivo. *PLoS Biol.* **2004**, *2*, e96. [CrossRef] [PubMed]

39. Winship, I.R.; Plaa, N.; Murphy, T.H. Rapid astrocyte calcium signals correlate with neuronal activity and onset of the hemodynamic response in vivo. *J. Neurosci. Off. J. Soc. Neurosci.* **2007**, *27*, 6268–6272. [CrossRef] [PubMed]

40. Ding, S.; Fellin, T.; Zhu, Y.; Lee, S.-Y.; Auberson, Y.P.; Meaney, D.F.; Coulter, D.A.; Carmignoto, G.; Haydon, P.G. Enhanced astrocytic Ca^{2+} signals contribute to neuronal excitotoxicity after status epilepticus. *J. Neurosci. Off. J. Soc. Neurosci.* **2007**, *27*, 10674–10684. [CrossRef] [PubMed]

41. Atkin, S.D.; Patel, S.; Kocharyan, A.; Holtzclaw, L.A.; Weerth, S.H.; Schram, V.; Pickel, J.; Russell, J.T. Transgenic mice expressing a cameleon fluorescent Ca^{2+} indicator in astrocytes and Schwann cells allow study of glial cell Ca^{2+} signals in situ and in vivo. *J. Neurosci. Methods* **2009**, *181*, 212–226. [CrossRef] [PubMed]

42. Ghosh, A.; Wyss, M.T.; Weber, B. Somatotopic astrocytic activity in the somatosensory cortex. *Glia* **2013**, *61*, 601–610. [CrossRef] [PubMed]

43. Kanemaru, K.; Sekiya, H.; Xu, M.; Satoh, K.; Kitajima, N.; Yoshida, K.; Okubo, Y.; Sasaki, T.; Moritoh, S.; Hasuwa, H.; et al. In vivo visualization of subtle, transient, and local activity of astrocytes using an ultrasensitive $Ca^{(2+)}$ indicator. *Cell Rep.* **2014**, *8*, 311–318. [CrossRef] [PubMed]

44. Otsu, Y.; Couchman, K.; Lyons, D.G.; Collot, M.; Agarwal, A.; Mallet, J.-M.; Pfrieger, F.W.; Bergles, D.E.; Charpak, S. Calcium dynamics in astrocyte processes during neurovascular coupling. *Nat. Neurosci.* **2015**, *18*, 210–218. [CrossRef] [PubMed]

45. Stobart, J.L.; Ferrari, K.D.; Barrett, M.J.P.; Glück, C.; Stobart, M.J.; Zuend, M.; Weber, B. Cortical Circuit Activity Evokes Rapid Astrocyte Calcium Signals on a Similar Timescale to Neurons. *Neuron* **2018**, *98*, 726–735.e4. [CrossRef] [PubMed]

46. Ding, F.; O'Donnell, J.; Thrane, A.S.; Zeppenfeld, D.; Kang, H.; Xie, L.; Wang, F.; Nedergaard, M. α1-Adrenergic receptors mediate coordinated Ca^{2+} signaling of cortical astrocytes in awake, behaving mice. *Cell Calcium* **2013**, *54*, 387–394. [CrossRef] [PubMed]

47. Stobart, J.L.; Ferrari, K.D.; Barrett, M.J.P.; Stobart, M.J.; Looser, Z.J.; Saab, A.S.; Weber, B. Long-term In Vivo Calcium Imaging of Astrocytes Reveals Distinct Cellular Compartment Responses to Sensory Stimulation. *Cereb. Cortex* **2018**, *28*, 184–198. [CrossRef] [PubMed]

48. Straub, S.V.; Bonev, A.D.; Wilkerson, M.K.; Nelson, M.T. Dynamic inositol trisphosphate-mediated calcium signals within astrocytic endfeet underlie vasodilation of cerebral arterioles. *J. Gen. Physiol.* **2006**, *128*, 659–669. [CrossRef] [PubMed]

49. Piet, R.; Jahr, C.E. Glutamatergic and purinergic receptor-mediated calcium transients in Bergmann glial cells. *J. Neurosci. Off. J. Soc. Neurosci.* **2007**, *27*, 4027–4035. [CrossRef] [PubMed]

50. Tanaka, M.; Shih, P.-Y.; Gomi, H.; Yoshida, T.; Nakai, J.; Ando, R.; Furuichi, T.; Mikoshiba, K.; Semyanov, A.; Itohara, S. Astrocytic Ca^{2+} signals are required for the functional integrity of tripartite synapses. *Mol. Brain* **2013**, *6*, 6. [CrossRef] [PubMed]

51. Haustein, M.D.; Kracun, S.; Lu, X.-E.; Shih, T.; Jackson-Weaver, O.; Tong, X.; Xu, J.; Yang, X.W.; O'Dell, T.J.; Marvin, J.S.; et al. Conditions and constraints for astrocyte calcium signaling in the hippocampal mossy fiber pathway. *Neuron* **2014**, *82*, 413–429. [CrossRef] [PubMed]

52. Rao, S.P.; Sikdar, S.K. Astrocytes in 17beta-estradiol treated mixed hippocampal cultures show attenuated calcium response to neuronal activity. *Glia* **2006**, *53*, 817–826. [CrossRef] [PubMed]

53. Suadicani, S.O.; Cherkas, P.S.; Zuckerman, J.; Smith, D.N.; Spray, D.C.; Hanani, M. Bidirectional calcium signaling between satellite glial cells and neurons in cultured mouse trigeminal ganglia. *Neuron Glia Biol.* **2010**, *6*, 43–51. [CrossRef] [PubMed]

54. Schulz, K.; Sydekum, E.; Krueppel, R.; Engelbrecht, C.J.; Schlegel, F.; Schröter, A.; Rudin, M.; Helmchen, F. Simultaneous BOLD fMRI and fiber-optic calcium recording in rat neocortex. *Nat. Methods* **2012**, *9*, 597–602. [CrossRef] [PubMed]

55. Volterra, A.; Liaudet, N.; Savtchouk, I. Astrocyte Ca^{2+} signalling: An unexpected complexity. *Nat. Rev. Neurosci.* **2014**, *15*, 327–335. [CrossRef] [PubMed]

56. Papanikolaou, M.; Lewis, A.; Butt, A.M. Store-operated calcium entry is essential for glial calcium signalling in CNS white matter. *Brain Struct. Funct.* **2017**, *222*, 2993–3005. [CrossRef] [PubMed]

57. Verkhratsky, A.; Parpura, V. Store-operated calcium entry in neuroglia. *Neurosci. Bull.* **2014**, *30*, 125–133. [CrossRef] [PubMed]

58. Gao, X.; Xia, J.; Munoz, F.M.; Manners, M.T.; Pan, R.; Meucci, O.; Dai, Y.; Hu, H. STIMs and Orai1 regulate cytokine production in spinal astrocytes. *J. Neuroinflamm.* **2016**, *13*, 126. [CrossRef] [PubMed]

59. Müller, M.S.; Fox, R.; Schousboe, A.; Waagepetersen, H.S.; Bak, L.K. Astrocyte glycogenolysis is triggered by store-operated calcium entry and provides metabolic energy for cellular calcium homeostasis. *Glia* **2014**, *62*, 526–534. [CrossRef] [PubMed]

60. Rungta, R.L.; Bernier, L.-P.; Dissing-Olesen, L.; Groten, C.J.; LeDue, J.M.; Ko, R.; Drissler, S.; MacVicar, B.A. Ca2+ transients in astrocyte fine processes occur via Ca^{2+} influx in the adult mouse hippocampus. *Glia* **2016**, *64*, 2093–2103. [CrossRef] [PubMed]

61. Zhang, Y.V.; Ormerod, K.G.; Littleton, J.T. Astrocyte Ca^{2+} Influx Negatively Regulates Neuronal Activity. *eNeuro* **2017**, *4*. [CrossRef] [PubMed]

62. Motiani, R.K.; Hyzinski-García, M.C.; Zhang, X.; Henkel, M.M.; Abdullaev, I.F.; Kuo, Y.-H.; Matrougui, K.; Mongin, A.A.; Trebak, M. STIM1 and Orai1 mediate CRAC channel activity and are essential for human glioblastoma invasion. *Pflugers Arch.* **2013**, *465*, 1249–1260. [CrossRef] [PubMed]

63. Vicidomini, C.; Ponzoni, L.; Lim, D.; Schmeisser, M.J.; Reim, D.; Morello, N.; Orellana, D.; Tozzi, A.; Durante, V.; Scalmani, P.; et al. Pharmacological enhancement of mGlu5 receptors rescues behavioral deficits in SHANK3 knock-out mice. *Mol. Psychiatry* **2017**, *22*, 689–702. [CrossRef] [PubMed]

64. Oh-hora, M.; Rao, A. Calcium signaling in lymphocytes. *Curr. Opin. Immunol.* **2008**, *20*, 250–258. [CrossRef] [PubMed]

65. Eder, P. Cardiac Remodeling and Disease: SOCE and TRPC Signaling in Cardiac Pathology. *Adv. Exp. Med. Biol.* **2017**, *993*, 505–521. [CrossRef] [PubMed]

66. Li, T.; Finch, E.A.; Graham, V.; Zhang, Z.-S.; Ding, J.-D.; Burch, J.; Oh-hora, M.; Rosenberg, P. STIM1-Ca^{2+} signaling is required for the hypertrophic growth of skeletal muscle in mice. *Mol. Cell. Biol.* **2012**, *32*, 3009–3017. [CrossRef] [PubMed]

67. Somasundaram, A.; Shum, A.K.; McBride, H.J.; Kessler, J.A.; Feske, S.; Miller, R.J.; Prakriya, M. Store-operated CRAC channels regulate gene expression and proliferation in neural progenitor cells. *J. Neurosci. Off. J. Soc. Neurosci.* **2014**, *34*, 9107–9123. [CrossRef] [PubMed]

68. Bernardinelli, Y.; Randall, J.; Janett, E.; Nikonenko, I.; König, S.; Jones, E.V.; Flores, C.E.; Murai, K.K.; Bochet, C.G.; Holtmaat, A.; et al. Activity-dependent structural plasticity of perisynaptic astrocytic domains promotes excitatory synapse stability. *Curr. Biol.* **2014**, *24*, 1679–1688. [CrossRef] [PubMed]

69. Cheung, G.; Sibille, J.; Zapata, J.; Rouach, N. Activity-Dependent Plasticity of Astroglial Potassium and Glutamate Clearance. *Neural Plast.* **2015**, *2015*, 109106. [CrossRef] [PubMed]

70. Pirttimaki, T.M.; Parri, H.R. Astrocyte plasticity: Implications for synaptic and neuronal activity. *Neurosci. Rev. J. Bring. Neurobiol. Neurol. Psychiatry* **2013**, *19*, 604–615. [CrossRef] [PubMed]

71. Otmakhov, N.; Khibnik, L.; Otmakhova, N.; Carpenter, S.; Riahi, S.; Asrican, B.; Lisman, J. Forskolin-induced LTP in the CA1 hippocampal region is NMDA receptor dependent. *J. Neurophysiol.* **2004**, *91*, 1955–1962. [CrossRef] [PubMed]

72. Oh, M.C.; Derkach, V.A.; Guire, E.S.; Soderling, T.R. Extrasynaptic membrane trafficking regulated by GluR1 serine 845 phosphorylation primes AMPA receptors for long-term potentiation. *J. Biol. Chem.* **2006**, *281*, 752–758. [CrossRef] [PubMed]

73. Lu, Y.; Jen, P.H. GABAergic and glycinergic neural inhibition in excitatory frequency tuning of bat inferior collicular neurons. *Exp. Brain Res.* **2001**, *141*, 331–339. [CrossRef] [PubMed]

74. Lu, W.; Man, H.; Ju, W.; Trimble, W.S.; MacDonald, J.F.; Wang, Y.T. Activation of synaptic NMDA receptors induces membrane insertion of new AMPA receptors and LTP in cultured hippocampal neurons. *Neuron* **2001**, *29*, 243–254. [CrossRef]

75. Behnisch, T.; Yuanxiang, P.; Bethge, P.; Parvez, S.; Chen, Y.; Yu, J.; Karpova, A.; Frey, J.U.; Mikhaylova, M.; Kreutz, M.R. Nuclear translocation of jacob in hippocampal neurons after stimuli inducing long-term potentiation but not long-term depression. *PLoS ONE* **2011**, *6*, e17276. [CrossRef]

76. Mizui, T.; Sekino, Y.; Yamazaki, H.; Ishizuka, Y.; Takahashi, H.; Kojima, N.; Kojima, M.; Shirao, T. Myosin II ATPase activity mediates the long-term potentiation-induced exodus of stable F-actin bound by drebrin A from dendritic spines. *PLoS ONE* **2014**, *9*, e85367. [CrossRef] [PubMed]

77. Sokal, D.M.; Mason, R.; Parker, T.L. Multi-neuronal recordings reveal a differential effect of thapsigargin on bicuculline- or gabazine-induced epileptiform excitability in rat hippocampal neuronal networks. *Neuropharmacology* **2000**, *39*, 2408–2417. [CrossRef]

78. Oltedal, L.; Haglerød, C.; Furmanek, T.; Davanger, S. Vesicular release of glutamate from hippocampal neurons in culture: An immunocytochemical assay. *Exp. Brain Res.* **2008**, *184*, 479–492. [CrossRef] [PubMed]

79. Porter, J.T.; McCarthy, K.D. Hippocampal astrocytes in situ respond to glutamate released from synaptic terminals. *J. Neurosci. Off. J. Soc. Neurosci.* **1996**, *16*, 5073–5081. [CrossRef]

80. Sun, W.; McConnell, E.; Pare, J.-F.; Xu, Q.; Chen, M.; Peng, W.; Lovatt, D.; Han, X.; Smith, Y.; Nedergaard, M. Glutamate-dependent neuroglial calcium signaling differs between young and adult brain. *Science* **2013**, *339*, 197–200. [CrossRef] [PubMed]

81. Ruffinatti, F.; Tapella, L.; Gregnanin, I.; Stevano, A.; Chiorino, G.; Canonico, P.L.; Distasi, C.; Genazzani, A.A.; Lim, D. Transcriptional remodeling in primary hippocampal astrocytes from an Alzheimer's disease mouse model. *Curr. Alzheimer Res.* **2018**. [CrossRef] [PubMed]

82. Lim, D.; Bertoli, A.; Sorgato, M.C.; Moccia, F. Generation and usage of aequorin lentiviral vectors for Ca^{2+} measurement in sub-cellular compartments of hard-to-transfect cells. *Cell Calcium* **2016**, *59*, 228–239. [CrossRef] [PubMed]

International Journal of
Molecular Sciences

MDPI

Article

Stroke-Like Episodes and Cerebellar Syndrome in Phosphomannomutase Deficiency (PMM2-CDG): Evidence for Hypoglycosylation-Driven Channelopathy

Mercè Izquierdo-Serra [1,†], Antonio F. Martínez-Monseny [2,†], Laura López [3],
Julia Carrillo-García [1], Albert Edo [1], Juan Darío Ortigoza-Escobar [4,5], Óscar García [6],
Ramón Cancho-Candela [7], M Llanos Carrasco-Marina [8], Luis G. Gutiérrez-Solana [3],
Daniel Cuadras [9], Jordi Muchart [4,5], Raquel Montero [4,5], Rafael Artuch [4,5], Celia Pérez-Cerdá [10],
Belén Pérez [10], Belén Pérez-Dueñas [4,5], Alfons Macaya [11], José M. Fernández-Fernández [1,*] and
Mercedes Serrano [2,4,5,*]

[1] Laboratori de Fisiologia Molecular, Departament de Ciències Experimentals i de la Salut, Universitat
    Pompeu Fabra, 08003 Barcelona, Spain; merce.izquierdo@upf.edu (M.I.-S.); julia.carrillo@upf.edu (J.C.-G.);
    albert.edo01@estudiant.upf.edu (A.E.)
[2] Genetic Medicine and Rare Diseases Pediatric Institute, Hospital Sant Joan de Déu, 08002 Barcelona, Spain;
    afmartinez@hsjdbcn.org
[3] Unit of Child Neurology, Department of Pediatrics, Hospital Infantil Universitario Niño Jesús de Madrid,
    28009 Madrid, Spain; laural.marin@hotmail.com (L.L.); lgutierrez.hnjs@salud.madrid.org (L.G.G.-S.)
[4] Neuropediatric, Radiology and Clinical Biochemistry Departments, Hospital Sant Joan de Déu,
    08002 Barcelona, Spain; jortigoza@sjdhospitalbarcelona.org (J.D.O.-E.); jmuchart@hsjdbcn.org (J.M.);
    rmontero@hsjdbcn.org (R.M.); rartuch@hsjdbcn.org (R.A.); bperez@hsjdbcn.org (B.P.-D.)
[5] U-703 Centre for Biomedical Research on Rare Diseases (CIBER-ER), Instituto de Salud Carlos III,
    08002 Barcelona, Spain
[6] Pediatric Department, Hospital Virgen de la Salud, 45004 Toledo, Spain; oscargarcam@hotmail.com
[7] Pediatric Neurology Unit, Pediatrics Department, Hospital Universitario Rio Hortega, 47012 Valladolid,
    Spain; rcanchoc@saludcastillayleon.es
[8] Neuropediatric Department, Pediatric Service, Hospital Universitario Severo Ochoa, Leganés, 28009 Madrid,
    Spain; maria.llanos@salud.madrid.org
[9] Statistics Department, Fundació Sant Joan de Déu, 08002 Barcelona, Spain; dcuadras@fsjd.org
[10] Centro de Diagnóstico de Enfermedades Moleculares (CEDEM), Universidad Autónoma de Madrid (UAM),
    U-746 Centre for Biomedical Research on Rare Diseases (CIBER-ER) Madrid, Instituto de Salud Carlos III,
    IdiPAZ, 28009 Madrid, Spain; cpcerda@cbm.csic.es (C.P.-C.); bperez@cbm.csic.es (B.P.)
[11] Grup de Recerca en Neurologia Pediàtrica, Institut de Recerca Vall d'Hebron, Universitat Autònoma de
    Barcelona, Secció de Neurologia Pediàtrica, Hospital Universitari Vall d'Hebron, 08002 Barcelona, Spain;
    amacaya@vhebron.net
*   Correspondence: jmanuel.fernandez@upf.edu (J.M.F.-F.); mserrano@hsjdbcn.org (M.S.);
    Tel.: +34-93-316-0854 (J.M.F.-F.); +34-93-253-2100 (M.S.);
    Fax: +34-93-316-0901 (J.M.F.-F.); +34-93-280-3626 (M.S.)
†   Both authors contributed equally to this work.

Received: 10 January 2018; Accepted: 18 February 2018; Published: 22 February 2018

Abstract: Stroke-like episodes (SLE) occur in phosphomannomutase deficiency (PMM2-CDG), and may complicate the course of channelopathies related to Familial Hemiplegic Migraine (FHM) caused by mutations in *CACNA1A* (encoding $Ca_V2.1$ channel). The underlying pathomechanisms are unknown. We analyze clinical variables to detect risk factors for SLE in a series of 43 PMM2-CDG patients. We explore the hypothesis of abnormal $Ca_V2.1$ function due to aberrant *N*-glycosylation as a potential novel pathomechanism of SLE and ataxia in PMM2-CDG by using whole-cell patch-clamp, *N*-glycosylation blockade and mutagenesis. Nine SLE were identified. Neuroimages showed no signs of stroke. Comparison of characteristics between SLE positive versus negative patients' group showed no differences. Acute and chronic phenotypes of patients with PMM2-CDG or *CACNA1A* channelopathies show similarities. Hypoglycosylation of both Ca$_V$2.1 subunits (α_{1A} and $\alpha_{2\alpha}$) induced gain-of-function effects on channel gating that mirrored those reported for pathogenic *CACNA1A* mutations linked to FHM and ataxia. Unoccupied *N*-glycosylation site N283 at α_{1A} contributes to a gain-of-function by lessening Ca$_V$2.1 inactivation. Hypoglycosylation of the $\alpha_2\delta$ subunit also participates in the gain-of-function effect by promoting voltage-dependent opening of the Ca$_V$2.1 channel. Ca$_V$2.1 hypoglycosylation may cause ataxia and SLEs in PMM2-CDG patients. Aberrant Ca$_V$2.1 *N*-glycosylation as a novel pathomechanism in PMM2-CDG opens new therapeutic possibilities.

Keywords: ataxia; cerebellum; congenital disorders of glycosylation; magentic resonance Imaging (MRI); stroke-like; Ca$_V$2.1 voltage-gated calcium channel

---

1. Introduction

Phosphomannomutase deficiency (PMM2-CDG), caused by mutations in *PMM2* (*601785OMIM), is the most frequent congenital disorder of *N*-linked glycosylation [1,2]. Patients with PMM2-CDG develop cerebellar atrophy and a cerebellar syndrome (axial and limb ataxia, abnormal eye movements, dysarthria, cognitive impairment, and long-term disability) [3,4].

Stroke-like episodes (SLE) are among the acute neurological complications that may occur in PMM2-CDG patients. SLE are typically triggered by infections and have been described in about 20–55% of patients with PMM2-CDG [5–8]. The term SLE was initially coined for MELAS (Mitochondrial myopathy, Encephalopathy, Lactic acidosis, and Strokelike episodes) syndrome to stress the non-ischemic origin of their events [9], but it may be used for different neurological diseases that are associated with focal deficits that mimic clinically, but not neuroradiologically, an ischemic injury.

SLE in PMM2-CDG patients are characterized by confusional status, mono- or hemiparesis, and sometimes, epileptic seizures. The underlying pathomechanisms are not fully understood, and clinical guidelines helping their diagnosis, and prevention and treatment are missing, resulting in a very stressful situation for families and doctors.

Although coagulation abnormalities are very frequent in PMM2-CDG, no significant correlations have been found between the presence of thrombophilia and the occurrence of SLE [5–8]. Besides, the majority of patients with PMM2-CDG and SLE did not show vascular occlusions on magnetic resonance angiography, the affected brain lesions did not reveal restricted diffusion, or did not follow well-defined vascular territories on MRI (magnetic resonance imaging) [5–11].

SLE may also complicate the course of other neurological diseases such as channelopathies related to Familial Hemiplegic Migraine (FHM), a paroxysmal neurological disease caused by mutations in *CACNA1A* (encoding the neuronal pore-forming Ca$_V$2.1 channel α_{1A} subunit), *ATP1A2* (encoding the Na$^+$, K$^+$-ATPase pump α_2 subunit) or *SCN1A* (encoding the neuronal Na$_V$1.1 channel α_1 subunit). As for PMM2-CDG, the pathogenesis of SLE in channelopathies is still not fully understood [12,13]. Interestingly, some clinical, neuroimaging and neurophysiological features of PMM2-CDG patients and *CACNA1A* mutated patients are similar [13], including not only SLE, but also ataxia, ocular motor disturbances, and cerebellar atrophy on MRI [14].

Ion channels are highly *N*-glycosylated and we are not aware of studies on the effect of an abnormal glycosylation of the α_{1A} channel subunit on Ca$_V$2.1 functional expression and gating. As other high-voltage activated (HVA) Ca^{2+} channels, Ca$_V$2.1 also contains at least the regulatory $\alpha_2\delta$ and β subunits [15]. Impaired glycosylation of the auxiliary $\alpha_2\delta$ subunit reduces the functional expression of different HVA channels (including Ca$_V$2.1) in the plasma membrane [16–18], but its relevance (if any) on Ca$_V$2.1 channel gating is unknown. Moreover, increased or decreased Ca$_V$2.1-mediated Ca^{2+} influx has been related to different clinical features. Thus, FHM relates to gain-of-function mutations on *CACNA1A*, while ataxia is associated with the alteration of a narrow window of Ca^{2+} homeostasis in Purkinje cells due to either *CACNA1A* loss- or gain-of-function mutations [19].

We report on 9 SLE in a series of 43 PMM2-CDG patients and analyze different clinical, laboratory, and neuroimaging variables to detect potential SLE risk factors. We demonstrate the clinical similarities between the acute and chronic phenotypes of patients with PMM2-CDG or patients carrying *CACNA1A* mutations, establishing a novel link between both diseases. With those evident clinical similarities and knowing that gain-of-function *CACNA1A* mutations can lead to both paroxysmal neurological symptoms and cerebellar symptoms including cerebellar atrophy (such as progressive or congenital ataxia) [19], we explore the hypothesis of increased $Ca_V2.1$ activity due to deficient *N*-glycosylation as an underlying biological reason, offering a potential novel pathomechanism of SLE and cerebellar syndrome in PMM2-CDG.

2. Results

Forty-three children and young adults with PMM2-CDG were included. Seven patients had nine SLE (mean age 7.5 ± 4.8 years, range 3.3–15.0 years) representing 17.9% of the patients (SLE positive group). Using our SLE definition, two clinical episodes were excluded: one caused by a true cerebrovascular event as confirmed by the MRI, and one caused by an epileptic seizure occurring at the onset of an infection process of the brain.

Thirty-two PMM2-CDG patients (mean age 16.2 ± 9.6, range 4.2–38 years) did not have SLE and were used as PMM2-CDG control group (SLE negative group).

2.1. Stroke-Like Episodes

Table 1 summarizes the age at the SLE, molecular studies, clinical presentation and follow-up, neurophysiological and neuroimaging findings of all patients.

In six SLE, a mild-to-moderate head trauma was identified as a potential trigger. SLE-related neurological symptoms started after a symptom-free interval ranging between 1 and 24 h. For three patients suffering from infections it was not possible to exactly quantify the symptom-free interval. Three out of nine episodes were characterized by focal neurological deficits and three patients presented with clinical seizures. All patients had abnormal consciousness (Supplementary Video 1).

Almost all patients developed hyperthermia at presentation leading to a laboratory work-up for a possible underlying infection or inflammatory process (Table 1). Acute phase proteins and complete blood counts were normal for all patients with SLE after head trauma.

In seven SLE a brain MRI was performed during the first 72 h. Neuroimaging showed the preexisting cerebellar atrophy with absence of new findings in all but Patient 1 (Table 1), who showed diffuse cortical swelling and FLAIR-hyperintense signal in the right hemisphere with no restricted diffusion, suggestive of vasogenic edema (Figure 1).

EEG revealed normal findings in six out of nine studies (Table 1). On the second day after SLE onset, the EEG of Patient 1 showed an asymmetric (right) slow background activity with moderately low voltage in temporal regions, and frontal intermittent rhythmic delta activity (FIRDA) in the right hemisphere (Figure 1).

Antiepileptic drugs were used in six of nine SLE with two aims: to treat irritability using midazolam (two out of four cases), and to control or prevent epileptic seizures using levetiracetam, midazolam, valproic acid, phenytoin, and diazepam (Table 1). Neither anticoagulants nor antiaggregants were used in eight of nine episodes, as there was no evidence of a procoagulant situation. Patient 4 developed a deep venous thrombosis during her second SLE. Her coagulation tests revealed abnormal findings that required fibrinogen substitution.

Table 1. Molecular, clinical, electrophysiological and neuroimaging details from stroke-like episodes.

Patient/ Episode	Sex/ Age	Molecular Findings	ICARS	Trigger	Free Symptom Period	Initial Clinical Presentation	Body Temperature (°C)	C-reactive Protein/PCT	EEG	Neuroimaging (<72 h)	Treatment	Duration of Symptoms (h)	Recovery
1	M 3 yr 11 mo	p.V44A p.Q33stop	NA	Head trauma	3 h	Lethargy, epileptic seizures, left hemiparesis	38.5	Normal	Asymmetric slow background, right hemisphere FIRDA epileptic activity	MRI: Asymmetric hemispheric vasogenic edema	LEV, PHE, MDZ	96	Tonic upgaze some days after discharge. Left hemiparesis until 4 months later
2	M 15 yr	p.V44A p.R123G	NA	Head trauma	24 h	Somnolence, vomiting	37.4	No data available	Normal	MRI: No significant changes	No	48	Complete recovery
3	F 5 yr	p.T237M IVS7-9T > G	46	Head trauma	4 h	Somnolence, vomiting	37.5	Normal	Normal	No significant changes	No	7	Complete recovery
4-I	F 3 yr 6 mo	p.L32R	8	Head trauma	1 h	Irritability and lethargy	38.0	Normal	Normal	No significant changes	No	3	Complete recovery
4-II	F 14 yr 5 mo	IVS-1G > C		Enterovirus	NA	Dysarthria, dysphasia, irritability, left hemiparesis	38.0	Normal	Normal	CT angiography: No significant changes	LEV	36	Complete recovery
5-I	F 7 yr 4 mo	p.D65Y	56	Head trauma	12 h	Irritability, lethargy, aphasia and dysphagia	38.6	No data available	Normal	MRI: No significant changes	Risperidone	168	Complete recovery
5-II	F 10 yr 2 mo	IVS7-9T > G		Head trauma	2 h	Irritability and lethargy	38.5	Normal	Normal	MRI: No significant changes	MDZ	24	Complete recovery
6	F 3 yr 3 mo	p.P113L p.F207S	NA	Influenza virus infection	NA	Epileptic seizures, irritability and lethargy	38.9	38.0 mg/L 1.04 ng/mL	Asymmetric slow background, epileptic activity	CT: No significant changes	LEV, PHE, MDZ	264	Tonic upgaze deviation and irritability until 1 month later. Fully recovered 7 months later.
7	M 5 yr 8 mo	p.D65Y p.R141H	65	Upper respiratory viral infection	NA	Lethargy, epileptic seizures, left arm monoparesis	38.5	11.4 mg/L 24.7 ng/mL	Asymmetric slow background, no epileptic activity	MRI: No significant changes	VPA, LEV, DZP, MDZ	144	Distal weakness left arm, fully recovered in 2 weeks
Mean values/SD	7 yr 7 mo/ 4 yr 10 mo			6 out of 9 Head trauma	13.0/8.4		38.2/0.5		3 out of 9 Abnormal EEG	1/9 Hemispheric edema		87.8 h/86.9 h	
Range	3 yr 3 mo/15 yr				3–24 h		37.4–38.9					3–264 h	

M: male; F: female; yr: years; mo: months; ICARS: International Cooperative Ataxia Rating Scale; NA: Not available, difficult to determine; PCT: procalcitonin; FIRDA: Frontal intermittent rhythmic delta activity; LEV: levetiracetam; PHE: phenytoin; MDZ: midazolam; VPA: valproic acid; DZP: diazepam.

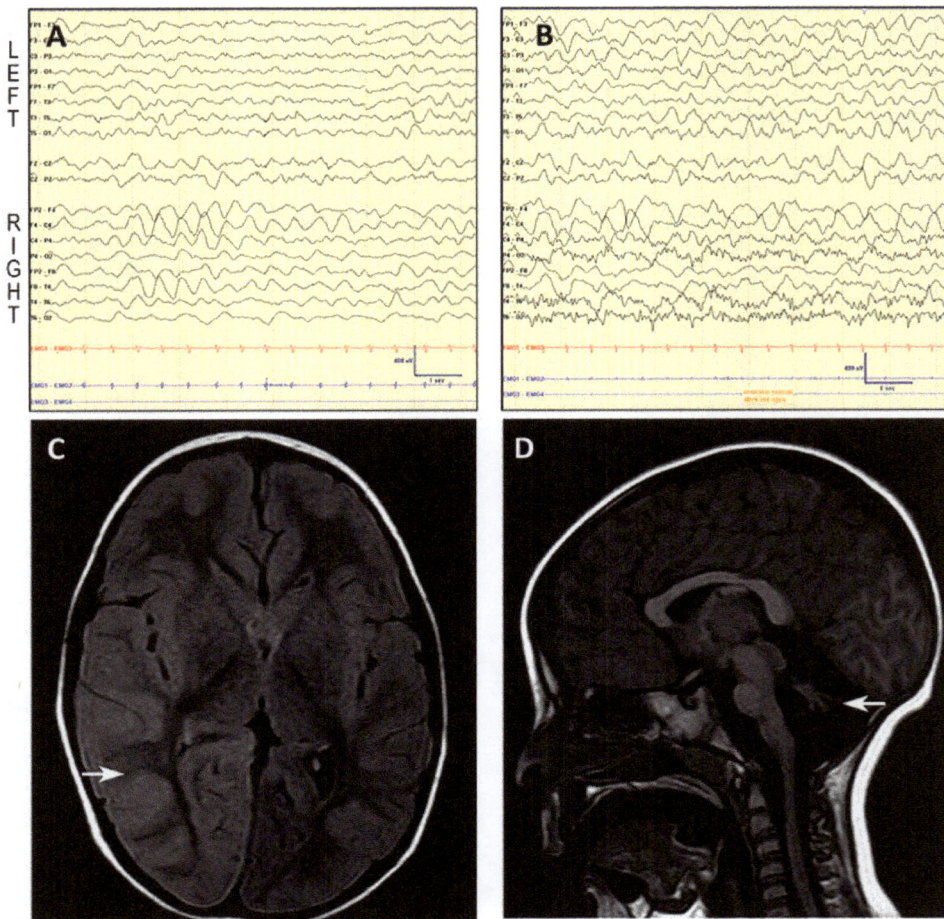

Figure 1. Patient 1: EEG and MRI. EEG shows asymmetric (right) slow background trace with moderately low voltage in temporal regions (**A**,**B**), and frontal intermittent rhythmic delta activity (FIRDA) in the right hemisphere, coherent with the edema found on the MRI and the left hemiparesis. Below, axial fluid attenuation inversion recovery (FLAIR) MR image reveals cortical diffuse edema in the right hemisphere, mainly in the parieto-occipital region (**C**, see arrow). Midsagittal T1-weighted MR image of the same patient shows a small cerebellar vermis with enlarged interfolial spaces representing atrophy and secondary enlargement of the fourth ventricle (**D**, see arrow).

Most of patients recovered completely within one week. Three patients, however, experienced a more delayed recovery (Table 1). After SLE onset, Patient 1 and Patient 6 developed episodic tonic upgaze that progressively decreased in frequency and disappeared three and five weeks after onset, respectively (Supplementary Video 2).

2.2. SLE Versus SLE Negative Group

Statistical analysis of clinical, laboratory, and neuroimaging variables showed no significant differences between PMM2-CDG patients with and without SLE. Epidemiological and clinical data are shown in Table 2.

Table 2. Comparison of clinical and epidemiological data between PMM2-CDG patients with and without SLE.

		SLE Positive (SD)	SLE Negative (SD)	*p*-Value
Number of patients/episodes		7/9	32	
Age (years)		13.7 (11.6) *	16.2 (9.6)	0.46
Sex (Male:Female)		3:4	21:11	0.19
Liver function	AST (UI/mL) Normal values 2–34 UI/mL	378 (317.3)	172 (235.5)	0.61
	ALT (UI/mL) Normal values 2–36 UI/mL	277 (220.6)	216 (327.1)	0.70
Coagulation	PT (%) Normal values 80–120%	95.4 (13.1)	95.9(24.0)	0.82
	aPTT (seconds) Normal values 23–35 seconds	33.3 (4.6)	31.5 (6.0)	0.53
	F IX (%) Normal values 50–120%	49.5 (14.9)	75.2 (25.0)	0.17
	F XI (%) Normal values 50–120%	35.9 (15.1)	59.3 (35.5)	0.28
	AT III (%) Normal values 60–120%	43.0 (14.1)	51.2 (26.8)	0.82
	Protein C (%) Normal values 60–140%	38.6 (17.2)	83.4 (129.9)	0.46
	Protein S (%) Normal values 60–140%	52.9 (22.6)	63.5 (17.5)	0.31
Vascular events	Positive personal history	1/7	3/32	0.51
	Positive familial history	1/7	2/32	0.41
Vermis Midsagittal Relative Diameter (MRI)		0.48	0.41	0.44
Seizures	Febrile seizures	2/7	4/32	0.23
	Epilepsy	2/7	6/32	0.58

* Age at the time of the present study; SD: Standard deviation; AST: aspartate transaminase; ALT: Alanine transaminase; PT: prothrombin time; aPTT: activated partial thromboplastin time; F IX: coagulation factor IX; F XI: coagulation factor XI; ATIII: antithrombin III. Included values for the statistical studies are for the SLE group, values during the episode, and for those "SLE negative" patients, most abnormal values among all the different liver function tests (maximal AST and ALT) and coagulation analysis (lowest coagulation factor activities).

2.3. Literature Review Comparing Clinical Presentations between Patients with PMM2-CDG and Patients with CACNA1A Mutations

Table 3 compares the acute and chronic clinical presentations that have been reported in patients with pathogenic variants in *CACNA1A* and *PMM2*, respectively. The similarities in acute and chronic neurological presentations suggest a possible common underlying pathomechanism for SLE.

Table 3. Findings in patients carrying mutations in *CACNA1A* compared to findings in PMM2-CDG patients, including our cohort.

Gene	CACNA1A	PMM2
Acute: stroke-like episodes		
Prevalence	21% [8], >70 cases reported [14].	18 to 40–50%, >49 cases reported [20].
Age at onset	At any age between 1 to 73 years old [21].	Any age, frequently between 3 to 6 years old [5,8,11].
Trigger/Onset of encephalopathy after trigger	Minor head trauma [22], infections, diagnostic procedures (cerebral or coronary angiography), physical activity [21]; Immediately or 2–3 h after the trigger [14].	Infections, head trauma, angiography, alcohol ingestion [5]; Minutes to hours after the trigger.
Prodromi	Severe headache, yawning, truncal unsteadiness [22].	None
Focal deficits	Hemiparesis, dysphasia, nystagmus, dyskinetic limb movement [12–14].	Hemiparesis, dysphasia, dysphagia, conjugated eye deviation, blindness [5,8,11].
Epilepsy/EEG findings	Hemi-clonic or hemi-tonic convulsions/Globally slow background activity, spikes, spike-and-slow wave complexes, slow wave bursts, photoparoxysmal response [14].	Clonic convulsions/Low voltage pattern in affected areas, asymmetric slow background activity with moderately low frontal intermittent rhythmic delta activity (FIRDA) in the affected hemisphere [8].
Autonomic signs	High fever, recurrent vomiting [14].	High fever, recurrent vomiting [11].
MRI findings during the episode	Normal, ischemic lesion with prominent perifocal edema, or panhemispheric edema [14,23].	Normal or asymmetric hemispheric cytotoxic edema [8,11].
Duration/number of episodes	Few minutes to 10 days/1–11 episodes [14,21].	1h to several months/1–2 episodes [5,11].
Recovery	24 h to months. Complete recovery in the majority of the patients, but residual hemiplegia possible [14,22].	1h to several months. Complete recovery in the majority of the patients, but exceptionally residual motor symptoms may persist [5].
Treatment	Analgesia, antiepileptic drugs, sleep, acetazolamide, verapamil [13,22–24].	IV hydration, antiepileptic drugs [5].
Chronic: time between stroke-like episodes		
Ataxia	Episodic attacks of cerebellar dysfunction lasting from 5 minutes to 5 h [23].	Persistent ataxia [1,2,4,5,11].
Abnormal eye movement	Saccadic eye movements, downgaze nystagmus, strabismus, tonic upgaze episodes [22].	Strabismus, nystagmus, tonic upgaze episodes [4,20,25].
Other neurologic manifestations	Developmental disability, dysarthria, migraine with aura, motor stereotypies, benign paroxysmal torticollis [14,21].	Developmental disability, hypotonia, dysarthria [4,20,25].
MRI	Cerebellar vermis atrophy, mild delay in white matter myelination [22].	Atrophy of the cerebellar hemispheres and vermis [25], pontine hypoplasia [11].

2.4. Deficient N-Glycosylation Alters $Ca_V2.1$ Functional Expression, Activation, and Inactivation

Whole-cell patch-clamp recordings obtained from HEK293 cells expressing $Ca_V2.1$ channels composed of α_{1A}, β_3 and $\alpha_2\delta_1$ subunits, showed that tunicamycin treatment (which specifically blocks the first step of N-glycosylation by inhibiting N-acetylglucosamine transferase in the endoplasmic reticulum, Figure 2), reduced $Ca_V2.1$ Ca^{2+} currents in a concentration-dependent manner.

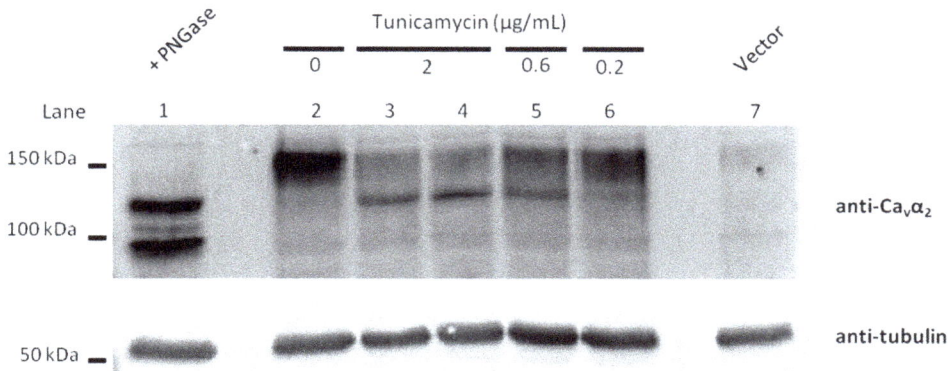

Figure 2. Hypoglycosylation levels of the Ca_V regulatory $\alpha_2\delta$ subunits under different concentrations of tumicamycin. Immunoblot analysis of glycosylated fragments of $\alpha_2\delta$ subunits heterologously expressed in HEK293 cells (along with the α_{1A} $Ca_V2.1$ pore-forming channel subunit and β_3 regulatory subunit), treated or not with increasing concentrations of tunicamycin (0.2, 0.6 and 2 μg/mL, as indicated). The extent of $\alpha_2\delta$ glycosylation was identified by using antibody anti-α_2 (1:500 dilution, Sigma D219, St. Louis, MO, USA) (top panels) and is shown as the difference in molecular weight between the glycosylated and unglycosylated forms. Molecular weight markers are indicated on the left. PNGase F was also added to protein extraction from tunicamycin-untreated cells in order to identify unglycosylated forms in vitro. Upper bands potentially corresponding to glycosylated $\alpha_2\delta$ progressively decrease as the concentration of tunicamycin increases, which in turn promotes the appearance of a lower band potentially corresponding to unglycosylated $\alpha_2\delta$ (lanes 2 to 6). In vitro PNGase F treatment (lane 1) shows two main bands potentially corresponding to unglycosylated $\alpha_2\delta$ (upper band) and unglycosylated α_2 (lower band), as previously reported [26].Protein extraction from HEK293 cells transfected only with the vector plasmid was included as negative control (lane 7). For each experimental condition, the protein sample was probed with anti-tubulin (1:2000 dilution, Sigma T6074) (bottom panel) as loading control.

Thus, 2 μg/mL tunicamycin almost abolished $Ca_V2.1$ Ca^{2+} currents (Figure 3A right panel, and B) compared to cells treated with the vehicle (DMSO) (Figure 3A left panel, and B). Such tunicamycin effect was mimicked by removal of the regulatory $\alpha_2\delta_1$ subunit (Figure 3B). Under these experimental conditions, maximal Ca^{2+} current density was reduced from -149.8 ± 14.9 pA/pF ($n = 27$) to -1.9 ± 1.8 pA/pF by tunicamycin ($n = 6$) ($p < 0.001$) and to -7.8 ± 2.9 pA/pF in the absence of the $\alpha_2\delta_1$ subunit ($n = 7$) ($p < 0.001$) (Kruskal-Wallis test followed by Dunn post hoc test) (Figure 3B).

At lower concentrations, a smaller decrease of $\alpha_2\delta_1$ glycosylation by tunicamycin was found (Figure 2) and the effect on the amplitude of $Ca_V2.1$ Ca^{2+} currents was less important, with reductions in peak current density to -107.03 ± 25.1 pA/pF ($n = 13$) ($p > 0.05$) and to -26.7 ± 7.9 pA/pF ($n = 16$) ($p < 0.001$) induced by 0.2 and 0.6 μg/mL tunicamycin, respectively (Kruskal-Wallis test followed by Dunn post hoc test) (Figure 4A,B). The $Ca_V2.1$ voltage-dependent activation curve was left-shifted and the potential for half-maximal channel activation ($V_{1/2}$ activation, directly related to the energy necessary to open the channel) was significantly decreased in response to both 0.2 and 0.6 μg/mL tunicamycin treatments (by ~3.5 mV and ~5 mV, respectively) (Figure 4C,D). Consistently, the maximum current amplitude was elicited by depolarizing pulses to +15 mV or +10 mV for control (vehicle) or tunicamycin treatments, respectively (Figure 4B,C).

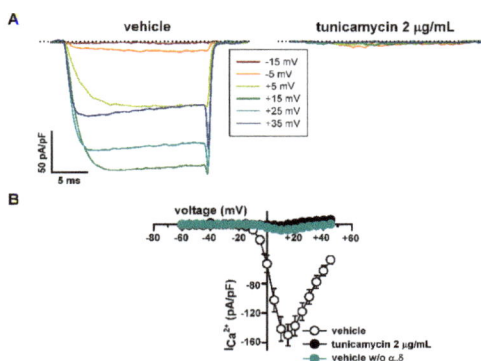

Figure 3. Inhibition of *N*-glycosylation strongly reduces functional expression of Ca$_V$2.1 channels containing β$_3$ and α$_2$δ$_1$ subunits, heterologously expressed in HEK293 cells. (**A**) Representative current traces elicited by 20 ms depolarizing pulses from −80 mV to the indicated voltages (inset), illustrating the significant (see Results) reduction in Ca^{2+} current amplitude through Ca$_V$2.1 channels induced by tunicamycin (2 μg/mL) versus vehicle (DMSO) treatment. The zero current level is indicated by dotted lines. (**B**) Average Ca^{2+} current density-voltage relationships for DMSO-treated cells in the presence (open circles, *n* = 27) and in the absence (*w/o*) (filled cyan circles, *n* = 7) of the α$_2$δ$_1$ subunit, and for tunicamycin-treated cells (filled black circles, *n* = 6). No significant differences on the maximal current density between cells treated with tunicamycin and cells lacking α$_2$δ$_1$ were observed (Kruskal-Wallis test followed by Dunn post hoc test).

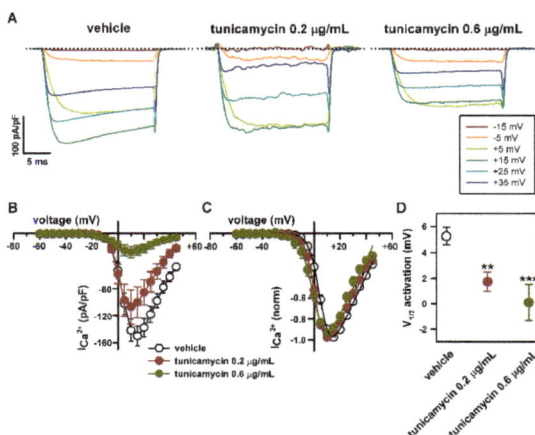

Figure 4. Reduction in the levels of *N*-glycosylation favors voltage-dependent activation of Ca$_V$2.1 channels heterologously expressed in HEK293 cells. (**A**) Current traces elicited by 20 ms depolarizing pulses from −80 mV to the indicated voltages (inset), illustrating the shift of Ca$_V$2.1 activation to lower depolarization, induced by 0.2 and 0.6 μg/mL tunicamycin. Dotted lines indicate the zero current level. Average Ca^{2+} current density-voltage relationships (**B**) and I-V curves normalized to peak Ca^{2+} current density (**C**) for Ca$_V$2.1 channels expressed in DMSO-treated cells (open circles, *n* = 27) and in cells treated with 0.2 μg/mL (filled red circles, *n* = 13) or 0.6 μg/mL tunicamycin (filled green circles, *n* = 16). (**D**) Reduction in V$_{1/2}$ for Ca$_V$2.1 channel activation (estimated from normalized I-V curves, as indicated in Materials and Methods) induced by 0.2 (*n* = 13) and 0.6 μg/mL (*n* = 16) tunicamycin. ** *p* < 0.01 and *** *p* < 0.001 versus the control condition (vehicle, *n* = 27; Kruskal-Wallis test followed by Dunn *post hoc* test).

Tunicamycin (0.6 μg/mL) also impaired the inactivation of $Ca_V2.1$ Ca^{2+} currents: compared to the control (vehicle) condition, the degree of inactivation at the end of 3s depolarizing pulses to +20 (Figure 5A,B) or 0 mV (Figure 5D,E) was lower (by ~14% and ~16%, respectively); and there was a tendency for slow $Ca_V2.1$ inactivation, which reached statistical significance for Ca^{2+} currents elicited by depolarization to 0 mV (Figure 5A,C,D,F).

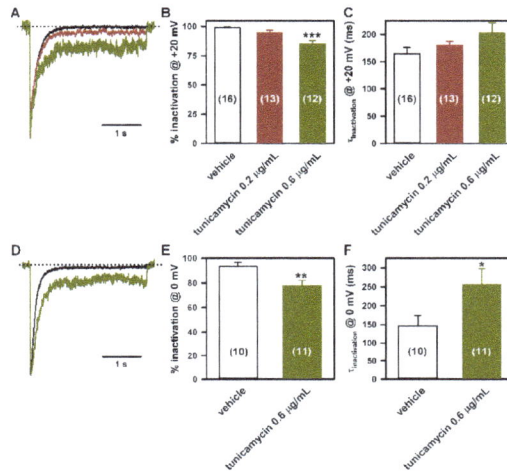

Figure 5. Inactivation of $Ca_V2.1$ channels heterologously expressed in HEK293 cells is impaired by lowering *N*-glycosylation. Representative current traces (normalized to the corresponding peak amplitude) illustrating the effects of 0.2 (red trace) and 0.6 μg/mL (green traces) tunicamycin on $Ca_V2.1$ inactivation when compared to DMSO-treated cells (black traces), in response to a 3 s depolarizing pulse to +20 mV (**A**) or 0 mV (**D**). The zero current level is indicated by dotted lines. (**B,E**) Average Ca^{2+} current inactivation (in %) at the end of these 3 s depolarizing pulses obtained from HEK293 cells expressing $Ca_V2.1$ channels and treated with DMSO (vehicle) or with tunicamycin (0.2 and 0.6 μg/mL). Data are expressed as the mean ± SEM of the number of experiments shown in brackets. *** $p < 0.001$ (Kruskal-Wallis test followed by Dunn post hoc test) and ** $p < 0.01$ (Student's *t*-test) versus the control (vehicle) condition. (**C,F**) Average τ inactivation values of Ca^{2+} currents through $Ca_V2.1$ channels expressed in HEK293 cells treated with DMSO (vehicle) or with tunicamycin (0.2 and 0.6 μg/mL), elicited by a 3 s depolarizing pulse to +20 mV or 0 mV, as indicated. Data are expressed as the mean ± SEM of the number of experiments shown in brackets. * $p < 0.05$ versus the control (vehicle) condition (Mann-Whitney *U*-test).

2.5. Mutation of α_{1A} Potential Glycosylation Site Mimics Tunicamycin Effect on $Ca_V2.1$ Inactivation

To evaluate whether abnormal glycosylation of the α_{1A} channel subunit may contribute to the effect produced by tunicamycin on $Ca_V2.1$ biophysical properties, we first searched for potential *N*-glycosylation sites in the sequence of the human α_{1A} subunit (GenBank No. O00555) by using Uniprot Knowledgebase (www.uniprot.org, access on 30 November 2017). One single glycosylation site at residue N283, located in the extracellular P-loop region at domain I (DI) of α_{1A}, was identified (Figure 6A). This asparagine residue is highly conserved through evolution (Figure 6B, top). In addition, potential glycosylation of a single N residue at P-loop-DI is a trait shared by all human high-voltage activated (HVA) Ca^{2+} channels (Figure 6B, bottom).

A

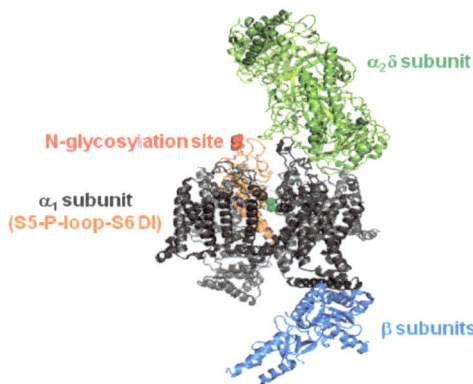

B

```
Human Ca_v2.1 (NP_001167551.1)    249 MGKFHTTCFEEGTDDIQGESPAPCGTEEPARTCPNGTKCQPYWEGPNNGITQFDNILFAVLTVFQCITMEGWTDLLYNSNDASGNTW 335
Bovin Ca_v2.1 (NP_001068597.1)    249 MGKFHTTCFEEGTDDIQGESPAPCGTEEPARTCPNGTKCQPYWEGPNNGITQFDNILFAVLTVFQCITMEGWTDLLYNSNDASGNTW 335
Rabbit Ca_v2.1 (P27884.1)         249 MGKFHTTCFEEGTDDIQGESPAPCGTEEPARTCPNGTRCQPYWEGPNNGITQFDNILFAVLTVFQCITMEGWTDLLYNSNDASGNTW 335
Mouse Ca_v2.1 (NP_037050.2)       251 MGKFHTTCFEEGTDDIQGESPAPCGTEEPARTCPNGTKCQPYWEGPNNGITQFDNILFAVLTVFQCITMEGWTDLLYNSNDASGNTW 337
Rat Ca_v2.1 (NP_031604.3)         251 MGKFHTTCFEEGTDDIQGESPAPCGTEEPARTCPNGTKCQPYWEGPNNGITQFDNILFAVLTVFQCITMEGWTDLLYNSNDASGNTW 337
Zebrafish Ca_v2.1 (XP_690548.5)   242 MGKFHTTCFDEITDEIREEF--PCGEEVPARICPNGTVCKKYWLGPNYGITQFDNILFAVLTVFQCITMEGWTDLLYYSNDAAGSAW 326
                                      *********::  **:*; *   *** * *** *****  *; ** *** ***************************** ****:*.:*
```

```
hCa_v2.1 (NP_001167551.1)  249 MGKFHTTCFEEGTDDIQ--GE-SPAPCGTE-EPARTCP-NGTKCQPYWEGPNNGITQFDNILFAVLTVFQCITMEGWTDLLYNSNDASGNTW 335
hCa_v2.2 (Q00975.1)        246 MGKFHKACFPNSTDA-E--PV-GDFPCCGKE-APARLCE-GDTECREYWPGPNFGITNFDNILFAILTVFQCITMEGWTDILYNTNDAAGNTW 331
hCa_v2.3 (Q15878.3)        245 -GKLHRACFMNNSGILE--GFDPPHPCGVQ-----GCP-AGYECKD-WIGPNDGITQFDNILFAVLTVFQCITMEGWTTVLYNTNDALGATW 326
hCa_v1.1 (Q13698.4)        219 KGKMHKTCYFIGTDIVATVENEEPSPCART-GSGRRCTINGSECRGGWPGPNHGITHFDNFGFSMLTVYQCITMEGWTDVLYWVNDAIGNEW 309
hCa_v1.2 (Q13936.4)        291 MGKMHKTCYNQEG-IADVPAEDDPSPCALETGHGRQCQ-NGTVCKPGWDGPKHGITNFDNFAFAMLTVFQCITMEGWTDVLYWVNDAVGRDW 380
hCa_v1.3 (Q01668.2)        294 IGKMHKTCFFAD---SDIVAEEDPAPCAFS-GNGRQCTANGTECRSGWVGPNGGITNFDNFAFAMLTVFQCITMEGWTDVLYWMNDAMGFEL 381
hCa_v1.4 (O60840.2)        260 LGRMHKTCYFLG---SDMEAEEDPSPCASS-GSGRACTLNQTECRGRWPGPNGGITNFDNFFFAMLTVFQCVTMEGWTDVLYWMQDAMGYEL 347
                               *::* :*:     **. *  *:  * **: ***:***; *::***;**:***** :** ;** *
```

Figure 6. Location and conservation of the N283 amino acid residue. (**A**) Illustration showing the location of the potential glycosylation site at residue N283 (in red) at the extracellular P-loop between S5 and S6 transmembrane segments in domain I of the pore forming α_{1A} subunit (in orange). The structure of the Ca$_V$1.1 complex, containing α_1 (in grey), β (in blue) and $\alpha_2\delta$ (in green) subunits, was used as model (PDB 5GJV). (**B**) Sequence alignment of P-loop regions at domain I (DI) of Ca$_V$2.1 channel α_{1A} subunits from different organisms (as indicated) (top), and sequence alignment of P-loop-DI α_1 regions of human high-voltage activated Ca^{2+} channels belonging to the Ca$_V$2.x and Ca$_V$1.x families (bottom). Alignments were performed with Clustal Omega (www.ebi.ac.uk/Tools/msa/clustalo/, access on 30 November 2017). Potential sites for *N*-glycosylation (including N283 at the human Ca$_V$2.1 channel) are shown in red. Asterisk means identical residues; colon and period indicates the existence of conservative and semi-conservative amino acid substitutions, respectively.

Hence, we compared the biophysical properties of Ca$_V$2.1 channels composed by wild-type (WT) or N283Q glycosylation mutant α_{1A}, β_3 and $\alpha_2\delta_1$ subunits, heterologously expressed in HEK293 cells. Peak Ca^{2+} current density for N283Q mutant Ca$_V$2.1 channels were ~72% lower than for WT channels (Figure 7A,B). N283Q did not alter Ca$_V$2.1 voltage-dependent activation (Figure 7C,D). Interestingly, tunicamycin (0.6 µg/mL) shifted $V_{1/2}$ activation for N283Q Ca$_V$2.1 channels to less depolarized potentials (by ~4.5 mV) (Figure 7E–F), in line with its gain-of-function effect on WT Ca$_V$2.1 activation (Figure 4A–D).

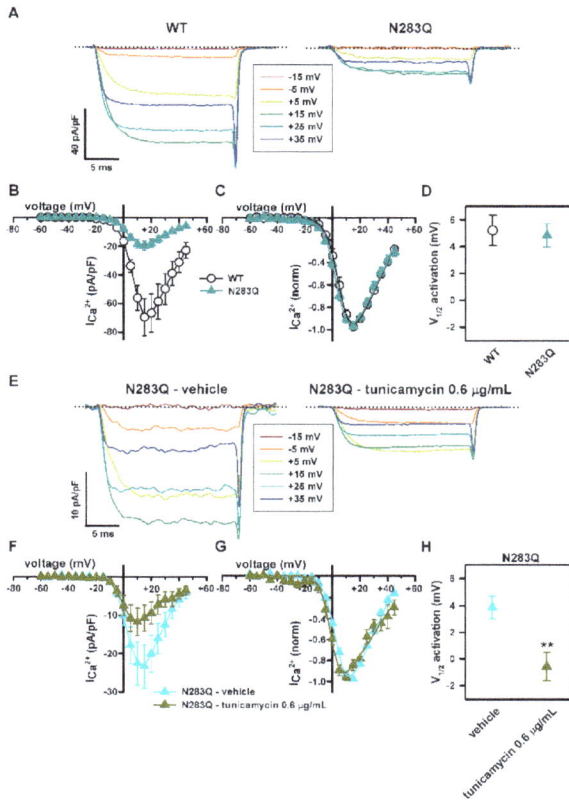

Figure 7. N283Q glycosylation site mutation reduces Ca^{2+} current density through $Ca_V2.1$ channels heterologously expressed in HEK293 cells, without altering their voltage-dependent activation. (**A**) Current traces elicited by 20 ms depolarizing pulses from -80 mV to the indicated voltages (inset) illustrating the decrease in Ca^{2+} current density through $Ca_V2.1$ channels containing the mutation at the α_{1A} glycosylation site N283, which exhibit voltage dependence of activation similar to that of WT channels. Dotted lines mark the zero current level. Average Ca^{2+} current density-voltage relationships (**B**) and normalized I-V curves (**C**) for WT (open circles, $n = 13$) and N283Q (filled dark cyan triangles, $n = 19$) $Ca_V2.1$ channels. N283Q reduces maximal Ca^{2+} current density (obtained by membrane depolarization to +15 mV) from -69.4 ± 12.9 pA/pF ($n = 13$) to -19.5 ± 3.2 pA/pF ($n = 19$) ($p < 0.001$, Mann-Whitney *U*-test). (**D**) N283Q mutation has no significant effect on the $V_{1/2}$ for $Ca_V2.1$ channel activation (estimated from normalized I-V curves shown in **C** as indicated in Materials and Methods, $p = 0.798$, Student's *t*-test). (**E**) Current traces elicited by 20 ms depolarizing pulses from -80 mV to the indicated voltages (inset) illustrating the shift of activation to lower depolarization induced by 0.6 μg/mL tunicamycin on N283Q mutant $Ca_V2.1$ channels containing β_3 and $\alpha_2\delta_1$ subunits. The zero current level is indicated by dotted lines. Average Ca^{2+} current density-voltage relationships (**F**) and normalized I-V curves (**G**) for N283Q mutant $Ca_V2.1$ channels expressed in DMSO-treated cells (filled cyan triangles, $n = 10$) and in cells treated with 0.6 μg/mL tunicamycin (filled green circles, $n = 8$). Peak Ca^{2+} current density through N283Q $Ca_V2.1$ channels after vehicle (DMSO) and tunicamycin treatments were -23.3 ± 5.6 pA/pF ($n = 10$) and -11.6 ± 3.6 pA/pF ($n = 8$), respectively ($p = 0.06$, Student's *t*-test). (**H**) Reduction in $V_{1/2}$ for activation of $Ca_V2.1$ channels composed by N283Q mutant α_{1A}, β_3 and $\alpha_2\delta_1$ subunits (estimated from normalized I-V curves shown in **G** as indicated in Materials and Methods) produced by 0.6 μg/mL tunicamycin ($n = 8$). ** $p < 0.01$ versus the control condition (vehicle, $n = 10$; Student's *t*-test).

Finally, Ca^{2+} currents through N283Q mutant Ca$_V$2.1 channels in response to 3 s depolarizing pulses to either +20 mV (Figure 8A–C) or 0 mV (Figure 8D–F), showed lower (~22% and ~35%, respectively) and slower inactivation than currents through WT channels.

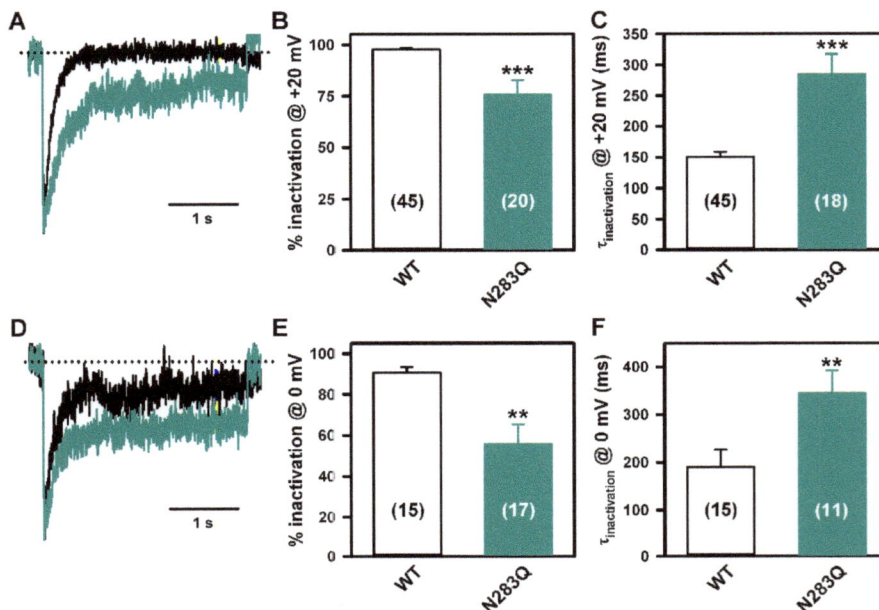

Figure 8. N283Q glycosylation site mutation lessens Ca$_V$2.1 channel inactivation. Ca^{2+} current traces (normalized to the corresponding peak amplitude) illustrating differential inactivation of WT (black traces) and N283Q (cyan traces) Ca$_V$2.1 channels, in response to a 3 s depolarizing pulse to +20 mV (**A**) or 0 mV (**D**). Dotted lines indicate the zero current level. (**B,E**) Average Ca^{2+} current inactivation (in %) at the end of these 3s depolarizing pulses obtained from HEK293 cells expressing either WT or N283Q Ca$_V$2.1 channels. Data are expressed as the mean ± SEM of the number of experiments shown in brackets (*** $p < 0.001$ and ** $p < 0.01$ versus WT, Mann-Whitney *U*-test). (**C,F**) Average τ inactivation values of Ca^{2+} currents through WT (open bars) and N283Q (cyan bars) Ca$_V$2.1 channels expressed in HEK293 cells, elicited by a 3 s depolarizing pulse to +20 mV or 0 mV, as indicated. Data are expressed as the mean ± SEM of the number of experiments shown in brackets (*** $p < 0.001$ and ** $p < 0.01$ versus WT, Mann-Whitney *U*-test).

3. Materials and Methods

3.1. Patients

Patients with a molecular diagnosis of PMM2-CDG were recruited from the hospitals collaborating in the Spanish PMM2-CDG Network, and were followed from May 2014 to December 2016.

We used the following SLE definition: "Acute event consisting of sudden onset of a focal neurological deficit, irritability or decreased consciousness that may associate with seizures, headache or other transient symptoms, in the absence of another diagnosis explaining these symptoms". The definition does not include neurophysiological and neuroimaging findings. Patients with "true" ischemic stroke on MRI were excluded.

We collected epidemiological and molecular data, potential triggers of SLE, clinical findings that occurred shortly before the episode, symptoms during the episode and the recovery process, and laboratory, neurophysiological (EEG), and neuroimaging (computed tomography (CT) and MRI) findings. Laboratory studies included blood cell counts, liver enzymes, acute phase reactants and coagulation parameters. Clinical severity of cerebellar syndrome was assessed through International Cooperative Ataxia Rating Scale (ICARS) [25].

To compare PMM2-CDG patients with SLE (SLE group) and without SLE (SLE negative group), we used epidemiological and molecular data, laboratory findings, personal history of epilepsy, personal or familial history of vascular events, and neuroimaging findings, using the midsagittal vermis relative diameter (MVRD) [25,27].

Biochemical studies were performed during the SLE. For the patients without SLE, the most abnormal liver function and coagulation laboratory values were included in the statistical analysis. For this comparison, four patients with early death in young childhood (before 2 years) were excluded, because of their severe systemic involvement.

Molecular studies had been carried out in all PMM2-CDG patients enrolled. Genetic analysis was performed in the "Centro de Diagnóstico de Enfermedades Moleculares" in Madrid. Total mRNA and genomic DNA were isolated from venous whole blood or patient-derived fibroblasts using a MagnaPure system following the manufacturer's protocol (Roche Applied Science, Indianapolis). Mutational analysis was performed by genomic DNA analysis both in patients' and parents' samples to assure that both changes are on different alleles and to rule out the presence of a large genomic rearrangement. In some cases, the effect on splicing was analyzed by cDNA profile analysis. The primers used for cDNA and genomic DNA amplifications were designed using the ENSEMBL database (http://www.ensembl.org/index.html, ENSG00000140650) and GenBank accession number NM_000303.2.

3.2. Literature Review

To compare the cerebellar syndrome and SLE in PMM2-CDG patients and the phenotype related to *CACNA1A* mutated patients, we searched PubMed for articles on *PMM2*, PMM2-CDG, CDG-Ia, congenital disorders of glycosylation, *CACNA1*, familial hemiplegic migraine, episodic ataxia type 2 and spinocerebellar ataxia type 6 that have been published between 1 January 1980, and 31 May 2017, and used different combinations of these terms.

3.3. cDNA Constructs

cDNA of the human voltage-gated Ca^{2+} ($Ca_V2.1$) channel α_{1A} wild-type (WT) subunit (originally cloned into a pCMV vector) was a gift from Professor J. Striessnig (University of Innsbruck, Austria). cDNAs of the rabbit $\alpha_2\delta_1$ and rat β_3 regulatory subunits (subcloned into a pcDNA3 expression vector) were gifts from Dr. L. Birnbaumer (National Institutes of Health, Durham, NC, USA). $Ca_V2.1$ N283Q mutant channel was generated by site-directed mutagenesis of the human α_{1A} cDNA (GenScript Corporation, Piscataway, NJ, USA). All cDNA clones used in this study were sequenced in full to confirm their integrity.

3.4. Heterologous Expression

HEK293 cells were transfected using a linear polyethylenimine (PEI) derivative, the polycation ExGen500 (Fermentas Inc., Hanover, MD, USA) as previously reported (eight equivalents PEI/3.3 µg DNA/dish) [28]. Transfection was performed using the ratio for α_{1A} (WT or N283Q), β_3, $\alpha_2\delta_1$, and EGFP (transfection marker) cDNA constructs of 1:1:1:0.3. When required, $\alpha_2\delta_1$ cDNA was replaced for pcDNA3 empty vector.

3.5. Inhibition of N-Glycosylation in Live Cells and Western Blot

One day after transfection, cells were grown for another day in the presence of tunicamycin (0.2, 0.6 or 2 µg/mL) or vehicle (dimethyl sulfoxide, DMSO). At the end of the treatment, cells were washed with $1\times$ PBS, returned to the incubator, and electrophysiological recordings were performed 4 h later at room temperature (22–24 °C). For Western Blot analysis of glycosylated fragments of heterologously expressed $\alpha_2\delta_1$ subunit, after 4 h from recovery of the above mentioned tunicamycin or vehicle (DMSO) treatments, HEK293 cells were lysed on ice in lysis buffer (50 mM Tris-HCl, pH 7.5, 150 mM NaCl, 0.5% *v/v* Nonidet P-40, 5 mM EDTA, 1 mM DTT, 10 mM β-glycerolphosphate, 0.1 mM Na_3VO_4, 1 μg/mL pepstatin, 2 μg/mL aprotinin, 0.1 mM phenylmethylsulfonyl fluoride) containing protease inhibitors (Complete Mini protease inhibitor cocktail, Roche). Lysates were vortexed for 15 min at 4 °C and centrifuged at 4 °C for 7 min at $13,500\times g$ to pellet debris. Next, protein concentrations of cleared lysates were determined using Pierce BCA protein assay kit (Thermo Scientific, Madrid, Spain). Cell lysates (20 μg of proteins) were first incubated under denaturing conditions (0.5% SDS and 40 mM DTT) at 100 °C for 10 min, and then incubated in presence or absence of 500 units of peptide-*N*-glycosidase F (PNGase F, New England Biolab, Ipswich, MA, USA) during 1 h at 37 °C, according to the manufacturer's instructions. Incubated lysates were denatured to inactivate PNGase by incubation for 10 min at 80 °C, with 4× LDS Sample Buffer (Life Technologies, Carlsbad, CA, USA) and 10× sample reducing agent (Life Technologies). Samples were electrophoresed on an 8% SDS-polyacrylamide denaturing gel, transferred to a nitrocellulose membrane with iBlot (Invitrogen, Madrid, Spain), and probed with anti-rabbit $Ca_V\alpha_2$ antibody (1:500 dilution, Sigma D219) and anti-tubulin (1:2000 dilution, Sigma T6074) as a loading control, and mouse secondary antibody (GE Healthcare, Piscataway, NJ, USA). Signal was detected with the SuperSignal West Pico Chemiluminescent Substrate (Thermo Scientific). Blots were visualized with the ChemiDoc XRS documentation system (Bio-Rad, Hercules, CA, USA).

3.6. Electrophysiology

Ca^{2+} currents ($I_{Ca}{}^{2+}$) through $Ca_V2.1$ channels were measured using the whole-cell configuration of the patch-clamp technique as described in detail previously [28]. In brief, pipettes had a resistance of 2–3 MΩ when filled with a solution containing (in mM): 140 CsCl, 1 EGTA, 4 Na_2ATP, 0.1 Na_3GTP, and 10 HEPES (pH 7.2–7.3 and 290–300 mOsmol/L). The external bath solution contained (in mM): 140 tetraethylammonium-Cl (TEACl), 3 CsCl, 2.5 $CaCl_2$, 1.2 $MgCl_2$, 10 HEPES and 10 D-glucose (pH 7.4 and 300–310 mOsmol/L). Recordings were obtained with a D-6100 Darmstadt (List Medical, Darmstadt-Eberstadt, Germany) or an Axopatch 200A amplifier (Molecular Devices, San Jose, CA, USA), and the pClamp8 or pClamp10.5 software (Molecular Devices) was used for pulse generation, data acquisition and subsequent analysis.

Maximal inward Ca^{2+} current ($I_{Ca}{}^{2+}$) densities in response to 20 ms depolarizing pulses were measured from cells clamped at −80 mV, as detailed in previous report [28]. To evaluate the effect of tunicamycin and mutation N283Q on $Ca_V2.1$ voltage-dependent activation, normalized current-voltage (I–V) relationships were individually fitted with the modified Boltzmann equation, as previously reported [28]:

$$I = \frac{Gmax(V - Vrev)}{1 + e^{\frac{-(V - V1/2\ act)}{kact}}} \tag{1}$$

where *I* is the peak current, G_{max} is the maximal conductance of the cell, *V* is the membrane potential, V_{rev} is the extrapolated reversal potential of $I_{Ca}{}^{2+}$, $V_{1/2\ act}$ is the voltage for half-maximal current activation, and k_{act} is the slope factor of the Boltzmann term.

Time constant for $Ca_V2.1$ inactivation ($\tau_{inactivation}$) under the different experimental conditions was calculated after single exponential fits of the $I_{Ca}{}^{2+}$ inactivation phase during a 3 s pulse from a holding potential of −80 mV to a test potential of +20 mV or 0 mV. The degree of $I_{Ca}{}^{2+}$ inactivation (in %) at the end of these 3 s depolarizing pulses was also measured.

3.7. Ethics Statement

This study was approved by the Research and Ethics Committee of the "Sant Joan de Déu Hospital (SJDH)" (Internal code PIC-108-14) (Internal code PIC-108-14). Parents gave their written informed consent. Samples and data were obtained in accordance with the Helsinki Declaration of 1964, as revised in October 2013 (Fortaleza, Brazil).

3.8. Statistical Analysis

Statistical analysis was performed using Program R 3.2 (Vienna, Austria). Numerical variables are compared between groups by means of Mann-Whitney's *U*-test (non-parametric), and categorical variables by Fisher's exact test. For electrophysiological analysis, data are presented as the means ± S.E.M. and statistical tests included Kruskal-Wallis test followed by Dunn post hoc test, Student's *t*-test or Mann-Whitney *U*-test, as appropriate. Differences were considered significant if $p < 0.05$.

4. Discussion and Conclusions

In this multicenter cohort of 43 PMM2-CDG patients, we found a SLE incidence of 16.2% (7/39), slightly lower than previously reported (20–55%) [5–7]. Underdiagnosis due to the lack of defined clinical criteria for SLE, might explain such lower incidence. This situation also makes the differentiation from other more common complications of PMM2-CDG—such as epileptic seizures and true ischemic events—very challenging. To overcome this problem, we proposed and applied a SLE clinical definition.

Our results suggest two main trigger factors for SLE in PMM2-CDG: head trauma and viral infection. The association between cranial traumatisms and the occurrence of SLE in PMM2-CDG patients has been occasionally reported [5]. However, the role of head trauma and hyperthermia in triggering encephalopathic episodes is also described in other inborn errors of metabolism and child neurology conditions [29–32]. Here we report on six SLE occurring after mild accidental head trauma, a common situation in children with ataxia and hypotonia. Why only a minority of PMM2-CDG patients experiencing mild head trauma during their life develop SLE is currently unknown.

In our cohort, MRI during SLE revealed diffuse cortical edema in the parieto-occipital region of right hemisphere in Patient 1 (Figure 1), whereas no acute brain injuries were observed in the other eight SLE, as reported in many patients [8,20].

In PMM2-CDG different underlying pathomechanisms have been suggested for SLE, being hypoperfusion or ischemia the most discussed. However, they do not completely explain the nature and temporal course of the neuronal dysfunction [8,10]. In our PMM2-CDG cohort, no vascular injury was seen on brain MRI. Besides, no differences in coagulation factors were found between PMM2-CDG patients with and without SLE (Table 2).

Three patients showed abnormal EEG findings, including asymmetrical background activity with FIRDA, a characteristic finding of severe encephalopathy, but nonspecific for SLE (Figure 1), supporting the need of EEG monitoring in SLE and the use of antiepileptic drugs to treat neuronal dysfunction [8].

Although rarely reported [20], our results show that hyperthermia without any sign of infection in blood or CSF is consistently present during SLE, presumably secondary to hypothalamic thermoregulatory dysregulation. Although contribution of hyperthermia to cerebral/cerebellar damage has not been documented in PMM2-CDG, aiming at normothermia appears as a reasonable goal in SLE, in line with general neuroprotective recommendations in stroke.

Analysis of epidemiological, clinical, laboratory, and neuroimaging findings did not reveal differences between PMM2-CDG patients with and without SLE. Regarding the genotype, all kind of pathogenic variants were distributed among both groups.

Similarities in clinical presentation encompass both the acute and chronic manifestations of patients with PMM2-CDG and *CACNA1A* channelopathies (Table 3). Those in the acute phenotype led to the not previously explored hypothesis of a similar underlying pathomechanism for FHM episodes (related to *CACNA1A* gain-of-function mutations) and SLE in PMM2-CDG. The characteristics of episodic encephalopathic crises after minor head trauma associated with seizures, hemiparesis, hyperthermia, altered consciousness and reversible hemispheric swelling on brain MRI were common features of both disorders [12,13]. Also, a chronic course including cerebellar ataxia, nystagmus, and episodes of tonic upgaze has been described in patients with $Ca_V2.1$ (*CACNA1A*) channelopathy [33], and in PMM2-CDG patients recovering from severe SLE as shown by this study (Video 2). Interestingly, in patients with *CACNA1A*-linked FHM, verapamil or acetazolamide may improve the symptoms and prevent recurrences [22–24,34]. This observation is of great importance because the similarity between the two diseases may be an argument to explore a potentially beneficial therapeutic use of these drugs in PMM2-CDG patients.

As a novel pathophysiological approach, we explored the hypothesis of abnormal $Ca_V2.1$ function due to aberrant *N*-glycosylation as the potential cause of SLE in PMM2-CDG. We studied the effect of hypoglycosylation of heterologously expressed $Ca_V2.1$ channels on their functional expression and gating. Our results suggest that $\alpha_2\delta_1$ hypoglycosylation induced by strong inhibition of *N*-glycosylation with 2 μg/mL tunicamycin may mediate the reduction in Ca^{2+} current density through $Ca_V2.1$ channels. This agree with previous reports showing that: (1) glycosylation of specific asparagine residues at $\alpha_2\delta$ increases cationic current density through distinct Ca_V channels (including $Ca_V2.1$) [16–18], (2) mutation of several $\alpha_2\delta$ consensus *N*-glycosylation sites disrupt its cell surface expression and diminish protein stability [18], which may reduce the number of functional Ca_V channels in the membrane by destabilizing the interaction between $\alpha_2\delta$ and pore-forming α_1 subunits [16–18]. At lower tunicamycin concentrations (0.2 and 0.6 μg/mL) the reduction of $Ca_V2.1$ Ca^{2+} current density is less important and a double gain-of-function effect is observed: 1) the $Ca_V2.1$ voltage-dependent activation is favored, as indicated by the tunicamycin-induced shift of the current activation curve towards less depolarized potentials (by ~3.5–5 mV); and 2) the $Ca_V2.1$ channel inactivation is impaired due to a lower degree of inactivation (that allows a persistent Ca^{2+} influx at the end of long (seconds) depolarizing pulses) and the slowing of inactivation kinetic.

A gain-of-function in the voltage-dependence of $Ca_V2.1$ activation is the common feature described for all FHM-linked *CACNA1A* mutations that have been functionally analyzed, both in heterologous expression systems and in excitatory neurons from FHM $Ca_V2.1$ knock-in mice: they reduce the voltage threshold of channel activation by 5.6–20.9 mV [28,35–41]. Besides, for some *CACNA1A* mutations, a relationship between impaired $Ca_V2.1$ inactivation and a greater severity in the clinical phenotype (including progressive cerebellar or congenital ataxia) has been described [41–44].

Despite the association with current density reduction, increase in $Ca_V2.1$ activity due to both favored voltage-dependent activation and impaired inactivation results in higher Ca^{2+} influx into the cell in response to stimuli of physiological relevance in neurons, such as either single or trains of action potential [41]. An augment in intracellular Ca^{2+} influx leads to the specific increase in cortical excitatory neurotransmission that promotes the generation and propagation of cortical spreading depression, which correlates with the neurological symptoms (auras) and triggers the headache phase that are typical for FHM [36–39,45–48]. In the cerebellum, enhanced $Ca_V2.1$ activity favors the generation of somatic action potentials and dendritic Ca^{2+} spikes in Purkinje cells (PCs). This induces PCs hyperexcitability and, hence, mild constant cerebellar ataxia in a FHM knock-in mouse model [19,46].

Our functional analysis of the N283Q mutant $Ca_V2.1$ channel α_{1A} subunit suggests that N283 glycosylation is essential for proper $Ca_V2.1$ inactivation. To our knowledge, this is the first study on the importance of *N*-glycosylation at residues located at the P-loop-DI region of pore-forming α_1 subunits in the biophysical properties of HVA Ca_V complexes formed by α_1, β and $\alpha_2\delta$. Only for low-voltage activated $Ca_V3.2$ channels it has been shown that α_1 subunit hypoglycosylation slowed inactivation kinetic [49,50].

Since α_{1A} mutation N283Q does not affect the voltage-dependence of $Ca_V2.1$ opening (Figure 7A–D), the tunicamycin-induced gain-of-function effect on channel activation seems to be independent of aberrant α_{1A} subunit glycosylation, and is most probably due to $\alpha_2\delta$ hypoglycosylation. Supporting this observation, tunicamycin treatment of cells expressing the $Ca_V2.1$ channel formed by the N283Q mutant α_{1A}, β_3 and $\alpha_2\delta_1$ subunits induces a reduction in the $V_{1/2}$ for channel activation of similar magnitude to that induced on the wild-type (WT) channel (Figures 4 and 7E–H). It is well known that $\alpha_2\delta$ subunits not only affect the surface expression of HVA Ca_V channels but also their function. $\alpha_2\delta$ effect on Ca_V activation most likely depends on the particular pore-forming α_1 subunit of the channel. $\alpha_2\delta_1$ favored the opening of $Ca_V1.2$ (α_{1C}) channels by shifting their activation 10 mV to more hyperpolarized voltages [51,52]. Accordingly, the presence of $\alpha_2\delta_1$ glycosylation mutants impaired Ca$_V$1.2 activation by voltage [18]. On the contrary, $\alpha_2\delta_1$ impaired the opening of Ca$_V$2.3 (α_{1E}) channels by depolarizing their voltage-dependent activation (by 7–12 mV) [53]. The latter observation matches our data showing that tunicamycin-induced hypoglycosylation favors the activation of another Ca$_V$2 family member (Ca$_V$2.1) containing the regulatory $\alpha_2\delta$ subunit along with either WT or N283Q glycosylation mutant α_{1A} subunit (Figures 4 and 7E–H).

In summary, we propose a clinical definition for SLE to avoid misclassification of acute events in PMM2-CDG patients. The lack of well-defined diagnostic criteria and the high level of suspicion required for the diagnosis may result in SLE underdiagnosis. Our data suggest that mild cranial trauma and infections may trigger SLE in PMM2-CDG patients. Similarities in the clinical presentation of SLE in PMM2-CDG and *CACNA1A*-related encephalopathies suggest a possible common underlying pathomechanism. Our findings show that hypoglycosylation of different Ca$_V$2.1 subunits promotes gain-of-function effects that are similar to those induced by *CACNA1A* pathogenic variants linked to FHM and different forms of ataxia. Hence, our results support the hypothesis that aberrant Ca$_V$2.1 *N*-glycosylation may cause not only a cerebellar syndrome, but also SLE in PMM2-CDG patients. This is not to the detriment of the possible contribution of other hypoglycosylated proteins in the phenotypic expression of this disease, as we are aware that is difficult to extrapolate total cell hypoglycosylation to the effects on specific glycosylation sites at Ca$_V$2.1 channel subunits. Clarifying SLE pathogenesis in PMM2-CDG may prove paramount to develop prophylactic or therapeutic strategies for this acute and stressful complication.

Supplementary Materials: Supplementary materials can be found at http://www.mdpi.com/1422-0067/19/2/619/s1.

Acknowledgments: In memory of Andrea Poretti. We thank the patients and their families for their kind collaboration in all our projects, and particularly in this study. We are grateful to the doctors and institutions from the Spanish national network for the study of glycosylation disorders for their collaboration. We thank J. Striessnig (University of Insbruck, Innsbruck, Austria) for the gift of human *CACNA1A* cDNA and L. Birnbaumer (National Institutes of Health, North CA, USA) for providing the cDNAs encoding rabbit $\alpha_2\delta_1$, and rat β_3 regulatory subunits. We also thank F. Rubio-Moscardó for excellent technical assistance. This work was supported by national grant PI14/00021 and PI17/00101 from the National Plan on I+D+I, cofinanced by ISC-III (Subdirección General de Evaluación y Fomento de la Investigación Sanitaria), the Spanish Ministry of Economy and Competitiveness (Grants IPT-2012-0561-010000, SAF2015-69762-R, MDM-2014-0370 through the "María de Maeztu" Programme for Units of Excellence in R&D to "Departament de Ciències Experimentals i de la Salut"), FEDER (Fondo Europeo de Desarrollo Regional), and the Migraine Research Foundation (New York, USA). Mercè Izquierdo-Serra holds a "Juan de la Cierva-Formación" Fellowship funded by the Spanish Ministry of Economy and Competitiveness.We are very grateful to Víctor de Diego, M Pilar Póo, Ramón Velázquez Fragua, Mª Concepción Sierra-Córcoles, Bernabé Robles, Pilar Quijada-Fraile, Eduardo López-Laso, Mª Concepción Miranda, Ana Felipe, Mª Teresa García-Silva, Mª Luz Couce, Francisco Carratalá, Sergio Aguilera-Albesa, collaborators in the CDG Spanish Consortium.

Author Contributions: Mercè Izquierdo-Serra and Antonio F. Martínez-Monseny contributed equally to the presented work. Conception and design of study: Mercè Izquierdo-Serra, Antonio F. Martínez-Monseny, Julia Carrillo-García, Albert Edo, Daniel Cuadras, Jordi Muchart, Alfons Macaya, Juan Darío Ortigoza-Escobar, Belén Pérez-Dueñas, José M. Fernández-Fernández and Mercedes Serrano. Acquisition and analysis of data: all authors. Critical revision of manuscript: all authors. Drafting manuscript and figure: Mercè Izquierdo-Serra, Antonio F. Martínez-Monseny, Julia Carrillo-García, Albert Edo, Laura López, Óscar García, Ramón Cancho-Candela, M Llanos Carrasco-Marina, Luis G. Gutiérrez-Solana, Raquel Montero, Rafael Artuch, Celia Pérez-Cerdá, Belén Pérez, Daniel Cuadras, Jordi Muchart, Alfons Macaya, Juan Darío Ortigoza-Escobar, Belén Pérez-Dueñas, José M. Fernández-Fernández and Mercedes Serrano.

Conflicts of Interest: The authors report no conflicts of interest. We did not have any sponsor in any of the phases of the study. None of us, the authors, has received any payment to produce the manuscript.

References

1. Freeze, H.H.; Eklund, E.A.; Ng, B.G.; Patterson, M.C. Neurology of inherited glycosylation disorders. *Lancet Neurol.* **2012**, *11*, 453–466. [CrossRef]
2. Grünewald, S.; Matthijs, G.; Jaeken, J. Congenital disorders of glycosylation: A review. *Pediatr. Res.* **2002**, *52*, 618–624. [CrossRef] [PubMed]

3. Feraco, P.; Mirabelli-Badenier, M.; Severino, M.; Alpigiani, M.G.; Di Rocco, M.; Biancheri, R.; Rossi, A. The shrunken, bright cerebellum: A characteristic MRI finding in congenital disorders of glycosylation type 1a. *AJNR Am. J. Neuroradiol.* **2012**, *33*, 2062–2067. [CrossRef] [PubMed]
4. Pérez-Dueñas, B.; García-Cazorla, A.; Pineda, M.; Poo, P.; Campistol, J.; Cusí, V.; Schollen, E.; Matthijs, G.; Grunewald, S.; Briones, P.; et al. Long-term evolution of eight Spanish patients with CDG type Ia: Typical and atypical manifestations. *Eur. J. Paediatr. Neurol.* **2009**, *13*, 444–451. [CrossRef] [PubMed]
5. Barone, R.; Carrozzi, M.; Parini, R.; Battini, R.; Martinelli, D.; Elia, M.; Spada, M.; Lilliu, F.; Ciana, G.; Burlina, A.; et al. A nationwide survey of PMM2-CDG in Italy: High frequency of a mild neurological variant associated with the L32R mutation. *J. Neurol.* **2015**, *262*, 154–164. [CrossRef] [PubMed]
6. Pearl, P.L.; Krasnewich, D. Neurologic course of congenital disorders of glycosylation. *J. Child Neurol.* **2001**, *16*, 409–413. [CrossRef] [PubMed]
7. Kjaergaard, S.; Schwartz, M.; Skovby, F. Congenital disorder of glycosylation type Ia (CDG-Ia): Phenotypic spectrum of the R141H/F119L genotype. *Arch. Dis. Child.* **2001**, *85*, 236–239. [CrossRef] [PubMed]
8. Dinopoulos, A.; Mohamed, I.; Jones, B.; Rao, S.; Franz, D.; deGrauw, T. Radiologic and neurophysiologic aspects of stroke-like episodes in children with congenital disorders of glycosylation type-1a. *Pediatrics* **2007**, *119*, 768–772. [CrossRef] [PubMed]
9. Sproule, D.M.; Kaufmann, P. Mitochondrial encephalopathy, lactic acidosis, and strokelike episodes: Basic concepts, clinical phenotype, and therapeutic management of MELAS syndrome. *Ann. N. Y. Acad. Sci.* **2008**, *1142*, 133–158. [CrossRef] [PubMed]
10. Van Baalen, A.; Stephani, U.; Rohr, A. Increased brain lactate during stroke-like episode in a patient with congenital disorder of glycosylation type Ia. *Brain Dev.* **2009**, *31*, 183. [CrossRef] [PubMed]
11. Ishikawa, N.; Tajima, G.; Ono, H.; Kobayashi, M. Different neuroradiological findings during two stroke-like episodes in a patient with a congenital disorder of glycosylation type Ia. *Brain Dev.* **2009**, *31*, 240–243. [CrossRef] [PubMed]
12. Freeman, J.L.; Coleman, L.T.; Smith, L.J.; Shield, L.K. Hemiconvulsion-hemiplegia-epilepsy syndrome: Characteristic early magnetic resonance imaging findings. *J. Child Neurol.* **2002**, *17*, 10–16. [CrossRef] [PubMed]
13. Hart, A.R.; Trinick, R.; Connolly, D.J.; Mordekar, S.R. Profound encephalopathy with complete recovery in three children with familial hemiplegic migraine. *J. Paediatr. Child Health* **2009**, *45*, 154–157. [CrossRef] [PubMed]
14. Blumkin, L.; Michelson, M.; Leshinsky-Silver, E.; Kivity, S.; Lev, D.; Lerman-Sagie, T. Congenital ataxia, mental retardation, and dyskinesia associated with a novel *CACNA1A* mutation. *J. Child Neurol.* **2010**, *25*, 892–987. [CrossRef] [PubMed]
15. Catterall, W.A. Voltage-gated calcium channels. *Cold Spring Harb. Perspect. Biol.* **2011**, *3*, a003947. [CrossRef] [PubMed]
16. Gurnett, C.A.; de Waard, M.; Campbell, K.P. Dual function of the voltage-dependent Ca^{2+} channel $\alpha_2\delta$ subunit in current stimulation and subunit interaction. *Neuron* **1996**, *16*, 431–440. [CrossRef]
17. Sandoval, A.; Oviedo, N.; Andrade, A.; Felix, R. Glycosylation of asparagines 136 and 184 is necessary for the $\alpha_2\delta$ subunit-mediated regulation of voltage-gated Ca^{2+} channels. *FEBS Lett.* **2004**, *576*, 21–26. [CrossRef] [PubMed]
18. Tétreault, M.P.; Bourdin, B.; Briot, J.; Segura, E.; Lesage, S.; Fiset, C.; Parent, L. Identification of glycosylation sites essential for surface expression of the $Ca_V\alpha_2\delta_1$ subunit and modulation of the cardiac $Ca_V1.2$ channel activity. *J. Biol. Chem.* **2016**, *291*, 4826–4843. [CrossRef] [PubMed]
19. Gao, Z.; Todorov, B.; Barrett, C.F.; van Dorp, S.; Ferrari, M.D.; van den Maagdenberg, A.M.; De Zeeuw, C.I.; Hoebeek, F.E. Cerebellar ataxia by enhanced $Ca_V2.1$ currents is alleviated by Ca^{2+}-dependent K^+-channel activators in Cacna1a S218L mutant mice. *J. Neurosci.* **2012**, *32*, 15533–15546. [CrossRef] [PubMed]
20. Arnoux, J.B.; Boddaertm, N.; Valayannopoulos, V.; Romano, S.; Bahi-Buisson, N.; Desguerre, I.; de Keyzer, Y.; Munnich, A.; Brunelle, F.; Seta, N.; et al. Risk assessment of acute vascular events in congenital disorder of glycosylation type Ia. *Mol. Genet. Metab.* **2008**, *93*, 444–449. [CrossRef] [PubMed]
21. Barros, J.; Damásio, J.; Tuna, A.; Alves, I.; Silveira, I.; Pereira-Monteiro, J.; Sequeiros, J.; Alonso, I.; Sousa, A.; Coutinho, P. Cerebellar ataxia, hemiplegic migraine, and related phenotypes due to a CACNA1A missense mutation: 12-year follow-up of a large Portuguese family. *JAMA Neurol.* **2013**, *70*, 235–240. [CrossRef] [PubMed]

22. Knierim, E.; Leisle, L.; Wagner, C.; Weschke, B.; Lucke, B.; Bohner, G.; Dreier, J.P.; Schuelke, M. Recurrent stroke due to a novel voltage sensor mutation in Ca$_V$2.1 responds to verapamil. *Stroke* **2011**, *42*, e14–e17. [CrossRef] [PubMed]

23. García Segarra, N.; Gautschi, I.; Mittaz-Crettol, L.; Kallay Zetchi, C.; Al-Qusairi, L.; Van Bemmelen, M.X.; Maeder, P.; Bonafé, L.; Schild, L.; Roulet-Perez, E. Congenital ataxia and hemiplegic migraine with cerebral edema associated with a novel gain of function mutation in the calcium channel CACNA1A. *J. Neurol. Sci.* **2014**, *342*, 69–78. [CrossRef] [PubMed]

24. Malpas, T.J.; Riant, F.; Tournier-Lasserve, E.; Vahedi, K.; Neville, B.G. Sporadic hemiplegic migraine and delayed cerebral oedema after minor head trauma: A novel de novo *CACNA1A* gene mutation. *Dev. Med. Child Neurol.* **2010**, *52*, 103–104. [CrossRef] [PubMed]

25. Serrano, M.; de Diego, V.; Muchart, J.; Cuadras, D.; Felipe, A.; Macaya, A.; Velázquez, R.; Poo, M.P.; Fons, C.; O'Callaghan, M.M.; et al. Phosphomannomutase deficiency (PMM2-CDG): Ataxia and cerebellar assessment. *Orphanet J. Rare Dis.* **2015**, *10*, 138. [CrossRef] [PubMed]

26. Douglas, L.; Davies, A.; Wratten, J.; Dolphin, A.C. Do voltage-gated calcium channel $\alpha_2\delta$ subunits require proteolytic processing into α_2 and δ to be functional? *Biochem. Soc. Trans.* **2006**, *34*, 894–898. [CrossRef] [PubMed]

27. De Diego, V.; Martínez-Monseny, A.F.; Muchart, J.; Cuadras, D.; Montero, R.; Artuch, R.; Pérez-Cerdá, C.; Pérez, B.; Pérez-Dueñas, B.; Poretti, A.; Serrano, M. Collaborators of the CDG Spanish-Consortium. Longitudinal volumetric and 2D assessment of cerebellar atrophy in a large cohort of children with phosphomannomutase deficiency (PMM2-CDG). *J. Inherit. Metab. Dis.* **2017**, *40*, 709–713. [CrossRef] [PubMed]

28. Serra, S.A.; Fernàndez-Castillo, N.; Macaya, A.; Cormand, B.; Valverde, M.A.; Fernández-Fernández, J.M. The Hemiplegic Migraine associated Y1245C mutation in CACNA1A results in a gain of channel function due to its effect on the voltage sensor and G-protein mediated inhibition. *Pflügers Arch.* **2009**, *458*, 489–502. [CrossRef] [PubMed]

29. Van der Knaap, M.S.; Barth, P.G.; Gabreëls, F.J.; Franzoni, E.; Begeer, J.H.; Stroink, H.; Rotteveel, J.J.; Valk, J. A new leukoencephalopathy with vanishing white matter. *Neurology* **1997**, *48*, 845–855. [CrossRef] [PubMed]

30. Kölker, S.; Sauer, S.W.; Hoffmann, G.F.; Müller, I.; Morath, M.A.; Okun, J.G. Pathogenesis of CNS involvement in disorders of amino and organic acid metabolism. *J. Inherit. Metab. Dis.* **2008**, *31*, 194–204. [CrossRef] [PubMed]

31. Iida, S.; Nakamura, M.; Asayama, S.; Kunieda, T.; Kaneko, S.; Osaka, H.; Kusaka, H. Rapidly progressive psychotic symptoms triggered by infection in a patient with methylenetetrahydrofolate reductase deficiency: A case report. *BMC Neurol.* **2017**, *17*, 47. [CrossRef] [PubMed]

32. Serrano, M.; Rebollo, M.; Depienne, C.; Rastetter, A.; Fernández-Álvarez, E.; Muchart, J.; Martorell, L.; Artuch, R.; Obeso, J.A.; Pérez-Dueñas, B. Reversible generalized dystonia and encephalopathy from thiamine transporter 2 deficiency. *Mov. Disord.* **2012**, *27*, 1295–1298. [CrossRef] [PubMed]

33. Tantsis, E.M.; Gill, D.; Griffiths, L.; Gupta, S.; Lawson, J.; Maksemous, N.; Ouvrier, R.; Riant, F.; Smith, R.; Troedson, C.; et al. Eye movement disorders are an early manifestation of CACNA1A mutations in children. *Dev. Med. Child Neurol.* **2016**, *58*, 639–644. [CrossRef] [PubMed]

34. Asghar, S.J.; Milesi-Hallé, A.; Kaushik, C.; Glasier, C.; Sharp, G.B. Variable manifestations of familial hemiplegic migraine associated with reversible cerebral edema in children. *Pediatr. Neurol.* **2012**, *47*, 201–204. [CrossRef] [PubMed]

35. Plomp, J.J.; van den Maagdenberg, A.M.; Molenaar, P.C.; Frants, R.R.; Ferrari, M.D. Mutant P/Q-type calcium channel electrophysiology and migraine. *Curr. Opin. Investig. Drugs* **2001**, *2*, 1250–1260. [PubMed]

36. Van den Maagdenberg, A.M.; Pietrobon, D.; Pizzorusso, T.; Kaja, S.; Broos, L.A.; Cesetti, T.; van de Ven, R.C.; Tottene, A.; van der Kaa, J.; Plomp, J.J.; et al. A Cacna1a knockin migraine mouse model with increased susceptibility to cortical spreading depression. *Neuron* **2004**, *41*, 701–710. [CrossRef]

37. Tottene, A.; Conti, R.; Fabbro, A.; Vecchia, D.; Shapovalova, M.; Santello, M.; van den Maagdenberg, A.M.; Ferrari, M.D.; Pietrobon, D. Enhanced excitatory transmission at cortical synapses as the basis for facilitated spreading depression in Ca$_V$2.1 knockin migraine mice. *Neuron* **2009**, *61*, 762–773. [CrossRef] [PubMed]

38. Pietrobon, D. Ca$_V$2.1 channelopathies. *Pflügers Arch.* **2010**, *460*, 375–393. [CrossRef] [PubMed]

39. Di Guilmi, M.N.; Wang, T.; Inchauspe, C.G.; Forsythe, I.D.; Ferrari, M.D.; van den Maagdenberg, A.M.; Borst, J.G.; Uchitel, O.D. Synaptic gain-of-function effects of mutant Ca$_V$2.1 channels in a mouse model of familial hemiplegic migraine are due to increased basal [Ca^{2+}]$_i$. *J. Neurosci.* **2014**, *34*, 7047–7058. [CrossRef] [PubMed]

40. Vecchia, D.; Tottene, A.; van den Maagdenberg, A.M.; Pietrobon, D. Mechanism underlying unaltered cortical inhibitory synaptic transmission in contrast with enhanced excitatory transmission in Ca$_V$2.1 knockin migraine mice. *Neurobiol. Dis.* **2014**, *69*, 225–234. [CrossRef] [PubMed]

41. Bahamonde, M.I.; Serra, S.A.; Drechsel, O.; Rahman, R.; Marcé-Grau, A.; Prieto, M.; Ossowski, S.; Macaya, A.; Fernández-Fernández, J.M. A Single amino acid deletion (ΔF1502) in the S6 segment of Ca$_V$2.1 domain III associated with congenital ataxia increases channel activity and promotes Ca^{2+} influx. *PLoS ONE* **2015**, *10*, e0146035. [CrossRef] [PubMed]

42. Fitzsimons, R.B.; Wolfenden, W.H. Migraine coma: Meningitic migraine with cerebral oedema associated with a new form of autosomal dominant cerebellar ataxia. *Brain* **1985**, *108*, 555–577. [CrossRef] [PubMed]

43. Ducros, A.; Denier, C.; Joutel, A.; Vahedi, K.; Michel, A.; Darcel, F.; Madigand, M.; Guerouaou, D.; Tison, F.; Julien, J.; et al. Recurrence of the T666M calcium channel CACNA1A gene mutation in familial hemiplegic migraine with progressive cerebellar ataxia. *Am. J. Hum. Genet.* **1999**, *64*, 89–98. [CrossRef] [PubMed]

44. Kors, E.E.; Terwindt, G.M.; Vermeulen, F.L.; Fitzsimons, R.B.; Jardine, P.E.; Heywood, P.; Love, S.; van den Maagdenberg, A.M.; Haan, J.; Frants, R.R.; et al. Delayed cerebral edema and fatal coma after minor head trauma: Role of the CACNA1A calcium channel subunit gene and relationship with familial hemiplegic migraine. *Ann. Neurol.* **2001**, *49*, 753–760. [CrossRef] [PubMed]

45. Bolay, H.; Reuter, U.; Dunn, A.K.; Huang, Z.; Boas, D.A.; Moskowitz, M.A. Intrinsic brain activity triggers trigeminal meningeal afferents in a migraine model. *Nat. Med.* **2002**, *8*, 136–142. [CrossRef] [PubMed]

46. Pietrobon, D.; Striessnig, J. Neurobiology of migraine. *Nat. Rev. Neurosci.* **2003**, *4*, 386–398. [CrossRef] [PubMed]

47. Pelzer, N.; Stam, A.H.; Haan, J.; Ferrari, M.D.; Terwindt, G.M. Familial and sporadic hemiplegic migraine: Diagnosis and treatment. *Curr. Treat. Options Neurol.* **2013**, *15*, 13–27. [CrossRef] [PubMed]

48. Pietrobon, D.; Moskowitz, M.A. Pathophysiology of migraine. *Annu. Rev. Physiol.* **2013**, *75*, 365–391. [CrossRef] [PubMed]

49. Weiss, N.; Black, S.A.G.; Bladen, C.; Chen, L.; Zamponi, G.W. Surface expression and function of Ca$_V$3.2 T-type calcium channels are controlled by asparagine-linked glycosylation. *Pflugers Arch.* **2013**, *465*, 1159–1170. [CrossRef] [PubMed]

50. Orestes, P.; Osuru, H.P.; McIntire, W.E.; Jacus, M.O.; Salajegheh, R.; Jagodic, M.M.; Choe, W.; Lee, J.; Lee, S.S.; Rose, K.E.; et al. Reversal of neuropathic pain in diabetes by targeting glycosylation of Ca$_V$3.2 T-Type calcium channels. *Diabetes* **2013**, *62*, 3828–3838. [CrossRef] [PubMed]

51. Felix, R.; Gurnett, C.A.; De Waard, M.; Campbell, K.P. Dissection of functional domains of the voltage-dependent Ca^{2+} channel α$_2$δ subunit. *J. Neurosci.* **1997**, *17*, 6884–6891. [PubMed]

52. Platano, D.; Qin, N.; Noceti, F.; Birnbaumer, L.; Stefani, E.; Olcese, R. Expression of the α$_2$δ subunit interferes with prepulse facilitation in cardiac L-type calcium channels. *Biophys. J.* **2000**, *78*, 2959–2972. [CrossRef]

53. Qin, N.; Olcese, R.; Stefani, E.; Birnbaumer, L. Modulation of human neuronal α$_{1E}$-type calcium channel by α$_2$δ-subunit. *Am. J. Physiol.* **1998**, *274*, 1324–1331. [CrossRef]

International Journal of
Molecular Sciences

MDPI

Article

ATP-Induced Increase in Intracellular Calcium Levels and Subsequent Activation of mTOR as Regulators of Skeletal Muscle Hypertrophy

Naoki Ito [1],*, Urs T. Ruegg [2] and Shin'ichi Takeda [1],*

[1]   Department of Molecular Therapy, National Institute of Neuroscience, National Center of Neurology and Psychiatry, Kodaira 187-8502, Japan

[2]   Pharmacology, Geneva-Lausanne School of Pharmaceutical Sciences, University of Geneva, CH 1211 Geneva, Switzerland; urs.ruegg@unige.ch

*   Correspondence: naoki.ito.198571@gmail.com (N.I.); takeda@ncnp.go.jp (S.T.)

Received: 21 August 2018; Accepted: 13 September 2018; Published: 18 September 2018

Abstract: Intracellular signaling pathways, including the mammalian target of rapamycin (mTOR) and the mitogen-activated protein kinase (MAPK) pathway, are activated by exercise, and promote skeletal muscle hypertrophy. However, the mechanisms by which these pathways are activated by physiological stimulation are not fully understood. Here we show that extracellular ATP activates these pathways by increasing intracellular Ca^{2+} levels ($[Ca^{2+}]_i$), and promotes muscle hypertrophy. $[Ca^{2+}]_i$ in skeletal muscle was transiently increased after exercise. Treatment with ATP induced the increase in $[Ca^{2+}]_i$ through the $P2Y_2$ receptor/inositol 1,4,5-trisphosphate receptor pathway, and subsequent activation of mTOR in vitro. In addition, the ATP-induced increase in $[Ca^{2+}]_i$ coordinately activated Erk1/2, p38 MAPK and mTOR that upregulated translation of JunB and interleukin-6. ATP also induced an increase in $[Ca^{2+}]_i$ in isolated soleus muscle fibers, but not in extensor digitorum longus muscle fibers. Furthermore, administration of ATP led to muscle hypertrophy in an mTOR- and Ca^{2+}-dependent manner in soleus, but not in plantaris muscle, suggesting that ATP specifically regulated $[Ca^{2+}]_i$ in slow muscles. These findings suggest that ATP and $[Ca^{2+}]_i$ are important mediators that convert mechanical stimulation into the activation of intracellular signaling pathways, and point to the P2Y receptor as a therapeutic target for treating muscle atrophy.

Keywords: skeletal muscle; muscle hypertrophy; muscle atrophy; ATP; Ca^{2+}; P2Y receptor; IP_3 receptor; mammalian target of rapamycin (mTOR); mitogen-activated protein kinase (MAPK)

1. Introduction

Skeletal muscle maintains an adequate volume that is commensurate with its surrounding environment. This is regulated by a balance between protein synthesis and degradation [1]. The insulin-like growth factor-1 (IGF-1) activates the mammalian target of rapamycin (mTOR) through the class I phosphatidylinositol 3-kinase (PI3K)/Akt pathway, which induces protein synthesis and subsequent muscle growth or hypertrophy [2]. Physiological stimulation, such as exercise and mechanical load, also promotes muscle hypertrophy by upregulating protein synthesis [3]. However, recent studies have revealed that exercise- or mechanical load-induced activation of mTOR is not mediated by the IGF-1/class I PI3K/Akt pathway, especially at an early stage of muscle hypertrophy [4–7]. This suggests the presence of another mediator that converts physiological stimulation into the activation of mTOR. We have previously demonstrated that mechanical load-induced activation of mTOR and subsequent muscle hypertrophy were regulated by a TRPV1-mediated increase in intracellular Ca^{2+} levels ($[Ca^{2+}]_i$) [8,9]. However, it is not clear whether TRPV1 is the only channel that converts mechanical load into intracellular signaling events that induce muscle hypertrophy. This motivated us to investigate the potential role of other Ca^{2+} channels involved in this process.

Release of ATP from muscles in response to contraction, exercise or electrical stimulation was observed both in humans and in rodents [10–13]. ATP and other purine- or pyrimidine-based agonists mediate a variety of biological functions by activating purinergic receptors, namely the ligand-gated P2X receptors or the G-protein-coupled P2Y receptors [14]. Activation of P2Y receptors induces the production of inositol 1,4,5-trisphosphate (IP_3) through the Gαq–mediated activation of phospholipase C (PLC), thereby causing Ca^{2+} to be released from the sarcoplasmic reticulum (SR)-localized inositol 1,4,5-trisphosphate receptors (IP_3R) [15,16]. However, to our knowledge, the role of ATP in the activation of mTOR and muscle hypertrophy has not been analyzed

2. Results

2.1. ATP Induces an Increase in [Ca^{2+}]$_i$ by Activating the P2Y$_2$ Receptor and Downstream IP$_3$R

Plasma ATP levels are known to increase following exercise or muscle contraction [10–13]. To investigate the functional consequences of this increase, we focused on ATP in regulating [Ca^{2+}]$_i$ and subsequent muscle hypertrophy. Using pCAGGS-GCaMP2 mice, which express a genetically encoded Ca^{2+} indicator [17], we found that [Ca^{2+}]$_i$ increased after exercise in both the soleus and plantaris muscles, which are mainly composed of slow and fast muscle fibers, respectively (Figure 1A). This increase was observed immediately after and during 30 min after exercise (Figure 1B), suggesting a causal relationship between exercise-induced increase of plasma ATP levels and exercise-induced increase in [Ca^{2+}]$_i$.

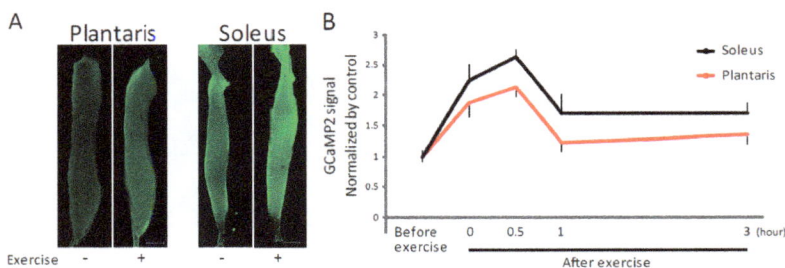

Figure 1. Exercise induces increased intracellular Ca^{2+} levels. (**A**) Representative fluorescence imaging of isolated plantaris and soleus muscles from pCAGGS-GCaMP2 mice before and after exercise. Scale bar: 1 mm. (**B**) Quantitative analysis of GCaMP2 signal intensity (*n* > 4). Plantaris and soleus muscles were isolated before or immediately, 0.5, 1 and 3 h after exercise. Error bars indicate S.E.M.

We next analyzed the ATP-induced Ca^{2+} signaling events in vitro. Similarly to previous reports [18–24], exogenous ATP treatment of C2C12 myotubes resulted in a transient increase in [Ca^{2+}]$_i$ (Figure 2A,B). This increase was also observed by the closely related pyrimidine triphosphate UTP, whereas GTP, CTP and TTP had no effect (Figure 2A,B). ATPγS or 2-MeS ATP also increased [Ca^{2+}]$_i$, whereas ADP, AMP or adenosine had no effect (Supplementary Figure S1A). ATP or UTP induced an increase in [Ca^{2+}]$_i$ in a concentration-dependent manner (Supplementary Figure S1B,C). To investigate which receptor was mediating the ATP- or UTP-induced increase in [Ca^{2+}]$_i$, we examined P2Y receptors and the subsequent PLC-mediated activation of IP$_3$R. Using mouse NG108-15 cells, Lustig et al. showed that the relative agonist potencies on the P2Y$_2$ receptor were ATP = UTP > ATPγS > 2-MeS ATP [25]. Our similar results suggest that the P2Y$_2$ receptor may be the target of ATP in C2C12 myotubes (Supplementary Figure S1A). In agreement with this, the ATP- or UTP-induced increase in [Ca^{2+}]$_i$ was prevented either by knockdown of the P2Y$_2$ receptor (Supplementary Figure S1D,E) or by treatment with the P2Y inhibitor, suramin (Figure S1F). This increase was also inhibited by the PLC inhibitor, U73122 (Supplementary Figure S1F). To determine the source of the intracellular Ca^{2+}, the effect of ATP,

UTP or ATPγS was assessed in the absence of extracellular Ca^{2+}. Removal of extracellular Ca^{2+} did not affect the increase in $[Ca^{2+}]_i$ (Figure S1G). However, this increase was blocked by pre-incubation with thapsigargin, which depletes intracellular Ca^{2+} stores (Figure S1G), indicating that ATP provoked Ca^{2+} release from the SR. Although other subtypes of P2Y and P2X receptors are known to be expressed in C2C12 cells [24], the facts that removal of extracellular Ca^{2+} had no effect on the increase in $[Ca^{2+}]_i$ (Supplementary Figure S1G), and that knockdown of the P2Y$_2$ receptor almost completely inhibited the ATP-induced increase in $[Ca^{2+}]_i$ (Figure S1E), suggest that other purinergic receptors were not involved in ATP- or UTP-induced increase in $[Ca^{2+}]_i$ in vitro. Furthermore, the increase in $[Ca^{2+}]_i$ was inhibited either by the overexpression of an IP$_3$-sponge (Figure S1H), a shortened and soluble form of the IP$_3$R containing its high-affinity IP$_3$-binding sequence [26,27], or by the IP$_3$R inhibitor, xestospongin C (XeC) (Figure S1I). These results indicate that the ATP-induced increase in $[Ca^{2+}]_i$ occurs by activating the P2Y$_2$ receptor and the downstream PLC/IP$_3$R pathway in C2C12 myotubes.

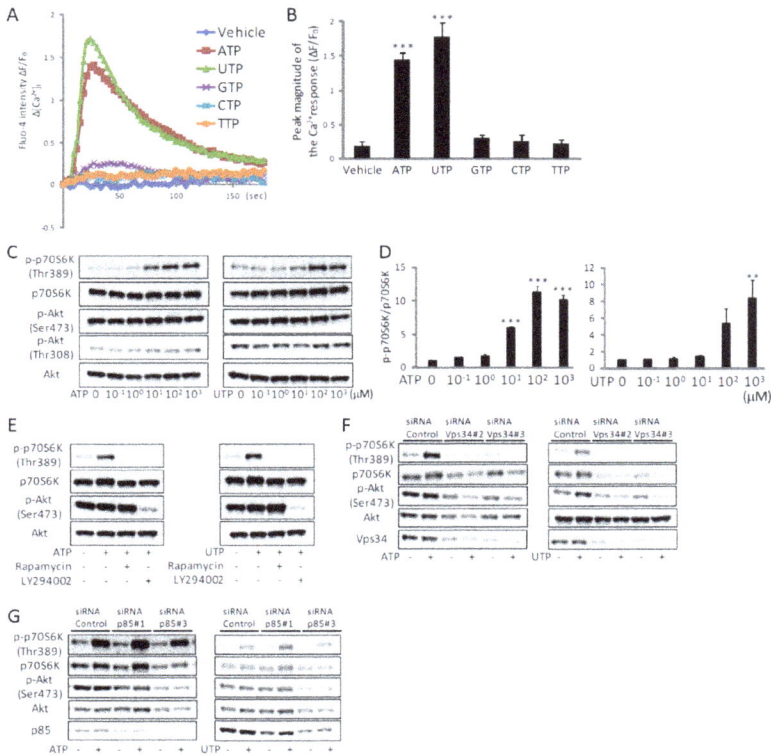

Figure 2. ATP induces an increase in $[Ca^{2+}]_i$ and activation of mTOR via Vps34 in C2C12 myotubes. (**A**) Traces of Fluo-4 fluorescence in C2C12 myotubes treated with nucleotide triphosphates at a concentration of 100 μM (*n* = 4). (**B**) Quantitative analysis of peak magnitudes of (**A**) (*n* = 4). (**C**) Representative Western blot analysis showing concentration-dependent effects of ATP (left) or UTP (right) on phosphorylation of p70S6K and Akt (*n* = 4). (**D**) Quantitative signal intensities of p-p70S6K/p70S6K in (**C**) (*n* = 4). (**E–G**) Representative Western blot analysis showing the effects of rapamycin or LY294002 (**E**), siRNA for Vps34 (**F**) or p85 (**G**) on ATP- or UTP-induced phosphorylation of p70S6K and Akt (*n* = 4). The concentration of ATP and UTP in (**E–G**) was 100 μM and the duration of treatment in all experiments shown in this figure was 30 min. ** $p < 0.01$, *** $p < 0.001$ by ANOVA with Tukey-Kramer test. Error bars indicate S.E.M.

2.2. ATP Activates mTOR via the P2Y₂ Receptor/PLC/IP₃R Pathway In Vitro

We have previously shown that the TRPV1-mediated increase of $[Ca^{2+}]_i$ activates mTOR, and promotes muscle hypertrophy [8,9]. Therefore, to investigate the downstream events of the ATP-induced increase in $[Ca^{2+}]_i$, we focused on the activation of mTOR. Treatment of C2C12 myotubes with ATP or UTP induced phosphorylation of p70S6K at Thr389, which is an mTOR-regulated phosphorylation site indispensable for p70S6K activation [28], in a concentration-dependent manner (Figure 2C,D). Furthermore, this phosphorylation was inhibited by the mTOR inhibitor, rapamycin (Figure 2E), indicating that mTOR was activated by ATP. It is worth noting that phosphorylation of Akt was not altered by ATP or UTP (Figure 2C). Nevertheless, phosphorylation of p70S6K was prevented by the PI3K inhibitor, LY294002 (Figure 2E), suggesting an involvement of PI3K in the activation of mTOR. Because LY294002 inhibits both, class I and class III PI3K [29], we investigated which PI3K was involved in the activation of mTOR by gene knockdown of Vps34, a member of class III PI3K, as well as p85, a member of class I PI3K [30]. Knockdown of Vps34 completely prevented the ATP- or UTP-induced phosphorylation of p70S6K (Figure 2F), whereas knockdown of p85 had no effect (Figure 2G). These results indicate that the ATP-induced activation of mTOR was mediated by class III PI3K, rather than class I PI3K/Akt, the major downstream pathway of IGF-1.

The ATP- or UTP-induced phosphorylation of p70S6K was prevented by the intracellular Ca^{2+} chelator, BAPTA-AM, indicating that the activation of mTOR was $[Ca^{2+}]_i$-dependent (Figure 3A). Removal of extracellular Ca^{2+} by EGTA partially prevented phosphorylation of p70S6K (Figure 3A). In line with the effects on the ATP- or UTP-induced increase of $[Ca^{2+}]_i$ (Figure S1), knockdown of P2Y₂ prevented the ATP- or UTP-induced phosphorylation of p70S6K (Figure 3B). Treatment with suramin, U73122, XeC, or overexpression of IP₃-sponge-eGFP, also prevented this phosphorylation (Figure 3C–F). Taken together, these results indicate that ATP induced activation of mTOR, and that this activation is mediated by the P2Y₂ receptor/PLC/IP₃R pathway.

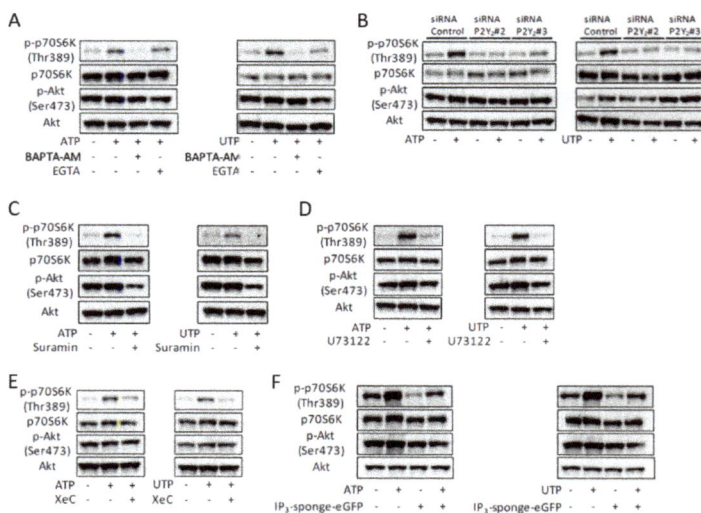

Figure 3. ATP-induced activation of mTOR is mediated by the P2Y₂ receptor and its downstream PLC/IP₃R pathway in C2C12 myotubes. (**A–E**) Representative Western blots showing the effects of BAPTA-AM or EGTA (**A**), siRNA for the P2Y₂ receptor (**B**), suramin (**C**), U73122 (**D**), XeC (**E**) on ATP- or UTP-induced phosphorylation of p70S6K and Akt (*n* = 4). (**F**) ATP- or UTP-induced phosphorylation of p70S6K and Akt in the IP₃-sponge-overexpressed C2C12 myotubes (*n* = 4). The concentration of ATP and UTP was 100 µM and the duration of treatment in all experiments shown in this figure was 30 min.

2.3. ATP Upregulates JunB and IL-6 by Activating MAPKs and mTOR

Treatment with ATP or UTP also induced phosphorylation of Erk1/2 and p38 mitogen-activated protein kinases (MAPK), whereas the degree of phosphorylation of AMP-activated protein kinase α (AMPKα) remained unchanged (Figure 4A). These results indicate that the ATP-induced increase in $[Ca^{2+}]_i$ activated both mTOR and MAPKs, suggesting a coordinated effect on subsequent events. To identify the potential downstream targets of Erk1/2 and p38 MAPK, we focused on JunB and interleukin-6 (IL-6), which are known to be regulated by Erk1/2 and p38 MAPK. These genes are upregulated by exercise or mechanical load, and are important for subsequent muscle hypertrophy [31–37]. Exposure of C2C12 myotubes to ATP increased the expression of JunB and IL-6 (Figure 4B,C, black), but this was prevented by BAPTA-AM (Figure 4B,C, white), indicating that their expression is $[Ca^{2+}]_i$-dependent. Treatment of the myotubes with the MEK inhibitor, U0126, blocked the upregulation of JunB by ATP, whereas treatment with the p38 MAPK inhibitor, SB203580, prevented the upregulation of IL-6 (Figure 4B,C, blue and yellow). These results indicate that ATP induced the transcription of JunB and IL-6 through the $[Ca^{2+}]_i$-dependent activation of Erk1/2 and p38 MAPK pathways, respectively. Treatment with rapamycin had no effect on the upregulation of JunB or IL-6 (Figure 4B,C, red). However, the ATP- or UTP-induced increase of JunB or IL-6 at the protein level was prevented by rapamycin (Figure 4D,E), suggesting that the ATP/UTP-induced activation of Erk1/2 and p38 MAPK contributed to the transcription of JunB and IL-6, and that the activation of mTOR promoted their translation.

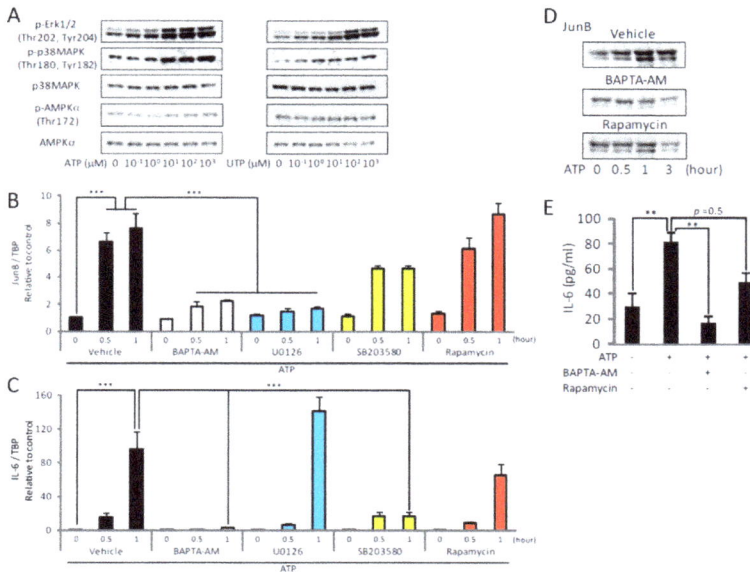

Figure 4. ATP upregulates JunB and IL-6 by activating Erk1/2, p38 MAPK and mTOR in C2C12 myotubes. (**A**) Concentration-dependent effects of ATP (left) or UTP (right) on phosphorylation of Erk1/2, p38 MAPK and AMPKα (*n* = 4). (**B,C**) Gene expression analysis showing the effects of BAPTA-AM (white), U0126 (blue), SB203580 (yellow) and rapamycin (red) on the ATP-induced expression of JunB (**B**) and IL-6 (**C**) (*n* = 4). (**D**) Representative Western blots showing the effects of BAPTA-AM or rapamycin on the ATP-induced increase of JunB (*n* = 4). (**E**) Effects of BAPTA-AM or rapamycin on the ATP-induced increase of IL-6 protein level (*n* = 4). The concentration of ATP in (**B–E**) was 100 μM and the duration of treatment in all experiments shown in this figure was 30 min. ** *p* < 0.01, *** *p* < 0.001 by ANOVA with Tukey-Kramer test. Error bars indicate S.E.M.

2.4. ATP Induces an Increase in $[Ca^{2+}]_i$ Specifically in Slow Muscles

To investigate the effects of ATP on mature muscle fibers, we isolated single fibers from the soleus and extensor digitorum longus (EDL) muscles. Similarly to the results of a previous report [38], ATP had no effect on $[Ca^{2+}]_i$ in EDL fibers (Figure 5A). However, treatment with ATP showed a slow, but significant increase in $[Ca^{2+}]_i$ in the soleus fibers, suggesting that ATP increases $[Ca^{2+}]_i$ only in slow fibers (Figure 5A,B). Similarly to C2C12 myotubes, the ATP-induced increase of $[Ca^{2+}]_i$ in soleus fibers was inhibited by suramin, U73122 and XeC, but not by removal of extracellular Ca^{2+} (Figure 5C), suggesting that the effect of ATP on single fibers was also mediated by the P2Y receptor and the downstream PLC/IP$_3$R pathway. However, different to C2C12 myotubes, the effects of 2MeS-ATP and UTP on $[Ca^{2+}]_i$ were significantly higher and lower than that of ATP, respectively (Figure 5D). These results suggest that the target P2Y receptor of ATP might not be the P2Y$_2$ receptor in mature muscle fibers.

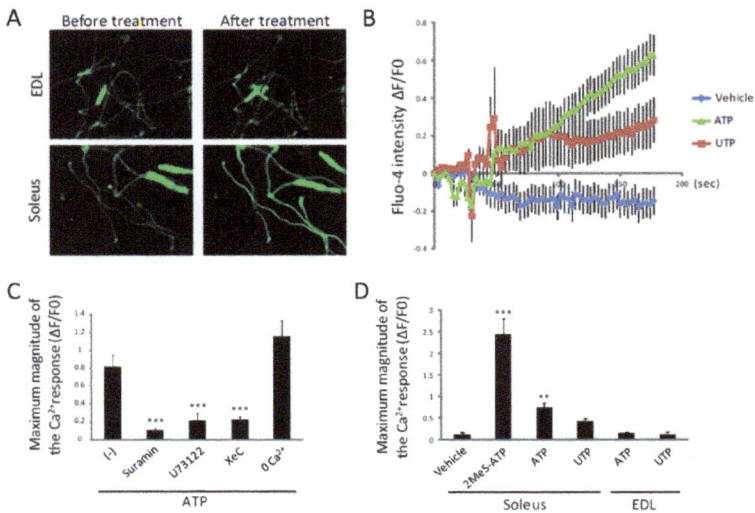

Figure 5. ATP induces an increase in $[Ca^{2+}]_i$ by the activation of the P2Y receptor and its downstream PLC/IP$_3$R pathway in single fibers isolated from soleus muscle. (**A**) Representative fluorescence imaging of Fluo-4-loaded single fibers isolated from EDL (upper) and soleus (lower) muscles before and after treatment with ATP for 3 min. (**B**) Traces of Fluo-4 fluorescence in single fibers isolated from soleus muscle. (**C**) Quantitative analysis for the maximum magnitudes of Fluo-4 intensity in suramin-, U73122-, XeC- or 0 Ca^{2+}-buffer-treated single fibers isolated from soleus muscle. (**D**) Quantitative analysis of the maximum Fluo-4 intensity in 2MeS-ATP-, ATP- or UTP-treated soleus or EDL single fibers at a concentration of 100 μM. At least 10 fibers were analyzed in B, C and D. ** $p < 0.01$, *** $p < 0.001$ by ANOVA with Tukey-Kramer test. Error bars indicate S.E.M.

2.5. ATP Induces Muscle Hypertrophy by Regulating $[Ca^{2+}]_i$

Finally, to investigate the effect of the ATP-induced increase in $[Ca^{2+}]_i$ on muscle hypertrophy in vivo, ATP was administered intramuscularly daily for 1 week. This treatment led to a slight but significant increase in muscle weight in both the soleus and gastrocnemius muscles (Figure 6A). Administration of ATP also increased fiber size in the soleus muscle (Figure 6B). Consistent with the slow fiber-specific effect of ATP (Figure 5), this treatment had no effect on plantaris muscle (Figure 6A), even though we did not observe significant differences in the expression of P2Y$_1$ or P2Y$_2$ receptors between soleus and plantaris muscles (Figure 6C). This increase was not observed when BAPTA-AM or rapamycin was co-administered (Figure 6A), indicating that ATP promotes muscle hypertrophy by regulating $[Ca^{2+}]_i$ and mTOR. Furthermore, administration of ATP during hindlimb-suspension or denervation attenuated the decrease of muscle weight (Figure 6D). Taken together, these results suggest the P2Y receptor to be a potential therapeutic target for treating muscle atrophy.

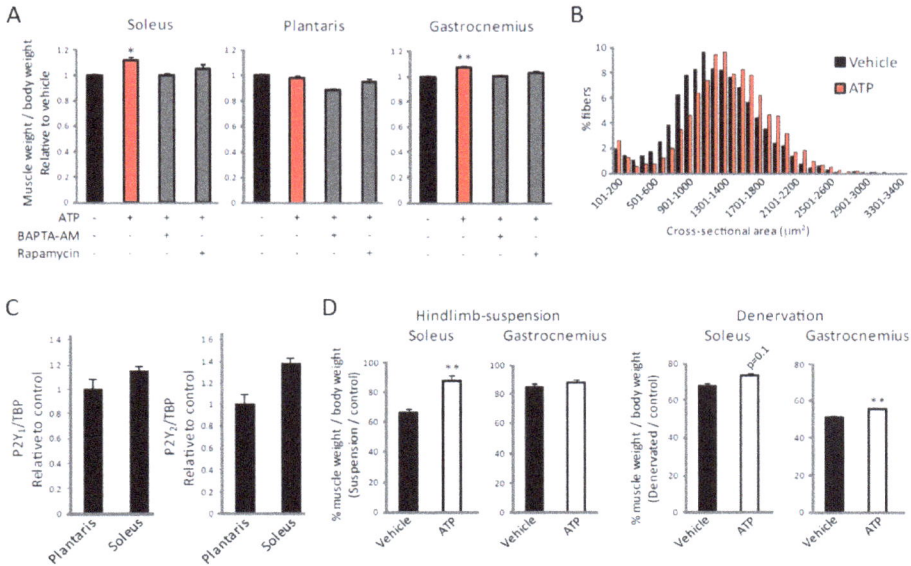

Figure 6. ATP induces muscle hypertrophy and alleviates muscle atrophy in an $[Ca^{2+}]_i$- and mTOR-dependent manner. (**A**) Effects of ATP with or without BAPTA-AM or rapamycin on the increase of muscle weight in soleus, plantaris and gastrocnemius muscles (*n* = 4). (**B**) The cross-sectional area distributions were plotted as frequency histograms. At least 2800 fibers from 4 different soleus muscles were counted. (**C**) Expression of $P2Y_1$ and $P2Y_2$ receptor mRNA in soleus and plantaris muscle (*n* = 8). (**D**) Effects of ATP on soleus and gastrocnemius muscles weights in hindlimb-suspended (left) or denervated (right) muscle shown as a percentage of the control group (*n* > 4). * *p* < 0.05, ** *p* < 0.01 by ANOVA with Tukey-Kramer test in (**A**), and ** *p* < 0.01 by Student's *t*-test in (**D**). Error bars indicate S.E.M.

3. Discussion

Extracellular ATP plays a crucial role in the regulation of a considerable number of biological processes [39,40]. Plasma ATP levels are increased by exercise or muscle contraction and can function in an autocrine or paracrine manner [10–13]. Of note, an earlier study showed that ATP was released from skeletal muscle in response to electrical stimulation through the pannexin channel, suggesting a regulated involvement of extracellular ATP in contraction- or exercise-induced events [39]. However, to our knowledge, the role of extracellular ATP on the induction of muscle hypertrophy has not been studied. Therefore, we investigated the ATP-dependent activation of the P2Y receptor and its downstream PLC/IP_3R pathway in the context of muscle hypertrophy. ATP activated the $P2Y_2$ receptor and initiated a subsequent PLC-mediated increase in $[Ca^{2+}]_i$ through IP_3R in vitro. In turn, this led to the activation of mTOR and MAPKs, and subsequent expression of JunB or IL-6. Furthermore, ATP affected $[Ca^{2+}]_i$ and muscle hypertrophy in soleus muscle, but not in EDL or plantaris muscles, suggesting that ATP regulates $[Ca^{2+}]_i$ specifically in slow muscles. This is the first report that shows the role of extracellular ATP in the activation of mTOR and subsequent muscle hypertrophy. Our discovery demonstrates an important role of extracellular ATP and intracellular Ca^{2+} signaling in transducing physical activity into activation of intracellular signaling pathways that subsequently lead to muscle hypertrophy.

The IGF-1-induced activation of Akt, which is mediated by class I PI3K, is essential for inducing muscle growth and hypertrophy [2]. Although the regulation of Akt by Ca^{2+} has been reported [41,42], our results show that the effect of $[Ca^{2+}]_i$ on mTOR was not mediated by class I PI3K or Akt (Figure 2G), suggesting that the IGF-1-induced muscle growth and the exercise-induced muscle hypertrophy are differently regulated. Consistent with these results, Gulati et al. demonstrated that amino acid-induced activation of mTOR was mediated by increased $[Ca^{2+}]_i$, and that the direct target of Ca^{2+} was hVps34, a component of class III PI3K in Hela cells [43]. Vps34 regulates intracellular protein trafficking and autophagy, and is known to be activated by a variety of physiological stimuli, such as exercise [44,45]. These results suggest that both, amino acid- and ATP-induced activation of mTOR share similar mechanisms that are mediated by $[Ca^{2+}]_i$ and Vps34. However, in contrast to our observation, Mercan et al. showed that leucine-induced increase of $[Ca^{2+}]_i$ and subsequent activation of mTOR was not mediated by hVps34 in C2C12 myoblasts [46]. Currently, the differences between Vps34-dependent and –independent activation of mTOR are still unclear. Future studies using adequate experimental models are required in order to elucidate the molecular links between $[Ca^{2+}]_i$ and mTOR as well as the role of mTOR phosphorylation in response to ATP.

Moreover, we showed JunB and IL-6 to be downstream targets of ATP-induced activation of MAPKs. Several lines of evidence suggest that the exercise-induced upregulation of IL-6 is controlled by $[Ca^{2+}]_i$ [31,47]. Our results indicate that this also applies to JunB. Raffaello et al. showed that overexpression of JunB was sufficient to increase protein synthesis and muscle hypertrophy without activating the Akt/mTOR pathway [33]. In view of our observation that the ATP-induced increase of JunB at the protein level was inhibited by rapamycin (Figure 4D), the increase of protein synthesis after physiological stimulation, which is considered to be due solely to mTOR, might be at least partially caused by JunB that was upregulated by mTOR.

Most of our results regarding signal transduction in C2C12 myotubes are similar to those in mature muscle fibers. However, it appears that the receptor activated by ATP was different in C2C12 myotubes as compared to mature muscle. Because the relative agonist potencies for the $P2Y_1$ receptor are 2MeS-ATP > ATP [48], we suspected the potential involvement of $P2Y_1$, or a combination of $P2Y_1$ and other P2Y receptors in the increase of $[Ca^{2+}]_i$ in soleus muscle fibers. Furthermore, and interestingly, the effects of ATP in vivo appeared to occur only in slow muscles, even though we did not detect a difference in expression levels of $P2Y_1$ and $P2Y_2$ receptor between fast and slow muscles (Figure 6C). A previous study showed that the expression level of IP_3R was much higher in slow muscle than in fast muscle [49], suggesting that the different levels of IP_3R, or a post-transcriptional regulation of P2Y receptors might be involved in the different responses to ATP between fast and slow fibers. Notably, our results do not exclude an involvement of other proteins, such as ATP transporters or pannexin channels in the exercise-induced events.

The concentration of ATP in the blood is thought to be kept below 1 µM [12]. Our results indicate that more than 1 µM of ATP was required to activate mTOR (Figure 2C). It is possible that ATPase activity keeps overall circulating ATP low, but cell-adjacent concentrations of ATP might be much higher than 1 µM. If this is the case, a physiological role for exercise-induced increase of extracellular ATP, activation of mTOR and subsequent muscle hypertrophy would exist. Furthermore, although the effect of ATP on muscle in vivo might be below what we observed in this study, it is likely that the repeated increase of extracellular ATP levels occurring in contracting muscle contributes to activate mTOR and muscle hypertrophy. Our results on ATP administration to mice showing an increase in soleus muscle weight (Figures 5 and 6) are in favor of such a role, at least for slow muscle. In addition, although the exercise-induced increase of $[Ca^{2+}]_i$ returned to a steady-state level within 1 h (Figure 1B), it is likely that the ATP-induced signaling cascade persists for a much longer period after exercise. In addition to the target P2Y receptor subtype of ATP in vivo, its role in the exercise-induced activation of mTOR and subsequent muscle hypertrophy should be analyzed in more detail in the future.

We have previously shown that mechanical load activates neuronal nitric oxide synthase (nNOS) and promotes muscle hypertrophy by regulating TRPV1 [8]. In the current study, the ATP-induced increase in $[Ca^{2+}]_i$ occurs exclusively in slow muscle (Figures 5 and 6). It is noteworthy that NOS activity is much greater in fast than slow muscle [50]. Considering that both plantaris and soleus muscle showed increase in $[Ca^{2+}]_i$ after exercise (Figure 1), the subsequent activation of mTOR might be differently regulated in different fiber types.

Finally, our results point to the P2Y receptor as a therapeutic target for the treatment of muscle atrophy, as it occurs in immobilized muscle or in cancer cachexia. The anti-cancer activity of purine-based agonists has already been demonstrated. In fact, intraperitoneal injection of ATP has been shown to inhibit cachexia and prolong survival in mice [51]. Although the administered ATP was thought to alleviate cachexia by inhibiting tumor growth [51], our results suggest that part of its effect may be mediated by an action on skeletal muscle. In conclusion, our results indicate that the ATP-induced activation of the P2Y receptor and the subsequent increase in $[Ca^{2+}]_i$ through IP_3R convert a mechanical load into activation of intracellular signaling pathways, subsequently leading to muscle hypertrophy.

4. Materials and Methods

4.1. Animals

Twelve to 14-week-old male C57BL/6 mice were purchased from Nihon CREA. pCAGGS-GCaMP2 mice were provided by Dr. Michael Kotlikoff [17]. All mice were housed at the animal facility at the National Center of Neurology and Psychiatry (NCNP, Tokyo, Japan). All of the animal procedures were approved by the Experimental Animal Care and Use Committee at the National Center of Neurology and Psychiatry (Approval number: 2016001, Approval date: 2/25/2016). All of the experimental methods were performed in accordance with approved guidelines.

4.2. Materials

Antibodies against Akt (#9272), phospho-Akt (Ser473) (#9271), phospho-Akt (Thr308) (#9275), p70S6K (#9202), phospho-p70S6K (Thr389) (#9205), p85 (#4257), Vps34 (#3358), AMPKα (#2603), phospho-AMPKα (Thr172) (#2535), phospho-ERK (Thr202/Tyr204) (#9101), p38 MAPK (#9212), and phospho-p38 MAPK (Thr180, Tyr182) (#9211) were purchased from Cell Signaling Technology. The $P2Y_2$ antibody (ab10270) came from Abcam. The JunB (sc-46) and α-tubulin (sc-12462) antibodies were obtained from Santa Cruz. The laminin α2 (Clone: 4H-8, ALX-804-190) antibody was supplied by Alexis Biochemicals (San Diego, CA, USA). The following compounds were obtained from the sources indicated and used at the final concentrations indicated: ATP (0.1–1000 μM, Sigma-Aldrich, Saint Louis, MO, USA), UTP (0.1–1000 μM, Sigma-Aldrich), BAPTA-AM (50 μM, Calbiochem, Sigma-Aldrich, Saint Louis, MO, USA), EGTA (2 mM, Sigma-Aldrich), thapsigargin (2 μM, Calbiochem), rapamycin (0.1 μM, Sigma-Aldrich), LY294002 (10 μM, Sigma-Aldrich), suramin (100 μM, Sigma-Aldrich), U73122 (10 μM, Sigma-Aldrich), XeC (2 μM, Sigma-Aldrich), GTP (100 μM, Sigma-Aldrich), CTP (100 μM, Sigma-Aldrich), TTP (100 μM, Sigma-Aldrich), ATPγS (100 μM, Sigma-Aldrich), 2MeS-ATP (100 μM, Sigma-Aldrich), ADP (100 μM, Sigma-Aldrich), AMP (100 μM, Sigma-Aldrich), adenosine (100 μM, Sigma-Aldrich), U0126 (10 μM, Sigma-Aldrich) and SB203580 (10 μM, Sigma-Aldrich).

4.3. C2C12 Cell Culture

C2C12 myoblasts were cultured in a growth medium (DMEM; (Wako) supplemented with 10% FBS and 1% penicillin-streptomycin) at 37 °C with 5% CO_2 as previously described [52]. Myogenic differentiation was induced by changing the growth medium to a differentiation medium (DMEM supplemented with 2% horse serum and 1% penicillin-streptomycin) at 37 °C with 5% CO_2. For Western blotting, myotubes were deprived of serum for 8 h, and then cultured for an additional 2 h in Earle's

Balanced Salt Solution (Sigma-Aldrich) to deprive both serum and amino acids. C2C12 myotubes were treated with the indicated reagents for 30 min.

4.4. Gene Knockdown by RNA Interference

The transfection of siRNA (100 nM, Sigma-Aldrich) into C2C12 myotubes was performed using jetPRIME (Polyplus). The following siRNA sequences were used: p85#1 5'-ggaauaugauagauuauautt-3', p85#3 5'-ggauuaugcauaaccaugatt-3', Vps34#2 5'-caucugaccacgaucucaatt-3', Vps34#3 5'-cuacaaggcguuu aguacatt-3', P2Y$_2$#1 5'-caaucauuuguacgugugatt-3', P2Y$_2$#2 5'-cuaaggacauucggcuauatt-3', P2Y$_2$#3 5'-gucuugacccgguacucuatt-3'. The control siRNA was purchased from Sigma-Aldrich.

4.5. Retrovirus-Mediated Gene Transfer

The red genetically encoded Ca^{2+} indicators for optimal imaging (R-GECO) [53] or the IP$_3$-sponge-eGFP cDNA were cloned into a pMXs-puro vector [54]. Viral particles were prepared by introducing the resultant pMXs-R-GECO and pMXs-IP$_3$-sponge-eGFP into PLAT-E retrovirus packaging cells using polyethylenimine (Polysciences, Warrington, PA, USA) [55]. Subsequently, the concentrated supernatant was added to the C2C12 myoblast culture containing 8 ng/mL polybrene (Sigma-Aldrich). The next day, cells stably expressing the viral vector were selected with puromycin (Invitrogen, Carlsbad, CA, USA).

4.6. Intracellular Ca^{2+} Level Measurement

For in vitro experiments, [Ca^{2+}]$_i$ was monitored using Fluo-4 or R-GECO as previously described [8]. Briefly, C2C12 myotubes were cultured in a serum-free medium for 6 to 9 h and then cultured for an additional 2 h in a buffer containing 140 mM NaCl, 5 mM KCl, 2.5 mM CaCl$_2$, 1 mM MgCl$_2$, 10 mM HEPES and 10 mM glucose (pH 7.0). Single fibers from soleus or EDL muscles were isolated by digestion using type 1 collagenase as previously described [9,52]. Myotubes or isolated single fibers were incubated in 4 μM Fluo-4 AM (Dojindo, Rockville, MA, USA) for 30 min at room temperature to allow homogenous intracellular distribution of the dye. After incubation at 37 °C, the Fluo-4 loaded cells were placed on the stage of an inverted microscope (Olympus, Tokyo, Japan), and fluorescent intensity changes were recorded every 3 s. Data were calculated as normalized fluorescence $\Delta F/F0$: $\Delta F/F0 = (Fmax - F0)/F0$, where Fmax was the maximum fluorescence and F0 was the fluorescence before exposure to a reagent. Where noted, thapsigargin was combined with Fluo-4 AM, and the extracellular Ca^{2+} was depleted using a Ca^{2+}-free buffer containing 2 mM EGTA. For in vivo Ca^{2+} imaging analysis, soleus or plantaris muscles were immediately dissected from pCAGGS-GCaMP2 mice before or after exercise, and the fluorescence was observed using a Biozero digital microscope (Keyence, Osaka, Japan).

4.7. Western Blot Analysis

Western blot analysis was performed as described previously [52] with minor modifications. Briefly, cells were frozen immediately by liquid nitrogen, and protein was extracted using a sample buffer containing 0.1% Triton X-100. 50 mM HEPES (pH 7.4), 4 mM EGTA, 10 mM EDTA, 15 mM Na$_4$P$_2$O$_7$, 100 mM glycerophosphate, 25 mM NaF, 5 mM Na$_2$VO$_4$ and a complete protease inhibitor cocktail (Roche, Basel, Switzerland). Protein concentrations were determined by Coomassie Brilliant Blue G-250 (Bio-Rad, Hercules, CA, USA). Just before SDS-PAGE, an aliquot of the extracted protein solution was mixed with an equal volume of sample loading buffer containing 30% glycerol, 5% 2-mercaptoethanol, 2.3% SDS, 62.5 mM Tris-HCl (pH 6.8) and 0.05% bromophenol blue. The mixture was heated at 60 °C for 10 min. Proteins were separated on an SDS-polyacrylamide gel and electrically transferred from the gel to a polyvinylidene difluoride membrane (Millipore, Burlington, MA, USA). The signals were detected using the ECLTM Western Blotting Detection system and ImageQant LAS 4000 (GE Healthcare, Chicago, IL, USA), quantified by scanning densitometry (ImageQuant TL) and expressed in arbitrary units.

4.8. RNA Preparation and RT-qPCR

RT-qPCR was performed as described previously [52] with minor modifications. Briefly, total RNA was isolated by TRIzol (Invitrogen). Single strand cDNA was synthesized by the QuantiTect Reverse Transcription Kit (Qiagen, Hilden, Germany). Expression levels of JunB, IL-6, P2Y$_1$ and P2Y$_2$ receptors were evaluated by quantitative RT-PCR using SYBR Premix Ex Taq II (Takara, Kyoto, Japan) on a MyiQ single-color system (Bio-Rad), and were normalized to the TATA-binding protein (TBP). Primer sequences for RT-qPCR were as follows: JunB forward: 5′-ctcaacctggcggatccctatc-3′, reverse: 5′-gtgtctgatccctgacccgaaa-3′; IL-6 forward: 5′-agatctactcggcaaacc-3′, reverse: 5′-cgtagagaacaacataagtcag-3′; P2Y$_1$ forward: 5′-cgtggctatctggatgttcgt-3′, reverse: 5′-agatgagggctggtagggt-3′; P2Y$_2$ forward: 5′-ctgggatacaagtgtcgtt-3′, reverse: 5′-atagagagccacgacgtt-3′; TBP forward: 5′-cagcctcagtacagcaatcaac-3′, reverse: 5′-taggggtcataggagtcattgg-3′.

4.9. Exercise Model

Exercise experiment was performed as described previously [8,9] with minor modifications. Mice were placed on a flat MK-680S treadmill (Muromachi Kikai, Tokyo, Japan) and forced to run at 5 m/min. After 5 min, the speed was gradually increased by 1 m/min every minute. After 30 min, the test was stopped, and gastrocnemius or plantaris muscle were immediately isolated.

4.10. ELISA

The release of IL-6 elicited by ATP (100 µM) from C2C12 myotubes was measured in both, the absence and the presence of BAPTA-AM (50 µM) or rapamycin (10 µM). Myotubes were deprived of serum for 8 h, and then cultured for an additional 2 h in Earle's Balanced Salt Solution to deprive both serum and amino acids. Four hours after stimulation by ATP, 50 µL samples were taken from each culture plate and IL-6 levels were measured using an ELISA kit (Thermo Scientific, Waltham, MA, USA).

4.11. Hindlimb-Suspension and Denervation Surgery

Hindlimb-suspension analysis and denervation surgery were performed as described previously [8]. For hindlimb-suspension analysis, mice were randomly assigned to control or hindlimb-suspension groups. To induce muscle atrophy, mice were suspended by their tail so that their hindlimbs were 2 mm off the floor for 14 days. For denervation surgery, the sciatic nerve was excised from a small incision made in the mid-lateral thigh under general anesthesia. Mice were sacrificed 14 days after surgery. Muscle weight was normalized to body weight and was presented as a percentage of the control group. ATP (500 µM; 150 µL) was injected into hindlimb muscles once a day for 13 days. Since the injected volume was 150 µL, and muscle weights were about 200–300 mg, a dilution of more than two-fold should result.

4.12. Histological and Immunohistological Analysis

Histological and immunohistological analysis were performed as described previously [52,56,57] with minor modifications. After cervical dislocation, soleus muscle was immediately isolated and frozen in isopentane cooled by liquid nitrogen. Eight-µm thick cryosections were cut across the middle part of the muscle. For immunohistochemistry, cryosections were fixed in cold acetone for 10 min, and incubated in PBS containing 1% bovine serum albumin and 5% goat serum for 15 min at room temperature. Anti-laminin-α2 antibody in PBS containing 1% bovine serum albumin was applied overnight at 4 °C. Following incubation with secondary antibodies, mounted sections were observed using a Biozero digital microscope (Keyence), and cross-sectional areas were determined.

4.13. Statistical Analysis

All values are expressed as mean ± S.E.M. Statistical differences were assessed by the Student's *t*-test or one-way analysis of variance, with differences among the groups assessed by a Tukey-Kramer post-hoc analysis. Probabilities less than 5% (*, $p < 0.05$), 1% (**, $p < 0.01$) or 0.1% (***, $p < 0.001$) were considered to be statistically significant.

Supplementary Materials: Supplementary materials can be found at http://www.mdpi.com/1422-0067/19/9/2804/s1.

Author Contributions: N.I. conceived, designed and performed the experiments. N.I. also wrote the manuscript. U.T.R. and S.T. supervised the study, and all authors discussed the results and commented on the manuscript.

Funding: This work was supported by JSPS (Japan Society for the Promotion of Science) Grant-in-Aid for Scientific Research (B) (Number 25282202), Grant-in-Aid for JSPS Research Fellow (Number 11J04790), and Intramural Research Grant (28-6) for Neurological and Psychiatric Disorders of NCNP.

Acknowledgments: We would like to thank Isao Kii and Takashi Nishiyama for contributing to valuable discussions. We also thank Michael Kotlikoff for providing pCAGGS-GCaMP2 mice. IP$_3$-sponge-eGFP and pMXs-puro vector were kindly provided by H. Llewelyn Roderick and Toshio Kitamura, respectively.

Conflicts of Interest: The authors declare no conflict of interest.

Abbreviations

AMPKα	AMP-activated protein kinase α
EDL	extensor digitorum longus
IP$_3$R	inositol 1,4,5-trisphosphate receptor
IGF-1	linear dichroism
IL-6	interleukin-6
[Ca^{2+}]$_i$	intracellular Ca^{2+} level
mTOR	mammalian target of rapamycin
MAPK	mitogen-activated protein kinase
nNOS	neuronal nitric oxide synthase
PI3K	phosphatidylinositol 3-kinase
PLC	phospholipase C
SR	sarcoplasmic reticulum
TBP	TATA-binding protein
XeC	xestospongin C

References

1. Egerman, M.A.; Glass, D.J. Signaling pathways controlling skeletal muscle mass. *Crit. Rev. Biochem. Mol. Biol.* **2014**, *49*, 59–68. [CrossRef] [PubMed]
2. Rommel, C.; Bodine, S.C.; Clarke, B.A.; Rossman, R.; Nunez, L.; Stitt, T.N.; Yancopoulos, G.D.; Glass, D.J. Mediation of IGF-1-induced skeletal myotube hypertrophy by Pl(3)K/Alt/mTOR and Pl(3)K/Akt/GSK3 pathways. *Nat. Cell Biol.* **2001**, *3*, 1009. [CrossRef] [PubMed]
3. Goodman, C.A. The role of mTORC1 in regulating protein synthesis and skeletal muscle mass in response to various mechanical stimuli. *Rev. Physiol. Biochem. Pharmacol.* **2014**, *166*, 43–95. [PubMed]
4. Spangenburg, E.E.; Le Roith, D.; Ward, C.W.; Bodine, S.C. A functional insulin-like growth factor receptor is not necessary for load-induced skeletal muscle hypertrophy. *J. Physiol.* **2008**, *586*, 283–291. [CrossRef] [PubMed]
5. Philp, A.; Hamilton, D.L.; Baar, K. Signals mediating skeletal muscle remodeling by resistance exercise: PI3-kinase independent activation of mTORC1. *J. Appl. Physiol.* **2011**, *110*, 561–568. [CrossRef] [PubMed]
6. Hornberger, T.A.; Stuppard, R.; Conley, K.E.; Fedele, M.J.; Fiorotto, M.L.; Chin, E.R.; Esser, K.A. Mechanical stimuli regulate rapamycin-sensitive signalling by a phosphoinositide 3-kinase-, protein kinase B- and growth factor-independent mechanism. *Biochem. J.* **2004**, *380*, 795–804. [CrossRef] [PubMed]

7. Miyazaki, M.; Mccarthy, J.J.; Fedele, M.J.; Esser, K.A. Early activation of mTORC1 signalling in response to mechanical overload is independent of phosphoinositide 3-kinase/Akt signalling. *J. Physiol.* **2011**, *589*, 1831–1846. [CrossRef] [PubMed]

8. Ito, N.; Ruegg, U.T.; Kudo, A.; Miyagoe-Suzuki, Y.; Takeda, S. Activation of calcium signaling through Trpv1 by nNOS and peroxynitrite as a key trigger of skeletal muscle hypertrophy. *Nat. Med.* **2013**, *19*, 101–106. [CrossRef] [PubMed]

9. Ito, N.; Ruegg, U.T.; Kudo, A.; Miyagoe-Suzuki, Y.; Takeda, S. Capsaicin mimics mechanical load-induced intracellular signaling events. *Channels* **2013**, *7*, 221–224. [CrossRef] [PubMed]

10. Buvinic, S.; Almarza, G.; Bustamante, M.; Casas, M.; López, J.; Riquelme, M.; Sáez, J.C.; Huidobro-Toro, J.P.; Jaimovich, E. ATP released by electrical stimuli elicits calcium transients and gene expression in skeletal muscle. *J. Biol. Chem.* **2009**, *284*, 34490–34505. [CrossRef] [PubMed]

11. Forrester, T.; Lind, A.R. Identification of adenosine triphosphate in human plasma and the concentration in the venous effluent of forearm muscles before, during and after sustained contractions. *J. Physiol.* **1969**, *204*, 347–364. [CrossRef] [PubMed]

12. Li, J.; King, N.C.; Sinoway, L.I. ATP concentrations and muscle tension increase linearly with muscle contraction. *J. Appl. Physiol.* **2003**, *95*, 577–583. [CrossRef] [PubMed]

13. Ogawa, K.; Seta, R.; Shimizu, T.; Shinkai, S.; Calderwood, S.K.; Nakazato, K.; Takahashi, K. Plasma adenosine triphosphate and heat shock protein 72 concentrations after aerobic and eccentric exercise. *Exerc. Immunol. Rev.* **2011**, *17*, 136–149. [PubMed]

14. Fumagalli, M.; Lecca, D.; Abbracchio, M.P.; Ceruti, S. Pathophysiological role of purines and pyrimidines in neurodevelopment: Unveiling new pharmacological approaches to congenital brain diseases. *Front. Pharmacol.* **2017**, *8*, 941. [CrossRef] [PubMed]

15. Burnstock, G.; Verkhratsky, A. Receptors for Purines and Pyrimidines. In *Purinergic Signalling and the Nervous System*; Springer: Berlin/Heidelberg, Germany, 2012; ISBN 978-3-642-28862-3.

16. Kania, E.; Roest, G.; Vervliet, T.; Parys, J.B.; Bultynck, G. IP₃ Receptor-Mediated Calcium Signaling and Its Role in Autophagy in Cancer. *Front. Oncol.* **2017**, *7*, 140. [CrossRef] [PubMed]

17. Tallini, Y.N.; Ohkura, M.; Choi, B.-R.; Ji, G.; Imoto, K.; Doran, R.; Lee, J.; Plan, P.; Wilson, J.; Xin, H.-B.; et al. Imaging cellular signals in the heart in vivo: Cardiac expression of the high-signal Ca^{2+} indicator GCaMP2. *Proc. Natl. Acad. Sci. USA* **2006**, *103*, 4753–4758. [CrossRef] [PubMed]

18. Henning, R.H.; Nelemans, A.; van den Akker, J.; den Hertog, A. The nucleotide receptors on mouse C2C12 myotubes. *Br. J. Pharmacol.* **1992**, *106*, 853–858. [CrossRef] [PubMed]

19. Henning, R.H.; Duin, M.; den Hertog, A.; Nelemans, A. Characterization of P2-purinoceptor mediated cyclic AMP formation in mouse C2C12 myotubes. *Br. J. Pharmacol.* **1993**, *110*, 133–138. [CrossRef] [PubMed]

20. Henning, R.H.; Duin, M.; den Hertog, A.; Nelemans, A. Activation of the phospholipase C pathway by ATP is mediated exclusively through nucleotide type P2-purinoceptors in C2C12 myotubes. *Br. J. Pharmacol.* **1993**, *110*, 747–752. [CrossRef] [PubMed]

21. Ling, K.K.Y.; Siow, N.L.; Choi, R.C.Y.; Tsim, K.W.K. ATP potentiates the formation of AChR aggregate in the co-culture of NG108-15 cells with C2C12 myotubes. *FEBS Lett.* **2005**, *579*, 2469–2474. [CrossRef] [PubMed]

22. Deli, T.; Tóth, B.I.; Czifra, G.; Szappanos, H.; Bíró, T.; Csernoch, L. Differences in purinergic and voltage-dependent signalling during protein kinase Cα overexpression- and culturing-induced differentiation of C2C12 myoblasts. *J. Muscle Res. Cell Motil.* **2006**, *27*, 617–630. [CrossRef] [PubMed]

23. Rigault, C.; Bernard, A.; Georges, B.; Kandel, A.; Pfützner, E.; Le Borgne, F.; Demarquoy, J. Extracellular ATP increases L-carnitine transport and content in C2C12 cells. *Pharmacology* **2008**, *81*, 246–250. [CrossRef] [PubMed]

24. Banachewicz, W.; Supłat, D.; Krzemiński, P.; Pomorski, P.; Barańska, J. P2 nucleotide receptors on C2C12 satellite cells. *Purinergic Signal.* **2005**, *1*, 249. [CrossRef] [PubMed]

25. Lustig, K.D.; Shiau, A.K.; Brake, A.J.; Julius, D. Expression cloning of an ATP receptor from mouse neuroblastoma cells. *Proc. Natl. Acad. Sci. USA* **1993**, *90*, 5113–5117. [CrossRef] [PubMed]

26. Uchiyama, T.; Yoshikawa, F.; Hishida, A.; Furuichi, T.; Mikoshiba, K. A novel recombinant hyperaffinity inositol 1,4,5-trisphosphate (IP₃) absorbent traps IP₃, resulting in specific inhibition of IP₃-mediated calcium signaling. *J. Biol. Chem.* **2002**, *277*, 8106–8113. [CrossRef] [PubMed]

27. Hanson, C.J.; Bootman, M.D.; Roderick, H.L. Cell signalling: IP₃ receptors channel calcium into cell death. *Curr. Biol.* **2004**, *14*, R933–R935. [CrossRef] [PubMed]

28. Shimobayashi, M.; Hall, M.N. Making new contacts: The mTOR network in metabolism and signalling crosstalk. *Nat. Rev. Mol. Cell Biol.* **2014**, *15*, 155. [CrossRef] [PubMed]

29. Mizushima, N.; Yoshimori, T.; Levine, B. Methods in Mammalian Autophagy Research. *Cell* **2010**, *140*, 313–326. [CrossRef] [PubMed]

30. Vanhaesebroeck, B.; Guillermet-Guibert, J.; Graupera, M.; Bilanges, B. The emerging mechanisms of isoform-specific PI3K signalling. *Nat. Rev. Mol. Cell Biol.* **2010**, *11*, 329. [CrossRef] [PubMed]

31. Bustamante, M.; Fernandez-Verdejo, R.; Jaimovich, E.; Buvinic, S. Electrical stimulation induces IL-6 in skeletal muscle through extracellular ATP by activating Ca^{2+} signals and an IL-6 autocrine loop. *AJP Endocrinol. Metab.* **2014**, *306*, E869–E882. [CrossRef] [PubMed]

32. Braun, T.; Gautel, M. Transcriptional mechanisms regulating skeletal muscle differentiation, growth and homeostasis. *Nat. Rev. Mol. Cell Biol.* **2011**, *12*, 349. [CrossRef] [PubMed]

33. Raffaello, A.; Milan, G.; Masiero, E.; Carnio, S.; Lee, D.; Lanfranchi, G.; Goldberg, A.L.; Sandri, M. JunB transcription factor maintains skeletal muscle mass and promotes hypertrophy. *J. Cell Biol.* **2010**, *191*, 101–113. [CrossRef] [PubMed]

34. Serrano, A.L.; Baeza-Raja, B.; Perdiguero, E.; Jardí, M.; Muñoz-Cánoves, P. Interleukin-6 Is an Essential Regulator of Satellite Cell-Mediated Skeletal Muscle Hypertrophy. *Cell Metab.* **2008**, *7*, 33–44. [CrossRef] [PubMed]

35. Mahoney, D.J.; Parise, G.; Melov, S.; Safdar, A.; Tarnopolsky, M.A. Analysis of global mRNA expression in human skeletal muscle during recovery from endurance exercise. *FASEB J.* **2005**, *19*, 1498–1500. [CrossRef] [PubMed]

36. Eferl, R.; Wagner, E.F. AP-1: A double-edged sword in tumorigenesis. *Nat. Rev. Cancer* **2003**, *3*, 859. [CrossRef] [PubMed]

37. Pedersen, B.K.; Akerström, T.C.A.; Nielsen, A.R.; Fischer, C.P. Role of myokines in exercise and metabolism. *J. Appl. Physiol.* **2007**. [CrossRef] [PubMed]

38. Blaauw, B.; Del Piccolo, P.; Rodriguez, L.; Hernandez Gonzalez, V.-H.; Agatea, L.; Solagna, F.; Mammano, F.; Pozzan, T.; Schiaffino, S. No evidence for inositol 1,4,5-trisphosphate-dependent Ca^{2+} release in isolated fibers of adult mouse skeletal muscle. *J. Gen. Physiol.* **2012**. [CrossRef] [PubMed]

39. Díaz-Vegas, A.; Campos, C.A.; Contreras-Ferrat, A.; Casas, M.; Buvinic, S.; Jaimovich, E.; Espinosa, A. ROS production via P2Y1-PKC-NOX2 is triggered by extracellular ATP after electrical stimulation of skeletal muscle cells. *PLoS ONE* **2015**. [CrossRef] [PubMed]

40. Casas, M.; Buvinic, S.; Jaimovich, E. ATP signaling in skeletal muscle: From fiber plasticity to regulation of metabolism. *Exerc. Sport Sci. Rev.* **2014**. [CrossRef] [PubMed]

41. Li, T.; Finch, E.A.; Graham, V.; Zhang, Z.-S.; Ding, J.-D.; Burch, J.; Oh-hora, M.; Rosenberg, P. STIM1-Ca^{2+} Signaling Is Required for the Hypertrophic Growth of Skeletal Muscle in Mice. *Mol. Cell. Biol.* **2012**. [CrossRef] [PubMed]

42. Zanou, N.; Schakman, O.; Louis, P.; Ruegg, U.T.; Dietrich, A.; Birnbaumer, L.; Gailly, P. Trpc1 ion channel modulates phosphatidylinositol 3-kinase/Akt pathway during myoblast differentiation and muscle regeneration. *J. Biol. Chem.* **2012**. [CrossRef] [PubMed]

43. Gulati, P.; Gaspers, L.D.; Dann, S.G.; Joaquin, M.; Nobukuni, T.; Natt, F.; Kozma, S.C.; Thomas, A.P.; Thomas, G. Amino Acids Activate mTOR Complex 1 via Ca^{2+}/CaM Signaling to hVps34. *Cell Metab.* **2008**. [CrossRef] [PubMed]

44. Mackenzie, M.G.; Hamilton, D.L.; Murray, J.T.; Taylor, P.M.; Baar, K. mVps34 is activated following high-resistance contractions. *J. Physiol.* **2009**. [CrossRef] [PubMed]

45. MacKenzie, M.G.; Hamilton, D.L.; Murray, J.T.; Baar, K. mVps34 is activated by an acute bout of resistance exercise. *Biochem. Soc. Trans.* **2007**. [CrossRef] [PubMed]

46. Mercan, F.; Lee, H.; Kolli, S.; Bennett, A.M. Novel Role for SHP-2 in Nutrient-Responsive Control of S6 Kinase 1 Signaling. *Mol. Cell. Biol.* **2013**. [CrossRef] [PubMed]

47. Febbraio, M.A.; Pedersen, B.K. Muscle-derived interleukin-6: Mechanisms for activation and possible biological roles. *FASEB J.* **2002**. [CrossRef] [PubMed]

48. von Kügelgen, I. Pharmacological profiles of cloned mammalian P2Y-receptor subtypes. *Pharmacol. Ther.* **2006**, *110*, 415–432. [CrossRef] [PubMed]

49. Moschella, M.C.; Watras, J.; Jayaraman, T.; Marks, A.R. Inositol 1,4,5-trisphosphate receptor in skeletal muscle: Differential expression in myofibers. *J. Muscle. Res. Cell. Motil.* **1995**, *16*, 390–400. [CrossRef] [PubMed]

50. Kobzik, L.; Reid, M.B.; Bredt, D.S.; Stamler, J.S. Nitric oxide in skeletal muscle. *Nature* **1994**, *372*, 546. [CrossRef] [PubMed]

51. White, N.; Burnstock, G. P2 receptors and cancer. *Trends Pharmacol. Sci.* **2006**, *27*, 211–217. [CrossRef] [PubMed]

52. Ito, N.; Kii, I.; Shimizu, N.; Tanaka, H.; Takeda, S. Direct reprogramming of fibroblasts into skeletal muscle progenitor cells by transcription factors enriched in undifferentiated subpopulation of satellite cells. *Sci. Rep.* **2017**, *7*. [CrossRef] [PubMed]

53. Zhao, Y.; Araki, S.; Wu, J.; Teramoto, T.; Chang, Y.F.; Nakano, M.; Abdelfattah, A.S.; Fujiwara, M.; Ishihara, T.; Nagai, T.; et al. An expanded palette of genetically encoded Ca^{2+} indicators. *Science* **2011**. [CrossRef] [PubMed]

54. Kitamura, T.; Koshino, Y.; Shibata, F.; Oki, T.; Nakajima, H.; Nosaka, T.; Kumagai, H. Retrovirus-mediated gene transfer and expression cloning: Powerful tools in functional genomics. *Exp. Hematol.* **2003**, *31*, 1007–1014. [CrossRef]

55. Morita, S.; Kojima, T.; Kitamura, T. Plat-E: An efficient and stable system for transient packaging of retroviruses. *Gene Ther.* **2000**. [CrossRef] [PubMed]

56. Ito, N.; Shimizu, N.; Tanaka, H.; Takeda, S. Enhancement of Satellite Cell Transplantation Efficiency by Leukemia Inhibitory Factor. *J. Neuromuscul. Dis.* **2016**. [CrossRef] [PubMed]

57. Hyzewicz, J.; Tanihata, J.; Kuraoka, M.; Ito, N.; Miyagoe-Suzuki, Y.; Takeda, S. Low intensity training of mdx mice reduces carbonylation and increases expression levels of proteins involved in energy metabolism and muscle contraction. *Free Radic. Biol. Med.* **2015**, *82*, 122–136. [CrossRef] [PubMed]

International Journal of
Molecular Sciences

MDPI

Article

The Amino-Terminal Domain of GRK5 Inhibits Cardiac Hypertrophy through the Regulation of Calcium-Calmodulin Dependent Transcription Factors

Daniela Sorriento [1], Gaetano Santulli [1,2], Michele Ciccarelli [3], Angela Serena Maione [4], Maddalena Illario [4], Bruno Trimarco [1] and Guido Iaccarino [3,*]

[1] Dipartimento di "Scienze Biomediche Avanzate", Università "Federico II" di Napoli, Via Pansini 5, 80131 Napoli, Italy; danisor@libero.it (D.S.); gsantulli001@gmail.com (G.S.); trimarco@unina.it (B.T.)
[2] Department of Medicine, Albert Einstein College of Medicine, Montefiore University Hospital, 1300 Morris Park Avenue, Bronx, NY 10461, USA
[3] Dipartimento di Medicina, Chirurgia e Odontoiatria "Scuola Medica Salernitana"/DIPMED, Università degli Studi di Salerno, Via S. Allende, 84081 Baronissi (SA), Italy; mciccarelli@unisa.it
[4] Dipartimento di "Scienze Mediche Traslazionali", Università "Federico II" di Napoli, Via Pansini 5, 80131 Napoli, Italy; mercoledi85@gmail.com (A.S.M.); illario@unina.it (M.I.)
*   Correspondence: giaccarino@unisa.it; Tel.: +39-089-965-021

Received: 24 January 2018; Accepted: 9 March 2018; Published: 15 March 2018

Abstract: We have recently demonstrated that the amino-terminal domain of G protein coupled receptor kinase (GRK) type 5, (GRK5-NT) inhibits NFκB activity in cardiac cells leading to a significant amelioration of LVH. Since GRK5-NT is known to bind calmodulin, this study aimed to evaluate the functional role of GRK5-NT in the regulation of calcium-calmodulin-dependent transcription factors. We found that the overexpression of GRK5-NT in cardiomyoblasts significantly reduced the activation and the nuclear translocation of NFAT and its cofactor GATA-4 in response to phenylephrine (PE). These results were confirmed in vivo in spontaneously hypertensive rats (SHR), in which intramyocardial adenovirus-mediated gene transfer of GRK5-NT reduced both wall thickness and ventricular mass by modulating NFAT and GATA-4 activity. To further verify in vitro the contribution of calmodulin in linking GRK5-NT to the NFAT/GATA-4 pathway, we examined the effects of a mutant of GRK5 (GRK5-NTPB), which is not able to bind calmodulin. When compared to GRK5-NT, GRK5-NTPB did not modify PE-induced NFAT and GATA-4 activation. In conclusion, this study identifies a double effect of GRK5-NT in the inhibition of LVH that is based on the regulation of multiple transcription factors through means of different mechanisms and proposes the amino-terminal sequence of GRK5 as a useful prototype for therapeutic purposes.

Keywords: cardiac hypertrophy; transcription factors; calmodulin; GRK

1. Introduction

Left ventricular hypertrophy (LVH) is an adaptive response of the heart to stress that eventually, and progressively, turns into maladaptive, evolving towards cardiac dysfunction and heart failure [1–4]. Available therapies targeting the pro-hypertrophic pathways, including angiotensin converting enzyme (ACE) inhibitors and β-adrenergic receptor (β-AR) blockers, can reduce hypertrophy significantly, but not completely [5]. During the past decade, scientific research has focused on the identification of new targets for the treatment of cardiac hypertrophy [5–7]. In particular, alterations in intracellular calcium fluxes activate intracellular pathways that are involved in the progression of LVH and myocardial remodeling [8]. Intracellular Calcium fluxes are the primary component of excitation-contraction coupling and maintain heart contractility [9,10], but can also ignite nuclear gene transcription. In particular, several transcription factors, including nuclear factor of activated T cells (NFAT), are calcium-dependent [11]. Calcium binds calmodulin (CaM) and activates the serine/threonine calcium-calmodulin phosphatase calcineurin. Activated calcineurin, in turn, dephosphorylates the transcription factor NFAT, which quickly moves from cytosol to the nucleus [12,13]. Here NFAT regulates the transcription of genes involved in the development of cardiac hypertrophy, including Atrial Natriuretic Factor, TNF-α, and Endothelin-1 [14]. NFAT works in association with other transcription factors, such as GATA-4. This latter is activated by ERK phosphorylation [15], but also Ca/CaM dependent pathway can enable it.

Previously, we demonstrated that the RH domain of G protein-coupled receptor kinase 5 (GRK5) inhibits NFκB transcriptional activity through binding of the inhibitory protein IκBα [16–18]. In vivo, the intra-myocardial injection of an adenovirus encoding for the amino-terminal domain of GRK5 (AdGRK5-NT), which comprises the RH domain, reduces LVH in spontaneously hypertensive rats (SHR) in a blood pressure-independent manner through the inhibition of NFκB transcription activity [19]. Interestingly, the same amino-terminal sequence of GRK5 that contains the RH domain also flanks a calmodulin binding site [20]. Therefore, it is possible to speculate that GRK5-NT is involved in the regulation of calcium-calmodulin dependent events of activation of transcription factors. This study aims to evaluate whether GRK5-NT can regulate the activation of transcription factors NFAT and GATA-4 through the interaction with calcium-calmodulin dependent signaling pathways.

2. Results

2.1. GRK5-NT Regulates the Activation of Calcium-Calmodulin Dependent Transcription Factors In Vitro

In cultured cardiomyoblasts, hypertrophy was induced by chronic PE stimulation. PE induced the activation of the transcription factors GATA-4, NFκB, and NFAT (Figure 1A). The overexpression of GRK5-NT inhibited the activation of these transcription factors in response to PE (Figure 1A). On the contrary, TAT-RH, which is a peptide that reproduces only the RH domain of GRK5 lacking the amino-terminal domain, did not affect GATA-4 and NFAT activation in response to PE, but was active on NFκB inhibition, as consistent with our previous findings [18]. These data suggest that GRK5-NT is able to regulate calcium-calmodulin-dependent transcription factors, through means of its amino-terminal domain and independently from the RH domain. To verify these findings, we assessed NFAT and GATA 4 nuclear translocation. Strikingly, GRK5-NT overexpression reduced their nuclear accumulation in response to PE, whereas TATRH had no effects (Figure 1B).

2.2. GRK5-NT Inhibits NFAT Activation by Competing for Binding to Calmodulin

To further clarify in vitro the molecular mechanisms by which GRK5-NT regulates the activation of calcium-calmodulin-dependent transcription factors, we tested the hypothesis that GRK5-NT regulates these factors by sequestrating calmodulin. To this aim, cardiomyoblasts were transfected with a plasmid encoding a mutated form of GRK5 in the calmodulin binding site (GRK5-NTPB), which enables the kinase to bind calmodulin [21]. Figure 2A shows that, when compared to GRK5-NT, GRK5-NTPB had no significant effect on PE-induced nuclear translocation of GATA-4 and NFAT, strongly suggesting that GRK5-NT regulates GATA-4 and NFAT activation by competing with calcineurin for binding to calmodulin. Therefore, we evaluated the effects of GRK5-NT and GRK5-NTPB on calmodulin/calcineurin interaction via immunoprecipitation and western blot. We observed that PE induces such interaction, which is not affected by GRK5-NTPB, but is markedly reduced by GRK5-NT (Figure 2B).

A

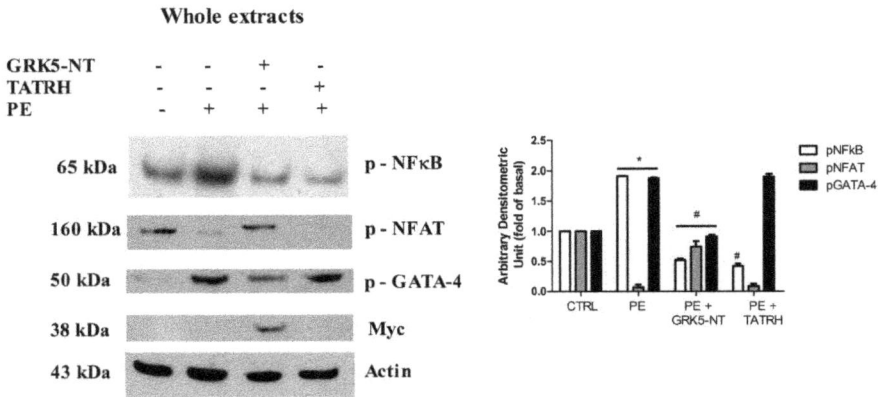

B

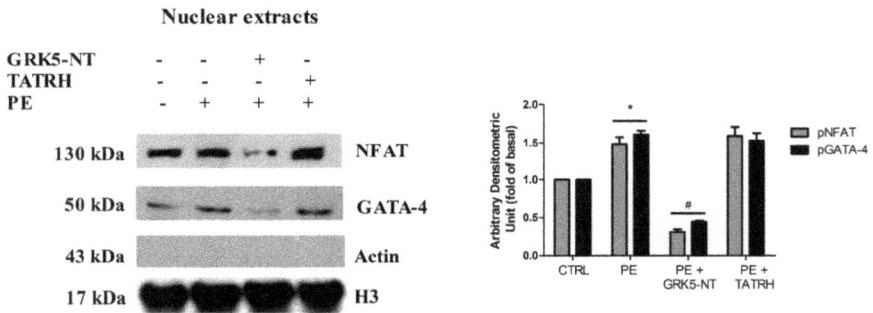

Figure 1. The amino-terminal domain of the G Protein Coupled Receptor Kinase 5 (GRK5-NT) regulates the activation of calcium-calmodulin dependent transcription factors in vitro. (**A**) In cultured cardiomyoblasts H9C2, hypertrophy was induced by chronic stimulation with Phenylephrine (PE) and the activation of NFκB, nuclear factor of activated T cells (NFAT), and GATA-4 in response to PE was evaluated by western blot. PE triggered the phosphorylation of GATA-4, NFκB and the dephosphorylation of NFAT. The overexpression of GRK5-NT reduces such phenomenon while TAT-RH, which has only the RH domain of GRK5, did not modify PE-induced GATA-4 and NFAT activation albeit being effective on NFκB activation. Images are representative of three independent experiments. Densitometric analysis is shown in bar graph as mean ± SD; * $p < 0.05$ vs. Control and # $p < 0.05$ vs. PE; (**B**) nuclear accumulation of NFAT and GATA-4 was evaluated by western blot in nuclear extracts from H9C2 cells. GRK5-NT reduced the PE-dependent nuclear translocation of these factors while TATRH is not able to exert the same effect. Actin was used as control of nuclear extracts purity. Images are representative of three independent experiments. Densitometric analysis is shown in bar graph as mean ± SD; * $p < 0.05$ vs. Control and # $p < 0.05$ vs. PE.

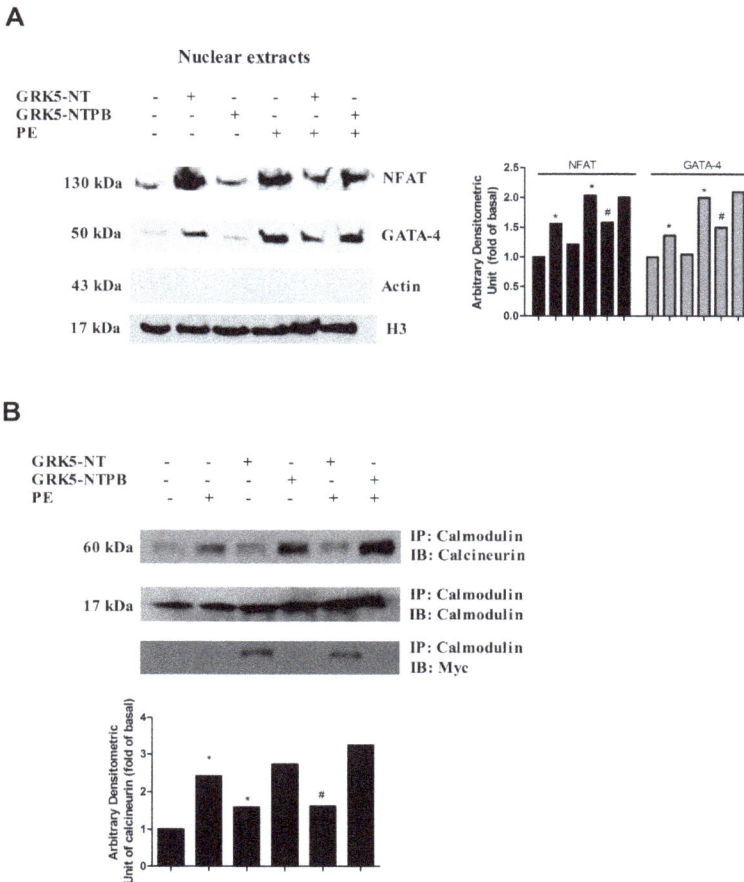

Figure 2. The amino-terminal domain of the G Protein Coupled Receptor Kinase 5 (GRK5-NT) inhibits NFAT activation by sequestrating calmodulin. (**A**) Cardiomyoblasts were transfected with a plasmid encoding GRK5-NT or GRK5-NTPB and NFAT and GATA-4 nuclear translocation was evaluated by western blot in presence and absence of phenylephrine (PE). GRK5-NTPB had no significant effect on nuclear translocation and activation of NFAT and GATA4; instead GRK5-NT inhibited such phenomenon. Actin was used as control of nuclear extracts purity and histone 3 (H3) was used as loading control. Densitometric analysis is shown in bar graph as mean $\pm$ SD, * $p < 0.05$ vs. Control and # $p < 0.05$ vs. PE; (**B**) cardiomyoblasts were transfected with a plasmid encoding GRK5-NT or GRK5-NTPB and calmodulin was precipitated in whole lysates from these cells. Calcineurin was evaluated by western blot. GRK5-NT reduced calmodulin/calcineurin interaction. GRK5-NTPB had no effect on such interaction. All images are representative of three independent experiments. Densitometric analysis is shown in bar graph as mean $\pm$ SD, * $p < 0.05$ vs. Control and # $p < 0.05$ vs. PE.

2.3. GRK5-NT Regulates the Activation of Calcium-Calmodulin Dependent Transcription Factors In Vivo

To confirm our in vitro data, we evaluated the effects of GRK5-NT on NFAT and GATA-4 activation in an animal model of hypertrophy. Spontaneously hypertensive rats (SHR) underwent intra-myocardial injections of an adenovirus encoding for GRK5-NT (AdGRK5-NT) or Lac-Z as control (AdLac-Z), as previously described [19]. Rats were monitored for three weeks by CUS to assess the effect of such treatment on LVH. After 21 days, LVH was significantly reduced, as underlined by the reduction of IVS (Figure 3A) and LVM/BW (Figure 3B). Moreover, cardiac function was recovered in treated rats (Figure 3C). Hearts were then collected and the nuclear translocation of NFAT and GATA-4 was assessed by immunoblot. In hypertrophic SHR rats, there was a significant accumulation of NFAT and GATA-4 in the nucleus as compared with WKY (Figure 4A). The treatment with AdGRK5-NT significantly inhibited such nuclear translocation (Figure 4A) thereby indicating that GRK5-NT can reduce NFAT and GATA-4 activation in response to hypertrophy in vivo. Importantly, this finding was confirmed by EMSA assay, which shows that GRK5-NT significantly reduces both NFAT (Figure 4B) and GATA-4 (Figure 4C) ability to bind DNA in an established model of cardiac hypertrophy (i.e., SHR).

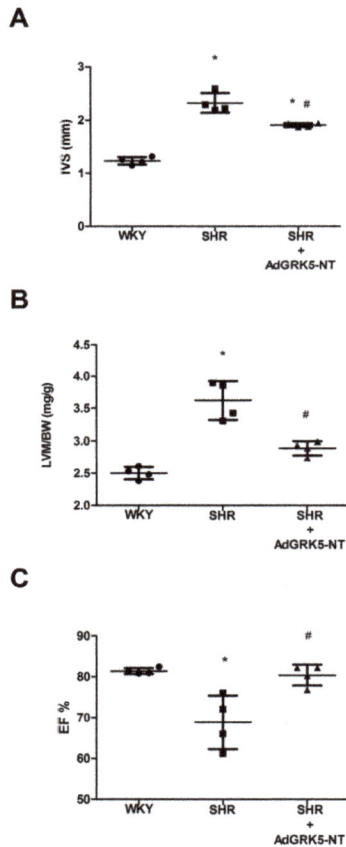

Figure 3. The amino-terminal domain of the G Protein Coupled Receptor Kinase 5 (GRK5-NT) inhibits calcium-calmodulin dependent transcription factors in vivo. (**A–C**) Spontaneously hypertensive and hypertrophic rats (SHR) were treated with an intra-myocardial injection of an adenovirus encoding for GRK5-NT (AdGRK5-NT) or Lac-Z (AdLac-Z), as described in methods, and cardiac hypertrophy was evaluated by echocardiography. 21 days after injection, a reduction of IVS (**A**) and LVM/BW (**B**), was found in AdGRK5-NT versus SHR controls. Cardiac function was recovered in treated SHR vs. SHR (**C**) * $p < 0.05$ vs. WKY and # $p < 0.05$ vs. SHR + AdLac-Z.

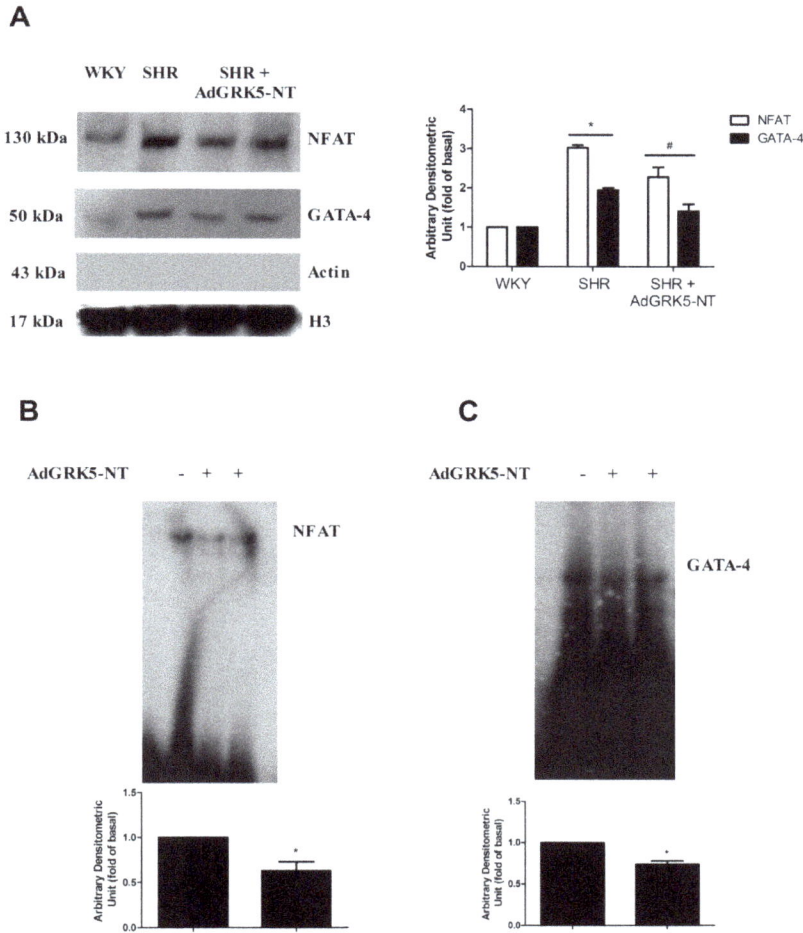

Figure 4. The amino-terminal domain of the G Protein Coupled Receptor Kinase 5 (GRK5-NT) regulates NFAT and GATA-4 activity in vivo. (**A**) The nuclear translocation of NFAT and GATA-4 was evaluated by immunoblot in rat hearts. In hypertrophic SHR rats, there was a significant accumulation of NFAT and GATA-4 in nuclear extracts respect to WKY. The treatment with AdGRK5-NT significantly reduced nuclear translocation of these factors. Densitometric analysis is shown in bar graph as mean ± SD. Actin was used as control of nuclear extracts purity; * $p < 0.05$ vs. WKY and # $p < 0.05$ vs. SHR + AdLac-Z; (**B,C**) to confirm the inhibition of transcription factors activity, we analyzed NFAT and GATA-4 ability to bind DNA by EMSA in control and treated SHR. GRK5-NT reduced both NFAT (**B**) and GATA-4 (**C**) activity in response to hypertrophy. All of the images are representative of at least three independent experiments. Densitometric analysis is shown in bar graph as mean ± SD; * $p < 0.05$ vs. SHR + AdLac-Z.

3. Discussion

Several transcription factors are involved in the regulation and development of LVH [11,22,23] by regulating the expression of critical hypertrophic genes in response to specific stimuli [24,25]. We have recently demonstrated that the treatment with GRK5-NT inhibits NFκB activity in hypertrophied hearts [19]. Here, we describe a novel level of inhibition of LVH by using the ability of the GRK5 sequence to regulate CaM signaling and to prevent NFAT and GATA-4 activation. Indeed, we show that GRK5-NT regulates the activation of the calcium-calmodulin-dependent transcription factor, NFAT, and its cofactor GATA-4, through binding to calmodulin (Figure 5). Other calcium-calmodulin dependent pathways are involved in the development of cardiac hypertrophy, such as CaMKs signaling, and we cannot exclude the possibility that they could be affected by GRK5-NT. However, in this study, we focused on NFAT and its co-factor GATA-4, since they are among the main cardiac transcription factors whose activation is strictly dependent on calcium-calmodulin interactions. Hence, our data indicate that within the GRK5-NT sequence, there are two regions with the potentiality to regulate cardiac hypertrophy in two different ways: by inhibiting NFκB through the binding of the RH domain to IκBα [19] and by inhibiting NFAT through the binding and sequestration of calmodulin in the amino-terminal domain. GRK5-NT can simultaneously affect these intracellular signaling pathways since our data show that it can inhibit both NFκB and NFAT when compared with TATRH.

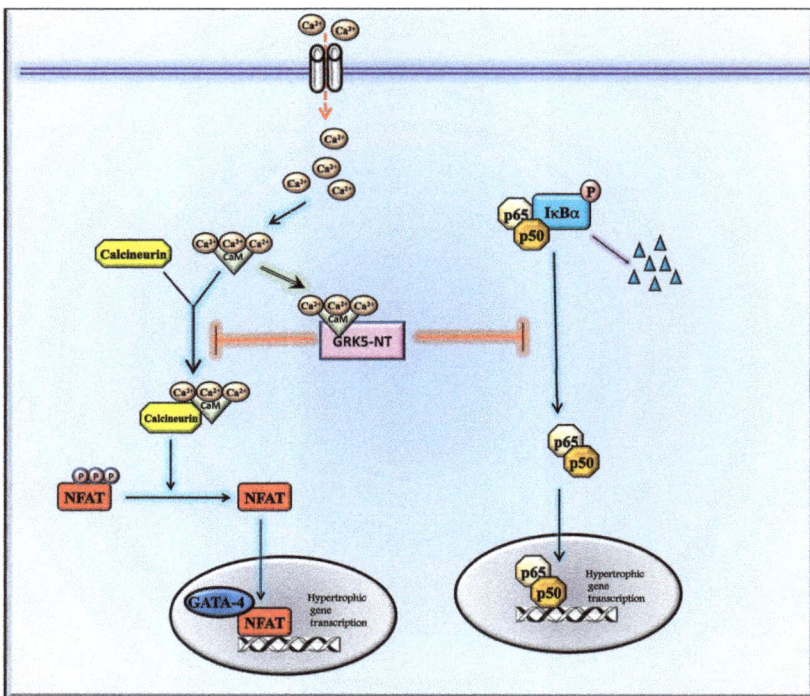

Figure 5. Schematic representation of the molecular mechanism. GRK5-NT regulates GATA-4 and NFAT activation by competing with calcineurin for binding to calcium-calmodulin. PE induces an increase of intracellular calcium levels. Calcium binds calmodulin and activates calcium-calmodulin dependent factors such as calcineurin. Calcineurin, in turn, dephosphorylates NFAT leading to NFAT nuclear translocation and activation of gene transcription. GRK5-NT binds calcium-calmodulin, through the calmodulin binding site (CBS), thus inhibiting its binding to calcineurin. Inactivated calcineurin is not anymore able to dephosphorylate NFAT and this causes the inhibition of NFAT dependent gene transcription and the inhibition of NFAT dependent GATA-4 activation. Besides this effect of GRK5-NT, the peptide is also able to inhibit IκB degradation, thus preventing NFκB transcription activity.

A direct association between NFAT and NFκB activity in cardiac myocytes has been shown to promote cardiac hypertrophy and ventricular remodeling [26]. In particular, since the inhibition of NFκB with IκBαM or the dominant negative of IKKβ reduces NFAT activity both in vitro and in vivo [26], it is likely that the inhibitory effect of GRK5-NT on NFAT activity could be a consequence of GRK5-NT dependent inhibition of NFκB. Actually, in our model, we did not find a mechanistic association between NFκB and NFAT activity. Indeed, while GRK5-NT regulates both NFκB and NFAT activity, TAT-RH—which specifically inhibits NFκB signaling—does not modify NFAT or GATA-4 activation. Such discrepancy could be attributable to the different ways of inhibition of NFκB transcriptional activity. Indeed, the inhibition of NFκB that was obtained via overexpressing IκBαM or a dominant negative IKKβ is mainly based on the inhibition of IκB phosphorylation. In our model, the effects of GRK5-NT are instead based on a protein-protein interaction without interfering with IκB phosphorylation [16]. Such binding leads to the generation of a macromolecular complex that could prevent the binding of NFAT. Furthermore, GRK5-NT regulates NFAT in different manner respect to NFκB that is based on the sequestration of calmodulin through the amino-terminal domain. These findings suggest that GRK5-NT dependent inhibition of NFAT is not due to NFκB regulation, but the peptide is able to regulate both intracellular signalings simultaneously.

Here, we show that GRK5-NT also regulates GATA-4, which is not itself a calcium-calmodulin dependent transcription factor. It has been shown that its activation induced by PE stimulation is coupled with serine phosphorylation by Extracellular signal–regulated kinase 2 (ERK2) [15]. However, besides phosphorylation, the transcriptional activity of GATA4 is also regulated through interaction with other cofactors such as p300, MEF2, SRF, and NFAT [11,24]. Among them, the interaction with NFAT is noteworthy since NFAT plays a critical role in activating the hypertrophic gene program, and this activity is partly dependent on its interaction with GATA4 [24]. These findings suggest the functional importance of calcium signaling also in the activation of GATA4 and support our finding on GRK5-NT dependent GATA-4 inhibition through the modulation of calcium-calmodulin signaling.

Previous reports show that GRK5 also induces cardiac hypertrophy [27,28] by activating NFAT-dependent gene transcription [29]. In this context, GRK5-NT exerts opposite effects compared with the full-length sequence of GRK5. This is because GRK5-NT includes the amino-terminal domain of GRK5 (first 170 aminoacidic sequence) and lacks the catalytic domain that is instead responsible for GRK5 effects on gene transcription. Indeed, GRK5 exacerbates cardiac hypertrophy by phosphorylation of different substrates (plasma membrane receptors and transcription factors) [27–30]. On the contrary, GRK5-NT acts by protein-protein interaction within the RH domain and calmodulin binding within the amino-terminal sequence. However, we cannot exclude the possibility that GRK5-NT could interfere with endogenous GRK5 signaling. Literature is quite discordant on the effects of calmodulin binding on GRK5 signaling. Indeed, some reports show that CaM binding to GRK5 inhibits its catalytic activity [20], while others suggest that this interaction inhibits the binding of GRK5 to plasma membrane favoring its nuclear translocation to induce HDAC phosphorylation [27]. Pitcher et al. which generate the mutant plasmid of GRK5 (GRK5-NTPB) showed that the binding of nuclear Calcium-Calmodulin to the amino-terminal CaM-binding site of GRK5 is required for nuclear export [21]. Thus, further studies are needed to better clarify this issue. Our data show that GRK5-NT in basal condition exerts a pro-hypertrophic effect inducing NFAT nuclear translocation, and this appears to be in contrast with the effect of the peptide in response to PE. Hypertrophic stimuli increase the intracellular calcium levels and enhance the affinity of calmodulin to its substrates. Thus the effect of GRK5-NT to compete with calcineurin for calmodulin binding could take place mainly in response to stimuli. In resting conditions, GRK5-NT may interfere with other physiological pathways and indirectly regulate NFAT nuclear translocation. Taken together, our data demonstrate the ability of GRK5-NT to control several fundamental cardiac transcription factors in response to hypertrophy by different mechanisms of action which involve different domains of GRK5. Specifically, here we demonstrate the ability of GRK5-NT to regulate calcium-calmodulin dependent transcription factors, proposing GRK5 as a potential regulator of calcium signaling through its calmodulin binding site. In conclusion, our data confirm the usefulness of GRK5-NT for the treatment of LVH, suggesting the GRK5-NT sequence as a prototype for the generation of small molecules that are to be used for therapeutic applications.

4. Materials and Methods

4.1. Cell Culture

A cell line of cardiac myoblasts (H9C2) was maintained in culture in Dulbecco Modified Eagle Medium (DMEM) supplemented with 10% FBS at 37 °C in 95% air-5% CO_2.

4.2. Plasmids

p-GRK5-NT is a pcDNA3.1 myc/his plasmid encoding the amino-terminal domain of GRK5, including the RH domain (aa 1-176), and was described previously [16]; p-GRK5-NTPB, which is a plasmid encoding a mutant form of GRK5 which lacks the ability to bind calmodulin, was a kind gift of Julie Pitcher (University College London, London, UK) and was used as template for cloning p-GRK5-NT mutant (GRK5-NT mut), as previously described [16]; TAT-RH was described previously [18]. All experiments were performed using TAT alone and empty pcDNA3.1 as controls. Transient transfection of the plasmids was performed, as previously described [31], using Lipofectamine 2000 from Invitrogen (Thermo Fisher Scientific, Waltham, MA, USA) in 70% confluent H9C2, accordingly to manufacturer instructions.

4.3. Western Blot

The experiments were performed as described previously [16,32]. H9C2 were treated with 10^{-7} M phenylephrine (PE, Sigma-Aldrich Corporation, St. Louis, MO, USA) for 24 h. In some experiments, cells were treated with the synthetic peptide TAT-RH, which only reproduce the RH domain of GRK5, as described previously [18] or transfected with plasmids encoding for GRK5-NT or its mutant in the amino-terminal calmodulin binding. At the end of the treatment, cells were lysed in RIPA/SDS buffer, and protein concentration was determined by using Pierce BCA assay kit (Thermo Fisher Scientific, Waltham, MA, USA) [33,34]. Total extracts were electrophoresed by SDS/PAGE and transferred to nitrocellulose [35]. The antibodies anti-NFATc4 (B-2) (SC-271597), GATA-4 (G-4) (sc-25310), p-NFATc4 (80.S168/170) (sc-135770), p-GATA-4 (H-4) (sc-377543), β-actin (C-4) (sc-47778), and histone H3 (FL-136) (sc-10809) were from Santa Cruz Biotechnology (Santa Cruz Biotechnology, Inc, Dallas, TX, USA). In some experiments, nuclear proteins were isolated from heart samples as previously described [16]. Densitometric analysis was performed using Image Quant 5.2 software (Molecular Dynamics Inc., Caesarea, Israel). Images are representative of at least three independent experiments quantified and corrected for appropriate loading control.

4.4. In Vivo Study

Experiments were carried out accordingly with the Federico II University Ethical Committee on 12-week–old spontaneously hypertensive male rats SHR (AdLac-Z $n = 6$, and AdGRK5-NT $n = 6$) and 4 normotensive Wistar-Kyoto (WKY) rats as control. The animals were obtained from Charles River (Wilmington, MA, USA) and had access to water and food ad libitum. Anesthesia was obtained through isoflurane (4%). After the induction of anesthesia, rats were orotracheally intubated, the inhaled concentration of isoflurane was reduced to 1.8%, and the lungs were mechanically ventilated (New England Medical Instruments Scientific, Inc., Chelmsford, MA, USA). The chest was opened under sterile conditions through a right parasternal mini-thoracotomy to expose the heart. Then, we performed four injections (50 µL each) of AdGRK5-NT (10^{10} pfu/mL) or AdLac-Z (10^{10} pfu/mL) as control, into the cardiac wall (anterior, lateral, posterior, and apical), as previously validated [19]. Finally, the chest wall was quickly closed in layers, and animals were observed and monitored until recovery.

Int. J. Mol. Sci. **2018**, *19*, 861

4.5. Cardiac Ultrasounds (CUS)

Transthoracic CUS was performed at days 0, 7, 14, and 21 after surgery using a dedicated small-animal high-resolution imaging system (VeVo 770, Visualsonics, Inc., Amsterdam, The Netherlands). The rats were anesthetized with isoflurane (4%) inhalation and maintained by mask ventilation, as described above. The chest was shaved with a depilatory cream (Veet, Reckitt-Benckiser, Milan, Italy). Left ventricular (LV) end-diastolic and LV end-systolic diameters (LVEDD and LVESD, respectively) were measured at the level of the papillary muscles from the parasternal short-axis view [36,37]. Intraventricular septal (IVS) and LV posterior wall thickness (PW) were measured at the end of the diastolic phase. LV mass (LVM) was obtained, as described and corrected by body weight [19,36]. All of the measurements were averaged on at least five consecutive cardiac cycles and were analyzed by investigators that were blinded to treatment.

4.6. EMSA

Electrophoretic mobility shift assay (EMSA) was performed on lysates from treated hearts as previously described [16].

4.7. Statistical Analysis

All data are presented as mean ± SEM. Two-way ANOVA with Bonferroni post hoc test was performed to compare the different parameters between groups. A p value < 0.05 was considered significant. Statistical analysis was performed using GraphPad Prism version 5.01 (GraphPad Software, Inc., San Diego, CA, USA).

5. Conclusions

This study identifies a double effect of GRK5-NT in the inhibition of LVH that is based on the regulation of multiple transcription factors through means of different mechanisms. Indeed, GRK5-NT is able to inhibit NFκB activity through the interaction of RH domain with IkBα and to inhibit NFAT activity by means of calcium-calmodulin sequestration. In conclusion, here we propose the amino-terminal sequence of GRK5 as a useful prototype for therapeutic purposes.

Acknowledgments: We are grateful to Julie Pitcher (University College London, London, UK) for providing GRK5-NTPB plasmid. Research was funded by "BEYONDSILOS, FP7-ICT-PCP" and "Campania Bioscience, PON03PE_00060_8" Grants to Guido Iaccarino.

Author Contributions: Daniela Sorriento and Guido Iaccarino conceived and designed the work; Daniela Sorriento, Gaetano Santulli, Michele Ciccarelli and Angela Serena Maione performed the experiments; Daniela Sorriento, Gaetano Santulli, Michele Ciccarelli, Maddalena Illario, Bruno Trimarco and Guido Iaccarino analyzed data and wrote the paper.

Conflicts of Interest: The Authors declare no conflicts of interest.

References

1. Devereux, R.B.; Roman, M.J. Left ventricular hypertrophy in hypertension: Stimuli, patterns, and consequences. *Hypertens. Res.* **1999**, *22*, 1–9. [CrossRef] [PubMed]
2. Schmieder, R.E.; Messerli, F.H. Hypertension and the heart. *J. Hum. Hypertens.* **2000**, *14*, 597–604. [CrossRef] [PubMed]
3. Santulli, G.; Iaccarino, G. Adrenergic signaling in heart failure and cardiovascular aging. *Maturitas* **2016**, *93*, 65–72. [CrossRef] [PubMed]
4. Vakili, B.A.; Okin, P.M.; Devereux, R.B. Prognostic implications of left ventricular hypertrophy. *Am. Heart J.* **2001**, *141*, 334–341. [CrossRef] [PubMed]
5. Hardt, S.E.; Sadoshima, J. Negative regulators of cardiac hypertrophy. *Cardiovasc. Res.* **2004**, *63*, 500–509. [CrossRef] [PubMed]
6. Sadoshima, J.; Izumo, S. The cellular and molecular response of cardiac myocytes to mechanical stress. *Annu. Rev. Physiol.* **1997**, *59*, 551–571. [CrossRef] [PubMed]

7. Molkentin, J.D.; Dorn, G.W., II. Cytoplasmic signaling pathways that regulate cardiac hypertrophy. *Annu. Rev. Physiol.* **2001**, *63*, 391–426. [CrossRef] [PubMed]

8. Heineke, J.; Molkentin, J.D. Regulation of cardiac hypertrophy by intracellular signalling pathways. *Nat. Rev. Mol. Cell Biol.* **2006**, *7*, 589–600. [CrossRef] [PubMed]

9. Gambardella, J.; Trimarco, B.; Iaccarino, G.; Santulli, G. New Insights in Cardiac Calcium Handling and Excitation-Contraction Coupling. *Adv. Exp. Med. Biol.* **2017**. [CrossRef]

10. Santulli, G.; Nakashima, R.; Yuan, Q.; Marks, A.R. Intracellular calcium release channels: An update. *J. Physiol.* **2017**, *595*, 3041–3051. [CrossRef] [PubMed]

11. Akazawa, H.; Komuro, I. Roles of cardiac transcription factors in cardiac hypertrophy. *Circ. Res.* **2003**, *92*, 1079–1088. [CrossRef] [PubMed]

12. Schulz, R.A.; Yutzey, K.E. Calcineurin signaling and NFAT activation in cardiovascular and skeletal muscle development. *Dev. Biol.* **2004**, *266*, 1–16. [CrossRef] [PubMed]

13. Yuan, Q.; Yang, J.; Santulli, G.; Reiken, S.R.; Wronska, A.; Kim, M.M.; Osborne, B.W.; Lacampagne, A.; Yin, Y.; Marks, A.R. Maintenance of normal blood pressure is dependent on IP3R1-mediated regulation of eNOS. *Proc. Nat. Acad. Sci. USA* **2016**, *113*, 8532–8537. [CrossRef] [PubMed]

14. Hogan, P.G.; Chen, L.; Nardone, J.; Rao, A. Transcriptional regulation by calcium, calcineurin, and NFAT. *Genes Dev.* **2003**, *17*, 2205–2232. [CrossRef] [PubMed]

15. Liang, Q.; Wiese, R.J.; Bueno, O.F.; Dai, Y.S.; Markham, B.E.; Molkentin, J.D. The transcription factor GATA4 is activated by extracellular signal-regulated kinase 1- and 2-mediated phosphorylation of serine 105 in cardiomyocytes. *Mol. Cell Biol.* **2001**, *21*, 7460–7469. [CrossRef] [PubMed]

16. Sorriento, D.; Ciccarelli, M.; Santulli, G.; Campanile, A.; Altobelli, G.G.; Cimini, V.; Galasso, G.; Astone, D.; Piscione, F.; Pastore, L.; et al. The G-protein-coupled receptor kinase 5 inhibits NFκB transcriptional activity by inducing nuclear accumulation of IκBα. *Proc. Nat. Acad. Sci. USA* **2008**, *105*, 17818–17823. [CrossRef] [PubMed]

17. Sorriento, D.; Illario, M.; Finelli, R.; Iaccarino, G. To NFκB or not to NFκB: The Dilemma on How to Inhibit a Cancer Cell Fate Regulator. *Transl. Med. UniSa* **2012**, *4*, 73–85. [PubMed]

18. Sorriento, D.; Campanile, A.; Santulli, G.; Leggiero, E.; Pastore, L.; Trimarco, B.; Iaccarino, G. A new synthetic protein, TAT-RH, inhibits tumor growth through the regulation of NFκB activity. *Mol. Cancer* **2009**, *8*, 97. [CrossRef] [PubMed]

19. Sorriento, D.; Santulli, G.; Fusco, A.; Anastasio, A.; Trimarco, B.; Iaccarino, G. Intracardiac injection of AdGRK5-NT reduces left ventricular hypertrophy by inhibiting NFκB-dependent hypertrophic gene expression. *Hypertension* **2010**, *56*, 696–704. [CrossRef] [PubMed]

20. Pronin, A.N.; Satpaev, D.K.; Slepak, V.Z.; Benovic, J.L. Regulation of G protein-coupled receptor kinases by calmodulin and localization of the calmodulin binding domain. *J. Biol. Chem.* **1997**, *272*, 18273–18280. [CrossRef] [PubMed]

21. Johnson, L.R.; Scott, M.G.; Pitcher, J.A. G protein-coupled receptor kinase 5 contains a DNA-binding nuclear localization sequence. *Mol. Cell Biol.* **2004**, *24*, 10169–10179. [CrossRef] [PubMed]

22. Frey, N.; Olson, E.N. Cardiac hypertrophy: The good, the bad, and the ugly. *Annu. Rev. Physiol.* **2003**, *65*, 45–79. [CrossRef] [PubMed]

23. Russell, B.; Motlagh, D.; Ashley, W.W. Form follows function: How muscle shape is regulated by work. *J. Appl. Physiol.* **2000**, *88*, 1127–1132. [CrossRef] [PubMed]

24. Molkentin, J.D.; Lu, J.R.; Antos, C.L.; Markham, B.; Richardson, J.; Robbins, J.; Grant, S.R.; Olson, E.N. A calcineurin-dependent transcriptional pathway for cardiac hypertrophy. *Cell* **1998**, *93*, 215–228. [CrossRef]

25. Gordon, J.W.; Shaw, J.A.; Kirshenbaum, L.A. Multiple facets of NF-κB in the heart: To be or not to NF-κB. *Circ. Res.* **2011**, *108*, 1122–1132. [CrossRef] [PubMed]

26. Liu, Q.; Chen, Y.; Auger-Messier, M.; Molkentin, J.D. Interaction between NFκB and NFAT coordinates cardiac hypertrophy and pathological remodeling. *Circ. Res.* **2012**, *110*, 1077–1086. [CrossRef] [PubMed]

27. Martini, J.S.; Raake, P.; Vinge, L.E.; DeGeorge, B.R., Jr.; Chuprun, J.K.; Harris, D.M.; Gao, E.; Eckhart, A.D.; Pitcher, J.A.; Koch, W.J. Uncovering G protein-coupled receptor kinase-5 as a histone deacetylase kinase in the nucleus of cardiomyocytes. *Proc. Nat. Acad. Sci USA* **2008**, *105*, 12457–12462. [CrossRef] [PubMed]

28. Belmonte, S.L.; Blaxall, B.C. G protein-coupled receptor kinase 5: Exploring its hype in cardiac hypertrophy. *Circ. Res.* **2012**, *111*, 957–958. [CrossRef] [PubMed]

29. Hullmann, J.E.; Grisanti, L.A.; Makarewich, C.A.; Gao, E.; Gold, J.I.; Chuprun, J.K.; Tilley, D.G.; Houser, S.R.; Koch, W.J. GRK5-mediated exacerbation of pathological cardiac hypertrophy involves facilitation of nuclear NFAT activity. *Circ. Res.* **2014**, *115*, 976–985. [CrossRef] [PubMed]

30. Dzimiri, N.; Muiya, P.; Andres, E.; Al-Halees, Z. Differential functional expression of human myocardial G protein receptor kinases in left ventricular cardiac diseases. *Eur. J. Pharmacol.* **2004**, *489*, 167–177. [CrossRef] [PubMed]

31. Sorriento, D.; Santulli, G.; Del Giudice, C.; Anastasio, A.; Trimarco, B.; Iaccarino, G. Endothelial cells are able to synthesize and release catecholamines both in vitro and in vivo. *Hypertension* **2012**, *60*, 129–136. [CrossRef] [PubMed]

32. Santulli, G.; Campanile, A.; Spinelli, L.; Assante di Panzillo, E.; Ciccarelli, M.; Trimarco, B.; Iaccarino, G. G protein-coupled receptor kinase 2 in patients with acute myocardial infarction. *Am. J. Cardiol.* **2011**, *107*, 1125–1130. [CrossRef] [PubMed]

33. Ciccarelli, M.; Sorriento, D.; Cipolletta, E.; Santulli, G.; Fusco, A.; Zhou, R.H.; Eckhart, A.D.; Peppel, K.; Koch, W.J.; Trimarco, B.; et al. Impaired neoangiogenesis in β2-adrenoceptor gene-deficient mice: Restoration by intravascular human β2-adrenoceptor gene transfer and role of NFκB and CREB transcription factors. *Br. J. Pharmacol.* **2011**, *162*, 712–721. [CrossRef] [PubMed]

34. Santulli, G.; Basilicata, M.F.; De Simone, M.; Del Giudice, C.; Anastasio, A.; Sorriento, D.; Saviano, M.; Del Gatto, A.; Trimarco, B.; Pedone, C.; et al. Evaluation of the anti-angiogenic properties of the new selective αVβ3 integrin antagonist RGDechiHCit. *J. Transl. Med.* **2011**, *9*, 7. [CrossRef] [PubMed]

35. Iaccarino, G.; Izzo, R.; Trimarco, V.; Cipolletta, E.; Lanni, F.; Sorriento, D.; Iovino, G.L.; Rozza, F.; De Luca, N.; Priante, O.; et al. β2-adrenergic receptor polymorphisms and treatment-induced regression of left ventricular hypertrophy in hypertension. *Clin. Pharmacol. Ther.* **2006**, *80*, 633–645. [CrossRef] [PubMed]

36. Santulli, G.; Cipolletta, E.; Sorriento, D.; Del Giudice, C.; Anastasio, A.; Monaco, S.; Maione, A.S.; Condorelli, G.; Puca, A.; Trimarco, B.; et al. CaMK4 Gene Deletion Induces Hypertension. *J. Am. Heart Assoc.* **2012**, *1*, e001081. [CrossRef] [PubMed]

37. Santulli, G.; Xie, W.; Reiken, S.R.; Marks, A.R. Mitochondrial calcium overload is a key determinant in heart failure. *Proc. Nat. Acad. Sci. USA* **2015**, *112*, 11389–11394. [CrossRef] [PubMed]

International Journal of
Molecular Sciences

MDPI

Article

HDAC Inhibition Improves the Sarcoendoplasmic Reticulum Ca^{2+}-ATPase Activity in Cardiac Myocytes

Viviana Meraviglia [1,†], Leonardo Bocchi [2,†], Roberta Sacchetto [3], Maria Cristina Florio [1], Benedetta M. Motta [1], Corrado Corti [1], Christian X. Weichenberger [1], Monia Savi [2], Yuri D'Elia [1], Marcelo D. Rosato-Siri [1], Silvia Suffredini [1], Chiara Piubelli [1], Giulio Pompilio [4,5], Peter P. Pramstaller [1], Francisco S. Domingues [1], Donatella Stilli [2,*,‡] and Alessandra Rossini [1,*,‡]

[1] Institute for Biomedicine, Eurac Research, 39100 Bolzano, Italy (affiliated institute of the University of Lübeck, 23562 Lübeck, Germany); viviana.meraviglia@eurac.edu (V.M.); cristiflorio@gmail.com (M.C.F.); benedetta.motta@eurac.edu (B.M.M.); corrado.corti@eurac.edu (C.C.); christian.weichenberger@eurac.edu (C.X.W.); yuri.delia@eurac.edu (Y.D.); Marcelo.RosatoSiri@eurac.edu (M.D.R.-S.); silvia.suffredini@alice.it (S.S.); chiara.piubelli@sacrocuore.it (C.P.); peter.pramstaller@eurac.edu (P.P.P.); francisco.domingues@eurac.edu (F.S.D.)
[2] Department of Chemistry, Life Sciences and Environmental Sustainability, University of Parma, 43124 Parma, Italy; leonardo.bocchi@unipr.it (L.B.); monia.savi@unipr.it (M.S.)
[3] Department of Comparative Biomedicine and Food Science, University of Padova, 35020 Legnaro (Padova), Italy; roberta.sacchetto@unipd.it
[4] Vascular Biology and Regenerative Medicine Unit, Centro Cardiologico Monzino, IRCCS, 20138 Milano, Italy; giulio.pompilio@ccfm.it
[5] Dipartimento di Scienze Cliniche e di Comunità, Università degli Studi di Milano, 20122 Milano, Italy
*   Correspondence: donatella.stilli@unipr.it (D.S.); alessandra.rossini@eurac.edu (A.R.); Tel.: +39-0521-906-117 (D.S.); +39-0471-055-504 (A.R.); Fax: +39-0521-905-673 (D.S.); +39-0471-055-599 (A.R.)
†   These authors contributed equally to this work.
‡   These authors contributed equally to this work.

Received: 14 December 2017; Accepted: 29 January 2018; Published: 31 January 2018

Abstract: SERCA2a is the Ca^{2+} ATPase playing the major contribution in cardiomyocyte (CM) calcium removal. Its activity can be regulated by both modulatory proteins and several post-translational modifications. The aim of the present work was to investigate whether the function of SERCA2 can be modulated by treating CMs with the histone deacetylase (HDAC) inhibitor suberanilohydroxamic acid (SAHA). The incubation with SAHA (2.5 μM, 90 min) of CMs isolated from rat adult hearts resulted in an increase of SERCA2 acetylation level and improved ATPase activity. This was associated with a significant improvement of calcium transient recovery time and cell contractility. Previous reports have identified K464 as an acetylation site in human SERCA2. Mutants were generated where K464 was substituted with glutamine (Q) or arginine (R), mimicking constitutive acetylation or deacetylation, respectively. The K464Q mutation ameliorated ATPase activity and calcium transient recovery time, thus indicating that constitutive K464 acetylation has a positive impact on human SERCA2a (hSERCA2a) function. In conclusion, SAHA induced deacetylation inhibition had a positive impact on CM calcium handling, that, at least in part, was due to improved SERCA2 activity. This observation can provide the basis for the development of novel pharmacological approaches to ameliorate SERCA2 efficiency.

Keywords: SERCA2; acetylation; HDAC inhibition; ATPase activity; calcium transients; cardiomyocyte mechanics

1. Introduction

The Sarco(Endo)plasmic Reticulum (SR) Ca^{2+} ATPase (SERCA) is a Ca^{2+} pump that uses the energy of ATP hydrolysis to translocate two Ca^{2+} ions from the cytosol to the SR lumen. It is composed of a single polypeptide weighing approximately 110 kDa and organized into four different domains: one transmembrane domain composed of ten alpha-helices, one nucleotide (ATP) binding site, one phosphorylation/catalytic domain that drives ATP hydrolysis, and one actuator domain involved in the gating mechanism that regulates calcium binding and release [1,2]. In vertebrates, three different SERCA genes have been identified so far, showing high degree of conservation among species: *ATP2A1*, *ATP2A2* and *ATP2A3*, encoding respectively for the SERCA1, SERCA2 and SERCA3 proteins. Thanks to alternative splicing mechanisms, they can give rise to different protein isoforms, whose expression is regulated during development and in a tissue specific manner [3]. SERCA2a is the most expressed isoform in cardiac muscle [3] and is the responsible for the level of SR Ca^{2+} load and the majority of the cytosolic calcium removal after contraction (70% in human, up to 90% in mouse and rats), thus representing a key player for cardiomyocyte (CM) relaxation [4]. Alterations in SERCA2a are recognized among the major factors contributing to ventricular dysfunction in several heart diseases, like diabetic cardiomyopathy [5,6] and heart failure [7,8].

Owing its essential role in cardiac physiology, the mechanisms regulating SERCA2a have been extensively studied [9]. Evidence has been provided that the pump activity can be directly modulated by several post-translational modifications [10]. Specifically, sumoylation [11] and glutathionylation [12] increase the activity, while glycosylation [13] and nitration [14] seem to decrease SERCA2a function. A recent study by Foster and coworkers [15] identified three acetylation sites within the structure of SERCA2a. However, the potential impact of SERCA2a acetylation on intracellular Ca^{2+} cycling has never been evaluated.

The enzymes responsible for protein acetylation and deacetylation are histone acetyltransferases (HATs) and histone deacetylases (HDACs), respectively. HATs and HDACs act not only on histones, but also on other nuclear and cytoplasmic proteins, such as transcription factors and structural proteins [16].

The acetylation balance carried out by HATs and HDACs plays a role in the physiology of adult CMs. For instance, it has been demonstrated that HDAC inhibition prevented ventricular arrhythmias and restored conduction defects of the heart in a model of dystrophic mice [17].

The specific aim of the present work was to investigate the impact of deacetylation inhibition on SERCA2 activity. Specifically, adult ventricular CMs isolated from normal rat hearts were exposed to the HDAC general inhibitor suberanilohydroxamic acid (SAHA, or Vorinostat). Additionally, three stable lines of HEK cells were produced to express the human SERCA2a either wild type or with mutations mimicking constitutive acetylation/deacetylation on a selected lysine residue.

In this manuscript, we demonstrate for the first time that HDAC inhibition promotes improved efficiency in cytosolic Ca^{2+} removal, along with increased SERCA2 acetylation and ATPase activity.

2. Results

2.1. Effect of SAHA Treatment on SERCA2a Acetylation and Function in Control Rat Cardiomyocytes

Cardiomyocytes (CMs) isolated from normal adult rat hearts were used as model to investigate whether SAHA-promoted HDAC inhibition can impact SERCA2a acetylation and function. Following enzymatic dissociation, CMs were either untreated (CTR) or incubated with 2.5 μM SAHA (CTR+SAHA) for 90 min. As expected [18], SAHA administration induced higher tubulin acetylation level (Figure 1a).

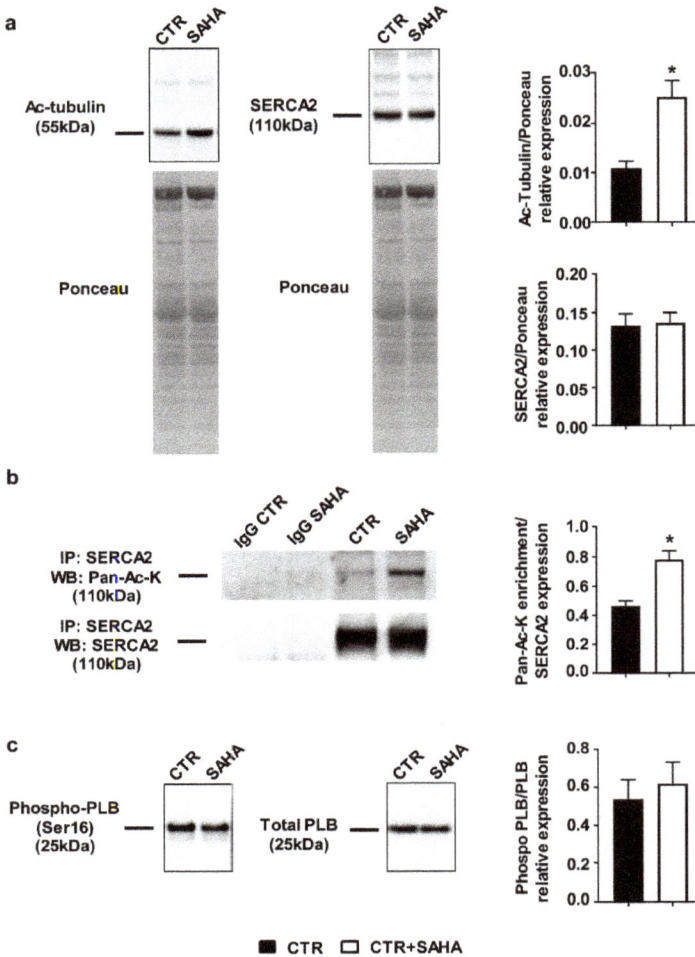

Figure 1. Effect of SAHA treatment on cardiomyocytes isolated from adult rat hearts. (**a**) Western blot panels (**on the left**) and densitometric analysis (**on the right**) showing SERCA2 and Ac-Tubulin protein expression after SAHA treatment ($n = 7$). Mann–Whitney U-test: * $p < 0.005$ vs. CTR; (**b**) Immunoprecipitation experiments and densitometric analysis (**on the right**) indicating higher acetylation level of SERCA2 after SAHA treatment ($n = 5$). Mann–Whitney U-test: * $p < 0.005$ vs. CTR; (**c**) Western blot analysis showing the expression of phosphorylated phospholamban (Phospho-PLB, Ser16) compared to total phospholamban (PLB) in adult rat CMs after SAHA treatment ($n = 7$).

Co-immunoprecipitation experiments revealed that SAHA increased SERCA2 acetylation level (Figure 1b and Figure S1), without inducing significant changes in SERCA2 protein expression (Figure 1a). The ratio of phosphorylated phospholamban (PLB)/total phospholamban (Figure 1c) remained unchanged.

To investigate whether increased acetylation could also affect SERCA2 functional properties, ATPase activity was measured on microsomes [19] isolated from both CTR and CTR+SAHA CMs. HDAC inhibition resulted in an increase of ATPase activity when microsomes were exposed to 10 μM calcium concentration (corresponding to pCa5) [20,21] (Figure 2a).

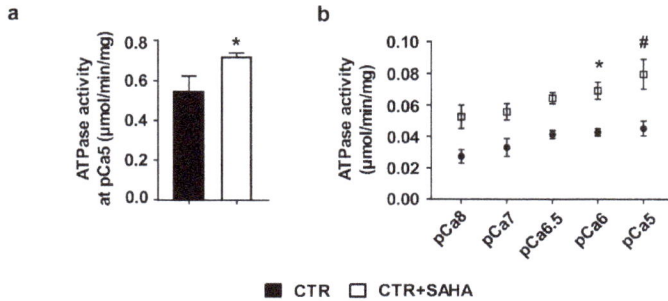

Figure 2. Effect of SAHA treatment on SERCA2 ATPase activity evaluated in cardiomyocytes isolated from adult rat hearts and HL-1 cells. (**a**) ATPase activity assay performed on microsomes isolated from adult rat CMs either untreated or treated with SAHA at pCa5. Experiments were performed on 3 independent CM sets per group and repeated twice. Unpaired Student's *t*-test: * $p < 0.005$ vs. CTR; (**b**) ATPase activity assay performed on microsomes isolated from HL-1 cells at different pCa. Each point represents the mean ± SEM of at least 4 independent experiments; Two-way ANOVA followed by Sidak's multiple comparison: * $p < 0.05$ vs. SAHA pCa6; # $p < 0.05$ vs. SAHA pCa5. All data are presented as mean ± SEM.

In order to confirm our result on SERCA2 functional properties, we decided to perform an additional set of experiments measuring ATPase activity on microsomes isolated from HL-1 cells, derived from the AT-1 mouse atrial cardiomyocyte tumor lineage. These cells partially maintain an adult cardiac phenotype and are able to contract [22]. The calcium-dependence of ATPase activity on HL-1 cells, either untreated (CTR) or treated for 90 min with 2.5 µM SAHA, was analyzed at different calcium concentration (from pCa8 to pCa5). ATPase activity of microsomes extracted from HL-1 cells was increased after SAHA treatment in comparison with CTR and the difference reached statistical significance when microsomes were exposed to 1 and 10 µM calcium concentration (corresponding to pCa6 and pCa5, respectively; Figure 2b).

2.2. Effect of SAHA Treatment on Calcium Transients and Cell Mechanics in CMs Isolated from Adult Rat Hearts

We then investigated whether SAHA treatment affected CM functional parameters that directly depend on SERCA2 activity, namely calcium transients and cell contractility. The amplitude and the time to peak (TTP) of the calcium transient were comparable in CTR and CTR+SAHA groups, while the rate of cytosolic calcium clearing was significantly higher in SAHA-treated cardiomyocytes (Figure 3a,c). Specifically, SAHA induced a 21% decrease in the time constant tau, as well as a significant reduction in the time to 10%, 50% and 90% of fluorescence signal decay (BL10, BL50, BL90; Figure 3c). Consistent with this finding, SAHA also affected CM mechanics during the re-lengthening phase, as documented by the significant increase in the maximal rate of re-lengthening (+dl/dt$_{max}$, approximately 16%) associated with a decrease in the time to 10%, 50% and 90% of re-lengthening (Figure 3b,d). Conversely, the average diastolic sarcomere length, the fraction of shortening and the maximal rate of shortening were comparable in CTR and CTR+SAHA cardiomyocytes (Figure 3d). Of note, SAHA exposure, at the conditions used in the present manuscript (90 min, 2.5 µM), did not affect cardiomyocytes diastolic nor systolic calcium concentration assessed by Fura-2 dye (Figure S2).

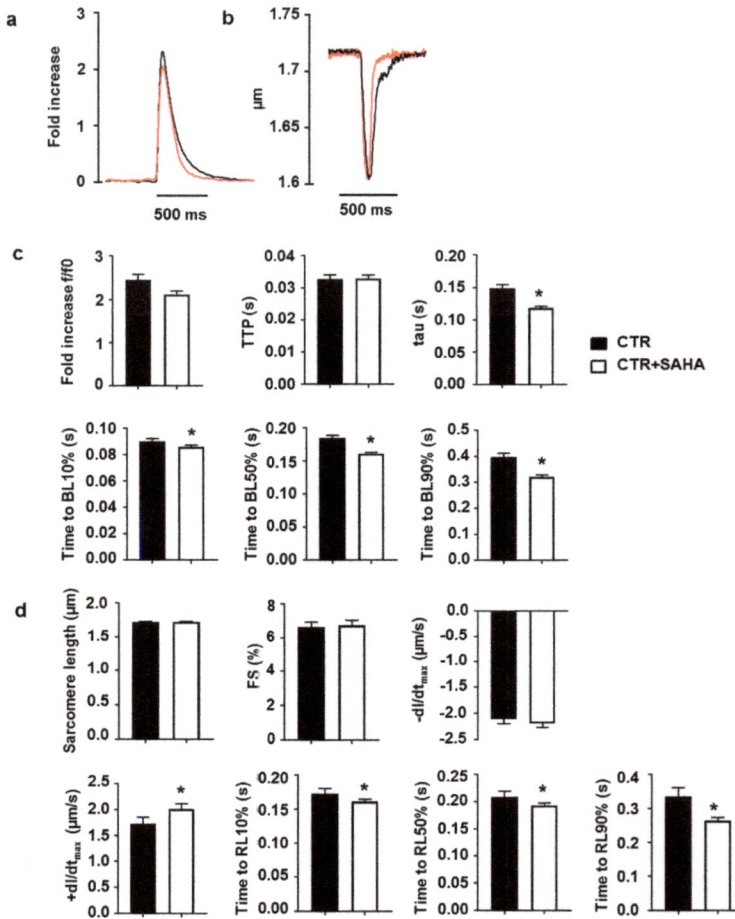

Figure 3. Effect of SAHA treatment on calcium transients and cell mechanics in CMs isolated from adult rat hearts. Representative examples of calcium transients (**a**) normalized traces: fold increase and sarcomere shortening (**b**) recorded from CTR (black line) and CTR+SAHA (red line) ventricular myocytes. (**c**) Calcium transient parameters: amplitude of calcium transient (expressed as calcium peak fluorescence normalized to baseline fluorescence, f/f0: fold increase), time to peak of the calcium transient (TTP), time constant of the rate of intracellular Ca^{2+} clearing (tau), and recovery phase of Ca^{2+} transients indicated as time to 10% (BL10), 50% (BL50) and 90% (BL90) of fluorescence signal decay. (**d**) Cell mechanics: diastolic sarcomere length and fraction of shortening (FS), maximal rates of shortening ($-dl/dt_{max}$) and re-lengthening ($+dl/dt_{max}$), time of re-lengthening at 10% (RL10), 50% (RL50) and 90% (RL90). All data are presented as mean $\pm$ SEM and analysed using General Linear Model (GLM) ANOVA for repeated measurements ($n = 45$ untreated CMs, $n = 70$ CMs treated with SAHA): * $p < 0.005$ vs. CTR.

2.3. Nε-Lysine Acetylation Sites of Human SERCA2a

To support the hypothesis that direct acetylation can boost SERCA2a function, we searched in PHOSIDA and PhosphoSitePlus for reported Nε-lysine acetylation sites in human SERCA2. PHOSIDA Post Translation Modification Database reports acetylation at K464, while PhosphoSitePlus reports acetylation at K31, K33, K218, K464, K476 and K514. All these residues are mostly conserved in mammals (Figure S3). Out of these candidates, site K464 was selected as the target for further analysis given two previous independent reports identifying K464 as an acetylation site in human SERCA2 [23,24]. Homology modeling results indicate that K464 is solvent accessible at the surface of the nucleotide binding-domain (Figure 4a and Figure S4).

To assess the impact of Nε-lysine acetylation on hSERCA2a function, two different mutants were generated where lysine in position 464 was changed into glutamine (Q) or arginine (R) to mimic constitutive acetylation or constitutive deacetylation, respectively [15]. The expression of either the hSERCA2a wild type form (WT) and the mutant forms (K464Q/R) was induced in HEK human cells (Figure 4b), expressing lower levels of the SERCA2 isoform compared to other commonly used human cell lines, such as HeLa. SERCA2 expression was significantly increased in all stable transfected HEK cells compared to non-transfected cells (NT; Figure 4b). However, although the observed difference did not reach statistical significance, the expression of SERCA2 appeared lower in both mutants compared to the HEK transfected with WT (Figure 4b).

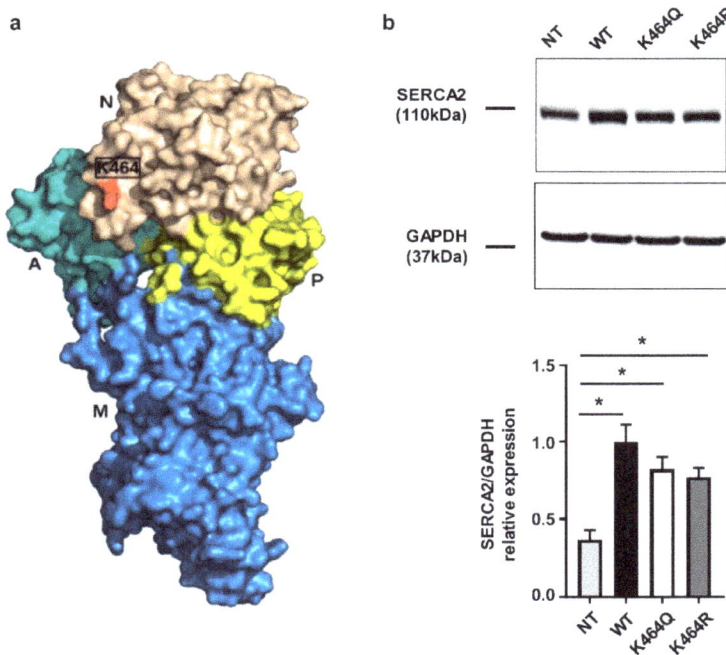

Figure 4. Analysis of human SERCA2a mutants transfected in HEK cells: bioinformatics prediction and protein expression. (**a**) Molecular surface view of predicted hSERCA2a. Domains are visualized in different colors: Actuator (A) in light blue, transmembrane (M) in dark blue, phosphorylation (P) in yellow and nucleotide-binding (N) in beige. K464 is highlighted in red; (**b**) Western blot analysis of SERCA2 expression in stable transfected HEK cells. Densitometry is reported in the bar graph ($n = 8$). Ordinary one-way ANOVA followed by Bonferroni's multiple comparison: $p < 0.005$ vs. NT.

2.4. Effects of K464 Mutation on hSERCA2a ATPase Activity and HEK Calcium Transients

The effects of mutagenesis were evaluated on ATPase activity measured in the microsomal fraction isolated from transfected HEK cells. In order to take into consideration the possible impact of different level of protein expression on hSERCA2a function, the values of ATPase activity recorded for each sample were normalized to the correspondent SERCA2 expression evaluated by Western blot on cells obtained from the same sample (Figure 4b). Taking into account the variable amount of SERCA2 expression in each cell line and for each experiment, we observed that SERCA2 function was significantly ameliorated in mutants where K464 were mutated into Q compared with WT (Figure 5a). Conversely, the K464R mutant did not show any differences in comparison with the WT (Figure 5a). Thus, constitutive acetylation of hSERCA2a K464 improved hSERCA2a ATPase activity, while K464 mutation into R (constitutive deacetylation) did not have a significant effect (Figure 5a). Importantly, untransfected HEK cells responded with a transient increase of intracellular calcium concentration to caffeine puff (1 s, 10 mM, Figure 5b). Therefore, we further evaluated the effects of SERCA2a mutagenesis on calcium transients induced by caffeine in HEK cells loaded with the calcium sensitive dye Fluo-4. Again, in order to consider the effects of different SERCA2 protein levels in HEK mutants, calcium transients were corrected on the mean SERCA2 expression detected by Western Blot in the correspondent cell line. The constitutive acetylation of SERCA2 K646Q promoted an acceleration of Ca^{2+} transient recovery phase at BL10%, BL50% compared to WT and K464R (Figure 5c), thus indicating that constitutive acetylation of K464 has a positive impact on hSERCA2a function.

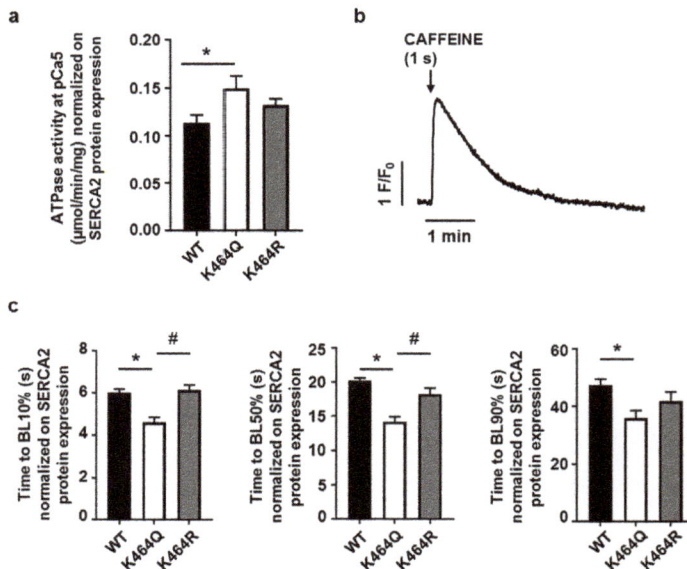

Figure 5. ATPase activity on microsomes and calcium transients evaluated in HEK cells transfected with human SERCA2a mutants. (a) ATPase activity assay performed on microsomes isolated from stable transfected HEK cells at pCa5 normalized on SERCA2 protein expression. Experiments were performed on 4 independent HEK sets per group and repeated twice. Ordinary one-way ANOVA followed by Bonferroni's multiple comparison: $p < 0.005$ WT vs. K464Q. All data are presented as mean ± SEM; (b) Representative trace of the rise in intracellular calcium concentration evoked by a caffeine pulse in untransfected HEK cells; (c) Calcium transients in SERCA2a K464 mutants normalized on SERCA2 protein expression. K464 is mutated either into Q or R, mimicking constitutive acetylation and deacetylation, respectively ($n = 99$ WT, $n = 69$ K464Q and $n = 36$ K464R). Time to 10% (BL10) and time to 90% (BL90) were analysed by Kruskal-Wallis followed by Dunn's multiple comparisons test: * $p < 0.005$ WT vs. K464Q, # $p < 0.005$ K464Q vs. K464R. Time to 50% (BL50) was analysed by ordinary one-way ANOVA followed by Bonferroni's multiple comparisons test: * $p < 0.005$ WT vs. K464Q, # $p < 0.005$ K464Q vs. K464R. All data are presented as mean ± SEM.

3. Discussion

SERCA2 activity has been recently suggested to be affected by acetylation/deacetylation mechanisms [11,15], although no experimental evidence has so far been provided. In the present work, we show that HDAC inhibition promotes SERCA2 acetylation. This is associated with a significant increase of microsomal ATPase activity (which is recognized to be mainly due to SERCA2 function [25]) and a significant amelioration of calcium clearing parameters measured in CMs isolated from normal adult rats. Our results indicate for the first time a relation between HDACs, SERCA2 acetylation and its function. In addition, lysine 464 has been mutated in the human SERCA2a (hSERCA2a) coding sequence transfected in a HEK stable expression system, in order to provide evidence for a causal association between SERCA2 deacetylation and the ATPase activity of the pump.

In our CM experimental model, SERCA2 acetylation was promoted by SAHA (Vorinostat), a general HDAC inhibitor, acting on class I and II HDACs [26]. SAHA use was approved by the US FDA in October 2006 for the treatment of refractory cutaneous T-cell lymphoma [27]. Although several in vitro studies [28,29] have reported that 2.5 μM SAHA is able to inhibit the growth of cancer cells, it has also been reported that 2 μM SAHA is able to reduce CM death after simulated ischaemia/reperfusion injury [30]. In our work, the decision to treat adult CMs with 2.5 μM SAHA for 90 min was taken on the basis of current literature showing that SAHA in the range of 1 to 5 μM increases histone acetylation in an in vitro model within one hour, reaching the plateau effect at 3 h [31]. Therefore, we assumed that 90 min are sufficient to achieve the acetylation increase of all the other HDAC potential targets, including cytoplasmic substrates and to preserve cell viability. We demonstrated higher levels of SERCA2 acetylation following SAHA treatment in adult CMs from CTR rats by immunoprecipitation and Western blot for Acetylated-Lysine. The SERCA2 ATPase activity has been then explored using isolated microsomes from normal adult rat CMs, where an increase has been shown after SAHA treatment. Although it is a common praxis to perform measures at pCa5 [21,32–34], we decided to test the calcium-dependence of ATPase activity in order to confirm our results on SERCA2 ATPase activity also using microsomes isolated from HL-1 cells, as additional in vitro model. Specifically, HL-1 cells, derived from mouse atrial cardiomyocyte tumor lineage [22], represent a more abundant source of cardiomyocytes compared to freshly isolated rat adult CMs.

The treatment with SAHA ameliorated intracellular Ca^{2+} dynamics and, accordingly, contractile performance of adult rat CMs. These findings support the hypothesis that the inhibition of SERCA2 deacetylation induced by SAHA plays a crucial role in promoting a more efficient SR calcium re-uptake, which in turn results in a higher rate of re-lengthening and shorter relaxation times [35].

In view of these results, one should also consider that Gupta and colleagues recently observed an enhanced calcium sensitivity of myofilaments following HDAC inhibition. This aspect can play a key role in explaining the improved contractile efficiency of SAHA treated CMs [36]. On the other hand, an increase of myofilament calcium sensitivity would be expected to delay SR Ca^{2+} uptake and consequently prolong twitch duration, thus having an opposite effect compared to our observation. Nevertheless, it is important to notice that Gupta and coworkers have demonstrated that acetylation increases myofilament Ca^{2+} sensitivity in a model of skinned cardiac papillary muscle fibers [36]. It is known that the muscle skinning process leads to the loss of all the other sensitizing and desensitizing intracellular components that synergistically act in determining the effective calcium sensitivity of myofilaments [37]. Therefore it is conceivable that the discrepancy between our and Gupta results depends, at least in part, on the use of a whole cell model instead of a reductionist model consisting only of free calcium ions and myofilaments.

It is now well recognized that Ca^{2+} handling not only governs contractile events, but can also directly or indirectly influence ion channel function in cardiomyocytes, resulting in electrophysiological effects. An important component of Na^+ channel regulation is due to Ca^{2+}, calmodulin (CaM) and CaM-dependent protein kinase II (CaMKII) pathway that affects channel function [38]. L-type Ca^{2+} (ICaL) and sodium(Na^+)-Ca^{2+} exchanger (NCX) currents are Ca^{2+}-sensitive, as well as the slowly activating delayed rectifier current (IKs), which plays an important role in regulating action potential duration [39,40]. In this work, we did not perform electrophysiological measurements on isolated ventricular myocytes that could help us in evaluating the impact on cardiac ion channels and pumps of SERCA2a activity amelioration induced by SAHA deacetylation inhibition. However, several findings reported in the present study can be useful to speculate on this topic. We observed that the amplitude of the calcium transient, which is dependent on the amplitude of Ca^{2+} entry, was comparable in control and SAHA-treated cardiomyocytes, suggesting that SAHA does not induce substantial changes in the calcium entry, through L-type Ca^{2-} channels and reverse activity of NCX. In addition, ryanodine receptors should not be affected by SAHA treatment, as indirectly suggested by the calcium transient rate of rise, measured as time to peak fluorescence, showing similar values in treated and untreated cells. Also, it should be considered that diastolic and systolic cytosolic Ca^{2+} levels were unchanged in SAHA treated cells as compared to control cardiomyocytes. This observation, besides confirming that calcium homeostasis is maintained even in the presence of a faster SERCA2 calcium re-uptake, should imply no changes in both the intracellular Na^+ concentration that is tightly coupled to calcium concentration regulation, via electrogenic Na^+/Ca^{2+} exchange, and Ca^{2+}-sensitive *I*Ks current.

Interestingly, it has been recently shown that SAHA administration in vivo in a mouse model of muscular dystrophy can reduce the appearance of cardiac arrhythmias by reverting the remodeling of connexins and sodium channels [17]. Although the mechanisms of action are still unclear, those findings, together with our results, support the positive effect of SAHA treatment on cardiac function.

In order to explore possible acetylation sites influencing human SERCA2 activity, 3D models of hSERCA2a were produced by bioinformatic prediction methods, identifying the residue 464 as the most promising candidate. Residue K464 is located at the surface of the N domain in hSERCA2a. Although homology modeling of the different conformational states does not reveal an obvious functional impact for the K464 acetylation, the site is expected to locate on the side of a large cleft in state E2P [1,41] at the interface between domains N, P and M, near a putative phospholamban binding site (Figure S4). The potential effect of the K464 acetylation in protein electrostatics, flexibility and phospholamban binding affinity should be further investigated. Indeed, this lysine, conserved in all mammals, has already been reported as acetylated in the guinea pig heart, although no evidence of its functional role has been provided [15].

We produced three mutant HEK lines expressing WT hSERCA2a, or hSERCA2a where K464 was mutated into Q or R, mimicking constitutive acetylated and deacetylated state of the residue, respectively [15]. Of note, HEK cells have already been used by other groups as a valuable model to overexpress mutant SERCA and to study the consequent effects on the protein pumping activity [11]. HEK cells overexpressing mutated SERCA have also been recently used as a tool to investigate intracellular calcium dynamics [42]. Importantly, despite the debate on the presence of functional RyRs in HEK cells [43–45], we found that the application of a pulse of caffeine caused the abrupt increase of intracellular calcium concentration. Therefore, we decided to expose HEK cells transfected with WT hSERCA2a or mutated hSERCA2a to caffeine in order to increase the intracellular calcium concentration and then compare the time of recovery in WT compared to mutants. It is important to notice that, although the recovery phase in HEK cells was considerably longer than that observed in adult cardiomyocytes, SERCA has been described as the major player in the recovery of calcium transients long up to 4 s [46]. We cannot exclude a possible involvement of other Ca^{2+} removal systems (such as mitochondria and Peripheral Membrane Calcium ATPase) as well as of Ca^{2+} buffering in restoring basal conditions. However, given that in our experimental conditions the hSERCA2a pump was overexpressed, we can reasonably assume that the recovery phase in (transfected) HEK cells at least partially depends on the activity of transfected hSERCA2a. Intriguingly, K464Q mutants exhibited a higher ATPase activity and an acceleration of Ca^{2+} transient recovery phase compared to WT, supporting the hypothesis that the acetylation is a positive regulator of SERCA2 function. However, it is important to notice that SAHA could inhibit the acetylation of different lysine residues at the same time. Thus, mutating only one lysine at a time might only partially mimic the global effect obtained after HDAC inhibition on CMs.

Interestingly, Hajjar and coworkers demonstrated that sumoylation of lysine residue 585 enhances SERCA2a stability and activity [11]. It has been recently shown that HDAC inhibition can stimulate protein sumoylation possibly through acetylation of the SUMO machinery [47]. Thus, our findings in isolated SAHA-treated cardiomyocytes sustain the idea that HDAC-inhibition could support other approaches, such as SUMO-1 overexpression [11], in the treatment of heart disease associated with SERCA2 impairment.

The main limitation of the present study is the fact that acetylation is only one of the factors contributing to SERCA2 function in health and disease. Additionally, we cannot exclude a possible action of SAHA on other proteins, such as SERCA2 interacting proteins or transporters involved in calcium dynamics, such as the sodium-calcium exchanger [48]. However, SERCA2 is recognized as the main player in calcium re-uptake during the relaxation/recovery phase in CMs, removing approximately 70–90% of calcium ions from the cytosol in species including humans, mice and rats [4]. This observation let us hypothesize that SERCA2 is the key factor in SR reuptake whose function is modulated by SAHA. The fact that the time to peak (TTP) of calcium transients recorded in isolated rat cardiomyocytes remained unchanged following SAHA administration seems also to indicate that RyRs are not affected by the treatment, at least under our experimental conditions. Further, our results indicate that phosphorylated and total PLB remain unchanged after treatment, thus suggesting that PKA activity might not be affected by SAHA. Nevertheless, deeper investigation are required in order to explain whether acetylation can directly or indirectly modulate other mechanisms of SERCA2 regulation, like CAMKII mediated phosphorylation. Importantly, it has been recently hypothesized that SERCA2 can be potentially regulated by microRNAs [49]. Concomitantly, experimental evidence has been provided that knock-out mice for miR-22 expression exhibit prolonged calcium transients and a decrease of SERCA transcript induced by indirect interaction [50]. More recently, it has been shown that miR-275 can directly target SERCA in mosquito gut [51], thus providing new basis for further studies on SERCA regulation by microRNA in vertebrates and humans. In this context, it should be reported that SAHA is well recognized as a microRNA modulator [52–54]. Further studies are required to understand whether HDAC inhibition can exert its action on SERCA and calcium dynamics by acting on microRNA target. Additionally, other potential mechanism of action should be taken into consideration for future studies, such as the role of the TGF-beta pathway, already shown to be involved in cardiac calcium signaling [55,56] and to be modulated by HDAC inhibitors in organs other than heart [57–59].

However, we would like to underline the fact that the aim of the present paper was to evaluate whether HDAC inhibition can modulate cardiac cell function by acting on SERCA2. SERCA2 is reasonably one of the multiple targets of SAHA action that can result in the enhancement or in the inhibition of many proteins, the net effect on EC coupling depending on the combination of the single contributions and on the model used.

Another question remaining unanswered is which enzymes are responsible for SERCA2a acetylation/deacetylation. It is known that SAHA at low concentration is a potent inhibitor of the class I HDAC isoforms 1 and 8 (IC50 < 20 nM), while it is active on other class I and class II HDACs at higher concentrations [60]. It is also well recognized that class II HDAC have both nuclear and cytoplasmic localization [61]. However, although traditionally considered as nuclear enzymes, class I HDACs have been recently demonstrated to be present in the endoplasmic reticulum [62]. Therefore, it is conceivable that both class of enzymes might act on SERCA2, contributing to the control of its acetylation status. Further studies are required to elucidate this point.

In conclusion, the results here demonstrate for the first time that SERCA2 acetylation and activity are increased following HDAC inhibition promoted by SAHA. Importantly, our findings suggest that the modulation of acetylation can be a novel strategy to treat diseases where calcium handling is affected, such as diabetic cardiomyopathy [6,63], dilated cardiomyopathy [64], ischemia/reperfusion injury [65], cardiac hypertrophy [66] and heart failure [67].

4. Materials and Methods

4.1. Rat Population

This study was carried out in strict accordance with the recommendations in the Guide for the Care and Use of Laboratory Animals (National Institute of Health, Bethesda, MD, USA, revised 1996). The investigation was approved by the Veterinary Animal Care and Use Committee of the University of Parma-Italy (Prot. No. 59/12; 9 May 2012) and conforms to the National Ethical Guidelines of the Italian Ministry of Health. All efforts were made to minimize animal suffering.

The study population consisted of 5 male Wistar rats (*Rattus norvegicus*) aged 12–14 weeks, weighing 379 ± 8 g, individually housed in a temperature-controlled room at 22–24 °C, with the light on between 7.00 AM and 7.00 PM.

4.2. Rat Cardiomyocyte Isolation

Individual left ventricular (LV) cardiomyocytes (CMs) were enzymatically isolated by collagenase perfusion from the adult rat hearts in accordance with a procedure previously described [68]. Briefly, rats were anesthetised with ketamine chloride (Imalgene, Merial, Milan, Italy; 40 mg/kg ip) plus medetomidine hydrochloride (Domitor, Pfizer Italia S.r.l., Latina, Italy; 0.15 mg/kg ip). The heart was then removed and rapidly perfused at 37 °C by means of an aortic cannula with the following sequence of solutions: (1) a calcium-free solution for 5 min to remove the blood. Calcium-free solution contains the following (in mM): 126 NaCl, 22 dextrose, 5 $MgCl_2$, 4.4 KCl, 20 taurine, 5 creatine, 5 sodium pyruvate, 1 NaH_2PO_4, and 24 HEPES (pH 7.4, adjusted with NaOH), and the solution was gassed with 100% O_2; (2) a low-calcium solution (0.1 mM) plus 1 mg/mL type 2 collagenase (Worthington Biochemical, NJ, USA) and 0.1 mg/mL type XIV protease (Sigma-Aldrich, Milan, Italy) for about 20 min; (3) an enzyme-free, low-calcium solution for 5 min. The left ventricle was then minced and shaken for 10 min. The cells were filtered through a nylon mesh, re-suspended in low-calcium solutions (0.1 mM) for 30 min, and then used for recording cell mechanics and calcium transients. Following enzymatic dissociation, CMs were either untreated or incubated with 2.5 µM SAHA for 90 min. A fraction of cells from each experimental group was washed three times with low-calcium solution and centrifuged (42× *g* for 5 min). After removing the supernatant, the pellet was stored at −80 °C for subsequent microsome isolation (see the Section 4.6). The remaining cardiomyocytes were used for recording cell mechanics and calcium transients, as described in the Section 4.8.

4.3. HL-1 Cardiomyocyte Culture

HL-1 cells, derived from the AT-1 mouse atrial cardiomyocyte tumor lineage [22], were kindly donated by Prof. W.C. Claycomb. Cells were grown as previously described [22] in Claycomb Medium (Sigma-Aldrich S.r.l., Milan, Italy) supplemented with 100 µM norepinephrine (Sigma-Aldrich), 10% FBS (Sigma-Aldrich), 1% *v/v* penicillin streptomycin (Invitrogen by Thermo Fisher Scientific, Monza, Italy) and 2 mM L-glutamine (Invitrogen). Cells were cultured at 37 °C in a humid atmosphere of 5% CO_2 and 95% air. Cells were plated onto gelatin/fibronectin-coated flasks at a density of 10,000 cells/cm^2. When cells reached 90% of confluency, culture medium was replaced and supplemented for 90 min with 2.5 µM SAHA.

4.4. Western Blot Analysis

Cells were washed in PBS and harvested in protein extraction buffer (10 mM Tris-HCl pH 7.4, 150 mM NaCl, 1% Igepal CA630, 1% sodium deoxycholate, 0.1% SDS (Sodium Dodecyl Sulphate) and 1% Glycerol supplemented with protease/phosphatase inhibitor mix (Roche Diagnostics S.p.A., Milan, Italy). Proteins were quantified by BCA protein kit (Thermo Fisher Scientific, Monza, Italy) following manufacturer's instructions. Then, 30 µg of protein extracts were separated by SDS-PAGE on precast gradient (4–12%) gels (Invitrogen) using MOPS running buffer (Invitrogen) and transferred onto nitrocellulose membranes (Bio-Rad, Segrate, Milan, Italy) in Transfer buffer (Life Technologies by

Thermo Fisher Scientific) supplemented with 10% (*v/v*) methanol (Sigma-Aldrich, Milan, Italy). After blocking in 5% BSA in PBS containing 0.05% Tween 20 (1 h at room temperature), membranes were incubated overnight at 4 °C with primary antibodies: SERCA2 (1:1000, SantaCruz, Dallas, TX, USA, sc-376235), Acetylated-tubulin (1:1000, Sigma-Aldrich # T7451), GAPDH (glyceraldehyde-3-phosphate dehydrogenase, 1:2000, SantaCruz, Dallas, TX, USA, sc-32233), Pan-Ac-K (Acetylated-Lysine, 1:1000, Cell Signaling, Danvers, MA, USA, #9441), Phospho-PLB (Ser16) (Phospho-Phospholamban (Ser16), 1:5000, Merck Millipore, Billerica, MA, USA, # 07-052) and PLB (Phospholamban, 1:8000, Abcam, Hong Kong, China, #ab2865). Blots were washed three times in PBS-Tween buffer and then incubated with appropriate horseradish peroxidase conjugated secondary antibody (SantaCruz) for 1 h at room temperature. Detection was performed by enhanced chemiluminescence system (Supersignal West Dura Extended Duration Substrate, Thermo Scientific). Results were quantified by Image Lab software 5.2.1 (BioRad, Segrate-Milan, Italy).

4.5. Immunoprecipitation

Co-immunoprecipitation experiments were performed using protein-Agarose accordingly to the manufacturer's protocol (Roche Diagnostics S.p.A., Milan, Italy). In particular, 1 mg of protein extracts was co-immunoprecipitated with 10 µg of anti-SERCA2 antibody (SantaCruz, Dallas, TX, USA). Negative controls were performed with the same amount of protein extract (derived from rat cardiomyocytes) and were immunoprecipitated with the corresponding purified IgG antisera (SantaCruz) in the absence of primary antibody. Immunoprecipitated samples were resolved by SDS–PAGE, transferred onto nitrocellulose membrane (Bio-Rad, Segrate-Milan; Italy) and western blots were performed as described in the Section 4.4.

4.6. Isolation of the Microsomal Fraction

Microsome isolation from rat cardiomyocytes and HL-1 cells was performed according to Maruyama and MacLennan (1988) with minor modifications [69]. Specifically, cells were washed twice with PBS and then homogenized with 30 strokes in a glass Dounce homogenizer in 10 mM Tris-HCl, pH 7.5, 0.5 mM $MgCl_2$. The homogenate was diluted with an equal volume of a solution of 10 mM Tris-HCl, pH 7.5, 0.5 M sucrose, 300 mM KCl. The suspension was centrifuged at $10,000\times g$ for 30 min at 4 °C to pellet nuclei and mitochondria. The pellet was discarded and the supernatant was further centrifuged at $100,000\times g$ for 150 min at 4 °C to sediment the microsomal fraction. The pellet was re-suspended in a solution containing 0.25 M sucrose and 10 mM MOPS. All solutions were enriched in protease inhibitors (Roche). The microsome concentration was measured using the BCA protein kit (Thermo Fisher Scientific) following the manufacturer's instructions.

4.7. ATP/NADH Coupled Assay for Calcium ATPase Activity

The ATPase activity of the microsomal fraction isolated from HL-1 cells (25 µg/mL), was measured by spectrophotometric determination of NADH oxidation coupled to an ATP regenerating system, as previously described [70]. We used a Beckman DU 640 spectrophotometer adjusted at a wavelength of 340 nm. The assay was performed at 37 °C in a final volume of 1 mL of a buffer containing 20 mM histidine pH 7.2, 5 mM $MgCl_2$, 0.5 mM EGTA, 100 mM KCl, 2 mM ATP, 0.5 mM phosphoenolpyruvate, 0.15 mM NADH, 1.4 units of pyruvate kinase/lactic dehydrogenase, in the presence of 2 µg/mL A23187 Ca^{2+}-ionophore. ATPase activity was obtained subtracting the basal activity (measured in the presence of EGTA without calcium added) from the maximal activity and was expressed as micromoles/min/mg of protein. For investigating the calcium-dependence of ATPase activity, the concentration of free calcium was varied from pCa8 to pCa5 using EGTA buffered solutions.

NADH coupled ATPase assay protocol was then adapted for use on a 96-well microplate reader, according to the method described by Kiianitsa and coworkers [71]. We used the microplate protocol to analyze the ATPase activity of the microsomal fraction isolated from adult rat ventricular cardiomyocytes (5 µg/well) and from HEK293 cells (20 µg/well) expressing either Wild Type or mutated human SERCA2a (see Section 4.10). The absorbance change at 340 nm was monitored using the EnVision (Perkin Elmer, Waltham, MA, USA) plate reader. The experiments were performed at 37 °C in a final volume of 200 μL at pCa5 [21] on the same buffer described above. Technical duplicates were performed for each experiment.

4.8. Rat Cardiomyocyte Mechanics and Calcium Transients

CM mechanical properties were assessed by using the IonOptix fluorescence and contractility systems (IonOptix, Milton, MA, USA). CMs were placed in a chamber mounted on the stage of an inverted microscope (Nikon-Eclipse TE2000-U, Nikon Instruments, Florence, Italy) and superfused (1 mL/min at 37 °C) with a Tyrode solution containing (in mM): 140 NaCl, 5.4 KCl, 1 $MgCl_2$, 5 HEPES, 5.5 glucose, and 1 $CaCl_2$ (pH 7.4, adjusted with NaOH). Only rod-shaped myocytes with clear edges and average sarcomere length ≥ 1.7 μm were selected for the analysis. All the selected myocytes did not show spontaneous contractions. The cells were field stimulated at a frequency of 0.5 Hz by constant current pulses (2 ms in duration, and twice diastolic threshold in intensity) delivered by platinum electrodes placed on opposite sides of the chamber, connected to a MyoPacer Field Stimulator (IonOptix). The stimulated myocyte was displayed on a computer monitor using an IonOptix MyoCam camera. Load-free contraction of myocytes was measured with the IonOptix system, which captures sarcomere length dynamics via a Fast Fourier Transform algorithm. Sampling rate was fixed at 1 KHz. A total of 115 control isolated ventricular myocytes (of which 45 were used as untreated control and 70 were exposed to SAHA) were analyzed to compute the following parameters: mean diastolic sarcomere length and fraction of shortening (FS), maximal rates of shortening and re-lengthening ($\pm dl/dt_{max}$), and time at 10%, 50% and 90% of re-lengthening (RL10, RL50 and RL90, respectively). Steady-state contraction of myocytes was achieved before data recording by means of a 10 s conditioning stimulation.

Calcium transients were measured simultaneously with cell motion. Ca^{2+} transients were detected by epifluorescence after loading the myocytes with Fluo-3 AM (10 μmol/L; Invitrogen, Carlsbad, CA, USA) for 30 min. Excitation length was 480 nm, with emission collected at 535 nm using a $40\times$ oil objective lens (NA: 1.3). Fluo-3 signals were expressed as normalized fluorescence (f/f0: fold increase). The time course of the fluorescence signal decay was described by a single exponential equation, and the time constant (tau) was used as a measure of the rate of intracellular calcium clearing [72]. The time to peak of the calcium transients (TTP) and the times to 10%, 50% and 90% of fluorescence signal decay were also measured (time to BaseLine fluorescence BL10, BL50 and BL90, respectively).

4.9. Bioinformatics Analysis

The post-translation modification databases PHOSIDA [73] and PhosphoSitePlus [74] were used for the identification of acetylation sites investigated in human SERCA2. The location of the lysine acetylation was investigated in human SERCA2a structural models derived by homology modelling based on rabbit homologue templates with 83% sequence identity to hSERCA2a, and using MODELLER version 9.11 [75].

4.10. Mutagenesis

SERCA2a mutants were generated by PCR-site directed mutagenesis (for a list of primers see Table S1) and verified by complete sequencing (Eurofins Genomics, Ebersberg, Germany). The SERCA2a wild-type coding sequence (NM_001681.3; SER-WT) as well as the SERCA2a coding sequence with Lys 464 residue mutated into glutamine (SER-K464Q) or arginine (SER-K464R) were subsequently cloned into the pCDNA3 vector carrying Geneticin resistance (Life Technologies) for subsequent cell transfection. Transfections were performed on HEK-MSR1 cells (Human Embryonic Kidney 293 cells) that stably express human Macrophage Scavenger Receptor 1, facilitating cell adhesion in blasticidin selection (5 μg/mL, Invitrogen) [76]. Lipofectamine LTX (Invitrogen) was used for transfection following manufacturer's instructions. Stable cell lines were selected 24 h after transfections in culture

Int. J. Mol. Sci. **2018**, *19*, 419

medium composed by: MEM (Invitrogen) supplemented with 10% FBS (Sigma-Aldrich), 1% *v/v* glutamine (Invitrogen), 1% *v/v* minimum non-essential amino acids (Invitrogen) and 450 µg/mL Geneticin (Invitrogen).

4.10.1. SERCA2 Mutant Calcium Transients

HEK-MSR1 stably transfected with K464 onto SERCA2 sequence mutated in Q or R were plated on the bottom of glass dishes (25 mm diameter) and grown for 2 days in MEM medium supplemented with FBS 10%. Cytosolic Ca^{2+} dynamics were evaluated in Fluo-4 AM loaded cells. Briefly, cells were loaded with 2.5 µM of Fluo-4 AM (Molecular Probes, Invitrogen) for 30 min at 37 °C and then washed with Tyrode's solution containing in mM: 10 D-(+)-glucose, 5.0 Hepes, 140 NaCl, 5.4 KCl, 1.2 $MgCl_2$, 1.8 $CaCl_2$, pH adjusted to 7.3 with NaOH. Dishes were mounted in a perfusion chamber and placed on the stage of an epifluorescence microscope (Nikon) equipped with a 75 W Xenon lamp and connected to a CCD camera (Cool SnapTM EZ Photometrics, Tucson, AZ, USA). Cells were excited at 488 nm wavelength, fluorescence emission was measured at 520 nm and recorded using the MetaFluor Software (Molecular Devices, Sunnyvale, CA, USA); imaging was scanned repeatedly at 20 Hz. This low temporal acquisition is justified by the longer calcium transient in HEK cells lasting on average 1 min and 20 s. Calcium transients were evoked by a 1 s caffeine-pulse (10 mM). Temperature was maintained at 36.5 ± 0.5 °C throughout the experiment.

The background fluorescence signal was fitted to a linear regression and then subtracted. Each trace was subsequently resampled and denoised with an Rbf interpolator (f = 20, e = 1, s = 1). The baseline was calculated per-trace using the median value of the first 250 ms before the excitation and then subtracted.

Traces which didn't reach a peak value of at least 2 times the baseline level, or dropped under 90% of the absolute value of the baseline were discarded. We then calculated the time to decay at the first intersection of 10%, 50% and 90% of the curves, thus defining time to BL10, BL50 and BL90. Only traces that successfully recovered to BL90 within the recording timeline (90 s) were considered.

4.10.2. Statistical Analysis

All data are reported as means ± standard error (SEM). Normality of the data was evaluated by D'Agostino & Pearson normality test. For two groups, significance was assessed by two-tailed unpaired *t*-test or non-parametric Mann–Whitney *U*-test when appropriate. For more than two groups either two-way ANOVA, parametric one-way ANOVA or Kruskal-Wallis non parametric test were used followed by appropriate post-hoc individual comparisons (GraphPad Prism software 6.03). General Linear Model (GLM) ANOVA for repeated measurements was used to compare cell contractility and calcium transient data (IBM-SPSS 24.0, SPSS Inc., Chicago, IL, USA). The details on the specific test used are reported in the figure legend of each specific experiment. A *p*-value < 0.05 was considered statistically significant.

Supplementary Materials: Supplementary materials can be found at www.mdpi.com/2079-6382/19/2/419/s1.

Acknowledgments: The authors would like to thank William C. Claycomb for providing the HL-1 cell line. This work was supported by the Department of Innovation, Research and Universities of the Autonomous Province of Bolzano-South Tyrol (Italy) and local research funding from University of Parma (Italy) (FIL2014 and FIL2016). The founding sponsors had no role in the design of the study, in the collection, analyses, or interpretation of data, in the writing of the manuscript, and in the decision to publish the results.

Author Contributions: Viviana Meraviglia, Leonardo Bocchi, Donatella Stilli and Alessandra Rossini conceived and designed the experiments; Viviana Meraviglia, Leonardo Bocchi, Roberta Sacchetto, Maria Cristina Florio, Benedetta M. Motta, Corrado Corti, Monia Savi, Marcelo D. Rosato-Siri, Silvia Suffredini, Chiara Piubelli performed the experiments; Viviana Meraviglia, Leonardo Bocchi, Roberta Sacchetto, Marcelo D. Rosato-Siri, Yuri D'Elia analyzed the data; Christian X. Weichenberger and Francisco S. Domingues performed the bioinformatics analysis; Viviana Meraviglia, Francisco S. Domingues, Donatella Stilli and Alessandra Rossini wrote the paper; Giulio Pompilio and Peter P. Pramstaller critically and substantively revised the paper; Donatella Stilli and Alessandra Rossini supervised and managed the project.

Conflicts of Interest: The authors declare no conflict of interest.

References

1. Toyoshima, C. Structural aspects of ion pumping by Ca^{2+}-ATPase of sarcoplasmic reticulum. *Arch. Biochem. Biophys.* **2008**, *476*, 3–11. [CrossRef] [PubMed]
2. Asahi, M.; Sugita, Y.; Kurzydlowski, K.; De Leon, S.; Tada, M.; Toyoshima, C.; MacLennan, D.H. Sarcolipin regulates sarco(endo)plasmic reticulum Ca^{2+}-ATPase (SERCA) by binding to transmembrane helices alone or in association with phospholamban. *Proc. Natl. Acad. Sci. USA* **2003**, *100*, 5040–5045. [CrossRef] [PubMed]
3. Periasamy, M.; Kalyanasundaram, A. SERCA pump isoforms: Their role in calcium transport and disease. *Muscle Nerve* **2007**, *35*, 430–442. [CrossRef] [PubMed]
4. Bers, D.M. Cardiac excitation-contraction coupling. *Nature* **2002**, *415*, 198–205. [CrossRef] [PubMed]
5. Sulaiman, M.; Matta, M.J.; Sunderesan, N.R.; Gupta, M.P.; Periasamy, M.; Gupta, M. Resveratrol, an activator of SIRT1, upregulates sarcoplasmic calcium ATPase and improves cardiac function in diabetic cardiomyopathy. *Am. J. Physiol. Heart Circ. Physiol.* **2010**, *298*, H833–H843. [CrossRef] [PubMed]
6. Savi, M.; Bocchi, L.; Mena, P.; Dall'Asta, M.; Crozier, A.; Brighenti, F.; Stilli, D.; Del Rio, D. In vivo administration of urolithin A and B prevents the occurrence of cardiac dysfunction in streptozotocin-induced diabetic rats. *Cardiovasc. Diabetol.* **2017**, *16*, 80. [CrossRef] [PubMed]
7. Del Monte, F.; Harding, S.E.; Schmidt, U.; Matsui, T.; Kang, Z.B.; Dec, G.W.; Gwathmey, J.K.; Rosenzweig, A.; Hajjar, R.J. Restoration of contractile function in isolated cardiomyocytes from failing human hearts by gene transfer of SERCA2a. *Circulation* **1999**, *100*, 2308–2311. [CrossRef]
8. Park, W.J.; Oh, J.G. SERCA2a: A prime target for modulation of cardiac contractility during heart failure. *BMB Rep.* **2013**, *46*, 237–243. [CrossRef] [PubMed]
9. Periasamy, M.; Bhupathy, P.; Babu, G.J. Regulation of sarcoplasmic reticulum Ca^{2+} ATPase pump expression and its relevance to cardiac muscle physiology and pathology. *Cardiovasc. Res.* **2008**, *77*, 265–273. [CrossRef] [PubMed]
10. Stammers, A.N.; Susser, S.E.; Hamm, N.C.; Hlynsky, M.W.; Kimber, D.E.; Kehler, D.S.; Duhamel, T.A. The regulation of sarco(endo)plasmic reticulum calcium-ATPases (SERCA). *Can. J. Physiol. Pharmacol.* **2015**, *93*, 843–854. [CrossRef] [PubMed]
11. Kho, C.; Lee, A.; Jeong, D.; Oh, J.G.; Chaanine, A.H.; Kizana, E.; Park, W.J.; Hajjar, R.J. SUMO1-dependent modulation of SERCA2a in heart failure. *Nature* **2011**, *477*, 601–605. [CrossRef] [PubMed]
12. Adachi, T.; Weisbrod, R.M.; Pimentel, D.R.; Ying, J.; Sharov, V.S.; Schoneich, C.; Cohen, R.A. S-Glutathiolation by peroxynitrite activates SERCA during arterial relaxation by nitric oxide. *Nat. Med.* **2004**, *10*, 1200–1207. [CrossRef] [PubMed]
13. Bidasee, K.R.; Zhang, Y.; Shao, C.H.; Wang, M.; Patel, K.P.; Dincer, U.D.; Besch, H.R., Jr. Diabetes increases formation of advanced glycation end products on Sarco(endo)plasmic reticulum Ca^{2+}-ATPase. *Diabetes* **2004**, *53*, 463–473. [CrossRef] [PubMed]
14. Knyushko, T.V.; Sharov, V.S.; Williams, T.D.; Schoneich, C.; Bigelow, D.J. 3-Nitrotyrosine modification of SERCA2a in the aging heart: A distinct signature of the cellular redox environment. *Biochemistry* **2005**, *44*, 13071–13081. [CrossRef] [PubMed]
15. Foster, D.B.; Liu, T.; Rucker, J.; O'Meally, R.N.; Devine, L.R.; Cole, R.N.; O'Rourke, B. The cardiac acetyl-lysine proteome. *PLoS ONE* **2013**, *8*, e67513. [CrossRef] [PubMed]
16. Minucci, S.; Pelicci, P.G. Histone deacetylase inhibitors and the promise of epigenetic (and more) treatments for cancer. *Nat. Rev. Cancer* **2006**, *6*, 38–51. [CrossRef] [PubMed]
17. Colussi, C.; Berni, R.; Rosati, J.; Straino, S.; Vitale, S.; Spallotta, F.; Baruffi, S.; Bocchi, L.; Delucchi, F.; Rossi, S.; et al. The histone deacetylase inhibitor suberoylanilide hydroxamic acid reduces cardiac arrhythmias in dystrophic mice. *Cardiovasc. Res.* **2010**, *87*, 73–82. [CrossRef] [PubMed]
18. McLendon, P.M.; Ferguson, B.S.; Osinska, H.; Bhuiyan, M.S.; James, J.; McKinsey, T.A.; Robbins, J. Tubulin hyperacetylation is adaptive in cardiac proteotoxicity by promoting autophagy. *Proc. Natl. Acad. Sci. USA* **2014**, *111*, E5178–E5186. [CrossRef] [PubMed]

19. Bianchini, E.; Testoni, S.; Gentile, A.; Cali, T.; Ottolini, D.; Villa, A.; Brini, M.; Betto, R.; Mascarello, F.; Nissen, P.; et al. Inhibition of ubiquitin proteasome system rescues the defective sarco(endo)plasmic reticulum Ca^{2+}-ATPase (SERCA1) protein causing Chianina cattle pseudomyotonia. *J. Biol. Chem.* **2014**, *289*, 33073–33082. [CrossRef] [PubMed]

20. Lytton, J.; Westlin, M.; Burk, S.E.; Shull, G.E.; MacLennan, D.H. Functional comparisons between isoforms of the sarcoplasmic or endoplasmic reticulum family of calcium pumps. *J. Biol. Chem.* **1992**, *267*, 14483–14489. [PubMed]

21. Kho, C.; Lee, A.; Jeong, D.; Oh, J.G.; Gorski, P.A.; Fish, K.; Sanchez, R.; DeVita, R.J.; Christensen, G.; Dahl, R.; et al. Small-molecule activation of SERCA2a SUMOylation for the treatment of heart failure. *Nat. Commun.* **2015**, *6*, 7229. [CrossRef] [PubMed]

22. Claycomb, W.C.; Lanson, N.A., Jr.; Stallworth, B.S.; Egeland, D.B.; Delcarpio, J.B.; Bahinski, A.; Izzo, N.J., Jr. HL-1 cells: A cardiac muscle cell line that contracts and retains phenotypic characteristics of the adult cardiomyocyte. *Proc. Natl. Acad. Sci. USA* **1998**, *95*, 2979–2984. [CrossRef] [PubMed]

23. Mertins, P.; Qiao, J.W.; Patel, J.; Udeshi, N.D.; Clauser, K.R.; Mani, D.R.; Burgess, M.W.; Gillette, M.A.; Jaffe, J.D.; Carr, S.A. Integrated proteomic analysis of post-translational modifications by serial enrichment. *Nat. Methods* **2013**, *10*, 634–637. [CrossRef] [PubMed]

24. Choudhary, C.; Kumar, C.; Gnad, F.; Nielsen, M.L.; Rehman, M.; Walther, T.C.; Olsen, J.V.; Mann, M. Lysine acetylation targets protein complexes and co-regulates major cellular functions. *Science* **2009**, *325*, 834–840. [CrossRef] [PubMed]

25. Lytton, J.; Westlin, M.; Hanley, M.R. Thapsigargin inhibits the sarcoplasmic or endoplasmic reticulum Ca-ATPase family of calcium pumps. *J. Biol. Chem.* **1991**, *266*, 17067–17071. [PubMed]

26. Munshi, A.; Tanaka, T.; Hobbs, M.L.; Tucker, S.L.; Richon, V.M.; Meyn, R.E. Vorinostat, a histone deacetylase inhibitor, enhances the response of human tumor cells to ionizing radiation through prolongation of gamma-H2AX foci. *Mol. Cancer Ther.* **2006**, *5*, 1967–1974. [CrossRef] [PubMed]

27. Witt, O.; Milde, T.; Deubzer, H.E.; Oehme, I.; Witt, R.; Kulozik, A.; Eisenmenger, A.; Abel, U.; Karapanagiotou-Schenkel, I. Phase I/II intra-patient dose escalation study of vorinostat in children with relapsed solid tumor, lymphoma or leukemia. *Klinische Pädiatrie* **2012**, *224*, 398–403. [CrossRef] [PubMed]

28. Butler, L.M.; Zhou, X.; Xu, W.S.; Scher, H.I.; Rifkind, R.A.; Marks, P.A.; Richon, V.M. The histone deacetylase inhibitor SAHA arrests cancer cell growth, up-regulates thioredoxin-binding protein-2, and down-regulates thioredoxin. *Proc. Natl. Acad. Sci. USA* **2002**, *99*, 11700–11705. [CrossRef] [PubMed]

29. Richon, V.M.; Sandhoff, T.W.; Rifkind, R.A.; Marks, P.A. Histone deacetylase inhibitor selectively induces p21WAF1 expression and gene-associated histone acetylation. *Proc. Natl. Acad. Sci. USA* **2000**, *97*, 10014–10019. [CrossRef] [PubMed]

30. Xie, M.; Kong, Y.; Tan, W.; May, H.; Battiprolu, P.K.; Pedrozo, Z.; Wang, Z.V.; Morales, C.; Luo, X.; Cho, G.; et al. Histone deacetylase inhibition blunts ischemia/reperfusion injury by inducing cardiomyocyte autophagy. *Circulation* **2014**, *129*, 1139–1151. [CrossRef] [PubMed]

31. Tiffon, C.; Adams, J.; van der Fits, L.; Wen, S.; Townsend, P.; Ganesan, A.; Hodges, E.; Vermeer, M.; Packham, G. The histone deacetylase inhibitors vorinostat and romidepsin downmodulate IL-10 expression in cutaneous T-cell lymphoma cells. *Br. J. Pharmacol.* **2011**, *162*, 1590–1602. [CrossRef] [PubMed]

32. Simonides, W.S.; van Hardeveld, C. An assay for sarcoplasmic reticulum Ca^{2+}-ATPase activity in muscle homogenates. *Anal. Biochem.* **1990**, *191*, 321–331. [CrossRef]

33. Kennedy, D.; Omran, E.; Periyasamy, S.M.; Nadoor, J.; Priyadarshi, A.; Willey, J.C.; Malhotra, D.; Xie, Z.; Shapiro, J.I. Effect of chronic renal failure on cardiac contractile function, calcium cycling, and gene expression of proteins important for calcium homeostasis in the rat. *J. Am. Soc. Nephrol.* **2003**, *14*, 90–97. [CrossRef] [PubMed]

34. Kennedy, D.J.; Vetteth, S.; Xie, M.; Periyasamy, S.M.; Xie, Z.; Han, C.; Basrur, V.; Mutgi, K.; Fedorov, V.; Malhotra, D.; et al. Ouabain decreases sarco(endo)plasmic reticulum calcium ATPase activity in rat hearts by a process involving protein oxidation. *Am. J. Physiol. Heart Circ. Physiol.* **2006**, *291*, H3003–H3011. [CrossRef] [PubMed]

35. Frank, K.F.; Bolck, B.; Erdmann, E.; Schwinger, R.H. Sarcoplasmic reticulum Ca^{2+}-ATPase modulates cardiac contraction and relaxation. *Cardiovasc. Res.* **2003**, *57*, 20–27. [CrossRef]

36. Gupta, M.P.; Samant, S.A.; Smith, S.H.; Shroff, S.G. HDAC4 and PCAF bind to cardiac sarcomeres and play a role in regulating myofilament contractile activity. *J. Biol. Chem.* **2008**, *283*, 10135–10146. [CrossRef] [PubMed]

37. Chung, J.H.; Biesiadecki, B.J.; Ziolo, M.T.; Davis, J.P.; Janssen, P.M. Myofilament Calcium Sensitivity: Role in Regulation of In vivo Cardiac Contraction and Relaxation. *Front. Physiol.* **2016**, *7*, 562. [CrossRef] [PubMed]

38. Chen-Izu, Y.; Shaw, R.M.; Pitt, G.S.; Yarov-Yarovoy, V.; Sack, J.T.; Abriel, H.; Aldrich, R.W.; Belardinelli, L.; Cannell, M.B.; Catterall, W.A.; et al. Na+ channel function, regulation, structure, trafficking and sequestration. *J. Physiol.* **2015**, *593*, 1347–1360. [CrossRef] [PubMed]

39. Kennedy, M.; Bers, D.M.; Chiamvimonvat, N.; Sato, D. Dynamical effects of calcium-sensitive potassium currents on voltage and calcium alternans. *J. Physiol.* **2017**, *595*, 2285–2297. [CrossRef] [PubMed]

40. Bers, D.M.; Chen-Izu, Y. Sodium and calcium regulation in cardiac myocytes: From molecules to heart failure and arrhythmia. *J. Physiol.* **2015**, *593*, 1327–1329. [CrossRef] [PubMed]

41. Olesen, C.; Picard, M.; Winther, A.M.; Gyrup, C.; Morth, J.P.; Oxvig, C.; Moller, J.V.; Nissen, P. The structural basis of calcium transport by the calcium pump. *Nature* **2007**, *450*, 1036–1042. [CrossRef] [PubMed]

42. Ying, J.; Tong, X.; Pimentel, D.R.; Weisbrod, R.M.; Trucillo, M.P.; Adachi, T.; Cohen, R.A. Cysteine-674 of the sarco/endoplasmic reticulum calcium ATPase is required for the inhibition of cell migration by nitric oxide. *Arterioscler. Thromb. Vasc. Biol.* **2007**, *27*, 783–790. [CrossRef] [PubMed]

43. Tong, J.; Du, G.G.; Chen, S.R.; MacLennan, D.H. HEK-293 cells possess a carbachol- and thapsigargin-sensitive intracellular Ca2+ store that is responsive to stop-flow medium changes and insensitive to caffeine and ryanodine. *Biochem. J.* **1999**, *343 Pt 1*, 39–44. [CrossRef] [PubMed]

44. Querfurth, H.W.; Haughey, N.J.; Greenway, S.C.; Yacono, P.W.; Golan, D.E.; Geiger, J.D. Expression of ryanodine receptors in human embryonic kidney (HEK293) cells. *Biochem. J.* **1998**, *334 Pt 1*, 79–86. [CrossRef] [PubMed]

45. Luo, D.; Sun, H.; Xiao, R.P.; Han, Q. Caffeine induced Ca2+ release and capacitative Ca2+ entry in human embryonic kidney (HEK293) cells. *Eur. J. Pharmacol.* **2005**, *509*, 109–115. [CrossRef] [PubMed]

46. Itzhaki, I.; Rapoport, S.; Huber, I.; Mizrahi, I.; Zwi-Dantsis, L.; Arbel, G.; Schiller, J.; Gepstein, L. Calcium handling in human induced pluripotent stem cell derived cardiomyocytes. *PLoS ONE* **2011**, *6*, e18037. [CrossRef] [PubMed]

47. Blakeslee, W.W.; Wysoczynski, C.L.; Fritz, K.S.; Nyborg, J.K.; Churchill, M.E.; McKinsey, T.A. Class I HDAC inhibition stimulates cardiac protein SUMOylation through a post-translational mechanism. *Cell Signal.* **2014**, *26*, 2912–2920. [CrossRef] [PubMed]

48. Bers, D.M.; Despa, S. Cardiac myocytes Ca2+ and Na+ regulation in normal and failing hearts. *J. Pharmacol. Sci.* **2006**, *100*, 315–322. [CrossRef] [PubMed]

49. Bostjancic, E.; Zidar, N.; Glavac, D. MicroRNAs and cardiac sarcoplasmic reticulum calcium ATPase-2 in human myocardial infarction: Expression and bioinformatic analysis. *BMC Genom.* **2012**, *13*, 552. [CrossRef] [PubMed]

50. Gurha, P.; Abreu-Goodger, C.; Wang, T.; Ramirez, M.O.; Drumond, A.L.; van Dongen, S.; Chen, Y.; Bartonicek, N.; Enright, A.J.; Lee, B.; et al. Targeted deletion of microRNA-22 promotes stress-induced cardiac dilation and contractile dysfunction. *Circulation* **2012**, *125*, 2751–2761. [CrossRef] [PubMed]

51. Zhao, B.; Lucas, K.J.; Saha, T.T.; Ha, J.; Ling, L.; Kokoza, V.A.; Roy, S.; Raikhel, A.S. MicroRNA-275 targets sarco/endoplasmic reticulum Ca2+ adenosine triphosphatase (SERCA) to control key functions in the mosquito gut. *PLoS Genet.* **2017**, *13*, e1006943. [CrossRef] [PubMed]

52. Lee, E.M.; Shin, S.; Cha, H.J.; Yoon, Y.; Bae, S.; Jung, J.H.; Lee, S.M.; Lee, S.J.; Park, I.C.; Jin, Y.W.; et al. Suberoylanilide hydroxamic acid (SAHA) changes microRNA expression profiles in A549 human non-small cell lung cancer cells. *Int. J. Mol. Med.* **2009**, *24*, 45–50. [PubMed]

53. Yang, H.; Lan, P.; Hou, Z.; Guan, Y.; Zhang, J.; Xu, W.; Tian, Z.; Zhang, C. Histone deacetylase inhibitor SAHA epigenetically regulates miR-17-92 cluster and MCM7 to upregulate MICA expression in hepatoma. *Br. J. Cancer* **2015**, *112*, 112–121. [CrossRef] [PubMed]

54. Poddar, S.; Kesharwani, D.; Datta, M. Histone deacetylase inhibition regulates miR-449a levels in skeletal muscle cells. *Epigenetics* **2016**, *11*, 579–587. [CrossRef] [PubMed]

55. Li, S.; Li, X.; Zheng, H.; Xie, B.; Bidasee, K.R.; Rozanski, G.J. Pro-oxidant effect of transforming growth factor-beta1 mediates contractile dysfunction in rat ventricular myocytes. *Cardiovasc. Res.* **2008**, *77*, 107–117. [CrossRef] [PubMed]

56. Mufti, S.; Wenzel, S.; Euler, G.; Piper, H.M.; Schluter, K.D. Angiotensin II-dependent loss of cardiac function: Mechanisms and pharmacological targets attenuating this effect. *J. Cell. Physiol.* **2008**, *217*, 242–249. [CrossRef] [PubMed]

57. Ammanamanchi, S.; Brattain, M.G. Restoration of transforming growth factor-beta signaling through receptor RI induction by histone deacetylase activity inhibition in breast cancer cells. *J. Biol. Chem.* **2004**, *279*, 32620–32625. [CrossRef] [PubMed]

58. Khan, S.; Jena, G. Sodium butyrate, a HDAC inhibitor ameliorates eNOS, iNOS and TGF-beta1-induced fibrogenesis, apoptosis and DNA damage in the kidney of juvenile diabetic rats. *Food Chem. Toxicol.* **2014**, *73*, 127–139. [CrossRef] [PubMed]

59. Xie, L.; Santhoshkumar, P.; Reneker, L.W.; Sharma, K.K. Histone deacetylase inhibitors trichostatin A and vorinostat inhibit TGFbeta2-induced lens epithelial-to-mesenchymal cell transition. *Investig. Ophthalmol. Vis. Sci.* **2014**, *55*, 4731–4740. [CrossRef] [PubMed]

60. Huber, K.; Doyon, G.; Plaks, J.; Fyne, E.; Mellors, J.W.; Sluis-Cremer, N. Inhibitors of histone deacetylases: Correlation between isoform specificity and reactivation of HIV type 1 (HIV-1) from latently infected cells. *J. Biol. Chem.* **2011**, *286*, 22211–22218. [CrossRef] [PubMed]

61. Clocchiatti, A.; Florean, C.; Brancolini, C. Class IIa HDACs: From important roles in differentiation to possible implications in tumourigenesis. *J. Cell. Mol. Med.* **2011**, *15*, 1833–1846. [CrossRef] [PubMed]

62. Kahali, S.; Sarcar, B.; Prabhu, A.; Seto, E.; Chinnaiyan, P. Class I histone deacetylases localize to the endoplasmic reticulum and modulate the unfolded protein response. *FASEB J.* **2012**, *26*, 2437–2445. [CrossRef] [PubMed]

63. Ligeti, L.; Szenczi, O.; Prestia, C.M.; Szabo, C.; Horvath, K.; Marcsek, Z.L.; van Stiphout, R.G.; van Riel, N.A.; Op den Buijs, J.; Van der Vusse, G.J.; et al. Altered calcium handling is an early sign of streptozotocin-induced diabetic cardiomyopathy. *Int. J. Mol. Med.* **2006**, *17*, 1035–1043. [CrossRef] [PubMed]

64. Pieske, B.; Kretschmann, B.; Meyer, M.; Holubarsch, C.; Weirich, J.; Posival, H.; Minami, K.; Just, H.; Hasenfuss, G. Alterations in intracellular calcium handling associated with the inverse force-frequency relation in human dilated cardiomyopathy. *Circulation* **1995**, *92*, 1169–1178. [CrossRef] [PubMed]

65. Saini, H.K.; Dhalla, N.S. Defective calcium handling in cardiomyocytes isolated from hearts subjected to ischemia-reperfusion. *Am. J. Physiol. Heart Circ. Physiol.* **2005**, *288*, H2260–H2270. [CrossRef] [PubMed]

66. Balke, C.W.; Shorofsky, S.R. Alterations in calcium handling in cardiac hypertrophy and heart failure. *Cardiovasc. Res.* **1998**, *37*, 290–299. [CrossRef]

67. Luo, M.; Anderson, M.E. Mechanisms of altered Ca^{2+} handling in heart failure. *Circ. Res.* **2013**, *113*, 690–708. [CrossRef] [PubMed]

68. Zaniboni, M.; Pollard, A.E.; Yang, L.; Spitzer, K.W. Beat-to-beat repolarization variability in ventricular myocytes and its suppression by electrical coupling. *Am. J. Physiol. Heart Circ. Physiol.* **2000**, *278*, H677–H687. [CrossRef] [PubMed]

69. Maruyama, K.; MacLennan, D.H. Mutation of aspartic acid-351, lysine-352, and lysine-515 alters the Ca^{2+} transport activity of the Ca^{2+}-ATPase expressed in COS-1 cells. *Proc. Natl. Acad. Sci. USA* **1988**, *85*, 3314–3318. [CrossRef] [PubMed]

70. Sacchetto, R.; Testoni, S.; Gentile, A.; Damiani, E.; Rossi, M.; Liguori, R.; Drogemuller, C.; Mascarello, F. A defective SERCA1 protein is responsible for congenital pseudomyotonia in Chianina cattle. *Am. J. Pathol.* **2009**, *174*, 565–573. [CrossRef] [PubMed]

71. Kiianitsa, K.; Solinger, J.A.; Heyer, W.D. NADH-coupled microplate photometric assay for kinetic studies of ATP-hydrolyzing enzymes with low and high specific activities. *Anal. Biochem.* **2003**, *321*, 266–271. [CrossRef]

72. Bassani, J.W.; Bassani, R.A.; Bers, D.M. Relaxation in rabbit and rat cardiac cells: Species-dependent differences in cellular mechanisms. *J. Physiol.* **1994**, *476*, 279–293. [CrossRef] [PubMed]

73. Gnad, F.; Gunawardena, J.; Mann, M. PHOSIDA 2011: The posttranslational modification database. *Nucleic Acids Res.* **2011**, *39*, D253–D260. [CrossRef] [PubMed]

74. Hornbeck, P.V.; Zhang, B.; Murray, B.; Kornhauser, J.M.; Latham, V.; Skrzypek, E. PhosphoSitePlus, 2014: Mutations, PTMs and recalibrations. *Nucleic Acids Res.* **2015**, *43*, D512–D520. [CrossRef] [PubMed]

75. Marti-Renom, M.A.; Stuart, A.C.; Fiser, A.; Sanchez, R.; Melo, F.; Sali, A. Comparative protein structure modeling of genes and genomes. *Annu. Rev. Biophys. Biomol. Struct.* **2000**, *29*, 291–325. [CrossRef] [PubMed]
76. Robbins, A.K.; Horlick, R.A. Macrophage scavenger receptor confers an adherent phenotype to cells in culture. *Biotechniques* **1998**, *25*, 240–244. [PubMed]

International Journal of
Molecular Sciences

MDPI

Article

STIM1 Knockout Enhances PDGF-Mediated Ca^{2+} Signaling through Upregulation of the PDGFR–PLCγ–STIM2 Cascade

Tzu-Yu Huang [1], Yi-Hsin Lin [1], Heng-Ai Chang [2], Tzu-Ying Yeh [1], Ya-Han Chang [1], Yi-Fan Chen [2], Ying-Chi Chen [1], Chun-Chun Li [3] and Wen-Tai Chiu [1,2,*]

[1] Department of Biomedical Engineering, National Cheng Kung University, Tainan 701, Taiwan;
     sowa010750@gmail.com (T.-Y.H.); wsp90020@gmail.com (Y.-H.L.); blackcatfantasy555@gmail.com (T.-Y.Y.);
     ovov665@hotmail.com.tw (Y.-H.C.); yingtzai@gmail.com (Y.-C.C.)
[2] Institute of Basic Medical Sciences, National Cheng Kung University, Tainan 701, Taiwan;
     lisa40723@gmail.com (H.-A.C.); fanny870017@hotmail.com (Y.-F.C.)
[3] Department of Life Sciences, National Cheng Kung University, Tainan 701, Taiwan; ccli@mail.ncku.edu.tw
*   Correspondence: wtchiu@mail.ncku.edu.tw; Tel.: +886-6-275-7575 (ext. 63435)

Received: 5 May 2018; Accepted: 14 June 2018; Published: 18 June 2018

Abstract: Platelet-derived growth factor (PDGF) has mitogenic and chemotactic effects on fibroblasts. An increase in intracellular Ca^{2+} is one of the first events that occurs following the stimulation of PDGF receptors (PDGFRs). PDGF activates Ca^{2+} elevation by activating the phospholipase C gamma (PLCγ)-signaling pathway, resulting in ER Ca^{2+} release. Store-operated Ca^{2+} entry (SOCE) is the major form of extracellular Ca^{2+} influx following depletion of ER Ca^{2+} stores and stromal interaction molecule 1 (STIM1) is a key molecule in the regulation of SOCE. In this study, wild-type and STIM1 knockout mouse embryonic fibroblasts (MEF) cells were used to investigate the role of STIM1 in PDGF-induced Ca^{2+} oscillation and its functions in MEF cells. The unexpected findings suggest that STIM1 knockout enhances PDGFR–PLCγ–STIM2 signaling, which in turn increases PDGF-BB-induced Ca^{2+} elevation. Enhanced expressions of PDGFRs and PLCγ in STIM1 knockout cells induce Ca^{2+} release from the ER store through PLCγ–IP3 signaling. Moreover, STIM2 replaces STIM1 to act as the major ER Ca^{2+} sensor in activating SOCE. However, activation of PDGFRs also activate Akt, ERK, and JNK to regulate cellular functions, such as cell migration. These results suggest that alternative switchable pathways can be observed in cells, which act downstream of the growth factors that regulate Ca^{2+} signaling.

Keywords: Ca^{2+}; STIM1; STIM2; SOCE; PDGF; PLCγ

1. Introduction

Ca^{2+} plays an important and ubiquitous role in intracellular signaling in response to several cellular functions, including cell proliferation, development, differentiation, migration, transcription factor activation, and apoptosis [1,2]. To coordinate these functions, Ca^{2+} signaling is regulated meticulously and is distinguished by several patterns of spatio–temporal parameters with altering amplitude, frequency, or duration [2–4]. In cells, Ca^{2+} can be released from the internal store, the endoplasmic reticulum (ER), to increase the intracellular Ca^{2+} concentration. Moreover, there is a common signal transduction pathway to increase cytosolic Ca^{2+} concentration. After the ligand binds to receptor tyrosine kinase (RTK), phospholipase C gamma (PLCγ) is phosphorylated and activated, which in turn hydrolyzes membrane-bound phosphatidylinositol 4,5-bisphosphate (PIP2) into 1,4,5-inositol trisphosphate (IP3) and diacylglycerol (DAG). IP3 can induce the diffusion of Ca^{2+} from the ER into the cytoplasm through the IP3 receptor (IP3R) on the ER membrane.

Subsequent exhaustion of ER Ca^{2+} activates a refill mechanism called the store-operated Ca^{2+} entry (SOCE), which also increases cytosolic Ca^{2+} concentration.

SOCE is the major mechanism that increases and replenishes intracellular Ca^{2+} concentration in most non-excitable and a few excitable cells and is triggered following the depletion of ER Ca^{2+} concentration via the store-operated channels on the plasma membrane [5–7]. Stromal interaction molecules (STIMs) are distributed on ER membranes as Ca^{2+} store sensors. Following the depletion of ER Ca^{2+} concentration, the activated STIM proteins aggregate near the plasma membrane and interact with two Ca^{2+}-permeable components of store-operated channels (SOCs), viz., Ca^{2+} release-activated calcium modulator 1 (CRAM1, also called Orai1) and transient receptor potential canonical 1 (TRPC1), which contribute to SOCE [8–10]. Extracellular Ca^{2+} can be passively transported into the cytoplasm through the STIM–Orai1–TRPC1 complex to regulate Ca^{2+}-sensitive enzymes and Ca^{2+}-dependent transcription factors, such as cAMP response element binding protein (CREB) and nuclear factor of activated T-cells (NFAT) [11–14].

The STIM family senses ER Ca^{2+} and activates store-operated channels (SOCs) to regulate cell proliferation [15,16] and migration [17,18]. There are two conserved isoforms of STIM proteins, STIM1 and STIM2, which differ from the ER-plasma membrane junction of the STIM structure [19–21]. STIM1 has been reported to play a key role in SOCE and regulates SOCE when ER Ca^{2+} store depletes through interacting with Ca^{2+} release-activated calcium modulator 1 (Orai1) in mouse embryonic fibroblasts (MEF) cells [8,22,23]. Previous studies have shown that STIM2 exhibits more sensitivity but lower affinity to Ca^{2+} in the ER lumen than STIM1. Thus, it has been suggested that STIM2 responds to minor elevation in ER luminal Ca^{2+} concentration [19,20,24]. However, the exact role of STIM2 in Ca^{2+} regulation is still unclear. There are three homologues of Orais, viz., Orai1, Orai2, and Orai3, and all three Orai proteins are expressed in plasma membranes and mediate SOCE [25–28]. TRPC is a non-selective cation channel that works as an SOCE channel. TRPC is classified into seven subtypes, namely TRPC1–TRPC7, all of which respond to PIP2 hydrolysis [9,29]. Of these, TRPC1 is the major subtype involved in STIM-mediated Ca^{2+} influx via the STIM1–Orai1–TRPC1 complex [30,31]. Previous studies have suggested that STIMs, Orais, and TRPC1 play key roles in SOCE, an important mechanism that regulates intracellular Ca^{2+} hemostasis [32,33].

Platelet-derived growth factor (PDGF) is a mitogen that is involved in cellular migration, proliferation, angiogenesis, and extracellular matrix generation [34–37]. The PDGF family consists of five proteins: PDGF-AA, PDGF-BB, PDGF-AB, PDGF-CC, and PDGF-DD, which are structurally and functionally related and contain a disulfide-linked dimer of two polypeptide chains that are conserved similarly to that in the endothelial growth factor (EGF) family [35,38]. The isoforms bind with different affinities to two receptor units, PDGFRα and PDGFRβ. PDGFRs are composed of two related isoforms of RTKs, namely PDGFRα and PDGFRβ, and form three different subtype receptors, viz., PDGFR-αα, PDGFR-ββ, and PDGFR-αβ, which react with different ligands to initiate various reactions [39–41]. Each subtype of PDGF activates different combinations of PDGFRs; however, only PDGF-BB has been verified to bind to all kinds of receptors, resulting in auto-phosphorylation of RTK. The active form of PDGFR activates signaling pathways, such as Akt, JNK, ERK, and STAT3, to regulate cell function and hydrolyzes PIP2 to DAG and IP3 to induce ER Ca^{2+} release [42,43]. Previous studies have shown that PDGF-BB activates PDGFRs (PDGFRα and PDGFRβ) and that the phosphorylation of PDGFR activates PLCγ, which hydrolyze PIP2 to DAG and IP3, leading to a depletion in ER Ca^{2+} store and Ca^{2+} re-entering from the extracellular solution through SOCs [44,45]. Recent studies have also shown that PDGF-BB induces SOCE through STIM1 and Orai1 to promote smooth muscle cell migration [46,47].

Mitogens, such as PDGF and EGF, are well recognized as inducing STIM1-mediated Ca^{2+} elevation including ER Ca^{2+} release and SOCE, for which STIM1 and Orai1, but not STIM2, are essential in PDGF-BB-induced Ca^{2+} elevation and cell migration [16,47,48]. However, STIM2 has reportedly responded to minor decreases in basal Ca^{2+} concentration and replenished Ca^{2+} concentration through the STIM–Orai puncta-like complex [20,24]. In most studies, PDGF-BB has been reported as the SOCE activator that leads to ER Ca^{2+} depletion through the PDGFR–PLCγ–IP3 signaling pathway. However, the role between STIM1 and PDGF-BB signaling is still unclear. Thus, in this study, we employed a particular cell line, mouse embryonic fibroblast-STIM1 knockout cells (MEF-STIM1$^{-/-}$), to study the effects of STIM1 knockout on PDGF-BB-stimulated Ca^{2+} response.

2. Results

2.1. Stromal Interaction Molecule 1 (STIM1) Knockout Represses Store-Operated Ca^{2+} Entry (SOCE) in Mouse Embryonic Fibroblasts (MEF) Cells

In order to investigate the effect of STIM1 on SOCE, wild-type (MEF-WT) and STIM1 knockout MEF (MEF-STIM1$^{-/-}$) cell lines were used and intracellular concentration of Ca^{2+} was monitored by using a ratiometric fluorescent dye, Fura-2/AM. Thapsigargin (TG), a sarco/endoplasmic reticulum Ca^{2+}-ATPase (SERCA) inhibitor and SOCE activator, is usually used to deplete ER Ca^{2+} store to activate SOCE. The cells were exposed to 2 μM TG in the absence of extracellular Ca^{2+}. An initial transient intracellular Ca^{2+} elevation was observed due to Ca^{2+} released from the ER. Next, 2 mM Ca^{2+} was reintroduced into the extracellular solution, leading to Ca^{2+} influx, called SOCE (Figure 1A). Quantification analysis revealed that the ER released Ca^{2+} was similar in both cell types (Figure 1B), but SOCE was significantly reduced in MEF-STIM1$^{-/-}$ cells compared to that in MEF-WT cells (Figure 1C). In addition, cells were exposed to 2 mM extracellular Ca^{2+} and stimulated with 2 μM TG to mimic normal physiological Ca^{2+} concentration. Representative traces indicate a quick two-fold increase in intracellular Ca^{2+} concentration, which then decreased by 1.4-fold in MEF-WT cells. The resultant Ca^{2+} concentration was higher than the baseline and was sustained for a long period. The initial peak indicated that this Ca^{2+} release from the ER was accompanied by Ca^{2+} influx from the extracellular solution, which sustained the higher Ca^{2+} concentration. In MEF-STIM$^{-/-}$ cells, the initial peak was 1.4-fold higher, which then quickly reverted to the baseline concentration (Figure 1D). These results suggest that TG-mediated Ca^{2+} elevation after extracellular 2 mM Ca^{2+} exposure showed an initial peak (Figure 1E) and that the total Ca^{2+} elevation (Figure 1F) in MEF-WT cells was more dominant than that in MEF-STIM1$^{-/-}$ cells. Thus, STIM1 knockout reduced Ca^{2+} elevation in MEF cells, particularly the Ca^{2+} influx.

2.2. PDGF-BB Induced Significant Ca^{2+} Elevation in STIM1 Knockout MEF Cells

PDGF promotes cell migration through regulation of Ca^{2+} signaling. Here, we attempted to investigate the effects of STIM1 knockout on Ca^{2+} elevation in MEF cells following PDGF-BB stimulation. Unexpectedly, PDGF-BB induced Ca^{2+} elevation dose-dependently in MEF-STIM1$^{-/-}$ cells (Figure 2C,D) but not in MEF-WT cells (Figure 2A,B). Moreover, quantification analysis of the initial Ca^{2+} revealed that PDGF-BB at 100 ng/mL exerted optimal stimulation in MEF-STIM1$^{-/-}$ cells (Figure 2C) and produced an initial peak that was 1.8-fold higher, which is similar to TG-mediated Ca^{2+} elevation in MEF-WT cells (Figure 1D). In contrast to the TG-mediated Ca^{2+} elevation shown previously, PDGF-BB induced a notable dose-dependent response in MEF-STIM1$^{-/-}$ cells and had no effect in MEF-WT cells. To clarify how PDGF-BB induces Ca^{2+} elevation, intracellularly or extracellularly, MEF-STIM1$^{-/-}$ cells were treated with PDGF-BB (0–200 ng/mL) in the absence of extracellular Ca^{2+}, after which 2 mM Ca^{2+} was added to the extracellular solution. There was a transient increase in intracellular Ca^{2+} concentration in MEF-STIM1$^{-/-}$ cells due to the release of ER Ca^{2+}, which then reverted to the baseline concentration. Following the addition of 2 mM Ca^{2+}, Ca^{2+} influx appeared to replenish the cytosol (Figure 2E). In addition, PDGF-BB at 50 ng/mL triggered substantial Ca^{2+} release from the ER, 100 ng/mL of which was sufficient to induce optimal Ca^{2+} elevation (Figure 2F). More specifically, Ca^{2+} influx analysis with different concentrations of PDGF-BB showed that treatment with 100 ng/mL PDGF-BB induced the highest Ca^{2+} elevation following ER Ca^{2+} depletion (Figure 2G). These data also show that PDGF-BB-mediated Ca^{2+} elevation occurred primarily through intracellular Ca^{2+} release and then through SOCE. Furthermore, knockdown of STIM2 resulted in the downregulation of SOCE upon TG or PDGF-BB treatment in MEF-STIM1$^{-/-}$ cells (Figure S2).

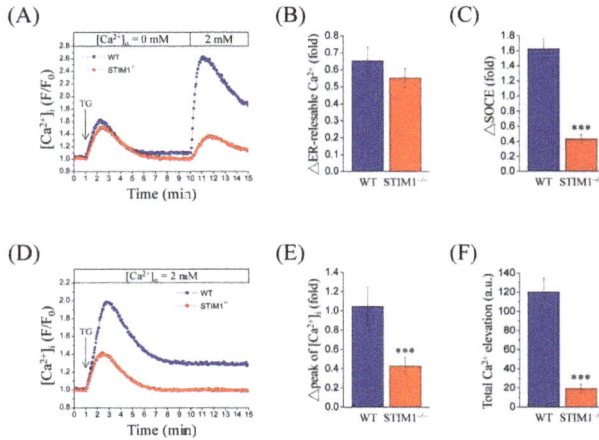

Figure 1. Thapsigargin (TG)-mediated store-operated Ca^{2+} entry (SOCE) is suppressed in mouse embryonic fibroblast-STIM1 knockout (MEF-STIM1$^{-/-}$) cells. (**A,D**) Representative tracings show the effect of 2 µM TG (arrow) on Fura-2/AM loaded MEF-WT (wild-type) and MEF-STIM1$^{-/-}$ cells (**A**) in absence of extracellular Ca^{2+} followed by addition of 2 mM Ca^{2+} to the extracellular buffer or (**D**) at 2 mM extracellular Ca^{2+}. Intracellular Ca^{2+} ($[Ca^{2+}]_i$) was monitored using a single-cell fluorimeter for 15 min. Each trace represents the mean of at least four independent experiments. The bar charts show (**B**) ER Ca^{2+} release, (**C**) SOCE, (**E**) initial Ca^{2+} peak (change of peak value), and (**F**) total Ca^{2+} elevation (area under the curve) following the addition of TG. Bars represent mean ± SEM. *** $p < 0.001$ by Student's *t*-test. TG, thapsigargin; a.u., arbitrary unit.

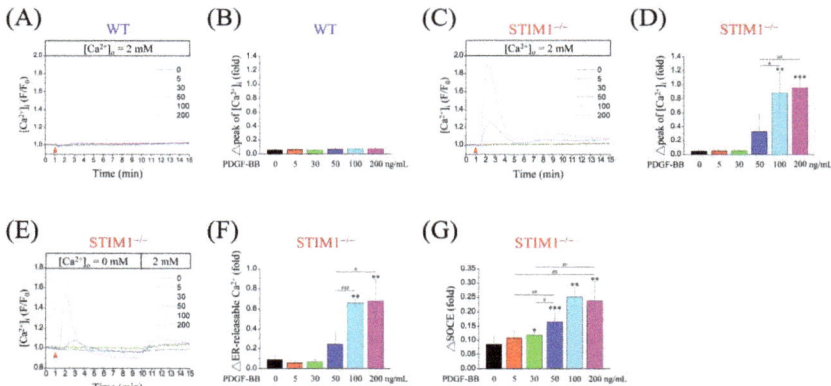

Figure 2. PDGF-BB induces Ca^{2+} elevation in MEF-STIM1$^{-/-}$ cells but not in MEF-WT cells. Representative tracings showing the effect of PDGF-BB (0–200 ng/mL, arrowhead) in Fura-2/AM-loaded, serum-starved (**A**) MEF-WT, (**C**) MEF-STIM1$^{-/-}$ cells at 2 mM extracellular Ca^{2+}, and (**E**) MEF-STIM1$^{-/-}$ cells in the absence of extracellular Ca^{2+} followed by addition of 2 mM Ca^{2+} to the extracellular buffer. Intracellular Ca^{2+} ($[Ca^{2+}]_i$) was monitored using a single-cell fluorimeter for 15 min. (**B,D,F,G**) Bar charts indicate (**B**) initial Ca^{2+} peak of MEF-WT and (**D**) initial Ca^{2+} peak, (**F**) ER Ca^{2+} release, and (**G**) SOCE of MEF-STIM1$^{-/-}$ cells following the addition of PDGF-BB. Bars represent mean ± SEM. *,#: $p < 0.05$; **,##: $p < 0.01$; ***,###: $p < 0.001$ by one-way ANOVA with Dunnett's post-hoc test.

2.3. Upregulation and Activation of PDGFRα, PDGFRβ, and Phospholipase C Gamma (PLCγ) in MEF-STIM1$^{-/-}$ Cells

Previous studies have shown that PDGF-BB activates PDGFRs (PDGFRα and PDGFRβ) and that PDGFR phosphorylation activates PLCγ to hydrolyze PIP2 into DAG and IP3, which leads to a depletion of the ER Ca^{2+} store. Therefore, we examined PDGF-BB-mediated signaling pathways. Immunoblotting showed that expressions of PDGFRα, PDGFRβ, and PLCγ were enhanced in MEF-STIM1$^{-/-}$ cells compared to those in MEF-WT cells (Figure 3A), indicating that the upregulation was due to PDGF-BB stimulation. Quantification analyses of the ratio of phosphorylated PDGFRβ:PDGFRβ (Figure 3B) and phosphorylated PLCγ:PLCγ (Figure 3C) also confirmed the results, because their activities following PDGF-BB treatment were evidently increased in MEF-STIM1$^{-/-}$ cells compared to those in MEF-WT cells. CREB activation by phosphorylation can be triggered by both PDGF and Ca^{2+} signal transduction pathways and inhibition of CREB expression or activation decreases PDGF-induced smooth muscle cell migration. Thus, we examined the phosphorylation of CREB in response to PDGF-BB stimulation. The results showed that CREB was phosphorylated in MEF-STIM1$^{-/-}$ cells and the phosphorylation levels were higher than those in MEF-WT cells (Figure 3D). STIM2 knockdown did not affect the expressions of PDGFRα and PDGFRβ and the PDGF-BB-induced PDGFRβ phosphorylation, whereas STIM1 overexpression downregulated the expressions of PDGFRα and PDGFRβ and the PDGF-BB-induced PDGFRβ phosphorylation (Figure 3E). We then sought to determine other non-Ca^{2+}-conducting PDGF-BB-induced downstream signaling molecules, including Akt, JNK, ERK and STAT3 (Figure 4A). Upon PDGF-BB stimulation, Akt phosphorylation increased within 3 min in MEF-STIM1$^{-/-}$ cells and was sustained for at least 10 min; however, in MEF-WT cells, Akt was activated within 5 min and then decreased quickly (Figure 4B). Although phosphorylation of JNK was triggered by PDGF-BB in both cell types, the levels of phosphorylation were higher in MEF-STIM1$^{-/-}$ cells than those in the MEF-WT cells (Figure 4C). In addition, PDGF-BB induced higher levels of ERK phosphorylation in MEF-STIM1$^{-/-}$ cells than that in MEF-WT cells (Figure 4D). Activation of STAT3 upon PDGF-BB stimulation was not significantly different between MEF-WT and MEF-STIM1$^{-/-}$ cells. Taken together, these findings support the responses of PDGF-BB-induced Ca^{2+} elevation in MEF-STIM1$^{-/-}$ cells due to the elevated protein levels of PDGFRs, resulting in higher activation of PLCγ, Akt, JNK, and ERK than in MEF-WT cells. PLCγ was the key regulator upstream of IP3-mediated ER Ca^{2+} release and SOCE. The PDGFR kinase inhibitor AG1295 was used to verify the role of PDGFR activation in PDGF-BB-induced PLCγ signaling (Figure S1A). Our results showed that inhibition of PDGFRβ activation using AG1295 (Figure S1B) resulted in cessation of PDGF-BB-mediated PLCγ signaling in MEF-STIM1$^{-/-}$ cells (Figure S1C). Moreover, PLC inhibitors U73122 and D609 were used to examine PDGF-BB-induced Ca^{2+} responses, such as CREB phosphorylation (Figure S3A). These results indicated that PLC inhibitors did not affect PDGFR phosphorylation upon PDGF-BB stimulation (Figure S3A) but inhibited phosphorylation of PLCγ and CREB (Figure S3B,C).

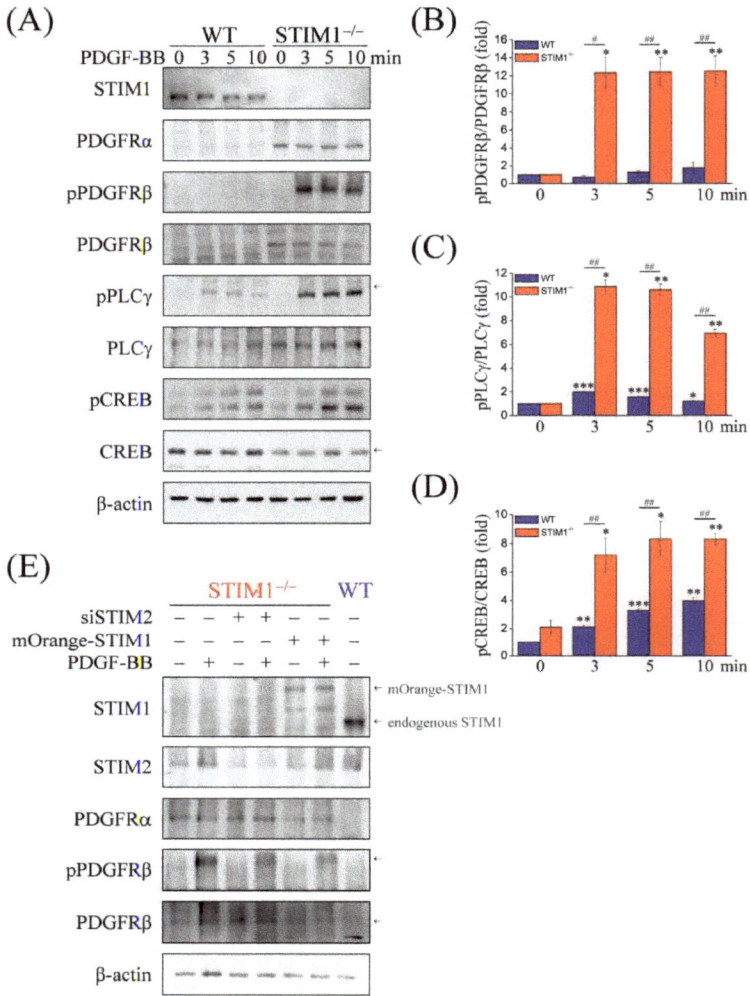

Figure 3. STIM1 knockout enhances the expression of PDGFRα and PDGFRβ, and increases phosphorylation of phospholipase C gamma (PLCγ) and cAMP response element binding protein (CREB). (**A**) MEF-WT and MEF-STIM1$^{-/-}$ cells were starved in a serum-free medium for 12 h and then stimulated with 100 ng/mL PDGF-BB for 3, 5, and 10 min. Immunoblotting analysis using antibodies against STIM1, PDGFRα, phospho-PDGFRβ (pPDGFRβ), PDGFRβ, phospho-PLCγ (pPLCγ), PLCγ, phospho-CREB (CREB), and CREB. β-actin served as the internal control; (**B–D**) Measurement of the relative intensities of protein phosphorylation are represented as mean ± SEM from three independent experiments for both MEF-WT and MEF-STIM1$^{-/-}$ cells. Bar charts show phosphorylation levels of (**B**) pPDGFRβ, (**C**) pPLCγ, and (**D**) pCREB, which were normalized to the total protein. *,#: $p < 0.05$; **,##: $p < 0.01$; ***: $p < 0.001$ by Student's *t*-test; (**E**) Knockdown of STIM2 or overexpression of STIM1 upon transient transfection with STIM2 siRNA (siSTIM2) and mOrange-tagged STIM1 plasmid (mOrange-STIM1) for 48 h in MEF-STIM1$^{-/-}$ cells, respectively. Cells were starved in a serum-free medium for 12 h and then stimulated with or without 100 ng/mL PDGF-BB for 5 min. Immunoblotting analysis using antibodies against STIM1, STIM2, PDGFRα, phospho-PDGFRβ, and PDGFRβ. β-actin served as the internal control. The MEF-WT cells were used as a control for MEF-STIM1$^{-/-}$ cells. Arrows indicate the target proteins.

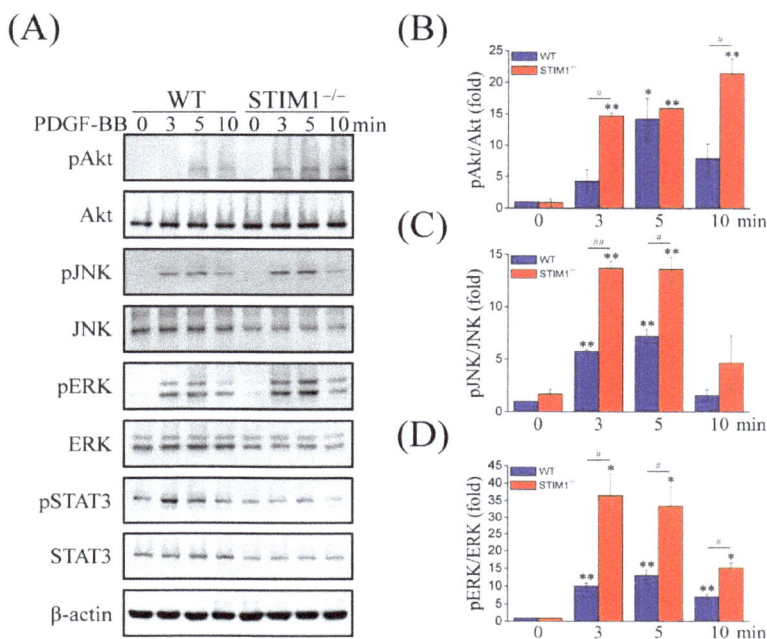

Figure 4. STIM1 knockout increases phosphorylation of Akt, JNK, and ERK but not STAT3 under PDGF-BB stimulation. (**A**) MEF-WT and MEF-STIM1$^{-/-}$ cells were starved in a serum-free medium for 12 h and then stimulated with 100 ng/mL PDGF-BB for 3, 5, and 10 min. Immunoblotting analysis using antibodies against phospho-Akt (pAkt), Akt, phospho-JNK (pJNK), JNK, phospho-ERK (pERK), ERK, phospho-STAT3 (pSTAT3), and STAT3. β-actin served as the internal control; (**B–D**) Measurement of the relative intensities of protein phosphorylation are represented as mean ± SEM from three independent experiments for both MEF-WT and MEF-STIM1$^{-/-}$ cells. Bar charts show the phosphorylation levels of (**B**) pAkt, (**C**) pJNK, and (**D**) pERK, which were normalized to the total protein. *,#: $p < 0.05$; **,##: $p < 0.01$ by Student's *t*-test.

2.4. PDGF-BB-Mediated ER Store-Depletion Activates STIM2 Translocation and Puncta Formation

STIMs, Orais, and TRPC1 play key roles in the regulation of SOCE. As shown in Figure 2, PDGF-BB induced SOCE in MEF-STIM1$^{-/-}$ cells but not in MEF-WT cells. To understand how PDGF-BB induced Ca^{2+} elevation in MEF-STIM1$^{-/-}$ cells, proteins that contribute to SOCE were examined by immunoblotting analysis. Although STIM1 knockout did not affect STIM2; Orai1, Orai2, Orai3, and TRPC1 were decreased in MEF-STIM1$^{-/-}$ cells (Figure 5A). These results supported the observation that Ca^{2+} influx following TG-mediated ER store depletion in MEF-STIM1$^{-/-}$ cells is lower than that in MEF-WT cells because of lower SOCE-associated protein levels in MEF-STIM1$^{-/-}$ cells (Figure 1). To investigate the role of STIM proteins in MEF cells in response to PDGF-BB stimulation, the subcellular localization of endogenous STIM1 and STIM2 was examined by immunostaining. The confocal images showed that STIM1 was not expressed in MEF-STIM1$^{-/-}$ cells (Figure 5B). Moreover, the STIM2 protein translocated near the plasma membrane and formed distinct intracellular puncta in MEF-STIM1$^{-/-}$ cells after PDGF-BB treatment from 3 to 10 min. Maximal activity was observed at 5 min, after which it decreased over time. In contrast, PDGF-BB did not induce puncta formation and plasma membrane translocation of STIMs in MEF-WT cells (Figures 5C and 6A,B). In addition, total internal reflection fluorescence microscopy (TIRFM) and high magnification laser scanning confocal microscopy were applied to verify plasma membrane translocation and intracellular puncta formation of activated STIM2, respectively. TIRFM is a technique used to record the fluorescence signals within 50 to 100 nm range of the basal plasma membrane. We observed a significant plasma membrane translocation of STIM2 after PDGF-BB stimulation for 5 min in MEF-STIM1$^{-/-}$ cells (Figure 6C). On the other hand, large size of the internal aggregation of the STIM2 puncta was presented in PDGF-BB stimulated MEF-STIM1$^{-/-}$ cells (Figure 6C). Quantitative results showed that PDGF-BB induced 5–6-fold increases of STIM2 aggregation (Figure 6D). We also demonstrated colocalization between SITM2 and Orai1, Orai2, Orai3, or TRPC1 after PDGF-BB treatment (Figure S4).

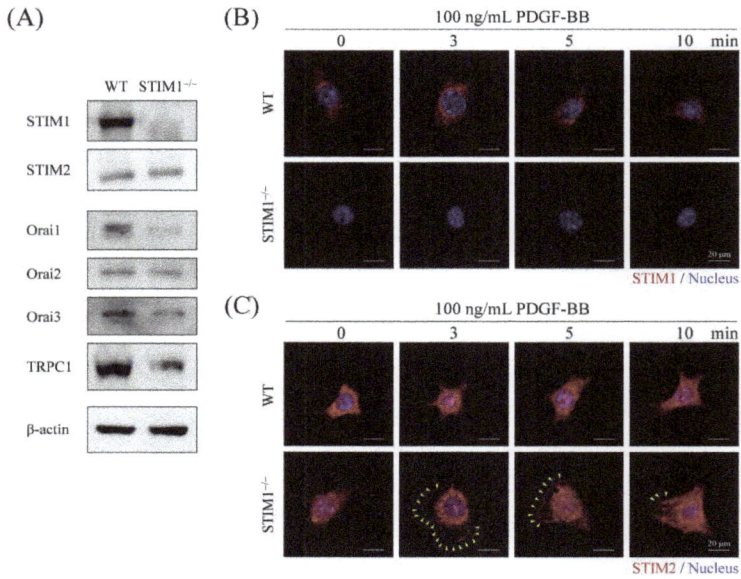

Figure 5. PDGF-BB induces STIM2 puncta formation and translocation in MEF-STIM1$^{-/-}$ cells. (**A**) STIM1, STIM2, Orai1, Orai2, Orai3 and transient receptor potential canonical 1 (TRPC1) were detected using immunoblotting in both MEF-WT and MEF-STIM1$^{-/-}$ cells. β-actin served as the internal control; (**B**,**C**) MEF-WT and MEF-STIM1$^{-/-}$ cells were starved in a serum-free medium and then stimulated with 100 ng/mL PDGF-BB for 3, 5, and 10 min. Then, cells were processed for immunofluorescence staining using (**B**) anti-STIM1 or (**C**) anti-STIM2 antibody, followed by adding a secondary antibody coupled to Alexa Fluor® 594 dye and Hoechst 33342 (blue) was used as the nuclear marker. Fluorescence images were captured using a laser scanning confocal microscope. Yellow arrowheads indicate plasma membrane translocation of the STIM2 protein. Scale bars = 20 μm.

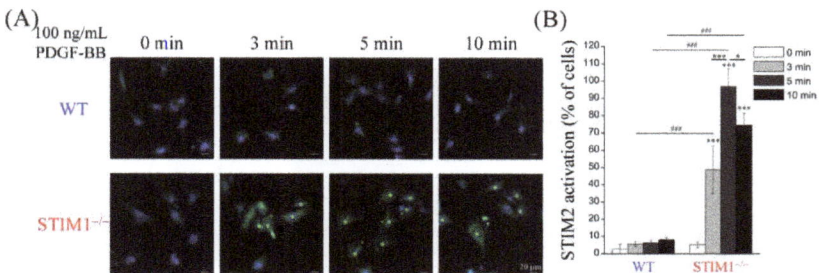

Figure 6. *Cont.*

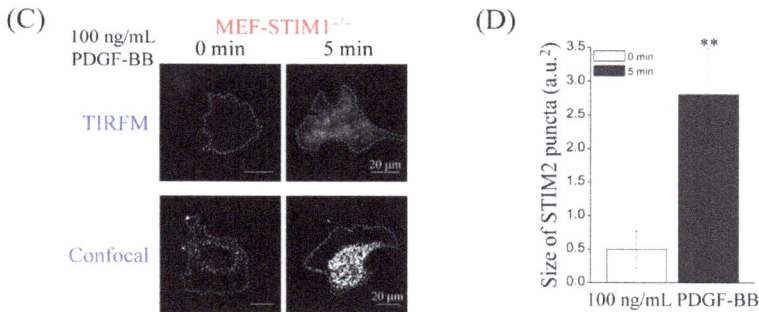

Figure 6. PDGF-BB induces STIM2 activation in MEF-STIM1$^{-/-}$ cells. (**A**) MEF-WT (WT) and MEF-STIM1$^{-/-}$ (STIM1$^{-/-}$) cells were starved in a serum-free medium and then stimulated with 100 ng/mL PDGF-BB for 3, 5, and 10 min. Then, cells were processed for immunofluorescence staining using anti-STIM2 antibody, followed by adding a secondary antibody coupled to Alexa Fluor textsuperscript® 488 dye. Fluorescence images were captured using a laser scanning confocal microscope. Yellow stars indicate cells with activated STIM2, presented as puncta formation and plasma membrane translocation. Scale bars = 20 μm; (**B**) Quantitative analysis of PDGF-BB-induced STIM2 activation that was assessed from the STIM2 plasma membrane translocated cells from three independent experiments. Bars represent mean ± SEM. *: $p < 0.05$; ***,###: $p < 0.001$ by Student's *t*-test; (**C**) MEF-STIM1$^{-/-}$ cells were starved in a serum-free medium and then stimulated with 100 ng/mL PDGF-BB for 5 min. Then, cells were processed for immunofluorescence staining using an anti-STIM2 antibody, followed by adding a secondary antibody coupled to Alexa Fluor® 488 dye. Fluorescence images were captured using a total internal reflection fluorescence microscope or laser scanning confocal microscope. Blue dashed lines indicate the periphery of cells. Scale bars = 20 μm; (**D**) Quantitative analysis of PDGF-BB-induced aggregation of STIM2 puncta from three independent experiments ($n > 30$ cells). Bars represent mean ± SEM. **: $p < 0.01$ by Student's *t*-test.

2.5. SOCE Inhibitors Decrease PDGF-BB-Induced Ca^{2+} Elevation in MEF-STIM1$^{-/-}$ Cells

Pharmacological inhibitors of SOCE, such as 2-APB (Figure 7A), SKF96365 (Figure 7B), YM-58483 (Figure 7C), La^{3+} (Figure 7D), and Gd^{3+} (Figure 7E) were used to examine the role of Ca^{2+} mobilization in PDGF-induced Ca^{2+} elevation in MEF-STIM1$^{-/-}$ cells exposed to 2 mM Ca^{2+} solution. Quantification analysis of the initial Ca^{2+} and total Ca^{2+} elevation following PDGF-BB stimulation showed that Ca^{2+} signaling was inhibited by these inhibitors in a dose-response manner (Figure 7).

2.6. PDGFR–PLCγ–STIM2 Signaling Induces Cell Migration in MEF-STIM1$^{-/-}$ Cells.

PDGF signaling regulates cell proliferation and migration. To determine the influence of PDGF-BB stimulation in MEF-STIM1$^{-/-}$ cells, we performed a wound healing assay and cell count with Hoechst 33342 nuclear staining to examine cell migration and proliferation, respectively. To minimize the effect of serum, we reduced the concentration of serum to 0.1%. By comparing MEF-WT and MEF-STIM1$^{-/-}$ cells, it was observed that the proliferation rate of MEF-WT cells was higher than that of MEF-STIM1$^{-/-}$ cells and that the number of MEF-STIM1$^{-/-}$ cells decreased slowly after 24 h (Figure S5A). To avoid cell proliferation interference, cells were treated with 0.1% fecal bovine serum (FBS) as a control and observed over 24 h. We found that 100 ng/mL PDGF-BB with 0.1% FBS did not affect the proliferation of MEF cells (Figure S5B). In addition, MEF-STIM1$^{-/-}$ cells treated with PDGF-BB led to a substantial increase in wound closure, wherein the wound healed completely after 18 h, unlike MEF-WT cells (Figure 8). These data indicated that STIM1 knockout enhances cell migration when compared with MEF-WT cells.

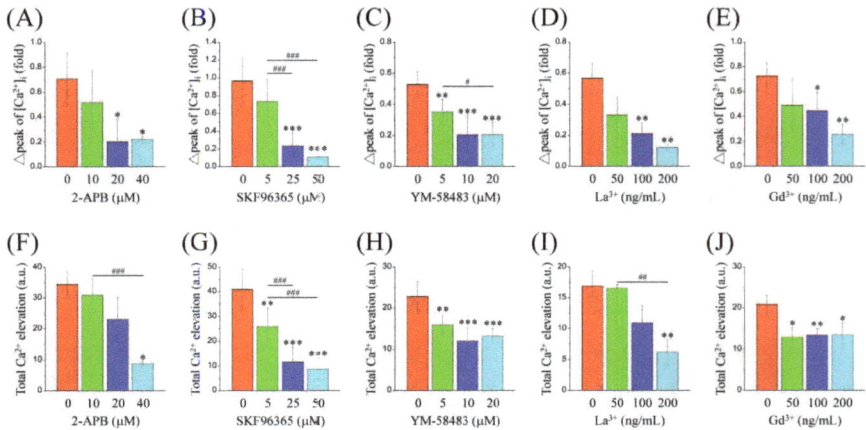

Figure 7. Dose-response inhibition of PDGF-BB-mediated Ca^{2+} elevation by the Ca^{2+} mobilization inhibitors. MEF-STIM1$^{-/-}$ cells were starved in a serum-free medium and then loaded with Fura-2/AM and co-treated with (**A,F**) 0–40 μM 2-APB, (**B,G**) 0–50 μM SKF96365, (**C,H**) 0–20 μM YM58483, (**D,I**) 0–200 ng/mL La^{3+}, and (**E,J**) 0–200 ng/mL Gd^{3+} for 30 min. Cells were then stimulated with 100 ng/mL PDGF-BB at 2 mM extracellular Ca^{2+}. Intracellular Ca^{2+} ([Ca^{2+}]$_i$) was monitored using a single-cell fluorimeter for 15 min for at least three independent experiments. The bar charts show (**A–E**) initial Ca^{2+} peak and (**F,J**) total Ca^{2+} elevation following the addition of PDGF-BB. Bars represent mean ± SEM. *,#: $p < 0.05$; **,##: $p < 0.01$; ***,###: $p < 0.001$ by one-way ANOVA with Dunnett's post-hoc test.

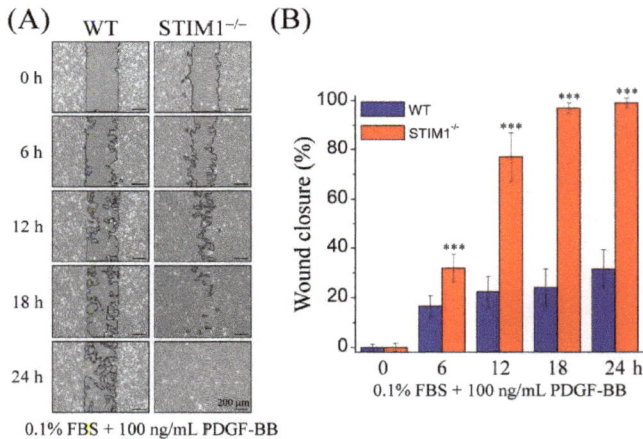

Figure 8. PDGF-BB induces cell migration in MEF-STIM1$^{-/-}$ cells. (**A**) MEF-WT and MEF-STIM1$^{-/-}$ cells were seeded into silicon inserts with 0.5% fecal bovine serum (FBS) medium. Following cell adhesion, inserts were removed, followed by the addition of Dulbecco's modified Eagle's medium (DMEM) with 0.1% FBS + PDGF-BB (100 ng/mL) for 24 h. Phase images were captured using inverted phase-contrast microscopy and wound spaces were analyzed using ImageJ; (**B**) Comparison of the effects of PDGF-BB-induced wound closure following insert removal in MEF-WT and MEF-STIM1$^{-/-}$ cells. Cellular migratory ability is presented as the percentages of wound closure. Bars represent mean ± SEM. ***: $p < 0.001$ by Student's *t*-test. Scale bars = 200 μm.

3. Discussion

STIM1 plays the major role in controlling Ca^{2+} homeostasis through regulation of SOCE in many non-excitable cells. In this study, we found the unexpected result that PDGF-BB induced higher intracellular Ca^{2+} elevation in MEF-STIM1$^{-/-}$ cells than that in the parental MEF-WT cells (Figure 2). The observed results were associated with STIM2-dependent SOCE (Figure S2). Immunoblotting revealed that PDGFRα, PDGFRβ, and PLCγ were overexpressed in MEF-STIM1$^{-/-}$ cells and higher PDGFR and PLCγ phosphorylation was observed after PDGF-BB stimulation (Figure 3). In contrast, the levels of both PDGFRα and PDGFRβ were very low in MEF-WT cells, and hence, PDGF-BB could not induce PDGFR phosphorylation and Ca^{2+} elevation (Figures 2 and 3). Increased PLCγ activation may increase the propensity of IP3 to activate the IP3 receptors and lead to Ca^{2+} leakage from the ER lumen, which is the major source of PDGF-BB-mediated Ca^{2+} elevation in MEF-STIM1$^{-/-}$ cells (Figure 2). Depletion of the ER Ca^{2+} store through PDGFR–PLCγ axis following extracellular Ca^{2+} influx is termed as SOCE. However, STIM1 was not expressed in MEF-STIM1$^{-/-}$ cells and Ca^{2+} influx following the re-introduction of 2 mM Ca^{2+} was lower than that of TG-induced SOCE in MEF-WT cells (Figures 1 and 2).

PDGFR (AG1295) and PLC (U73122 and D609) inhibitors were employed to verify whether the PDGFR–PLCγ axis could be activated by PDGF-BB stimulation in MEF-STIM1$^{-/-}$ cells [49–51]. Immunoblotting results supported the essential roles of PDGFRs and PLCγ in PDGF-BB-mediated activation (Figures S1 and S3). In addition, the Ca^{2+}-activated transcription factor CREB was also observed in correspondence to activation of PDGFR–PLCγ axis-mediated signaling. These results indicated that PDGF-BB increased Ca^{2+} elevation through the activation of PLCγ.

SOCE inhibitors manifestly lowered intracellular Ca^{2+} elevation in a dose-response manner (Figure 7). Accordingly, SOCE may contribute to PDGF-BB-mediated Ca^{2+} elevation upon PDGF-BB stimulation. Here, we confirmed that STIM2 activation was observed in MEF-STIM1$^{-/-}$ cells but not in MEF-WT cells (Figures 5 and 6). PDGF-BB induced interaction between STIM2 and SOCE-related channel proteins in MEF-STIM1$^{-/-}$ cells (Figure S4). However, the level of PDGF-BB-mediated SOCE were lower than ER released Ca^{2+} in MEF-STIM1$^{-/-}$ cells (Figures 1 and 2), which may be due to the downregulation of SOCs Orais and TRPC1 (Figures 2C and 5A). Our results demonstrated that enhancement of Ca^{2+} signaling in MEF-STIM1$^{-/-}$ cells was mediated by the PDGFR–PLCγ–STIM2 pathway.

Proliferation assay of MEF cells also indicated that MEF-STIM1$^{-/-}$ cells proliferated more slowly than MEF-WT cells [22]. Moreover, we showed that MEF-WT cells proliferated more quickly than MEF-STIM1$^{-/-}$ cells, even in a low-serum concentration containing 0.1% FBS, whereas PDGF-BB did not affect MEF proliferation in the same condition (Figure S5). Furthermore, our results showed that the ability of cell migration with PDGF-BB treatment in MEF-STIM1$^{-/-}$ cells was higher than that in MEF-WT cells, regardless of the low cell proliferation rate. It may be that STIM1 knockout evoked migration via PDGFR–PLCγ–STIM2 enhancement, as demonstrated by the wound healing assay (Figure 8). PDGF-BB-induced Ca^{2+} elevation may promote MEF-STIM1$^{-/-}$ cell migration as well as activation of Akt, ERK, and JNK, which could also participate in cell migration regulation. Both EGF and PDGF share similar signal pathways, regulate many similar functions of the cell, and induce SOCE following ER Ca^{2+} depletion through PLCγ signaling [47,52]. We also observed EGFR downregulation in MEF-STIM1$^{-/-}$ cells. Moreover, EGF induced higher activity of EGFR, PLCγ, Akt, JNK, and ERK in MEF-WT cells compared to those in MEF-STIM1$^{-/-}$ cells (Figure S6). PDGF-BB induced higher response in MEF-STIM1$^{-/-}$ cells, whereas EGF induced higher response in MEF-WT cells. These contradictions resulted from a switch in these different expressions of EGFR and PDGFR.

STIM1 has been vigorously studied and is well-established in response to high Ca^{2+} leakage from the ER lumen, which leads to huge and constitutive Ca^{2+} influx through interacting with Orai1 and TRPC1 in the plasma membrane [8,53]. Without STIM1, STIM2 replaces the role of STIM1 and induces lower Ca^{2+} influx [54]. A recent study shows the essential role of STIM proteins in Ca^{2+} homeostasis and their crucial role in controlling multiple functions of smooth muscle cells [12]. The critical role of STIM1 can be partially rescued by STIM2. STIM1 and Orai1, but not STIM2, are essential for PDGF-BB-induced Ca^{2+} elevation and cell migration in human airway smooth muscle cells, which express both STIM proteins [47]. Our results also showed that STIM1 played a curial role in response to store depletion after TG treatment (Figure 1). Interestingly, we found that PDGF-BB induced Ca^{2+} elevation in MEF-STIM1$^{-/-}$ cells but not in MEF-WT cells. STIM2 seemed to partially replace the regulatory role of SOCE by STIM1. To our knowledge, this is the first report that shows enhanced Ca^{2+} elevation in MEF-STIM1$^{-/-}$ cells compared to that in MEF-WT cells because of overexpression of PDGFRs and PLCγ in MEF-STIM1$^{-/-}$ cells.

4. Materials and Methods

4.1. Cell Culture and Reagents

Wild-type (MEF-WT) and STIM1 knockout (MEF-STIM1$^{-/-}$) mouse embryonic fibroblasts were maintained in high-glucose Dulbecco's modified Eagle's medium (DMEM; Caisson, Smithfield, VA, USA) supplemented with 5% fetal bovine serum (FBS; GIBCO, Big Cabin, OK, USA), 100 IU/mL penicillin, 100 μg/mL streptomycin, non-essential amino acid (NEAA; GIBCO), and 1 mM sodium pyruvate (NEAA; GIBCO) under 5% CO_2 at 37 °C. Recombinant mouse PDGF-BB and EGF were purchased from R&D Systems (Minneapolis, MN, USA).

4.2. Single-Cell Ca^{2+} Measurement

Intracellular Ca^{2+} was measured by the ratiometric Fura-2 fluorescence method using a single-cell fluorimeter (Till Photonics, Grafelfing, Germany). Fura-2 were excited alternatively between 340 nm and 380 nm using the Polychrome IV monochromator (Till Photonics), which is highly sensitive and can detect rapid changes in Ca^{2+} concentration with high temporal and spatial resolution and images were captured using an Olympus IX71 inverted microscope equipped with a xenon illumination system and an IMAGO CCD camera (Till Photonics). Emission fluorescence intensity of 510 nm was used to calculate intracellular Ca^{2+} levels using the TILLvisION 4.0 software (Till Photonics).

4.3. Immunoblotting Analyses

Cell lysates were harvested in radioimmunoprecipitation assay (RIPA) buffer (150 mM NaCl, 1 mM EGTA, 50 mM Tris at pH 7.4, 10% glycerol, 1% Triton X-100, 1% sodium deoxycholate, 0.1% SDS, and CompleteTM), and the lysates were analyzed by Western blotting using antibodies against STIM1 (BD, Franklin Lakes, NJ, USA), Orai1, Orai3 (ProSci, Poway, CA, USA), Orai2 (Enzo, Farmingdale, NY, USA), TRPC1 (Proteintech, Rosemont, IL, USA), PDGFRα, JNK, phospho-EGFR, EGFR (Santa Cruz, Santa Cruz, CA, USA), phospho-PDGFRβ, PDGRβ (Abnova, San Francisco, CA, USA), STIM2, phospho-PLCγ, PLCγ, phospho-Akt, Akt, phospho-JNK, phospho-ERK, ERK, phospho-STAT3, STAT3, phospho-CREB, CREB (Cell Signaling, Beverly, MA, USA), and β-actin (Sigma, Saint Louis, MO, USA). The immune complexes were then detected with horseradish peroxidase-conjugated IgG (Jackson ImmunoResearch Laboratories, West Grove, PA, USA), and the reaction was developed using an enhanced chemiluminescence (ECL) detection kit under an ImageQuant LAS 4000 system (GE Healthcare Life Sciences, Pittsburgh, PA, USA).

4.4. Immunofluorescence Staining and Confocal Microscopy

MEF cells (3×10^4 cells) were seeded in a 3-cm glass-bottom dish (Alpha Plus, Taoyuan, Taiwan). After adhesion, cells were starved in a serum-free medium for 12 h and then treated with PDGF-BB (100 ng/mL) for 3, 5, and 10 min. Then, cells were fixed with 4% paraformaldehyde (Alfa Aesar, Haverhill, MA, USA) for 10 min; incubated in PBS washing buffer containing 0.1% Triton X-100 for 10 min; and then in CAS-Block solution (Invitrogen, Carlsbad, CA, USA) for 1 h with gentle shaking. Subsequently, the cells were loaded with primary antibody, anti-STIM1 (Cell Signaling) and anti-STIM2 (Alomone Labs, Jerusalem, Israel) diluted to 1:100 in filtered PBS. Finally, cells were stored overnight at 4°C. Alexa Fluor-conjugated secondary antibodies (Invitrogen) were used to counteract primary antibodies, and Hoechst 33342 was loaded as the nuclear marker. The fluorophores were excited by laser at 405 and 594 nm and detected by a laser scanning confocal microscope (FV-1000, Olympus, Tokyo, Japan). The FV10-ASW software (version 4.2a, Tokyo, Japan) was used to access puncta formation and translocation of the STIM proteins.

4.5. Wound Healing Assay

Culture inserts (ibidi) were applied to assess MEF cell migration. The insert consisted of two wells separated by a 500 μm-thick silicon wall. The culture insert was placed into a 3-cm culture dish and slightly pressed on the top of the insert to ensure adhesion on the dish. MEF-WT and MEF-STIM1$^{-/-}$ cells were seeded at an equal density (3×10^4 cells in 100 μL) with 0.5% FBS medium and incubated at 37 °C with 5% CO_2 overnight. The insert was removed after cells were well-attached and formed a monolayer. The cells were then incubated in DMEM containing 0.1% FBS with 100 ng/mL PDGF-BB. Cell migrating into the gap (initially ~500 μm) was recorded every 6 h via phase-contrast microscopy. The data were collected from three independent experiments and analyzed as wound closure (%) by the ImageJ software (version 1.47, Bethesda, MD, USA).

4.6. Cell Proliferation Analysis

To assess MEF cell proliferation, cells were stained using Hoechst 33342. MEF cells (WT and STIM1$^{-/-}$) were seeded into a 3-cm culture dish with 0.5% FBS medium for cell starvation and incubated at 37 °C and 5% CO_2 overnight. After cells had attached properly to the dish, the medium was changed (0.1% FBS with 100 ng/mL PDGF-BB). The cells were collected and fixed with 4% paraformaldehyde at 0, 24, 48, and 72 h. Next, the cells were stained with Hoechst 33342 for 30 min. Fluorescence images were captured randomly and counted using fluorescence microscopy and ImageJ software, respectively.

4.7. Statistical Analysis

All data were reported as mean ± SEM (standard error of the mean). For statistical analysis, Student's *t*-test or one-way ANOVA with Dunnett's post-hoc test were processed using the Origin software, for which differences were considered significant when *p*-value was <0.05.

Supplementary Materials: Supplementary materials can be found at www.mdpi.com/xxx/s1.

Author Contributions: T.-Y.H. and W.-T.C. conceived and supervised the study; T.-Y.H., T.-Y.Y., Y.-H.L., Y.-H.C., Y.-F.C. and W.-T.C. designed and performed experiments; T.-Y.H., Y.-C.C., C.-C.L. and W.-T.C. analyzed and interpreted data; W.-T.C. wrote the manuscript; Y.-H.L., H.-A.C. and W.-T.C. made manuscript revisions. All authors have read and approved the manuscript.

Funding: The funding was provided by the Ministry of Science and Technology of Taiwan under Grant 105-2628-B-006-003-MY3.

Acknowledgments: We thank Axel Methner at Johannes Gutenberg University Mainz, Germany for kindly providing the wild-type and STIM1$^{-/-}$ MEF cell lines. We also thank the "Bioimaging Core Facility of the National Core Facility Program for Biopharmaceuticals, Ministry of Science and Technology, Taiwan" for their technical services.

Conflicts of Interest: The authors declare no conflict of interest.

Abbreviations

MEF	Mouse embryonic fibroblasts
PDGF	Platelet-derived growth factor
PDGFR	Platelet-derived growth factor receptor
SOCE	Store-operated Ca^{2+} entry
STIM1	Stromal interaction molecule 1
Orai1	Ca^{2+} release-activated calcium modulator 1
TRPC1	Transient receptor potential canonical 1
PLCγ	Phospholipase C gamma
RTK	Receptor tyrosine kinase
CREB	cAMP response element binding protein

References

1. Berchtold, M.W.; Villalobo, A. The many faces of calmodulin in cell proliferation, programmed cell death, autophagy, and cancer. *Biochim. Biophys. Acta* **2014**, *1843*, 398–435. [CrossRef] [PubMed]
2. Berridge, M.J.; Bootman, M.D.; Lipp, P. Calcium—A life and death signal. *Nature* **1998**, *395*, 645–648. [CrossRef] [PubMed]
3. Boulware, M.J.; Marchant, J.S. Timing in cellular Ca^{2+} signaling. *Curr. Biol.* **2008**, *18*, R769–R776. [CrossRef] [PubMed]
4. Li, L.; Stefan, M.I.; Le Novère, N. Calcium input frequency, duration and amplitude differentially modulate the relative activation of calcineurin and CaMKII. *PLoS ONE* **2012**, *7*, e43810. [CrossRef] [PubMed]
5. Putney, J.W. Pharmacology of store-operated calcium channels. *Mol. Interv.* **2010**, *10*, 209. [CrossRef] [PubMed]
6. Smyth, J.T.; Hwang, S.Y.; Tomita, T.; DeHaven, W.I.; Mercer, J.C.; Putney, J.W. Activation and regulation of store-operated calcium entry. *J. Cell. Mol. Med.* **2010**, *14*, 2337–2349. [CrossRef] [PubMed]
7. Targos, B.; Barańska, J.; Pomorski, P. Store-operated calcium entry in physiology and pathology of mammalian cells. *Acta Biochim. Pol.* **2005**, *52*, 397–409. [PubMed]
8. Cheng, K.T.; Ong, H.L.; Liu, X.; Ambudkar, I.S. Contribution of TRPC1 and Orai1 to Ca^{2+} entry activated by store depletion. *Adv. Exp. Med. Biol.* **2011**, *704*, 435–449. [PubMed]
9. Ambudkar, I.S.; Ong, H.L.; Liu, X.; Bandyopadhyay, B.C.; Cheng, K.T. TRPC1: The link between functionally distinct store-operated calcium channels. *Cell Calcium* **2007**, *42*, 213–223. [CrossRef] [PubMed]
10. Park, C.Y.; Hoover, P.J.; Mullins, F.M.; Bachhawat, P.; Covington, E.D.; Rauser, S.; Walz, T.; Garcia, K.C.; Dolmetsch, R.E.; Lewis, R.S. STIM1 clusters and activates CRAC channels via direct binding of a cytosolic domain to Orai1. *Cell* **2009**, *136*, 876–890. [CrossRef] [PubMed]
11. Daskoulidou, N.; Zeng, B.; Berglund, L.M.; Jiang, H.; Chen, G.L.; Kotova, O.; Bhandari, S.; Ayoola, J.; Griffin, S.; Atkin, S.L.; et al. High glucose enhances store-operated calcium entry by upregulating ORAI/STIM via calcineurin-NFAT signalling. *J. Mol. Med.* **2015**, *93*, 511–521. [CrossRef] [PubMed]
12. Mancarella, S.; Potireddy, S.; Wang, Y.; Gao, H.; Gandhirajan, R.K.; Autieri, M.; Scalia, R.; Cheng, Z.; Wang, H.; Madesh, M.; et al. Targeted STIM deletion impairs calcium homeostasis, NFAT activation, and growth of smooth muscle. *FASEB J.* **2013**, *27*, 893–906. [CrossRef] [PubMed]
13. Pulver, R.A.; Rose-Curtis, P.; Roe, M.W.; Wellman, G.C.; Lounsbury, K.M. Store-operated Ca^{2+} entry activates the CREB transcription factor in vascular smooth muscle. *Circ. Res.* **2004**, *94*, 1351–1358. [CrossRef] [PubMed]
14. Srikanth, S.; Gwack, Y. Orai1-NFAT signalling pathway triggered by T cell receptor stimulation. *Mol. Cells* **2013**, *35*, 182–194. [CrossRef] [PubMed]
15. Somasundaram, A.; Shum, A.K.; McBride, H.J.; Kessler, J.A.; Feske, S.; Miller, R.J.; Prakriya, M. Store-operated CRAC channels regulate gene expression and proliferation in neural progenitor cells. *J. Neurosci.* **2014**, *34*, 9107–9123. [CrossRef] [PubMed]
16. Umemura, M.; Baljinnyam, E.; Feske, S.; De Lorenzo, M.S.; Xie, L.H.; Feng, X.; Oda, K.; Makino, A.; Fujita, T.; Yokoyama, U.; et al. Store-operated Ca^{2+} entry (SOCE) regulates melanoma proliferation and cell migration. *PLoS ONE* **2014**, *9*, e89292. [CrossRef] [PubMed]
17. Bisaillon, J.M.; Motiani, R.K.; Gonzalez-Cobos, J.C.; Potier, M.; Halligan, K.E.; Alzawahra, W.F.; Barroso, M.; Singer, H.A.; Jourd'heuil, D.; Trebak, M. Essential role for STIM1/Orai1-mediated calcium influx in PDGF-induced smooth muscle migration. *Am. J. Physiol. Cell Physiol.* **2010**, *298*, C993–C1005. [CrossRef] [PubMed]
18. Michaelis, M.; Nieswandt, B.; Stegner, D.; Eilers, J.; Kraft, R. STIM1, STIM2, and Orai1 regulate store-operated calcium entry and purinergic activation of microglia. *Glia* **2015**, *63*, 652–663. [CrossRef] [PubMed]
19. Bhardwaj, R.; Müller, H.M.; Nickel, W.; Seedorf, M. Oligomerization and Ca^{2+}/calmodulin control binding of the ER Ca^{2+}-sensors STIM1 and STIM2 to plasma membrane lipids. *Biosci. Rep.* **2013**, *33*, e00077. [CrossRef] [PubMed]
20. Rana, A.; Yen, M.; Sadaghiani, A.M.; Malmersjö, S.; Park, C.Y.; Dolmetsch, R.E.; Lewis, R.S. Alternative splicing converts STIM2 from an activator to an inhibitor of store-operated calcium channels. *J. Cell Biol.* **2015**, *209*, 653–669. [CrossRef] [PubMed]

21. Zhou, Y.; Mancarella, S.; Wang, Y.; Yue, C.; Ritchie, M.; Gill, D.L.; Soboloff, J. The short N-terminal domains of STIM1 and STIM2 control the activation kinetics of Orai1 channels. *J. Biol. Chem.* **2009**, *284*, 19164–19168. [CrossRef] [PubMed]

22. Chen, Y.W.; Chen, Y.F.; Chen, Y.T.; Chiu, W.T.; Shen, M.R. The STIM1–Orai1 pathway of store-operated Ca^{2+} entry controls the checkpoint in cell cycle G1/S transition. *Sci. Rep.* **2016**, *6*, 22142. [CrossRef] [PubMed]

23. Henke, N.; Albrecht, P.; Pfeiffer, A.; Toutzaris, D.; Zanger, K.; Methner, A. Stromal interaction molecule 1 (STIM1) is involved in the regulation of mitochondrial shape and bioenergetics and plays a role in oxidative stress. *J. Biol. Chem.* **2012**, *287*, 42042–42052. [CrossRef] [PubMed]

24. Brandman, O.; Liou, J.; Park, W.S.; Meyer, T. STIM2 is a feedback regulator that stabilizes basal cytosolic and endoplasmic reticulum Ca^{2+} levels. *Cell* **2007**, *131*, 1327–1339. [CrossRef] [PubMed]

25. DeHaven, W.I.; Smyth, J.T.; Boyles, R.R.; Putney, J.W., Jr. Calcium inhibition and calcium potentiation of Orai1, Orai2, and Orai3 calcium release-activated calcium channels. *J. Biol. Chem.* **2007**, *282*, 17548–17556. [CrossRef] [PubMed]

26. Parekh, A.B. Store-operated CRAC channels: Function in health and disease. *Nat. Rev. Drug Discov.* **2010**, *9*, 399–410. [CrossRef] [PubMed]

27. Soboloff, J.; Rothberg, B.S.; Madesh, M.; Gill, D.L. STIM proteins: Dynamic calcium signal transducers. *Nat. Rev. Mol. Cell Biol.* **2012**, *13*, 549–565. [CrossRef] [PubMed]

28. Rosado, J.A.; Diez, R.; Smani, T.; Jardín, I. STIM and Orai1 variants in store-operated calcium entry. *Front. Pharmacol.* **2015**, *6*, 325. [CrossRef] [PubMed]

29. Cheng, K.T.; Ong, H.L.; Liu, X.; Ambudkar, I.S. Contribution and regulation of TRPC channels in store-operated Ca^{2+} entry. *Curr. Top. Membr.* **2013**, *71*, 149–179. [PubMed]

30. De Souza, L.B.; Ong, H.L.; Liu, X.; Ambudkar, I.S. Fast endocytic recycling determines TRPC1–STIM1 clustering in ER–PM junctions and plasma membrane function of channel. *Biochim. Biophys. Acta* **2015**, *1853*, 2709–2721. [CrossRef] [PubMed]

31. Salido, G.M.; Jardín, I.; Rosado, J.A. The TRPC ion channels: Association with Orai1 and STIM1 proteins and participation in capacitative and non-capacitative calcium entry. *Adv. Exp. Med. Biol.* **2011**, *704*, 413–433. [PubMed]

32. Ma, G.; Wei, M.; He, L.; Liu, C.; Wu, B.; Zhang, S.L.; Jing, J.; Liang, X.; Senes, A.; Tan, P.; et al. Inside-out Ca^{2+} signalling prompted by STIM1 conformational switch. *Nat. Commun.* **2015**, *6*, 7826. [CrossRef] [PubMed]

33. Ong, H.L.; Cheng, K.T.; Liu, X.; Bandyopadhyay, B.C.; Paria, B.C.; Soboloff, J.; Pani, B.; Gwack, Y.; Srikanth, S.; Singh, B.B.; et al. Dynamic assembly of TRPC1-STIM1-Orai1 ternary complex is involved in store-operated calcium influx. Evidence for similarities in store-operated and calcium release-activated calcium channel components. *J. Biol. Chem.* **2007**, *282*, 9105–9116. [CrossRef] [PubMed]

34. Bowen-Pope, D.F.; Raines, E.W. History of discovery: Platelet-derived growth factor. *Arterioscler. Thromb. Vasc. Biol.* **2011**, *31*, 2397–2401. [CrossRef] [PubMed]

35. Cao, Y. Multifarious functions of PDGFs and PDGFRs in tumor growth and metastasis. *Trends Mol. Med.* **2013**, *19*, 460–473. [CrossRef] [PubMed]

36. Raica, M.; Cimpean, A.M. Platelet-derived growth factor (PDGF)/PDGF receptors (PDGFR) axis as target for antitumor and antiangiogenic therapy. *Pharmaceuticals* **2010**, *3*, 572–599. [CrossRef] [PubMed]

37. Yang, P.S.; Wang, M.J.; Jayakumar, T.; Chou, D.S.; Ko, C.Y.; Hsu, M.J.; Hsieh, C.Y. Antiproliferative activity of hinokitiol, a tropolone derivative, is mediated via the inductions of p-JNK and p-PLCγ1 signaling in PDGF-BB-stimulated vascular smooth muscle cells. *Molecules* **2015**, *20*, 8198–8212. [CrossRef] [PubMed]

38. Andrae, J.; Gallini, R.; Betsholtz, C. Role of platelet-derived growth factors in physiology and medicine. *Genes Dev.* **2008**, *22*, 1276–1312. [CrossRef] [PubMed]

39. Donovan, J.; Shiwen, X.; Norman, J.; Abraham, D. Platelet-derived growth factor alpha and beta receptors have overlapping functional activities towards fibroblasts. *Fibrogenesis Tissue Repair* **2013**, *6*, 10. [CrossRef] [PubMed]

40. Fretto, L.J.; Snape, A.J.; Tomlinson, J.E.; Seroogy, J.J.; Wolf, D.L.; LaRochelle, W.J.; Giese, N.A. Mechanism of platelet-derived growth factor (PDGF) AA, AB, and BB binding to alpha and beta PDGF receptor. *J. Biol. Chem.* **1993**, *268*, 3625–3631. [PubMed]

41. Sachinidis, A.; Locher, R.; Hoppe, J.; Vetter, W. The platelet-derived growth factor isomers, PDGF-AA, PDGF-AB and PDGF-BB, induce contraction of vascular smooth muscle cells by different intracellular mechanisms. *FEBS Lett.* **1990**, *275*, 95–98. [CrossRef]

42. Bornfeldt, K.E.; Raines, E.W.; Graves, L.M.; Skinner, M.P.; Krebs, E.G.; Ross, R. Platelet-derived growth factor. Distinct signal transduction pathways associated with migration versus proliferation. *Ann. N. Y. Acas. Sci.* **1995**, *766*, 416–430. [CrossRef]

43. Pinzani, M. PDGF and signal transduction in hepatic stellate cells. *Front. Biosci.* **2002**, *7*, d1720–d1726. [CrossRef] [PubMed]

44. Heldman, A.W.; Kandzari, D.E.; Tucker, R.W.; Crawford, L.E.; Fearon, E.R.; Koblan, K.S.; Goldschmidt-Clermont, P.J. EJ-Ras inhibits phospholipase C gamma 1 but not actin polymerization induced by platelet-derived growth factor-BB via phosphatidylinositol 3-kinase. *Circ. Res.* **1996**, *78*, 312–321. [CrossRef] [PubMed]

45. Yao, H.; Duan, M.; Yang, L.; Buch, S. Platelet-derived growth factor-BB restores human immunodeficiency virus Tat-cocaine-mediated impairment of neurogenesis: Role of TRPC1 channels. *J. Neurosci.* **2012**, *32*, 9835–9847. [CrossRef] [PubMed]

46. Spinelli, A.M.; González-Cobos, J.C.; Zhang, X.; Motiani, R.K.; Rowan, S.; Zhang, W.; Garrett, J.; Vincent, P.A.; Matrougui, K.; Singer, H.A. Airway smooth muscle STIM1 and Orai1 are upregulated in asthmatic mice and mediate PDGF-activated SOCE, CRAC currents, proliferation, and migration. *Pflugers Arch.* **2012**, *464*, 482–491. [CrossRef] [PubMed]

47. Suganuma, N.; Ito, S.; Aso, H.; Kondo, M.; Sato, M.; Sokabe, M.; Hasegawa, Y. STIM1 regulates platelet-derived growth factor-induced migration and Ca^{2+} influx in human airway smooth muscle cells. *PLoS ONE* **2012**, *7*, e45056. [CrossRef] [PubMed]

48. Ma, R.; Sansom, S.C. Epidermal growth factor activates store-operated calcium channels in human glomerular mesangial cells. *J. Am. Soc. Nephrol.* **2001**, *12*, 47–53. [PubMed]

49. Banai, S.; Wolf, Y.; Golomb, G.; Pearle, A.; Waltenberger, J.; Fishbein, I.; Schneider, A.; Gazit, A.; Perez, L.; Huber, R.; et al. PDGF-receptor tyrosine kinase blocker AG1295 selectively attenuates smooth muscle cell growth in vitro and reduces neointimal formation after balloon angioplasty in swine. *Circulation* **1998**, *97*, 1960–1969. [CrossRef] [PubMed]

50. Cecchetti, S.; Bortolomai, I.; Ferri, R.; Mercurio, L.; Canevari, S.; Podo, F.; Miotti, S.; Iorio, E. Inhibition of phosphatidylcholine-specific phospholipase C interferes with proliferation and survival of tumor initiating cells in squamous cell carcinoma. *PLoS ONE* **2015**, *10*, e0136120. [CrossRef] [PubMed]

51. Mauban, J.R.; Zacharia, J.; Fairfax, S.; Wier, W.G. PC-PLC/sphingomyelin synthase activity plays a central role in the development of myogenic tone in murine resistance arteries. *Am. J. Physiol. Heart Circ. Physiol.* **2015**, *308*, H1517–H1524. [CrossRef] [PubMed]

52. Vandenberghe, M.; Raphaël, M.; Lehen'kyi, V.; Gordienko, D.; Hastie, R.; Oddos, T.; Rao, A.; Hogan, P.G.; Skryma, R.; Prevarskaya, N. Orai1 calcium channel orchestrates skin homeostasis. *Proc. Natl. Acad. Sci. USA* **2013**, *110*, E4839–E4848. [CrossRef] [PubMed]

53. Numaga-Tomita, T.; Putney, J.W. Role of STIM1- and Orai1-mediated Ca^{2+} entry in Ca^{2+}-induced epidermal keratinocyte differentiation. *J. Cell Sci.* **2013**, *126*, 605–612. [CrossRef] [PubMed]

54. Hooper, R.; Samakai, E.; Kedra, J.; Soboloff, J. Multifaceted roles of STIM proteins. *Pflugers Arch.* **2013**, *465*, 1383–1396. [CrossRef] [PubMed]

International Journal of
Molecular Sciences

MDPI

Article

Dandelion Root Extract Induces Intracellular Ca^{2+} Increases in HEK293 Cells

Andrea Gerbino [1],*, Daniela Russo [2], Matilde Colella [1], Giuseppe Procino [1], Maria Svelto [1], Luigi Milella [2] and Monica Carmosino [2],*

[1] Department of Biosciences, Biotechnologies and Biopharmaceutics, University of Bari, 70126 Bari, Italy; matilde.colella@uniba.it (M.C.); giuseppe.procino@uniba.it (G.P.); maria.svelto@uniba.it (M.S.)

[2] Department of Sciences, University of Basilicata, 85100 Potenza, Italy; daniela.russo@unibas.it (D.R.); luigi.milella@unibas.it (L.M.)

* Correspondence: andrea.gerbino@uniba.it (A.G.); monica.carmosino@unibas.it (M.C.); Tel.: +39-080-544-3334 (A.G.); +39-335-6302642 (M.C.)

Received: 6 March 2018; Accepted: 4 April 2018; Published: 7 April 2018

Abstract: Dandelion (*Taraxacum officinale* Weber ex F.H.Wigg.) has been used for centuries as an ethnomedical remedy. Nonetheless, the extensive use of different kinds of dandelion extracts and preparations is based on empirical findings. Some of the tissue-specific effects reported for diverse dandelion extracts may result from their action on intracellular signaling cascades. Therefore, the aim of this study was to evaluate the effects of an ethanolic dandelion root extract (DRE) on Ca^{2+} signaling in human embryonic kidney (HEK) 293 cells. The cytotoxicity of increasing doses of crude DRE was determined by the Calcein viability assay. Fura-2 and the fluorescence resonance energy transfer (FRET)-based probe ERD1 were used to measure cytoplasmic and intraluminal endoplasmic reticulum (ER) Ca^{2+} levels, respectively. Furthermore, a green fluorescent protein (GFP)-based probe was used to monitor phospholipase C (PLC) activation (pleckstrin homology [PH]–PLCδ–GFP). DRE (10–400 µg/mL) exposure, in the presence of external Ca^{2+}, dose-dependently increased intracellular Ca^{2+} levels. The DRE-induced Ca^{2+} increase was significantly reduced in the absence of extracellular Ca^{2+}. In addition, DRE caused a significant Ca^{2+} release from the ER of intact cells and a concomitant translocation of PH–PLCδ–GFP. In conclusion, DRE directly activates both the release of Ca^{2+} from internal stores and a significant Ca^{2+} influx at the plasma membrane. The resulting high Ca^{2+} levels within the cell seem to directly stimulate PLC activity.

Keywords: Ca^{2+} signaling; Ca^{2+} influx; plasma membrane; endoplasmic reticulum; phospholipase C; Fura-2; Ca^{2+} fluorescent sensors; herbal extract; bioactive compounds

---

1. Introduction

To date, natural products from medicinal plants, microorganisms, marine organisms (e.g., *Acmella oleracea* [1], *Sclerocarya birrea* [2,3], *Agelas clathrodes* [4], and *Tedania ignis* [5]) are in the spotlight of many research and industrial laboratories for (i) biomedical applications, (ii) the recovery of bioactives, and (iii) the development of pharmacological drugs. *Taraxacum officinale* Weber, known as dandelion, is a perennial weed that has been used for hundreds of years (starting from the 10th and 11th centuries) as a traditional medical remedy for several diseases [6]. As a matter of fact, dandelion is nowadays commercialized as a healthy food because of its health-promoting ethno-medical properties, including anti-inflammatory, anti-rheumatic, anti-oxidant, anti-carcinogenic, diuretic, choleretic and cholagogue, laxative, and hypoglycemic activities [6,7]. However, only some of these empirical effects have been validated by a proper scientific investigation [8], therefore more experimental evidence is needed to justify such extensive use of dandelion as a natural therapeutic remedy.

A possible explanation of such a wide panel of physiological effects might be found in the chemical composition of the different dandelion preparations. Dandelion's extract might be composed by the whole plant or by different parts of it (e.g., roots, leaves, stem, flowers), alone or in combination. In addition, its chemical composition strictly depends on both the extraction protocol and the solvents used (ethanol, acetone, water, or methanol) that have to be properly designed and identified to efficiently produce extracts containing the desired bioactive compounds. For example, the phytochemical composition of dandelion roots extracts (DRE) reported the presence of sesquiterpenes, various triterpenes, phytosterols, and phenolic compounds [6]. Among the phenolic compounds, hydroxycinnamic acid derivatives (chlorogenic, caffeic, 4-coumaric, 3-coumaric, ferulic acids) are the main represented class, whereas a small amount of flavonoids and hydroxybenzoic acid derivatives is generally reported [9–11]. Compared to the roots, dandelion leaves and flowers are more enriched in flavonoids (luteolin and its glycoside derivatives, chrysoeriol) and coumarins (cichoriin and aesculin), but also hydroxycinnamic acid derivatives (caffeic, chlorogenic, chicoric, and monocaffeoyltartaric acids) were reported to be present [10,12].

These various chemical components may act individually, additively, or in synergy to regulate different tissue-specific functions. Still, the surprising amount of physiological functions that are somehow related to the medical use of *T. officinale* suggests the ability of cells to respond to its herbal extracts through a basic and widespread intracellular mechanism.

Therefore, in this study, we evaluated whether the acute exposure of HEK293 cells to an ethanolic dandelion root extract (DRE) might impact intracellular Ca^{2+} homeostasis. Among the more general transduction pathways of a cell we decided to analyze specifically Ca^{2+} signaling for two main reasons. First, even though there is no information available in the literature about the effects of dandelion extracts on intracellular Ca^{2+} dynamics, numerous reports regarding the effects of single bioactive components, especially hydroxycinnamic acid derivatives, have been published. Caffeic acid (CA) and its derivatives are natural phenolic compounds that affect cellular Ca^{2+} homeostasis in different experimental models, such as human gastric cancer (SCM1) cells [13], T lymphocytes [14], and Jurkat cells [15]. Chlorogenic (CGA) and ferulic acids (FA) are polyphenols that protect different cell models (human umbilical vein endothelial cells and rat cortical neurons) from Ca^{2+}-mediated insults by reducing, through cell-specific mechanisms, Ca^{2+} entry at the plasma membrane [16–18].

Second, most of the physiological functions that are regulated by dandelion extracts are finely modulated by intracellular Ca^{2+} events (e.g., inflammation, proliferation, diuresis, just to name a few). This implies either that the two players (namely, Ca^{2+} and DRE) share common cellular targets or that Ca^{2+} is a dandelion-induced mediator within the cells.

Therefore, we used Fura-2 and the fluorescence resonance energy transfer (FRET)-based probe ERD1 [19] to measure cytoplasmic and intraluminal Ca^{2+} levels within the endoplasmic reticulum, respectively, in HEK293 cells in response to acute exposure to DRE. In addition, a green fluorescent protein (GFP)-based probe [20] was used to monitor the time course of phospholipase C (PLC) activation (pleckstrin homology [PH]–PLCδ–GFP). All the experiments performed depict a scenario in which DRE induced a rapid and reversible increase in intracellular Ca^{2+} levels. Such rise in cytosolic Ca^{2+} levels was due to both Ca^{2+} release from the endoplasmic reticulum and Ca^{2+} entry at the plasma membrane, likely via store-operated channels. DRE exposure also induced the activation of a downstream signaling element such as PLC, which is elicited, even though not exclusively, by G-protein-coupled receptors [21].

The precise dissection of DRE-induced intracellular Ca^{2+} events at the cellular and molecular level will pave the way for the proper use of this promising extract as a natural pharmaceutical tool.

2. Results

2.1. High-Performance Liquid Chromatography (HPLC) Analysis of the Ethanolic DRE and Its Cytotoxic Activity

Dandelion dried roots were extracted with ethanol by exhaustive maceration. Ethanol is an appropriate solvent for both phenols and other bioactive compounds because of its intermediate polarity. It is well known that solvents affect the extraction of bioactive compounds and the phytochemical profiles of extracts from various plant parts [22]. Consequently, the phytochemical constituents and their amount in different plant extracts reflect different biological properties of a plant.

The chemical composition of our dandelion ethanolic root extract (DRE) was evaluated by reversed-phase high-performance liquid chromatography (RP-HPLC). Figure 1a shows the HPLC profile of the ethanolic DRE. The results of the quali–quantitative analysis reported the presence of hydroxybenzoic acid derivatives and hydroxycinnamic acids. In particular, chlorogenic acid (retention time, hereafter Rt: 7.75 min) and caffeic acid (Rt: 10.17 min) were identified in the extract as the main components (see Table 1).

Table 1. HPLC quantitative analysis of the ethanolic extract from dandelion root. The values are expressed as mean ± standard deviation (SD) of milligram (mg) of compound per gram (g) of extract ($n = 3$).

Compound	Rt (min)	Average ± SD (mg/g)
Gallic acid (1)	4.72	0.164 ± 0.018
Chlorogenic acid (2)	7.75	1.256 ± 0.012
Caffeic acid (3)	10.17	0.776 ± 0.063
Vanillic acid (4)	10.83	0.148 ± 0.012
Syringic acid (5)	11.32	0.234 ± 0.009
p-Coumaric acid (6)	14.72	0.041 ± 0.005
Ferulic acid (7)	17.71	0.273 ± 0.070

We also identified, although present in lower amount, gallic acid (Rt: 4.72 min), vanillic acid (Rt: 10.83 min), syringic acid (Rt: 11.32 min), *p*-coumaric acid (Rt: 14.72 min), and ferulic acid (Rt: 17.71 min). All compounds were identified comparing their retention time and UV spectrum with those of reference standards. The other peaks of the extract could not be identified because of a lack of validated reference standards to compare with. All these compounds had already been identified in dandelion root extracts [9]. In this study we first aimed at identifying the non-toxic concentration of the DRE by performing a Calcein cytotoxic assay in HEK293 cells. As shown in Figure 1b, compared with vehicle control (ethanol), DRE induced significant cell death only at 750 µg/mL.

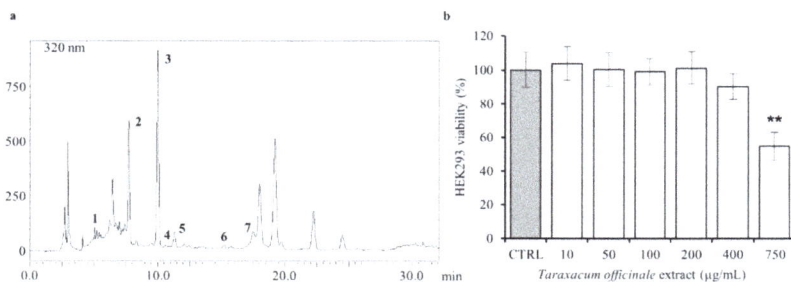

Figure 1. HPLC analysis of dandelion root extract (DRE) and evaluation of its cytotoxicity. (**a**) Chromatogram obtained by HPLC analysis of a *Taraxacum officinale* root extract at 320 nm. The compounds identified are gallic acid (1), chlorogenic acid (2), caffeic acid (3), vanillic acid (4), syringic acid (5), p-coumaric (6), and ferulic acid (7). (**b**) Viability of HEK293 cells cultured in the absence (ethanol 0.6% as control, CTRL) and presence of diverse doses of *T. officinale* root extract for 24 h by Calcein viability assay. The reported values are means ± SEM of three independent experiments (** $p < 0.01$ vs. CTRL).

On the basis of this result, in the following functional studies, we used DRE at 400 µg/mL (or lower).

2.2. Exposure to DRE Increased Intracellular Ca²⁺ Levels

The results collected in Figure 2 show that DRE causes an increase of intracellular Ca^{2+} levels in HEK293 cells. In the presence of 1.2 mM extracellular Ca^{2+}, 400 µg/mL DRE elicited cytosolic Ca^{2+} transients that were not significantly larger (118.88 ± 7.96% of control peak; $p = 0.4$) than those produced by a saturating dose of the Ca^{2+}-mobilizing agonist ATP (100 µM) in the same cells (Figure 2a; $n = 5$ experiments, HEK293 cells).

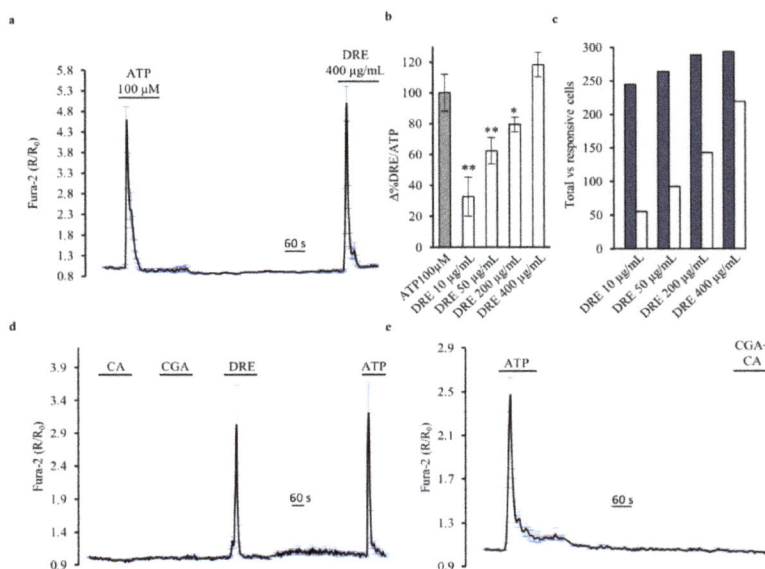

Figure 2. DRE increases intracellular Ca^{2+} levels, while caffeic and chlorogenic acids do not, as measured by Fura-2 in single HEK293 cells. (**a**) Response of Fura-2-loaded HEK293 cells to the Ca^{2+}-mobilizing agonist ATP (100 µM) compared with the Ca^{2+} increase elicited by DRE (400 µg/mL). The reported values are means ± SE from all the responsive cells on a single coverslip. (**b**) Changes in the Fura-2 ratio in Ca^{2+}-containing solutions in response to increasing doses of DRE (10, 50, 200, and 400 µg/mL) normalized to the response induced by ATP. The reported values are means ± SE from all the responsive cells of all experiments performed (** $p < 0.01$ vs. ATP, * $p < 0.05$ vs. ATP). (**c**) Number of DRE-treated (10, 50, 200, and 400 µg/mL) Ca^{2+}-responsive cells versus the total number of cells analyzed. The bar graph summarizes the number of cells of all the experiments performed for each concentration. (**d**) Response of Fura-2-loaded HEK293 cells to caffeic acid (CA, 150 µM), chlorogenic acid (CGA, 200 µM), and DRE (400 µg/mL) compared with the effect induced by the Ca^{2+}-mobilizing agonist ATP (100 µM). The reported values are means ± SE from all the responsive cells on a single coverslip. (**e**) Response of Fura-2-loaded HEK293 cells to the simultaneous exposure to caffeic (CA, 150 µM) and chlorogenic acid (CGA, 200 µM) compared with the effect induced by the Ca^{2+}-mobilizing agonist ATP (100 µM). The reported values are means ± SE from all the responsive cells on a single coverslip.

The Ca^{2+} response induced by DRE was dose-dependent. The smallest, albeit significant, increase in intracellular Ca^{2+} was detected at a concentration of DRE as low as 10 µg/mL (32.7 ± 12.6% of control peak; $n = 3$, $m = 149$ cells, $p < 0.01$). Concentrations of 50 and 200 µg/mL DRE induced linear increases in intracellular Ca^{2+} that were, however, significantly smaller than those induced by ATP (Figure 2b, 62.4 ± 8.60% of control peak; $n = 3$, m = 129 cells, $p < 0.01$ and 79.4 ± 4.75% of control peak; $n = 4$, $m = 187$ cells, $p < 0.05$, respectively). In addition, the higher was the concentration of DRE used, the larger was the number of responsive cells on a single coverslip (Figure 2c), clearly indicating the activation of a cellular mechanism that was strictly dependent on DRE concentration.

Were these Ca^{2+} transients induced by a specific component of the DRE? To answer this question, we analyzed the effects of both caffeic and chlorogenic acids (CA and CGA, respectively) on the intracellular Ca^{2+} levels. Both compounds, enriched in our DRE (see Table 1), are known to impact in vitro Ca^{2+} homeostasis [13–17]. We exposed HEK293 cells to different concentrations of CA and CGA, starting with the concentration at which these compounds were present in our extract, but no intracellular Ca^{2+} changes were recorded (data not shown). Also, when we exposed HEK293 cells to the highest investigated concentrations of either caffeic acid (150 μM) or chlorogenic acid (200 μM) we did not record intracellular Ca^{2+} changes (Figure 2d). On the other hand, the exposure of cells on the same coverslip to DRE induced a significant increase in the intracellular Ca^{2+} levels that was similar to that induced by ATP.

The same effect, namely, no intracellular Ca^{2+} increase, was also obtained when both components were perfused at the same time (Figure 2e), indicating that the response elicited by DRE is induced by other components (even though present in a smaller dose) or by the extract as a whole.

The Ca^{2+}-mediated response induced by 400 μg/mL DRE was only slightly, but significantly, decreased, when compared to ATP, in the absence of external Ca^{2+} (Figure 3a; 81.69 ± 7.17% of control response to ATP; $n = 5$, $m = 235$ cells, $p < 0.001$). These results suggest that DRE was able to release Ca^{2+} from internal stores. Indeed, Figure 3b shows that, when the cells were treated with the reversible sarco/endoplasmic reticulum Ca^{2+}-ATPase (SERCA) pump inhibitor cyclopiazonic acid (CPA) in free extracellular Ca^{2+} to deplete intracellular Ca^{2+} stores, the Ca^{2+} response induced by DRE in control conditions was completely prevented. Note that this experimental approach also revealed that the acute exposure to 400 μg/mL DRE did not interfere with the mechanisms underlining CPA-induced endoplasmic reticulum (ER) Ca^{2+} release and with the capacitative Ca^{2+} entry induced by perfusion with 5 mM Ca^{2+} after store emptying. *T. officinale* has been usually associated with diuretic effects, thus renal cells represent putative physiological targets. In separate experiments, we showed that exposure to DRE (400 μg/mL) induced Ca^{2+} transients similar to those obtained in HEK293 cells either in the presence or absence of extracellular Ca^{2+} also in a renal model of epithelial cells (MCD4, mouse collecting duct cells) (data not shown).

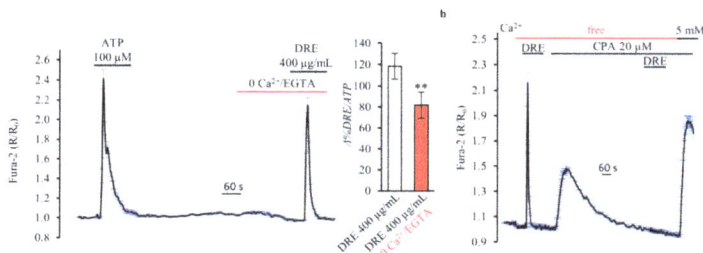

Figure 3. DRE increases intracellular Ca^{2+} levels in a Ca^{2+}-free extracellular solution as measured by Fura-2 in single HEK293 cells. (**a**) Response of Fura-2-loaded HEK293 cells to the Ca^{2+}-mobilizing agonist ATP (100 μM) compared with the Ca^{2+} increase elicited by DRE (400 μg/mL) in the absence of extracellular Ca^{2+}. The reported values are means ± SE from all the responsive cells on a single coverslip. Right inset: Changes in the Fura-2 ratio in response to DRE (400 μg/mL) in the presence or absence of extracellular Ca^{2+}, normalized to the response induced by ATP. The reported values are means ± SE from all the responsive cells of all experiments performed (** $p < 0.01$ vs. DRE with Ca^{2+}). (**b**) Real-time measurements of intracellular Ca^{2+} levels in response to DRE (400 μg/mL) before and after application of 20 μM cyclopiazonic acid (CPA) in the absence or presence of extracellular Ca^{2+} (5.0 mM) in Fura-2-loaded HEK293. The reported values are means ± SE from all the responsive cells on a single coverslip.

2.3. DRE Exposure Induced Ca^{2+} Release from CPA-Sensitive Stores

We next examined the ER Ca^{2+} release process in more detail in intact HEK293 cells transfected with the FRET-based probe ERD1. Figure 4a shows that exposure to 400 µg/mL DRE in extracellular Ca^{2+}-free solution for 4 min decreased the intraluminal free Ca^{2+} levels by 46.12 ± 6.6% ($n = 4$, $m = 32$ cells, $p < 0.001$). DRE-induced intraluminal reduction was normalized with respect to the maximal ER-emptying effect elicited by 5 µM ionomycin in the absence of Ca^{2+}. Exposure of cells to the Ca^{2+}-mediated agonist ATP (Figure 4b) was able to induce a decrease in ER Ca^{2+} levels, corresponding to 29.84 ± 8.1% of the maximal ionomycin response ($n = 4$, $m = 41$ cells, $p < 0.001$).

It is widely recognized that the emptying of the endoplasmic reticulum activates store-operated Ca^{2+} entry at the plasma membrane [23]. As showed in Figure 4c, re-addition of 2 mM extracellular Ca^{2+} in a Ringer's solution nominally free of Ca^{2+}, in the continuous presence of DRE, induced an increase in the intracellular Ca^{2+} levels likely through store-operated channels (SOCs). Comparable results were obtained when we used, in the same experimental protocol, ATP instead of DRE.

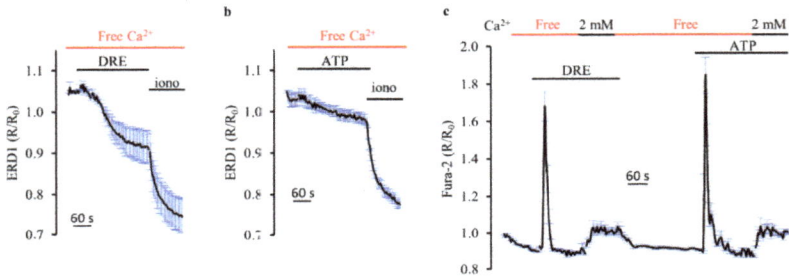

Figure 4. DRE induces both Ca^{2+} release from the endoplasmic reticulum and Ca^{2+} entry at the plasma membrane. Response of ERD1-transfected HEK-293 cells to (**a**) DRE (400 µg/L) or (**b**) ATP (100 µM) in extracellular Ca^{2+}-free solution. Both responses were compared with the maximal ER Ca^{2+} release elicited by ionomycin 5 µM in extracellular Ca^{2+}-free solution. The reported values are means ± SE from all the responsive cells on a single coverslip (**c**) Response of Fura-2-loaded HEK293 cells to either DRE (400 µg/mL), in the absence or presence (2 mM) of extracellular Ca^{2+}, or ATP (100 µM), in the absence or presence (2 mM) of extracellular Ca^{2+}. The reported values are means ± SE from all the responsive cells on a single coverslip.

2.4. Dissecting DRE Effect on Ca^{2+} Signaling

The results collected so far suggest that DRE can either cross the plasma membrane and exert direct actions on internal Ca^{2+} stores (e.g., like a Ca^{2+}-ionophore such as ionomycin) or activate a plasma membrane receptor coupled to a Gq alpha subunit (Gαq). However, when we used ionomycin in a protocol similar to that depicted in Figure 4c, we found completely different results.

Figure 5a shows that ionomycin elicited a rapid and reversible increase in cytosolic Ca^{2+}, considered in this protocol as the control response. The quantal increase (1.2, 3.0, 10 mM) of the extracellular Ca^{2+} concentration rapidly raised the cytosolic amount of Ca^{2+} to about 160% in the presence of 10 mM extracellular Ca^{2+} (164.84 ± 10.4%, $n = 3$, versus the control response). When we repeated the same protocol with DRE instead of ionomycin, we recorded, in the presence of 10 mM extracellular Ca^{2+}, smaller increases in cytosolic Ca^{2+} (Figure 5b, 20.15 ± 8.8%, $n = 3$, versus the control response) that suggested a capacitative Ca^{2+} entry rather than a ionophore-mediated influx through the plasma membrane.

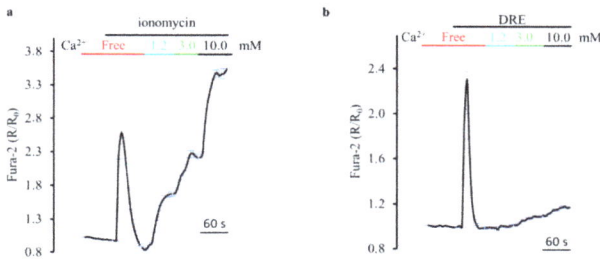

Figure 5. DRE does not act as an ionophore-like compound such as ionomycin. (**a**) Response of Fura-2-loaded HEK293 cells to the ionophore ionomycin (5 μM) in the absence or presence (1.2, 3.0, and 10 mM) of extracellular Ca^{2+}. The reported values are means ± SE from all the responsive cells on a single coverslip. (**b**) Response of Fura-2-loaded HEK293 cells to DRE (400 μg/mL) in the absence or presence (1.2, 3.0, and 10 mM) of extracellular Ca^{2+}. The reported values are means ± SE from all the responsive cells on a single coverslip.

Thus, is there a G-protein coupled receptor (GPCR) involved? It is pharmacologically complicated to specifically block Gαq. This is the reason why we decided to characterize features that are generally related to the increase in cytosolic Ca^{2+} mediated by GPCR.

First, the Ca^{2+} response induced by DRE was significantly decreased when repeated, at the same concentration, after 10–15 min of DRE washout (Figure 6a, 65.75 ± 7.17%, $n = 4$, $p < 0.01$). When we increased the interval between two DRE stimulations, we found a significant recovery of the second Ca^{2+} response that resulted non-significantly changed compared to the first one after 45 min (105.70 ± 8.26%, $n = 4$, $p = 0.19$). One possible explanation could be that it takes time to refill the ER that has been challenged after the first DRE stimulation. However, when we repeated the same experimental protocol with ATP, we found a complete recovery of the Ca^{2+} response after 15–20 min (Figure 6b). Indeed, these results can be explained considering the receptor desensitization of a putative GPCR involved. Considering that in HEK293 cells (i) the cytosolic amount of Ca^{2+} recorded under both agonists (DRE and ATP) stimulation is of similar extent and (ii) the Ca^{2+} homeostatic mechanisms are the same for the two agonists, we conclude that the hypothesis that DRE is activating a GPCR has to be taken into account.

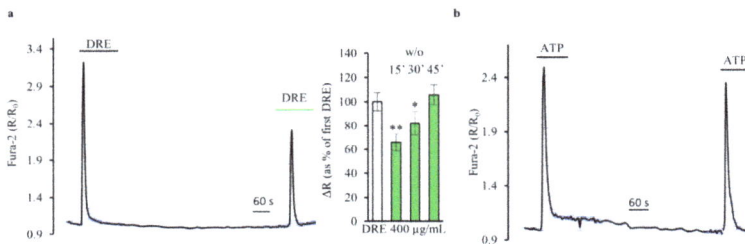

Figure 6. Does DRE activate plasma membrane receptors? (**a**) Response of Fura-2-loaded HEK293 cells to two repeated exposures to DRE (400 μg/mL) separated by a 15 min washout (*w/o*). The reported values are means ± SE from all the responsive cells on a single coverslip. Right inset: Changes in the Fura-2 ratio in Ca^{2+}-containing solutions in response to repeated exposures to DRE (400 μg/mL) separated by washouts of different durations (15′, 30′, and 45′). The second response to DRE is normalized to the first one. The reported values are means ± SE from all the responsive cells of all experiments performed (** $p < 0.01$ vs. first DRE, * $p < 0.05$ vs. first DRE) (**b**) Response of Fura-2-loaded HEK293 cells to two repeated exposures to ATP (100 μM) separated by a 15 min washout (*w/o*). The reported values are means ± SE from all the responsive cells on a single coverslip.

Under this scenario, the Ca^{2+} release from the ER would have been the result of the activation of the $G\alpha q$/PLC/IP3 pathway. Therefore, we used 10 µM U73122 in order to inhibit PLC. Figure 7a shows that DRE (400 µg/mL) was unable to increase intracellular Ca^{2+} levels both in the presence of U73122 and 10 min after its washout. However, these results are significantly affected by the fact that U73122 likely increased intracellular Ca^{2+} levels by emptying the endoplasmic reticulum, as already shown for baby hamster kidney (BHK21) cells [24]. This secondary and non-specific effect of U73122 fully explains its inhibitory action on the DRE-induced Ca^{2+} increase that we observed in our study. Therefore, we cannot use these experiments to extrapolate information about the putative activation of PLC following exposure to DRE.

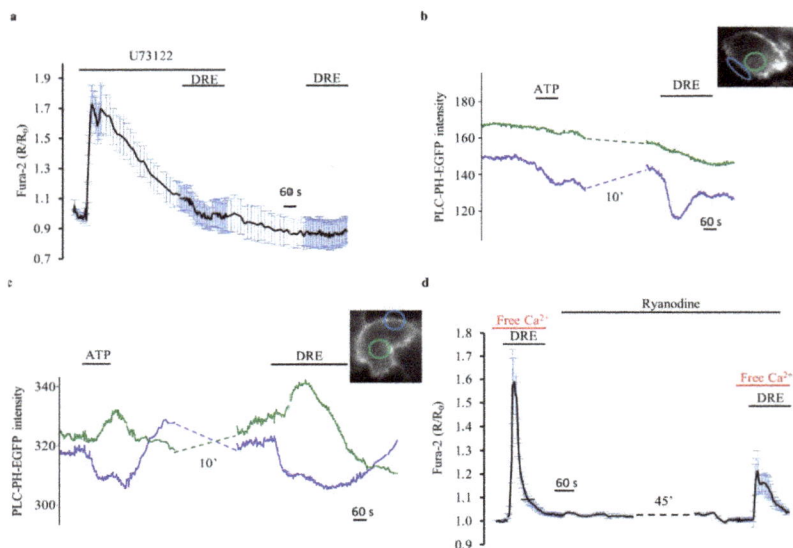

Figure 7. DRE exposure activates the downstream signaling effector phospholipase C. (**a**) Response of Fura-2-loaded HEK293 cells to DRE (400 µg/mL) in the presence of or after exposure to U73122 (10 µM). The reported values are means ± SE from all the responsive cells on a single coverslip. (**b,c**) Representative traces of the measurement of PLC activation in response to ATP and DRE, as measured by translocation of PH–PLCδ–EGFP from the plasma membrane (blue trace) to the cytosol (green trace). Inset image: Fluorescence of a representative cell excited at 480 nm showing, at rest, PH–PLCδ–EGFP associated mostly at the plasma membrane (blue ROI) but also weakly present in the peripheral cytoplasm (green ROI). (**d**) Response of Fura-2-loaded HEK293 cells to DRE (400 µg/mL) in extracellular Ca^{2+}-free solution before and after a prolonged (about 50 min) exposure to 100 µM Ryanodine. The reported values are means ± SE from all the responsive cells on a single coverslip.

Thus, as an alternative strategy, we used a GFP-based indicator, PH–PLCδ–EGFP [20,24] that allowed us to directly examine in real time the action of DRE on PLC activation in single HEK293 cells. In resting conditions, the PH domain of this probe interacts with PIP2 at the plasma membrane, so that it is predominantly visible at the periphery of the cells. Upon PLC activation and PIP2 hydrolysis, the PH domain translocates toward the cytosol with inositol trisphosphate (InsP3).

The two typical patterns of response observed in single cell epifluorescence experiments in PH–PLCδ–EGFP-expressing HEK293 cells upon stimulation with ATP and DRE are shown in Figure 7b,c. Most probably because of the morphology and the large volume occupied by the nucleus in HEK293 cells plated at high confluence, the expected antiparallel signal of PH–PLCδ–EGFP [9] was not always appreciable, both upon agonist and upon DRE stimulation ($n = 4$, $m = 63$). In order to have a semiquantitative appreciation of the responses, as for the previous experiments, we analyzed only cells which clearly responded to ATP with either a decrease of the plasma membrane signal or the classical antiparallel behavior of the membrane and cytosolic fluorescence intensity.

Figure 7b shows a representative trace in which, albeit only a small increase in the cytosolic signal is appreciable upon agonist stimulation, a clear and reversible reduction of PLC fluorescence intensity is apparent at the membrane, indicative of PIP2 hydrolysis. In the cells which showed this behavior upon ATP stimulation, a clear response was also recordable upon DRE perfusion (36/36 cells). Importantly, in half of the ATP responsive cells (18/36 cells), also a clear translocation of the probe to the cytosol was appreciable upon ATP stimulation, and again an analogous behavior was recordable with DRE (18/18 cells) (Figure 7c). All these data clearly indicated a similar mechanism of action of DRE and ATP on PLC, as measured by PH–PLCδ–EGFP.

Thus, DRE-induced activation of PLC might induce Ca^{2+} release from the endoplasmic reticulum through the formation of IP3 and the activation of IP3R. Ca^{2+} released from the ER might, in turn, activate the ryanodine receptors (RyRs) showed to be expressed and functional in HEK293 cells [25]. Of note, as shown in Figure 7d, DRE-induced increase in intracellular Ca^{2+} was significantly reduced (by about 60%, $n = 3$, $p < 0.01$) under prolonged (at least 50 min) RyRs blockade with high concentrantions of ryanodine (100 μM), clearly indicating Ca^{2+}-induced Ca^{2+} release as an additional mechanism in DRE-elicited ER Ca^{2+} release.

3. Discussion

The use of plant extracts to treat diseases is ancient, and popular observations regarding their use and efficacy prompt the investigation of their therapeutic properties [26]. The identification of bioactive compounds and their action mechanisms against diseases are the assumptions for the potential use of plant preparations as healthy sources [3]. *T. officinale* (dandelion) is an ethnomedical herb used as anti-inflammatory, antioxidant, diuretic, choleretic, laxative, and to treat arthritis and liver disorders [6]. However, despite dandelion incredible popularity, scientific research regarding the mechanisms of action of its extracts is still limited [8].

In this study, we showed, for the first time, that acute exposure (2–5 min, 400 μg/mL) to an ethanolic dandelion root extract (DRE) induced a dose-dependent and reversible Ca^{2+} increase in HEK293 cells (Figure 2). DRE-mediated increase in intracellular Ca^{2+} was characterized by using both classical experimental manoeuvres (e.g., removal of extracellular Ca^{2+}, inhibition of the SERCA pump, blockers of mechanisms regulating Ca^{2+} homeostasis) and direct comparison with drugs (e.g., ATP and ionomycin) that have a well-known action on Ca^{2+} homeostasis. We found that the rapid Ca^{2+} transient that follows DRE exposure results from a significant Ca^{2+} release from the endoplasmic reticulum and Ca^{2+} entry at the plasma membrane probably through store-operated Ca^{2+} channels. Of note, DRE stimulation induced a significant activation of PLC. This important observation was obtained considering the subcellular localization of a GFP-tagged PH–PLCδ construct [20,24]. In the absence of DRE, the probe was properly located at the plasma membrane whilst, in the presence of DRE, the PH domain moved to the cytosol with InsP3 (IP3) [20,24], thus providing both a direct indication of PLC activation and indirect measurements of IP3 formation [27].

Collectively, these results draw two possible scenarios as to how the presence of the extract is sensed and transduced by the cells (see Figure 8). In the first one, DRE (or one of its components) crosses the plasma membrane ("cross theory") leading to Ca^{2+} entry and "in situ" Ca^{2+} release from the ER and activation of PLC that seconds the cytosolic Ca^{2+} increase. In the second scenario, DRE (or one of its components) activates a G-protein coupled receptor at the plasma membrane and PLC and IP3 production which, in turn, induces Ca^{2+} release from the endoplasmic reticulum potentiated by Ca^{2+}-induced Ca^{2+} release from RyRs (Figure 7d) and the consequent store-operated Ca^{2+} entry. In addition, the fact that the activity of DRE, in terms of activation of Ca^{2+} transients and number of responsive cells, is strictly dose-dependent, is also in line with the "GPCR theory". Finally, in the same direction goes the evidence that close, repeated stimulations with DRE seemed associated with an apparent desensitization of the "putative" receptor, which, on the other hand, did not affect the receptor for ATP.

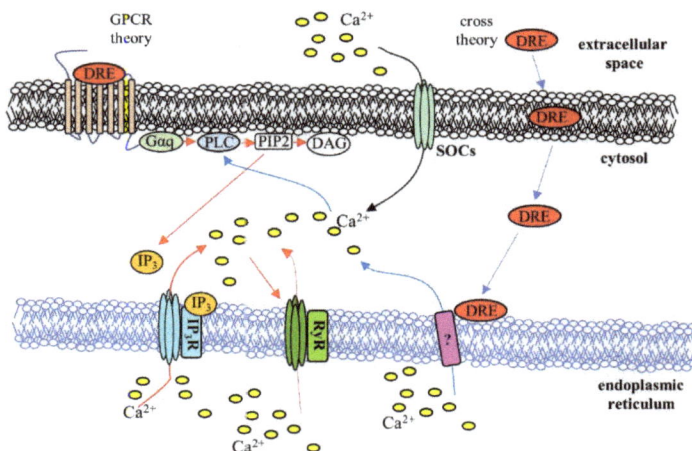

Figure 8. Schematic illustration of the putative mechanisms for DRE-induced Ca^{2+} signaling. The "cross theory" (blue arrows) and the "GPCR theory" (red arrows) have been proposed to explain the mechanisms underlying the Ca^{2+} increase induced by exposure to DRE: See the text for a detailed description. DRE, dandelion root extract; IP3, Inositol trisphosphate; IP3R, Inositol trisphosphate receptor; RyR, Ryanodine receptor; DAG, diacylglycerol, PIP2, phosphatidylinositol 4,5-bisphosphate; PLC, phospholipase C; Gαq, alpha subunit of Gq; SOCs, store-operated Ca^{2+} channels, "?" indicates unknown mechanisms.

Under the same scenario, Chang et al. showed that 800 μM caffeic acid evoked Ca^{2+} release from thapsigargin-sensitive stores in SCM1 human gastric cancer cells, speculating the activation of plasma membrane receptors [13]. In addition, the authors proved that Ca^{2+} release was induced by a phospholipase C-dependent mechanism, since it was blocked when phospholipase C activity was inhibited by U73122. However, in our hands, caffeic acid did not activate Ca^{2+} transients neither in HEK293 (Figure 3) nor in MCD4 cells (data not shown). This difference is likely related to the lower concentration used in our Fura-2 experiment that, in turn, matched the range of concentration of caffeic measured in the DRE (see Table 1). In our experimental protocols, also chlorogenic acid, previously reported to regulate Ca^{2+} homeostasis even though via different cellular mechanisms (e.g., inhibition of agonists-stimulated Ca^{2+} entry), was ineffective when used at a concentration (200 μM) proportional to its presence in our DRE. Therefore, regardless of which theory is correct, it is currently unknown if the Ca^{2+} effects measured in response to DRE are mediated by the whole extract or by individual components diverse from caffeic and chlorogenic acids.

Whatever the specific mechanism of action at the plasma or intracellular membrane, the direct observation that DRE impacts Ca^{2+} homeostasis at different subcellular locations is an important piece of the puzzle for understanding the wide array of biological functions (of different organs and systems) reported to be modulated by *T. officinale*. As a matter of fact, and as suggested for *Taraxacum* [28–31], Ca^{2+} signaling finely regulates processes like cellular proliferation [32], bile secretion [33], diuresis [1,34], inflammation [35], and oxidative stress [35], just to name a few. Thus, being aware of this *Taraxacum*–Ca^{2+} interplay can be considered as a starting point to better project and more precisely target the use of *T. officinale* extracts as pharmacological tools.

Then, it is also important to note that both the composition and the effects induced by these herbal extracts strictly depend on which part of the plant has been used to produce them. In the last years, reports on the phytochemical composition of *T. officinale* extracts are increasing, and it has been shown that dandelion root extracts are mainly rich in hydroxycinnamic acids, whereas hydroxybenzoic acids derivatives and flavonoids have been reported in lower amounts. In this study, the chemical composition of an ethanol extract of dandelion roots was investigated by RP-HPLC analysis (Figure 1) and chlorogenic and caffeic acids were found to be the most abundant compounds (Table 1). This result is in agreement with a previous study [9] where chlorogenic acid was the most abundant chemical compound in a dandelion root extract, followed by caffeic acid. Also, ferulic, syringic, vanillic, *p*-coumaric, and gallic acids were identified. Overall, the root extract of dandelion can be considered a good natural product with antioxidant activity, immunostimulatory capacity, anticancer, anti-inflammatory, and antimicrobial properties [7,36–39]. On the other hand, dandelion leaves and flowers are more enriched in flavonoids (luteolin and its glycoside derivatives, chrysoeriol) and coumarins (cichoriin and aesculin) but also contain hydroxycinnamic acid derivatives (caffeic, chlorogenic, chicoric, and monocaffeoyltartaric acids) [8,10]. Because of this specific composition, extracts produced using different portions of the plant have different biological effects. For example, a dandelion leaves extract has been extensively used as diuretic [40], that is why in Italy it goes under the name of "piscialetto" (bedwetter). Interestingly, we have in vitro evidence that argue against the putative diuretic effect of our DRE. Accordingly, the British Herbal Medicine Association indicated the leaves as the part of the *Taraxacum* associated with a diuretic action.

DRE is not the first herbal extract tested for an effect on the mechanisms underlying Ca^{2+} homeostasis. We recently showed that a methanolic extract of *A. oleracea* (flowers, leaves, and stems) induced a significant increase in cytosolic Ca^{2+} levels in HEK293 and MCD4 cells. We proved that spilanthol, the main component of the *Acmella* extract, acted like a potent and natural diuretic in mouse by activating a mechanism based on Ca^{2+}-induced inhibition of cytosolic cyclic adenosine monophosphate (cAMP) [1]. More examples are available in the literature about olive leaf extracts [41], *Scrophularia orientalis* [42], and *Echinacea* extracts [43,44]. For this latter extract, it has been shown that different bioactive components of *Echinacea*, such as alkamides (via activation of cannabinoid receptors) [43] and unknown lipophilic fractions (via direct activation of IP3 receptor) [44], are able to induce intracellular Ca^{2+} peaks.

In conclusion, this report highlights the effect of an ethanolic dandelion root extract on Ca^{2+} homeostasis in HEK293 cells. The mechanism proposed, although not completely characterized, indicates Ca^{2+} signaling as the cellular mediator used by the cells to transduce the effects of the exposure to the bioactive components of the *T. officinale* extract. Of note, this important evidence has a number of perspectives. First, researchers know how the mechanisms underlying Ca^{2+} homeostasis work, how to modulate them pharmacologically, and how they impact diverse biological activities. Thus, it will be possible to finely tune *T. officinale* effects on Ca^{2+}-signaling in order to improve the biological outcome. Second, Ca^{2+} signaling intersects the activity of a huge number of other signaling molecules (e.g., cAMP [45]) and/or effectors (e.g., mitogen-activated protein (MAP) kinases [46]). Thus, it is likely that *T. officinale*-induced Ca^{2+} signaling activates other pathways, which have to be brought to light by tissue-specific research projects. Third, a deeper comparative analysis of *Taraxacum* extracts (from roots, leaves, flowers) in terms of their relative composition of bioactive compounds will help to link the presence of specific components to the activation of precise signaling mechanisms. This latter perspective might have an enormous pharmaceutical impact.

4. Materials and Methods

4.1. Plant Extraction

T. officinale was collected in Basilicata region (40°34′50.9″ N 15°49′38.2″ E) in October 2017; the roots were separated from the aerial parts and then dried at room temperature in the dark.

The dried roots (15 g) were extracted by maceration with EtOH 95% at 1:10 w/v (plant material–solvent ratio) under continuous stirring for 3 days, changing the solvent every 24 h. The extract was filtered, and the solvent was removed in a rotary evaporator. The dried crude extract was kept in the dark until use.

4.2. High-Performance Liquid Chromatography (HPLC) Analysis

The re-dissolved crude extract (100 mg/mL) was analyzed on an analytical HPLC–DAD (Shimadzu Corp., Kyoto, Japan) unit using an EC 250/4.6 Nucleodur 100-5 C18ec column (Macherey-nagel, Düren, Germany). The mobile phase involved two solvents: water–formic acid (5%) (A) and acetonitrile (B). The gradient elution method has already been described by our group [47]. All peaks were collected in the range of 200–400 nm, and chromatograms were recorded at 280 nm for hydroxybenzoic acids, 320 nm for hydroxycinnamic acids, and 350 nm for flavonoids. Standards, such as caffeic acid, chlorogenic acid, syringic acid, ferulic acid, vanillic acid, *p*-coumaric acid, and gallic acid, were used (Sigma Aldrich, Milan, Italy). The quantification of phenolic compounds was realized by the comparison between chromatogram recorded absorbances and external calibration standards. All experiments were carried out in triplicate.

4.3. Cell Culture

HEK293 cells were grown in Dulbecco's modified Eagle medium (DMEM) Glutamax supplemented with 10% fetal bovine serum and 1% penicillin/streptomycin (all these products were from Thermo Fisher Scientific, Waltham, MA, USA) and were maintained in a humidified incubator at 37°C in the presence of 5% CO_2/95% air. MCD4 cells, a clone of M-1 cells stably transfected with human-Aquaporin (AQP) 2, were cultured as described elsewhere [48]. Trypsin was used to subculture cells that were used for no longer than ten/twelve passages after thawing. The cells were seeded on glass coverslips (at a density of 60–70%) and used 24/48 h later for imaging evaluations of intracellular Ca^{2+} levels with Fura-2-AM (Thermo Fisher Scientific, Waltham, MA, USA). Alternatively, 24 h after plating, the cells were transfected with ERD1 [19] or the PLC GFP-reporter [20] and used the next day for fluorescence imaging experiments.

4.4. In Vitro Cytotoxic Assay

The cytotoxic effect of the crude DRE was determined by the Calcein-AM viability assay [2]. HEK293 and MCD4 cells were seeded into 96-well black plates for 24 h and thereafter exposed to different concentrations of DRE (750, 400, 200, 100, 50, 10, 5 µg/mL) or vehicle (ethanol 0.6%) as a control, at 37°C for 24 h. Afterward, the medium was removed, and 100 µL of 1µM Calcein-AM (Thermo Fisher Scientific, Waltham, MA, USA) in phosphate buffered saline (PBS) was added for 30 min at 37°C. The fluorescence was detected by FLEX STATION 3 (Molecular Devices, San Jose, CA, USA) plate reader using a blue filter (Ex 490 nm, Em 510–570 nm).

4.5. Intracellular Ca^{2+} Measurements

For cytosolic Ca^{2+} recordings, the cells were seeded on 25 (or 18) mm Ø glass coverslips. HEK293 cells were loaded with 2–4 µM Fura-2-AM (Thermo Fisher Scientific, Waltham, MA, USA) for 30 min at 37°C in DMEM Glutamax, followed by 15 min in an extracellular solution to allow Fura-2 de-esterification. The coverslips with dye-loaded cells were mounted in an open perfusion chamber (a modified version of FCS2 Closed Chamber System, Biopthechs, Butler, PA, USA), and recordings were carried out using an inverted Eclipse TE2000-S microscope (Nikon, Shinagawa, Tokyo, Japan) equipped for single cell fluorescence evaluations and imaging analysis. Each Fura-2-AM loaded sample was illuminated every 5 s through a 40× oil immersion objective (numerical aperture = 1.30) at 340 and 380 nm. The emitted fluorescence was passed through a dichroic mirror, filtered at 510 nm (Omega Optical, Brattleboro, VT, USA), and captured by a cooled CCD CoolSNAP HQ camera (Photometrics,

Tucson, AZ, USA). Additional technical information about the setup used in our imaging facility is available in other publications from our group [34,49,50]. Fluorescence measurements were carried out using the MetaFluor Fluorescence Ratio Imaging Software (Version 7.7.3.0, Molecular Devices, San Jose, CA, USA). The ratio of the fluorescence signal acquired upon excitation at 340 and 380 nm was normalized to the basal fluorescence ratio obtained in the absence of the stimulus (reported as R/R0). Bar graphs show the averaged rate of fluorescence ratio changes normalized to those induced by maximal ATP stimulation, the latter used in each experiment as an internal positive control. The data from at least 30 cells were summarized in a single run and averaged in a plot ± SE; at least three independent experiments were conducted. *n* indicates the number of experiments performed for each protocol, *m* the number of cells analyzed in *n* experiments.

4.6. Measurement of Intraluminal ER Ca^{2+} Levels

The same imaging setup and perfusion apparatus were used for the measurements of Ca^{2+} levels within the endoplasmic reticulum of HEK293 cells transfected with ERD1. Real-time FRET experiments were carried out using the MetaFluor Fluorescence Ratio Imaging Software (Version 7.7.3.0, Molecular Devices, San Jose, CA, USA). FRET from cyan fluorescent protein (CFP) to yellow fluorescent protein (YFP) was evaluated by excitation of CFP (435 nm) and measurement of the fluorescence emitted by YFP. The results are presented as the emission ratio 485/535 nm collected every 5 s in control conditions and after stimulation with the agonists (DRE, ATP, or ionomycin). The emission ratio was normalized to the maximal store emptying induced by 5 μM ionomycin in an extracellular Ca^{2+}-free solution (R/R0). The data from 5 to 10 transfected cells were summarized in a single run and averaged in a plot ± SE, and at least four independent experiments for each agonist were conducted. *n* indicates the number of experiments performed for each protocol, *m* the number of cells analyzed in *n* experiments.

4.7. Fluorescence Imaging of GFP-Based Reporters

The synthesis of intracellular InsP3/PLC activation was evaluated with the GFP-tagged pleckstrin homology domain of PLCδ1 (PH–PLCδ1–EGFP) [20,27,51]. HEK293 cells were transfected transiently with the GFP-based indicator using Lipofectamine 2000 (Thermo Fisher Scientific, Waltham, MA, USA). Fluorescence images of cells expressing the probe were acquired every 5 s using the MetaFluor Fluorescence Ratio Imaging Software (Version 7.7.3.0, Molecular Devices, San Jose, CA, USA) and the setup and the perfusion apparatus described in [52,53]. GFP was excited at 485 nm, and the emission was collected above 530 nm. *n* indicates the number of experiments performed for each protocol, *m* the number of cells analyzed in *n* experiments.

4.8. Solutions and Materials

Most of the chemicals were obtained from Sigma (Sigma-Aldrich, Saint Louis, MO, USA). The experiments were carried out with an extracellular (Ringer's) solution containing (in mmol/L): 140 NaCl, 5 KCl, 1 CaCl$_2$, 1 MgCl$_2$, 5 glucose, 10 HEPES, adjusted to pH 7.40 with NaOH.

Ionomycin and Fura-2-AM were from Thermo Fisher Scientific (Waltham, MA, USA). When dimethyl sulfoxide or ethanol were used as a vehicle, the final solvent concentration was always below 0.01% or 0.1%, respectively.

4.9. Data Analysis

Whenever possible, responses to DRE and ATP (as internal control) were compared in the same cell (paired data), thus eliminating concerns about the variability of the starting Fura-2 ratio. Whenever appropriate, paired data were assessed for statistical significance using the Student's *t* test. The data were expressed as means ± SE with n equal to the number of experimental runs. For Fura-2 and ERD1 ratio imaging experiments, $p < 0.05$ was considered statistically significant.

Acknowledgments: The authors would like to thank Stephen Ferguson, University of Ottawa, Canada for providing the pleckstrin-homology [PH]–PLCδ–GFP construct. The work of Andrea Gerbino and Daniela Russo was supported by funding from the CLUSTER TECNOLOGICO REGIONALE "DICLIMAX" (project # MTJU9H8) to Maria Svelto. This research was also supported by the Master in "Structural Osteopathy" of the University of Basilicata to Monica Carmosino. The founding sponsors had no role in the design of the study, in the collection, analyses, or interpretation of data, in the writing of the manuscript, and in the decision to publish the results. Andrea Gerbino would like to thank Giuseppe Cassano, University of Bari, for his comments and suggestions regarding the Ca^{2+}-imaging experiments.

Author Contributions: Andrea Gerbino conceived and designed the experiments; Andrea Gerbino, Daniela Russo, and Matilde Colella performed the experiments and analyzed the data; Daniela Russo and Luigi Milella collected, extracted, and performed quali–quantitative HPLC analysis. Andrea Gerbino and Monica Carmosino wrote the paper; Giuseppe Procino and Maria Svelto critically and substantively revised the paper; Andrea Gerbino and Monica Carmosino supervised and managed the project.

Conflicts of Interest: The authors declare no conflict of interest.

Abbreviations

DRE	Dandelion root extract
ATP	Adenosine triphosphate
SERCA	Sarco/endoplasmic reticulum Ca^{2+}-ATPase
CPA	Cyclopiazonic acid

References

1. Gerbino, A.; Schena, G.; Milano, S.; Milella, L.; Barbosa, A.F.; Armentano, F.; Procino, G.; Svelto, M.; Carmosino, M. Spilanthol from *Acmella oleracea* lowers the intracellular levels of camp impairing NKCC2 phosphorylation and water channel AQP2 membrane expression in mouse kidney. *PLoS ONE* **2016**, *11*, e0156021. [CrossRef] [PubMed]

2. Armentano, M.F.; Bisaccia, F.; Miglionico, R.; Russo, D.; Nolfi, N.; Carmosino, M.; Andrade, P.B.; Valentão, P.; Diop, M.S.; Milella, L. Antioxidant and proapoptotic activities of *Sclerocarya birrea* [(A. Rich.) hochst.] methanolic root extract on the hepatocellular carcinoma cell line HepG2. *Biomed. Res. Int.* **2015**, *2015*, 561589. [CrossRef] [PubMed]

3. Russo, D.; Miglionico, R.; Carmosino, M.; Bisaccia, F.; Andrade, P.B.; Valentão, P.; Milella, L.; Armentano, M.F. A comparative study on phytochemical profiles and biological activities of *Sclerocarya birrea* (A. Rich.) Hochst leaf and bark extracts. *Int. J. Mol. Sci.* **2018**, *19*, 186. [CrossRef] [PubMed]

4. Costantino, V.; Fattorusso, E.; Imperatore, C.; Mangoni, A. Glycolipids from sponges. Part 17. Clathrosides and isoclathrosides, unique glycolipids from the caribbean sponge *Agelas clathrodes*. *J. Nat. Prod.* **2006**, *69*, 73–78. [CrossRef] [PubMed]

5. Costantino, V.; Fattorusso, E.; Mangoni, A.; Perinu, C.; Teta, R.; Panza, E.; Ianaro, A. Tedarenes A and B: Structural and stereochemical analysis of two new strained cyclic diarylheptanoids from the marine sponge *Tedania ignis*. *J. Org. Chem.* **2012**, *77*, 6377–6383. [CrossRef] [PubMed]

6. Schütz, K.; Carle, R.; Schieber, A. *Taraxacum*—A review on its phytochemical and pharmacological profile. *J. Ethnopharmacol.* **2006**, *107*, 313–323. [CrossRef] [PubMed]

7. Wirngo, F.E.; Lambert, M.N.; Jeppesen, P.B. The physiological effects of dandelion (*Taraxacum officinale*) in type 2 diabetes. *Rev. Diabet. Stud.* **2016**, *13*, 113–131. [CrossRef] [PubMed]

8. Martinez, M.; Poirrier, P.; Chamy, R.; Prüfer, D.; Schulze-Gronover, C.; Jorquera, L.; Ruiz, G. *Taraxacum officinale* and related species—An ethnopharmacological review and its potential as a commercial medicinal plant. *J. Ethnopharmacol.* **2015**, *169*, 244–262. [CrossRef] [PubMed]

9. Kenny, O.; Smyth, T.; Hewage, C.; Brunton, N. Antioxidant properties and quantitative UPLC-MS/MS analysis of phenolic compounds in dandelion (*Taraxacum officinale*) root extracts. *Free Radic. Antioxid.* **2014**, *4*, 7. [CrossRef]

10. Williams, C.A.; Goldstone, F.; Greenham, J. Flavonoids, cinnamic acids and coumarins from the different tissues and medicinal preparations of *Taraxacum officinale*. *Phytochemistry* **1996**, *42*, 121–127. [CrossRef]

11. Ivanov, I. Polyphenols content and antioxidant activities of *Taraxacum officinale* F.H. Wigg (dandelion) leaves. *Int. J. Pharmacogn. Phytochem. Res.* **2014**, *6*, 889–893.

12. Budzianowski, J. Coumarins, caffeoyltartaric acids and their artifactual methyl esters from *Taraxacum officinale* leaves. *Planta Med.* **1997**, *63*, 288. [CrossRef] [PubMed]

13. Chang, H.T.; Chen, I.L.; Chou, C.T.; Liang, W.Z.; Kuo, D.H.; Shieh, P.; Jan, C.R. Effect of caffeic acid on Ca^{2+} homeostasis and apoptosis in SCM1 human gastric cancer cells. *Arch. Toxicol.* **2013**, *87*, 2141–2150. [CrossRef] [PubMed]

14. Nam, J.H.; Shin, D.H.; Zheng, H.; Kang, J.S.; Kim, W.K.; Kim, S.J. Inhibition of store-operated Ca^{2+} entry channels and K^+ channels by caffeic acid phenethylester in T lymphocytes. *Eur. J. Pharmacol.* **2009**, *612*, 153–160. [CrossRef] [PubMed]

15. Bose, J.S.; Gangan, V.; Jain, S.K.; Manna, S.K. Novel caffeic acid ester derivative induces apoptosis by expressing fasl and downregulating NF-KappaB: Potentiation of cell death mediated by chemotherapeutic agents. *J. Cell. Physiol.* **2009**, *218*, 653–662. [CrossRef] [PubMed]

16. Mikami, Y.; Yamazawa, T. Chlorogenic acid, a polyphenol in coffee, protects neurons against glutamate neurotoxicity. *Life Sci.* **2015**, *139*, 69–74. [CrossRef] [PubMed]

17. Jung, H.J.; Im, S.S.; Song, D.K.; Bae, J.H. Effects of chlorogenic acid on intracellular calcium regulation in lysophosphatidylcholine-treated endothelial cells. *BMB Rep.* **2017**, *50*, 323–328. [CrossRef] [PubMed]

18. Lin, T.Y.; Lu, C.W.; Huang, S.K.; Wang, S.J. Ferulic acid suppresses glutamate release through inhibition of voltage-dependent calcium entry in rat cerebrocortical nerve terminals. *J. Med. Food* **2013**, *16*, 112–119. [CrossRef] [PubMed]

19. Palmer, A.E.; Jin, C.; Reed, J.C.; Tsien, R.Y. Bcl-2-mediated alterations in endoplasmic reticulum Ca^{2+} analyzed with an improved genetically encoded fluorescent sensor. *Proc. Natl. Acad. Sci. USA* **2004**, *101*, 17404–17409. [CrossRef] [PubMed]

20. Dale, L.B.; Babwah, A.V.; Bhattacharya, M.; Kelvin, D.J.; Ferguson, S.S. Spatial-temporal patterning of metabotropic glutamate receptor-mediated inositol 1,4,5-triphosphate, calcium, and protein kinase C oscillations: Protein kinase C-dependent receptor phosphorylation is not required. *J. Biol. Chem.* **2001**, *276*, 35900–35908. [CrossRef] [PubMed]

21. Berridge, M.J.; Lipp, P.; Bootman, M.D. The versatility and universality of calcium signalling. *Nat. Rev. Mol. Cell Biol.* **2000**, *1*, 11–21. [CrossRef] [PubMed]

22. Altemimi, A.; Lakhssassi, N.; Baharlouei, A.; Watson, D.G.; Lightfoot, D.A. Phytochemicals: Extraction, isolation, and identification of bioactive compounds from plant extracts. *Plants* **2017**, *6*, 42. [CrossRef] [PubMed]

23. Hofer, A.M.; Fasolato, C.; Pozzan, T. Capacitative Ca^{2+} entry is closely linked to the filling state of internal Ca^{2+} stores: A study using simultaneous measurements of ICRAC and intraluminal $[Ca^{2+}]$. *J. Cell Biol.* **1998**, *140*, 325–334. [CrossRef] [PubMed]

24. Lau, B.W.; Colella, M.; Ruder, W.C.; Ranieri, M.; Curci, S.; Hofer, A.M. Deoxycholic acid activates protein kinase C and phospholipase C via increased Ca^{2+} entry at plasma membrane. *Gastroenterology* **2005**, *128*, 695–707. [CrossRef] [PubMed]

25. Querfurth, H.W.; Haughey, N.J.; Greenway, S.C.; Yacono, P.W.; Golan, D.E.; Geiger, J.D. Expression of ryanodine receptors in human embryonic kidney (HEK293) cells. *Biochem. J.* **1998**, *334*, 79–86. [CrossRef] [PubMed]

26. Mezrag, A.; Malafronte, N.; Bouheroum, M.; Travaglino, C.; Russo, D.; Milella, L.; Severino, L.; De Tommasi, N.; Braca, A.; Dal Piaz, F. Phytochemical and antioxidant activity studies on *Ononis angustissima* L. Aerial parts: Isolation of two new flavonoids. *Nat. Prod. Res.* **2017**, *31*, 507–514. [CrossRef] [PubMed]

27. Okubo, Y.; Kakizawa, S.; Hirose, K.; Iino, M. Visualization of IP_3 dynamics reveals a novel AMPA receptor-triggered IP_3 production pathway mediated by voltage-dependent Ca^{2+} influx in Purkinje cells. *Neuron* **2001**, *32*, 113–122. [CrossRef]

28. Rácz-Kotilla, E.; Rácz, G.; Solomon, A. The action of *Taraxacum officinale* extracts on the body weight and diuresis of laboratory animals. *Planta Med.* **1974**, *26*, 212–217. [CrossRef] [PubMed]

29. Yasukawa, K.; Akihisa, T.; Oinuma, H.; Kasahara, Y.; Kimura, Y.; Yamanouchi, S.; Kumaki, K.; Tamura, T.; Takido, M. Inhibitory effect of Di- and trihydroxy triterpenes from the flowers of compositae on 12-O-tetradecanoylphorbol-13-acetate-induced inflammation in mice. *Biol. Pharm. Bull.* **1996**, *19*, 1329–1331. [CrossRef] [PubMed]

30. Hu, C.; Kitts, D.D. Dandelion (*Taraxacum officinale*) flower extract suppresses both reactive oxygen species and nitric oxide and prevents lipid oxidation in vitro. *Phytomedicine* **2005**, *12*, 588–597. [CrossRef] [PubMed]

31. Takasaki, M.; Konoshima, T.; Tokuda, H.; Masuda, K.; Arai, Y.; Shiojima, K.; Ageta, H. Anti-carcinogenic activity of *Taraxacum* plant. I. *Biol. Pharm. Bull.* **1999**, *22*, 602–605. [CrossRef] [PubMed]

32. Humeau, J.; Bravo-San Pedro, J.M.; Vitale, I.; Nuñez, L.; Villalobos, C.; Kroemer, G.; Senovilla, L. Calcium signaling and cell cycle: Progression or death. *Cell Calcium* **2018**, *70*, 3–15. [CrossRef] [PubMed]

33. Amaya, M.J.; Nathanson, M.H. Calcium signaling and the secretory activity of bile duct epithelia. *Cell Calcium* **2014**, *55*, 317–324. [CrossRef] [PubMed]

34. Procino, G.; Gerbino, A.; Milano, S.; Nicoletti, M.C.; Mastrofrancesco, L.; Carmosino, M.; Svelto, M. Rosiglitazone promotes AQP2 plasma membrane expression in renal cells via a Ca^{2+}-dependent/cAMP-independent mechanism. *Cell. Physiol. Biochem.* **2015**, *35*, 1070–1085. [CrossRef] [PubMed]

35. Zierler, S.; Hampe, S.; Nadolni, W. TRPM channels as potential therapeutic targets against pro-inflammatory diseases. *Cell Calcium* **2017**, *67*, 105–115. [CrossRef] [PubMed]

36. Dekdouk, N.; Malafronte, N.; Russo, D.; Faraone, I.; De Tommasi, N.; Ameddah, S.; Severino, L.; Milella, L. Phenolic compounds from *Olea europaea* L. Possess antioxidant activity and inhibit carbohydrate metabolizing enzymes in vitro. *Evid. Based Complement. Altern. Med.* **2015**, *2015*, 684925. [CrossRef] [PubMed]

37. Ou, S.; Kwok, K.-C. Ferulic acid: Pharmaceutical functions, preparation and applications in foods. *J. Sci. Food Agric.* **2004**, *84*, 1261–1269. [CrossRef]

38. Chong, K.P.; Rossall, S.; Atong, M. In vitro antimicrobial activity and fungitoxicity of syringic acid, caffeic acid and 4-hydroxybenzoic acid against *Ganoderma boninense*. *J. Agric. Sci.* **2009**, *1*, 6. [CrossRef]

39. Dos Santos, M.D.; Almeida, M.C.; Lopes, N.P.; de Souza, G.E. Evaluation of the anti-inflammatory, analgesic and antipyretic activities of the natural polyphenol chlorogenic acid. *Biol. Pharm. Bull.* **2006**, *29*, 2236–2240. [CrossRef] [PubMed]

40. Clare, B.A.; Conroy, R.S.; Spelman, K. The diuretic effect in human subjects of an extract of *Taraxacum officinale* folium over a single day. *J. Altern. Complement. Med.* **2009**, *15*, 929–934. [CrossRef] [PubMed]

41. Marchetti, C.; Clericuzio, M.; Borghesi, B.; Cornara, L.; Ribulla, S.; Gosetti, F.; Marengo, E.; Burlando, B. Oleuropein-enriched olive leaf extract affects calcium dynamics and impairs viability of malignant mesothelioma cells. *Evid. Based Complement. Altern. Med.* **2015**, *2015*, 908493. [CrossRef] [PubMed]

42. Lange, I.; Moschny, J.; Tamanyan, K.; Khutsishvili, M.; Atha, D.; Borris, R.P.; Koomoa, D.L. *Scrophularia orientalis* extract induces calcium signaling and apoptosis in neuroblastoma cells. *Int. J. Oncol.* **2016**, *48*, 1608–1616. [CrossRef] [PubMed]

43. Raduner, S.; Majewska, A.; Chen, J.Z.; Xie, X.Q.; Hamon, J.; Faller, B.; Altmann, K.H.; Gertsch, J. Alkylamides from *Echinacea* are a new class of cannabinomimetics. Cannabinoid type 2 receptor-dependent and -independent immunomodulatory effects. *J. Biol. Chem.* **2006**, *281*, 14192–14206. [CrossRef] [PubMed]

44. Wu, L.; Rowe, E.W.; Jeftinija, K.; Jeftinija, S.; Rizshsky, L.; Nikolau, B.J.; McKay, J.; Kohut, M.; Wurtele, E.S. Echinacea-induced cytosolic Ca^{2+} elevation in HEK293. *BMC Complement. Altern. Med.* **2010**, *10*, 72. [CrossRef] [PubMed]

45. Gerbino, A.; Ruder, W.C.; Curci, S.; Pozzan, T.; Zaccolo, M.; Hofer, A.M. Termination of cAMP signals by Ca^{2+} and $G\alpha_i$ via extracellular Ca^{2+} sensors: A link to intracellular Ca^{2+} oscillations. *J. Cell Biol.* **2005**, *171*, 303–312. [CrossRef] [PubMed]

46. Colella, M.; Gerbino, A.; Hofer, A.M.; Curci, S. Recent advances in understanding the extracellular calcium-sensing receptor. *F1000Res* **2016**, *5*. [CrossRef] [PubMed]

47. Russo, D.; Valentão, P.; Andrade, P.B.; Fernandez, E.C.; Milella, L. Evaluation of antioxidant, antidiabetic and anticholinesterase activities of *Smallanthus sonchifolius* landraces and correlation with their phytochemical profiles. *Int. J. Mol. Sci.* **2015**, *16*, 17696–17718. [CrossRef] [PubMed]

48. Procino, G.; Barbieri, C.; Tamma, G.; De Benedictis, L.; Pessin, J.E.; Svelto, M.; Valenti, G. AQP2 exocytosis in the renal collecting duct—Involvement of SNARE isoforms and the regulatory role of Munc18b. *J. Cell Sci.* **2008**, *121*, 2097–2106. [CrossRef] [PubMed]

49. Carmosino, M.; Gerbino, A.; Hendy, G.N.; Torretta, S.; Rizzo, F.; Debellis, L.; Procino, G.; Svelto, M. NKCC2 activity is inhibited by the bartter's syndrome type 5 gain-of-function CaR-A843E mutant in renal cells. *Biol. Cell* **2015**, *107*, 98–110. [CrossRef] [PubMed]

50. Carmosino, M.; Gerbino, A.; Schena, G.; Procino, G.; Miglionico, R.; Forleo, C.; Favale, S.; Svelto, M. The expression of Lamin A mutant R321X leads to endoplasmic reticulum stress with aberrant Ca^{2+}. *J. Cell. Mol. Med.* **2016**, *20*, 2194–2207. [CrossRef] [PubMed]

51. Stauffer, T.P.; Ahn, S.; Meyer, T. Receptor-induced transient reduction in plasma membrane Ptdins(4,5)P_2 concentration monitored in living cells. *Curr. Biol.* **1998**, *8*, 343–346. [CrossRef]

52. Gerbino, A.; Maiellaro, I.; Carmone, C.; Caroppo, R.; Debellis, L.; Barile, M.; Busco, G.; Colella, M. Glucose increases extracellular [Ca^{2+}] in rat insulinoma (INS-1E) pseudoislets as measured with Ca^{2+}-sensitive microelectrodes. *Cell Calcium* **2012**, *51*, 393–401. [CrossRef] [PubMed]

53. Cardone, A.; Lopez, F.; Affortunato, F.; Busco, G.; Hofer, A.M.; Mallamaci, R.; Martinelli, C.; Colella, M.; Farinola, G.M. An aryleneethynylene fluorophore for cell membrane staining. *Biochim. Biophys. Acta* **2012**, *1818*, 2808–2817. [CrossRef] [PubMed]

International Journal of
Molecular Sciences

MDPI

Article

Opposing Roles of Calcium and Intracellular ATP on Gating of the Purinergic P2X2 Receptor Channel

Milos B. Rokic [1], Patricio Castro [2,3], Elias Leiva-Salcedo [1,4,5], Melanija Tomic [1,†], Stanko S. Stojilkovic [1] and Claudio Coddou [1,2,*]

[1] Section on Cellular Signaling, National Institutes of Child Health and Human Development, NIH, Bethesda, MD 20892, USA; milos.rokic@nih.gov (M.B.R.); eliasleiva@gmail.com (E.L.-S.); stojilks@mail.nih.gov (S.S.S.)

[2] Departamento de Ciencias Biomédicas, Facultad de Medicina, Universidad Católica del Norte, Coquimbo 1781421, Chile; pacastrom@gmail.com

[3] Laboratory of Developmental Physiology, Department of Physiology, Faculty of Biological Sciences, Universidad de Concepción, Concepción 4030000, Chile

[4] Departamento de Biología, Facultad de Química y Biología, Universidad de Santiago de Chile, Santiago 9170022, Chile

[5] Centro para el Desarrollo de Nanociencias y Nanotecnología (CEDENNA), Santiago 9170022, Chile

* Correspondence: ccoddou@ucn.cl; Tel.: +56-512205937

† Deceased on March 2017.

Received: 6 March 2018; Accepted: 3 April 2018; Published: 11 April 2018

Abstract: P2X2 receptors (P2X2R) exhibit a slow desensitization during the initial ATP application and a progressive, calcium-dependent increase in rates of desensitization during repetitive stimulation. This pattern is observed in whole-cell recordings from cells expressing recombinant and native P2X2R. However, desensitization is not observed in perforated-patched cells and in two-electrode voltage clamped oocytes. Addition of ATP, but not ATPγS or GTP, in the pipette solution also abolishes progressive desensitization, whereas intracellular injection of apyrase facilitates receptor desensitization. Experiments with injection of alkaline phosphatase or addition of staurosporine and ATP in the intracellular solution suggest a role for a phosphorylation-dephosphorylation in receptor desensitization. Mutation of residues that are potential phosphorylation sites identified a critical role of the S363 residue in the intracellular ATP action. These findings indicate that intracellular calcium and ATP have opposing effects on P2X2R gating: calcium allosterically facilitates receptor desensitization and ATP covalently prevents the action of calcium. Single cell measurements further revealed that intracellular calcium stays elevated after washout in P2X2R-expressing cells and the blockade of mitochondrial sodium/calcium exchanger lowers calcium concentrations during washout periods to basal levels, suggesting a role of mitochondria in this process. Therefore, the metabolic state of the cell can influence P2X2R gating.

Keywords: purinergic receptor channels; desensitization; ATP; calcium; allosteric; covalent

1. Introduction

Extracellular ATP-gated P2X receptor channels (P2XRs) are composed of three homologous or heterologous subunits; in mammals seven subunits, designated P2X1-7, have been identified. The gating of these channels is very complex and usually consists of three phases: a rapid rise in inward current caused by binding of ATP, termed activation; a progressively developing decay of the current in the continued presence of ATP, termed desensitization; and a relatively rapid decay in the current after washout of the ligand, termed deactivation [1]. Desensitization kinetics is receptor-specific, from the fast desensitizing P2X1R and P2X3R to the slow desensitizing P2X2R and P2X4R, and even the non-desensitizing P2X7R, which exhibits increased currents upon sustained activation [2].

Numerous studies were focused on structural elements accounting for receptor-specific desensitization pattern. Briefly, the cysteine-rich head domain located above the ATP-binding cleft of P2X1R [3] and the negatively charged aspartate residues located below the ATP-binding site of P2X3R [4] play important roles in the transition from the open to the desensitized state. The transmembrane tyrosine residues stabilize the desensitized state [5]. Interactions between the ectodomain and regions around the transmembrane domains have also been implicated as being relevant in receptor desensitization [6]. It is also well established that the N- and C-terminal domains provide the structural elements involved in controlling the kinetics of receptor desensitization. These studies included work with splice forms of P2X2Rs [7–9], and mutagenesis studies focused on identification of the residues that account for those effects [10,11]. Experiments with chimeric P2X2R and P2X1R also suggested a role of intracellular N-terminus in receptor desensitization [12]. In particular, the most attention has been devoted to the conserved threonine and tyrosine residues located in the N-terminus [13–18].

P2XRs are highly regulated channels and can respond to changes in their extracellular and intracellular environment, which further contributes to the complexity in their gating properties [2]. This includes the roles of bath and intracellular calcium in receptor gating; its main effect is observed in a millimolar concentration range causing inhibition of agonist-induced currents [19], but it has also been suggested that Ca^{2+} facilitates recovery from desensitization of P2X3Rs [20]. Recently, we identified a novel aspect of P2X2R desensitization, termed use-dependent desensitization (UDD), which is manifested by a progressive increase in rates of P2X2R desensitization during repetitive application of ATP [21]. This process was observed in whole-cell recordings from HEK293 cells expressing both splice forms of the rat, mouse, and human receptors, termed P2X2aR and P2X2bR [22]. We also found that this process depends on calcium influx and that domain calcium is sufficient to establish UDD in the whole-cell recording [22]. However, the fact that nanomolar calcium concentrations are sufficient to develop UDD strongly suggest that other molecules present in the intracellular environment are also important to counteract calcium effects and therefore to regulate P2X2R gating.

Here, we utilized electrophysiology and mutagenesis to test the relevance of intracellular ATP and calcium in P2XRs gating, an issue that has been previously addressed for other plasma membrane channels [23–26], but not for P2XRs, that are gated by extracellular ATP. We found that P2X2R gating is dependent on both intracellular ATP and calcium, having these ligands opposite effects in P2X2R desensitization, with calcium accelerating and ATP impeding desensitization. The action of intracellular calcium is allosteric, whereas intracellular ATP acts covalently, as the phosphate source for receptor phosphorylation. We also identified the relevance of the C-terminal S363 residue for intracellular ATP regulation. These data suggest that P2X2Rs represent an important functional crosslink between extracellular and intracellular ATP and calcium signaling and provide integration of cellular responses to these ligands.

2. Results

2.1. The P2X2R Current Response Pattern Depends on Recording Conditions

We recently reported that P2X2R undergoes a progressive calcium-dependent desensitization during repetitive ATP application, a phenomenon termed UDD. This process was observed in whole-cell recordings from HEK293 cells expressing both splice forms P2X2aR and P2X2bR of the rat, mouse, and human receptors [22]. Figure 1A shows an example of UDD using rat P2X2aR. The cells were stimulated four times for 40 s with 100 µM ATP followed by 4 min washout periods in calcium-containing (2.5 mM) or calcium-deficient (0.09 mM) medium. For the cells bathed in calcium-containing medium (top), but not in calcium-deficient medium (bottom), there was pronounced receptor desensitization during the first ATP application and a further increase in the rate of receptor desensitization with each subsequent ATP application. In those experiments, the peak amplitude of currents using calcium-containing or calcium-deficient medium were not significantly different (5.4 ± 0.3 μA vs. 5.0 ± 0.7 nA; *n* = 8–11). UDD was independent of the method of transfection; it was observed in both the polymer-based and lipid-based transfection experiments.

However, when we used perforated patch-clamp recording in HEK293 cells, no difference in the rate of receptor desensitization was observed during five consecutive agonist applications (Figure 1B) in cells bathed in calcium-containing (top) or calcium-deficient (bottom) medium. The lack of the ability of repetitive ATP application to generate UDD in calcium-containing medium was also observed in two-electrode voltage clamp TEVC recording in P2X2aR-expressing *Xenopus* oocytes. Figure 1C shows that the profiles of the currents induced by four consecutive ATP applications in oocytes bathed in calcium-containing (top) and calcium-deficient (bottom) medium were comparable. These results indicated that calcium influx affects the rate of P2X2aR desensitization in the whole-cell patch clamp configuration, but not in the recording configurations that preserve the intracellular cytosol.

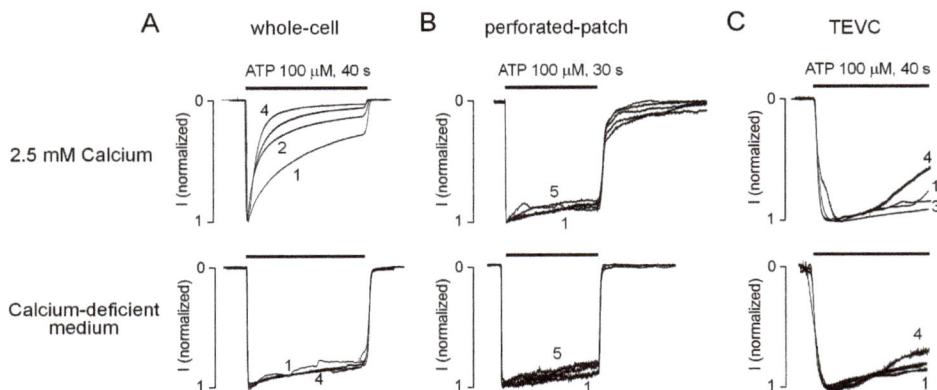

Figure 1. Dependence on recording conditions of the pattern of rat P2X2a receptor (P2X2aR) desensitization to repetitive ATP application. (**A**) Whole-cell recording in HEK293 cells. Traces of currents induced by repetitive application of 100 μM ATP for 40 s with 4 min washout periods for cells bathed in calcium-containing medium (2.5 mM; top traces) or in calcium-deficient medium (0.09 mM calcium; bottom traces) at a holding potential of −60 mV are shown. The progressive increase in the rates of receptor desensitization is termed use-dependent desensitization (UDD); (**B**) Perforated patch-clamp recording in HEK293 cells. Notice that there is no difference in the rate of receptor desensitization during repetitive agonist application (100 μM ATP for 30 s with 4 min washing periods) in cells bathed in calcium-containing (top) or calcium-deficient (bottom) medium; (**C**) Two-electrode voltage clamp (TEVC) recording in *Xenopus* oocytes. The currents induced by four consecutive ATP applications in oocytes bathed in calcium-containing (top) or calcium-deficient (bottom) medium are shown. In (**A**,**C**), the data shown are normalized representative recordings. In all panels, traces shown are representative of at least five similar experiments. In this and the following figures, horizontal black bars indicate duration of ATP application, and the numbers indicate the order of each ATP application.

The experiments shown in Figure 1A, top panels, were performed using an intrapipette solution containing 142 mM NaCl, 10 mM EGTA, and 10 mM HEPES and extracellular medium containing 142 mM NaCl, 3 mM KCl, 1 mM MgCl, 2.5 mM CaCl$_2$, 10 mM glucose, and 10 mM HEPES. To examine whether the equimolar concentrations of Na$^+$ in extracellular and intracellular media specifically account for development of UDD, in additional experiments, KCl (140 mM) or CsCl (154 mM) was substituted for the intracellular NaCl. Figure 2A shows that UDD also developed in the presence of the K$^+$- and Cs$^+$-containing intracellular solutions. Furthermore, Figure 2B shows that the rates of receptor desensitization were quantitatively comparable in all three experimental conditions. Similar to the results of the experiments using the Na$^+$-containing intrapipette solution (Figure 1A, bottom panel). Thus, Ca^{2+} influx is critical for development of UDD independent of the nature of the major intrapipette cation and independent of the expression system and/or gene transfection technique.

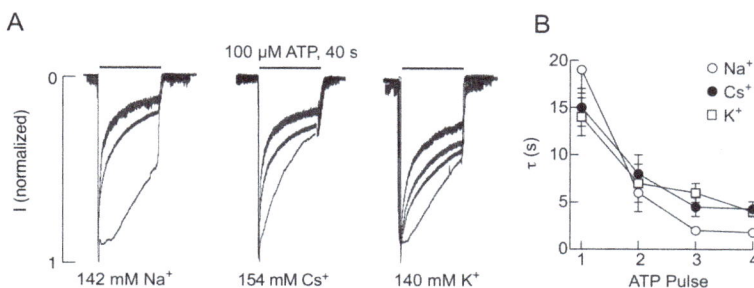

Figure 2. UDD is present and comparable in various intracellular ionic environments. (**A**) The traces shown are representative of whole-cell recordings in HEK293 cells with the intrapipette medium containing sodium (left), cesium (middle), and potassium (right) chloride. In all cases, 2.5 mM calcium was present in the extracellular medium, and the holding potential was −60 mV; (**B**) Mean ± SEM values of the rate of receptor desensitization (τ) derived from a monoexponential fit for sodium, cesium, and potassium-based intracellular solutions (*n* = 10–14 per group).

2.2. Endogenous P2X2R-Mediated Currents also Exhibit UDD

The experiments of UDD shown so far are all performed in heterologous systems that overexpress the recombinant rat P2X2R. We next tested if this phenomenon is also observed in cells that endogenously express the P2X2R. For that purpose we used undifferentiated PC12 cells, the source of the first rat P2X2R cDNA cloned [27]. In whole-cell recordings, these cells exhibited robust, slow-desensitizing ATP currents, resembling the P2X2R desensitization profile (Figure 3A), with an average amplitude of 302 ± 43 pA, that is more than 10-fold smaller that the amplitudes obtained in P2X2R-expressing HEK293 cells. The use of an intracellular solution with 10 mM EGTA resulted in slow desensitizing currents with no further increases in receptor desensitization (Figure 3A), but when EGTA was lowered to 0.05 mM EGTA the endogenous currents immediately started to develop UDD (Figure 3B). This observation was confirmed when we analyzed the changes in desensitization rates in both conditions (Figure 3C,D). The variations on intracellular EGTA did not affect the current densities obtained in each condition (Figure 3E).

2.3. Intracellularly Applied ATP Abolishes UDD

Performing whole-cell recording is associated with dilution of numerous intracellular factors, including ATP. In order to test the potential role of intracellular ATP on P2X2R gating, the whole-cell recording was done in HEK293 cells with the potassium-containing intrapipette solution without and with 5 mM ATP. In contrast to the control conditions, which showed typical UDD, there was a loss of the receptor UDD in the cells filled with 5 mM ATP-containing medium (Figure 4). In the presence of ATP, there was no difference in the pattern of currents in the cells bathed in calcium-containing and calcium–deficient medium, further confirming that calcium-dependent UDD does not occur in cells filled with 5 mM ATP.

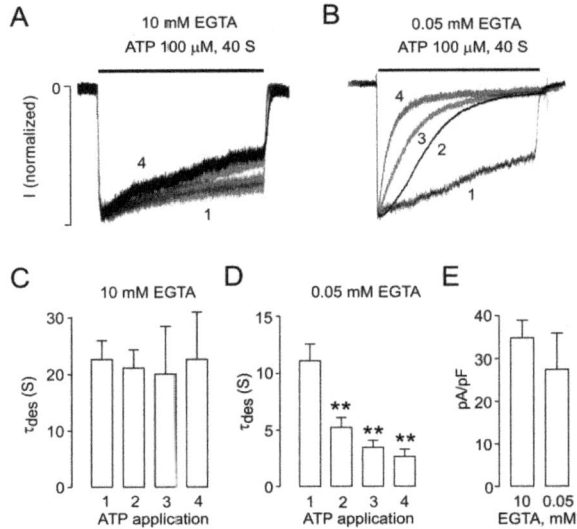

Figure 3. Use-dependent desensitization on PC12 cells. (**A**) Representative tracings of a single PC12 cell challenged with four successive applications of 100 μM ATP and the intracellular solution containing 10 mM EGTA; (**B**) The same experiment in other PC12 cells in which the intracellular solution includes 0.05 mM EGTA; UDD is developed under such conditions; (**C**,**D**) Summary of the desensitization constants in four consecutive ATP applications in cells clamped with an intracellular solution containing 10 mM (**C**) or 0.05 mM (**D**) EGTA; (**E**) Current densities for both conditions. ** $p < 0.01$, test. Mann-Whitney test. $n = 4$–8.

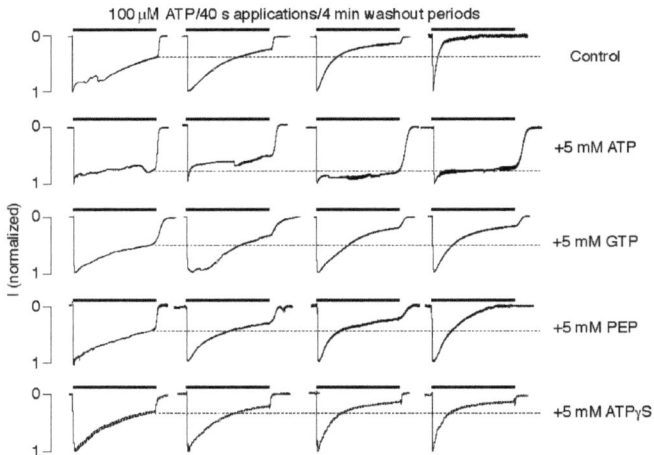

Figure 4. Intracellular ATP, but not GTP, phosphoenolpyruvate (PEP), or ATPγS, blocks receptor desensitization in the whole-cell recordings performed in HEK293 cells expressing P2X2aR. The cells were dialyzed with ATP, GTP, PEP, or ATPγS for 7 min before the start of the extracellular ATP application, and the holding potential was −60 mV. The horizontal dotted lines illustrate the peak in the current response at the end of the first ATP application. Notice that UDD is preserved in all cases except for the cells filled with 5 mM ATP.

We also examined whether GTP, phosphoenolpyruvate, or ATPγS could substitute for ATP in blocking the calcium-dependent desensitization of P2X2R in the whole-cell recording. GTP is a source of energy and an activator of substrates in metabolic reactions, similar to ATP, but is more specific due to the difference in the purine ring structure. However, GTP could not substitute for ATP in blocking UDD (Figure 4). The phosphoenolpyruvate anion has the phosphate bond with the highest free energy found in living organisms and is involved in glycolysis and gluconeogenesis but was unable to block development of UDD (Figure 4). ATPγS is a non-hydrolyzable ATP analog that has agonist activity, i.e., can substitute for ATP in the activation of P2XRs, but this analog could not mimic the regulatory role of ATP in the channel gating (Figure 4). Together, these results indicate that ATP does not affect P2X2R gating directly by binding to an intracellular binding site but, presumably, by the ATP-dependent and kinase-mediated phosphorylation of the channels.

To address this hypothesis, we examined the effects of intracellular ATP on the pattern of current signaling in the absence and presence of 10 μM staurosporine, a prototypical ATP-competitive kinase inhibitor that binds to many kinases with high affinity but very little selectivity [28]. In contrast to the results for ATP alone, in the presence of ATP and staurosporine, there was pronounced receptor desensitization, but UDD was lost (Figure 5), which supported the hypothesis. Intracellular calcium also affects the status of phosphoinositides through various pathways [29], and these compounds also contribute to the control of P2XR gating [30–34]. We assessed the potential role of phosphoinositides in P2X2R desensitization in the absence of intracellular ATP by adding the phosphoinositides PI(4,5)P, PI(3,5)P (Figure 6A,B) to the intracellular solution. Additionally, we inhibited PI3K and PIK4K with wortmannin (wort, Figure 6C). The potential role of protein phosphatases in the absence of intracellular ATP was also tested by adding the PP2B and calcineurin inhibitors, okadaic acid and cyclosporin A, respectively (Figure 6D,E). In all conditions tested, P2X2R-mediated currents preserved the Ca^{2+}-induced increase in receptor desensitization. As expected, the effects of staurosporine shown in Figure 5 were not observed in experiments lacking intracellular ATP (Figure 6F).

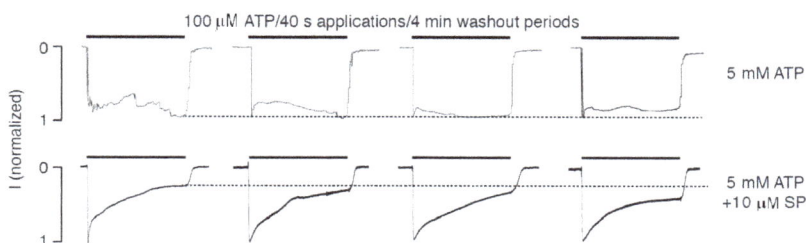

Figure 5. Potential role of protein kinases in the intracellular ATP effects on receptor desensitization; whole-cell recordings in HEK293 cells expressing P2X2aR at a holding potential of −60 mV. (**Top panels**) The loss of receptor desensitization in cells filled with 5 mM ATP-containing medium; (**Bottom panels**) Restoration of receptor desensitization in cells filled with a medium containing 5 mM ATP plus 10 μM staurosporine (SP). Notice the lack of UDD in the presence of SP. In both experiments, the cells were dialyzed for 7 min before the extracellular ATP was applied. The horizontal dotted lines indicate the peak amplitude of the current at the end of the first pulse of extracellular ATP.

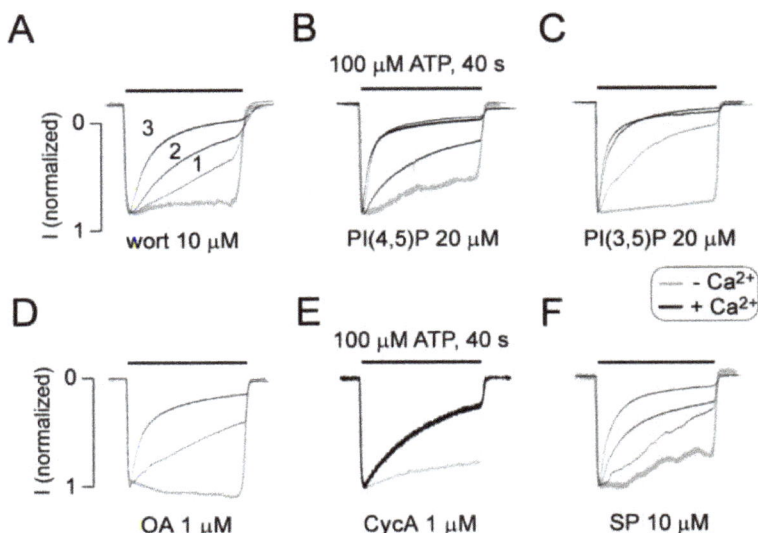

Figure 6. Independence of UDD of the status of kinases, phosphatases, and phosphoinositides in the absence of intracellular ATP. The results show whole-cell recordings from P2X2aR-expressing HEK293 cells clamped at −60 mV and containing kinase and phosphatase inhibitors or activators in the absence (gray traces) or in the presence (black traces) of 2 mM extracellular calcium. (**A–F**) The lack of effects of wortmannin (WT), an inhibitor of PI_3 and PI_4 kinases (**A**), the phosphoinositides $PI(4,5)P_2$ (**B**) and $PI(3,5)P_2$ (**C**), the phosphatase PP2B inhibitor okadaic acid (OA) (**D**), the calcineurin inhibitor cyclosporin A (CycA) (**E**) and staurosporine (SP), a nonselective kinase inhibitor at the concentration used (**F**), on the development of UDD during repetitive agonist application. The cells were dialyzed for 7 min before the start of the experiment, and the washout periods between ATP applications were 4 min each. The recordings shown are representative of at least 3 different experiments.

Next, we tested different intracellular ATP concentrations to elucidate the range at which ATP can prevent UDD. The experiments were performed in both HEK293 cells overexpressing the P2X2R and in PC12 that endogenously express this receptor. In both systems 5 mM of intracellular ATP completely prevented the develop of UDD, 4 mM was able to partially block UDD and 3 mM ATP induced a modest UDD, as compared to 1 mM ATP (Figure 7). These results suggest that there are differences in the gating of the P2X2R in the 3–5 mM range, suggesting that this receptor can be regulated by changes in cell metabolism.

In further experiments, we examined the dependence of the rates of receptor desensitization on the status of the intracellular ATP concentration using TEVC recording in oocytes. To manipulate the intracellular levels of ATP, the oocytes were injected with 2.5 U/µL apyrase, a calcium-activated enzyme that catalyzes the hydrolysis of ATP to yield AMP and inorganic phosphate [35]. Figure 8A shows that injection of this enzyme into the oocytes significantly increased the level of receptor desensitization. Injection of 1 U/µL alkaline phosphatase, a hydrolytic enzyme responsible for removing phosphate groups from many types of molecules [36], also facilitated the desensitization of receptors (Figure 8B). Thus, in intact cells, manipulation of the intracellular ATP levels and the status of kinase/phosphatase activity also influenced the P2X2R desensitization.

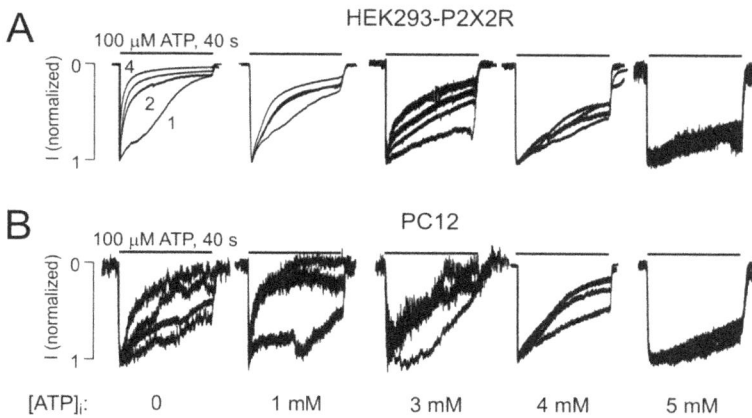

Figure 7. Intracellular ATP concentration-response experiments. (**A,B**) Traces of currents induced by repetitive application of 100 μM ATP for 40 s with 4 min washout periods for cells bathed in calcium-containing medium with the intracellular solution containing 0, 1, 3, 4 or 5 mM ATP in HEK293 cells transfected with the P2X2R cDNA (**A**) or in PC12 cells (**B**) that endogenously express the P2X2R. Recordings were performed in at least three different cells for each condition.

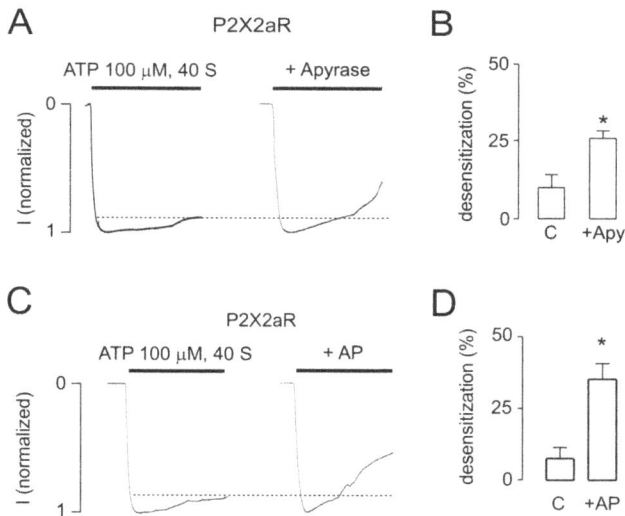

Figure 8. Dependence of the rates of receptor desensitization on the status of intracellular ATP concentration and constitutive phosphorylation of channels. The experiments were performed in oocytes expressing P2X2aR. (**A,C**) Representative recordings from oocytes expressing P2X2aR either uninjected (left panels) or injected with 2.5 U/μL apyrase, an ATPase (**A**, right), or 1 U/μL alkaline phosphatase (AP), a hydrolase enzyme responsible for removing phosphate groups from many types of molecules (**C**, right). In both cases, the currents were generated by 100 μM ATP; (**B,D**) The mean ± SEM values of the percentage of desensitization after 40 s of ATP application in uninjected cells or those cells injected with apyrase (**B**, $n = 8$) or AP (**D**, $n = 4$); * $p < 0.01$ compared to control columns.

2.4. The S363 Residue is Critical for the Intracellular ATP Effects on the Channel Gating

The intracellular N- and C-terminal domains of P2X2R contain several residues that could account for constitutive and/or regulated phosphorylation. We utilized the NetPhosK 1.0 Server to predict the residues of P2X2R that could potentially be phosphorylated. Those residues are as follows: Y16, T18, and T372 by MAP kinases; T354, Y362, and S363 by protein kinase C; S377 and S378 by protein kinases C and A; and Y398 by protein kinase A. Because T372, S377, S378, and Y398 are all located on the region of the protein that is absent in the splice variant P2X2bR, and this splice variant also shows UDD and calcium-dependent desensitization [22] (Table 1), we excluded these residues as potential targets for UDD-related phosphorylation and/or dephosphorylation. We then generated mutants of all remaining residues to examine their potential contribution to the development of UDD. We also mutated the D15 and E17 residues because of their potential role in directly coordinating calcium. Finally, we also generated a truncated receptor in which the C-terminal Val^{439}-Leu^{472} sequence, a sequence that is absent in the fast-desensitizing variant expressed in mouse P2X2eR [22], was deleted. All of the mutants were expressed in HEK293 cells, and the patterns of extracellular ATP-induced currents were examined using the whole-cell recording mode with the potassium-containing intracellular solution. Table 1 summarizes the results obtained with these mutants and with the wild-type P2X2bR.

Table 1. Rates of desensitization and maximal currents after repetitive stimulation of the P2X2aR, P2X2bR and mutant P2X2aRs by 100 μM ATP for 40 s followed by 4-min washout periods. The data are presented as the mean ± standard error values, with n of 3–17 per receptor. The P2X2bR is a splice variant lacking the Val^{370}–Gln^{438} sequence in the C terminal domain. ΔCR, Truncated receptor with the C-terminal sequence Val^{444}–Leu^{472} deleted; n.d. not defined; τ_1–τ_4, rates of receptor desensitization and I$_1$–I$_4$, the peak amplitude of current responses during the four ATP applications. All results are expressed as the mean ± SEM. Nonparametric Mann Whitney-test ($p < 0.01$) was utilized to compare the wild type and mutant receptor amplitude responses and desensitization times. Asterisk (*) denotes significant differences.

Receptor	τ_1 (s)	τ_2 (s)	τ_3 (s)	τ_4 (s)	I$_1$ (nA)	I$_2$ (nA)	I$_3$ (nA)	I$_4$ (nA)
P2X2a	14.6 ± 1.7	7.4 ± 1.8	6.0 ± 1.6	4.3 ± 0.1	2.6 ± 0.5	2.1 ± 0.3	1.4 ± 0.3	0.7 ± 0.1
D15N	14.1 ± 3.2	17.5 ± 1.4 *	6.8 ± 2.5	2.8 ± 0.5	1.4 ± 0.3 *	1.2 ± 0.1 *	0.8 ± 0.1 *	0.56 ± 0.04
Y16D	13.2 ± 2.6	3.6 ± 1.04 *	1.7 ± 0.1 *	1.2 ± 0.4 *	2.1 ± 0.3	0.4 ± 0.1 *	0.2 ± 0.1 *	0.2 ± 0.1 *
E17A	15.2 ± 2.4	5.3 ± 3.0	1.9 ± 0.1 *	n.d.	9.1 ± 0.6	5.7 ± 1.4 *	2.9 ± 0.2 *	n.d.
T18V	0.7 ± 0.1 *	n.d.	n.d.	n.d.	0.2 ± 0.1 *	n.d.	n.d.	n.d.
T354A	12.03 ± 4.6	10.5 ± 1.3	9.7 ± 2.2 *	5.9 ± 1.9	2.2 ± 0.4	0.9 ± 0.3 *	0.8 ± 0.2 *	0.4 ± 0.1 *
Y362A	4.6 ± 1.2	3.9 ± 0.8 *	2.6 ± 0.5 *	0.9 ± 0.4 *	2.1 ± 0.3	0.9 ± 0.1 *	0.6 ± 0.1 *	0.1 ± 0.03 *
S363A	8.4 ± 2.6 *	5.3 ± 1.1	5.1 ± 1.2	4.5 ± 0.9	3.8 ± 0.8	2.8 ± 0.3	2.2 ± 0.6 *	2.4 ± 0.4 *
S363C	3.8 ± 0.5 *	3.1 ± 0.6 *	3.9 ± 0.6	4.8 ± 0.8	1.4 ± 0.3 *	0.4 ± 0.01 *	0.2 ± 0.02 *	0.1 ± 0.01 *
S363K	8.1 ± 2.2 *	7.2 ± 1.8	7.0 ± 1.6	6.9 ± 2.2	1.9 ± 0.6	1.0 ± 0.3 *	0.8 ± 0.3	0.6 ± 0.2
S363D	5.2 ± 1.1 *	4.2 ± 0.9 *	3.4 ± 0.8 *	2.3 ± 0.5	1.9 ± 0.2	1.0 ± 0.2 *	0.9 ± 0.2	0.6 ± 0.2
S363G	7.6 ± 0.9 *	9.6 ± 1.7	9.5 ± 2.9 *	8.2 ± 1.3 *	1.2 ± 0.3 *	1.0 ± 0.2 *	0.7 ± 0.1 *	0.5 ± 0.1
S363Y	8.2 ± 2.8 *	8.7 ± 2.8	6.9 ± 1.7	5.3 ± 1.4	0.7 ± 0.2 *	0.5 ± 0.2 *	0.3 ± 0.1 *	0.2 ± 0.1 *
P2X2b	6.8 ± 1.1	3.5 ± 1.3	2.6 ± 0.5	n.d.	4.8 ± 0.6	4.1 ± 0.5	3.3 ± 0.8	n.d.
ΔCR	22.3 ± 6.9 *	14.5 ± 5.4 *	8.3 ± 0.5	1.5 ± 0.4 *	1.9 ± 0.4	1.5 ± 0.4	0.5 ± 0.1	0.17 ± 0.04

Substitution of the T18 residue with valine generated a channel that was practically nonfunctional, whereas all other mutants were functional. Several mutants showed significant variations in receptor function compared to the wild type receptor, but UDD was clearly visible in most of the mutants tested, which argued against their roles in intracellular ATP-dependent receptor gating (Table 1). The only exception was the S363 mutant. So next, we studied this residue in more detail.

We generated five additional S363 mutants: S363C, S363D, S363G, S363K, and S363Y. These mutants in the naïve state showed variable rates of receptor desensitization during the 40 s application of extracellular ATP, but none of them responded with a clearly observable UDD (Table 1, Figure 9). In contrast to the wild-type receptor, addition of intracellular ATP had no influence on UDD. Figure 9 shows the profiles of the current during repetitive 40 s applications of extracellular ATP in the cells expressing S363D, S363K, and S363Y mutants in the absence (left panels) and presence (right panels) of 5 mM intrapipette ATP. The S363D mutant desensitized rapidly in the naïve state, whereas the two other mutants desensitized more slowly, and the rates of receptor desensitization were highly comparable in the presence or absence of intracellular ATP. These results indicate a critical role of the S363 residue in intracellular ATP-dependent regulation of P2X2 and confirm its relevance in UDD.

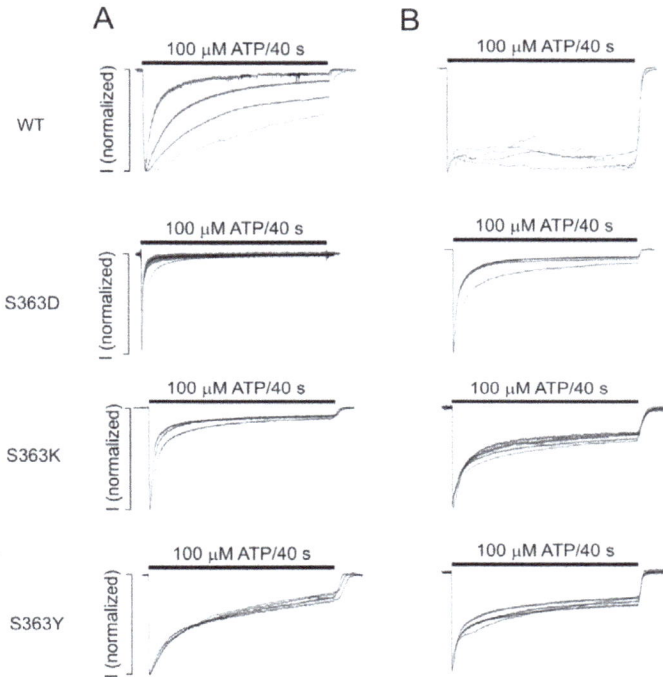

Figure 9. The patterns of extracellular ATP-induced current responses in cells expressing the wild type (WT) and the S363D, S363K, and S363Y mutant receptors of P2X2aR, in the absence (**A**) or in the presence of 5 mM intracellular ATP (**B**). The experiments were performed in HEK393 cells expressing P2X2R at a holding potential of −60 mV. Notice the loss of UDD in all S363-P2X2aR mutants in the absence and presence of intracellular ATP.

2.5. Role of Mitochondria in P2X2R Desensitization

It is well established that mitochondria accumulate cytosolic Ca^{2+} at elevated (above 300 nM) concentrations and release it slowly when $[Ca^{2+}]_i$ drops below this value [37]. Therefore, we further tested the hypothesis that such redistribution of intracellular Ca^{2+} during ATP pulses and interpulse intervals determine basal $[Ca^{2+}]_i$, can influence in P2X2Rs desensitization. To test this hypothesis, we first performed Ca^{2+} recordings on single intact P2X2aR-expressing cells. In cells loaded with the low-affinity Ca^{2+} indicator Fura-FF, application of 3 ATP pulses caused both UDD and a small but consistent increase in basal $[Ca^{2+}]_i$ (Figure 10A). To have a better estimation of basal $[Ca^{2+}]_i$, we next used cells loaded with the more sensitive Fura-2 indicator; in those experiments we consistently observed an increase in basal $[Ca^{2+}]_i$ during the washout period (as long as 15 min) in cells expressing the P2X2aR (Figure 10B). Next, we used CGP 37157, a blocker of the mitochondrial Na^+-Ca^{2+} exchanger (NCX) [38], but also a blocker of plasma membrane NCXs at concentrations >10 μM [38,39]. Application of 4 μM CGP 37157 prevented and reverted the increase in basal $[Ca^{2+}]_i$ after repetitive ATP applications, indicating that mitochondrial NCXs are responsible for the sustained increase in intracellular Ca^{2+}. As shown in Figure 10C, the addition of 4 μM CGP 37157 to P2X2aR-expressing GT1-7 cells completely reverted the increase in basal $[Ca^{2+}]_i$ observed after an ATP pulse. Moreover, when CGP 37157 was present during the whole experiment no increase in $[Ca^{2+}]_i$ was observed (Figure 10D). In patch-clamp experiments, pre and co-application of CGP 37157 prevented UDD when P2X2aR-expressing HEK293 cells where stimulated with 10 μM ATP, in the presence of extracellular Ca^{2+} (Figure 10E). However, when 100 μM ATP was used to trigger the currents, CGP 37157 lost its ability to prevent UDD (Figure 10F).

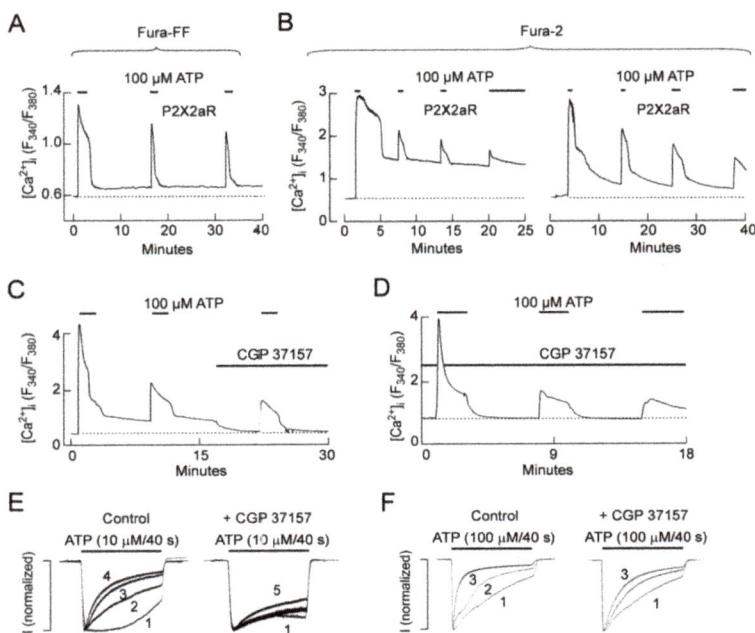

Figure 10. Repetitive ATP application increases basal $[Ca^{2+}]_i$. (**A,B**) Representative $[Ca^{2+}]_i$ recordings from a P2X2aR-expressing GT1-7 cell in which 3 consecutive ATP application were done (horizontal black bars) with washout period of 15 min between pulses. Recordings were done using Fura-FF (**A**) or Fura-2 (**B**) as Ca^{2+}-indicator; (**C,D**) Representative $[Ca^{2+}]_i$ recordings from a P2X2aR-expressing GT1-7 cells in which 4 μM CGP 37157 was applied in the middle (**C**) or during the whole experiment (**D**). Horizontal black bars represent 100 μM ATP applications and dotted lines indicate the basal $[Ca^{2+}]_i$ before to the first ATP application; (**E,F**) Representative recordings from P2X2aR-HEK293 cells in the absence (left) and in the presence (right) of 10 μM CGP 37157; currents were gated with 10 μM (**E**) or 100 μM (**F**) ATP.

3. Discussion

This study summarizes our progress in understanding the regulation of P2X2Rs desensitization by ATP and calcium, and strongly suggests that their gating depends on the metabolic state of the cell. In general, ATP acts extracellularly as an orthosteric agonist for P2XRs, whereas extracellular calcium inhibits or stimulates these channels depending on the subtype of receptors [2]. Several groups observed a strong inhibitory effect of bath calcium on P2X2R, with patterns typical for allosteric mode of regulation [40–42]. Calcium also regulates P2XR function by acting as an intracellular messenger. For P2X2R, an interaction with the neuronal calcium sensor VILIP1 was reported [43]. Our recent work with this receptor is focused on UDD [21], which depends on calcium influx and its intracellular action, with domain calcium being sufficient for development of this form of receptor desensitization [22].

The intracellular effect of calcium on receptor desensitization could be allosteric or through activation of some intracellular signaling pathway. For example, in the case of P2X7R, extracellular calcium is an allosteric regulator [44] whereas intracellular calcium acts through CaM [45]. However, CaM was not involved in the intracellular action of calcium on P2X2aR, as indicated by the experiments with the CaM inhibitory peptide and CaM-KII inhibitor KN-93. Additionally, it has been shown that CAM-KII regulates membrane insertion of the P2X3R through direct phosphorylation of the intracellular T388 [46]. It has also been suggested that the protein kinases A and C signaling pathways contribute to the control of P2XR gating [13,47]. Our experiments also suggest that phosphorylation/dephosphorylation have an important role in P2X2aR gating. First, we did not observe UDD in perforated-patch clamp recording, although there is a substantial rise in intracellular calcium concentration under these experimental conditions [48]. UDD was also not present in the TEVC recordings in oocytes, which further suggests that cytosol dilution during whole-cell recordings accounts for the calcium-dependent facilitation of receptor desensitization. Consistent with this, we found that UDD did not occur in cells containing 5 mM ATP in the intracellular solution during whole-cell recordings. The dependence of the receptor desensitization rates on the intracellular ATP concentration was also confirmed using TEVC recording in oocytes. First, injection of apyrase, a calcium-activated enzyme that catalyzes the hydrolysis of ATP to yield AMP and inorganic phosphate [35], significantly increased the level of receptor desensitization. Second, injection of alkaline phosphatase, a hydrolase enzyme responsible for removing phosphate groups from many types of molecules [36], also facilitated receptor desensitization.

The role of intracellular ATP in P2XR gating has not been studied previously, but such mode of regulation is well established for other channels. One of the first channels that are identified as intracellular ATP regulated channel is the CFTR channel involved in pathogenesis of cystic fibrosis [23]. This channel is regulated by phosphorylation of intracellular domains and by allosteric action of ATP [24]. Transient receptor potential (TRP) channels are also strongly influenced by intracellular ATP; TRPV1 is rescued from desensitized state by long action at high concentration ATP [25], whereas TRPC5 is inhibited by both intracellular ATP and its non-hydrolizable analogue AMP-PNP, indicating the allosteric nature of ATP action [49]. The desensitization of TRPV subfamily of channels is also influenced by intracellular calcium [50]. Intracellular ATP also controls gating of inwardly rectifying potassium $K_{ir}6$ channels [26].

Given that ATPγS is a synthetic non-hydrolyzable analog of ATP in which the terminal gamma phosphate cannot be dissociated from the molecule and transferred to substrates, any potential effect of this agonist could only be allosteric. However, this effect was lacking, in our experiments. We also utilized phosphoenolpyruvate, which has the highest thermodynamic propensity to release and transfer its phosphate group to different substrates among the many phosphate-containing species [51], to verify whether a non-enzymatically mediated phosphorylation plays a role in P2X2aR desensitization. UDD did develop in the presence of intrapipette phosphoenolpyruvate, excluding non-enzymatically mediated phosphorylation as a mechanism for the development of UDD. Furthermore, the presence of intrapipette GTP did not interfere with UDD, indicating that ATP selectively regulates this process. The receptor was desensitized in the presence of 5 mM ATP when kinase activity was blocked. Together, these results indicate that ATP does not affect P2X2aR gating directly by binding to an intracellular binding site but, presumably, by the ATP-dependent and kinase-mediated phosphorylation of the channels. Furthermore, a dilution of cytosolic ATP during whole-cell recording, but not in perforated patch-clamp recording, leads to the abolition of metabolic regulation of P2X2R and the development of allosteric effects of calcium.

We can infer from our experiments with PC12 cells that the calcium-induced increase in P2X2Rs desensitization could be of physiological significance. These cells that endogenously express the P2X2R have been extensively used as a model for chromaffin cell function specially in vesicle trafficking and hormone release, processes that are calcium dependent. In undifferentiated PC12 cells P2X2Rs are the main subtype expressed, this has been deducted both from the expression profile [52] and by the functional responses evoked by extracellular ATP [53]. In contrast NGF-differentiated PC12 cells can express different P2XR subtypes [52]. Thus, the rate of P2X2R desensitization will finally have an impact on the total amount of calcium that is entering through this receptor and will finally affect the vesicular release. In this context, the metabolic state of the cell will regulate this release, by affecting P2X2R gating properties. Interestingly, we found that this regulation mechanism can work at intracellular ATP concentrations between 3 and 5 mM, concentrations that could reflect a cell with a low or high metabolism, respectively.

The NetPhosK 1.0 Server (available online: http://www.cbs.dtu.dk/services/NetPhosK/), which predicts protein phosphorylation sites on the basis of eukaryotic kinase specificity [54], suggested the following potential residues of P2X2R that could be phosphorylated: Y16, T18, T354, Y362, S363, S377, S378 and Y398. As previously mentioned, the P2X2bR that lacks the *C*-terminal Val^{370}–Gln^{438} sequence can develop UDD, similar to P2X2aR (Table 1) [22]. For that reason, we excluded the residues T372, S377, S378 and Y398 as functionally relevant for the development of UDD. However, this does not rule out the putative role of these P2X2aR-specific residues in other desensitization-related processes; different splice variants of the P2X2R are known to exhibit different desensitization properties [2], and future experiments could help to clarify whether phosphorylation/dephosphorylation could be involved in these differences. Our mutagenesis and electrophysiological studies also investigated residues T18, Y362, and S363 as being potentially relevant for receptor desensitization. The T18V mutant was essentially nonfunctional, precluding further analysis. Moreover, this residue is conserved among all P2XRs subtypes, and because UDD and calcium-dependent desensitization are primarily observed on the P2X2R, it is more plausible that residues specific to this subtype are responsible for UDD. On the other hand, UDD was lost only in S363 mutants, and this effect was independent of intracellular ATP. These data strongly support the hypothesis that the S363 residue is constitutively phosphorylated at high intracellular ATP concentrations, and its dephosphorylation leads to an increase in P2X2R desensitization.

The discovery that the P2X2R gating can be regulated by intracellular calcium and ATP suggests that mitochondria, the source of intracellular ATP and an important regulator of intracellular calcium, could have a key role in this process. For that we run experiments to evaluate the potential role of this organelle in calcium handling during repetitive ATP applications and the potential involvement of such regulation in the desensitization properties of P2X2R using a blocker of mitochondrial calcium transport. CGP 37157 is a preferential blocker of the mitochondrial Na^+–Ca^{2+} exchanger (NCX) [38,55]. It is well known that mitochondria can regulate $[Ca^{2+}]_i$ by the action of several transporter proteins [37]. For example, Ca^{2+} uptake is mainly driven by the mitochondrial Ca^{2+} uniporter, whereas Ca^{2+} release to the cytosol is driven by NCX. In this context, the inhibition of mitochondrial NCX by CGP 37157 will cause Ca^{2+} accumulation in this organelle, thereby lowering $[Ca^{2+}]_i$ and then reverting and/or preventing UDD, as we observed when the currents were gated by low ATP concentrations. Interestingly, when 100 μM ATP was used to gate P2X2aR-currents CGP 37157 was not able to revert/prevent UDD. We infer that the higher calcium influx combined with the leak of intracellular ATP as result of the whole-cell configuration can account for these differential effects. Likewise, in untreated cells, the normal functioning of mitochondrial Ca^{2+} uniporter and NCX will provide a constant and long lasting source of Ca^{2+} to the cytosol, that will lead to increased basal $[Ca^{2+}]_i$ and therefore UDD will persist under these conditions.

In summary, these data demonstrate the opposite roles of intracellular calcium and ATP in the regulation of P2X2R gating. Intracellular calcium facilitates receptor desensitization in the absence of ATP, whereas in the presence of high ATP, this effect is suppressed, and the receptor operates practically as a non-desensitizing receptor. It is reasonable to conclude that UDD reflects the kinetics of the decay of the intracellular ATP concentration in whole-cell recording. The preservation of the UDD in the presence of various blockers of calcium-dependent enzymes supports the hypothesis that calcium acts allosterically. The loss of UDD in the presence of ATP further suggests that allosteric calcium binding site(s) is (are) not accessible when intracellular ATP is not depleted. In contrast to calcium, intracellular ATP does not act allosterically but is required for receptor phosphorylation. Mutagenesis studies indicated a critical role for the S363 residue as a putative phosphorylation site. Thus, constitutive phosphorylation of P2X2R blocks intracellular calcium-dependent facilitation of receptor desensitization, probably by blocking the allosteric action of calcium. Phosphorylated receptors, a consequence of higher intracellular ATP concentration, are non-desensitizing channels with a high capacity to elevate intracellular calcium for prolonged periods. In contrast, the dephosphorylated receptors with occupied calcium-binding sites, a reflection of a decrease in intracellular ATP concentrations, are rapidly desensitizing channels with limited capacity to maintain elevated intracellular calcium concentrations. Such a dual regulation of P2X2R could be of physiological relevance because the desensitization profiles of the receptors have direct and indirect (through voltage-gated calcium channels) impacts on calcium influx and calcium-dependent cellular processes.

4. Material and Methods

4.1. P2XR Constructs Used in Experiment

Experiments were performed using the wild type and mutant rat P2X2aRs, and wild-type P2X2bR, which were subcloned into the pIRES2-EGFP vector (Clontech, Mountain View, CA, USA). To generate the mutants, oligonucleotides (synthesized by Integrated DNA Technologies, Rockville, MD, USA) containing specific point mutations were introduced into the rP2X2a/pIRES2-EGFP template using Pfu Ultra DNA polymerase (Agilent Technologies, Santa Clara, CA, USA). A QIAprep Spin Miniprep Kit (QIAGEN, Venlo, The Netherlands) was used to isolate the plasmids for transfection. To identify and verify the presence of the mutations, sequencing was performed by MACROGEN Inc. (Rockville, MD, USA). The P2X2Rs cDNAs were inserted in the pIRES2-EGFP plasmid and expressed in HEK293 cells by polymer-based transfection with the JetPrime transfection reagent (PolyPlus-transfection, Illkirch, France), lipofectamine 2000 (Invitrogen, Carlsbad, CA, USA) or injected into *Xenopus* oocytes.

4.2. Receptor Transfection and Cell Culture

The HEK293 cells were routinely maintained in DMEM (Invitrogen, Carlsbad, CA, USA) containing 10% (v/v) fetal bovine serum (Gibco, Grand Island, NY, USA) and 1% (v/v) penicillin-streptomycin (Invitrogen, Carlsbad, CA, USA). For the electrophysiological experiments, the cells were grown on 35-mm Nunclon Surface dishes (NUNC, Rochester, NY, USA) at a density of 500,000 cells per 35 mm culture dish. The wild type and mutated receptors were transiently expressed in HEK293 cells by transfection using the JetPrime transfection reagent. Briefly, 2 μL of JetPrime and 2 μL of DNA were mixed in the manufacturer-provided buffer and incubated for 10 min at room temperature. The mixture was then added to the medium covering the cells in a 2 mL Nunclon Surface (NUNC, Rochester, NY, USA) cell culture dish. Alternatively, the transfection was conducted using 2 μg of DNA and 5 μL of Lipofectamine 2000 in 2 mL of serum-free Opti-MEM medium according to the manufacturer's instructions (Invitrogen, UK or Carlsbad, CA, USA). Experiments were performed 24–48 h after transfection, and the P2X2R-expressing cells were identified by GFP fluorescence.

PC12 cells (derived from pheochromocytoma of rat adrenal medulla, ATCC# CRL-1721) were grown in DMEM containing 10% FBS and penicillin-streptomycin (Invitrogen, Carlsbad, CA, USA). Cells were routinely subcultured and plated on 12-mm poly-L-lysine-coated glass coverslips 1–5 days before electrophysiological recordings.

4.3. Patch-Clamp Measurements

Electrophysiological experiments were performed on cells at room temperature using whole-cell patch-clamp recording techniques. The currents were recorded using an Axopatch 200B patch-clamp amplifier (Molecular Devices, Sunnyvale, CA, USA) and were filtered at 2 kHz using a low-pass Bessel filter. Patch electrodes, fabricated from borosilicate glass (type 1B150F-3; World Precision Instruments, Sarasota, FL, USA) using a Flaming Brown horizontal puller (P-87; Sutter Instruments, Novato, CA, USA), were heat polished to a final tip resistance of 2–4 MΩ. All current recordings were captured and stored using the pCLAMP 9 software package in conjunction with the Digidata 1322A analog-to-digital converter (Molecular Devices, Sunnyvale, CA, USA). The bath solution contained (in mM) 142 NaCl, 3 KCl, 1 $MgCl_2$, 2 $CaCl_2$, 10 glucose, and 10 HEPES (4-(2-hydroxyethyl)-1-piperazineethanesulfonic acid). The Ca^{2+}-deficient solution contained (in mM) 145 NaCl, 3 KCl, 10 glucose, and 10 HEPES. To discard the ocurrence of Mg^{2+}-dependent unspecific conductances we performed experiments using the Ca^{2+}-deficient solution in the absence and in the presence of 2 mM Mg^{2+} (Figure S1). We defined as Ca^{2+}-deficient solution because trace amounts of Ca^{2+} (90 μM) were detected in our ultra-purified water by atomic absorption spectrometry using a calcium lamp (AAnalyst 200 Atomic Absorption Spectrophotometer; PerkinElmer, Hongkong, China). In all cases, the pH was adjusted to 7.35, and the osmolarity of these solutions was 295–305 mOsm. Various intracellular solutions were prepared in order to perform these experiments. The sodium-based intracellular solution contained (in mM) 142 NaCl, 10 EGTA, and 10 HEPES. The potassium-based intracellular solution contained (in mM) 140 KCl, 3 $MgCl_2$, and 10 HEPES; its pH was adjusted to 7.2 with KOH. Finally, the cesium-based intracellular solution contained (in mM) 154 mM CsCl, 11 mM EGTA, and 10 mM HEPES, adjusted to pH 7.2 with 1.6 M CsOH. ATP was prepared daily in the bath buffer and was applied using the rapid solution changer system RSC-200 (Biologic Science Instruments, Claix, France). Stock solutions of the kinase or phosphatase activators/inhibitors and phosphoinositides were prepared in dimethylsulfoxide, and aliquots were stored at −20 °C. The current responses were recorded from single cells clamped at −60 mV. Perforated-patch recordings were conducted by including 300 μg/mL nystatin in the potassium-containing pipette solution. For intracellular ATP concentration-response experiments we used sodium-based intracellular solutions with 10 mM (HEK293) or 0.05 mM (PC12) EGTA in addition to 0–5 mM ATP and performed UDD protocols.

4.4. Oocyte Injection and Current Measurements

Experiments with *Xenopus* oocytes were conducted following the National Institutes of Health guidelines for the care and use of experimental animals and approved (FM N°17, 21 June 2016) by the Bioethics Committee of the Catholic University of the North, Chile. A segment of the ovary was surgically removed from *Xenopus laevis* female frogs under anesthesia (benzocaine, 2 g/L); the oocytes were manually defolliculated and then incubated with collagenase as previously detailed [56]. The nuclei of the oocytes were injected with 3–5 ng of cDNA coding for the various rat receptors using a NanojectII nanoliter injector (Drummond Scientific, Broomall, PA, USA). After 24–48 h incubation in Barth's solution (in mM: 88 NaCl, 1 KCl, 2.4 NaHCO$_3$, 10 HEPES, 0.82 MgSO$_4$, 0.33 Ca(NO$_3$)$_2$, and 0.91 $CaCl_2$; pH 7.5; supplemented with 10 IU/L penicillin/10 mg streptomycin and 2 mM pyruvate), additional $CaCl_2$ was added to obtain a final 2.5 mM calcium concentration; the Ca^{2+}-deficient solution was similar to that used in patch-clamp experiments, but included 1 mM $MgCl_2$ to avoid the develop unspecific cationic currents. The oocytes were clamped at −70 mV using the two-electrode voltage clamp (TEVC) configuration with an OC-725C clamper (Warner Instruments, Hamden, CT, USA). The ATP-gated currents were recorded after regular ATP applications repeated every 10 or more minutes depending on ATP concentration. For the experiments with apyrase and alkaline phosphatase, previously P2X2aR-injected oocytes were injected 30 min before the start of recordings with 40 nL of the ectonucleotidase apyrase (2.5 U/μL) or with alkaline phosphatase (1 U/μL). In all cases, we calculated the percentage of desensitization 40 s after the peak amplitude was induced by 100 μM ATP.

4.5. Intracellular Calcium Recordings

Transfected GT1-7 cells plated on coverslips were bathed in KR-like medium containing 2.5 μM Fura-FF or Fura-2 AM (Invitrogen) for 1 h at room temperature. After that, the coverslips were washed in KR-like media and they were mounted on the stage of an inverted Observer-D1 microscope (Carl Zeiss, Oberkochen, Germany) attached to an ORCA-ER camera (Hamamatsu Photonics, Hamamatsu City, Japan) and a Lambda DG-4 wavelength switcher (Sutter, Novato, CA, USA). Hardware control and image analysis was performed using Metafluor software (Molecular Devices, Downingtown, PA, USA). Experiments were done under a 40× oil-immersion objective during exposure to alternating 340- and 380-nm excitation beams, and the intensity of light emission at 520 nm (F340 and F380) was followed simultaneously in several single cells.

4.6. Data Analysis

Each experiment was repeated in at least four separate cells. Curve fitting and statistical analyses, including nonparametric Mann–Whitney tests, were performed using GraphPad software (San Diego, CA, USA) and Sigmastat (San Jose, CA, USA). Statistical significance was determined using a $p < 0.01$. The desensitization constant (τ) was calculated for each recording by curve fitting of the desensitization phase with a predefined monoexponential function ($f(t) = B \exp(-t/\tau)$) using the Clampfit 10.0 software (Molecular Devices, Sunnyvale, CA, USA). Data are reported as mean ± SEM unless stated otherwise.

Supplementary Materials: Supplementary materials can be found at http://www.mdpi.com/1422-0067/19/4/1161/ s1.

Acknowledgments: Funded by Intramural Research Program of the National Institute of Child Health and Human Development, National Institutes of Health (Stanko S. Stojilkovic, Milos B. Rokic, Claudio Coddou, Elias Leiva-Salcedo), FONDECYT Initiation Grant #11121302 and Regular Grant #1161490 (Claudio Coddou) and FONDECYT Regular Grant 1181814 and Financiamiento Basal para Centros Científicos y Tecnológicos de Excelencia grant FB0807 (Elias Leiva-Salcedo). We are thankful to Shuo Li for ΔCR construct. We thank Marcelo Catalán for help in flame emission spectrometry measurements.

Author Contributions: Claudio Coddou and Stanko S. Stojilkovic designed research; Milos B. Rokic, Patricio Castro, Elias Leiva-Salcedo, Melanija Tomic and Claudio Coddou performed research; Claudio Coddou, Milos B. Rokic, Patricio Castro and Stanko S. Stojilkovic analyzed data; Milos B. Rokic, Patricio Castro, Elias Leiva-Salcedo, Stanko S. Stojilkovic and Claudio Coddou wrote the paper.

Conflicts of Interest: The authors declare no competing financial interests.

Abbreviations

CaM	Calmodulin
P2XRs	P2X receptor channels
TEVC	two-electrode voltage clamp
TRP	Transient receptor potential

References

1. Samways, D.S.; Li, Z.; Egan, T.M. Principles and properties of ion flow in P2X receptors. *Front. Cell. Neurosci.* **2014**, *8*, 6. [CrossRef] [PubMed]
2. Coddou, C.; Yan, Z.; Obsil, T.; Huidobro-Toro, J.P.; Stojilkovic, S.S. Activation and regulation of purinergic P2X receptor channels. *Pharmacol. Rev.* **2011**, *63*, 641–683. [CrossRef] [PubMed]
3. Lorinczi, E.; Bhargava, Y.; Marino, S.F.; Taly, A.; Kaczmarek-Hajek, K.; Barrantes-Freer, A.; Dutertre, S.; Grutter, T.; Rettinger, J.; Nicke, A. Involvement of the cysteine-rich head domain in activation and desensitization of the P2X$_1$ receptor. *Proc. Natl. Acad. Sci. USA* **2012**, *109*, 11396–11401. [CrossRef] [PubMed]
4. Fabbretti, E.; Sokolova, E.; Masten, L.; D'Arco, M.; Fabbro, A.; Nistri, A.; Giniatullin, R. Identification of negative residues in the P2X$_3$ ATP receptor ectodomain as structural determinants for desensitization and the Ca^{2+}-sensing modulatory sites. *J. Biol. Chem.* **2004**, *279*, 53109–53115. [CrossRef] [PubMed]

5. Jindrichova, M.; Khafizov, K.; Skorinkin, A.; Fayuk, D.; Bart, G.; Zemkova, H.; Giniatullin, R. Highly conserved tyrosine 37 stabilizes desensitized states and restricts calcium permeability of ATP-gated P2X$_3$ receptor. *J. Neurochem.* **2011**, *119*, 676–685. [CrossRef] [PubMed]

6. He, M.L.; Zemkova, H.; Stojilkovic. S.S. Dependence of purinergic P2X receptor activity on ectodomain structure. *J. Biol. Chem.* **2003**, *278*, 10182–10188. [CrossRef] [PubMed]

7. Brandle, U.; Spielmanns, P.; Osteroth, R.; Sim, J.; Surprenant, A.; Buell, G.; Ruppersberg, J.P.; Plinkert, P.K.; Zenner, H.P.; Glowatzki, E. Desensitization of the P2X$_2$ receptor controlled by alternative splicing. *FEBS Lett.* **1997**, *404*, 294–298. [CrossRef]

8. Koshimizu, T.; Tomic, M.; Van Goor, F.; Stojilkovic, S.S. Functional role of alternative splicing in pituitary P2X$_2$ receptor-channel activation and desensitization. *Mol. Endocrinol.* **1998**, *12*, 901–913. [CrossRef] [PubMed]

9. Koshimizu, T.A.; Kretschmannova, K.; He, M.L.; Ueno, S.; Tanoue, A.; Yanagihara, N.; Stojilkovic, S.S.; Tsujimoto, G. Carboxyl-terminal splicing enhances physical interactions between the cytoplasmic tails of purinergic P2X receptors. *Mol. Pharmacol.* **2006**, *69*, 1588–1598. [CrossRef] [PubMed]

10. Koshimizu, T.; Tomic, M.; Koshimizu, M.; Stojilkovic, S.S. Identification of amino acid residues contributing to desensitization of the P2X$_2$ receptor channel. *J. Biol. Chem.* **1998**, *273*, 12853–12857. [CrossRef] [PubMed]

11. Fountain, S.J.; North, R.A. A C-terminal lysine that controls human P2X$_4$ receptor desensitization. *J. Biol. Chem.* **2006**, *281*, 15044–15049. [CrossRef] [PubMed]

12. Allsopp, R.C.; Evans, R.J. The intracellular amino terminus plays a dominant role in desensitization of ATP-gated P2X receptor ion channels. *J. Biol. Chem.* **2011**, *286*, 44691–44701. [CrossRef] [PubMed]

13. Boue-Grabot, E.; Archambault, V.; Seguela, P. A protein kinase C site highly conserved in P2X subunits controls the desensitization kinetics of P2X$_2$ ATP-gated channels. *J. Biol. Chem.* **2000**, *275*, 10190–10195. [CrossRef] [PubMed]

14. Liu, G.J.; Brockhausen, J.; Bennett, M.R. P2X$_1$ receptor currents after disruption of the PKC site and its surroundings by dominant negative mutations in HEK293 cells. *Auton. Neurosci.* **2003**, *108*, 12–16. [CrossRef]

15. Franklin, C.; Braam, U.; Eisele, T.; Schmalzing, G.; Hausmann, R. Lack of evidence for direct phosphorylation of recombinantly expressed P2X$_2$ and P2X$_3$ receptors by protein kinase C. *Purinergic Signal.* **2007**, *3*, 377–388. [CrossRef] [PubMed]

16. Vial, C.; Tobin, A.B.; Evans, R.J. G-protein-coupled receptor regulation of P2X$_1$ receptors does not involve direct channel phosphorylation. *Biochem. J.* **2004**, *382*, 101–110. [CrossRef] [PubMed]

17. Brown, D.A.; Yule, D.I. Protein kinase C regulation of P2X$_3$ receptors is unlikely to involve direct receptor phosphorylation. *Biochim. Biophys. Acta* **2007**, *1773*, 1661–1675.

18. Yan, Z.; Li, S.; Liang, Z.; Tomic, M.; Stojilkovic, S.S. The P2X$_7$ receptor channel pore dilates under physiological ion conditions. *J. Gen. Physiol.* **2008**, *132*, 563–573. [CrossRef] [PubMed]

19. North, R.A. Molecular physiology of P2X receptors. *Physiol. Rev.* **2002**, *82*, 1013–1067. [CrossRef] [PubMed]

20. Cook, S.P.; Rodland, K.D.; McCleskey, E.W. A memory for extracellular Ca^{2+} by speeding recovery of P2X receptors from desensitization. *J. Neurosci.* **1998**, *18*, 9238–9244. [CrossRef] [PubMed]

21. Khadra, A.; Yan, Z.; Coddou, C.; Tomic, M.; Sherman, A.; Stojilkovic, S.S. Gating properties of the P2X2a and P2X2b receptor channels: Experiments and mathematical modeling. *J. Gen. Physiol.* **2012**, *139*, 333–348. [CrossRef] [PubMed]

22. Coddou, C.; Yan, Z.; Stojilkovic, S.S. Role of domain calcium in purinergic P2X$_2$ receptor channel desensitization. *Am. J. Physiol. Cell Physiol.* **2015**, *308*, C729–C736. [CrossRef] [PubMed]

23. Gadsby, D.C.; Vergani, P.; Csanady, L. The ABC protein turned chloride channel whose failure causes cystic fibrosis. *Nature* **2006**, *440*, 477–483. [CrossRef] [PubMed]

24. Sheppard, D.N.; Welsh, M.J. Structure and function of the CFTR chloride channel. *Physiol. Rev.* **1999**, *79* (Suppl. 1), S23–S45. [CrossRef] [PubMed]

25. Liu, B.; Zhang, C.; Qin, F. Functional recovery from desensitization of vanilloid receptor TRPV1 requires resynthesis of phosphatidylinositol 4,5-bisphosphate. *J. Neurosci. Off. J. Soc. Neurosci.* **2005**, *25*, 4835–4843. [CrossRef] [PubMed]

26. Babenko, A.P.; Aguilar-Bryan, L.; Bryan, J. A view of sur/KIR6.X, KATP channels. *Ann. Rev. Physiol.* **1998**, *60*, 667–687. [CrossRef] [PubMed]

27. Brake, A.J.; Wagenbach, M.J.; Julius. D. New structural motif for ligand-gated ion channels defined by an ionotropic ATP receptor. *Nature* **1994**, *371*, 519–523. [CrossRef] [PubMed]

28. Karaman, M.W.; Herrgard, S.; Treiber, D.K.; Gallant, P.; Atteridge, C.E.; Campbell, B.T.; Chan, K.W.; Ciceri, P.; Davis, M.I.; Edeen, P.T.; et al. A quantitative analysis of kinase inhibitor selectivity. *Nat. Biotechnol.* **2008**, *26*, 127–132. [CrossRef] [PubMed]

29. Balla, T. Phosphoinositides: Tiny lipids with giant impact on cell regulation. *Physiol. Rev.* **2013**, *93*, 1019–1137. [CrossRef] [PubMed]

30. Fujiwara, Y.; Kubo, Y. Regulation of the desensitization and ion selectivity of ATP-gated P2X$_2$ channels by phosphoinositides. *J. Physiol.* **2006**, *576*, 135–149. [CrossRef] [PubMed]

31. Mo, G.; Bernier, L.P.; Zhao, Q.; Chabot-Dore, A.J.; Ase, A.R.; Logothetis, D.; Cao, C.Q.; Seguela, P. Subtype-specific regulation of P2X3 and P2X2/3 receptors by phosphoinositides in peripheral nociceptors. *Mol. Pain.* **2009**, *5*, 47. [CrossRef] [PubMed]

32. Bernier, L.P.; Ase, A.R.; Tong, X.; Hamel, E.; Blais, D.; Zhao, Q.; Logothetis, D.E.; Seguela, P. Direct modulation of P2X$_1$ receptor-channels by the lipid phosphatidylinositol 4,5-bisphosphate. *Mol. Pharmacol.* **2008**, *74*, 785–792. [CrossRef] [PubMed]

33. Bernier, L.P.; Ase, A.R.; Chevallier, S.; Blais, D.; Zhao, Q.; Boue-Grabot, E.; Logothetis, D.; Seguela, P. Phosphoinositides regulate P2X$_4$ ATP-gated channels through direct interactions. *J. Neurosci.* **2008**, *28*, 12938–12945. [CrossRef] [PubMed]

34. Ase, A.R.; Bernier, L.P.; Blais, D.; Pankratov, Y.; Seguela, P. Modulation of heteromeric P2X1/5 receptors by phosphoinositides in astrocytes depends on the P2X$_1$ subunit. *J. Neurochem.* **2010**, *113*, 1676–1684. [CrossRef] [PubMed]

35. Clark, G.; Roux, S.J. Apyrases, extracellular ATP and the regulation of growth. *Curr. Opin. Plant Biol.* **2011**, *14*, 700–706. [CrossRef] [PubMed]

36. Fernandez, N.J.; Kidney, B.A. Alkaline phosphatase: Beyond the liver. *Vet. Clin. Pathol. Am. Soc.* **2007**, *36*, 223–233. [CrossRef]

37. Babcock, D.F.; Hille, B. Mitochondrial oversight of cellular Ca^{2+} signaling. *Curr. Opin. Neurobiol.* **1998**, *8*, 398–404. [CrossRef]

38. Cox, D.A.; Conforti, L.; Sperelakis, N.; Matlib, M.A. Selectivity of inhibition of Na$^+$-Ca^{2+} exchange of heart mitochondria by benzothiazepine CGP-37157. *J. Cardiovasc. Pharmacol.* **1993**, *21*, 595–599. [CrossRef] [PubMed]

39. Czyz, A.; Kiedrowski, L. Inhibition of plasmalemmal Na$^+$/Ca^{2+} exchange by mitochondrial Na$^+$/Ca^{2+} exchange inhibitor 7-chloro-5-(2-chlorophenyl)-1,5-dihydro-4,1-benzothiazepin-2(3H)-one (CGP-37157) in cerebellar granule cells. *Biochem. Pharmacol.* **2003**, *66*, 2409–2411. [CrossRef] [PubMed]

40. Virginio, C.; Church, D.; North, R.A.; Surprenant, A. Effects of divalent cations, protons and calmidazolium at the rat P2X$_7$ receptor. *Neuropharmacology* **1997**, *36*, 1285–1294. [CrossRef]

41. Evans, R.J.; Lewis, C.; Virginio, C.; Lundstrom, K.; Buell, G.; Surprenant, A.; North, R.A. Ionic permeability of, and divalent cation effects on, two ATP-gated cation channels (P2X receptors) expressed in mammalian cells. *J. Physiol.* **1996**, *497*, 413–422. [CrossRef] [PubMed]

42. Ding, S.; Sachs, F. Ion permeation and block of P2X$_2$ purinoceptors: Single channel recordings. *J. Membr. Biol.* **1999**, *172*, 215–223. [CrossRef] [PubMed]

43. Chaumont, S.; Compan, V.; Toulme, E.; Richler, E.; Housley, G.D.; Rassendren, F.; Khakh, B.S. Regulation of P2X$_2$ receptors by the neuronal calcium sensor VILIP1. *Sci. Signal.* **2008**, *1*, ra8. [CrossRef] [PubMed]

44. Yan, Z.; Khadra, A.; Sherman, A.; Stojilkovic, S.S. Calcium-dependent block of P2X$_7$ receptor channel function is allosteric. *J. Gen. Physiol.* **2011**, *138*, 437–452. [CrossRef] [PubMed]

45. Roger, S.; Pelegrin, P.; Surprenant, A. Facilitation of P2X$_7$ receptor currents and membrane blebbing via constitutive and dynamic calmodulin binding. *J. Neurosci.* **2008**, *28*, 6393–6401. [CrossRef] [PubMed]

46. Chen, X.Q.; Zhu, J.X.; Wang, Y.; Zhang, X.; Bao, L. CaMKIIalpha and caveolin-1 cooperate to drive ATP-induced membrane delivery of the P2X3 receptor. *J. Mol. Cell Biol.* **2014**, *6*, 140–153. [CrossRef] [PubMed]

47. Chow, Y.W.; Wang, H.L. Functional modulation of P2X$_2$ receptors by cyclic AMP-dependent protein kinase. *J. Neurochem.* **1998**, *70*, 2606–2612. [CrossRef] [PubMed]

48. Stojilkovic, S.S.; Tomic, M.; Van Goor, F.; Koshimizu, T. Expression of purinergic P2X$_2$ receptor-channels and their role in calcium signaling in pituitary cells. *Biochem. Cell. Biol.* **2000**, *78*, 393–404. [CrossRef] [PubMed]

49. Dattilo, M.; Penington, N.J.; Williams, K. Inhibition of TRPC5 channels by intracellular ATP. *Mol. Pharmacol.* **2008**, *73*, 42–49. [CrossRef] [PubMed]

50. Mercado, J.; Gordon-Shaag, A.; Zagotta, W.N.; Gordon, S.E. Ca^{2+}-dependent desensitization of TRPV2 channels is mediated by hydrolysis of phosphatidylinositol 4,5-bisphosphate. *J. Neurosci. Off. J. Soc.* **2010**, *30*, 13338–13347. [CrossRef] [PubMed]

51. Berg, J.M.; Tymoczko, J.L.; Stryer, L. Biochemistry. 5th Edition. Section 14.1: Metabolism is Composed of Many Coupled, Interconnecting Reactions. 2002. Available online: http://www.ncbi.nlm.nih.gov/books/NBK22439/ (accessed on 26 February 2018).

52. Sun, J.H.; Cai, G.J.; Xiang, Z.H. Expression of P2X purinoceptors in PC12 phaeochromocytoma cells. *Clin. Exp. Pharmacol. Physiol.* **2007**, *34*, 1282–1386. [CrossRef] [PubMed]

53. Michel, A.D.; Grahames, C.B.; Humphrey, P.P. Functional characterisation of P2 purinoceptors in PC12 cells by measurement of radiolabelled calcium influx. *Naunyn Schmiedebergs Arch. Pharmacol.* **1996**, *354*, 562–571. [CrossRef] [PubMed]

54. Blom, N.; Sicheritz-Ponten, T.; Gupta, R.; Gammeltoft, S.; Brunak, S. Prediction of post-translational glycosylation and phosphorylation of proteins from the amino acid sequence. *Proteomics* **2004**, *4*, 1633–1649. [CrossRef] [PubMed]

55. Babcock, D.F.; Herrington, J.; Goodwin, P.C.; Park, Y.B.; Hille, B. Mitochondrial participation in the intracellular Ca^{2+} network. *J. Cell Biol.* **1997**, *136*, 833–844. [CrossRef] [PubMed]

56. Acuna-Castillo, C.; Morales, B.; Huidobro-Toro, J.P. Zinc and copper modulate differentially the $P2X_4$ receptor. *J. Neurochem.* **2000**, *74*, 1529–1537. [CrossRef] [PubMed]

International Journal of
Molecular Sciences

MDPI

Article

P2X4 Receptor-Dependent Ca^{2+} Influx in Model Human Monocytes and Macrophages

Janice A. Layhadi and Samuel J. Fountain *

School of Biological Sciences, University of East Anglia, Norwich Research Park, Norwich NR4 7TJ, UK;
j.layhadi@uea.ac.uk
* Correspondence: s.j.fountain@uea.ac.uk; Tel.: +44(0)-160-359-7326

Received: 28 September 2017; Accepted: 18 October 2017; Published: 27 October 2017

Abstract: Monocytes and macrophages express a repertoire of cell surface P2 receptors for adenosine 5′-triphosphate (ATP) a damage-associated molecular pattern molecule (DAMP), which are capable of raising cytoplasmic calcium when activated. This is achieved either through direct permeation (ionotropic P2X receptors) or by mobilizing intracellular calcium stores (metabotropic P2Y receptors). Here, a side-by-side comparison to investigate the contribution of P2X4 receptor activation in ATP-evoked calcium responses in model human monocytes and macrophages was performed. The expression of P2X1, P2X4, P2X5 and P2X7 was confirmed by qRT-PCR and immunocytochemistry in both model monocyte and macrophage. ATP evoked a concentration-dependent increase in intracellular calcium in both THP-1 monocyte and macrophages. The sarco/endoplasmic reticulum Ca^{2+}-ATPase inhibitor thasigargin (Tg) responses to the maximal ATP concentration (100 μM) in THP-1 monocytes, and responses in macrophage were significantly attenuated. Tg-resistant ATP-evoked calcium responses in the model macrophage were dependent on extracellular calcium, suggesting a requirement for calcium influx. Ivermectin (IVM) potentiated the magnitude of Tg-resistant component and slowed the decay of response in the model macrophage. The Tg-resistant component was attenuated by P2X4 antagonists 5-BDBD and PSB-12062 but not by the P2X1 antagonist Ro0437626 or the P2X7 antagonist A438079. shRNA-mediated P2X4 knockdown resulted in a significant reduction in Tg-resistant ATP-evoked calcium response as well as reduced sensitivies towards P2X4-specific pharmacological tools, IVM and PSB-12062. Inhibition of endocytosis with dynasore significantly reduced the magnitude of Tg-resistant component but substantially slowed decay response. Inhibition of calcium-dependent exocytosis with vacuolin-1 had no effect on the Tg-resistant component. These pharmacological data suggest that P2X4 receptor activation contributed significantly towards the ionotropic calcium response evoked by ATP of the model human macrophage.

Keywords: P2X4; ATP; calcium; monocyte; macrophage; purinergic receptor

1. Introduction

The release of adenosine 5′-triphosphate (ATP) as a signaling molecule by activated leukocytes, platelets and apoptotic cells is crucial for the induction of a variety of physiological responses across different cell types [1,2]. ATP is able to mediate these biological processes through the activation of purinergic P2 receptors: P2X (ligand-gated ion channels) or P2Y (G-protein coupled receptors) [3,4]. To date, seven mammalian P2X receptor subunits (P2X1–P2X7) have been identified with widespread distribution throughout neuronal and non-neuronal tissues [5]. The expression of P2X receptors has been reported in immune cells such as monocytes and macrophages, with P2X4 and P2X7 being the pre-dominant subunits expressed [6]. Although significant research has been performed in the past decade to elucidate a role of P2X7 receptor in the immune system, the same cannot be said for the P2X4 receptor. Research on the P2X4 receptor has been greatly hampered by the lack of selective

P2X4 receptor antagonists as well as the possible complicating presence of the P2X7 receptor [7]. In macrophages and microglia, P2X4 receptors have been shown to co-express with P2X7 receptors, although they do not appear to share the same downstream signaling pathways [8,9].

Although the role of P2X4 receptors in monocytes and macrophages is not clearly defined within the literature, P2X4 expression has been shown to be upregulated following nerve damage in microglia [10] and, in macrophages, P2X4 has been reported to mediate prostaglandin E2 (PGE2) release [11]. Studies by Li and Fountain (2012) in THP-1 monocytes also illustrated that P2X4 contributes towards the ATP-evoked Ca^{2+} response and that its activity is suppressed by cholesterol lowering drug, fluvastatin [12].

In the present study, we have performed a side-by-side comparison of the expression of P2X receptors in model human THP-1 monocytes and macrophages through the use of qRT-PCR and immunocytochemistry. Though evidence of functional P2X4 can be revealed by whole-cell patch-clamps, studies evaluating the role of P2X receptors in intact cells are limited. To this end, we investigated the contribution of P2X4 receptor activation towards ATP-evoked intracellular Ca^{2+} response, and this was assessed in the two cell models through intracellular Ca^{2+} measurements. Our findings allowed the identification of P2X4 contribution of towards ATP-evoked Ca^{2+} influx in model human monocytes and macrophages, while also assessing the reliability of commercially available pharmacological tools for the study of P2X4.

2. Results

P2X1, P2X4, P2X5 and P2X7 are expressed at the mRNA and protein level in THP-1 monocytes and macrophages. Expression of P2X receptors in human THP-1 monocytes and THP-1 differentiated macrophages were investigated at the mRNA and protein level. qRT-PCR analysis identified mRNA expression of all P2X genes with the exception of P2X2 in THP-1 and P2X2 and P2X3 in THP-1 differentiated macrophages (Figure 1A). Fold change in mRNA expression was also quantified and showed that, across three replicates, P2X1 and P2X2 mRNA expression was found to be downregulated (P2X1: 0.38 ± 0.058 fold; $n = 5$; $p < 0.001$ and P2X2: 0.13 ± 0.074 fold; $n = 5$; $p < 0.001$), while P2X4 were found to be upregulated (2.61 ± 0.52 fold; $n = 5$; $p < 0.05$) in THP-1-differentiated macrophages compared to THP-1 monocytes (Figure 1B). Using immunocytochemistry, protein expression of P2X receptors was also studied. P2X1, P2X4, P2X5 and P2X7 were all expressed in both THP-1 monocytes and THP-1-differentiated macrophages (Figure 1C,D, respectively).

2.1. ATP-Evoked Intracellular Ca^{2+} Responses in THP-1 Monocytes and Macrophage Models

ATP evoked a concentration-dependent increase in intracellular Ca^{2+} response in both THP-1 monocytes (EC50 = 0.174 ± 0.01 µM with extracellular Ca^{2+} and EC50 = 0.051 ± 0.005 µM without extracellular Ca^{2+}) (Figure 2A,B) and THP-1-differentiated macrophages (EC50 = 6.25 ± 1.92 µM with extracellular Ca^{2+} and EC50 = 14.9 ± 0.88 µM without extracellular Ca^{2+}; $n = 6$) (Figure 2C,D). Response to ATP (100 µM) was biphasic; in both cell types, the response involved an initial rapid increase in intracellular Ca^{2+} response, which peaked and was followed by a decay phase. While the response decayed back to the baseline level in THP-1-differentiated macrophages, the decay response remained at an elevated phase in THP-1 monocytes (Figure 2B,D). Two striking differences were observed in the phenotypic features of the ATP-evoked Ca^{2+} response in the two cell models: (1) peak magnitude was significantly higher in THP-1 monocytes compared to THP-1-differentiated macrophages (F ratio 3.24 ± 0.127 THP-1 monocytes vs. 1.09 ± 0.077 THP-1-differentiated macrophages; $n = 12$ and 8; $p < 0.001$; Figure 2B) and (2) decay response. The decay time was quantified for both cells and illustrated a much faster decaying response in THP-1-differentiated macrophages (84.0 ± 5.28 s THP-1 monocytes vs. 18.64 ± 2.18 s THP-1-differentiated macrophages; Figure 2E). Cell quantification throughout the intracellular Ca^{2+} assay using Hoechst stain H33342 showed no significant differences in the number of cells used throughout the experiments, eliminating the possibility that the difference in peak magnitude in the two cell types were due to difference in cell number (Figure 2F).

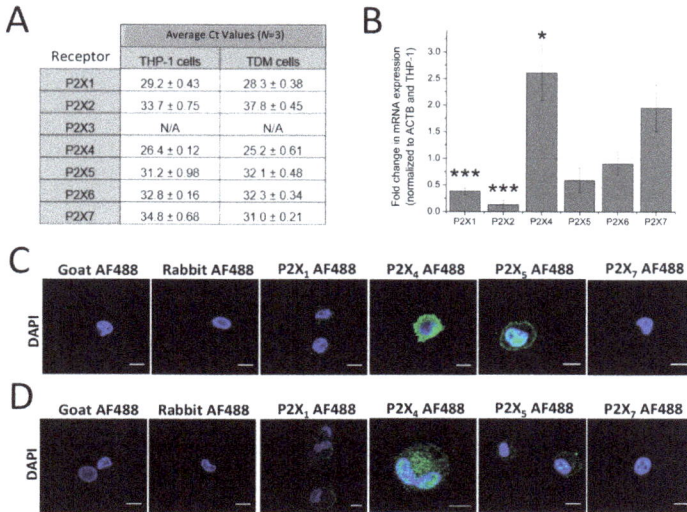

Figure 1. Expression of P2X receptors in THP-1 monocytes and THP-1-differentiated macrophages. (**A**) table showing qRT-PCR Ct values to identify which P2X genes are expressed in THP-1 cells and THP-1 differentiated macrophages. Ct values above 35.0 were considered absent ($n = 3$). Undetected genes are represented as N/A; (**B**) fold change in mRNA expression of P2X receptor genes in THP-1 differentiated macrophages, normalized to β-actin and THP-cells ($n = 3$). Asterisks include significant changes towards control (*** $p < 0.001$, * $p < 0.05$); (**C,D**) immunocytochemistry of P2X receptor expression in THP-1 monocytes and THP-1 differentiated macrophages, respectively, as observed under confocal microscopy. Blue represents 2-4(amidinophenyl)-1H-indole-6-carboxamidine (DAPI) nuclear stain while green represents human P2X receptor conjugated with AF488. Scale bar represents 10 μm. Images are representative of $n = 3$ replicates. TDM: THP-1 differentiated macrophages.

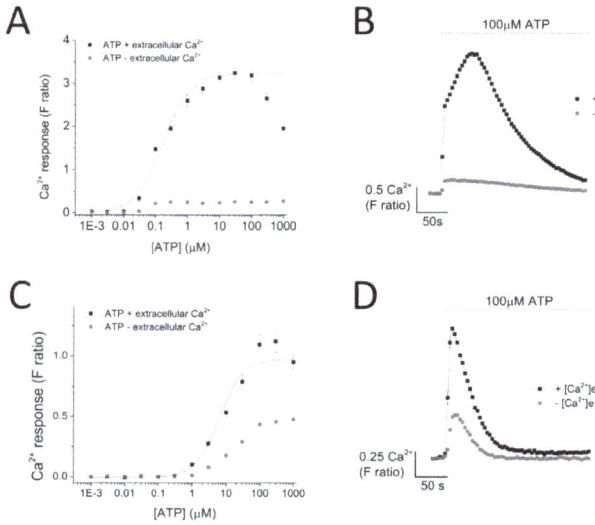

Figure 2. *Cont.*

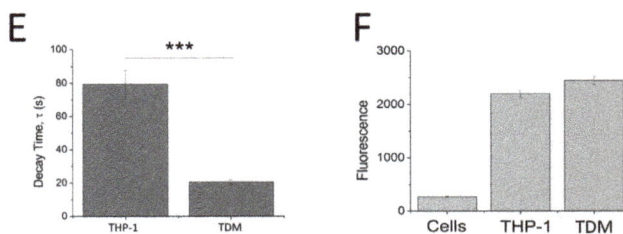

Figure 2. ATP-evoked intracellular Ca^{2+} responses in THP-1 monocytes and THP-1 differentiated macrophages. Concentration response curve of ATP (0.01–1000 μM) in the presence and absence of extracellular Ca^{2+} and its representative traces at 100 μM ATP in: (**A,B**) THP-1 cells ($n = 3$) and (**C,D**) THP-1 differentiated macrophages ($n = 3$); (**C**) bar chart representing decay kinetics of THP-1 cells ($n = 12$) and THP-1 differentiated macrophages ($n = 6$); (**D,E**) difference in decay time of 100 μM-evoked Ca^{2+} response in THP-1 versus TDM cells. Tau (τ) is the time constant calculated from fitting a single exponential decay to the falling phase of Ca^{2+} response ($n = 5$); (**F**) cell quantification assay using nuclear stain H-33342. Asterisks include significant changes towards control (*** $p < 0.001$, $n = 4$). TDM: THP-1 differentiated macrophages.

2.2. Dependency of ATP-Evoked Ca^{2+} Response on ER Calcium Store

As ATP evoked a global intracellular Ca^{2+} response, mediated by both the activation of P2X and P2Y receptors, a sarco/endoplasmic reticulum Ca^{2+}-ATPase inhibitor, Thapsigargin (Tg), was employed to eliminate metabotropic responses. Tg depletes intracellular stores of Ca^{2+} resulting in a raise in cytosolic Ca^{2+} level (F ratio 1.35 ± 0.0082 without Tg vs. 2.24 ± 0.065 with Tg; $n = 6$; $p < 0.001$; Figure 3). Pre-treatment of cells with 5 μM Tg completely attenuated ATP-evoked intracellular Ca^{2+} response in THP-1 monocytes (Figure 3A,B) while significantly inhibiting ATP-evoked intracellular Ca^{2+} response in THP-1-differentiated macrophages (88.7 ± 3.29% inhibition; $n = 6$; $p < 0.001$) (Figure 3C,D). Despite a significant inhibition with Tg, the results showed that 11.3 ± 3.29% ($n = 6$) of the ATP-evoked Ca^{2+} response appeared to be resistant to Tg. In the absence of extracellular Ca^{2+}, this Tg-resistant Ca^{2+} response was completely abolished, implying that it is dependent on Ca^{2+} influx (Figure 3E).

2.3. P2X4, but not P2X1 or P2X7, Receptor Activation Mediates the Ionotropic Calcium Response in Model Human Macrophage

The contribution of P2X4 receptor activation towards Tg-resistant ATP-evoked Ca^{2+} response was studied using a combination of pharmacological tools. Pre-treatment of cells with P2X4-specific positive allosteric modulator (ivermectin (IVM); 3 μM) resulted in the potentiation of peak magnitude of Tg-resistant Ca^{2+} response by 2.1 fold (F ratio 0.42 ± 0.025 without IVM to 0.90 ± 0.042 with IVM; $n = 6$; $p < 0.001$) (Figure 4A,B). In addition to this, IVM also significantly delayed the decay kinetics by 3.67 ± 0.37 fold (14.4 ± 0.78 s without IVM to 49.9 ± 3.8 s with IVM; $n = 6$; $p < 0.001$) (Figure 4C). Selective P2X4 receptor antagonists, 5-BDBD and PSB-12062, were also used to investigate the contribution of P2X4. In the presence of Tg, both 5-BDBD and PSB-12062 (both at 10 μM) caused a significant inhibition towards the ATP-evoked Ca^{2+} response by 52.9 ± 1.99% ($n = 3$; $p < 0.01$; Figure 5A) and 78.5 ± 3.23% ($n = 3$; $p < 0.001$; Figure 5C), respectively. In addition to inhibiting the peak magnitude of the Tg-resistant Ca^{2+} response, both antagonists also abolished the second slower response observed in the presence of Tg (Figure 5B,D). Finally, pre-treatment of THP-1-differentiated macrophages with P2X1 (Ro0437626; 30 μM) or P2X7 (A438079; 5 μM) antagonists had no inhibitory effect on Tg-resistant Ca^{2+} responses (Figure 5E,F, respectively).

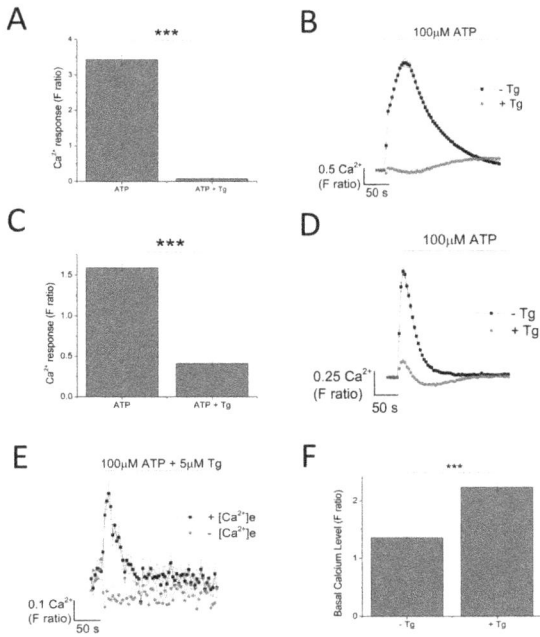

Figure 3. Dependency of ATP-evoked Ca^{2+} response on the endoplasmic reticulum Ca^{2+} store. The effect of 100 μM ATP in the presence and absence of 5 μM Tg in: (**A**,**B**) THP-1 monocytes ($n = 3$) and (**C**,**D**) THP-1 differentiated macrophages ($n = 6$); (**E**) time-response curve of the effect of Tg on 100 μM ATP in the presence and absence of extracellular Ca^{2+} ($n = 6$); (**F**) effect of 5 μM Tg on basal Ca^{2+} level in THP-1 differentiated macrophages ($n = 6$). Asterisks include significant changes towards control (*** $p < 0.001$).

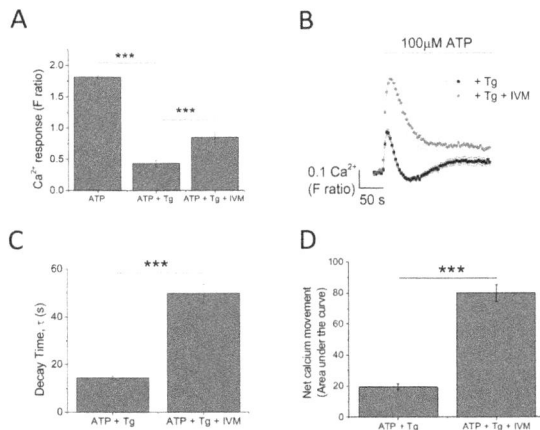

Figure 4. Investigating the contribution of P2X$_4$ receptor using the positive allosteric modulator ivermectin. (**A**,**B**) effect of 3 μM IVM on Tg-resistant Ca^{2+} response represented as bar chart and time-response curve, respectively ($n = 6$). The effect of 3 μM ivermectin (IVM) on: (**C**) decay kinetics ($n = 6$) and (**D**) net calcium movement as quantified by area under the curve ($n = 6$) of THP-1 differentiated macrophages. Asterisks include significant changes towards control (*** $p < 0.001$).

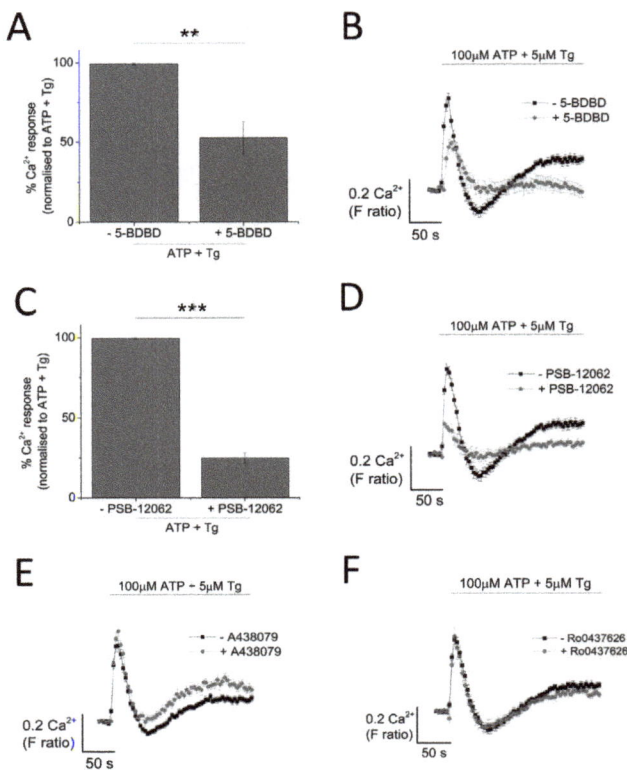

Figure 5. Effect of selective P2X receptor antagonists on Tg-resistant ATP-evoked Ca^{2+} response in THP-1 differentiated macrophages. (**A,B**) effect of $P2X_4$ receptor antagonist (10 μM 5-BDBD) on magnitude of Ca^{2+} response and net calcium movement on Tg-resistant Ca^{2+} response ($n = 3$); (**C,D**) effect of $P2X_4$ receptor antagonist (10 μM PSB-12062) on magnitude of Ca^{2+} response and net calcium movement on Tg-resistant Ca^{2+} response ($n = 3$); (**E**) effect of $P2X_7$ receptor antagonist (5 μM A438079) represented as time response traces ($n = 3$); (**F**) effect of $P2X_1$ receptor antagonist (30 μM Ro0437626) represented as time response traces ($n = 3$). Asterisks include significant changes towards control (*** $p < 0.001$, ** $p < 0.01$).

2.4. Silencing of P2X4 Abolished Tg-Resistant Calcium Response and Reduced Sensitivities Towards IVM and PSB-12062

The data so far illustrated that P2X4 receptors contribute towards the Tg-resistant ATP-evoked intracellular Ca^{2+} level. To provide further confirmation, a molecular-based approach to silence the P2X4 gene was employed. The success of the gene silencing technique was confirmed by the qRT-PCR approach ($48.8 \pm 9.64\%$ knockdown; $n = 3$; $p < 0.01$; Figure 6A) and flow cytometry ($57.7 \pm 4.37\%$ reduction in cells expressing extracellular P2X4; $n = 3$; $p < 0.01$; Figure 6B). Despite a reduction in cell surface expression of P2X4 receptor, no significant changes in 100 μM ATP-evoked intracellular Ca^{2+} responses were observed in the P2X4-shRNA cells when compared to scrambled cells (Figure 6C). However, silencing of P2X4 significantly reduced the Tg-resistant ATP-evoked Ca^{2+} response in THP-1-differentiated macrophages (F ratio 0.14 ± 0.02 ATP + Tg scrambled shRNA vs. F ratio 0.02 ± 0.014 ATP + Tg P2X4 shRNA; $n = 3$; $p < 0.001$; Figure 6D). The effect of IVM and PSB-12062 was assessed on the P2X4 shRNA cells. When compared to scrambled negative control cells (Figure 7A,B), P2X4 shRNA cells were found to be less sensitive towards both IVM and PSB-12062, as quantified using area under the curve (Figure 7B). While IVM caused a potentiation towards the net calcium movement of scrambled shRNA cells by 2.06 ± 0.23 fold ($n = 3$; $p < 0.01$; Figure 7A,B), IVM was only able to potentiate the net calcium movement of P2X4 shRNA cells by 1.53 ± 0.055 fold ($n = 3$; $p < 0.01$; Figure 7C,D). These data provide further evidence that P2X4 receptor contributes towards Ca^{2+} influx in THP-1-differentiated macrophages and that both IVM and PSB-12062 are useful tools for the study of P2X4.

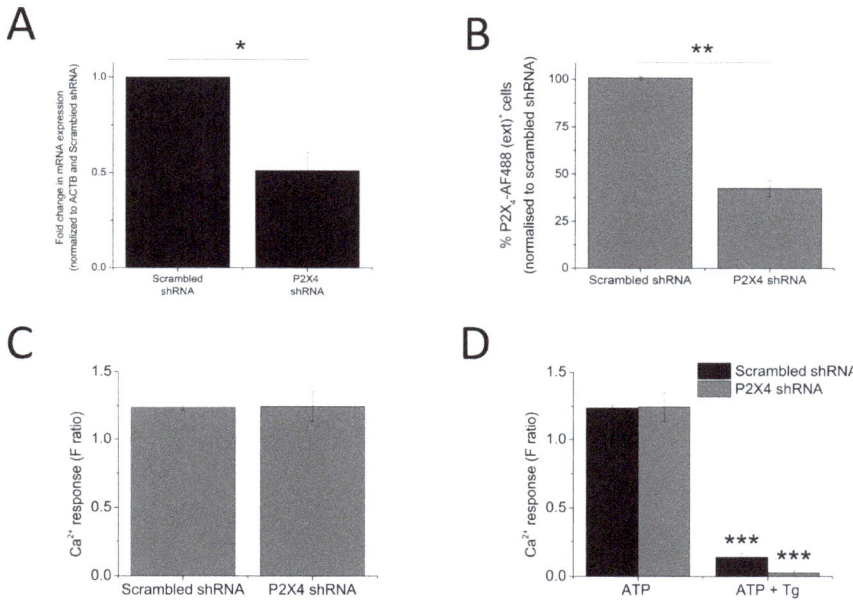

Figure 6. Loss of Tg-resistant ATP-evoked Ca^{2+} response in $P2X_4$ knockdown THP-1 differentiated macrophages. (**A,B**) quantification of $P2X_4$ knockdown level in THP-1 differentiated macrophages as measured using qRT-PCR for mRNA expression ($n = 3$) and flow cytometry for extracellular $P2X_4$ protein expression ($n = 3$), respectively. Knockdown level was quantified against negative control scrambled cells; (**C**) ATP-evoked Ca^{2+} response at 100 μM on scrambled shRNA and $P2X_4$ shRNA THP-1 differentiated macrophages ($n = 4$); (**D**) effect of Tg on 100 μM ATP response in scrambled versus $P2X_4$ shRNA THP-1 differentiated macrophages ($n = 3$). Asterisks include significant changes towards control (* $p < 0.05$, ** $p < 0.01$, *** $p < 0.001$).

2.5. Targeting Receptor Trafficking to Investigate P2X4 Contribution towards Tg-Resistant Calcium Response

Since P2X4 receptors have been previously described to be predominantly localized within lysosomal compartments [13], targeting trafficking mechanism (dynasore to block endocytosis and vacuolin-1 to block exocytosis) would provide further evidence towards the contribution of P2X4 towards ATP-evoked Ca^{2+} response. Pre-treatment of THP-1-differentiated macrophages with dynasore (80 μM) resulted in a significant reduction of the magnitude of Tg-resistant ATP-evoked Ca^{2+} response (0.42 ± 0.17 vs. 0.17 ± 0.25; $n = 3$; $p < 0.001$; Figure 8A). However, dynasore caused a significant delay in the decay kinetics (16.2 ± 1.20 s vs. 81.5 ± 7.03 s; $n = 3$; $p < 0.001$; Figure 8B) and a significant increase in the net calcium movement as quantified by area under the curve (AUC 19.9 ± 2.02 without dynasore vs. 33.36 ± 5.56 with dynasore; $n = 3$; $p < 0.05$; Figure 8C). Pre-treatment of cells with 1 μM vacuolin-1 had no effect on the magnitude of the Tg-resistant ATP-evoked Ca^{2+} response in THP-1-differentiated macrophages (Figure 8D).

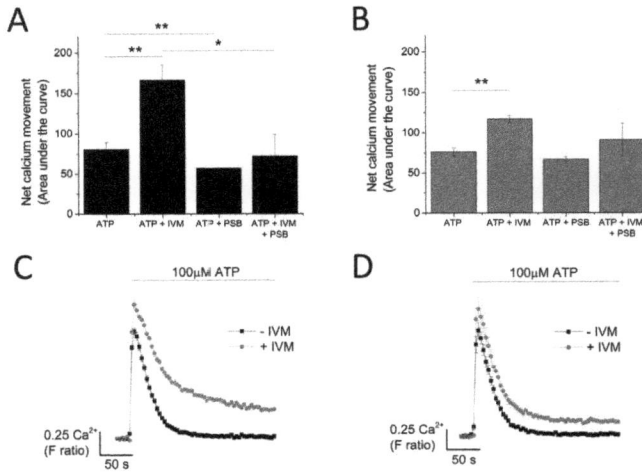

Figure 7. Knocking down P2X$_4$ reduced the sensitivity of THP-1 differentiated macrophages to IVM and PSB-12062. The effect of 10 μM PSB-12062 on ATP response in the presence and absence of IVM in: (**A**) scrambled shRNA ($n = 4$) vs. (**B**) P2X$_4$ shRNA ($n = 4$). Representative time-response curves of the effect of IVM on ATP response in: (**C**) scrambled shRNA ($n = 4$) vs. (**D**) P2X$_4$ shRNA ($n = 4$). Asterisks include significant changes towards control (*** $p < 0.001$, ** $p < 0.01$, * $p < 0.05$).

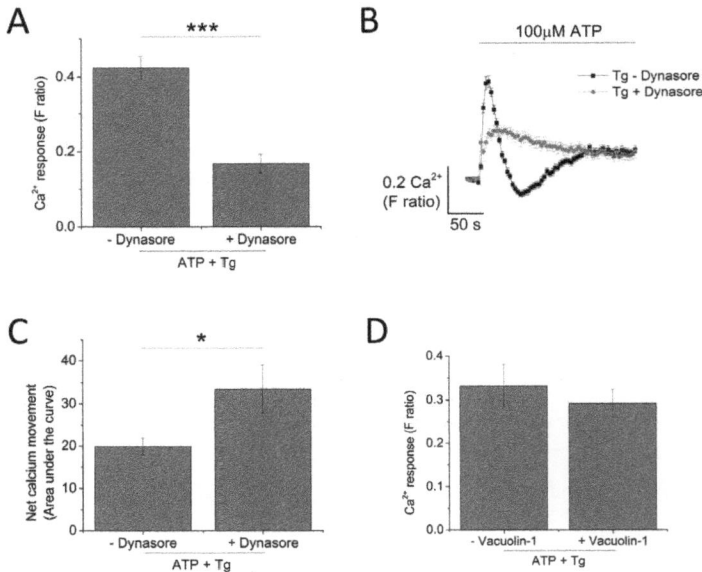

Figure 8. Targeting receptor trafficking to investigate contribution of P2X$_4$ receptor towards ATP-evoked Ca^{2+} response in THP-1 differentiated macrophages. (**A–C**) effect of blocking endocytosis (80 μM dynasore) on magnitude and net calcium movement ($n = 3$) of Tg-resistant ATP-evoked Ca^{2+} response; (**D**) effect of blocking lysosomal exocytosis (1 μM vacuolin-1) on magnitude of Tg-resistant ATP-evoked Ca^{2+} response ($n = 3$). Asterisks include significant changes towards control (*** $p < 0.001$, * $p < 0.05$).

3. Discussion

The expression of P2X receptors is widespread within the immune system, in particular P2X4 and P2X7 [6]. Despite this observation, there has been no clear evidence of the contribution of P2X4 receptor activation towards ATP-evoked Ca^{2+} influx in monocytes and macrophages. Since THP-1 monocytes and THP-1 differentiated macrophages are frequently used as an experimentally amenable model of human monocytes and macrophages [14–16], we utilized these model systems to investigate the aforementioned issue. Through qRT-PCR analysis, the present study illustrated that both THP-1 and THP-1 differentiated macrophages expresse all P2X receptors at the mRNA level with the exception of P2X2 for THP-1 and P2X2 and P2X3 for THP-1 differentiated macrophages. It was also interesting to observe that PMA differentiation of THP-1 into the model macrophage caused a downregulation of P2X1 and P2X2 mRNA expression while an upregulation of P2X4 mRNA expression was observed. To date, there is no reported comparison of mRNA expression level between THP-1 monocytes and THP-1 differentiated macrophages. However, our findings corroborated previously reported evidence by Myrtek et al. (2008) illustrating the expression of all P2X receptors except for P2X2, P2X3 and P2X6 [17]. Protein expression of P2X1, P2X4, P2X5 and P2X7 was further confirmed by immunocytochemistry analysis for both cell models.

To investigate the contribution of P2X4 receptor activation towards ATP-evoked intracellular Ca^{2+} responses in THP-1 monocytes and THP-1 differentiated macrophages, we employed intracellular Ca^{2+} measurements with the aid of various pharmacological tools. ATP elicited a concentration-dependent increase in intracellular Ca^{2+} level in both cells but with significantly different response characteristics. First, the peak magnitude of ATP-evoked Ca^{2+} responses was significantly larger in THP-1 monocytes, compared to THP-1 differentiated macrophages. Secondly, decay kinetic response in THP-1 cells was found to be significantly slower than that observed in THP-1 differentiated macrophages. Although the factors underlying these differences were not investigated further, several speculations can be made. The difference in peak magnitude may be a result of the different expression of P2Y receptors expressed on the cell surface of the two cell types. This is a plausible explanation as P2Y receptor activation has previously been reported as being responsible for mediating the first initial intracellular Ca^{2+} peak response [18,19]. A second plausible explanation would be the difference in ectoenzyme CD39 expression, responsible for the breakdown of ATP into ADP and AMP. Although monocytes and macrophages are reported to express ectoenzymes CD39 and CD73, there has been no direct comparison on the level of their expression [20,21]. On the other hand, the difference in decay kinetics of the two cells in response to ATP stimulation may be a result of the different buffering capacity of Ca^{2+} of the two cells. Previous studies revealed that antibody/FcR generated phagocytosis in macrophages is a process that relies on increase in cytosolic Ca^{2+} concentration [22,23]. Therefore, as macrophages are professional phagocytes, they may be more equipped to adapt to the changes in Ca^{2+} level.

As ATP elicits global P2 responses constituting of P2X and P2Y receptors, we utilized a SERCA inhibitor, Thapsigargin (Tg), to deplete the intracellular stores of Ca^{2+} to eliminate mobilization of Ca^{2+} from intracellular stores by P2Y receptors [24,25]. Interestingly, pre-treatment of THP-1 monocytes with 5 µM Tg completely abolished the ATP-evoked Ca^{2+} response, suggesting that it is entirely dependent on metabotropic P2Y receptor activation. The same treatment in THP-1 differentiated macrophages resulted in a significant inhibition of the ATP-evoked Ca^{2+} response, leaving a small Tg-resistant component, which was dependent on Ca^{2+} influx. Due to the lack of Tg-resistant Ca^{2+} response, we did not pursue investigation into THP-1 monocytes and focused on THP-1 differentiated macrophages instead. Our study provides evidence regarding the contribution of P2X4 receptor towards ATP-evoked Ca^{2+} response in THP-1 differentiated macrophages. This was evident from the effect observed by the use of positive allosteric modulator ivermectin (IVM) [26] and selective P2X4 receptor antagonists 5-BDBD and PSB-12062 [27–30]. Pre-treatment of THP-1 differentiated macrophages with 3 µM IVM caused a significant potentiation towards the peak magnitude of Tg-resistant Ca^{2+} response, accompanied by a significant delay in decay kinetics. It has been previously reported that the main characteristics of the IVM effect on P2X4 receptors include potentiation in magnitude and delay in

145

decay current, although studies by Norenberg et al. (2012) illustrated that only the effect on decay kinetics is specific to P2X4 receptors [26,28]. The effects of 5-BDBD and PSB-12062 on Tg-resistant Ca^{2+} response was also consistent, in which they both significantly reduced the peak magnitude as well as abolished the second slower response. Despite similar effect in inhibition, PSB-12062 appeared to be more potent in causing an inhibition towards the Tg-resistant Ca^{2+} response, and this may be due to the nature of the antagonists. While PSB-12062 has been described as an allosteric antagonist [28], 5-BDBD is thought to act competitively, which implies that its effect may be masked due to the high agonist concentration used throughout the study (100 µM ATP). Finally, the contribution of P2X4 receptor activation towards ATP-evoked Ca^{2+} response was confirmed with a knockdown approach, which illustrated that P2X4 knockdown THP-1 differentiated macrophages lacked a Tg-resistant ATP-evoked Ca^{2+} response as well as reduced sensitivities towards IVM and PSB-12062.

As P2X1, P2X4 and P2X7 are all expressed in the immune cells [31,32], we also investigated the contribution of P2X1 or P2X7 towards the Tg-resistant Ca^{2+} response. Selective P2X1 antagonist Ro0437626 and P2X7 antagonist A438079 had no significant inhibitory effect towards the Tg-resistant ATP-evoked Ca^{2+} response in THP-1 differentiated macrophages. The lack of contribution of P2X1 in ATP-evoked Ca^{2+} response in these cells contradict previous studies reported by Sim et al. and Wareham et al. illustrating functional evidence of P2X1 receptor in mouse macrophages [33] and human lung mast cells [31], respectively. Our qRT-PCR data illustrated that mRNA expression of P2X1 receptors in THP-1 differentiated macrophages is downregulated compared to THP-1 monocytes. It may therefore be interesting to investigate if this downregulation translates at the protein level, therefore being responsible for the lack of P2X1 contribution in these cells. Unlike P2X1, the lack of P2X7 receptor contribution towards ATP-evoked Ca^{2+} response in THP-1 differentiated macrophages is not as surprising. It is well established that 100 µM ATP used through this study is significantly below the activation threshold for P2X7 receptors (>500 µM) [34–36].

Altogether, the data provides evidence that P2X4 is functionally expressed in THP-1 differentiated macrophages as reflected from their contribution towards ATP-evoked Ca^{2+} response, but their functional evidence in THP-1 monocyte is lacking. Studies by Li et al. have previously reported functional P2X4 in THP-1 monocytes, although, in this case, PLC inhibitor U-73122 was used instead of Tg [12]. Furthermore, through intracellular Ca^{2+} measurements, we have showed that IVM and PSB-12062 are reliable tools for the study of P2X4, supported with findings from P2X4 knockdown cells.

4. Experimental and Materials

4.1. Chemicals and Reagents

The following reagents were used: PSB-12062 and ATP (Sigma-Aldrich, St. Louis, MO, USA), A438079 (Abcam, Cambridge, UK), Ivermectin, 5-BDBD, Ro0437626 and Thapsigargin (all from Tocris, Bristol, UK), and Fura 2-AM (TefLabs, Austin, TX, USA).

4.2. Cells

THP-1 cells were obtained from the European Collection of Cell Cultures (ECACC). Human THP-1 cells were cultured in Roswell Park Memorial Institute (RPMI) 1640 medium with 2 mM L-glutamine, 10% foetal bovine serum (FBS) and 50 IU/mL penicillin and 50 µg/mL streptomycin. Cells were maintained at 37 °C with 5% CO_2 in a humidified incubator. To generate THP-1 differentiated macrophages, cells were stimulated with 320 nM of phorbol 12-myristate 13-acetate (PMA) for 48 h at 37 °C, 5% CO_2 in a humidified incubator.

4.3. Intracellular Ca^{2+} Measurements

Cells at 2×10^5/well were loaded for 1 h at 37 °C with 2 µM Fura-2 AM. Cells were washed twice with physiological saline (SBS) buffer. Measurements were performed at 37 °C on a 96-well plate reader (FlexStation III, Molecular Devices, Sunnyvale, CA, USA). The change in intracellular

Ca^{2+} concentration is indicated as a ratio of fura-2 emission intensities for 340- and 380-nm excitation (F ratio). SBS buffer contained (mM): 130 NaCl, 5 KCl, 1.2 $MgCl_2$, 1.5 $CaCl_2$, 8 D-glucose, 10 HEPES; pH 7.4. Ca^{2+}-free SBS was prepared by excluding $CaCl_2$ and supplemented with 2 mM ethylene glycol-bis(β-aminoethyl ether)-N,N,N',N'-tetraacetic acid (EGTA). Loading of cells with Fura 2-AM was performed in SBS buffer supplemented with 0.01% (*w/v*) pluronic acid. All compounds were pre-treated onto cells for 30 min before Ca^{2+} measurements.

4.4. RNA Extraction and cDNA Synthesis

Total RNA was isolated from cells using TRI Reagent (Sigma Aldrich, St. Louis, MO, USA) and contaminating genomic DNA was eliminated using DNA-free kit (Ambion, Foster City, CA, USA). cDNA was synthesized from 0.5 μg of total RNA using SuperScript II reverse transcriptase kit (Invitrogen, Waltham, MA, USA).

4.5. Quantitative Real-Time PCR

Taqman primer probes sets for human P2X1 (Hs00175686_m1), P2X2 (Hs04176268_g1), P2X3 (Hs01125554_m1), P2X4 (Hs00602442 _m1), P2X5 (Hs01112471_m1), P2X6 (Hs01003997_m1), P2X7 (Hs00175721 _m1), β-actin (Hs01060665_g1) were obtained pre-designed from Applied Biosystems (Foster City, California, USA). qRT-PCR was performed in a 7500 Fast Real-Time PCR instrument (Applied Biosystems). Target gene expression was normalized to β-actin endogenous control and relative quantification was done by the $\Delta\Delta Ct$ method.

4.6. Generating Knockdown P2X4 THP-1 Cells

THP-1 cells (2.5 × 10^5) were transduced with either scrambled shRNA or P2X4 shRNA lentiviral particles (Thermo Scientific, Loughborough, UK) in the presence of 8 μg/mL hexadimethrine bromide and Lipofectamine 2000 (Thermo Fisher, Loughborough, UK).

4.7. Immunocytochemistry

Cells at 2.5 × 10^4/mL were fixed with 4% paraformaldehyde (15 min, RT) and permeabilized with 0.25% Triton X-100 (10 min, RT). Cells were blocked with 1% bovine serum albumin (30 min, RT) and incubated overnight at 4 °C with primary antibodies (rabbit polyclonal P2X4 (Alomone, Jerusalem, Israel), goat polyclonal P2X1,5,7 (Santa-Cruz biotechnology, Dallas, TX, USA). Cells were stained with secondary antibody goat anti-rabbit (Life Technologies, Waltham, MA, USA) or rabbit anti-goat (Abcam, Cambridge, UK) AF488. Nuclear staining was performed with Vectashield Antifade containing DAPI (Vectorlabs, Peterborough, UK). Cell imaging was performed using laser-scanning confocal microscope Zeiss LSM510 META (Zeiss, Oberkochen, Germany).

Flow cytometry—Cells (100 μL at 1 × 10^6/mL) were incubated for 10 min at room temperature (RT) with Fc block (BD, San Jose, CA, USA) and immunostained with anti-human P2X4 (Alomone, Jerusalem, Israel) or hIgG isotype control (Life Invitrogen, Waltham, MA, USA) followed by AlexaFluor 488 (AF488) (Life Invitrogen, Waltham, MA, USA). Cells were acquired on the Cytoflex instrument (Beckman Coulter, Franklin Lakes, NJ, USA). Analyses were performed on CytExpert software (Beckman Coulter, Franklin Lakes, NJ, USA).

Statistical analysis—Data were analyzed using Origin Pro 9.0 software (Origin Lab, Northampton, MA, USA). Concentration-response curves were fitted assuming a Hill coefficient of 1. Hypothesis testing for experiments with paired datasets were performed by means of paired Student's *t*-test using Origin Pro 9.0. Data are expressed as mean ± SEM of at least three independent experiments.

Acknowledgments: All sources of funding of the study should be disclosed. Please clearly indicate grants that you have received in support of your research work. Clearly state if you received funds for covering the costs to publish in open access.

Author Contributions: Janice A. Layhadi and Samuel J. Fountain designed the experiments. Janice A. Layhadi carried out all experimental work. Janice A. Layhadi and Samuel J. Fountain co-wrote the manuscript.

Conflicts of Interest: The authors declare no conflict of interest.

References

1. Elliott, M.R.; Chekeni, F.B.; Trampont, P.C.; Lazarowski, E.R.; Kadl, A.; Walk, S.F.; Park, D.; Woodson, R.I.; Ostankovich, M.; Sharma, P.; et al. Nucleotides released by apoptotic cells act as a find-me signal to promote phagocytic clearance. *Nature* **2009**, *461*, 282–286. [CrossRef] [PubMed]

2. Campwala, H.; Fountain, S.J. Constitutive and agonist stimulated ATP secretion in leukocytes. *Commun. Integr. Biol.* **2013**, *6*, e23631. [CrossRef] [PubMed]

3. Ralevic, V.; Burnstock, G. Receptors for purines and pyrimidines. *Pharmacol. Rev.* **1998**, *50*, 413–492. [PubMed]

4. Burnstock, G. Physiology and pathophysiology of purinergic neurotransmission. *Physiol. Rev.* **2007**, *87*, 659–797. [CrossRef] [PubMed]

5. Burnstock, G.; Knight, G.E. Cellular distribution and functions of P2 receptor subtypes in different systems. *Int. Rev. Cytol.* **2004**, *240*, 31–304. [PubMed]

6. Boumechache, M.; Masin, M.; Edwardson, J.M.; Górecki, D.C.; Murrell-Lagnado, R. Analysis of assembly and trafficking of native P2X4 and P2X7 receptor complexes in rodent immune cells. *J. Biol. Chem.* **2009**, *284*, 13446–13454. [CrossRef] [PubMed]

7. Surprenant, A.; North, R.A. Signaling at purinergic P2X receptors. *Annu. Rev. Physiol.* **2009**, *71*, 333–359. [CrossRef] [PubMed]

8. Tsuda, M.; Shigemoto-Mogami, Y.; Koizumi, S.; Mizokoshi, A.; Kohsaka, S.; Salter, M.W.; Inoue, K. P2X4 receptors induced in spinal microglia gate tactile allodynia after nerve injury. *Nature* **2003**, *424*, 778–783. [CrossRef] [PubMed]

9. Inoue, K.; Koizumi, S.; Tsuda, M.; Shigemoto-Mogami, Y. Signaling of ATP receptors in glia-neuron interaction and pain. *Life Sci.* **2003**, *74*, 189–197. [CrossRef] [PubMed]

10. Ulmann, L.; Hatcher, J.P.; Hughes, J.P.; Chaumont, S.; Green, P.J.; Conquet, F.; Buell, G.N.; Reeve, A.J.; Chessell, I.P.; Rassendren, F. Up-regulation of P2X4 receptors in spinal microglia after peripheral nerve injury mediates BDNF release and neuropathic pain. *J. Neurosci.* **2008**, *28*, 11263–11268. [CrossRef] [PubMed]

11. Ulmann, L.; Hirbec, H.; Rassendren, F. P2X4 receptors mediate PGE2 release by tissue-resident macrophages and initiate inflammatory pain. *EMBO J.* **2010**, *29*, 2290–2300. [CrossRef] [PubMed]

12. Li, J.; Fountain, S.J. Fluvastatin suppresses native and recombinant human P2X4 receptor function. *Purinergic Signal.* **2012**, *8*, 311–316. [CrossRef] [PubMed]

13. Qureshi, O.S.; Paramasivam, A.; Yu, J.C.; Murrell-Lagnado, R.D. Regulation of P2X4 receptors by lysosomal targeting, glycan protection and exocytosis. *J. Cell Sci.* **2007**, *120*, 3838–3849. [CrossRef] [PubMed]

14. Daigneault, M.; Preston, J.A.; Marriott, H.M.; Whyte, M.K.; Dockrell, D.H. The identification of markers of macrophage differentiation in PMA-stimulated THP-1 cells and monocyte-derived macrophages. *PLoS ONE* **2010**, *5*, e8668. [CrossRef] [PubMed]

15. Bosshart, H.; Heinzelmann, M. THP-1 cells as a model for human monocytes. *Ann. Transl. Med.* **2016**, *4*, 438. [CrossRef] [PubMed]

16. Qin, Z. The use of THP-1 cells as a model for mimicking the function and regulation of monocytes and macrophages in the vasculature. *Atherosclerosis* **2012**, *221*, 2–11. [CrossRef] [PubMed]

17. Myrtek, D.; Muller, T.; Geyer, V.; Derr, N.; Ferrari, D.; Zissel, G.; Durk, T.; Sorichter, S.; Luttmann, W.; Kuepper, M.; et al. Activation of human alveolar macrophages via P2 receptors: Coupling to intracellular Ca^{2+} increases and cytokine secretion. *J. Immunol.* **2008**, *181*, 2181–2188. [CrossRef] [PubMed]

18. Yu, D.; Buibas, M.; Chow, S.-K.; Lee, I.; Singer, Z.; Silva, G. Characterization of calcium-mediated intracellular and intercellular signaling in the rmc-1 glial cell line. *Cell. Mol. Bioeng.* **2009**, *2*, 144–155. [CrossRef] [PubMed]

19. Chow, J.; Liton, P.B.; Luna, C.; Wong, F.; Gonzalez, P. Effect of cellular senescence on the P2Y-receptor mediated calcium response in trabecular meshwork cells. *Mol. Vis.* **2007**, *13*, 1926–1933. [PubMed]

20. Hamidzadeh, K.; Mosser, D.M. Purinergic signaling to terminate TLR responses in macrophages. *Front. Immunol.* **2016**, *7*, 74. [CrossRef] [PubMed]

21. Antonioli, L.; Pacher, P.; Vizi, E.S.; Hasko, G. CD39 and CD73 in immunity and inflammation. *Trends Mol. Med.* **2013**, *19*, 355–367. [CrossRef] [PubMed]

22. Diler, E.; Schicht, M.; Rabung, A.; Tschernig, T.; Meier, C.; Rausch, F.; Garreis, F.; Brauer, L.; Paulsen, F. The novel surfactant protein SP-H enhances the phagocytosis efficiency of macrophage-like cell lines U937 and MH-S. *BMC Res. Notes* **2014**, *7*, 851. [CrossRef] [PubMed]

23. Lew, D.; Andersson, T.; Hed, J.; Di Virgilio, F.; Pozzan, T.; Stendahl, O. Ca^{2+}-dependent and Ca^{2+}-independent phagocytosis in human neutrophils. *Nature* **1985**, *315*, 509–511. [CrossRef] [PubMed]

24. Mori, M.; Hosomi, H.; Nishizaki, T.; Kawahara, K.; Okada, Y. Calcium release from intracellular stores evoked by extracellular ATP in a Xenopus renal epithelial cell line. *J. Physiol.* **1997**, *502*, 365–373. [CrossRef] [PubMed]

25. Desai, B.; Leitinger, N. Purinergic and calcium signalling in macrophage function and plasticity. *Front. Immunol.* **2014**, *5*, 580. [CrossRef] [PubMed]

26. Norenberg, W.; Sobottka, H.; Hempel, C.; Plotz, T.; Fischer, W.; Schmalzing, G.; Schaefer, M. Positive allosteric modulation by ivermectin of human but not murine P2X7 receptors. *Br. J. Pharmacol.* **2012**, *167*, 48–66. [CrossRef] [PubMed]

27. Fountain, S.J.; North, R.A. A C-terminal lysine that controls human P2X4 receptor desensitization. *J. Biol. Chem.* **2006**, *281*, 15044–15049. [CrossRef] [PubMed]

28. Hernandez-Olmos, V.; Abdelrahman, A.; El-Tayeb, A.; Freudendahl, D.; Weinhausen, S.; Muller, C.E. N-substituted phenoxazine and acridone derivatives: Structure-activity relationships of potent P2X4 receptor antagonists. *J. Med. Chem.* **2012**, *55*, 9576–9588. [CrossRef] [PubMed]

29. Balazs, B.; Danko, T.; Kovacs, G.; Koles, L.; Hediger, M.A.; Zsembery, A. Investigation of the inhibitory effects of the benzodiazepine derivative, 5-BDBD on P2X4 purinergic receptors by two complementary methods. *Cell Physiol. Biochem.* **2013**, *32*, 11–24. [CrossRef] [PubMed]

30. Donnelly-Roberts, D.; McGaraughty, S.; Shieh, C.C.; Honore, P.; Jarvis, M.F. Painful purinergic receptors. *J. Pharmacol. Exp. Ther.* **2008**, *324*, 409–415. [CrossRef] [PubMed]

31. Wareham, K.; Vial, C.; Wykes, R.C.E.; Bradding, P.; Seward, E.P. Functional evidence for the expression of P2X1, P2X4 and P2X7 receptors in human lung mast cells. *Br. J. Pharmacol.* **2009**, *157*, 1215–1224. [CrossRef] [PubMed]

32. Sluyter, R.; Watson, D. T cells continue to play on the ATP circuit. *J. Immunol. Res.* **2014**, *1*, 2.

33. Sim, J.A.; Park, C.K.; Oh, S.B.; Evans, R.J.; North, R.A. P2X1 and P2X4 receptor currents in mouse macrophages. *Br. J. Pharmacol.* **2007**, *152*, 1283–1290.

34. Dubyak, G. Go it alone no more—P2X7 joins the society of heteromeric ATP-gated receptor channels. *Mol. Pharmacol.* **2007**, *72*, 1402–1405. [CrossRef] [PubMed]

35. Khakh, B.S.; North, R.A. P2X receptors as cell-surface ATP sensors in health and disease. *Nature* **2006**, *442*, 527–532. [CrossRef] [PubMed]

36. North, R.A. Molecular physiology of P2X receptors. *Physiol. Rev.* **2002**, *82*, 1013–1067. [CrossRef] [PubMed]

International Journal of
Molecular Sciences

MDPI

Article

Beta-Estradiol Regulates Voltage-Gated Calcium Channels and Estrogen Receptors in Telocytes from Human Myometrium

Adela Banciu [1,2,†], Daniel Dumitru Banciu [1,†], Cosmin Catalin Mustaciosu [3,4], Mihai Radu [3], Dragos Cretoiu [5,6], Junjie Xiao [7], Sanda Maria Cretoiu [5], Nicolae Suciu [6,8] and Beatrice Mihaela Radu [1,9,*]

[1] Department of Anatomy, Animal Physiology and Biophysics, Faculty of Biology, University of Bucharest, Splaiul Independentei 91-95, 050095 Bucharest, Romania; adela.banciu79@gmail.com (A.B.); danieldumitrubanciu@gmail.com (D.D.B.)

[2] Faculty of Medical Engineering, University Politehnica of Bucharest, Gheorge Polizu Street 1-7, 011061 Bucharest, Romania

[3] Department of Life and Environmental Physics, Horia Hulubei National Institute of Physics and Nuclear Engineering, Reactorului 30, P.O. Box MG-6, 077125 Magurele, Romania; cosmin@nipne.ro (C.C.M.); mradu@nipne.ro (M.R.)

[4] Faculty of Applied Chemistry and Materials Science, University Politehnica of Bucharest, 011061 Bucharest, Romania

[5] Department of Cell and Molecular Biology and Histology, Carol Davila University of Medicine and Pharmacy, 050474 Bucharest, Romania; dragos@cretoiu.ro (D.C.); sanda@cretoiu.ro (S.M.C.)

[6] Alessandrescu-Rusescu National Institute of Mother and Child Health, Fetal Medicine Excellence Research Center, 020395 Bucharest, Romania; nsuciu54@yahoo.com

[7] Cardiac Regeneration and Ageing Lab, Experimental Center of Life Sciences, School of Life Science, Shanghai University, Shanghai 200444, China; junjiexiao@shu.edu.cn

[8] Department of Obstetrics and Gynecology, Polizu Clinical Hospital, 011062 Bucharest, Romania

[9] Life, Environmental and Earth Sciences Division, Research Institute of the University of Bucharest (ICUB), 91-95 Splaiul Independenței, 050095 Bucharest, Romania

* Correspondence: beatrice.radu@bio.unibuc.ro; Tel./Fax: +40-21-3181573

† These authors contributed equally to this work.

Received: 20 March 2018; Accepted: 2 May 2018; Published: 9 May 2018

Abstract: Voltage-gated calcium channels and estrogen receptors are essential players in uterine physiology, and their association with different calcium signaling pathways contributes to healthy and pathological conditions of the uterine myometrium. Among the properties of the various cell subtypes present in human uterine myometrium, there is increasing evidence that calcium oscillations in telocytes (TCs) contribute to contractile activity and pregnancy. Our study aimed to evaluate the effects of beta-estradiol on voltage-gated calcium channels and estrogen receptors in TCs from human uterine myometrium and to understand their role in pregnancy. For this purpose, we employed patch-clamp recordings, ratiometric Fura-2-based calcium imaging analysis, and qRT-PCR techniques for the analysis of cultured human myometrial TCs derived from pregnant and non-pregnant uterine samples. In human myometrial TCs from both non-pregnant and pregnant uterus, we evidenced by qRT-PCR the presence of genes encoding for voltage-gated calcium channels (Cav3.1, Ca3.2, Cav3.3, Cav2.1), estrogen receptors (ESR1, ESR2, GPR30), and nuclear receptor coactivator 3 (NCOA3). Pregnancy significantly upregulated Cav3.1 and downregulated Cav3.2, Cav3.3, ESR1, ESR2, and NCOA3, compared to the non-pregnant condition. Beta-estradiol treatment (24 h, 10, 100, 1000 nM) downregulated Cav3.2, Cav3.3, Cav1.2, ESR1, ESR2, GRP30, and NCOA3 in TCs from human pregnant uterine myometrium. We also confirmed the functional expression of voltage-gated calcium channels by patch-clamp recordings and calcium imaging analysis of TCs from pregnant human myometrium by perfusing with BAY K8644, which induced calcium influx through these channels. Additionally, we demonstrated that beta-estradiol (1000 nM) antagonized the effect of BAY K8644 (2.5 or 5 μM) in the same preparations. In conclusion, we evidenced the presence of voltage-gated calcium channels and estrogen receptors in TCs from non-pregnant and pregnant human uterine myometrium and their gene expression regulation by beta-estradiol in pregnant conditions. Further exploration of the calcium signaling in TCs and its modulation by estrogen hormones will contribute to the understanding of labor and pregnancy mechanisms and to the development of effective strategies to reduce the risk of premature birth.

Keywords: beta-estradiol; human uterine myometrium; telocytes; calcium signaling; voltage-gated calcium channels; estrogen receptors

1. Introduction

Uterine contractions represent a key point throughout pregnancy. Considered to be normal during pregnancy, uterine contractions may also be responsible for triggering premature or dysfunctional labor. During pregnancy, the myometrium suffers morphological and adaptive changes which enable it to become a forceful organ necessary for delivery. Myometrium contractility involves, among other physiological mechanisms that produce excitation in the uterus, changes in Ca^{2+} signals. Smooth myofiber contraction requires actin and myosin myofilaments and their interaction. The cross-bridge formation and contraction are mediated by elevated levels of intracellular Ca^{2+} and myosin light-chain phosphorylation.

Calcium signaling plays an essential role in uterine contractility and involves several key players, including voltage-gated calcium channels, calcium-activated chloride channels, large conductance Ca^{2+}-activated K^+ channels (BK(Ca)), calcium-sensing receptors, and transient receptor potential channels [1–6]. On the basis of their activation electrophysiological features, voltage-gated calcium channels have been classified in high-voltage-activated Ca^{2+} channels (HVA) and low-voltage-activated Ca^{2+} channels (LVA). Among the HVA channels, only the L-type Ca^{2+} channels are expressed in uterus. Their family includes Cav1.1 (*CACNA1S* gene), Cav1.2 (*CACNA1C* gene), Cav1.3 (*CACNA1D* gene), and Cav1.4 (*CACNA1F* gene) channels. Meanwhile, in uterus, the LVA channels are represented by T-type calcium channels, classified as Cav3.1 (*CACNA1G* gene), Cav3.2 (*CACNA1H* gene), and Cav3.3 channel (*CACNA1I* gene).

In uterine contractility, intracellular Ca^{2+} comes from two sources: entry across the sarcolemma through voltage-gated L-type Ca^{2+} channels and release from the sarcoplasmic reticulum [7]. In the uterus, the major Ca^{2+} source for contraction comes from the extracellular space through voltage-gated L-type Ca^{2+} channels [8]. Indeed, calcium influx through the L-type Ca^{2+} channels was demonstrated to be essential for labor and uterine contractility [1]. Moreover, uterine phasic contractions were abolished if the L-type Ca^{2+} channels were blocked [1]. In addition, the T-type Ca^{2+} channels might also contribute to calcium entry in smooth myocytes in human myometrium [9].

The estrogen 17beta-estradiol was demonstrated in different types of tissue (e.g., neuronal, uterine etc.) to modulate multiple components of the calcium transport pathways, including BK(Ca) [10], small conductance Ca^{2+}-activated potassium channel subtype 3 (SK3, [11]), HVA channels [12,13], LVA channels [14], Na^+/Ca^{2+} exchanger (NCX, [15]), transient receptor potential channels [16,17], etc. In particular, beta-estradiol and its receptors play an essential role in pregnancy and labor [18–20] and were shown to act on uterine myometrium by means of various calcium signaling pathways.

Telocytes (TCs) have been described by our team as a new cellular type in human myometrium [21]. Characterized by extremely long telopodes, TCs were described in the interstitial space of numerous organs, as 3-D network-forming cells by homocellular or heterocellular contacts [22,23]. It was shown that TCs release extracellular vesicles, such as exosomes and ectosomes, to regulate the functions of the surrounding cells in non-pregnant and pregnant myometrium [24]. The gene profiles, proteome and secretome features of TCs were recently revealed [25–27]. Moreover, our early studies showed that

TCs express estrogen and progesterone receptors, which indicates them as potentially responsible for myogenic contractility modulation under hormonal control [28].

There are some extensive reviews describing the main morphological features and possible functions of TCs in reproductive organs [29–32]. One of the most important aspects relates to the TCs stemness capacity which might contribute to regeneration and repair processes, as it has been previously shown [33]. In particular, a recent review highlights the contribution of calcium signaling in interstitial cells (including TCs, interstitial cells of Cajal, interstitial Cajal-like cells) to uterine, cardiac, and urinary physiology and pathology [34].

Recently, we demonstrated, by immunofluorescence and electrophysiology, the presence of the T-type calcium channels in human uterine myometrial TCs [35]. Our studies also confirmed the antagonistic pharmacological effect of mibefradil on voltage-gated calcium currents in TCs from human uterine myometrium [35] and its modulatory effect on telopodes growth after stimulation with near-infrared low-level lasers [36]. We also evidenced, by patch-clamp recordings, the presence of HVA calcium currents in TCs from human non-pregnant myometrium [35]. To date, no studies have yet quantified the mRNA levels encoding the voltage-gated calcium channels in human uterine myometrial TCs. Moreover, despite extensive studies describing the effect of beta-estradiol on human uterine myometrium, no description of its mechanism of action on TCs was done.

TCs were recently detected in human uterine leiomyoma in higher number compared with the areas of adjacent fibrotic and normal myometrium, suggesting that it may provide a scaffold for newly formed myocytes or control important downstream signaling pathways [37]. By contrast, another study claims that TCs are absent inside leiomyomas and represent ~2% of the cells in the normal myometrium [38]. Since leiomyomas are considered estrogen-dependent tumors, and leiomyoma tissue is more sensitive to estradiol (has more estrogen receptors in comparison to normal myometrium), authors suggested that the loss of the TC network might lead to myocytes taking up the hormonal sensor function, which can determine an uncontrollable proliferation of myocytes [38]. However, in our opinion, both studies have a major weakness since they do not demonstrate the presence of TCs with the aid of their most specific immunohistochemical markers CD34 and platelet-derived growth factor receptor (PDGFR) α or β [39,40], but use c-kit for the detection. TCs' c-kit positivity is also used by the authors to suggest the role of TC progenitor cells for the development of leiomyoma. Moreover, on the basis of the assumption that TCs are involved in angiogenesis [41,42], they also suggest that TCs loss may be responsible for decreased vessel formation within the myometrium and subsequent shifting from aerobic to anaerobic metabolism in smooth muscle cells [38]. The subsequent hypoxia and the decreased angiogenesis represent crucial factors for leiomyoma development [43,44].

Uterine TCs have been reported to be present in different reproductive states; while endometrial TCs have been hypothesized to be involved in glandular support and stromal cell communication, myometrial TCs were considered to be responsible for the initiation and propagation of contractile activity [45]. Calcium is a key player in uterine physiological activity, and multiple calcium signaling pathways have been evidenced to be activated in pregnancy and labor. In particular, voltage-gated calcium channels (e.g., HVA and LVA calcium channels) [46,47] or estrogen receptors [48] have been described to be involved in the spontaneous contractile activity in both pregnant and non-pregnant uterus. Therefore, our results in human myometrial TCs from pregnant and non-pregnant uterus are clinically relevant.

Our goal was to determine the level of expression of voltage-gated calcium channels and estrogen receptors in human myometrial TCs and to evidence potential differences in pregnant versus non-pregnant uterus. Additionally, we focused our attention on the modulatory effect of female sex hormones (e.g., beta-estradiol) exerted on the voltage-gated calcium channels and estrogen receptors in human myometrial TCs from pregnant uterine samples.

Our study is trying to highlight some of the calcium signaling mechanisms associated with voltage-gated calcium channels and estrogen receptors activation and to understand their involvement in the pregnancy state.

Int. J. Mol. Sci. **2018**, *19*, 1413

2. Results

2.1. Pregnancy Induces Changes in mRNA Levels Encoding the Voltage-Gated Calcium Channels Compared to the Non-Pregnant Condition in Human Uterine Myometrial TCs

In a previous study, we demonstrated the immunopositivity for T-type calcium channels (i.e., Cav3.1 and Cav3.2) in human myometrial TCs from pregnant and non-pregnant uterus [35]. We also evidenced the presence of T-type calcium currents and HVA currents in human myometrial TCs by employing patch-clamp recordings [35]. In the present study, we wanted to quantify the mRNA levels for voltage-gated calcium channels by qRT-PCR in both types of uterine samples. Indeed, we confirmed the presence of the followings genes: *CACNA1G* (encoding Cav3.1 T-type calcium channel), *CACNA1H* (encoding Cav3.2 T-type calcium channel) > *CACNA1I* (encoding Cav3.3 T-type calcium channel), and *CACNA1C* (encoding Cav1.2 L-type calcium channel) in TCs from both pregnant and non-pregnant uterine myometrium. The ranking of the mRNA levels encoding the voltage-gated calcium channels relative to *GAPDH* was: *CACNA1G* (Cav3.1) > *CACNA1C* (Cav1.2) > *CACNA1H* (Cav3.2) > *CACNA1I* (Cav3.3) in TCs from both pregnant (Figure 1C,D) and non-pregnant (Figure 1A,B) uterine myometrium. The same mRNA level ranking was obtained relative to the other housekeeping gene *18S rRNA*.

We also evaluated if there were any changes in the mRNA levels encoding the voltage-gated calcium channels due to the pregnancy condition (Figure 1E,F). When considering *GAPDH* as a reference gene and the non-pregnant condition as a calibrator, *CACNA1G* was upregulated 3.74-fold ($p < 0.01$), while *CACNA1H*, *CACNA1I*, and *CACNA1C* were downregulated 0.87-fold ($p < 0.05$), 0.81-fold ($p < 0.05$), and 0.56-fold (not significant), respectively, in pregnant samples versus non-pregnant samples. On the other hand, when considering *18S rRNA* as a reference gene and the non-pregnant condition as a calibrator, *CACNA1G* was upregulated 6.48-fold ($p < 0.01$), while *CACNA1H*, *CACNA1I*, and *CACNA1C* were downregulated 0.82-fold ($p < 0.05$), 0.71-fold ($p < 0.05$), and 0.29-fold (not significant), respectively, in pregnant samples versus non-pregnant samples.

2.2. Pregnancy Induces Changes in mRNA Levels Encoding the Estrogen Receptors Compared to the Non-Pregnant Condition

Estrogens (i.e., beta-estradiol) are essential hormones in uterine physiology. They may act on multiple cellular targets, including estrogen receptors (ESR1 and ESR2), G protein-coupled receptors (GPR30), and nuclear receptors (NCOA3). The role of estrogen receptors in the uterine myometrium physiology was demonstrated in different species, including human, canine, equine, rat etc. [49–53]. We might expect that these genes are present in different subtypes of cells of the human uterine myometrium, including TCs and smooth muscle cells. Therefore, we analyzed the presence of *ESR1*, *ESR2*, *GPR30*, and *NCOA3* genes in human myometrial TCs from non-pregnant and pregnant uterine samples. We confirmed the presence of all four analyzed genes, with the following ranking: *NCOA3* > *ESR1* > *GPR30* > *ESR2*. Their presence was not dependent on the pregnancy state or the used housekeeping gene (Figure 2A–D).

We also demonstrated a significant downregulation of the mRNA levels of all four genes in human myometrial TCs from pregnant uterine samples compared to non-pregnant uterine samples (Figure 2E,F). When considering *GAPDH* as a reference gene and the non-pregnant condition as a calibrator, the estrogen receptors were downregulated in pregnant samples 0.89-fold ($p < 0.05$, *ESR1*), 0.77-fold (not significant, *ESR2*), 0.67-fold ($p < 0.05$, *GPR30*), 0.86-fold ($p < 0.05$, *NCOA3*) versus non-pregnant samples. Meanwhile, when considering *18S rRNA* as a reference gene and the non-pregnant condition as a calibrator, the estrogen receptors were downregulated in pregnant samples 0.84-fold ($p < 0.05$, *ESR1*), 0.68-fold ($p < 0.05$, *ESR2*), 0.54-fold ($p < 0.01$, *GPR30*), 0.8-fold ($p < 0.05$, *NCOA3*) versus non-pregnant samples.

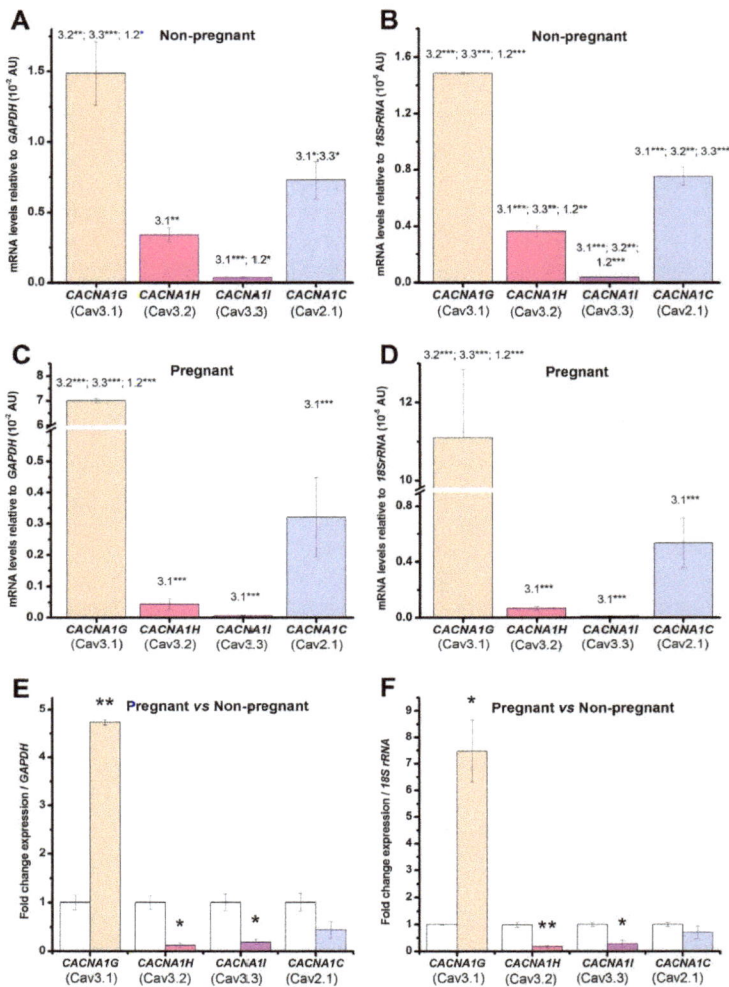

Figure 1. mRNA levels encoding the voltage-gated calcium channels in human myometrial telocytes (TCs) from non-pregnant (**A,B**) and pregnant (**C,D**) uterine samples. The mRNA levels for *CACNA1G* (Cav3.1 channel), *CACNA1H* (Cav3.2 channel), *CACNA1I* (Cav3.3 channel), and *CACNA1C* (Cav1.2 channel) were normalized against two different reference genes, *GAPDH* (**A,C**) and *18S rRNA* (**B,D**), and were plotted as mean ± SD (*N* = 3), corresponding to cell batches extracted from either pregnant or non-pregnant uterine myometrium. One-way ANOVA analysis, applied for the samples derived either from pregnant uterus or from non-pregnant uterus, indicated: (**A**) F = 78, *p* < 0.001; (**B**) F = 545, *p* < 0.001; (**C**) F = 2580, *p* < 0.001; (**D**) F = 77, *p* < 0.001. The post-hoc Bonferroni test was applied, and the statistically significant pairs are indicated by the numbers corresponding to the channels symbols, as follows: Cav3.1, 3.2, 3.3, and 1.2; (**E,F**) Fold-change for each voltage-gated calcium channel between non-pregnant and pregnant uterine samples, where non-pregnant samples were defined as calibrator, considering as housekeeping gene *GAPDH* (**E**) and *18S rRNA* (**F**), respectively. Statistical significance is indicated with asterisks (* 0.01 < *p* < 0.05; ** 0.001 < *p* < 0.01; *** *p* < 0.001).

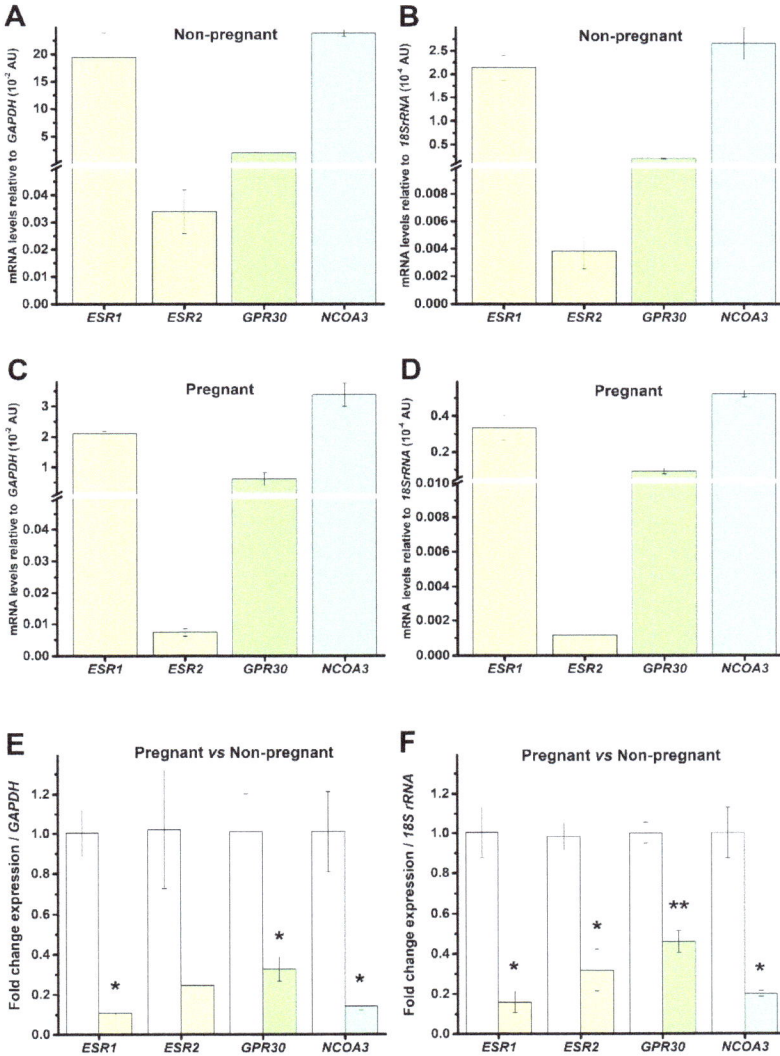

Figure 2. mRNA levels encoding the estrogen receptors and nuclear receptor co-activator in human myometrial TCs from non-pregnant (**A**,**B**) and pregnant (**C**,**D**) uterine samples. The mRNA levels of *ESR1* (estrogen receptor 1), *ESR2* (estrogen receptor 2), *GPR30* (G protein-coupled receptor 30), and *NCOA3* (nuclear receptor coactivator 3), were normalized against two different reference genes, *GAPDH* (**A**,**C**) and *18S rRNA* (**B**,**D**) and were plotted as mean ± SD (N = 3), corresponding to cell batches extracted from either pregnant or non-pregnant uterine myometrium. (**E**,**F**) Fold-change for each estrogen receptor in pregnant uterine samples versus the non-pregnant condition, where non-pregnant samples were defined as calibrator, considering as housekeeping gene *GAPDH* (**E**) and *18S rRNA* (**F**), respectively. Statistical significance is indicated with asterisks (* $0.01 < p < 0.05$; ** $0.001 < p < 0.01$).

2.3. Beta-Estradiol Downregulates Voltage-Gated Calcium Channels in TCs from Human Pregnant Uterine Myometrial Cultures

We considered only the pregnant condition for further testing the effect of beta-estradiol. Primary cultures of TCs from human pregnant myometrium were treated for 24 h with increasing concentrations of beta-estradiol (10, 100, and 1000 nM). The treatment protocol with beta-estradiol was established on the basis of previous studies [54]. We quantified by qRT-PCR the mRNA levels encoding the Cav3.1, Cav3.2, Cav3.3, and Cav1.2 channels at different doses of beta-estradiol, considering *GAPDH* or *18S rRNA* as housekeeping genes.

Beta-estradiol treatment downregulated the mRNAs encoding the voltage-gated calcium channels, with distinct patterns for each channel subtype. Considering the untreated cells as calibrator and *GAPDH* as the reference gene, *CACNA1G* (One way ANOVA, F = 13, $p < 0.05$), *CACNA1GH* (One way ANOVA, F = 14, $p < 0.05$), and *CACNA1GI* (One way ANOVA, F = 37, $p < 0.01$) channels were significantly downregulated (Figure 3A). The Bonferroni post-hoc analysis indicated that 100 nM beta-estradiol significantly downregulated *CACNA1G* by 0.67-fold ($p < 0.05$), *CACNA1GH* by 0.75-fold ($p < 0.05$), *CACNA1GH* by 0.86-fold ($p < 0.05$), versus non-treated samples. Considering the untreated cells as calibrator and *18S rRNA* as the reference gene, *CACNA1G* (One way ANOVA, F = 29, $p < 0.01$), *CACNA1GH* (One way ANOVA, F = 12.5, $p < 0.05$), *CACNA1GI* (One way ANOVA, F = 24, $p < 0.01$), and *CACNA1C* (One way ANOVA, F = 103, $p < 0.001$) channels were significantly downregulated (Figure 3B). The Bonferroni post-hoc analysis indicated that 100 and 1000 nM beta-estradiol significantly downregulated *CACNA1G* by 0.77-fold ($p < 0.01$) and 0.74-fold ($p < 0.05$), *CACNA1GH* by 0.73-fold ($p < 0.05$) and 0.82-fold ($p < 0.05$), *CACNA1GH* by 0.68-fold ($p < 0.05$) and 0.89-fold ($p < 0.01$), *CACNA1GC* by 0.49-fold ($p < 0.01$) and 0.36-fold ($p < 0.01$), respectively, versus the non-treated samples. It should be mentioned that the downregulation of *CACNA1GC* mRNA upon treatment with beta-estradiol was statistically significantly only relative to *18S rRNA* and not relative to *GAPDH*.

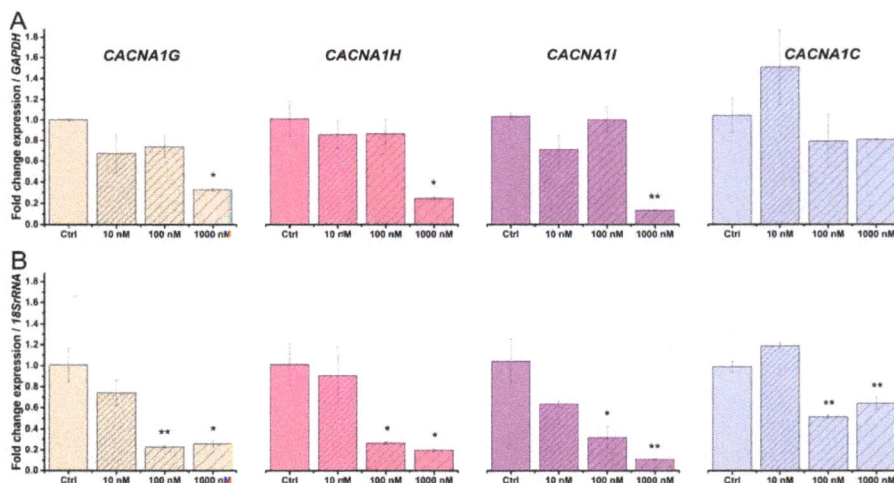

Figure 3. Effect of beta-estradiol treatment (10, 100, and 1000 nM) on the mRNA levels encoding the voltage-gated calcium channels in human myometrial TCs from pregnant uterine samples. Fold-change for each voltage-gated calcium channel upon treatment with beta-estradiol, where the non-treated samples were considered as calibrator. The data were normalized against *GAPDH* (**A**) and 18S rRNA (**B**) and were plotted as mean ± SD ($N = 3$). Statistical significance is indicated with asterisks (* $0.01 < p < 0.05$; ** $0.001 < p < 0.01$).

2.4. Beta-Estradiol Downregulates Estrogen Receptors in TCs from Human Pregnant Uterine Myometrial Cultures

Previous studies indicated that beta-estradiol was able to regulate the mRNA levels of estrogen receptors in different tissues [55–57]. Therefore, we analyzed the effect of a 24 h treatment with beta-estradiol (10,100, 1000 nM) on the mRNA levels for *ESR1* (estrogen receptor 1), *ESR2* (estrogen receptor 2), *GPR30* (G protein-coupled receptor 30), and *NCOA3* (nuclear receptor coactivator 3) in human myometrial TCs from pregnant uterus samples and normalized the data against the housekeeping genes *GAPDH* (Figure 4A) and *18S rRNA* (Figure 4B).

Considering the untreated cells as calibrator and *GAPDH* as the reference gene, beta-estradiol significantly downregulated *ESR1* (One way ANOVA, F = 31, $p < 0.01$), *ESR2* (One way ANOVA, F = 12, $p < 0.05$), *GPR30* (One way ANOVA, F = 49, $p < 0.01$), and *NCOA3* (One way ANOVA, F = 49, $p < 0.01$) (Figure 4A). The Bonferroni post-hoc analysis indicated that: 100 nM beta-estradiol significantly downregulated *ESR1* by 0.55-fold ($p < 0.05$) and *ESR2* by 0.89-fold ($p < 0.05$); 10, 100, and 1000 nM beta-estradiol significantly downregulated *GPR30* by 0.81-fold ($p < 0.01$), 0.97-fold ($p < 0.01$), and 0.72-fold ($p < 0.01$), respectively; 10 and 100 nM beta-estradiol significantly downregulated *NCAO3* by 0.89-fold ($p < 0.01$) and 0.96-fold ($p < 0.01$), respectively, versus the non-treated samples (Figure 4A).

Considering the untreated cells as calibrator and *18S rRNA* as the reference gene, *ESR1* (One way ANOVA, F = 27, $p < 0.01$), *ESR2* (One way ANOVA, F = 14, $p < 0.05$), *GPR30* (One way ANOVA, F = 46, $p < 0.01$), and *NCOA3* (One way ANOVA, F = 92, $p < 0.001$) were downregulated (Figure 4B). The Bonferroni post-hoc analysis indicated that: 100 nM beta-estradiol significantly downregulated *ESR1* by 0.37-fold ($p < 0.05$) and *ESR2* by 0.46-fold ($p < 0.05$); 10, 100, and 1000 nM beta-estradiol significantly downregulated *GPR30* by 0.8-fold ($p < 0.01$), 0.96-fold ($p < 0.01$), and 0.78-fold ($p < 0.01$), respectively; 10 and 100 nM beta-estradiol significantly downregulated *NCAO3* by 0.87-fold ($p < 0.01$) and 0.95-fold ($p < 0.01$), respectively, versus the non-treated samples (Figure 4B).

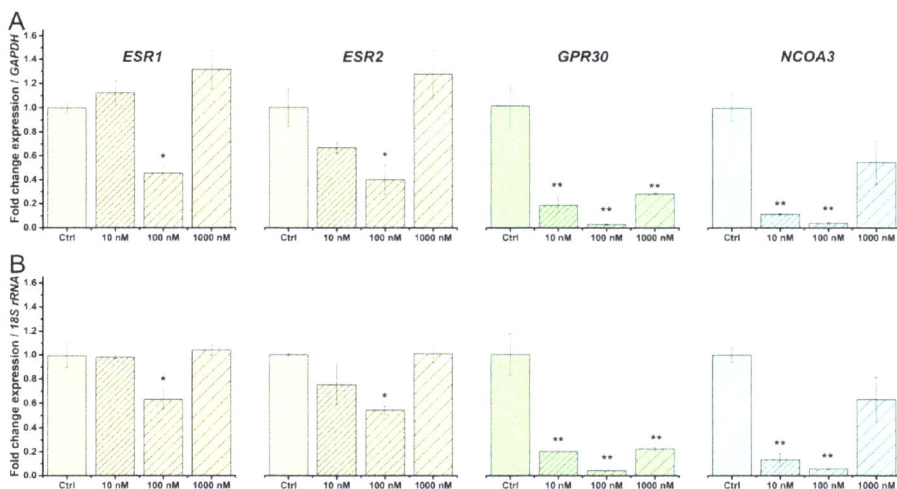

Figure 4. Effect of beta-estradiol treatment (10, 100, and 1000 nM) on the mRNA levels encoding the estrogen receptors and nuclear receptor co-activator in human myometrial TCs from pregnant uterine samples. Fold-change for each receptor upon treatment with beta-estradiol, where the non-treated samples were considered as calibrator. The data were normalized against *GAPDH* (**A**) and *18S rRNA* (**B**) and were plotted as mean ± SD ($N = 3$). Statistical significance is indicated with asterisks (* $0.01 < p < 0.05$; ** $0.001 < p < 0.01$).

2.5. Beta-Estradiol Partly Blocks Bay K8644-Induced Calcium Transients in Human Myometrial Uterine TCs

Bay K8644 was demonstrated to specifically activate HVA calcium channels [46]. We applied this agonist in order to test the functional activation of voltage-gated calcium channels in human myometrial TCs from pregnant myometrium by Fura-2AM calcium imaging (see loaded cells in Figure 5A). Consequently, we recorded calcium transients activated by 5 µM Bay K8644 (Figure 5B).

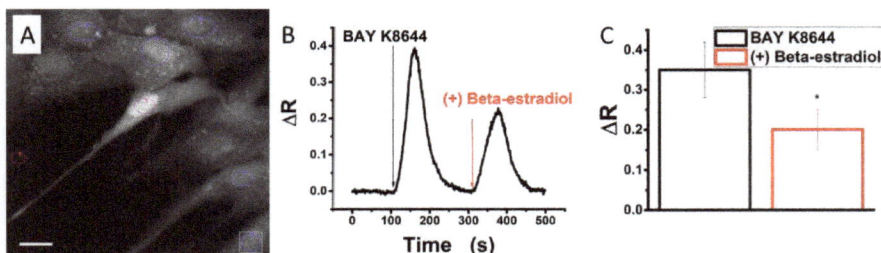

Figure 5. Beta-estradiol effect on voltage-gated calcium channels in TCs from pregnant uterine myometrium by Fura-2-based calcium imaging. (**A**) Cells in primary culture of human uterine myometrium loaded with Fura-2AM. The regions of interest (ROIs) are marked with circles corresponding to: red (reference field without cells), violet (TC), blue (other cells); Scale bar 20 µm (**B**) representative traces of fluorescence ratio (ΔR) changes induced by Bay K8644 (10 µM) in the absence and presence of beta-estradiol (1000 nM) in human myometrial TCs during calcium imaging recordings; (**C**) mean fluorescence ratio (ΔR) ± SD, $N = 5$, in the same conditions as (**B**). The statistical analysis was done by the paired Student's t test, and the level of significance is indicated with asterisks (* $p < 0.05$).

Previous studies indicated that beta-estradiol counteracts the activation exerted by Bay K8644 on cardiac calcium channels [58] and on rat cortical neurons [12,13,59]. In particular, 1000 nM beta-estradiol inhibited the HVA but not the LVA calcium currents in rat sensory neurons via a non-genomic mechanism [12,13,59]. The next step was to test the effect of beta-estradiol (1000 nM) on the calcium transients elicited by Bay K8644 (5 µM) in human myometrial TCs from pregnant uterine samples. Indeed, we evidenced the same partial inhibitory effect of beta-estradiol on calcium influx through voltage-gated calcium channels in TCs (Figure 5B). The mean fluorescence ratio of the calcium signal induced by Bay K8644 was significantly diminished from 0.35 ± 0.07 (in the absence of beta-estradiol) to 0.20 ± 0.05, $N = 5$, $p < 0.05$, paired Student's t test (in the presence of beta-estradiol) Figure 5C.

2.6. Beta-Estradiol Partly Inhibits Bay K8644-Induced Calcium Current in Human Myometrial Uterine TCs

We have also demonstrated that Bay K8644 was able to activate calcium currents in human myometrial uterine TCs (Figure 6A). The membrane capacitance had a mean of 115 ± 11 pF. Moreover, by adding beta-estradiol (1000 nM) in the presence of Bay K8644, we were able to partly inhibit Bay K8644-induced calcium current (Figure 6A). Beta-estradiol blocked Bay K8644-induced current from 85.12 ± 20.03 to 25.91 ± 10.34 pA, $N = 6$, $p < 0.05$, paired Student's t-test (Figure 6B). This inhibitory effect was present in all human myometrial uterine TCs in which Bay K8644 activated voltage-gated calcium currents.

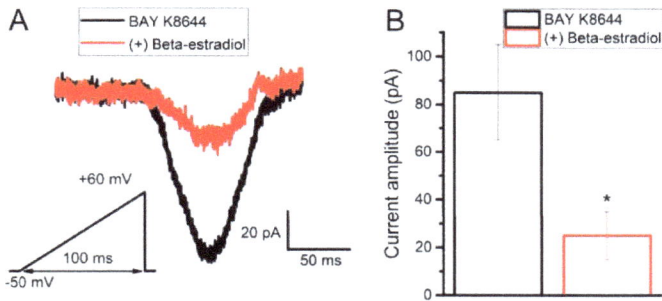

Figure 6. Beta-estradiol partly blocks voltage-gated calcium currents in TCs from pregnant uterine myometrium. (**A**) Representative patch-clamp recording in voltage-clamp mode on a human myometrial TC. Brief depolarizing ramp protocols from −50 to +60 mV with a duration of 100 ms were applied. HVA currents were elicited by perfusion with Bay K8644 (2.5 μM) for 1 min followed by the ramp protocol. To block the calcium current, beta-estradiol (1000 nM) was applied in the presence of Bay K8644; (**B**) amplitude of the voltage-gated calcium currents induced by Bay K8644 in the presence or absence of beta-estradiol. The statistical analysis was done by the paired Student's *t* test, and the level of significance is indicated with asterisks (* $p < 0.05$).

3. Discussion

3.1. Beta-Estradiol Regulates Voltage-Gated Calcium Channels in Human Myometrial TCs from Pregnant Uterus

We have previously evidenced HVA calcium currents induced by a brief ramp depolarization protocol in TCs from non-pregnant uterine myometrium [35]. In this study, we obtained Bay K8644-induced currents in TCs from pregnant uterine myometrium with lower amplitude compared to those recorded in TCs from non-pregnant uterine myometrium. This difference might be due to the pregnancy condition and is correlated with the lower expression of *CACNA1C* mRNA (without statistical significance) in pregnant versus non-pregnant uterine samples. While immunofluorescence indicated that the protein expression of the LVA calcium channels (e.g., Cav3.1 and Cav3.2) was upregulated in human myometrial TCs from pregnant versus non-pregnant uterine samples [35], this study showed that gene expression was upregulated for Cav3.1 and downregulated for Cav3.2 and Cav3.3.

Previous reports demonstrated that beta-estradiol regulates the mRNA expression of T-type calcium channel subunits, but the up- or downregulation was dependent on the type of tissue. To date, in the medial preoptic area and the arcuate nucleus, beta-estradiol has been shown to upregulate Cav3.1 and Cav3.2, but not Cav3.3 [60]. Meanwhile, in the pituitary, beta-estradiol has been shown to upregulate Cav3.1 and downregulate Cav3.2 and Cav3.3 [60]. By comparison, our study on human pregnant myometrial TCs indicated that chronic beta-estradiol treatment downregulated Cav3.1, Cav3.2, Cav3.3, and Cav1.2 at higher concentrations. We have also observed that the mRNA encoding Cav3.1 was upregulated in myometrial TCs in the untreated pregnant samples compared to the chronically estrogen-treated pregnant samples. Possible explanations might include the absence of some physiological factors in our preparations, including the bursting release of hormones during pregnancy and labor compared to the continuous chronic exposure (24 h) to estrogen, and the complexity and the interplay of other steroid hormones (e.g., progesterone, oxytocin etc.), prostaglandins, cytokines, nitric oxide, released in parturition [61].

Bay K8644 is a well-known agonist of the L-type calcium channels [38]. Previous studies demonstrated that beta-estradiol (1 μM) inhibited HVA, but not LVA, calcium currents in rat sensory neurons via a non-genomic mechanism [12,13,59]. This inhibition was done in a rapid, reversible and concentration-dependent manner and determined the hyperpolarization shift of the steady-state inactivation curve [12]. Our data are in accordance with these studies and demonstrate the blocking effect of beta-estradiol (1 μM) on high-voltage-activated calcium currents (Bay K8644-activated currents) in TCs from human uterine myometrium. However, 10-fold higher concentrations of beta-estradiol (10 μM) reduced not only the HVA Ba^{2+} current but also the T-type Ca^{2+} current in vascular smooth muscle cells [14]. This effect might be attributed to a non-specific antagonistic effect of beta-estradiol at higher concentrations on both types of voltage-gated calcium channels.

Interestingly, we should mention that beta-estradiol exerted different functional modulatory action on the HVA channels upon acute (1 min) or chronic (24 h) exposure. Thus, acute exposure determined a reduction in the current amplitude by partly antagonizing BAY K8644, while chronic exposure triggered gene expression downregulation.

It is of particular interest to understand if beta-estradiol regulates voltage-gated calcium channels by means of estrogen receptors activation. On this topic, there are controversial reports in the literature regarding the regulation of L-type calcium channels upon beta-estradiol treatment via estrogen receptors. To date, beta-estradiol was demonstrated to inhibit HVA calcium currents in rat cortical neurons without involving the estrogen receptors [12]. In the hypothalamus, beta-estradiol-induced upregulation in Cav3.1 was dependent on ERα, while its effect on Cav3.2 was dependent on both ERα and ERβ [60]. In the pituitary, beta-estradiol-induced effects were only dependent on the expression of ERα [60]. In our study, we did not evaluate if the beta-estradiol effect on voltage-gated calcium channels was mediated by the estrogen receptors. However, corroborating the modulatory effect of beta-estradiol exerted on voltage-gated calcium channels and on estrogen receptors, we might consider that there is an interplay between these key players that involves the activation of calcium signaling pathways. Other hormones (e.g., progesterone, oxytocin) have also been described to block voltage-gated calcium channels [62–64], and we might suppose a similar inhibition in TCs human myometrial preparations, but our study was focused only on estrogen effects.

3.2. Beta-Estradiol Regulates Estrogen Receptors in Human Myometrial TCs from Pregnant Uterus

Estrogens act specifically on estrogen receptors and mediate distinct roles in pregnancy and labor. Indeed, ERα plays a more prominent role than ERβ in mediating estrogen action in the induction of uterine oxytocin receptors before labor [65]. Alterations of the estrogen–estrogen receptors signaling have been associated with different pathologies of the reproductive system (e.g., preeclampsia, endometrial cancer, etc.). For example, low GPR30 expression levels, increased apoptosis, and reduced proliferation were associated with preeclampsia, and the pharmacological targeting of GPR30 might have clinical relevance [66]. Estrogen-induced PI3K–AKT signaling activated by GPR30 is involved in the regulation of endometrial cancer cell proliferation [67].

Previous studies on human pregnant myometrium samples indicated the expression of ERα and GPR30 mRNAs, while ERβ mRNA was not detectable [53]. By comparison, we retrieved ERα and GPR30 mRNAs expression in TCs from human pregnant myometrium samples, but we also detected low levels of ERβ mRNA. These findings are in accordance with our previous immunocytochemical studies on TCs cultured from human myometrium during and outside pregnancy [28,68]. Beside estrogen receptors, we also demonstrated, in an earlier study, the expression of progesterone receptors in TCs from human myometrial cell cultures [28], but an extensive analysis of progesterone receptors expression in pregnant and nonpregnant conditions was beyond the purpose of the present study.

Multiple studies have demonstrated that beta-estradiol regulates the mRNA levels encoding the estrogen receptors in various tissues and species, including mouse Sertoli cells [41], female rat brain [48], rat uterus [69], etc. In particular, 3 and 6 h after beta-estradiol subcutaneous injection, the downregulation of ERα and ERβ mRNA levels was demonstrated in rat uterus [69]. We showed that beta-estradiol downregulates ESR1, ESR2, GPR30, and NCOA3 in TCs from human pregnant uterine myometrial cultures. Our current data are in agreement with the in vivo downregulation of these receptors in rat uterus.

NCOA3, also known as steroid receptor coactivator-3 (src-3), was described as being associated with uterine endometrial cancer or other cancers, e.g., breast cancer [70,71]. Despite the great interest for this coregulator, little is known about its physiological role in uterine myometrium [71]. Our study documents for the first time NCOA3 expression in uterine TCs and its downregulation in pregnancy and upon beta-estradiol treatment.

Calcium was demonstrated to play an important role in the activation of estrogen receptors in different cell types. Interestingly, stimuli inducing the release of intracellular calcium determine the recruitment of estrogen receptors 1 and stimulate the expression of estrogen-responsive genes [72]. Several studies indicated that beta-estradiol regulates the expression of different receptors associated with calcium signaling mechanisms via estrogen receptors. In mouse N2A and human SK-N-SH neural cells, beta-estradiol upregulated the expression of the α and β subunits of BK channels via estrogen receptor β, in a concentration-dependent manner [10]. Alterations of calcium homeostasis may determine breast cancer progression by affecting various signaling pathways including estrogen receptors [73]. Both intra- and extracellular calcium was shown to influence ERα transcriptional activity in breast cancer cells [74].

In pregnancy, beta-estradiol exerts tissue-dependent regulatory actions. In particular, beta-estradiol and estrogens-specific agonists have been shown to differentially modulate the tone of uterine versus placental arteries, and their contribution to the regulation of human uteroplacental blood flow might be tissue-specific [75]. In rats ovariectomized on day 18 of pregnancy, estrogen treatment caused upregulation of oxytocin receptors [65].

In conclusion, our study brings novel highlights into the modulatory effects of beta-estradiol on the uterine myometrium and shows that voltage-gated calcium channels and estrogen receptors expressed in TCs are essential regulatory targets.

4. Materials and Methods

4.1. Human Uterine Samples

The myometrial tissue of non-pregnant women was obtained from a total of 10 menstruating, parous, pre-menopausal women during the proliferative phase, which were undergoing hysterectomy for benign gynecological reasons. The median age of the non-pregnant women was 42 years (range 35–49 years). The inclusion criteria used in the selection of the research subjects were indications for surgery for menometrorrhagia/dysmenorrhea as a consequence of localized or diffuse leiomyomas and no other associated medical conditions or treatment history. Hysterectomized uteri were biopsied in the pathology department where they were examined macroscopically, and apparently normal myometrial areas were chosen. The pregnant women included in the study (*n* = 8) were aged between 30 and 35 years, and the gestational age was between 38 and 40 weeks. The patients underwent cesarean surgery in labor because of dystocia and fetal distress. A small strip of myometrium was carefully dissected from the upper margin (in the midline) of the lower segment transverse incision of each patient.

All tissue samples were obtained in accordance with the protocol no. 12290 (18.08.2017) approved by the Ethics Committee of Alessandrescu-Rusescu National Institute of Mother and Child Health. The patients included in the study were enrolled from the Department of Obstetrics and Gynecology, "Polizu" Clinical Hospital, Alessandrescu-Rusescu National Institute of Mother and Child Health, Bucharest, Romania. All patients donating uterine tissue samples that were included in the study signed the informed written consent. The inclusion criteria requested that none of the patients included in the groups received any regular medication for chronic diseases.

4.2. Myometrial Cell Cultures

For the qRT-PCR experiments, primary cultures were prepared from human uterine myometrial biopsies as previously described [35]. On the basis of the immunopositivity for CD34 and platelet-derived growth factor receptor-α (PDGFRα) that was already described in human uterine myometrial TCs [35], we employed a double-labeling technique, using goat polyclonal anti-CD34 (#sc-7045, Santa Cruz

Biotechnology, Santa Cruz, CA, USA) and rabbit polyclonal anti-PDGFRα (#sc-338, Santa Cruz Biotechnology, USA) antibodies. The primary antibodies were detected by a secondary goat anti-rabbit antibody conjugated to AlexaFluor 488 and a donkey anti-goat antibody conjugated to AlexaFluor 546, from Invitrogen Molecular Probes, Eugene, OR, USA. An enriched TC culture ($CD34^+$/$PDGFR\alpha^+$ cells) was obtained by sorting the double-labeled cells using a BD FACSCanto II - Becton Dickinson flow cytometer (BD Biosciences, Waltham, MA, USA), as previously described for cardiac TCs [69]. For the qRT-PCR experiments, the cells were chronically exposed for 24 h to beta-estradiol (#E2758, Sigma-Aldrich, St. Louis, MO, USA) at concentrations of 10, 100, and 1000 nM in serum-free medium, in order to avoid the non-specific binding of beta-estradiol to the proteins in the serum.

For the patch-clamp and calcium imaging experiments, primary cultures were prepared from human uterine myometrial biopsies as previously described [35]. The cells were used without further sorting in order to maintain the physiological environment and to prevent any alteration in the calcium signaling due to the absence of human myometrial muscle cells in the cell culture. The human myometrial cells (between first and fourth passages) were plated at a density of 5×10^4 cells/cm^2 on 24 mm Petri dishes for the patch-clamp experiments or on 24 mm glass coverslips for the calcium imaging experiments. On the basis of their morphological features (e.g., long and moniliform telopodes), only TCs were selected under the microscope for patch-clamp or calcium imaging recordings. TCs were acutely exposed for 1 min to beta-estradiol in the patch-clamp and calcium-imaging experiments, in the presence or absence of BAY K8644.

4.3. qRT-PCR

To quantify the mRNA expression levels encoding for different genes (Table 1) in primary cell cultures from human uterine myometrium, total RNA was extracted using the GenElute Mammalian Total RNA MiniPrep Kit (RTN70, Sigma), according to the manufacturer's instructions. RNA concentrations were determined by spectrophotometric measurements of absorption at 260 and 280 nm (Beckman Coulter DU 730, Carlsbad, CA, USA), and DNase I treatment was applied in order to remove contaminating genomic DNA. In agreement with the manufacturer's guidelines (Sigma-Aldrich, USA), in our experiments, the A260:A280 ratio was 2.04 ± 0.05. Reverse transcription was done using the High-Capacity cDNA Archive Kit (Applied Biosystems, Waltham, MA, USA). The human primers and TaqMan probes (Life Technologies, Carlsbad, CA, USA) used in our experiments are listed in Table 1 and were used in accordance with the manufacturer's guidelines. *GAPDH* and *18S rRNA* were used as reference genes. The relative abundance of gene transcripts was assessed via qRT-PCR, using the TaqMan methodology and the ABI Prism 7300 Sequence Detection System (Applied Biosystems, USA). The reactions were carried out in triplicate for 50 cycles.

Table 1. Panel of primers (Life Technologies, Carlsbad, CA, USA).

Primer	Gene	Protein Encoded
Hs00167681_m1	*CACNA1C*	Cav1.2, L-type calcium channel
Hs00367969_m1	*CACNA1G*	Cav3.1, T-type calcium channel
Hs00234934_m1	*CACNA1H*	Cav3.2, T-type calcium channel
Hs00184168_m1	*CACNA1I*	Cav3.3, T-type calcium channel
Hs01922715_s1	*GPR30, GPER1*	G protein-coupled estrogen receptor 1
Hs01105253_m1	*NCOA3*	Nuclear receptor coactivator 3
Hs01100353_m1	*ESR2*	Estrogen receptor 2
Hs00174860_m1	*ESR1*	Estrogen receptor 1
Hs99999905_m1	*GAPDH*	Glyceraldehyde 3-phosphate dehydrogenase
Hs99999901_s1	*18S rRNA*	18S ribosomal RNA

4.4. Intracellular Calcium Imaging

Human myometrial TCs plated on 24 mm coverglasses were incubated for 45 min at room temperature in a dark chamber with 4 μM Fura-2 acetoxymethyl ester (Fura-2 AM; #F1221, Thermo Fisher Scientific,

Waltham, MA, USA) and 0.25% pluronic Pluronic™ F-127 (#P3000MP, Thermo Fisher Scientific, USA) in Dulbecco's Modified Eagle's Medium (DMEM; Thermo Fisher Scientific, USA). The cells were rinsed three times with Ringer solution (in mM: NaCl 126, HEPES 5, CaCl$_2$ 2, MgCl$_2$ 2, glucose 10, pH 7.4 (with Tris base), let recover for 15 min, and then imaged with an IX-71 Olympus microscope. The excitation was performed by a Xe lamp with monochromator Polychrome V (Till Photonics GmbH, Gräfelfing, Germany) at 340 ± 5 and 380 ± 5 nm, while the emission was collected by a filter at 510 ± 20 nm. The images were acquired by a cooled CCD camera iXON+ EM DU 897 (Andor, Belfast, Northern Ireland) controlled by iQ 1.8 software package, with a frequency of one pair of images per 2 s. Ca^{2+} transients were triggered by 1 min application of 5 μM (±)-Bay K8644 (#B112, Sigma-Aldrich, St. Louis, MO, USA), an L-type calcium channel agonist [46]. Beta-estradiol was applied for 1 min at 1000 nM in the presence of Bay K8644, and Ca^{2+} transients were recorded. The perfusion was performed with an MPS-2 system (World Precision Instruments, Sarasota, FL, USA). Each coverglass bearing human myometrial TCs was used for a single experimental variant.

4.5. Patch-Clamp Recordings on TCs

TCs from pregnant uterus were recorded in whole-cell configuration under the voltage-clamp mode, using an AxoPatch 200B amplifier (Molecular Devices, San Jose, CA, USA). The electrodes were pulled from borosilicate glass capillaries (GC150F; Harvard Apparatus, Edenbridge, Kent, UK) and heat polished. The final resistance of the pipette, when filled with internal solution, was 3–4 MΩ. The perfusion was performed with an MPS-2 (World Precision Instruments, Sarasota, FL, USA) system, with the tip placed at approximately 100 μm from the cell. Membrane currents were low-pass filtered at 3 kHz (−3 dB, three pole Bessel) and sampled with an Axon Digidata 1440 data acquisition system (Molecular Devices, USA), using pClamp 10 software in gap-free mode. All electrophysiological experiments were performed at room temperature (25 °C). The bath and pipette solutions were used as previously described [35,76]. We adapted the previously described protocol [35] and we applied brief depolarizing ramp protocols from −50 to +60 mV with a duration of 100 ms in order to elicit HVA currents. Uterine TCs were perfused with (±)-Bay K8644 (#B112, Sigma-Aldrich, USA) for 1 min at 2.5 μM concentration. Beta-estradiol was applied for 1 min at 1000 nM in the presence of Bay K8644.

4.6. Data Analysis

Quantitative RT-PCR data were obtained by normalizing the mRNA levels encoding for different genes to those of *GAPDH* mRNA level using the 2(-Delta C(T)) method, as previously described [77]. Statistical analysis comparing the mRNA expression for each subtype of voltage-gated calcium channel relative to the reference gene (*GAPDH* or *18S rRNA*) in pregnant and in non-pregnant myometrial TCs, was carried out by one-way ANOVA analysis followed by post-hoc Bonferroni test. The effect of pregnancy on the mRNA levels of the voltage-gated calcium channels or the estrogen receptors was analyzed by considering the unpregnant samples as calibrator and performing the one-way ANOVA analysis followed by the post-hoc Bonferroni test. The beta-estradiol effect on the mRNA levels of the voltage-gated calcium channels or the estrogen receptors was analyzed by considering the non-treated samples as calibrator and performing the one-way ANOVA analysis followed by the post-hoc Bonferroni test.

Quantitative calcium imaging results were expressed by means of emission ratio R = I$_{340}$/I$_{380}$, I being the emission intensity for excitation at 340 and 380 nm, which is proportional to the free cytosolic Ca^{2+} concentration ([Ca^{2+}]$_i$). The mean fluorescence ratios (ΔR) of the calcium transients induced by BAY K8644 in the absence or presence of beta-estradiol were compared by unpaired Student's *t* test.

The amplitudes of the HVA calcium currents recorded by patch-clamp in the absence or presence of beta-estradiol were compared by unpaired Student's *t* test.

All data analysis and data plotting were performed using OriginPro 8 (OriginLab Corporation, Northampton, MA, USA).

Author Contributions: S.M.C., N.S., and B.M.R. designed the study; S.M.C., D.C., and N.S. selected the patients included in the study and taken the human uterine samples; J.X. and A.D. planned the protocols for patch-clamp recordings and A.D. performed the patch-clamp experiments; A.D. and D.D.B. performed the qRT-PCR experiments; C.C.M performed the calcium-imaging experiments; A.D., M.R. and B.M.R analyzed the data; B.M.R and M.R. wrote the paper.

Acknowledgments: This work was supported by a grant of the Romanian Ministry of Research and Innovation, CCCDI - UEFISCDI, project number PN-III-P1-1.2-PCCDI-2017-0833/ 68/2018, within PNCDI III.

Conflicts of Interest: The authors declare no conflict of interest.

References

1. Wray, S.; Jones, K.; Kupittayanant, S.; Li, Y.; Matthew, A.; Monir-Bishty, E.; Noble, K.; Pierce, S.J.; Quenby, S.; Shmygol, A.V. Calcium signaling and uterine contractility. *J. Soc. Gynecol. Investig.* **2003**, *10*, 252–264. [CrossRef]
2. Herington, J.L.; Swale, D.R.; Brown. N.; Shelton, E.L.; Choi, H.; Williams, C.H.; Hong, C.C.; Paria, B.C.; Denton, J.S.; Reese, J. High-Throughput Screening of Myometrial Calcium-Mobilization to Identify Modulators of Uterine Contractility. *PLoS ONE* **2015**, *10*, e0143243. [CrossRef] [PubMed]
3. Bernstein, K.; Vink, J.Y.; Fu, X.W.; Wakita, H.; Danielsson, J.; Wapner, R.; Gallos, G. Calcium-activated chloride channels anoctamin 1 and 2 promote murine uterine smooth muscle contractility. *Am. J. Obstet. Gynecol.* **2014**, *211*, 688.e1–688.e10. [CrossRef] [PubMed]
4. Pistilli, M.J.; Petrik, J.J.; Holloway, A.C.; Crankshaw, D.J. Immunohistochemical and functional studies on calcium-sensing receptors in rat uterine smooth muscle. *Clin. Exp. Pharmacol. Physiol.* **2012**, *39*, 37–42. [CrossRef] [PubMed]
5. Ying, L.; Becard, M.; Lyell, D.; Han, X.; Shortliffe, L.; Husted, C.I.; Alvira, C.M.; Cornfield, D.N. The transient receptor potential vanilloid 4 channel modulates uterine tone during pregnancy. *Sci. Transl. Med.* **2015**, *7*, 319ra204. [CrossRef] [PubMed]
6. Wakle-Prabagaran, M.; Lorca, R.A.; Ma, X.; Stamnes, S.J.; Amazu, C.; Hsiao, J.J.; Karch, C.M.; Hyrc, K.L.; Wright, M.E.; England, S.K. BKCa channel regulates calcium oscillations induced by alpha-2-macroglobulin in human myometrial smooth muscle cells. *Proc. Natl. Acad. Sci. USA* **2016**, *113*, E2335–E2344. [CrossRef] [PubMed]
7. Wray, S. Insights from physiology into myometrial function and dysfunction. *Exp. Physiol.* **2015**, *100*, 1468–1476. [CrossRef] [PubMed]
8. Wray, S.; Burdyga, T.; Noble, D.; Noble, K.; Borysova, L.; Arrowsmith, S. Progress in understanding electro-mechanical signalling in the myometrium. *Acta Physiol.* **2015**, *213*, 417–431. [CrossRef] [PubMed]
9. Lee, S.E.; Ahn, D.S.; Lee, Y.H. Role of T-type Ca Channels in the Spontaneous Phasic Contraction of Pregnant Rat Uterine Smooth Muscle. *Korean J. Physiol. Pharmacol.* **2009**, *13*, 241–249. [CrossRef] [PubMed]
10. Li, X.T.; Qiu, X.Y. 17β-Estradiol Upregulated Expression of α and β Subunits of Larger-Conductance Calcium-Activated K(+) Channels (BK) via Estrogen Receptor β. *J. Mol. Neurosci.* **2015**, *56*, 799–807. [CrossRef] [PubMed]
11. Rahbek, M.; Nazemi, S.; Odum, L.; Gupta, S.; Poulsen, S.S.; Hay-Schmidt, A.; Klaerke, D.A. Expression of the small conductance Ca^{2+}-activated potassium channel subtype 3 (SK3) in rat uterus after stimulation with 17β-estradiol. *PLoS ONE* **2014**, *9*, e87652. [CrossRef] [PubMed]
12. Wang, Q.; Ye, Q.; Lu, R.; Cao, J.; Wang, J.; Ding, H.; Gao, R.; Xiao, H. Effects of estradiol on high-voltage-activated Ca(2+) channels in cultured rat cortical neurons. *Endocr. Res.* **2014**, *39*, 44–49. [CrossRef] [PubMed]
13. Sánchez, J.C.; López-Zapata, D.F.; Pinzón, O.A. Effects of 17beta-estradiol and IGF-1 on L-type voltage-activated and stretch-activated calcium currents in cultured rat cortical neurons. *Neuroendocrinol. Lett.* **2014**, *35*, 724–732. [PubMed]
14. Zhang, F.; Ram, J.L.; Standley, P.R.; Sowers, J.R. 17 beta-Estradiol attenuates voltage-dependent Ca^{2+} currents in A7r5 vascular smooth muscle cell line. *Am. J. Physiol.* **1994**, *266*, C975–C980. [CrossRef] [PubMed]
15. Sánchez, J.C.; López-Zapata, D.F.; Francis, L.; De Los Reyes, L. Effects of estradiol and IGF-1 on the sodium calcium exchanger in rat cultured cortical neurons. *Cell. Mol. Neurobiol.* **2011**, *31*, 619–627. [CrossRef] [PubMed]
16. Choi, Y.; Seo, H.; Kim, M.; Ka, H. Dynamic expression of calcium-regulatory molecules, TRPV6 and S100G, in the uterine endometrium during pregnancy in pigs. *Biol. Reprod.* **2009**, *81*, 1122–1130. [CrossRef] [PubMed]

17. Pohóczky, K.; Kun, J.; Szalontai, B.; Szőke, É.; Sághy, É.; Payrits, M.; Kajtár, B.; Kovács, K.; Környei, J.L.; Garai, J.; et al. Estrogen-dependent up-regulation of TRPA1 and TRPV1 receptor proteins in the rat endometrium. *J. Mol. Endocrinol.* **2016**, *56*, 135–149. [CrossRef] [PubMed]

18. Tulchinsky, D.; Korenman, S.G. The plasma estradiol as an index of fetoplacental function. *J. Clin. Investig.* **1971**, *50*, 1490–1497. [CrossRef] [PubMed]

19. Hatthachote, P.; Gillespie, J.I. Complex interactions between sex steroids and cytokines in the human pregnant myometrium: Evidence for an autocrine signaling system at term. *Endocrinology* **1999**, *140*, 2533–2540. [CrossRef] [PubMed]

20. Wu, J.J.; Geimonen, E.; Andersen, J. Increased expression of estrogen receptor beta in human uterine smooth muscle at term. *Eur. J. Endocrinol.* **2000**, *142*, 92–99. [CrossRef] [PubMed]

21. Cretoiu, S.M.; Cretoiu, D.; Popescu, L.M. Human myometrium—The ultrastructural 3D network of telocytes. *J. Cell. Mol. Med.* **2012**, *16*, 2844–2849. [CrossRef] [PubMed]

22. Cretoiu, S.M.; Popescu, L.M. Telocytes revisited. *Biomol. Concepts* **2014**, *5*, 353–369. [CrossRef] [PubMed]

23. Song, D.; Cretoiu, D.; Cretoiu, S.M.; Wang, X. Telocytes and lung disease. *Histol. Histopathol.* **2016**, *31*, 1303–1314. [PubMed]

24. Cretoiu, S.M.; Cretoiu, D.; Marin, A.; Radu, B.M.; Popescu, L.M. Telocytes: Ultrastructural, immunohistochemical and electrophysiological characteristics in human myometrium. *Reproduction* **2013**, *145*, 357–370. [CrossRef] [PubMed]

25. Song, D.; Cretoiu, D.; Zheng, M.; Qian, M.; Zhang, M.; Cretoiu, S.M.; Chen, L.; Fang, H.; Popescu, L.M.; Wang, X. Comparison of Chromosome 4 gene expression profile between lung telocytes and other local cell types. *J. Cell. Mol. Med.* **2016**, *20*, 71–80. [CrossRef] [PubMed]

26. Albulescu, R.; Tanase, C.; Codrici, E.; Popescu, D.I.; Cretoiu, S.M.; Popescu, L.M. The secretome of myocardial telocytes modulates the activity of cardiac stem cells. *J. Cell. Mol. Med.* **2015**, *19*, 1783–1794. [CrossRef] [PubMed]

27. Zheng, Y.; Cretoiu, D.; Yan, G.; Cretoiu, S.M.; Popescu, L.M.; Fang, H.; Wang, X. Protein profiling of human lung telocytes and microvascular endothelial cells using iTRAQ quantitative proteomics. *J. Cell. Mol. Med.* **2014**, *18*, 1035–1059. [CrossRef] [PubMed]

28. Cretoiu, S.M.; Cretoiu, D.; Simionescu, A.; Popescu, L.M. Telocytes in human fallopian tube and uterus express estrogen and progesterone receptors. In *Sex Steroids*; Kahn, S., Ed.; InTech: Rijeka, Croatia, 2012; pp. 91–114.

29. Roatesi, I.; Radu, B.M.; Cretoiu, D.; Cretoiu, S.M. Uterine Telocytes: A Review of Current Knowledge. *Biol. Reprod.* **2015**, *93*, 10. [CrossRef] [PubMed]

30. Cretoiu, D.; Xu, J.; Xiao, J.; Cretoiu, S.M. Telocytes and Their Extracellular Vesicles-Evidence and Hypotheses. *Int. J. Mol. Sci.* **2016**, *17*, 1322. [CrossRef] [PubMed]

31. Cretoiu, D.; Cretoiu, S.M. Telocytes in the reproductive organs: Current understanding and future challenges. *Semin. Cell Dev. Biol.* **2016**, *55*, 40–49. [CrossRef] [PubMed]

32. Cretoiu, D.; Radu, B.M.; Banciu, A.; Banciu, D.D.; Cretoiu, S.M. Telocytes heterogeneity: From cellular morphology to functional evidence. *Semin. Cell Dev. Biol.* **2017**, *64*, 26–39. [CrossRef] [PubMed]

33. Rusu, M.C.; Cretoiu, D.; Vrapciu, A.D.; Hostiuc, S.; Dermengiu, D.; Manoiu, V.S.; Cretoiu, S.M.; Mirancea, N. Telocytes of the human adult trigeminal ganglion. *Cell Biol. Toxicol.* **2016**, *32*, 199–207. [CrossRef] [PubMed]

34. Radu, B.M.; Banciu, A.; Banciu, D.D.; Radu, M.; Cretoiu, D.; Cretoiu, S.M. Calcium Signaling in Interstitial Cells: Focus on Telocytes. *Int. J. Mol. Sci.* **2017**, *18*, 397. [CrossRef] [PubMed]

35. Cretoiu, S.M.; Radu, B.M.; Banciu, A.; Banciu, D.D.; Cretoiu, D.; Ceafalan, L.C.; Popescu, L.M. Isolated human uterine telocytes: Immunocytochemistry and electrophysiology of T-type calcium channels. *Histochem. Cell Biol.* **2015**, *143*, 83–94. [CrossRef] [PubMed]

36. Campeanu, R.A.; Radu, B.M.; Cretoiu, S.M.; Banciu, D.D.; Banciu, A.; Cretoiu, D.; Popescu, L.M. Near-infrared low-level laser stimulation of telocytes from human myometrium. *Lasers Med. Sci.* **2014**, *29*, 1867–1874. [CrossRef] [PubMed]

37. Othman, E.R.; Elgamal, D.A.; Refaiy, A.M.; Abdelaal, I.I.; Abdel-Mola, A.F.; Al-Hendy, A. Identification and potential role of telocytes in human uterine leiomyoma. *Contracep. Reprod. Med.* **2016**, *20*, 12. [CrossRef] [PubMed]

38. Varga, I.; Klein, M.; Urban, L.; Danihel, L., Jr.; Polak, S.; Danihel, L., Sr. Recently discovered interstitial cells "telocytes" as players in the pathogenesis of uterine leiomyomas. *Med. Hypotheses* **2018**, *110*, 64–67. [CrossRef] [PubMed]

39. Cretoiu, S.M. Immunohistochemistry of Telocytes in the Uterus and Fallopian Tubes. *Adv. Exp. Med. Biol.* **2016**, *913*, 335–357. [PubMed]

40. Vannucchi, M.G.; Faussone-Pellegrini, M.S. The Telocyte Subtypes. *Adv. Exp. Med. Biol.* **2016**, *913*, 115–126. [PubMed]

41. Yang, J.; Li, Y.; Xue, F.; Liu, W.; Zhang, S. Exosomes derived from cardiac telocytes exert positive effects on endothelial cells. *Am. J. Transl. Res.* **2017**, *9*, 5375–5387. [PubMed]

42. Rusu, M.C.; Hostiuc, S.; Vrapciu, A.D.; Mogoantă, L.; Mănoiu, V.S.; Grigoriu, F. Subsets of telocytes: Myocardial telocytes. *Ann. Anat.* **2017**, *209*, 37–44. [CrossRef] [PubMed]

43. Uluer, E.T.; Inan, S.; Ozbilgin, K.; Karaca, F.; Dicle, N.; Sancı, M. The role of hypoxia related angiogenesis in uterine smooth muscle tumors. *Biotech. Histochem.* **2015**, *90*, 102–110. [CrossRef] [PubMed]

44. Sajewicz, M.; Konarska, M.; Wrona, A.N.; Aleksandrovych, V.; Bereza, T.; Komnata, K.; Solewski, B.; Maleszka, A.; Depukat, P.; Warchoł, Ł. Vascular density, angiogenesis and pro-angiogenic factors in uterine fibroids. *Folia Med. Cracov.* **2016**, *56*, 27–32. [PubMed]

45. Salama, N. Immunohistochemical characterization of telocytes in rat uterus in different reproductive states. *Egypt J. Histol.* **2013**, *36*, 85–194.

46. Kim, Y.H.; Chung, S.; Lee, Y.H.; Kim, E.C.; Ahn, D.S. Increase of L-type Ca^{2+} current by protease-activated receptor 2 activation contributes to augmentation of spontaneous uterine contractility in pregnant rats. *Biochem. Biophys. Res. Commun.* **2012**, *418*, 167–172. [CrossRef] [PubMed]

47. Seda, M.; Pinto, F.M.; Wray, S.; Cintado, C.G.; Noheda, P.; Buschmann, H.; Candenas, L. Functional and molecular characterization of voltage-gated sodium channels in uteri from nonpregnant rats. *Biol. Reprod.* **2007**, *77*, 855–863. [CrossRef] [PubMed]

48. Tica, A.A.; Dun, E.C.; Tica, O.S.; Gao, X.; Arterburn, J.B.; Brailoiu, G.C.; Oprea, T.I.; Brailoiu, E. G protein-coupled estrogen receptor 1-mediated effects in the rat myometrium. *Am. J. Physiol. Cell Physiol.* **2011**, *301*, C1262–C1269. [CrossRef] [PubMed]

49. Silva, E.S.; Scoggin, K.E.; Canisso, I.F.; Troedsson, M.H.; Squires, E.L.; Ball, B.A. Expression of receptors for ovarian steroids and prostaglandin E2 in the endometrium and myometrium of mares during estrus, diestrus and early pregnancy. *Anim. Reprod. Sci.* **2014**, *151*, 169–181. [CrossRef] [PubMed]

50. Kautz, E.; Gram, A.; Aslan, S.; Ay, S.S.; Selçuk, M.; Kanca, H.; Koldaş, E.; Akal, E.; Karakaş, K.; Findik, M.; et al. Expression of genes involved in the embryo-maternal interaction in the early-pregnant canine uterus. *Reproduction* **2014**, *147*, 703–717. [CrossRef] [PubMed]

51. Ilicic, M.; Butler, T.; Zakar, T.; Paul, J.W. The expression of genes involved in myometrial contractility changes during ex situ culture of pregnant human uterine smooth muscle tissue. *J. Smooth Muscle Res.* **2017**, *53*, 73–89. [CrossRef] [PubMed]

52. Vodstrcil, L.A.; Shynlova, O.; Westcott, K.; Laker, R.; Simpson, E.; Wlodek, M.E.; Parry, L.J. Progesterone withdrawal, and not increased circulating relaxin, mediates the decrease in myometrial relaxin receptor (RXFP1) expression in late gestation in rats. *Biol. Reprod.* **2010**, *83*, 825–832. [CrossRef] [PubMed]

53. Welsh, T.; Johnson, M.; Yi, L.; Tan, H.; Rahman, R.; Merlino, A.; Zakar, T.; Mesiano, S. Estrogen receptor (ER) expression and function in the pregnant human myometrium: Estradiol via ERα activates ERK1/2 signaling in term myometrium. *J. Endocrinol.* **2012**, *212*, 227–238. [CrossRef] [PubMed]

54. Chandran, S.; Cairns, M.T.; O'Brien, M.; Smith, T.J. Transcriptomic effects of estradiol treatment on cultured human uterine smooth muscle cells. *Mol. Cell. Endocrinol.* **2014**, *393*, 16–23. [CrossRef] [PubMed]

55. Lin, J.; Zhu, J.; Li, X.; Li, S.; Lan, Z.; Ko, J.; Lei, Z. Expression of genomic functional estrogen receptor 1 in mouse sertoli cells. *Reprod. Sci.* **2014**, *21*, 1411–1422. [CrossRef] [PubMed]

56. Yamaguchi, N.; Yuri, K. Estrogen-dependent changes in estrogen receptor-β mRNA expression in middle-aged female rat brain. *Brain Res.* **2014**, *1543*, 49–57. [CrossRef] [PubMed]

57. Murata, T.; Narita, K.; Ichimaru, T. Rat uterine oxytocin receptor and estrogen receptor α and β mRNA levels are regulated by estrogen through multiple estrogen receptors. *J. Reprod.* **2014**, *60*, 55–61. [CrossRef]

58. Bechem, M.; Hoffmann, H. The molecular mode of action of the Ca agonist (-) BAY K 8644 on the cardiac Ca channel. *Pflugers Arch.* **1993**, *424*, 343–353. [CrossRef] [PubMed]

59. Lee, D.Y.; Chai, Y.G.; Lee, E.B.; Kim, K.W.; Nah, S.Y.; Oh, T.H.; Rhim, H. 17Beta-estradiol inhibits high-voltage-activated calcium channel currents in rat sensory neurons via a non-genomic mechanism. *Life Sci.* **2002**, *70*, 2047–2059. [CrossRef]

60. Bosch, M.A.; Hou, J.; Fang, Y.; Kelly, M.J.; Rønnekleiv, O.K. 17Beta-estradiol regulation of the mRNA expression of T-type calcium channel subunits: Role of estrogen receptor alpha and estrogen receptor beta. *J. Comp. Neurol.* **2009**, *512*, 347–358. [CrossRef] [PubMed]

61. Ravanos, K.; Dagklis, T.; Petousis, S.; Margioula-Siarkou, C.; Prapas, Y.; Prapas, N. Factors implicated in the initiation of human parturition in term and preterm labor: A review. *Gynecol. Endocrinol.* **2015**, *31*, 679–683. [CrossRef] [PubMed]

62. Luoma, J.I.; Kelley, B.G.; Mermelstein, P.G. Progesterone inhibition of voltage-gated calcium channels is a potential neuroprotective mechanism against excitotoxicity. *Steroids* **2011**, *76*, 845–855. [CrossRef] [PubMed]

63. Sun, J.; Moenter, S.M. Progesterone treatment inhibits and dihydrotestosterone (DHT) treatment potentiates voltage-gated calcium currents in gonadotropin-releasing hormone (GnRH) neurons. *Endocrinology* **2010**, *151*, 5349–5358. [CrossRef] [PubMed]

64. Liu, B.; Hill, S.J.; Khan, R.N. Oxytocin inhibits T-type calcium current of human decidual stromal cells. *J. Clin. Endocrinol. Metab.* **2005**, *90*, 4191–4197. [CrossRef] [PubMed]

65. Murata, T.; Narita, K.; Honda, K.; Matsukawa, S.; Higuchi, T. Differential regulation of estrogen receptor alpha and beta mRNAs in the rat uterus during pregnancy and labor: Possible involvement of estrogen receptors in oxytocin receptor regulation. *Endocr. J.* **2003**, *50*, 579–587. [CrossRef] [PubMed]

66. Li, J.; Chen, Z.; Zhou, X.; Shi, S.; Qi, H.; Baker, P.N.; Zhang, H. Imbalance between proliferation and apoptosis-related impaired GPR30 expression is involved in preeclampsia. *Cell Tissue Res.* **2016**, *366*, 499–508. [CrossRef] [PubMed]

67. Wei, Y.; Zhang, Z.; Liao, H.; Wu, L.; Wu, X.; Zhou, D.; Xi, X.; Zhu, Y.; Feng, Y. Nuclear estrogen receptor-mediated Notch signaling and GPR30-mediated PI3K/AKT signaling in the regulation of endometrial cancer cell proliferation. *Oncol. Rep.* **2012**, *27*, 504–510. [PubMed]

68. Cretoiu, D.; Ciontea, S.M.; Popescu, L.M.; Ceafalan, L.; Ardeleanu, C. Interstitial Cajal-like cells (ICLC) as steroid hormone sensors in human myometrium: Immunocytochemical approach. *J. Cell. Mol. Med.* **2006**, *10*, 789–795. [CrossRef] [PubMed]

69. Li, Y.Y.; Zhang, S.; Li, Y.G.; Wang, Y. Isolation, culture, purification and ultrastructural investigation of cardiac telocytes. *Mol. Med. Rep.* **2016**, *14*, 1194–1200. [CrossRef] [PubMed]

70. Sakaguchi, H.; Fujimoto, J.; Sun, W.S.; Tamaya, T. Clinical implications of steroid receptor coactivator (SRC)-3 in uterine endometrial cancers. *J. Steroid Biochem. Mol. Biol.* **2007**, *104*, 237–240. [CrossRef] [PubMed]

71. Szwarc, M.M.; Kommagani, R.; Lessey, B.A.; Lydon, J.P. The p160/steroid receptor coactivator family: Potent arbiters of uterine physiology and dysfunction. *Biol. Reprod.* **2014**, *91*, 122. [CrossRef] [PubMed]

72. Divekar, S.D.; Storchan, G.B.; Sperle, K.; Veselik, D.J.; Johnson, E.; Dakshanamurthy, S.; Lajiminmuhip, Y.N.; Nakles, R.E.; Huang, L.; Martin, M.B. The role of calcium in the activation of estrogen receptor-alpha. *Cancer Res.* **2011**, *71*, 1658–1668. [CrossRef] [PubMed]

73. Tajbakhsh, A.; Pasdar, A.; Rezaee, M.; Fazeli, M.; Soleimanpour, S.; Hassanian, S.M.; FarshchiyanYazdi, Z.; Younesi Rad, T.; Ferns, G.A.; Avan, A. The current status and perspectives regarding the clinical implication of intracellular calcium in breast cancer. *J. Cell. Physiol.* **2018**. [CrossRef] [PubMed]

74. Leclercq, G. Calcium-induced activation of estrogen receptor alpha—New insight. *Steroids* **2012**, *77*, 924–927. [CrossRef] [PubMed]

75. Corcoran, J.J.; Nicholson, C.; Sweeney, M.; Charnock, J.C.; Robson, S.C.; Westwood, M.; Taggart, M.J. Human uterine and placental arteries exhibit tissue-specific acute responses to 17β-estradiol and estrogen-receptor-specific agonists. *Mol. Hum. Reprod.* **2014**, *20*, 433–441. [CrossRef] [PubMed]

76. Comunanza, V.; Carbone, E.; Marcantoni, A.; Sher, E.; Ursu, D. Calcium-dependent inhibition of T-type calcium channels by TRPV1 activation in rat sensory neurons. *Pflügers Arch. Eur. J. Physiol.* **2011**, *462*, 709–722. [CrossRef] [PubMed]

77. Livak, K.J.; Schmittgen, T.D. Analysis of relative gene expression data using real-time quantitative PCR and the $2^{-\Delta\Delta Ct}$. *Methods* **2001**, *25*, 402–408. [CrossRef] [PubMed]

International Journal of
Molecular Sciences

MDPI

Article

NR1 and NR3B Composed Intranuclear *N*-methyl-D-aspartate Receptor Complexes in Human Melanoma Cells

Tibor Hajdú [1], Tamás Juhász [1], Csilla Szűcs-Somogyi [1], Kálmán Rácz [2] and Róza Zákány [1,*]

[1] Faculty of Medicine, Department of Anatomy, Histology and Embryology, University of Debrecen, H-4032 Debrecen, Hungary; hajdu.tibor@anat.med.unideb.hu (T.H.); juhaszt@anat.med.unideb.hu (T.J.); somogyics@anat.med.unideb.hu (C.S.-S.)

[2] Department of Forensic Medicine, University of Debrecen, Faculty of Medicine, H-4032 Debrecen, Hungary; raczkalman@hotmail.com

* Correspondence: roza@anat.med.unideb.hu; Tel.: +36-52-255-567

Received: 8 June 2018; Accepted: 27 June 2018; Published: 30 June 2018

Abstract: Heterotetrameric *N*-methyl-D-aspartate type glutamate receptors (NMDAR) are cationic channels primarily permeable for Ca^{2+}. NR1 and NR3 subunits bind glycine, while NR2 subunits bind glutamate for full activation. As NR1 may contain a nuclear localization signal (NLS) that is recognized by importin-α, our aim was to investigate if NMDARs are expressed in the nuclei of melanocytes and melanoma cells. A detailed NMDAR subunit expression pattern was examined by RT-PCRs (reverse transcription followed by polymerase chain reaction), fractionated western blots and immunocytochemistry in human epidermal melanocytes and in human melanoma cell lines A2058, HT199, HT168M1, MEL35/0 and WM35. All kind of NMDAR subunits are expressed as mRNAs in melanocytes, as well as in melanoma cells, while NR2B protein remained undetectable in any cell type. Western blots proved the exclusive presence of NR1 and NR3B in nuclear fractions and immunocytochemistry confirmed NR1-NR3B colocalization inside the nuclei of all melanoma cells. The same phenomenon was not observed in melanocytes. Moreover, protein database analysis revealed a putative NLS in NR3B subunit. Our results support that unusual, NR1-NR3B composed NMDAR complexes are present in the nuclei of melanoma cells. This may indicate a new malignancy-related histopathological feature of melanoma cells and raises the possibility of a glycine-driven, NMDA-related nuclear Ca^{2+}-signalling in these cells.

Keywords: melanoma; melanocyte; NMDAR; NLS; nuclear Ca^{2+}-signalling

1. Introduction

Glutamate is the main excitatory neurotransmitter of the mammalian central nervous system (CNS) acting on a broad range of ionotropic (iGluR) and metabotropic glutamate receptors (mGluR). Roles of mGluRs in physiological functions [1] or tumorigenesis [2] of various peripheral tissues, including melanocytes and melanoma have been recognized, but less is known about iGluRs in these contexts. According to their pharmacological properties (ligand binding) and sequence analogy iGluRs are divided into four subtypes: *N*-methyl-D-aspartate (NMDA), alpha-amino-3-hydroxy-5-methyl-4-isoxazolepropionic acid (AMPA), kainate and δ type glutamate receptors [3]. *N*-methyl-D-aspartate type glutamate receptors (NMDAR) are nonselective cation channels, with high permeability for Ca^{2+} [4] and require glycine coagonist for full activation. NMDARs are mostly known for playing role in synaptic plasticity, memory and learning [5]. NMDARs assemble as di- or triheterotetrameric complexes comprising three types of subunits (NR1, NR2 and NR3) with each subunit having several isoforms and in certain cases, multiple splice variants [6]. Two NR1 subunits possessing glycine binding sites are essential for channel formation, whilst NR2 subunits contain the glutamate binding site [6]. The NR1-NR2 diheterotetrameric constellation is the most common in the CNS and is sensitized by membrane depolarization, exhibit channel blockade by Mg^{2+} and allows high level, mainly Ca^{2+}-selective conductance, although these parameters vary depending on the NR2 subtype [6]. The third subunit (NR3), discovered in the developing brain [7,8], does not bind glutamate, but similarly to NR1 contains a glycine binding site. Subsequent functional studies concluded that the rare variant of the NMDARs consisting of NR1-NR3 subunits are glycine gated, excitatory ion channels, that are unresponsive to glutamate or NMDA and insensitive to membrane depolarization as they do not have Mg^{2+}-blockade [9]. In addition, Ca^{2+}-influx generated by channels consisting of NR1-NR3 subunits has a smaller amplitude in comparison to conventional NMDARs and probably leads to intracellular signals others than those evoked by activation of NR1-NR2 channels [10].

Emerging evidence revealed the presence of glutamate receptors in various tissues other than the brain [11] and functional receptors have been reported (over)expressed in various neoplastic cells (gastric, colorectal, hepatocellular, prostate, lung, breast, ovarian cancers, etc.) [12]. Cutaneous melanoma only makes 3% of all skin tumors, but is responsible for 65% of skin-tumor-related deaths [13]. High mortality rates and increasing incidence highlight [14] the importance of understanding melanocyte and melanoma biology, and glutamatergic signalling seems to be a significant part of that. Melanoma cells can release elevated amounts of glutamate through a possible autocrine loop [15] and wide range of data has been gathered on mGluRs in this tumor type [2]. On the other hand, less is known about iGluRs in melanoma cells and melanocytes, albeit a recent study showed in vitro and in vivo that iGluR inhibition by MK-801 resulted in decrease of migration and loss of tumor growth [16]. In terms of NMDARs and their subunits, mutations frequently appear in the NR2A gene, *GRIN2A* [17], correlating with decreased survival of melanoma patients [18]. Furthermore, *GRIN2A* was shown to be a tumor suppressor in melanoma as loss of its activity resulted in uncontrolled cell proliferation [19]. NMDARs are also present on melanocytes [20] and regulate cell morphology and melanosome transfer [21]. Despite the NR2(A) subunit having been more often in the spotlight in melanoma cells and melanocytes, functions of the NR1 and NR3 subunits in this context remain elusive.

The NR1 subunit has a particular molecular attribute, insofar as it may contain a nuclear localization signal (NLS). It was demonstrated that the NR1-1a splice variant bears a functionally relevant intracellular domain: a bipartite NLS close to the C terminus, in the C1 cassette [22]. The NLS is a sequence of basic amino acids that may be on the surface of proteins allowing them to bind importins, thus permitting translocation to the nucleus. Unusual subcellular localization may give rise to novel functions, especially in the pathophysiological context. However, it is not clear whether NR1 undergoes regulated intramembrane proteolysis or appears in full sequence in the nucleus. Besides NR1, no other NMDAR subunits have been shown to possess a NLS.

In our present study, we aimed to examine the detailed subcellular expression pattern of NMDAR subunits in melanoma cells and melanocytes. As the most striking novel observation, we found that cells all of the investigated melanoma cell lines possessed full size nuclear NR1 and NR3B, which phenomenon was not observed in normal human epidermal melanocytes (NHEM). Immunocytochemistry of the melanoma cells proved that NR1-NR3B form heteromer complexes in the nucleus of melanoma cells. This finding raises the possibility of the existence of a malignant transformation related, glycine driven nuclear Ca^{2+}-signalling in melanoma cells and may open new perspectives in melanoma therapy.

2. Results

2.1. Melanocytes and Melanoma Cells Express NMDAR Subunit mRNAs

NMDA receptor subunits NR1 (*GRIN1*), NR2A (*GRIN2A*), NR2B (*GRIN2B*), NR2C (*GRIN2C*), NR2D (*GRIN2D*), NR3A (*GRIN3A*) and NR3B (*GRIN3B*) mRNA expression was examined by RT-PCR (reverse transcription followed by polymerase chain reaction) in five different melanoma cell lines

(A2058, HT169M1, HT199, M35/01, WM35), as well as in melanocytes. In general, mRNAs of all NR subunits were detected either in the samples of melanoma cell lines or in that of melanocytes (Figure 1). Human brain sample was used as a positive control for detection of NMDAR subunit mRNAs (See Supplementary Information Figure S1). Glycerol aldehyde phosphate dehydrogenase (GAPDH) mRNA expression was the internal control of the reaction.

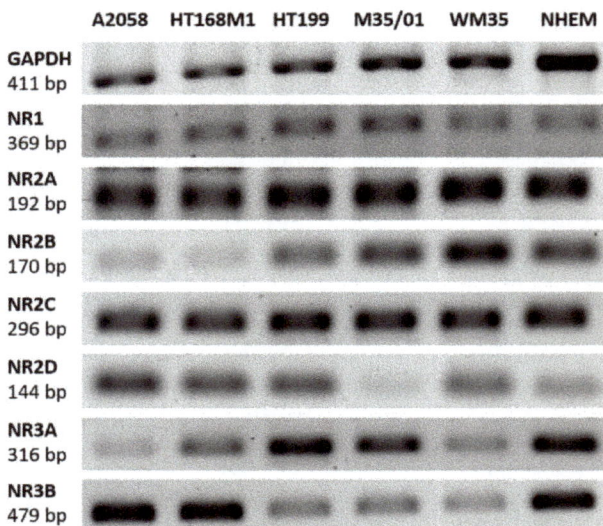

Figure 1. RT-PCR (reverse transcription followed by polymerase chain reaction) detection of NMDAR (*N*-methyl-D-aspartate receptor) subunit mRNA expression in melanoma cells and melanocytes. RT-PCRs proved the presence of the essential NR1, the widely expressed NR2 and also the rare NR3 subunits in A2058, HT169M1, HT199, M35/01 and WM35 melanoma cell lines, as well as in NHEMs (normal human epidermal melanocytes). GAPDH (Glycerol aldehyde phosphate dehydrogenase) mRNA expression served as internal control to the reaction. The positive control for NMDAR subunit detection was human brain tissue sample (see Supplementary Information Figure S1). Normalization of the relative optical density parameters of the subunit mRNA expressions to GAPDH is shown in a graphical format in the Supplementary Information Figure S2.

2.2. NR1 and NR3 Subunits Appear in Melanoma

Western blot analysis of melanocyte whole cell lysates corroborated the presence of NR2A. Although the NR1-1a subunit showed immunoblot signals, NR1, NR2B, NR3A and NR3B subunits were under the level of detection in NHEM (Figure 2A). Microphthalmia-associated transcription factor (MITF) was used as a marker for melanocyte differentiation. Actin and GAPDH were used as internal controls for western blots of melanocyte whole cell lysates.

Western blots of melanoma cellular fractions showed a detailed picture of NMDAR subunit expression in the cytosolic, nuclear and membrane compartments (Figure 2B). There was no difference found in the NMDAR subunit protein expression patterns of the radial growth phase (RGP) melanoma derived WM35 and the other, metastasizing melanoma-derived A2058, HT168M1, HT199 and M35/01 cell lines. The expression of NR1, NR2A, NR3A and NR3B subunits was confirmed in the cytosol and membranes of melanoma cells (Figure 2B). Immunoblot signals of the NLS containing NR1-1a splice variant were also detected. We found evidence of the presence of NR1, NR1-1a and NR3B subunits in the nuclei of melanoma cells, while NR2A and NR3A subunits were completely absent in the nuclear fractions (Figure 2B). NR2B signals remained undetectable in any of the examined cellular fractions, which confirmed lack or extremely weak protein levels of NR2B in melanoma cells. Multiple controls were used to prove clarity of fractionated samples. Actin was the internal control for the comparison of the cytosolic fraction to the membrane and to the nuclear fraction. TATA box binding protein (TATABP) was an internal control for nuclear fraction analysis. Human brain lysate was used as positive control for detection of NMDAR subunit protein expression (see Figure S3 in Supplementary Information).

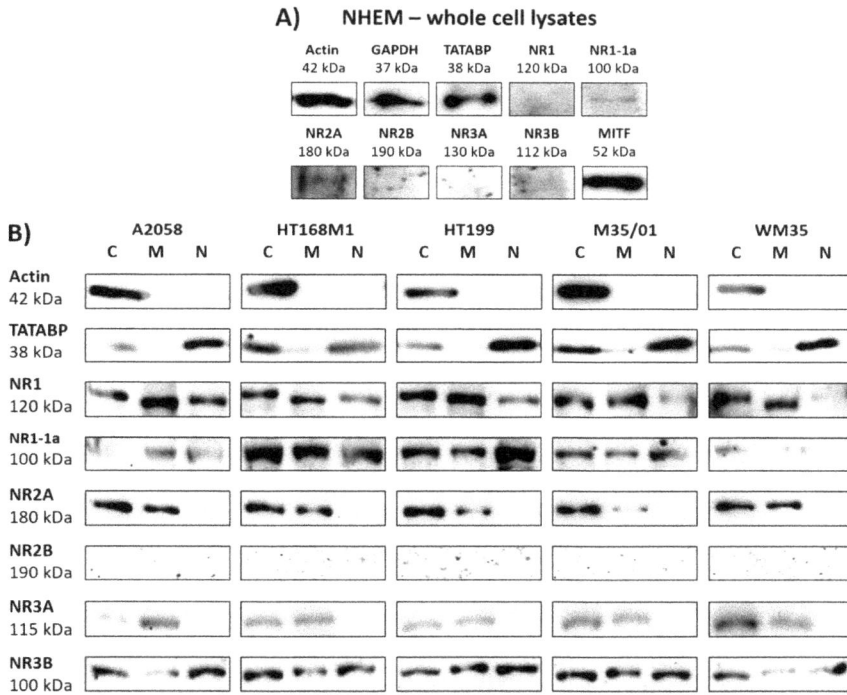

Figure 2. Western blot analysis gave detailed picture on NMDAR subunit protein expression. NR1-1a and NR2A subunits were expressed in melanocytes, while other subunits were undetectable by western blots (**A**). Fractionated melanoma samples revealed that NR1 and NR3B subunits are expressed in every cellular compartment (**B**). In each melanoma cell line NR2A and NR3A were present in the cytosol and membranes, but in the nucleus (**B**). NR2B was undetectable in each compartment (**B**). Immunoblot signals of NR1-1a were detected in the cytosol and membranes, but more importantly also in the nuclei of melanoma cells (**B**). Controls used in the experiments: MITF (microphthalmia associated transcription factor) for melanocyte differentiation control; GAPDH as an internal control for melanocyte whole cell lysates; actin as internal controls for the comparison of the cellular fractions; TATABP (TATA box binding protein) expression as internal control for nuclear fraction analysis. The positive control for detection of NMDAR subunit protein expression was human brain tissue lysate (see Supplementary Information Figure S3). Whole membrane pictures of western blots are presented as Supplementary Figure S4. Normalization of the relative optical density parameters of the protein subunit expressions to the respective control is shown in a graphical format in the Supplementary Information Figures S5–S7. (C = cytosolic, M = membrane, N = nuclear fractions of melanoma cells).

2.3. Unusual Nuclear Colocalization of Subunits in Cultured Melanoma Cells, But Not in Melanocytes

NR1-NR3B and NR1-1a-NR3B immunocytochemistry reactions were performed to reveal the subcellular (co)localization of the subunits. NR1 appeared in the cytosol of cells of every melanoma cell line, along with a less intense appearance of NR3B in the cytoplasm (Figure 3). Nevertheless, the nuclei showed remarkable patterns as NR1 and NR3B subunits colocalized in every investigated melanoma cell line (Figure 3). The prominent nuclear colocalization appeared diffusely inside the nuclei of melanoma cells rather than in the nuclear envelope or in perinuclear regions. We detected a similar expression pattern after the NR1-1a-NR3B immunocytochemistry reactions: the NLS containing NR1 splice variant specific antibody showed mostly nuclear localization and/or colocalizations with NR3B, while less cytoplasmic signals of NR1-1a were detected compared to NR1 (Figure 4). In another constellation of the experiments which was performed with an NR3B antibody produced in goat (a whole-membrane western blot picture confirming antibody specificity is shown as Figure S9), a more intense colocalization was observed, as seen in Supplementary Information Figure S10A,B. In contrast to these observations, melanocyte immunocytochemistry showed cytoplasmic localization of NR1 and NR3B subunits and confocal microscope revealed that the two subunits were undetectable in the nuclei (Figure 3). NR1-1a was also expressed in the cytoplasm of melanocytes but similarly to NR1, colocalization with NR3B could not be detected, although a very weak nuclear appearance of NR1-1a was observed (Figure 4).

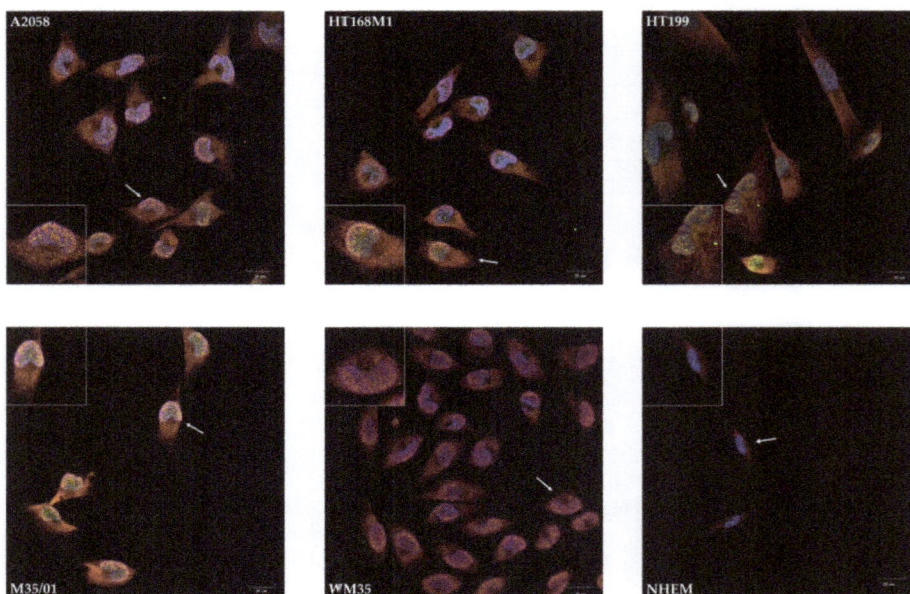

Figure 3. Immunocytochemistry demonstrating colocalization of NR1 and NR3B in the nuclei of melanoma cells, but not in the cell membrane in melanocytes. Confocal microscopic evaluation revealed that NR1 (red) was present in the cytosol of every melanoma cell line, while NR3B (green) showed weaker and less diffuse signals in the cytoplasm. Unambiguous plasma membrane immunofluorescent signals were not visible, but NR1 and NR3B colocalized inside the nuclei of melanoma cells. NR1 and NR3B colocalization was absent or undetected in the nuclei of NHEM. That 1-μm thick optical section was selected for presentation which specifically went through the majority of nuclei to show that all nuclei were immunopositive. One representative cell pointed by **white arrow** was selected for the inserts at each part of the panel. Scale bar: 20 μm. Positive controls for the secondary antibodies and special controls for the primary antibodies were also performed (see Supplementary Information Figure S8A,B).

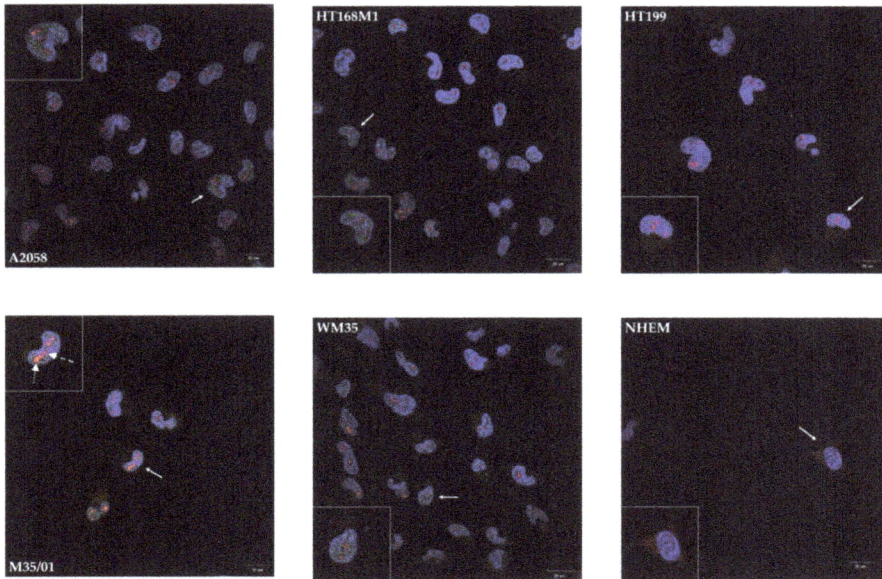

Figure 4. Immunocytochemistry demonstrating nuclear colocalization of NR1-1a and NR3B in melanoma cells. Confocal microscopic analysis demonstrated that NR1-1a (red) shows similar nuclear colocalization in melanoma cells with NR3B (green) as NR1. Nuclei of NHEM showed very weak signals of NR1-1a. NR1-1a was present in the cytoplasmic region of NHEM but unequivocal colocalization with NR3B was not detected at any cellular compartment. As an example, two extra arrows are used in the insert that shows M35/01 cells: the dashed arrow points at an area where the immunofluorescent signal of NR1-1a (red) colocalizes with DAPI (blue), resulting in purple tone whereas the spotted arrow points at an area where NR1-1a colocalizes with DAPI and NR3B (green), resulting in orange colour. Photomicrographs of 1 μm thick optical sections passing through the majority of nuclei in the area of the interest are shown. One representative cell pointed by **white arrow** with numerous (nuclear) colocalizing signals was selected on each picture and presented in the insert. Scale bar: 20 μm. Positive controls for the secondary antibodies and special controls for the primary antibodies were also performed (See Supplementary Information Figure S8A,B).

3. Discussion

Here we presented proofs for the presence of full size NR1 and NR3B NMDAR subunits, forming heteromer complexes in the nuclei of human melanoma cells, raising the possibility of a glycine sensitive autonomous nuclear Ca^{2+}-signalling of these cells. In the past 20 years, the nucleus has become a new field in the interpretation of cellular Ca^{2+}-homeostasis as it emerged as a subcellular compartment to have its own Ca^{2+}-related signalling [23–25]. Although the origin of Ca^{2+} is still a controversial question, several studies have been performed on isolated cell nuclei and showed that nucleoplasmic $[Ca^{2+}]$ changes are independent from the restricting second messengers (InsP$_3$, cyclic ADP ribose) permeability of the nuclear membrane [26,27].

Cytosolic Ca^{2+} transients can 'swamp' Ca^{2+}-changes of nuclear origin [28], making difficult to identify original nuclear transients. Nonetheless, certain observations showed that the nucleoplasmic reticulum and the nucleoplasm can generate autonomous Ca^{2+}-signals [29]. In addition, other findings showed that the nuclear interior is rich in Ca^{2+}-signalling components, such as inositol trisphosphate receptors (InsP$_3$Rs), cyclic adenosine diphosphate ribose receptors (cADPRs), nicotinic acid adenine diphosphate (NAADP) receptors and even sarco/endoplasmic reticulum Ca^{2+}-ATPase (SERCA) pumps and Na^+/Ca^{2+}-exchangers [23], supporting a concept of independent nuclear Ca^{2+}-signalling.

Recent reviews reported that metabotropic receptors, especially a broad range of G protein coupled receptors are present in the nuclei of various cell types, such as cardiomyocytes, HEK293 and HeLa cells [30]. Evidence has been gathered on nuclear ionotropic receptors as well: proteomic analysis revealed R-type Ca^{2+}-channels in the nuclear membrane, but K^+ and Cl^- channels were also proved to be active by nuclear patch clamping [31,32].

Presence of functional metabotropic glutamate receptors (mGluR5) were reported in the nuclear membranes of HEK293 cells and cortical neurons. mGluR5 presumably activates nuclear phosphatidyl-inositol-phospholipase C (PtdIns-PLCs), resulting in $InsP_3$ generation and consequent Ca^{2+}-signals in the nucleoplasm [33,34]. But so far, no evidence of functional nuclear NMDARs or other nuclear ionotropic glutamate receptors has been published yet.

Our results give pieces of evidence supporting the presence of ionotropic glutamate receptors in the nuclei of melanoma cells and afford valuable findings on NMDAR subunit expression in melanocytes. While a previous microarray analysis [20] revealed expression of NR2C subunit and confirmed absence of NR2B in melanocytes, our PCR results showed that all types of NMDAR subunits were present at mRNA level in NHEM. Nonetheless, western blots on NHEM could only prove the expression of NR1-1a and NR2A and very weak NR1 and NR3B signals were seen parallel to other subunits (NR2B, 2C and 3A) remaining undetectable. NR2A and NR3A were expressed in the cytosolic and membrane fractions of melanoma cells, but not in the nucleus, while NR2B was completely absent in melanoma cells, as in NHEM. This suggests that NR2 subunits do not play role in the possible nuclear NMDAR complex formation of melanoma cells. Therefore, detection of the essential NR1, and the unconventional NR3B subunits in every examined cellular compartments of melanoma cells may draw attention.

Confocal microscopic evaluation of immunocytochemistry reactions on melanoma cells demonstrated colocalization of NR1 and NR3B subunits suggesting the presence of molecular heteromers consisting of these subunits primarily inside the nuclei. We noticed spotted appearance of the colocalizing signals, which could be explained by the similar pattern in $InsP_3R$ expression [35]. That suggests NMDARs consisting of NR1 and NR3B subunits are likely to appear in the inner nuclear membrane, or in deep invaginations of the nuclear membrane or on the surface of nuclear Ca^{2+}-storages. Colocalization of NR1 or NR1-1 with NR3B was not observed in the nuclei of melanocytes and remained weakly detectable in other parts of the cells. Taken together these findings further strengthen the possibility that NR1-1a and NR3B composed NMDARs locating in the nuclear compartment may be connected to malignancy.

Besides the presumable Ca^{2+}-signalling role of these NMDAR complexes, another possible function (i.e., playing a direct role in transcription regulation) was also considered to explain nuclear presence of NMDAR subunits. We analysed if RNA, DNA or histone binding sequences are present either in NR1 or NR3B proteins by corresponding protein sequence databases (www.uniprot.org). As this screening carried out negative results we excluded the opportunity of an unusual function of these heteromers that is a role influencing transcriptional events or chromatin structure via direct molecular interactions with chromatin.

Conventional NMDARs are composed of two glycine binding NR1 subunits and two glutamate binding NR2 subunits each and are efficiently activated upon simultaneous binding of both glutamate and glycine. As neither NR1 nor NR3 subunits bind glutamate but glycine, these unorthodox NMDARs may function as alternative glycine receptors [36] which are permeable for Ca^{2+}. Nonetheless, glutamate insensitivity and the fact that previous membrane depolarization is unnecessary for channel opening [9] make difficult to investigate the function of NR1-NR3 heteromers by pharmacological approaches. Of note, other electrophysiological properties of this channel constellation also differ from usual NR1-NR2 or NR1-NR2-NR3 channels in terms of Ca^{2+}-currents [10].

As previously discussed, NR1-1a is the only splice variant of NR1 that contains NLS [22]. There has been no information published whether other NMDAR subunit splice variants possess NLS or not. As NLS in general is not a constant sequence that could be simply given, therefore, we decided to choose the NLSs described in NR1-1a [22] and based on sequence homology between human NR1 and NR3B. We matched the NLSs with the full-protein sequence of NR3B with the help of internet databases (www.uniprot.org). With this method, NLS-like motifs were found close to the intracellular C-terminus of NR3B (between amino acids 922 and 933), similar to NR1-1a. Therefore, we hypothesize the possibility of NR3B to have a functional NLS that could explain its nuclear localization. However, the NLS-suspicious area of NR3B needs thorough examination and functional confirmation [37].

Enhancing or modulating nuclear Ca^{2+}-signalling could already be useful for treatments of neurodegenerative diseases, anxiety and post-traumatic stress disorders [38], moreover successful in vitro studies have reported the efficacy of NMDAR (and iGluR) antagonists on cancer and melanoma cells [39]. However, novel melanoma (and cancer) therapies involving nuclear Ca^{2+}-signalling would still raise many questions, but the possibility is given for further examination and clarification as the nuclear Ca^{2+}-signalling machinery and according to our results NMDARs are present in melanoma cells.

4. Materials and Methods

4.1. Melanoma and Melanocyte Cell Cultures

The human cutaneous melanoma cell line A2058 was derived from lymph node metastasis of amelanotic melanoma; HT168M1 was established from the spleen of HT168-injected immunosuppressed mice; HT199 was isolated from a metastasizing nodular melanoma; M35/01 was selected from a vertical growth phase (VGP) superficial spreading melanoma; and WM35 was derived from nonmetastasizing superficial spreading melanoma. A2058 and WM35 cell lines were acquired from ATCC (ATCC® CRL-1661™, Manassas, VA, USA). HT168M1, HT199 and M35/01 cell lines were kind gifts of Andrea Ladányi (National Institute of Oncology, Budapest, Hungary). NHEM (PromoCell GmbH, Heidelberg, Germany) were isolated from the epidermis of juvenile or adult skin from different locations (face, breasts, abdomen, or thigh). Melanoma cells were cultured in RPMI-1640 culture medium (Sigma-Aldrich, St. Louis, MO, USA) supplemented with 10% FBS (Gibco, Gaithersburg, MD, USA), 4.1 g/L glucose, 2 mmol/L L-glutamine (Gibco), penicillin (100 units/mL) and streptomycin (100 μg/mL), and incubated at 37 °C, in a 5% CO_2 atmosphere and 80% humidity to approximately 80% confluence. NHEM cell line was cultured in Melanocyte Medium (PromoCell GmbH) according to the instructions of the manufacturer. Every cell line was routinely screened for a possible *Mycoplasma* infection.

4.2. mRNA Expression Analysis Using Reverse Transcription Followed by PCR (RT-PCR)

After reaching the expected confluence melanoma and melanocyte cell cultures were washed three times with physiological NaCl, dissolved in TRIzol (Applied Biosystems, Foster City, CA, USA), and following addition of 20% chloroform (Sigma-Aldrich) samples were centrifuged at $10,000 \times g$ for 15 min at 4 °C. Samples were incubated in 500 μL RNase-free isopropanol at -20 °C for 1 h, then total RNA was dissolved in nuclease-free water (Promega, Madison, WI, USA) and stored at -70 °C. The assay mixture for reverse transcriptase (RT) reactions was composed of 2000 ng RNA, 2 μL $10\times$ RT random primers; 0.8 μL $25\times$ deoxynucleotide triphosphate (dNTP) Mix (100 mM); 50 units (1 μL) of MultiScribe™ RT in 2 μL $10\times$ RT buffer (High Capacity RT kit; Applied Biosystems, Foster City, CA, USA). DNA was transcribed at 37 °C for 2 h.

Amplifications of specific cDNA sequences were carried out using specific primer pairs that were designed by Primer Premier 5.0 software (Premier Biosoft, Palo Alto, CA, USA) based on human nucleotide sequences published in GenBank and purchased from Integrated DNA Technologies, Inc. (IDT; Coralville, IA, USA). The specificity of custom-designed primer pairs was confirmed in silico by using the Primer-BLAST service of NCBI (Available online: http://www.ncbi.nlm.nih.gov/tools/primer-blast/). Nucleotide sequences of forward and reverse primers and reaction conditions are shown in Table 1. PCR reactions were carried out in a final volume of 21 µL containing 1 µL forward and 1 µL reverse primers (10 µM), 0.5 µL cDNA, 0.5 µL dNTP Mix (200 µM), and 0.625 unit (0.125 µL) of GoTaq® DNA polymerase in 1× Green GoTaq® Reaction Buffer (Promega) in a programmable thermal cycler (Labnet MultiGene™ 96-well Gradient Thermal Cycler; Labnet International, Edison, NJ, USA) with the following protocol: 2 min at 95 °C for initial denaturation followed by 35 repeated cycles of denaturation at 94 °C for 1 min, primer annealing for 1 min at an optimised temperature for each primer pair (see Table 1), and extension at 72 °C for 90 s. After the final cycle, further extension was allowed to proceed for another 10 min at 72 °C. PCR products were analysed using horizontal gel electrophoresis in 1.2% agarose gel containing ethidium bromide (Amresco Inc., Solon, OH, USA) at 120 V constant voltage. Signals were developed with a gel imaging system (Fluorchem E, Protein Simple, San Jose, CA, USA). Optical densities of signals were measured by using ImageJ 1.46R bundled with Java 1.8.0_112 freeware (https://imagej.nih.gov/ij/), and results were normalized to the internal control.

Table 1. Nucleotide sequences, amplification sites, GenBank accession numbers, amplimer sizes and PCR reaction conditions for each primer pair are shown.

Gene	Primer	Nucleotide Sequence (5′→3′)	GenBank ID	Annealing Temperature	Amplimer Size (bp)
NR1 (GRIN1)	sense	CCA CGC TGA GTG ATG GGA CA (1568–1587)	NM_007327.3	56.5 °C	369
	antisense	GGT ACT TGA AGG GCT TGG AAA A (1936–1915)			
NR2A (GRIN2A)	sense	TGC CAC AAC GAG AAG AAC (2946–2963)	NM_001134407.2	57 °C	192
	antisense	GAT GGA GAA GAG CAA CCC (3137–3120)			
NR2B (GRIN2B)	sense	CAA GAG GCG TAA GCA GC (3413–3429)	NM_000834.3	55 °C	170
	antisense	CCA GGT AGA AGT CCC GTA G (3582–3564)			
NR2C (GRIN2C)	sense	ACG TCC ACG GCA TTG TCT (676–693)	NM_000835.4	61 °C	296
	antisense	GGA TGT CAT TCC CAT AAC CA (952–971)			
NR2D (GRIN2D)	sense	TGG CAA GCA CGG AAA GAA GAT C (1618–1639)	NM_000836.2	62.5 °C	144
	antisense	TCC ACG AAG GGG ACG GAG AAG T (1761–1740)			
NR3A (GRIN3A)	sense	ACA CAA AAC CCA CTT CCA ACA TCC (2107–2130)	NM_133445.2	58 °C	316
	antisense	TGC TCC ATA CTT TCC ATC CCC TAC (2422–2399)			
NR3B (GRIN3B)	sense	CGC AAG TGC TGC TAC GGC TAC (1436–1456)	NM_138690.2	61 °C	479
	antisense	ACG GTG CGT CTG AAG AGG ATG (1914–1894)			
GAPDH	sense	CCA GAA GAC TGT GGA TGG CC (740–759)	NM_002046.5	54 °C	411
	antisense	CTG TAG CCA AAT TCG TTG TC (1150–1131)			

4.3. SDS-PAGE and Western Blot Analysis

Melanoma and melanocyte cell cultures were washed with physiological NaCl solution then harvested. After centrifugation at 2000× *g* for 10 min at room temperature, cell pellets were suspended in 100 µL of RIPA (Radio Immuno Precipitation Assay) homogenization buffer containing 150 mM NaCl, 1.0% NP40, 0.5% sodium deoxycholate, 50 mM Tris, 0.1% SDS (pH 8.0), supplemented with protein inhibitors as follows: aprotinin (10 µg/mL), 5 mM benzamidine, leupeptin (10 µg/mL), trypsin inhibitor

(10 µg/mL), 1 mM PMSF, 5 mM EDTA, 1 mM EGTA, 8 mM Na-fluoride and 1 mM Na-orthovanadate. All components were purchased from Sigma-Aldrich. Samples were stored at −70 °C.

Due to the availability of limited number of cells, melanocytes' whole cell lysates were only used for further steps. NHEM suspensions were sonicated by pulsing burst for three times 30 s by 50 cycles using an ultrasonic homogeniser (Cole-Parmer, Vernon Hills, IL, USA). Melanoma cell fractions were separated as described below.

Samples for SDS-PAGE (sodium-dodecyl-sulphate polyacrylamide gel electrophoresis) were prepared by adding Laemmli's electrophoresis sample buffer (4% SDS, 10% 2-mercaptoethanol, 20% glycerol, 0.004% bromophenol blue, 0.125 mM Tris–HCl; pH 6.8) to cell lysates to set equal protein concentrations, and heated at 95 °C for 10 min. 20 µL of NHEM lysate (2 µg/µL final concentration), 20 µL of cytosolic fraction (0.5 µg/µL), 20 µL of nuclear fraction (0.5 µg/µL) and 20 µL of membrane fraction (0.5 µg/µL) samples of melanoma cells were separated by 7.5% SDS-PAGE gel for immunological detection of NMDAR subunit proteins. Separated proteins were transferred to nitrocellulose membranes (Bio-Rad Trans Blot Turbo Midi Nitrocellulose Transfer Packs) by using a Bio-Rad Trans-Blot Turbo system (Bio-Rad Laboratories, Hercules, CA, USA). After blocking in 5% nonfat dry milk in PBS for 1 h, membranes were incubated with primary antibodies overnight at 4 °C as shown in Table 2. After washing for 30 min in PBS with 1% Tween-20 (Amresco) (PBST (phosphate buffered saline supplemented with 1% Tween-20)), membranes were incubated with the HRP (horse radish peroxidase)-conjugated secondary antibodies, anti-rabbit or anti-mouse IgG (Bio-Rad Laboratories, Hercules, CA, USA) in 1:1500 dilution. Membranes were developed by enhanced chemiluminescence (Advansta Inc., Menlo Park, CA, USA) according to the instructions of the manufacturer. Signals were developed with a gel imaging system (Fluorchem E, Protein Simple, San Jose, CA, USA). Optical densities of western blot signals were measured by using ImageJ 1.46R bundled with Java 1.8.0_112 freeware (https://imagej.nih.gov/ij/), and results were normalized to the internal control.

Table 2. Tables of antibodies used in the experiments.

Antibody	Host Animal	Dilution for Western Blot	Dilution for Immunocytochemistry	Distributor, Cat. Num.
Anti-NR1	rabbit, monoclonal	1:250	1:50	Cell Signaling Technology, Danvers, MO, USA #5704S lot:2
Anti-NR1-1a	rabbit, polyclonal	1:200	1:50	Merck-Millipore, Billerica, MA, USA AB5046P
Anti-NR2A	rabbit, polyclonal	1:200	-	Cell Signaling Technology, Danvers, MO, USA #4205S lot:1
Anti-NR2B	rabbit, polyclonal	1:200	-	Cell Signaling Technology, Danvers, MO, USA #4207S lot:2
Anti-NR3A	rabbit, polyclonal	1:300	-	Merck-Millipore, Billerica, MA, USA 07-356
Anti-NR3A	rabbit, polyclonal	1:500	-	Alomone Labs, Jerusalem, Israel AGC-030
Anti-NR3B	rabbit, polyclonal	1:600	-	Abcam, Cambridge, UK ab35677-100
Anti-NR3B	rabbit, polyclonal	1:250	1:50	Alomone Labs, Jerusalem, Israel AGC-031
Anti-TATABP	mouse, monoclonal	1:500	-	Abcam, Cambridge, UK ab51841
Anti-GAPDH	rabbit, polyclonal	1:500	-	Abcam, Cambridge, UK Ab9485
Anti-β-Actin	mouse, monoclonal	1:10.000	-	Sigma-Aldrich, St. Louis, MO, USA A5441

4.4. Subcellular Fractionation of Melanoma Cells

Cytosolic, membrane and nuclear fractions of melanoma cells were separated. Samples used for cytosol and membrane isolation were sonicated, and then centrifuged at $50,000 \times g$ for 90 min at 4 °C. Supernatants containing cytosolic fractions were separated. Pellets were resuspended in 60 µL of 1% Triton X-100 (Sigma-Aldrich) in RIPA buffer and triturated on ice for 1 h. Afterwards, samples were centrifuged again at $50,000 \times g$ for 55 min at 4 °C. Supernatants containing membrane fractions were carefully separated.

For nuclear fraction isolation, melanoma cells were suspended in 1 mL buffer-A (10 mM HEPES, 1.5 mM $MgCl_2$, 10 mM KCl, 0.1 mM EDTA, 0.1 mM EGTA, 1 mM DTT, supplemented with protein inhibitors as described above) and stored at -70 °C. Samples were not sonicated. After the addition of 50 µL Igepal CA-630 (Sigma-Aldrich) cells were homogenized by Dounce homogenizer. Samples were centrifuged at $770 \times g$ for 10 min at 4 °C and pellets were resuspended in 1 mL 2.2 mM sucrose-buffer-A solution. This was followed by a second step of centrifugation at $40,000 \times g$ for 90 min at 4 °C. Pellets were washed twice by 0.25 mM sucrose-buffer-A solution and suspended in 200 µL of the same solution.

4.5. Immunocytochemistry and Confocal Microscopy

Immunocytochemistry was performed on cells cultured on the surface of coverslips to visualize intracellular localizations of NR1, NR1-1a and NR3B. Cultures were fixed in 4% paraformaldehyde (Sigma-Aldrich) solution for 1 h and washed in distilled water. After rinsing in PBS (pH 7.4), nonspecific binding sites were blocked with PBS supplemented with 1% BSA (Amresco) for 30 min at 37 °C. After that samples were washed again three times in PBS, and cultures were incubated with rabbit polyclonal anti-NR3B antibody (Alomone Labs, Jerusalem, Israel), at a dilution of 1:50 in PBST, at 4 °C overnight. On the following day after washing three times with PBS, biotinylated goat anti-rabbit antibodies were added onto the samples (Vector Laboratories, Burlingame, CA, USA), at a dilution of 1:1000 in PBST, at room temperature for 2 h. After washing, cultures were incubated with monoclonal anti-NR1 antibody (Cell Signaling Technology, Danvers, MA, USA) or polyclonal anti-NR1-1a antibody (Merck-Millipore, Billerica, MA, USA), both produced in rabbit, at a dilution of 1:50 at 4 °C overnight. On the third day the biotinylated goat antibody was visualized with Streptavidin Alexa Fluor 488 conjugate, while the NR1/NR1-1a antibodies were visualized with anti-rabbit Alexa Fluor 555 secondary antibody (Life Technologies Corporation, Carlsbad, CA, USA) at a dilution of 1:1000 in PBST. Cultures were mounted in Vectashield mounting medium (Vector Laboratories, Peterborough, England) containing DAPI for nuclear DNA staining. Control experiments for these immunocytochemistry reactions are described in Supplementary Material and presented on Supplementary Figure S8A,B, S9, S10.

Immunocytochemistry reactions were repeated three times with NHEM and each melanoma cell line, and five recordings were examined with each constellation of the reaction.

Photomicrographs of the cells were taken with an Olympus FV3000 confocal microscope (Olympus Co., Tokyo, Japan) using a 60x PlanApo N oil-immersion objective (NA: 1.42) and FV31S-SW software (Olympus Co., Tokyo, Japan). Z image series of 1-µm optical thickness were recorded in sequential scan mode. For excitation 488 and 543 nm laser beams were used. The average pixel time was 4 µsec.

Supplementary Materials: Supplementary materials can be found at http://www.mdpi.com/1422-0067/19/7/1929/s1.

Author Contributions: Conceived and designed the experiments: T.H., T.J., R.Z. Performed the experiments: T.H., T.J. Analyzed the data: T.H., T.J., C.S.-S., R Z. Contributed materials: K.R. Wrote the paper: T.H., R.Z.

Funding:: This research was funded by OTKA Bridging Funds to Róza Zákány and Tamás Juhász from University of Debrecen.

Acknowledgments: The authors thank Krisztina Bíró for her excellent technical assistance and Sinyoung Cho medical student for her valuable contribution to the experimental work. We acknowledge Csaba Matta for his useful comments and suggestions during preparation of the manuscript. Our grateful thanks are extended to Andrea Ladányi (National Institute of Oncology, Hungary) for providing HT168M1, HT199 and M35/01 melanoma cell lines at the initial stage of our study. Tibor Hajdú and Tamás Juhász were supported by the Eötvös Loránd and Magyary Zoltán Funds of the Hungarian Academy of Sciences and by the European Union and the State of Hungary, co-financed by the European Social Fund in the framework of TÁMOP 4.2.4. A/2-11-1-2012-0001 'National Excellence Program'. Tamás Juhász was supported by the Bolyai János Research Scholarship and the Szodoray Lajos Fund. The project is cofinanced by the University of Debrecen (RH/751/2015).

Conflicts of Interest: The authors declare no conflict of interest. The founding sponsors had no role in any part of the study.

Abbreviations

CNS	central nervous system
iGluR	ionotropic glutamate receptor
mGluR	metabotropic glutamate receptor
NMDA	*N*-methyl-D-aspartate
NMDAR	*N*-methyl-D-aspartate receptor
NLS	Nuclear localization signal
NHEM	normal human epidermal melanocytes
RT-PCR	reverse transcription followed by polymerase chain reaction
dNTP	deoxynucleotide triphosphate
SDS-PAGE	sodium-dodecyl-sulphate polyacrylamide gel electrophoresis
RIPA	radio immuno precipitation assay
EDTA	ethylene diamine tetra-acetic acid
EGTA	ethylene glycol tetra-acetic acid
PBS	phosphate buffered saline
PBST	phosphate buffered saline supplemented with 1% Tween-20
HRP	horse radish peroxidase
HEPES	hydroxyethyl-piperazine-ethanesulfonic acid
DTT	dithiothreitol
GAPDH	glycerine aldehyde phosphate dehydrogenase
MITF	microphthalmia associated transcription factor
TATABP	TATA box binding protein
InsP$_3$R	inositol trisphosphate receptor

References

1. Hogan-Cann, A.D.; Anderson, C.M. Physiological Roles of Non-Neuronal NMDA Receptors. *Trends Pharmacol. Sci.* **2016**, *37*, 750–767. [CrossRef] [PubMed]

2. Teh, J.L.; Chen, S. Glutamatergic signaling in cellular transformation. *Pigment Cell Melanoma Res.* **2012**, *25*, 331–342. [CrossRef] [PubMed]

3. Traynelis, S.F.; Wollmuth, L.P.; McBain, C.J.; Menniti, F.S.; Vance, K.M.; Ogden, K.K.; Hansen, K.B.; Yuan, H.; Myers, S.J.; Dingledine, R. Glutamate receptor ion channels: Structure, regulation, and function. *Pharmacol. Rev.* **2010**, *62*, 405–496. [CrossRef] [PubMed]

4. Mayer, M.L.; Westbrook, G.L. Permeation and block of *N*-methyl-D-aspartic acid receptor channels by divalent cations in mouse cultured central neurones. *J. Physiol.* **1987**, *394*, 501–527. [CrossRef] [PubMed]

5. Malenka, R.C.; Nicoll, R.A. Long-term potentiation—A decade of progress? *Science* **1999**, *285*, 1870–1874. [CrossRef] [PubMed]

6. Paoletti, P.; Bellone, C.; Zhou, Q. NMDA receptor subunit diversity: Impact on receptor properties, synaptic plasticity and disease. *Nat. Rev. Neurosci.* **2013**, *14*, 383–400. [CrossRef] [PubMed]

7. Ciabarra, A.M.; Sullivan, J.M.; Gahn, L.G.; Pecht, G.; Heinemann, S.; Sevarino, K.A. Cloning and characterization of chi-1: A developmentally regulated member of a novel class of the ionotropic glutamate receptor family. *J. Neurosci.* **1995**, *15*, 6498–6508. [CrossRef] [PubMed]

8. Nishi, M.; Hinds, H.; Lu, H.P.; Kawata, M.; Hayashi, Y. Motoneuron-specific expression of NR3B, a novel NMDA-type glutamate receptor subunit that works in a dominant-negative manner. *J. Neurosci.* **2001**, *21*, RC185. [CrossRef] [PubMed]

9. Chatterton, J.E.; Awobuluyi, M.; Premkumar, L.S.; Takahashi, H.; Talantova, M.; Shin, Y.; Cui, J.; Tu, S.; Sevarino, K.A.; Nakanishi, N.; et al. Excitatory glycine receptors containing the NR3 family of NMDA receptor subunits. *Nature* **2002**, *415*, 793–798. [CrossRef] [PubMed]

10. Smothers, C.T.; Woodward, J.J. Pharmacological characterization of glycine-activated currents in HEK 293 cells expressing N-methyl-D-aspartate NR1 and NR3 subunits. *J. Pharmacol. Exp. Ther.* **2007**, *322*, 739–748. [CrossRef] [PubMed]

11. Hinoi, E.; Takarada, T.; Ueshima, T.; Tsuchihashi, Y.; Yoneda, Y. Glutamate signaling in peripheral tissues. *Eur. J. Biochem.* **2004**, *271*, 1–13. [CrossRef] [PubMed]

12. Stepulak, A.; Rola, R.; Polberg, K.; Ikonomidou, C. Glutamate and its receptors in cancer. *J. Neural Transm.* **2014**, *121*, 933–944. [CrossRef] [PubMed]

13. Cummins, D.L.; Cummins, J.M.; Pantle, H.; Silverman, M.A.; Leonard, A.L.; Chanmugam, A. Cutaneous malignant melanoma. *Mayo Clin. Prcc.* **2006**, *81*, 500–507. [CrossRef] [PubMed]

14. Erdei, E.; Torres, S.M. A new understanding in the epidemiology of melanoma. *Expert Rev. Anticancer Ther.* **2010**, *10*, 1811–1823. [CrossRef] [PubMed]

15. Namkoong, J.; Shin, S.S.; Lee, H.J.; Marin, Y.E.; Wall, B.A.; Goydos, J.S.; Chen, S. Metabotropic glutamate receptor 1 and glutamate signaling in human melanoma. *Cancer Res.* **2007**, *67*, 2298–2305. [CrossRef] [PubMed]

16. Song, Z.; He, C.D.; Liu, J.; Sun, C.; Lu, P.; Li, L.; Gao, L.; Zhang, Y.; Xu, Y.; Shan, L.; et al. Blocking glutamate-mediated signalling inhibits human melanoma growth and migration. *Exp. Dermatol.* **2012**, *21*, 926–931. [CrossRef] [PubMed]

17. Wei, X.; Walia, V.; Lin, J.C.; Teer, J.K.; Prickett, T.D.; Gartner, J.; Davis, S.; Program, N.C.S.; Stemke-Hale, K.; Davies, M.A.; et al. Exome sequencing identifies GRIN2A as frequently mutated in melanoma. *Nat. Genet.* **2011**, *43*, 442–446. [CrossRef] [PubMed]

18. D'Mello, S.A.; Flanagan, J.U.; Green, T.N.; Leung, E.Y.; Askarian-Amiri, M.E.; Joseph, W.R.; McCrystal, M.R.; Isaacs, R.J.; Shaw, J.H.; Furneaux, C.E.; et al. Evidence That GRIN2A Mutations in Melanoma Correlate with Decreased Survival. *Front. Oncol.* **2014**, *3*, 333. [CrossRef] [PubMed]

19. Prickett, T.D.; Zerlanko, B.J.; Hill, V.K.; Gartner, J.J.; Qutob, N.; Jiang, J.; Simaan, M.; Wunderlich, J.; Gutkind, J.S.; Rosenberg, S.A.; et al. Somatic mutation of GRIN2A in malignant melanoma results in loss of tumor suppressor activity via aberrant NMDAR complex formation. *J. Investig. Dermatol.* **2014**, *134*, 2390–2398. [CrossRef] [PubMed]

20. Hoogduijn, M.J.; Hitchcock, I.S.; Smit, N.P.; Gillbro, J.M.; Schallreuter, K.U.; Genever, P.G. Glutamate receptors on human melanocytes regulate the expression of MiTF. *Pigment Cell Res.* **2006**, *19*, 58–67. [CrossRef] [PubMed]

21. Ni, J.; Wang, N.; Gao, L.; Li, L.; Zheng, S.; Liu, Y.; Ozukum, M.; Nikiforova, A.; Zhao, G.; Song, Z. The effect of the NMDA receptor-dependent signaling pathway on cell morphology and melanosome transfer in melanocytes. *J. Dermatol. Sci.* **2016**, *84*, 296–304. [CrossRef] [PubMed]

22. Holmes, K.D.; Mattar, P.; Marsh, D.R.; Jordan, V.; Weaver, L.C.; Dekaban, G.A. The C-terminal C1 cassette of the N-methyl-D-aspartate receptor 1 subunit contains a bi-partite nuclear localization sequence. *J. Neurochem.* **2002**, *81*, 1152–1165. [CrossRef] [PubMed]

23. Resende, R.R.; Andrade, L.M.; Oliveira, A.G.; Guimaraes, E.S.; Guatimosim, S.; Leite, M.F. Nucleoplasmic calcium signaling and cell proliferation: Calcium signaling in the nucleus. *Cell Commun. Signal. CCS* **2013**, *11*, 14. [CrossRef] [PubMed]

24. Bootman, M.D.; Fearnley, C.; Smyrnias, I.; MacDonald, F.; Roderick, H.L. An update on nuclear calcium signalling. *J. Cell Sci.* **2009**, *122*, 2337–2350. [CrossRef] [PubMed]

25. Alonso, M.T.; Garcia-Sancho, J. Nuclear Ca(2+) signalling. *Cell Calcium* **2011**, *49*, 280–289. [CrossRef] [PubMed]

26. Adebanjo, O.A.; Biswas, G.; Moonga, B.S.; Anandatheerthavarada, H.K.; Sun, L.; Bevis, P.J.; Sodam, B.R.; Lai, F.A.; Avadhani, N.G.; Zaidi, M. Novel biochemical and functional insights into nuclear Ca(2+) transport through IP(3)Rs and RyRs in osteoblasts. *Am. J. Physiol. Renal Physiol.* **2000**, *278*, F784–F791. [CrossRef] [PubMed]

27. Gerasimenko, O.V.; Gerasimenko, J.V.; Tepikin, A.V.; Petersen, O.H. ATP-dependent accumulation and inositol trisphosphate- or cyclic ADP-ribose-mediated release of Ca^{2+} from the nuclear envelope. *Cell* **1995**, *80*, 439–444. [CrossRef]

28. Shirakawa, H.; Miyazaki, S. Spatiotemporal analysis of calcium dynamics in the nucleus of hamster oocytes. *J. Physiol.* **1996**, *494 Pt 1*, 29–40. [CrossRef] [PubMed]

29. Luo, D.; Yang, D.; Lan, X.; Li, K.; Li, X.; Chen, J.; Zhang, Y.; Xiao, R.P.; Han, Q.; Cheng, H. Nuclear Ca^{2+} sparks and waves mediated by inositol 1,4,5-trisphosphate receptors in neonatal rat cardiomyocytes. *Cell Calcium* **2008**, *43*, 165–174. [CrossRef] [PubMed]

30. Campden, R.; Audet, N.; Hebert, T.E. Nuclear G protein signaling: New tricks for old dogs. *J. Cardiovasc. Pharmacol.* **2015**, *65*, 110–122. [CrossRef] [PubMed]

31. Matzke, A.J.; Weiger, T.M.; Matzke, M. Ion channels at the nucleus: Electrophysiology meets the genome. *Mol. Plant* **2010**, *3*, 642–652. [CrossRef] [PubMed]

32. Bkaily, G.; Avedanian, L.; Jacques, D. Nuclear membrane receptors and channels as targets for drug development in cardiovascular diseases. *Can. J. Physiol. Pharmacol.* **2009**, *87*, 108–119. [CrossRef] [PubMed]

33. Jong, Y.J.; Kumar, V.; Kingston, A.E.; Romano, C.; O'Malley, K.L. Functional metabotropic glutamate receptors on nuclei from brain and primary cultured striatal neurons. Role of transporters in delivering ligand. *J. Biol. Chem.* **2005**, *280*, 30469–30480. [CrossRef] [PubMed]

34. Kumar, V.; Jong, Y.J.; O'Malley, K.L. Activated nuclear metabotropic glutamate receptor mGlu5 couples to nuclear Gq/11 proteins to generate inositol 1,4,5-trisphosphate-mediated nuclear Ca^{2+} release. *J. Biol. Chem.* **2008**, *283*, 14072–14083. [CrossRef] [PubMed]

35. Leite, M.F.; Thrower, E.C.; Echevarria, W.; Koulen, P.; Hirata, K.; Bennett, A.M.; Ehrlich, B.E.; Nathanson, M.H. Nuclear and cytosolic calcium are regulated independently. *Proc. Natl. Acad. Sci. USA* **2003**, *100*, 2975–2980. [CrossRef] [PubMed]

36. Low, C.M.; Wee, K.S. New insights into the not-so-new NR3 subunits of N-methyl-D-aspartate receptor: Localization, structure, and function. *Mol. Pharmacol.* **2010**, *78*, 1–11. [CrossRef] [PubMed]

37. Cokol, M.; Nair, R.; Rost, B. Finding nuclear localization signals. *EMBO Rep.* **2000**, *1*, 411–415. [CrossRef] [PubMed]

38. Bading, H. Nuclear calcium signalling in the regulation of brain function. *Nat. Rev. Neurosci.* **2013**, *14*, 593–608. [CrossRef] [PubMed]

39. Ribeiro, M.P.; Custodio, J.B.; Santos, A.E. Ionotropic glutamate receptor antagonists and cancer therapy: Time to think out of the box? *Cancer Chemother. Pharmacol.* **2017**, *79*, 219–225. [CrossRef] [PubMed]

International Journal of
Molecular Sciences

MDPI

Article

Cell Cycle Regulation by Ca^{2+}-Activated K$^+$ (BK) Channels Modulators in SH-SY5Y Neuroblastoma Cells

Fatima Maqoud [1], Angela Curci [1], Rosa Scala [1], Alessandra Pannunzio [2], Federica Campanella [2], Mauro Coluccia [2], Giuseppe Passantino [3], Nicola Zizzo [3] and Domenico Tricarico [1,*]

[1]   Section of Pharmacology, Department of Pharmacy-Pharmaceutical Sciences, University of Bari,
     Via Orabona 4, 70125 Bari, Italy; fatima.maqoud@uniba.it (F.M.); angela.curci@uniba.it (A.C.);
     rosa.scala@uniba.it (R.S)
[2]   Section of Anatomy Pathology, Department of Pharmacy-Pharmaceutical Sciences, University of Bari,
     Via Orabona 4, 70125 Bari, Italy; Alessandra.pannunzio@uniba.it (A.P.); federica.campanella@uniba.it (F.C.);
     mauro.coluccia@uniba.it (M.C.)
[3]   Section of Anatomy Pathology, Department of Veterinary Medicine, University of Bari, Via Orabona 4,
     70125 Bari, Italy; giuseppe.passantino@uniba.it (G.P.); nicola.zizzo@uniba.it (N.Z.)
*   Correspondence: domenico.tricarico@uniba.it or domenico.tricarico@fastwebnet.it;
     Tel.: +39-08-0544-2795; Fax: +39-08-0523-1744

Received: 16 July 2018; Accepted: 13 August 2018; Published: 18 August 2018

Abstract: The effects of Ca^{2+}-activated K$^+$ (BK) channel modulation by Paxilline (PAX) (10^{-7}–10^{-4} M), Iberiotoxin (IbTX) (0.1–1×10^{-6} M) and Resveratrol (RESV) (1–2×10^{-4} M) on cell cycle and proliferation, AKT1p^{Ser473} phosphorylation, cell diameter, and BK currents were investigated in SH-SY5Y cells using Operetta-high-content-Imaging-System, ELISA-assay, impedentiometric counting method and patch-clamp technique, respectively. IbTX (4×10^{-7} M), PAX (5×10^{-5} M) and RESV (10^{-4} M) caused a maximal decrease of the outward K$^+$ current at +30 mV (Vm) of $-38.3 \pm 10\%$, $-31.9 \pm 9\%$ and $-43 \pm 8\%$, respectively, which was not reversible following washout and cell depolarization. After 6h of incubation, the drugs concentration dependently reduced proliferation. A maximal reduction of cell proliferation, respectively of $-60 \pm 8\%$ for RESV (2×10^{-4} M) (IC50 = 1.50×10^{-4} M), $-65 \pm 6\%$ for IbTX (10^{-6} M) (IC50 = 5×10^{-7} M), $-97 \pm 6\%$ for PAX (1×10^{-4} M) (IC50 = 1.06×10^{-5} M) and AKT1p^{ser473} dephosphorylation was observed. PAX induced a G1/G2 accumulation and contraction of the S-phase, reducing the nuclear area and cell diameter. IbTX induced G1 contraction and G2 accumulation reducing diameter. RESV induced G2 accumulation and S contraction reducing diameter. These drugs share common actions leading to a block of the surface membrane BK channels with cell depolarization and calcium influx, AKT1p^{ser473} dephosphorylation by calcium-dependent phosphatase, accumulation in the G2 phase, and a reduction of diameter and proliferation. In addition, the PAX action against nuclear membrane BK channels potentiates its antiproliferative effects with early apoptosis.

Keywords: Ca^{2+}-activated K$^+$(BK) channel; calcium ions; cell cycle; AKT; cell volume

1. Introduction

The large conductance voltage and Ca^{2+}-activated K$^+$ (BK) channel is known to regulate cell excitability, proliferation, and migration in response to a variety of stimulus [1–3]. Calcium signaling is involved in the release of neurotransmitters, cell excitability, and muscle contractility. Physiological low intracellular calcium levels induce cell proliferation and survivals, while an abnormal enhancement of intracellular calcium associated with the pathophysiological conditions induces cell death apoptosis. The BK channel is a calcium sensor able to regulate different cell functions including the release of neurotransmitters in the Central Nervous System (CNS), the modulation of the vascular tone, and Excitation-Contraction Coupling (EC) coupling in skeletal muscle fibers being involved in different pathophysiological functions and diseases [4–6]. The BK channel is unique among the superfamily of K^+ channels; indeed, they are indeed tetramers characterized by a pore-forming α subunit containing seven transmembrane segments (instead of the six found in voltage-dependent K^+ channels) and a large C terminus composed of two K^+ conductance domains (RCK domains), where the Ca^{2+}-binding sites reside. The pore alpha subunit, coded by a single gene (*Slo1*), is activated by depolarizing potentials and by a rise in intracellular Ca^{2+} concentration. BK channels can be associated with accessory β subunits that play a role in the modulation of BK channel gating. Four β subunits have currently been identified (β1, β2, β3, and β4) and despite the fact that they all share the same topology, it has been shown that every β subunit has a specific tissue distribution and that they modify channel kinetics as well as their pharmacological properties and the apparent Ca^{2+} sensitivity of the α subunit in different ways [5]. Splicing isoforms of the *Slo1* gene and accessory gamma subunits are also important in regulating channel function [7].

It was demonstrated that BK channel is a target for a large variety of toxins and modulators; especially, the pore forming alpha subunit represents the binding-site of these compounds whereas the associated beta 1–4 subunits play a critical role in regulating their binding affinity to the pore [5]. Among these toxins, Iberiotoxin (IbTX) is a minor fraction of the crude venom of Buthus tamulus discovered by Galvez et al. in 1990 [8]. It is a relatively impermanent external channel pore blocker of the BK channel, largely used in structural and functional studies [8,9]. Also, IbTX is characterized by an amino acid chain of the same length than Charybdotoxin (ChTX), consisting of 37 residues that possesses 68% of the sequence identity associated with it. Despite their structural similarities, a multitude of functional studies have demonstrated that IbTX binds to the external mouth of the BK channel with higher affinity than ChTX, as indicated by the lower dissociation rate of IbTX compared with ChTX. The binding of these toxins to the BK channel is very sensitive to the electrostatic interactions, involving several basic residues of toxins and negative charges in the outer vestibule of the channel pore [10,11]. Thus, the surface charge distributions and the three-dimensional structures of toxins are important determinants of their recognition and interactions with BK channels [12].

Instead, the tremorgenic mycotoxin paxilline (PAX) is an extremely potent but non-peptide BK channel blocker [13]. It is characterized by a selectivity and specificity for the BK channel so high, comparable with that of IbTX, that different authors reported a very low nM Kd when it is applied from the internal side in an excised patch [13,14]. More recently, it has been reported that the IC50 for PAX may shift from nM values, when channels are closed, to a value of 10 μM, as maximal Po is approached. Then, these findings suggest a mechanism of inhibition in which the allosteric binding of a single molecule may alter the intrinsic L(0), favoring the occupancy of closed states, with an affinity for the closed conformation greater than the affinity for the open one [15].

Both these toxins are reported to inhibit cell migration and proliferation in a variety of cell lines. For instance, chronic exposure of human malignant glioma cells for 72 h with IbTX induces S phase accumulation, reducing cell proliferation [16]. PAX reduces cell proliferation of the human breast cancer MDA-MB-453 following 72 h of incubation time [17] and it is reported to inhibit cell migration in the micromolar concentration range in the malignant pleural mesothelioma [3]. Moreover, in human cardiac c-kit$^+$ progenitor cells, this toxin inhibits cell proliferation and leads to accumulation of the cells in G0/G1 phase leading to the inhibition of migration and proliferation following 42–74 h of incubation [18].

Other than IbTX and PAX, the unselective Kv/BK channel blocker tetraethylammonium (TEA) and the potassium channel modulator resveratrol (RESV) showed antiproliferative effects. TEA at 10 mM concentration inhibits cell proliferation, leading to the accumulation of oligodendrocyte progenitor cells in the G1 phase [19]. Instead, RESV is an unselective modulator of BK channels, leading to channel activation in skeletal muscle and neurons but inhibiting the BK channels in human neuroblastoma SH-SY5Y cells [7,20–23]. Its effects have been extensively investigated in vivo on breast, colorectal, liver, pancreatic, and prostate cancers [24]. Thus, it was shown that RESV affects cell growth, apoptosis, angiogenesis and invasion through a variety of death signaling cascades [25]; for example, it is reported to exert cytotoxic action in neuroblastoma cells, down-regulating the AKT signaling [26].

The C-terminus domain of the pore-forming α subunit of the BK channel contains several protein kinase and phosphatase binding motifs associated with various proteins involved in the regulation of channel gating and signaling pathways. Among these, the attention has been focused on RAC-a serine/threonine-protein kinase 1, also named AKT-1. AKT has a central role in cell survival, since it is implicated in the inhibition of apoptosis by phosphorylating and inactivating targets such as BAD and caspase-9 [27]. Recent data suggest a direct physical interaction between AKT and BK channel via AKT binding sites that overlap at the C terminus [28]. In addition, knockdown of AKT-1 increases BK expression, suggesting that BK is kept in check by these kinases. These results also suggest that both these proteins belong to a complex interacting network that regulates cell death or survival [28].

To date, the relationship between BK channel blocking mechanisms and the biological effects associated with cell survivals are poorly described, particularly after a short-term incubation time. In our experiments, we used the high affinity BK channel blockers IbTX and PAX to investigate on the relationship between their blocking mechanisms and biological effects, such as cell cycle progression and the associated intracellular signaling, morphological changes, and cell proliferation. The acute effects of these compounds were investigated after a short incubation time of 6 h, which was the minimal time required to observe the antiproliferative effects of these compounds in SH-SY5Y cells in our experiments. Finally, the effects of RESV and TEA on cell cycle progression and proliferation were also investigated. The effects of these BK channel modulators were investigated in the SH-SY5Y neuroblastoma cells because they express a large BK channel current representing 50% of the outward K^+ currents. The BK channel also regulates cell proliferation in this cell line [29,30]. The role of the BK channel in this cell is supported by the fact that the downregulation of the gene expression by gene-specific mRNA silencing using RNA interference (RNAi) significantly reduced *Slo1* mRNA expression and cell proliferation after three days of the transfection when compared to non-transfected cells [23].

We found that the early response of BK channels to the external channel pore blocker IbTX induces a contraction of the G1 phase and an accumulation in G2; whereas the allosteric blocker PAX induces the S Phase contraction and accumulation in G1/G2 phase. The two different blockers of the BK channels IbTX and PAX as well as the BK channel modulator RESV share a common action, leading to accumulation in G2 phase, AKT1p^{ser473} dephosphorylation, and a reduction of cell diameter. This can be related with the block of surface membrane BK channels. In addition, the action of PAX against intracellular BK channels and/or the interaction with the allosteric side of the pore can be associated with G1 accumulation, nuclear area reduction, and the potentiation of its antiproliferative effects with early apoptosis.

2. Results

2.1. Effects of the Application of Kv/BK Channels Targeting Compounds on the Whole-Cell K^+ Current of SH-SY5Y Cells

As previously shown, the SH-SY5Y cells showed an outward current carried by K^+ ions [30]. The I/V relationship intercepted the voltage axes at -49 ± 4 mV (Vm) (Figure 1) in the control condition. This value represented the resting membrane potential characterizing these cells, according to the well-described experimental observation that they show a more depolarized resting membrane potential than their normal counterpart, because of their cancerous nature [31]. These values represented the resting membrane potential characterizing these cells, according to the well-described experimental observation that they show a more depolarized resting membrane potential than their normal counterpart, because of their cancerous nature [31]. IbTX at 4×10^{-7} M concentration (*N* of cells = 7) and PAX at 5×10^{-5} M concentration (*N* of cells = 8) caused a comparable decrease of the whole-cell K^+ current at +30 mV (Vm) of $-38.3 \pm 10\%$ for IbTX and $-31.9 \pm 9\%$ for PAX, respectively, and were significantly different vs controls as evaluated by the Student's *t*-test (n.s. PAX data vs IbTX data; $p < 0.05$ PAX and IbTX vs controls). No further decrease of the current was observed at a higher toxin concentration. The maximal inhibitory response was $-38.6 \pm 8\%$ for PAX at 1×10^{-4} M concentration (N of cells = 5) and $-45.8 \pm 8\%$ for IbTX at 6×10^{-7} M concentration (N of cells = 5) significantly different vs controls as evaluated by the Student's *t*-test (n.s. PAX data vs IbTX data; $p < 0.05$ PAX and IbTX vs controls). The unselective BK modulator RESV also induced a marked reduction of the whole-cell K$^+$ current of $-63.7 \pm 15\%$ (N of cells = 5) at 2×10^{-6} M concentration in this cell line. The IbTX (6×10^{-7} M), PAX (1×10^{-4} M), and RESV (2×10^{-4} M) also shifted the I/V relationships to the right on the voltage axis; for instance, the resting membrane potential was -29 ± 4, -31 ± 5, and -27 ± 3 mV for IbTX, PAX and RESV, respectively (Figure 1). The addition of the broad-spectrum K$^+$ channel blocker TEA at the saturating concentration of 10^{-3} M caused in all experiments a full reduction of the total K$^+$ current of the $-93 \pm 11\%$, allowing the definition of the leak current. At +100 mV (Vm) the current reduction by both toxins was more marked than that calculated at +30 mV (Vm), supporting the idea that the BK channel current component represented about 35–50% of the total whole-cell K$^+$ current in this cell line.

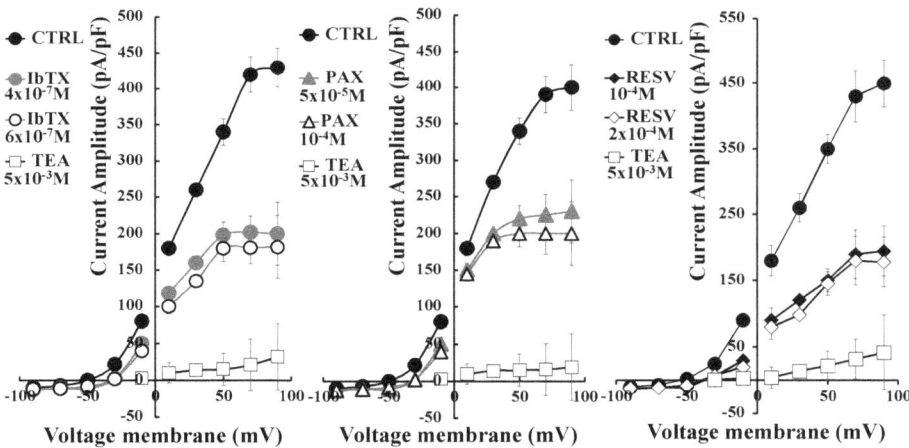

Figure 1. Current-voltage relationship of the whole-cell K$^+$ current recorded in SH-SY5Y cells: characterization of the maximal inhibitory effect of the application of the Ca^{2+}-activated K$^+$ (BK) channel blockers paxilline (PAX) and iberiotoxin (IbTX). A sigmoid I/V relationship of the K$^+$ current in a single SH-SY5Y cell recorded in the control condition (CTRL) and in the presence of the specific BK channel blockers PAX at 5×10^{-5} and 1×10^{-4} M concentrations, IbTX at 4×10^{-7} and 6×10^{-7} M and RESV at 10^{-4} and 2×10^{-4} M concentrations. The experiments were performed in asymmetrical K$^+$ ion concentrations (int K$^+$: 1.32×10^{-1} M; ext K$^+$: 2.8×10^{-3} M) and in the presence of an internal concentration of free Ca^{2+} ions of 1.6×10^{-6} M. Membrane currents were activated by 500 ms voltage steps between -90 to +90 mV, starting from an HP of -60 mV (Vm). The leak was obtained through the application of saturating concentration of the K$^+$ channel blocker tetraethylammonium (TEA) (5×10^{-3} M). Cells characterized by the same size were selected for patch clamp experiments. Each point represented the mean $\pm$ SEM of 5–8 different experimental points.

The inhibitory effect of TEA was fully reversible within 30 s of washout of the bath solution (Figure 2A). While, the effects of the application of the BK blockers IbTX and PAX on the whole-cell K$^+$-current were not reversible within 30 s of washout of the bath solution (Figure 2B,C). A washout period with bath solution >4 min failed to restore the K$^+$-currents to the control values. Similarly, the inhibitory effect of the not selective BK modulator RESV was not reversible following washout (Figure 2D, Table 1).

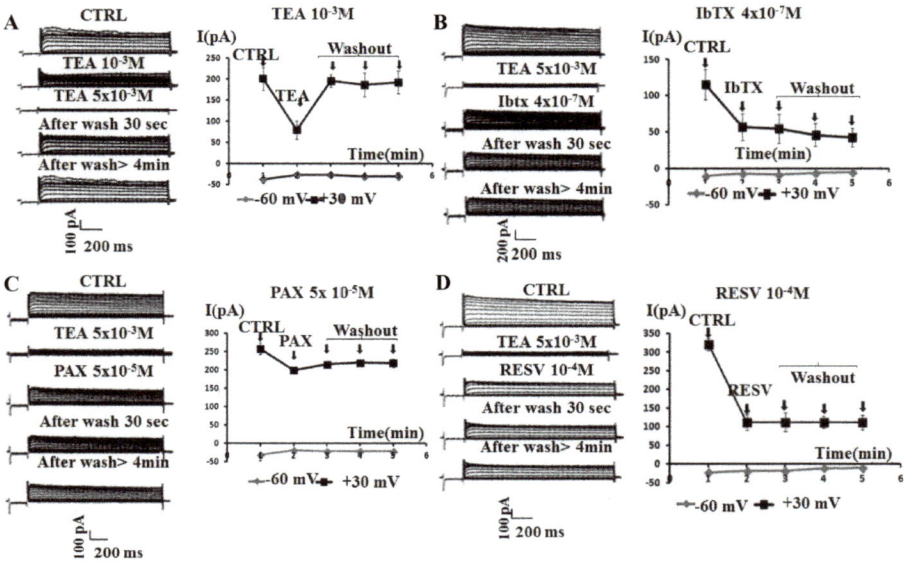

Figure 2. Sample traces of the whole-cell K$^+$-current and time courses recorded in SH-SY5Y cells in the CTRL condition (CTRL) and following the addition of the Kv/BK channel blockers TEA, IbTX, PAX, and resveratrol (RESV) to the bath solution. The current was recorded in asymmetrical K$^+$ ions concentrations (int K$^+$: 1.32×10^{-1} M; ext K$^+$: 2.8×10^{-3} M) and in the presence of an internal concentration of free Ca^{2+} ions of 1.6×10^{-6} M. Membrane currents were activated by 500 ms voltage steps between -100 to $+100$ mV, starting from an HP of -60 mV (Vm). The leak current was recorded in the presence of a saturating concentration of TEA (5×10^{-3} M). The four panels describe the effects of the application of the Kv/BK channel blockers under investigation on cells, according to the following scheme: current traces recorded in the control condition (CTRL); current traces recorded in the presence of TEA (5×10^{-3} M); current traces recorded after the application of specific concentration of Kv/BK channel targeting compounds under investigation: (**A**) TEA 5×10^{-3} M; (**B**) IbTX 4×10^{-7} M; (**C**) PAX 5×10^{-5} M; (**D**) RESV 10^{-4} M); current traces recorded following two washout periods with the bath solution of 30 s and >4 min. Time courses of the K$^+$ current in the presence or absence of the compounds under investigation recorded at physiological Vm of $+30$ and -60 mV. Each point is the mean + SEM of 3 cells. The incubation of the cells with TEA induced almost a total reduction of the current, which was fully reversible within 30 s of washout with bath solution. In contrast, the incubation of cells with IbTX, RESV, and PAX induced a partial reduction of the total K$^+$ current, which was not reversible within 4 min of washout with bath solution.

Table 1. Summary of the effects of the BK channel modulator on cell cycle progression, morphology, and proliferation in SH-SY5Y cells after 6 h of incubation.

Drugs	Mechanism of BK Channel Modulation	Cell Cycle Phase	Nuclear/Cell Morphology	AKT1p^{ser473} Phosporylation	Cell Proliferation
IbTX	Membrane impermeable; BK selective blocker; Open channel blocker; Not reversible action	G2 accumulation; G1 contraction; S not affected	No effects/ Diameter reduction	Dephosphorylation	Moderate reduction
PAX	Membrane permeable; BK selective blocker; Allosteric modulator and closed channel blocker; Not reversible action	G2 accumulation; S contraction; G1 accumulation	Nuclear area shrinking/ Diameter reduction	Marked dephosphorylation	Marked reduction
RESV	Membrane permeable; BK unselective modulator; Not reversible action	G2 accumulation; S contraction	Nuclear area enlargement/ Diameter reduction	Dephosphorylation	Moderate reduction
TEA	Kv/BK unselective blocker with reversible action	No effects	No Effects/ No Effects	No effects	No effects

In addition, we tested the possible effects of the patch perfusion with the bath solution on the recorded K$^+$ currents, reducing or increasing it by influencing the seal stability. In the control condition (*N* of cells = 15), the K$^+$-currents were stable all over the observation period (8 min).

The application of DMSO ($8.4 \times 10^{-10}\%$–$1.7 \times 10^{-1}\%$) to SH-SY5Y cells tested at the percent concentrations (*v*/*v*) corresponding to the amount used as a co-solvent caused a not significant decrease of K$^+$-currents with respect to that of the controls recorded at +30 mV (Vm) (*I* mean recorded after DMSO washout vs *I* mean recorded in control condition, Student's *t*-test *p* > 0.05).

2.2. Cell Proliferation and Cell Cycle Progression in SH-SY5Y Cell Following 6 h Incubation of the Cells with Kv/BK Channel Targeting Drugs

Cell proliferation evaluated using the impedentiometric technique (Scepter cell counter) revealed no effects of the compounds under investigation within 1–4 h of incubation time. These compounds failed to affect the cell cycle after 4 h of incubation. After 6 h of incubation the cell proliferation-time relationship for this cell line linearly increased with time. We therefore investigated the effects of these compounds after 6 h of incubation on cell cycle progression and proliferation. The cell-cycle changes induced by PAX, IbTX, and RESV tested in the range of concentrations inhibiting K$^+$ currents on SH-SY5Y cells have been investigated by the Operetta high-content imaging system, using a multiparametric approach to quantify DNA content and nuclear morphological properties (NuclearMask™ Blue Stain), DNA synthesis (EdU-positive cells), and mitotic cells (pHH3-positive cells). The vehicle DMSO ($8.4 \times 10^{-10}\%$–$1.7 \times 10^{-1}\%$) did not significantly affect the investigated parameters (Figure 3A–D). The effect of PAX, IbTX, and RESV on cell cycle and proliferation was investigated and compared by identifying, G1, S, G2, and M cell populations (Figure 3E–I). PAX, IbTX, and RESV reduced cell number in a concentration-dependent way and are characterized by a different inhibitory potency (Figure 3E); assuming that the application of increasing drug concentrations would be capable to induce a total reduction of cell number, the calculated IC$_{50}$ values were 1.06×10^{-5}, 5×10^{-7} and 1.5×10^{-4} M, respectively. IbTX (4×10^{-7} M) (*N* of experiments = 5) and PAX (5×10^{-5} M) (*N* of experiments = 5) respectively caused $-40 \pm 8\%$ and $-65 \pm 4\%$ significant reduction of cell proliferation vs vehicle (Ve) treated cells as evaluated by Student's *t*-test (PAX vs IbTX *p* < 0.05; PAX and IbTX vs vehicle, *p* < 0.05). A maximal reduction of cell proliferation respectively of $-62 \pm 7\%$ and $-65 \pm 6\%$ and for IbTX at 6×10^{-7} and 10^{-6} M concentrations, and of $-97 \pm 6\%$ for PAX (1×10^{-4} M); PAX data appeared to be statistically different vs IbTX data as determined by Student's *t*-test (*p* < 0.05). RESV at 2×10^{-4}M caused a reduction of cell proliferation of $-60 \pm 8\%$ (*N* of cells = 5). The effect on cell number was accompanied by modifications of nuclear morphological features in the case of RESV and PAX, whereas IbTX did not produce significant modifications. RESV and PAX produced opposite effects, showing a concentration-dependent nuclear area increase and decrease, respectively. The nuclear shrinkage induced by PAX at 10^{-4} M concentration was associated to an increase of nuclear fragmentation index, calculated from the coefficient of variance (CV) of nuclear stain intensity, indicative of apoptosis induction. All compounds induced cell cycle progression alterations, with a different pattern. In detail, the unselective permeable BK channel modulator RESV induced a G2 phase accumulation accompanied by a contraction of S-phase cells at 10^{-4} and 2×10^{-4} M. The selective membrane permeable BK channel blocker PAX also induced G2 accumulation and S contraction, and additionally, a G1-phase accumulation at 10^{-4} M. Finally, the selective external membrane impermanent BK channel blocker IbTX at 10^{-6} M concentration induced a G2-phase accumulation along with a slight contraction of G1 cells. Notably, IbTX is the only one for whom the fraction of cells actively synthesizing DNA was not reduced by treatment, thus suggesting a distinct mechanism of action (Figure 3G–I, Table 1).

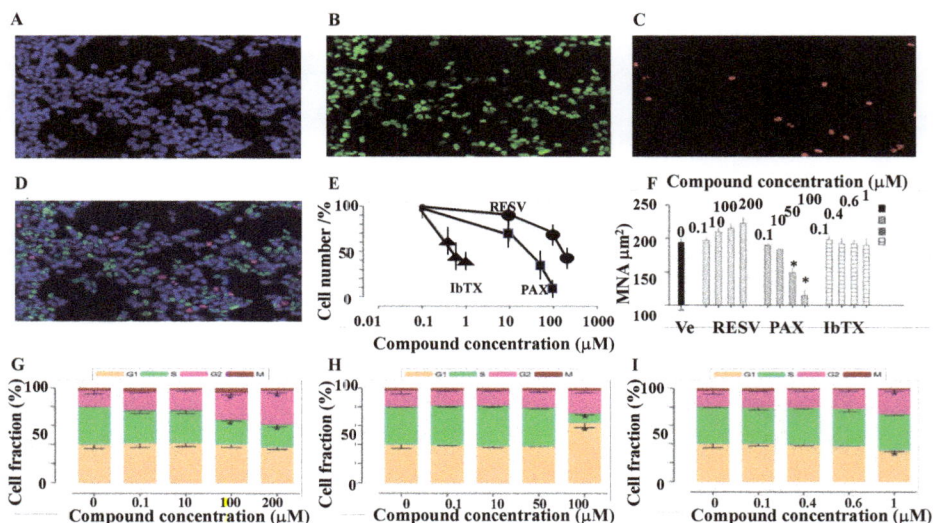

Figure 3. High-content cell-cycle analysis. (**A–D**) Example images of vehicle (DMSO 1.7×10^{-1}%) -treated SH-SY5Y cells stained with (**A**) NuclearMask Blue Stain, and (**B**) Alexa Fluor 488 (green, S-phase cells), or (**C**) anti-pHH3 (red, M-phase cells) antibodies are shown. (**D**) Merged image from panels (**A–C**). The effects of RESV, PAX, and IbTX treatment upon cell number and nuclear area are shown in panels (**E**) and (**F**), respectively. Concentration-response cell-cycle modifications by RESV, PAX, and IbTX determined using the multi-parametric approach are shown in panels (**G–I**). Values are the mean ± SD of 5 biological replicates; * $p < 0.05$ versus vehicle-treated controls (Student's *t*-test, two-tailed distribution, two sample equal variance). Operetta high-content imaging system (Perkin Elmer Life Sciences, Boston, MA, USA) was performed using a 20× long working distance objective. Typically, 10 fields per well were imaged which equated to approximately 6000 cells/well.

2.3. Cell Volume Changes Induced by Kv/BK Channels Targeting Compounds

To test the hypothesis that the observed effects on cell proliferation and cell cycle progression observed after 6 h of incubation may be associated to cell volume changes, the effects of the Kv/BK channel blockers on cell volume were evaluated using an impedentiometric method (Scepter cell counter). The distribution of SH-SY5Y cells as a function of their diameter size was well described by a Gaussian curve. In the CTRL condition, the majority of cells showed a diameter size in the range between 13–18 μm, which can be considered, for this cell type, the diameter range of normal cell dimensions (Figure 4). The application of the Kv/BK channels targeted compounds under investigation altered this cell distribution (Figure 4). In particular, the application of increasing concentrations of PAX (10^{-7}–10^{-4} M) reduced the number of cells in the range between 13–36 μm and determined the appearance of an additional peak in the range of diameter size between 6 and 12 μm. PAX induced a concentration-dependent increase of the cell population showing a diameter size between 6 and 12 μm vs controls. The same effects were observed following the incubation of SH-SY5Y cells with increasing a concentration of IbTX (10^{-7}–6 × 10^{-7} M) and RESV (10^{-7}–10^{-4} M).

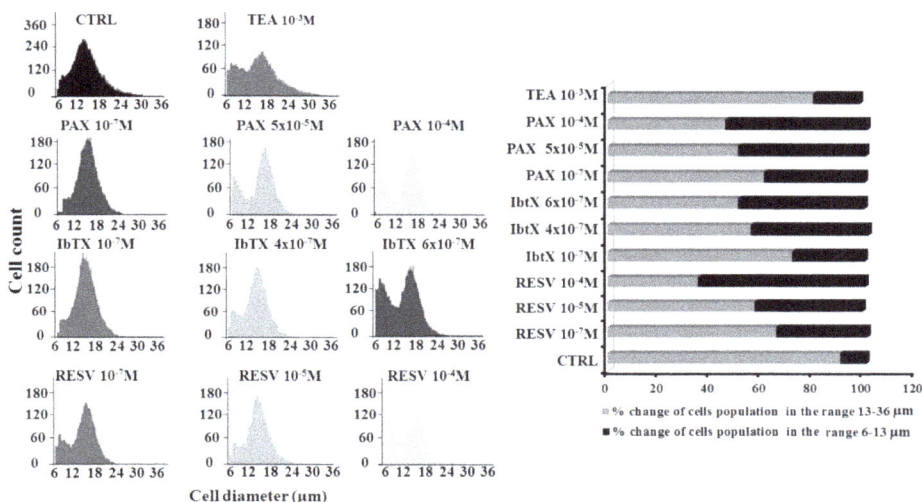

Figure 4. Gaussian type distribution of SH-SY5Y cells as a function of cell diameter size in the presence or absence of Kv/BK channel blockers: effects of the drugs on cell diameter after 6 h of incubation time. The effects of PAX (10^{-7}, 5 × 10^{-5} and 10^{-4} M), IbTX (10^{-7}, 4 × 10^{-7} and 6 × 10^{-7} M), RESV (10^{-7}, 10^{-5} and 10^{-4} M) and TEA (10^{-3} M) on cell distribution according to their diameter size were investigated in SH-SY5Y cells. The SH-SY5Y cell distribution follows a Gaussian curve when plotted as a function of their diameter size. Each graph is representative of cell distribution in a specific experimental condition. A biphasic distribution was observed after 6 h incubation time of cells with PAX, IbTX, and RESV. All compounds induced a concentration-dependent reduction of the cell numbers in the physiological diameter range of 13–36 μm for this cell type vs controls excepts TEA, while increasing the proportion of cell population in the diameter range of 6–13 μm.

We compared the effects of the BK channel targeting compounds under investigation with that of the unspecific K⁺ channel blocker TEA to underline differences in the behavior of cells exposed to a BK selective *vs* a broad spectrum Kv/BK channels blocker TEA. The application of the TEA (10^{-3} M) did not reduce the total number of cells and did not significantly affect their diameter size vs controls (Table 1).

2.4. AKT1p^{ser473}a Putative Node in the Signaling Pathways Regulating Cell Proliferation in Response to the Modulation of the BK Channel Activity

Finally, we tested the hypothesis that the downstream signaling pathway activated by the BK channel targeting compounds under study and resulting in the observed modulation of SH-SY5Y cell viability, may involve the serine-threonine kinase AKT-1. The incubation of the SH-SY5Y cells over a 6 h period with the Kv/BK channels targeting compounds under investigation affect the PhosphoAKT (Ser473) (AKT1p^{ser473})/ Pan AKT ratio. In particular, the incubation of cells with RESV

(10^{-7}, 10^{-5} and 10^{-4} M) caused a reduction of the PhosphoAKT (Ser473)/ Pan AKT ratio at the higher tested concentration. Also IbTX (4×10^{-7}, 6×10^{-7} M) and PAX (10^{-7}, 5×10^{-5} and 10^{-4} M) at the higher tested concentrations caused a significant reduction of PhosphoAKT (Ser473)/Pan AKT ratio; statistical analysis performed by One way ANOVA (p value = 0.0215, F value = 6.37) and Bonferroni test to assess the experimental groups statistically different from the CTRL condition, ($p < 0.05$) (Figure 5A).

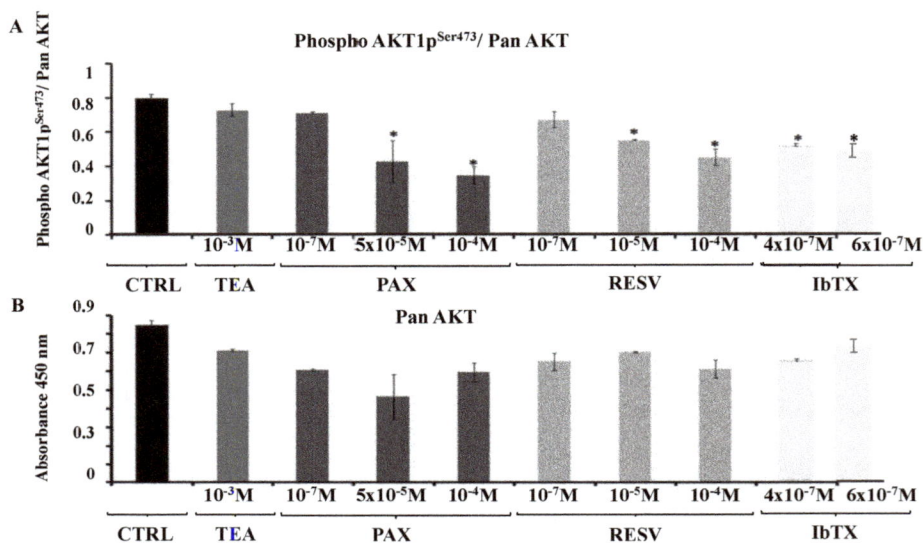

Figure 5. Effects of the Kv/BK channel blockers on the total AKT expression (Pan AKT) and PhosphoAKT (Ser473)/ Pan AKT ratio in SH-SY5Y neuroblastoma cells. The PhosphoAKT (Ser473) and the Pan AKT levels were evaluated in the cell lysates using the ELISA kit for PhosphoAKT (Ser473)-AKT pan (SIGMA Chemical Co., Milan, Italy) in the CTRL condition and after 6 h of incubation with PAX (10^{-7}, 5×10^{-5} and 10^{-4} M), IbTX (4×10^{-7} M), RESV (10^{-7}, 10^{-5} and 10^{-4} M) and TEA (10^{-3} M). (**A**) The incubation of cells with PAX, IbTX, and RESV significantly reduced the PhosphoAKT (Ser473) levels. (**B**) No effects were observed on the pan AKT levels reducing the PhosphoAKT (Ser473)/pan AKT ratio. The represented bars correspond to the mean of three different experimental data ± SEM. * Data significantly different with respect to the CTRL condition for $p < 0.05$ as determined by one-way ANOVA and Bonferroni test.

The Pan AKT values were not significantly affected by PAX, IbTX, and RESV, indicating that the observed reduction of the PhosphoAKT (Ser473)/Pan AKT ratio was due to the reduction of the PhosphoAKT (Ser473) value (Figure 5B). In contrast, the incubation with TEA (10^{-3} M), the only compound that was capable to induce a reversible decrease of the K⁺ current, failed to affect both Pan-AKT and the PhosphoAKT (Ser473)/ Pan AKT ratio (Figure 5A,B, Table 1).

IbTX (4×10^{-7} M) and PAX (1×10^{-5} M) at concentrations that fully blocked membrane surface BK channels induced AKT1p^{Ser473} dephosphorylation of $33 \pm 5\%$ and $50 \pm 8\%$ vs controls, respectively. A mild potentiation of the dephosphorylation is observed at higher toxins concentrations.

3. Discussion

In our experiments we used the high affinity BK channel blockers PAX and IbTX to investigate on the relationship between their blocking mechanisms and biological effects such as cell cycle progression and the associated intracellular signaling, the morphological changes, and cell proliferation after a short incubation time. The incubation time of 6 h was the minimal time required to observe the antiproliferative effects of these compounds in this cell line. No effects were indeed observed following 1–4 h of incubation of the cells with IbTX, PAX, and RESV.

IbTX is considered a not permeable BK channel blocker interacting on the external binding site of the pore and acting as an open channel blocker; tested at 4×10^{-7} and 6×10^{-7} M concentrations it caused a partial whole-cell K^+ current reduction and a full block of the membrane BK channels, leading to a partial reduction of cell proliferation and AKT1p^{ser473} dephosphorylation. No further effects were observed with IbTX at the highest concentration tested of 10^{-6} M.

PAX is instead considered a membrane permeable and allosteric blocker of BK channels; tested at concentrations of 5.0×10^{-4} and 10^{-4} M it caused a partial reduction of the whole-cell K^+ current and a full block of membrane BK channels, leading to a fully reduced cell proliferation and to a marked AKT1p^{ser473} dephosphorylation. These findings suggest that additional targets are involved in the PAX action other than the membrane BK channels.

The differences in the IC50 values of IbTX and PAX in inducing antiproliferative effects can be explained by the different binding affinity of the toxins to the BK channels site. Indeed, the placement of the binding site of these compounds plays a key role, since it is located on the intracellular side of the channel for PAX and on the extracellular portion of the pore for IbTX [9,15,32]. In this way, IbTX interaction with its binding site is expected to be faster than that of PAX, considering that the access to the binding site of PAX can be limited by the transmembrane flux, and consequently explaining why smaller concentrations of IbTX are required for exerting its own effects on cell proliferation.

Further insights into the toxins interaction with the BK channels may include the molecular composition of the channel. In fact, the β4 subunit leads to a BK channel phenotype that is characterized by a very low sensitivity to ChTX and IbTX [5]. For instance, the presence of the auxiliary β4 subunit dramatically decreases the BK channel sensitivity to IbTX by inducing a marked rise in the apparent half-maximal inhibitory concentration (IC$_{50}$) of the BK channel and a decrease of the IbTX association rate to it [33]. The PAX-BK channel interaction observed in our experiments could probably be affected by the presence of the auxiliary β4 subunit, which could modify channel properties, such as its kon and koff kinetics and the pharmacological features.

Actually, the β4 subunit seemed to be involved also in the modification of the toxins' dissociation from the BK channel complex [5]. In fact, it has been demonstrated that IbTX block of the BK channel takes place through a reversible bimolecular mechanism by which it can bind the external portion of the channel [9,32]. Moreover, also the mechanism by which PAX binds to the BK channel has been well characterized as inversely related to the open probability (Po) of the BK channel and completely reversible when applied from the internal side in excised-patch experiments [15]. On the contrary, in our experiments, the binding of both toxins as well as of RESV to the BK channel is irreversible. This effect can be likely explained by the presence of the modulatory β4 subunit in the SH-SY5Y cells as previously proposed [5].

Instead, the mechanism of interaction of RESV against the BK channel is not known. This natural compound was found to increase the BK channel activity in human vascular endothelial cells, when tested in whole cell patch, and in rat skeletal muscle fibers, in excised patch, thereby proposing this compound in neuromuscular disorders as a BK channel opener [6,7,34–37]. Others found a significant reduction of both the inward and the outward K^+ currents in pancreatic β cells by RESV [38–40].

Toxins and RESV induced also a concentration dependent reduction of the number of the cell population in the mean diameter range of 13–36 μm, which is the physiological range for healthy cells, accompanied by an increased number of suffering cells showing a mean diameter range of 6–12 μm. Either the selective BK channel blockers IbTX and PAX induced early an accumulation in the G2 phase and AKT1p^{ser473} dephosphorylation after 6 h of incubation time with a reduction of cell proliferation. Similarly, RESV, which is a membrane permeant but not selective BK channel modulator, induced G2 accumulation and AKT1p^{ser473} dephosphorylation reducing cell proliferation. These findings suggest that G2 accumulation is associated with AKT1p^{ser473} dephosphorylation following the acute but not reversible membrane channel block.

PAX and RESV also induced the contraction of the S phase, which is not observed with IbTX in our experimental condition. A possible interaction with intracellular BK channels located on the mitochondrial membrane and nuclear membrane can be associated with a reduced synthetic S phase. An additional G1 accumulation is observed with PAX, which is also capable to induce nuclear fragmentation and a reduction of the nuclear area, a well-known index of apoptosis. This effect can be associated with the interaction of PAX with the intracellular binding site located on the BK channel. In the brain for instance the BK channel interacts with proteins placed in cellular compartments other than in the plasma membrane. Recent studies revealed that the majority of BK channel interactome belongs to proteins localized, in a decreasing order, respectively in the cytoplasm, plasma membrane, nucleus, mitochondria, extracellular region, Golgi apparatus and endoplasmic reticulum [41]. PAX was therefore the most effective compound in our experiments showing G1/G2 accumulation and leading to −95% reduction of cell proliferation. The accumulation in G1 at the end of the first checkpoint is critical in making the decision if a cell should enter the S phase and divide, delay division, or enter G0. The accumulation in G2 at the second checkpoint limits the number of cells to progress into the mitosis. Several compounds acting as antiproliferative agents accumulate cells in G1 and G2 phases; for example 5-flurouracile and rapamicine induces accumulation in G1 phase whereas paclitaxel and other antimitotic agents accumulate in the G2/M phase [42].

G1/S transition is associated with upregulation of several types of K$^+$ channels that leads to the hyperpolarization of the cell membrane whereas intracellular blocking action may lead to G1 accumulation, as observed in our experiments with PAX [43]. The accumulation in G2 induced in our experiments by potassium channel blocking drugs like IbTX, PAX, and RESV is expected since the transition from G2 to M is associated with depolarization. During the M phase an elevated K$^+$ conductance is reported in several cell types [31]. Consequently, this is in line with the proposed role of the BK channel on the cell cycle. The BK channel can regulate cell volume, provide a driving force for Ca^{2+} entry, hyperpolarize the cell at the G1/S transition, depolarize it towards G2/M one and finally sustain the K$^+$ conductance during the M phase [31]. Calcium ions are indeed a key regulator of cell cycle progression and are required in the S phase, leading to cell proliferation at physiological concentrations. Abnormal enhancement of intracellular calcium may however induce calcium dependent apoptosis. Additionally, non-canonical, permeation-independent mechanisms may be involved, where BK channels recruit or modulate signaling cascades via protein–protein interactions. It is tempting to assume that signaling cascades activated by such interactions could link the nuclear clock control with its cytoplasmic counterpart [43].

Potassium channels regulate cell volume along the cell cycle [44]. Indeed, the elevated intracellular K$^+$ concentration increases the turgor pressure and cell diameter, which is required for cell growth. So, this is achieved by the activity of the inward rectifier K$^+$ channels. Instead, mitosis is associated to a decrease in turgor pressure owing to the K$^+$ efflux through the Kv/BK channels [43]. An increase in cell volume is expected very early following the drug blocking action of potassium channels that precede apoptosis [45]. In our experiments, toxins and RESV reduced the cell volume after 6 h of incubation and a lower intracellular K$^+$ ion concentration is expected when the apoptotic signaling are ongoing.

We focused on the capability of the drugs under investigation to alter the activity of AKT-1, a serine/threonine kinase that has emerged as a critical signaling node in the pathways that regulate principal cellular functions and activities such as cell survival, growth, proliferation, angiogenesis, metabolism, and migration in several types of cells [27]. This Ca^{2+} regulated kinase belongs to the PI3K/AKT pathway involved in the neuroblastoma development [46,47] for both primary neuroblastoma cells and cell lines, included also SH-SY5Y cells. This signaling pathway promotes cell survival and proliferation by activating survival associated proteins and by inhibiting the apoptotic pathway [48–50]. Moreover, a recently published BK channel sequence scanning, aiming to point out conserved BK channel interactions involving apoptosis, Ca^{2+} binding, and trafficking, in chicks, mice, and humans [28], revealed that a potential AKT-1 substrate binding site is present on the BK channel sequence and is highly conserved among species. AKT-1 mediated phosphorylation of the BK channel alpha subunit leads to a downregulation of the membrane of the BK channel [28]. We investigated if the observed effects induced by the BK channel targeting drugs under study on cell cycle and proliferation may be related with a modulation of AKT-1 activity. Interestingly, PAX, IbTX and RESV but not TEA induced a significant reduction of PhosphoAKT (Ser473), which represented the active form of this kinase [51]. The AKT-1 activation can stimulate cell proliferation through multiple downstream targets impinging on cell-cycle regulation, including the cyclin-dependent kinase inhibitors p27 and p21 that once phosphorylated and are sequestered in the cytosolic compartment, which avoids their nuclear localization and the possibility of exerting their inhibitory effects on the cell cycle [52–55]. Other AKT target proteins are represented by GSK3, TSC2, and PRAS40, which are involved in cell-cycle entry and progression [27].

We therefore propose the following cascade of events to explain the antiproliferative effects of the BK channel openers (Graphical abstract). The irreversible blocking action of the surface membrane BK channels by PAX, IbTX, and RESV induce persistent cell depolarization and influx of cations including Ca^{2+} ions. Cell depolarization is observed after acute exposure of the cells to the BK channel modulators in the presence of a micromolar concentration of internal calcium ions. In the presence of lower Ca^{2+} ions concentrations the contribution of the BK channel to the total currents and the effects of the channel blockers on resting potentials is expected to be proportionally reduced. Cell depolarization and cation influx was recently observed in glioma cells following exposure of the cells to Penitrem A, a BK channel blocker [56]. After 6 h of incubation, we believe that the cell depolarization would activate the death calcium dependent signaling with early apoptosis and a reduction of the cell diameter in the case of IbTX, RESV, and PAX. The increase of the intracellular Ca^{2+} ions activate phosphatase such as pleckstrin homology domain leucine-rich repeat protein phosphatase PHLPP that specifically dephosphorylate AKT-1 on Ser473 and with less extended protein phosphatases PP2A that may also dephosphorylate AKT-1 at Ser473 [57]. The cytosolic AKT-1 dephosphorylation is per se responsible for G2 phases accumulation [58]. The nuclear AKT-1 dephosphorylation can be responsible for G1 phases accumulation induced by PAX following interaction with the nuclear membrane BK channel and a reduction of the nuclear membrane area as observed in our experiments.

4. Materials and Methods

4.1. Drugs

The drugs and toxins were purchased from Sigma (SIGMA Chemical Co., Milan, Italy). Stock solutions of the drugs under investigation were prepared dissolving the drugs in dimethylsulfoxide (DMSO) at the concentration of 1.186×10^{-1} M (PAX and RESV) and 1×10^{-1} M (IbTX). Microliter amounts of the stock solutions were then added to the bath solutions, for patch clamp experiments, or to DMEM+, for cell proliferation, cell volume and AKT ELISA assays, as needed. DMSO did not exceed 0.017% in the bath or DMEM+; at this concentration the solvent does not normally affect the parameters under study.

4.2. Patch Clamp Solutions

Pipette (intracellular) solution composition (10^{-3} M): 132 K^+-glutamate, 1 ethylenegly-col bis (β-aminoethylether)-N,N,N,N-tetraaceticacid (EGTA), 10 NaCl, 2 $MgCl_2$, 10 HEPES, 1 Na_2ATP, 0.3 Na_2GTP, pH = 7.2 with KOH. $CaCl_2$ was added to the pipette solutions to give free Ca^{2+} ion concentration of 1.6×10^{-6} M in whole cell experiments. The calculation of the free Ca^{2+} ion concentration in the pipette was performed using the Maxchelator software (Stanford University, Stanford, CA, USA).

Bath (extracellular) solution composition (10^{-3} M): 142 NaCl, 2.8 KCl, 1 $CaCl_2$, 1 $MgCl_2$, 11 glucose, 10 HEPES, pH = 7.4 with NaOH.

4.3. Whole-Cell K+ Current Recordings in the SH-SY5Y Cells

The native K+-currents were recorded in the human neuroblastoma cell line SH-SY5Y through the application of the voltage protocol by 10 mV and 500 ms voltage steps, in the range of potentials going from −100 mV (Vm) to +100 (Vm) and starting from HP = −60 mV. The experiments were performed in asymmetrical K+ ion concentrations (int K+: 1.32×10^{-1} M; ext K+: 2.8×10^{-3} M), in the presence of free Ca^{2+} ions, using a patch-clamp technique in the whole cell configuration [59,60]. The resulting K+-currents were leak subtracted. The leak current was measured after adding the saturating concentration of an external solution containing TEA (1 or 5×10^{-3} M), which caused a full block of the Kv and BK channels. Drug effects were investigated in a physiological range of potentials going from −80 mV (Vm) to +30 mV (Vm) for all drugs. The K+-currents were recorded at 20 °C and sampled at 2 kHz (filter = 1 kHz) using an Axopatch-1D amplifier equipped with a CV-4 headstage (Axon Instruments, Foster City, CA, USA) as previously described [4,61,62].

Current analysis was performed using pCLAMP 10 software package (Axon Instruments, Foster City, CA, USA). The criteria for accepting the data entering were based on the stability of the seal evaluated by observing the noise levels not exceeding 0.6 pA. The pipettes resistance was 7 ± 0.2 MΩ. The cells were exposed to the drug solutions for 1 min. before recordings. In patch clamp experiments, the bath solution was tested against K+-currents in the absence of the control, in the presence of DMSO, and in the presence of the DMSO + drug. Increasing concentrations of drug solutions were applied to the cells by the fast perfusion system (AutoMate, Sci. Berkeley, CA, USA). No more than six different solutions were applied to the same cell. Seal resistance was continuously monitored during the patch solutions exchange. The reversibility of the drug response of the currents was evaluated in the presence or absence of compounds plotting raw data vs time in cells of different size. This type of experiment required the continuous monitoring of seal resistance for 5 min before the application of the compound. Cells not showing a stable seal during monitoring were disregarded and were not selected for the further analysis. The data reported in Figure 2 were pooled from three cells for each compound that reached a stable seal for at least 5 min before the application of the compound. By using this protocol we had a very low S.E.M. for each time point.

4.4. Cell Culture

The human neuroblastoma cell line SH-SY5Y (ATCC® CRL2266™) were purchased from American Type Culture Collection (ATCC, Manasass, VA, USA). The cells were cultured in Dulbecco's Modified Eagle's Medium (DMEM) supplemented with 10% fetal bovine serum (FBS), 2 mM L-glutamine and 1% (100 IU/mL penicillin and 100 µg/mL streptomycin) antibiotics. The cultures were maintained at 37 °C in a humidified atmosphere containing 95% air and 5% CO_2 in a humidified atmosphere containing 5% CO_2. All culture medium components were purchased from EuroClone (EuroClone S.p.A, Pero, Italy). The experiments were performed on undifferentiated cells at a passage comprised between 10 and 14 during the experiments.

4.5. High-Content Cell Cycle Analysis

Treatment-induced modifications of cell cycle phases were investigated using the high-content imaging system Operetta (Perkin Elmer Life Sciences, Boston, MA, USA), along with Harmony software and PhenoLOGIC machine learning, according to a method recently described by Massey [63]. Briefly, SH-SY5Y human neuroblastoma cells at passage 10 were seeded (6500 cells/well) in DMEM (Dulbecco's modified Eagle's medium) complete culture medium with 10% fetal bovine serum, L-glutamine and Penicillin/Streptomicin (cell biology reagents from EuroClone S.p.A, Pero, Italy) in 96-well collagen-coated plates (Perkin Elmer). Cells were allowed to attach for 24 h in a normal humidified atmosphere supplemented with 5% CO_2, and then treated in the same conditions for 6 h with different concentrations of freshly dissolved compounds or vehicles only. After compound removal and gentle cell washing, complete fresh medium was added in each well and the cells were incubated in the same conditions for a further 24 h. Cells were labelled with 10 mm (final concentration per well) 5-ethynyl-2'-deoxyuridine (EdU) for 30 min. After EdU incubation, media was removed immediately prior to fixation of adherent cells with formaldehyde 3.7% (*v*/*v*) in PBS at room temperature for 15 min. Cells were washed twice with PBS and permeabilized with Triton X-100 (Promega, Madison, MI, USA) for 15 min at room temperature. After two washes with PBS, incorporated EdU was detected with Alexa Fluor 488 by a Click-iT® EdU HCS Assay (C10350, Life Technologies, Grand Island, NY, USA) according to the manufacturer's recommended procedure. For phospho-histone H3 Ser28 detection, cells were incubated with 3% (*w*/*v*) bovine serum albumin (BSA) overnight at 4 °C, and then incubated for 1 h at room temperature in the dark with the anti-Phospho-Histone H3 primary antibody (Anti-Phospho-Histone H3 pSer28, clone HTA28, SIGMA), 2 µg/mL final concentration in BSA 3%. After three washes with 0.05% Tween® 20, cells were incubated with the secondary antibody (Alexa Fluor® 647 Goat anti-Rat IgG, ThermoFisher Scientific, Waltham, MA, USA), 10 µg/mL final concentration in BSA 3%, for 1 h at room temperature in the dark. Following additional three washes with 0.05% Tween® 20, nuclear staining was performed for 15 min at room temperature in the dark by NuclearMask™ Blue Stain (Life Technologies), 100 µL/well. Finally, following two PBS washes, cells were imaged with an Operetta high-content imaging system (Perkin Elmer) using a 20× long working distance objective. Typically, 10 fields per well were imaged which equated to approximately 6000 cells/well.

4.6. Impedentiometric Cell Volume Assay

The reduction of diameter and cell size is a well-known marker of apoptosis in different cell types and tissues [64–66]. Cytoplasmic condensation is per se capable to induce apoptosis [67] while cells in active proliferation often show an enhanced diameter size [44]. In our experiments the cells were sized electronically using the Scepter™ 2.0 cell counter (MERK-Millipore, Billerica, MA, USA). Because the Scepter™ cell counter measures volume using the Coulter Principle, it can quantify cells based on size and will discriminate larger cells from smaller debris, unlike vision-based techniques, which rely on object recognition software and cannot reliably detect small cells. In detail, precise cell volumes are drawn into the Scepter™ sensor. As cells flow through the aperture in the sensor, resistance increases. This increase in resistance causes a subsequent increase in voltage. Voltage changes are recorded as spikes with each passing cell and these are proportional to the cell volume. The spikes of the same size are bucketed into a histogram and counted. This histogram gives the quantitative data on cell morphology that can be used to examine the quality and health of the cell culture. The Scepter™ 2.0 cell counter (MERK-Millipore) is compatible with 60 and 40 µm sensors; in our experiments we used the 60 µm sensor for particles between 6 and 36 µm.

4.7. Phospo AKT (pSer^{473})/Pan AKT ELISA Assay

The pan AKT and phsphoAKT were measured by an enzyme-linked immunosorbent (ELISA) assay purchased by Sigma-Aldrich (Catalog number RAB0012, SIGMA Chemical Co., Milan, Italy). A pan- AKT antibody has been coated onto a 96-well plate. The samples that were constituted by cell lysates were pipetted into the wells and AKT present in a sample that was bound to the wells by the immobilized antibody. The wells were washed and anti- Phospo AKT (pSer^{473}) or pan- AKT were used to detect phosphorylated or pan- AKT, respectively. For statistical results, the assays were run in triplicate. The data from the crude protein preparations were compared with the data obtained from the standard calibration curves performed using the lyophilized positive control sample.

Cell lysates used as samples were prepared according to the instructions reported in the kit datasheet at the end of the 6 h incubation time of SH-SY5Y cells with drugs under investigation, as described in the experimental protocol of the cell volume assay.

4.8. Data Analysis and Statistics

Image acquisition and data analysis were performed by using the Harmony® Image Analysis Software with PhenoLOGIC (Perkin Elmer Life Sciences, Boston, MA, USA). Automated analysis was performed to obtain as readout the total number of cells (for information and quality control) and DNA content on a single-cell basis (nuclear mask intensity sum). From the single cell results a histogram of nuclear mask sum was generated (Excel MAC 16.15 software, Microsoft Corp., Redmond, WA, USA) and the intensity corresponding to the center of the G1 peak determined. Based upon this value, a histogram of normalized DNA (relative DNA content) was obtained from vehicle-treated wells on a per-plate basis and applied to all wells to identify G1, S, and G2/M cell populations. Multiparametric analysis of cell-cycle was then performed by using relative DNA content values along with mean marker intensity values of EdU-positive and PHH3-positive cells to specifically identify G1, S, G2, and M cells. In addition, the Harmony Software was used to calculate the nuclear morphological properties (NuclearMask™ Blue Stain), and the nuclear area and fragmentation index of the treated and control cells were then obtained on a single cell level. The fragmentation index was calculated on the basis of the coefficient of variance (CV) of the nuclear stain intensity. Healthy nuclei have a homogeneous intensity, resulting in a low CV value whereas fragmented nuclei have a non-homogeneously distributed intensity resulting in a higher CV value, namely the "fragmentation index".

The maximal inhibitory effect (E_{max}) of the Kv/BK channels targeting compounds on the whole cell K^+ current was calculated by the following equation:

$$E_{max} = (I\ drug - I\ leak)/(I\ CTRL - I\ leak) \times 100 - 100 \tag{1}$$

Where *I drug* is the outward current recorded at +30 mV in the presence of the specific drug concentration at which the maximal effect of inhibition of the total K^+ current was observed; *I leak* is the current recorded at +30 mV in the presence of a saturating concentration of TEA (1×10^{-3} M); *I CTRL* is the outward current recorded at +30 mV in the bath solution (without drugs).

The data were collected and analyzed using Excel software (Microsoft Office 2010). The statistical results are presented as mean ± SEM. The number of replicates relative to each experimental data set is reported in the description of the correspondent plot. The Student's *t*-test and one-way ANOVA were used as appropriate, to evaluate the significance of differences between the means of two or multiple groups, respectively. The one-way ANOVA analysis was completed by the Bonferroni test to assess the experimental groups significantly different from the CTRL. *p* values < 0.05 were considered to indicate a statistical significance.

5. Conclusions

In conclusion, the two different blockers of the BK channels IbTX and PAX as well as the BK channel modulator RESV share common actions leading to accumulation in the G2 phase, AKT1p[ser473] dephosphorylation, and a reduction of cell diameter. This can be related with the block of the surface membrane BK channels. In addition, the action of PAX against intracellular BK channels and/or the interaction with the allosteric side of the pore is associated with G1 accumulation, nuclear area reduction, and the potentiation of its antiproliferative effects with early apoptosis.

Author Contributions: Participated in research design: M.C. and D.T.; Conducted experiments: F.C., A.C., F.M.; Contributed new reagents or analytic tools: D.T. and M.C.; Performed data analysis: G.P., N.Z., F.M, A.P. and A.C.; Wrote or contributed to the writing of the manuscript: D.T. and R.S.

Funding: This research was funded by Contributi di ricerca D Tricarico "Fondi Ateneo 2012-14 Univ. of Bari, Italy; and leftover funds from: Incarico di ricerca al D Tricarico da parte del "Consorzio Interuniversitario di Ricerca in Chimica dei Metalli nei Sistemi Biologici" Sede legale Piazza Umberto I, Bari 1-70121; Incarico di ricerca al D Tricarico nell'ambito del Cluster Tecnologici Regionali.

Acknowledgments: The authors thanks funding supporters.

Conflicts of Interest: The authors declare no conflict of interest.

References

1. Roger, S.; Potier, M.; Vandier, C.; Le Guennec, J.Y.; Besson, P. Description and role in proliferation of iberiotoxin-sensitive currents in different human mammary epithelial normal and cancerous cells. *Biochim. Biophys. Acta* **2004**, *1667*, 190–199. [CrossRef] [PubMed]
2. Oeggerli, M.; Tian, Y.; Ruiz, C.; Wijker, B.; Sauter, G.; Obermann, E.; Guth, U.; Zlobec, I.; Sausbier, M.; Kunzelmann, K.; et al. Role of KCNMA1 in breast cancer. *PLoS ONE* **2012**, *7*, e41664. [CrossRef] [PubMed]
3. Cheng, Y.Y.; Wright, C.M.; Kirschner, M.B.; Williams, M.; Sarun, K.H.; Sytnyk, V.; Leshchynska, I.; Edelman, J.J.; Vallely, M.P.; McCaughan, B.C.; et al. KCa1.1, a calcium-activated potassium channel subunit α 1, is targeted by miR-17-5p and modulates cell migration in malignant pleural mesothelioma. *Mol. Cancer* **2016**, *15*, 44. [CrossRef] [PubMed]
4. Tricarico, D.; Capriulo, R.; Conte Camerino, D. Involvement of KCa^{2+} channels in the local abnormalities and hyperkalemia following the ischemia-reperfusion injury of rat skeletal muscle. *Neuromuscul. Disord.* **2002**, *12*, 258–265. [CrossRef]
5. Torres, Y.P.; Granados, S.T.; Latorre, R. Pharmacological consequences of the coexpression of BK channel alpha and auxiliary β subunits. *Front. Physiol.* **2014**, *5*, 383. [CrossRef] [PubMed]
6. Imbrici, P.; Liantonio, A.; Camerino, G.M.; de Bellis, M.; Camerino, C.; Mele, A.; Giustino, A.; Pierno, S.; de Luca, A.; Tricarico, D.; et al. Therapeutic Approaches to Genetic Ion Channelopathies and Perspectives in Drug Discovery. *Front. Pharmacol.* **2016**, *7*, 121. [CrossRef] [PubMed]
7. Dinardo, M.M.; Camerino, G.; Mele, A.; Latorre, R.; Conte Camerino, D.; Tricarico, D. Splicing of the rSlo gene affects the molecular composition and drug response of Ca^{2+}-activated K^+ channels in skeletal muscle. *PLoS ONE* **2012**, *7*, e40235. [CrossRef] [PubMed]
8. Galvez, A.; Gimenez-Gallego, G.; Reuben, J.P.; Roy-Contancin, L.; Feigenbaum, P.; Kaczorowski, G.J.; Garcia, M.L. Purification and characterization of a unique, potent, peptidyl probe for the high conductance calcium-activated potassium channel from venom of the scorpion Buthus tamulus. *J. Biol. Chem.* **1990**, *265*, 11083–11090. [PubMed]
9. Candia, S.; Garcia, M.L.; Latorre, R. Mode of action of iberiotoxin, a potent blocker of the large conductance Ca^{2+}-activated K^+ channel. *Biophys. J.* **1992**, *63*, 583–590. [CrossRef]
10. MacKinnon, R.; Miller, C. Mechanism of charybdotoxin block of the high-conductance, Ca^{2+}-activated K^+ channel. *J. Gen. Physiol.* **1988**, *91*, 335–349. [CrossRef] [PubMed]
11. MacKinnon, R.; Latorre, R.; Miller, C. Role of surface electrostatics in the operation of a high-conductance Ca^{2+}-activated K^+ channel. *Biochemistry* **1989**, *28*, 8092–8099. [CrossRef] [PubMed]
12. Yu, M.; Liu, S.; Sun, P.; Pan, H.; Tian, C.; Zhang, L. Peptide toxins and small-molecule blockers of BK channels. *Acta Pharmacol. Sin.* **2016**, *37*, 56–66. [CrossRef] [PubMed]
13. Knaus, H.G.; McManus, O.B.; Lee, S.H.; Schmalhofer, W.A.; Garcia-Calvo, M.; Helms, L.M.; Sanchez, M.; Giangiacomo, K.; Reuben, J.P.; Smith, A.B. Tremorgenic indole alkaloids potently inhibit smooth muscle high-conductance calcium-activated potassium channels. *Biochemistry* **1994**, *33*, 5819–5828. [CrossRef] [PubMed]
14. Imlach, W.L.; Finch, S.C.; Dunlop, J.; Dalziel, J.E. Structural determinants of lolitrems for inhibition of BK large conductance Ca^{2+}-activated K^+ channels. *Eur. J. Pharmacol.* **2009**, *605*, 36–45. [CrossRef] [PubMed]
15. Zhou, Y.; Lingle, C.J. Paxilline inhibits BK channels by an almost exclusively closed-channel block mechanism. *J. Gen. Physiol.* **2014**, *144*, 415–440. [CrossRef] [PubMed]
16. Weaver, A.K.; Liu, X.; Sontheimer, H. Role for calcium-activated potassium channels (BK) in growth control of human malignant glioma cells. *J. Neurosci. Res.* **2004**, *78*, 224–234. [CrossRef] [PubMed]
17. Khatun, A.; Fujimoto, M.; Kito, H.; Niwa, S.; Suzuki, T.; Ohya, S. Down-Regulation of Ca^{2+}-Activated K^+ Channel KCa1.1 in Human Breast Cancer MDA-MB-453 Cells Treated with Vitamin D Receptor Agonists. *Int. J. Mol. Sci.* **2016**, *17*, 2083. [CrossRef] [PubMed]
18. Zhang, Y.Y.; Li, G.; Che, H.; Sun, H.Y.; Xiao, G.S.; Wang, Y.; Li, G.R. Effects of BKCa and Kir2.1 Channels on Cell Cycling Progression and Migration in Human Cardiac c-kit$^+$ Progenitor Cells. *PLoS ONE* **2015**, *10*, e0138581. [CrossRef] [PubMed]

19. Chittajallu, R.; Chen, Y.; Wang, H.; Yuan, X.; Ghiani, C.A.; Heckman, T.; McBain, C.J.; Gallo, V. Regulation of Kv1 subunit expression in oligodendrocyte progenitor cells and their role in G1/S phase progression of the cell cycle. *Proc. Natl. Acad. Sci. USA* **2002**, *99*, 2350–2355. [CrossRef] [PubMed]

20. Robb, E.L.; Stuart, J.A. trans-Resveratrol as a neuroprotectant. *Molecules* **2010**, *15*, 1196–1212. [CrossRef] [PubMed]

21. Rieder, S.A.; Nagarkatti, P.; Nagarkatti, M. Multiple anti-inflammatory pathways triggered y resveratrol lead to amelioration of staphylococcal enterotoxin B-induced lung injury. *Br. J. Pharmacol.* **2012**, *167*, 1244–1258. [CrossRef] [PubMed]

22. Wang, Y.J.; Chan, M.H.; Chen, L.; Wu, S.N.; Chen, H.H. Resveratrol attenuates cortical neuron activity: Roles of large conductance calcium-activated potassium channels and voltage-gated sodium channels. *J. Biomed. Sci.* **2016**, *23*, 47. [CrossRef] [PubMed]

23. Curci, A.; Maqoud, F.; Mele, A.; Cetrone, M.; Angelelli, M.; Zizzo, N.; Tricarico, D. Antiproliferative effects of neuroprotective drugs targeting big Ca^{2+}-activated K^+ (BK) channel in the undifferentiated neuroblastoma cells. *Curr. Top. Pharmacol.* **2016**, *20*, 113–131.

24. Carter, L.G.; D'Orazio, J.A.; Pearson, K.J. Resveratrol and cancer: Focus on in vivo evidence. *Endocr. Relat. Cancer* **2014**, *21*, 209–225. [CrossRef] [PubMed]

25. Athar, M.; Back, J.H.; Kopelovich, L.; Bickers, D.R.; Kim, A.L. Multiple molecular Targets of Resveratrol: Anti-carcinogenic Mechanisms. *Arch. Biochem. Biophys.* **2009**, *486*, 95–102. [CrossRef] [PubMed]

26. Graham, R.M.; Hernandez, F.; Puerta, N.; de Angulo, G.; Webster, K.A.; Vanni, S. Resveratrol augments ER stress and the cytotoxic effects of glycolytic inhibition in neuroblastoma by downregulating AKT in a mechanism independent of SIRT1. *Exp. Mol. Med.* **2016**, *48*, e210-12. [CrossRef] [PubMed]

27. Manning, B.D.; Toker, A. AKT/PKB Signaling: Navigating the Network. *Cell* **2017**, *169*, 381–405. [CrossRef] [PubMed]

28. Sokolowski, B.; Orchard, S.; Harvey, M.; Sridhar, S.; Sakai, Y. Conserved BK Channel-Protein Interactions Reveal Signals Relevant to Cell Death and Survival. *PLoS ONE* **2011**, *6*, e28532. [CrossRef] [PubMed]

29. Park, J.H.; Park, S.J.; Chung, M.K.; Jung, K.H.; Choi, M.R.; Kim, Y.; Chai, Y.G.; Kim, S.J.; Park, K.S. High expression of large-conductance Ca^{2+}-activated K^+ channel in the CD133$^+$ subpopulation of SH-SY5Y neuroblastoma cells. *Biochem. Biophys. Res. Commun.* **2010**, *396*, 637–642. [CrossRef] [PubMed]

30. Curci, A.; Mele, A.; Camerino, G.M.; Dinardo, M.M.; Tricarico, D. The large conductance Ca^{2+}-activated K^+ (BKCa) channel regulates cell proliferation in SH-SY5Y neuroblastoma cells by activating the staurosporine-sensitive protein kinases. *Front. Physiol.* **2014**, *5*, 476. [CrossRef] [PubMed]

31. Yang, M.; Brackenbury, W.J. Membrane potential and cancer progression. *Front. Physiol.* **2013**, *4*, 185. [CrossRef] [PubMed]

32. Giangiacomo, K.M.; Garcia, M.L.; McManus, O.B. Mechanism of iberiotoxin block of the large-conductance calcium-activated potassium channel from bovine aortic smooth muscle. *Biochemistry* **1992**, *31*, 6719–6727. [CrossRef] [PubMed]

33. Meera, P.; Wallner, M.; Toro, L. A neuronal β subunit (KCNMB4) makes the large conductance, voltage- and Ca^{2+}-activated K^+ channel resistant to charybdotoxin and iberiotoxin. *Proc. Natl. Acad. Sci. USA* **2000**, *97*, 5562–5567. [CrossRef] [PubMed]

34. Li, H.F.; Chen, S.A.; Wu, S.N. Evidence for the stimulatory effect of resveratrol on Ca^{2+}-activated K^+ current in vascular endothelial cells. *Cardiovasc. Res.* **2000**, *45*, 1035–1045. [CrossRef]

35. Tricarico, D.; Mele, A.; Conte Camerino, D. Carbonic anhydrase inhibitors ameliorate the symptoms of hypokalaemic periodic paralysis in rats by opening the muscular Ca^{2+}-activated-K^+channels. *Neuromuscul. Disord.* **2006**, *16*, 39–45. [CrossRef] [PubMed]

36. Tricarico, D.; Lovaglio, S.; Mele, A.; Rotondo, G.; Mancinelli, E.; Meola, G.; Conte Camerino, D. Acetazolamide prevents vacuolar myopathy in skeletal muscle of K^+-depleted rats. *J Pharmacol* **2008**, *154*, 183–190. [CrossRef] [PubMed]

37. Tricarico, D.; Mele, A.; Liss, B.; Ashcroft, F.M.; Lundquist, A.L.; Desai, R.R.; George, A.L.J.; Conte Camerino, D. Reduced expression of Kir6.2/SUR2A subunits explains KATP deficiency in K^+-depleted rats. *Neuromuscul. Disord.* **2008**, *18*, 74–80. [CrossRef] [PubMed]

38. Orsini, F.; Verotta, L.; Lecchi, M.; Restano, R.; Curia, G.; Redaelli, E.; Wanke, E. Resveratrol derivatives and their role as potassium channels modulators. *J. Nat. Prod.* **2004**, *67*, 421–426. [CrossRef] [PubMed]

39. Gao, Z.B.; Hu, G.Y. Trans-resveratrol, a red wine ingredient, inhibits voltage-activated potassium currents in rat hippocampal neurons. *Brain Res.* **2005**, *1056*, 68–75. [CrossRef] [PubMed]

40. Chen, W.P.; Chi, T.C.; Chuang, L.M.; Su, M.J. Resveratrol enhances insulin secretion by blocking K_{ATP} and K_V channels of β cells. *Eur. J. Pharmacol.* **2007**, *568*, 269–277. [CrossRef] [PubMed]

41. Singh, H.; Li, M.; Hall, L.; Chen, S.; Sukur, S.; Lu, R.; Caputo, A.; Meredith, A.L.; Stefani, E.; Toro, L. MaxiK channel interactome reveals its interaction with GABA transporter 3 and heat shock protein 60 in the mammalian brain. *Neuroscience* **2016**, *317*, 76–107. [CrossRef] [PubMed]

42. Hoose, S.A.; Duran, C.; Malik, I.; Eslamfam, S.; Shasserre, S.C.; Downing, S.S.; Hoover, E.M.; Dowd, K.E.; Smith, R.; Polymenis, M. Systematic analysis of cell cycle effects of common drugs leads to the discovery of a suppressive interaction between gemfibrozil and fluoxetine. *PLoS ONE* **2012**, *7*, e36503. [CrossRef] [PubMed]

43. Urrego, D.; Tomczak, A.P.; Zahed, F.; Stuhmer, W.; Pardo, L.A. Potassium channels in cell cycle and cell proliferation. *Philos. Trans. R. Soc. Lond. B. Biol. Sci.* **2014**, *369*, 20130094. [CrossRef] [PubMed]

44. Dubois, J.M.; Rouzaire-Dubois, B. The influence of cell volume changes on tumour cell proliferation. *Eur. Biophys. J.* **2004**, *33*, 227–232. [CrossRef] [PubMed]

45. Dubois, J.M.; Rouzaire-Dubois, B. Roles of cell volume in molecular and cellular biology. *Prog. Biophys. Mol. Biol.* **2012**, *108*, 93–97. [CrossRef] [PubMed]

46. Johnsen, J.I.; Segerstrom, L.; Orrego, A.; Elfman, L.; Henriksson, M.; Kagedal, B.; Eksborg, S.; Sveinbjornsson, B.; Kogner, P. Inhibitors of mammalian target of rapamycin downregulate MYCN protein expression and inhibit neuroblastoma growth in vitro and in vivo. *Oncogene* **2008**, *27*, 2910–2922. [CrossRef] [PubMed]

47. Abid, M.R.; Guo, S.; Minami, T.; Spokes, K.C.; Ueki, K.; Skurk, C.; Walsh, K.; Aird, W.C. Vascular endothelial growth factor activates PI3K/AKT/forkhead signaling in endothelial cells. *Arterioscler. Thromb. Vasc. Biol.* **2004**, *24*, 294–300. [CrossRef] [PubMed]

48. Ho, R.; Minturn, J.E.; Hishiki, T.; Zhao, H.; Wang, Q.; Cnaan, A.; Maris, J.; Evans, A.E.; Brodeur, G.M. Proliferation of human neuroblastomas mediated by the epidermal growth factor receptor. *Cancer Res.* **2005**, *65*, 9868–9875. [CrossRef] [PubMed]

49. Wang, L.; Yang, H.J.; Xia, Y.Y.; Feng, Z.W. Insulin-like growth factor 1 protects human neuroblastoma cells SH-EP1 against MPP^+-induced apoptosis by AKT/GSK-3β/JNK signaling. *Apoptosis* **2010**, *15*, 1470–1479. [CrossRef] [PubMed]

50. Kwon, S.H.; Hong, S.I.; Kim, J.A.; Jung, Y.H.; Kim, S.Y.; Kim, H.C.; Lee, S.Y.; Jang, C.G. The neuroprotective effects of *Lonicera japonica* THUNB. against hydrogen peroxide-induced apoptosis via phosphorylation of MAPKs and PI3K/AKT in SH-SY5Y cells. *Food Chem. Toxicol.* **2011**, *49*, 1011–1019. [CrossRef] [PubMed]

51. Bayascas, J.R.; Alessi, D.R. Regulation of AKT/PKB Ser473 phosphorylation. *Mol. Cell* **2005**, *18*, 143–145. [CrossRef] [PubMed]

52. Zhou, B.P.; Liao, Y.; Xia, W.; Spohn, B.; Lee, M.H.; Hung, M.C. Cytoplasmic localization of p21Cip1/WAF1 by AKT-induced phosphorylation in HER-2/neu-overexpressing cells. *Nat. Cell Biol.* **2001**, *3*, 245–252. [CrossRef] [PubMed]

53. Viglietto, G.; Motti, M.L.; Bruni, P.; Melillo, R.M.; D'Alessio, A.; Califano, D.; Vinci, F.; Chiappetta, G.; Tsichlis, P.; Bellacosa, A.; et al. Cytoplasmic relocalization and inhibition of the cyclin-dependent kinase inhibitor p27(Kip1) by PKB/AKT-mediated phosphorylation in breast cancer. *Nat. Med.* **2002**, *8*, 1136–1144. [CrossRef] [PubMed]

54. Shin, I.; Yakes, F.M.; Rojo, F.; Shin, N.Y.; Bakin, A.V.; Baselga, J.; Arteaga, C.L. PKB/AKT mediates cell-cycle progression by phosphorylation of p27(Kip1) at threonine 157 and modulation of its cellular localization. *Nat. Med.* **2002**, *8*, 1145–1152. [CrossRef] [PubMed]

55. Liang, J.; Zubovitz, J.; Petrocelli, T.; Kotchetkov, R.; Connor, M.K.; Han, K.; Lee, J.H.; Ciarallo, S.; Catzavelos, C.; Beniston, R.; et al. PKB/AKT phosphorylates p27, impairs nuclear import of p27 and opposes p27-mediated G1 arrest. *Nat. Med.* **2002**, *8*, 1153–1160. [CrossRef] [PubMed]

56. Niklasson, M.; Maddalo, G.; Sramkova, Z.; Mutlu, E.; Wee, S.; Sekyrova, P.; Schmidt, L.; Nicolas Fritz, N.; Dehnisch, I.; Kyriatzis, G.; et al. Membrane-Depolarizing Channel Blockers Induce Selective Glioma Cell Death by Impairing Nutrient Transport and Unfolded Protein/Amino Acid Responses. *Cancer Res.* **2017**, *77*, 1741–1752. [CrossRef] [PubMed]

57. Liao, Y.; Hung, M.C. Physiological regulation of AKT activity and stability. *Am. J. Transl. Res.* **2010**, *2*, 19–42. [PubMed]

58. Duggal, S.; Jailkhani, N.; Midha, M.K.; Agrawal, N.; Rao, K.V.S.; Kumar, A. Defining the AKT1 interactome and its role in regulating the cell cycle. *Sci. Rep.* **2018**, *8*, 1303. [CrossRef] [PubMed]

59. Mele, A.; Buttiglione, M.; Cannone, G.; Vitiello, F.; Conte Camerino, D.; Tricarico, D. Opening/blocking actions of pyruvate kinase antibodies on neuronal and muscular KATP channels. *Pharmacol. Res.* **2012**, *66*, 401–408. [CrossRef] [PubMed]

60. Tricarico, D.; Mele, A.; Calzolaro, S.; Cannone, G.; Camerino, G.M.; Dinardo, M.M.; Latorre, R.; Conte Camerino, D. Emerging Role of Calcium-Activated Potassium Channel in the Regulation of Cell Viability Following Potassium Ions Challenge in HEK293 Cells and Pharmacological Modulation. *PLoS ONE* **2013**, *8*, e69551. [CrossRef] [PubMed]

61. Tricarico, D.; Barbieri, M.; Laghezza, A.; Tortorella, P.; Loiodice, F.; Conte Camerino, D. Dualistic actions of cromakalim and new potent 2*H*-1,4-benzoxazine derivatives on the native skeletal muscle K_{ATP} channel. *Br. J. Pharmacol.* **2003**, *139*, 255–262. [CrossRef] [PubMed]

62. Tricarico, D.; Montanari, L.; Conte Camerino, D. Involvement of $3Na^+/2K^+$ ATP-ase and Pi-3 kinase in the response of skeletal muscle ATP-sensitive K^+ channels to insulin. *Neuromuscul. Disord.* **2003**, *13*, 712–719. [CrossRef]

63. Massey, A.J. Multiparametric Cell Cycle Analysis Using the Operetta High-Content Imager and Harmony Software with PhenoLOGIC. *PLoS ONE* **2015**, *10*, e0134306. [CrossRef] [PubMed]

64. Tricarico, D.; Mele, A.; Camerino, G.M.; Bottinelli, R.; Brocca, L.; Frigeri, A.; Svelto, M.; George, A.L.J.; Conte Camerino, D. The KATP channel is a molecular sensor of atrophy in skeletal muscle. *J. Physiol.* **2010**, *588*, 773–784. [CrossRef] [PubMed]

65. Cetrone, M.; Mele, A.; Tricarico, D. Effects of the antidiabetic drugs on the age-related atrophy and sarcopenia associated with diabetes type II. *Curr. Diabetes Rev.* **2014**, *10*, 231–237. [CrossRef] [PubMed]

66. Mele, A.; Camerino, G.M.; Calzolaro, S.; Cannone, M.; Conte Camerino, D.; Tricarico, D. Dual response of the KATP channels to staurosporine: A novel role of SUR2B, SUR1 and Kir6.2 subunits in the regulation of the atrophy in different skeletal muscle phenotypes. *Biochem. Pharmacol.* **2014**, *91*, 266–275. [CrossRef] [PubMed]

67. Ernest, N.J.; Habela, C.W.; Sontheimer, H. Cytoplasmic condensation is both necessary and sufficient to induce apoptotic cell death. *J. Cell Sci.* **2008**, *121*, 290–297. [CrossRef] [PubMed]

International Journal of
Molecular Sciences

MDPI

Article

Phosphatidylethanolamine Induces an Antifibrotic Phenotype in Normal Human Lung Fibroblasts and Ameliorates Bleomycin-Induced Lung Fibrosis in Mice

Luis G. Vazquez-de-Lara [1,*], Beatriz Tlatelpa-Romero [1], Yair Romero [1], Nora Fernández-Tamayo [1], Fernando Vazquez-de-Lara [1,†], Jaime M. Justo-Janeiro [2], Mario Garcia-Carrasco [1], René de-la-Rosa Paredes [2], José G. Cisneros-Lira [3], Criselda Mendoza-Milla [3], Francesco Moccia [4] and Roberto Berra-Romani [1]

[1] Facultad de Medicina, Benemérita Universidad Autónoma de Puebla, Puebla 72410, Mexico; beatlarom@outlook.com (B.T.-R.); yair12@hotmail.com (Y.R.); norafertamayo@gmail.com (N.F.-T.); fvazdela@gmail.com (F.V.-d.-L.); mgc30591@yahoo.com (M.G.-C.); rberra001@hotmail.com (R.B.-R.)
[2] Hospital General del Sur, Puebla 72490, Mexico; jaime_justo@hotmail.com (J.M.J.-J.); rdelarosa2000@hotmail.com (R.d.-l.-R.P.)
[3] Instituto Nacional de Enfermedades Respiratorias "Ismael Cosío Villegas", México City 14080, Mexico; jgclira@yahoo.com.mx (J.G.C.-L.); criselda.mendoza@gmail.com (C.M.-M.)
[4] Laboratory of General Physiology, Department of Biology and Biotechnology "Lazzaro Spallanzani", University of Pavia, 27100 Pavia, Italy; francesco.moccia@unipv.it
* Correspondence: luis.vazquezdelara@correo.buap.mx; Tel.: +52-222-229-5500 (ext. 6074)
† Current address: Mount Sinai West Hospital, New York, NY 10019, USA.

Received: 28 August 2018; Accepted: 3 September 2018; Published: 14 September 2018

Abstract: Lung surfactant is a complex mixture of phospholipids and specific proteins but its role in the pathogenesis of interstitial lung diseases is not established. Herein, we analyzed the effects of three representative phospholipid components, that is, dipalmitoilphosphatidylcoline (DPPC), phosphatidylglycerol (PG) and phosphatidylethanolamine (PE), on collagen expression, apoptosis and Ca^{2+} signaling in normal human lung fibroblasts (NHLF) and probed their effect in an experimental model of lung fibrosis. Collagen expression was measured with RT-PCR, apoptosis was measured by using either the APOPercentage assay kit (Biocolor Ltd., Northern Ireland, UK) or the Caspase-Glo 3/7 assay (Promega, Madison, WI, USA) and Ca^{2+} signaling by conventional epifluorescence imaging. The effect in vivo was tested in bleomycin-induced lung fibrosis in mice. DPPC and PG did not affect collagen expression, which was downregulated by PE. Furthermore, PE promoted apoptosis and induced a dose-dependent Ca^{2+} signal. PE-induced Ca^{2+} signal and apoptosis were both blocked by phospholipase C, endoplasmic reticulum pump and store-operated Ca^{2+} entry inhibition. PE-induced decrease in collagen expression was attenuated by blocking phospholipase C. Finally, surfactant enriched with PE and PE itself attenuated bleomycin-induced lung fibrosis and decreased the soluble collagen concentration in mice lungs. This study demonstrates that PE strongly contributes to the surfactant-induced inhibition of collagen expression in NHLF through a Ca^{2+} signal and that early administration of Beractant enriched with PE diminishes lung fibrosis in vivo.

Keywords: pulmonary surfactant; phosphatidylethanolamine; lung fibrosis; bleomycin model; lung fibroblasts; Ca^{2+} signaling

1. Introduction

A common pathological feature in acute lung injury and diverse interstitial lung diseases, such as idiopathic pulmonary fibrosis (IPF), is the migration of fibroblasts and myofibroblasts from the interstitium to the alveolar spaces [1–3]. When the lung alveolar epithelium becomes injured and the basement membrane loses its integrity, an extravasation of plasma-derived fluid takes place and activated alveolar epithelial cells chemoattract mesenchymal cells to the alveolar regions [4]. In the process of migration through partially disrupted and denuded epithelial basement membranes, fibroblasts are exposed to the components of alveolar spaces including surfactant lipids and proteins.

The lipids of lung surfactant mainly include phospholipids (PL) (~90–95 wt %) and a small amount of neutral lipids (~5–10 wt %), such as cholesterol. Phosphatidylcholine (PC) accounts for ~80% of the total PL content, whereas the remaining 20% consists primarily of unsaturated anionic PL (e.g., lyso-bis-phosphatidic acid, phosphatidylinositol and phosphatidylglycerol (PG)), which account for ~15% of the total PL and by small amounts of non-PC zwitterionic PL (e.g., sphingomyelin and phosphatidylethanolamine (PE)) [5]. Beractant (Survanta™, Abbvie Inc., North Chicago, IL, USA) is a modified bovine pulmonary surfactant organic extract, to which synthetic dipalmitoilphosphatidylcoline (DPPC), triacylglycerol and palmitic acid are added. Beractant is widely used for the treatment of respiratory distress syndrome in premature newborns [6]. In a previous study, we showed that pulmonary surfactant promotes programmed cell death of normal human lung fibroblasts and induces the upregulation of matrix metalloproteinase-1 and the downregulation of type I collagen [7]. In that study, we used Beractant, a natural bovine lung extract containing phospholipids, neutral lipids, fatty acids and surfactant-associated hydrophobic proteins B and C. In other study [8], we found that Beractant promoted apoptosis and reduced the expression levels of type I collagen through an increase in intracellular Ca^{2+} concentration ($[Ca^{2+}]_i$) that was initiated by the recruitment of phospholipase C (PLC). Intriguingly, neither DPPC nor PG were able to induce an increase in $[Ca^{2+}]_i$, which suggests that other surfactant components induce apoptosis and regulate gene expression in normal lung fibroblasts (NHLF).

The aim of this work was to test the effect of three representative phospholipids on the expression of collagen in NHLF: DPPC (PC class), PG (anionic PL) and PE (non-PC zwitterionic PL) and to test this effect in vivo in an experimental model of lung fibrosis. Evidence is presented that physiological concentrations of PE downregulate collagen expression and induce apoptosis in NHLF. We also provide evidence that PE elicits an increase in $[Ca^{2+}]_i$ and that the pharmacological blockade of this Ca^{2+} signal attenuates type I collagen expression and apoptosis. Finally, we also show that Beractant enriched with PE ameliorates lung fibrosis in bleomycin-induced lung injury in mice.

2. Results

2.1. Effect of Different Phospholipid Components of Surfactant on Collagen mRNA Expression

We have previously demonstrated that surfactant reduced the expression of collagen by human lung fibroblasts [7]. To elucidate which components of surfactant participate in this process, we evaluated the effect of three representative phospholipids that are found in normal pulmonary surfactant: DPPG (PC class), PG (anionic PL) and PE (non-PC zwitterionic PL).

In the previous work, we saw that Beractant at 500 mg mL^{-1} decreases collagen expression in normal human lung fibroblasts. Considering the proportions of lipids reported in lung surfactant [5], DPPC was used at a dose of 200 µg·mL^{-1} and PG and PE were added at a concentration of 50 µg·mL^{-1}. Lung fibroblasts from passages 5–10 were placed in 6 well plates and when confluence was reached, culture medium was substituted with serum-free F-12 for 24 h. The phospholipids were added in triplicates and cells were incubated for 48 h. Culture medium was aspirated and RNA was isolated. As illustrated in Figure 1, no effect was observed with DPPC and PG. In contrast, collagen expression decreased significantly (cell line 1: 146.8 ± 28.3 vs. 68.3 ± 18.7; cell line 2: 12.9 ± 9.6 vs. 1.6 ± 1.9) with

PE at 50 µg·mL^{-1} ($p < 0.05$, ANOVA with post-hoc Dunnett Test). As shown elsewhere [7], Beractant (500 µg·mL^{-1}) also reduced collagen mRNA expression in NHLF.

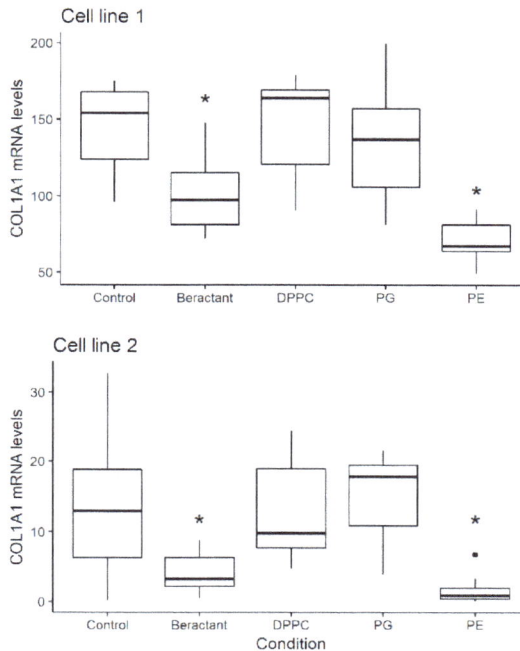

Figure 1. Effect of dipalmitoilphosphatidylcoline (DPPC), phosphatidylglycerol (PG) and phosphatidylethanolamine (PE) on collagen expression in lung fibroblasts. Cells from passages 5–10 were placed in 6 well plates and when confluence was reached, culture medium was substituted with serum-free F-12 for 24 h. Beractant (500 µg·mL^{-1}) and corresponding phospholipids were diluted in F-12 without serum and cells were incubated for 48 h. Controls were grown for the same time in F-12 without serum. DPPC: 200 µg·mL^{-1}; PG: 50 µg·mL^{-1}; PE: 50 µg·mL^{-1}. The figure represents the pooled data of two independent experiments in two cell lines. Asterisks denote significant differences ($p < 0.05$, ANOVA with post-hoc Dunnett test against the control).

2.2. Treatment of NHLF with PE Causes a Dose- and Time-Dependent Decrease of Collagen Expression

Since PE produced a significant and consistent reduction in the expression of collagen in relation to the other phospholipids tested, we performed a dose response experiment within physiologic ranges. Fibroblasts were incubated with PE at doses of 5, 20 and 50 µg·mL^{-1} for 48 h. At a dose of 5 µg·mL^{-1}, PE did not have a significant effect. However, at doses of 20 and 50 µg·mL^{-1}, a significant decrease was observed in collagen expression ($p < 0.05$; Figure 2). We also tested different incubation times with PE at 50 µg·mL^{-1} and observed that a significant decrease in collagen expression occurred at 24 h ($p < 0.05$, ANOVA with post-hoc Dunnett test, Figure 3).

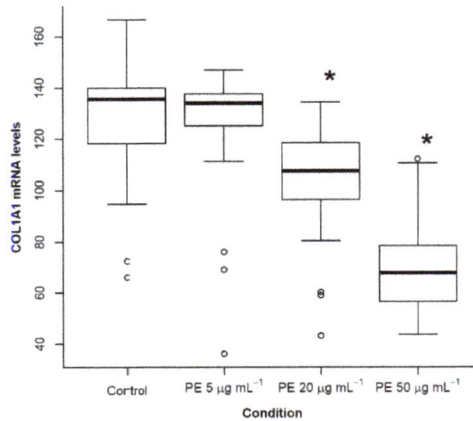

Figure 2. Effect of different concentrations of phosphatidylethanolamine (PE) on collagen expression in normal human lung fibroblasts. Cells from passages 5–10 were placed in 6 well plates and when confluence was reached, culture medium was substituted with serum-free medium for 24 h. PE was diluted in F-12 without serum at 5, 20 and 50 $\mu g \cdot mL^{-1}$ and cells were incubated for 48 h. Controls were grown for the same time in F-12 without serum. The figure represents the pooled data of three independent experiments. Asterisks denote significant differences ($p < 0.05$, ANOVA with post-hoc Dunnett test against the control).

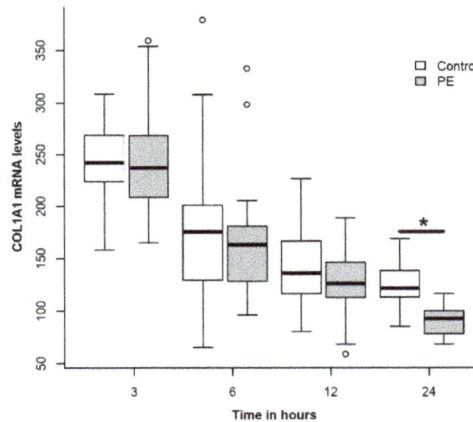

Figure 3. Effect of incubation time of phosphatidylethanolamine (PE) on collagen expression in normal human lung fibroblasts. Cells from passages 5–10 were placed in 6 well plates and when confluence was reached, culture medium was substituted with serum-free medium for 24 h. PE was diluted in F-12 without serum at 50 $\mu g \cdot mL^{-1}$ and cells were incubated for 3, 6, 12 and 48 h. Controls were grown for the same time in F-12 without serum. Asterisk indicates a significant difference when compared with the corresponding control ($p < 0.05$, Student *t*-test).

2.3. Effect of PE on Fibroblast Apoptosis

We have previously observed that Beractant induces fibroblast apoptosis [7], which represents an additional strategy to hamper collagen deposition and ameliorate pulmonary fibrosis [9]. Accordingly, we tested if this effect could be reproduced by PE. Figure 4 shows the effect of different concentrations of PE at 24 and 48 h. Apoptosis was not different from control when cells were incubated for 24 h with

25, 50 or 100 µg·mL^{-1} (Figure 4A); when fibroblasts were incubated for 48 h (Figure 4B), a significant difference in apoptosis was observed starting from a PE concentration of 25 µg·mL^{-1} ($p < 0.05$, ANOVA with post-hoc Dunnett test). The figure represents the pooled data of three independent experiments.

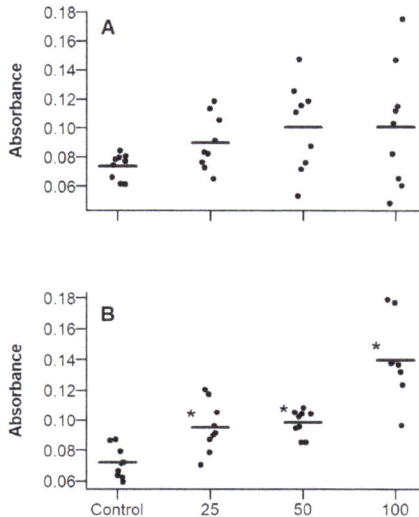

Figure 4. Effect of phosphatidylethanolamine (PE) on apoptosis of normal human lung fibroblasts. Forty-eight well plates were seeded with 40,000 cells per well and incubated for 24 (**A**) or 48 h (**B**) with 25, 50 or 100 µg·mL^{-1} PE in free-serum medium. Apoptosis was measured with the APOPercentage apoptosis assay (Biocolor). Lines represent the mean of the groups. Asterisks indicate significant differences ($p < 0.05$, ANOVA with post-hoc Dunnett test against the control).

2.4. A Calcium Signal Is Elicited by PE and a Phospholipase C Inhibitor Decreases PE Effect on Collagen Expression and Fibroblast Apoptosis

We have previously shown that Beractant elicits an intracellular Ca^{2+} signal in NHLF and that this Ca^{2+} response was not triggered either by DPPC or PG [8]. Thus, we tested if PE mimics this effect. NHLF were loaded with the Ca^{2+}-sensitive fluorochrome, Fura-2 and then stimulated with PE at increasing doses and in the absence and in the presence of specific inhibitors of the Ca^{2+} signaling toolkit. PE induced an increase in [Ca^{2+}]$_i$ similar to that evoked by Beractant (Figure 5) [8]. At 50 µg· mL^{-1}, around 20% of cells responded with a transient Ca^{2+} spike (Figure 5A, left); 500 µg·mL^{-1} PE induced a biphasic Ca^{2+} signal in all the cells, which consisted in an initial Ca^{2+} spike followed by a plateau phase of intermediate amplitude (Figure 5A, right) or Ca^{2+} oscillations (Figure 5B, left). The Ca^{2+} response to Beractant was mediated by the interaction between inositol-1,4,5-trisphosphate (InsP$_3$)-dependent Ca^{2+} release from the endoplasmic reticulum (ER) and store-operated Ca^{2+} entry (SOCE) [8]. Likewise, upon removal of extracellular Ca^{2+} (0Ca^{2+}) PE induced a transient increase in [Ca^{2+}]$_i$ which lacked the plateau or oscillation phase (Figure 5C). This finding confirms that, similar to Beractant, the Ca^{2+} response to PE is initiated by endogenous Ca^{2+} release and sustained by SOCE during the following plateau or oscillation phase. To assess this hypothesis, we pre-incubated the cells with the aminosteroid U73122 (10 µM, 30 min), a rather selective PLC blocker [5], or with 2-aminoethoxydiphenyl borate (2APB) (50 µM, 30 min), which selectively targets InsP$_3$ receptors (InsP$_3$Rs) under 0Ca^{2+} conditions [10]. As shown in Figure 5D and in Figure 5E, respectively, either U73122 or 2APB blocked Ca^{2+} response to PE. Furthermore, PE-induced increase in [Ca^{2+}]$_i$ was prevented by cyclopiazonic acid (CPA; 10 µM), a specific inhibitor of Sarco-Endoplasmic Reticulum Ca^{2+}-ATPase (SERCA) activity, which depletes the ER Ca^{2+} content in NHLF [8], Figure 5F. The statistical analysis is reported in Figure 6. To evaluate the participation of SOCE on the Ca^{2+} signaling evoked by PE, we applied 5 μM of Gd^{3+}, a concentration which selectively hinders store-operated channels [8,11]. Similar to what was reported for Beractant [8], Gd^{3+} interrupt both the prolonged plateau phase (Figure 7A) and the repetitive Ca^{2+} oscillations (Figure 7B) evoked by PE. Collectively, these finding strongly suggest that PE is the phospholipid component that actually triggers the Ca^{2+} response so surfactant in NHLF and that this Ca^{2+} signal arises downstream of PLC activation.

To test if there is a relationship between the Ca^{2+} signal and the PE effect on collagen expression, cells were incubated in the presence of PE with or without U73122 (10 μM, 30 min) for 48 h to prevent both $InsP_3$-dependent Ca^{2+} release and SOCE activation. The decrease in collagen expression provoked by PE was significantly attenuated in the presence of U73122 ($p < 0.05$, Student *t* test, Figure 8), thereby confirming our previous data with Beractant [8].

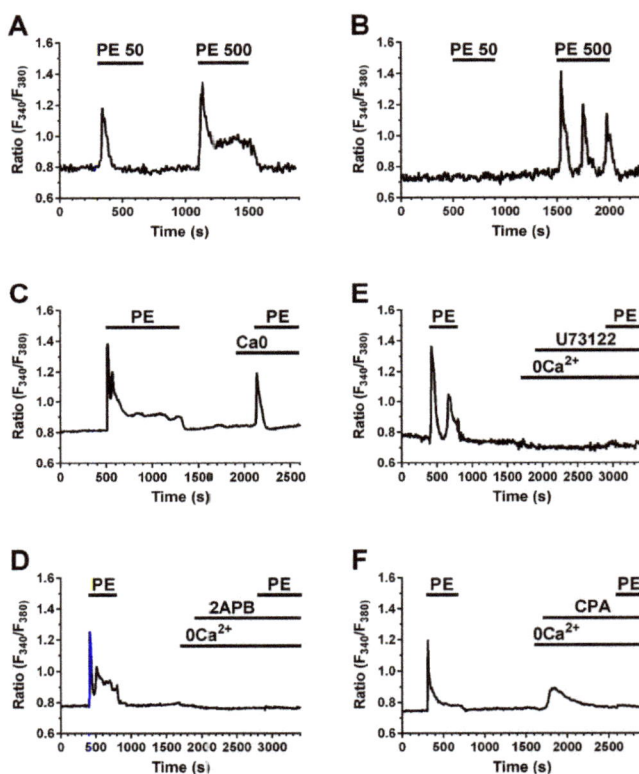

Figure 5. Phosphatidylethanolamine (PE) induced Ca^{2+} signals in normal human lung fibroblasts. (**A**) Representative tracing of a cell that displayed a Ca^{2+} signal in response to both 50 and 500 μg·mL^{-1} of PE; (**B**) Representative tracing of a cell that did not respond to PE 50 μg·mL^{-1}; (**C**) Representative Ca^{2+} tracing induced by PE 500 μg·mL^{-1} in the presence and absence of extracellular Ca^{2+} (0Ca^{2+}); (**D**) U73122 10 μM prevents the Ca^{2+} response to PE 500 μg·mL^{-1}; (**E**) 2APB 50 μM prevents the Ca^{2+} response to PE 500 μg·mL^{-1}; (**F**) CPA 10 μM, abolished the Ca^{2+} response to PE 500 μg·mL^{-1} under 0Ca^{2+} conditions.

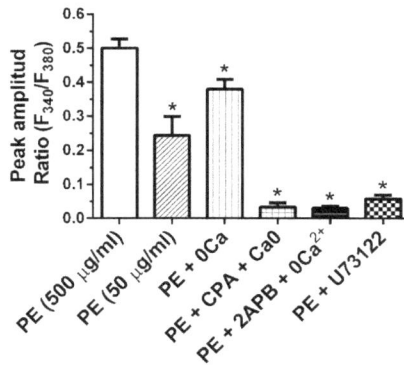

Figure 6. Statistical analysis of Ca^{2+} imaging experiments. Mean ± SE of the amplitude of the Ca^{2+} response to PE 500 $\mu g \cdot mL^{-1}$ under the designated treatments. Asterisks denote significant differences ($p < 0.05$, ANOVA with *post-hoc* Dunnett test against the control).

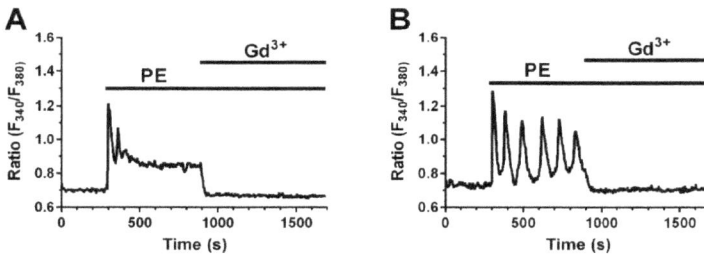

Figure 7. Gd^{3+} blocks phosphatidylethanolamine (PE)-induced Ca^{2+} plateau and Ca^{2+} oscillations in NHLF. (**A**) Addition of Gd^{3+} (5 μM) inhibited the sustained plateau phase of the Ca^{2+} signal induced by PE (500 $\mu g \cdot mL^{-1}$); (**B**) Application of La^{3+} (5 μM) inhibited PE-elicited Ca^{2+} oscillations (500 $\mu g \cdot mL^{-1}$).

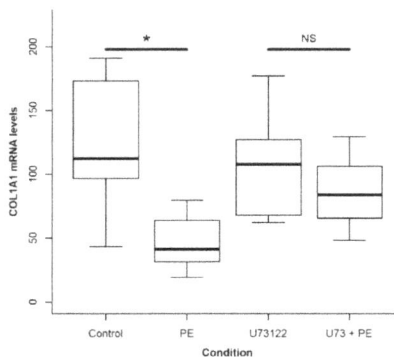

Figure 8. Influence of U73122 on phosphatidylethanolamine (PE)-induced collagen expression. Cells grown in six well plates were incubated for 48 h in serum-free culture medium with PE 50 $\mu g \ mL^{-1}$, either in the presence or absence of U73122 10 μM in dimethyl sulfoxide (DMSO). Controls were incubated with the same amount of DMSO used in the experimental wells. Individual comparisons were made for each pair of control and experimental groups with Student *t* test. Asterisk indicates a $p < 0.05$, NS: non-significant.

Finally, we assessed whether there is a relationship between the Ca^{2+} signal and the PE effect on apoptosis. We found that PE induced apoptosis when administered at 100 $\mu g \cdot mL^{-1}$ to human lung fibroblasts under control conditions for 48 h (Figure 9). However, the apoptotic process was significantly ($p < 0.05$) inhibited when the accompanying Ca^{2+} signal was impaired by using any of the following drugs: U73122 10 μM, CPA 5 μM and Gd^{3+} 5 μM. These data, therefore, endorse the view that PE uses intracellular Ca^{2+} signaling to promote apoptosis in human lung fibroblasts.

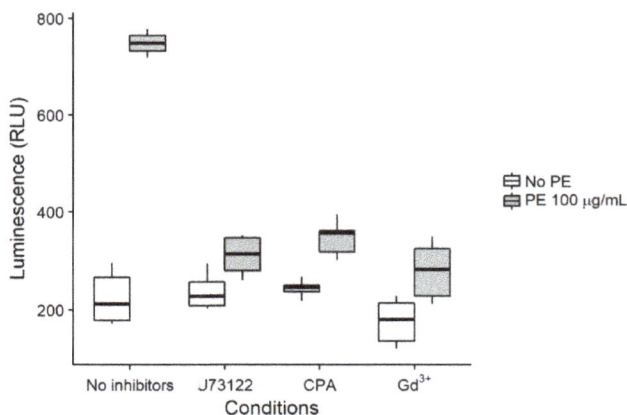

Figure 9. Inhibition of calcium signaling prevents phosphatidylethanolamine (PE)-induced apoptosis. Ninety-six well plates were seeded with 20,000 cells and incubated for 48 h in free serum medium with 100 $\mu g \cdot mL^{-1}$ PE in the absence or in the presence of U73122 (10 μM), CPA (5 μM) and gadolinium (5 μM). Apoptosis was measured by using the Caspase-Glo 3/7 assay (Promega). Asterisks indicate significant differences ($p < 0.05$, ANOVA with *post-hoc* Dunnett test against the control).

2.5. Phosphatidylethanolamine Mitigates Bleomycin-Induced Lung Fibrosis

Since PE strongly reduces the expression of collagen in fibroblast in vitro, we examined whether this phospholipid may have an effect on the development of lung fibrosis in vivo. For this purpose, mice were challenged with bleomycin and the lung fibrotic response was analyzed at 21 days. Five groups of animals were studied. Number of animals and treatment protocols are summarized in Table 1. All treatments were administered every 48 h for 6 doses, starting 2 h after bleomycin instillation via aerosol as described in Methods. Beractant was enriched with PE sonicated in normal saline at a final concentration of 0.75 $mg \cdot mL^{-1}$. The rationale was that PE comprises approximately 3% of surfactant phospholipids [5] and the intention was to duplicate the amount of PE. Twenty-one days after bleomycin instillation lungs were obtained, the right lung was used to measure soluble collagen content and the left lung for histological analysis.

Table 1. Treatment protocol of C57/BL mice.

Group	Intratracheal Bleomycin	Treatment	*n*
1	No (normal saline)	Normal saline (1 mL)	7
2	Yes (4 units Kg^{-1})	Normal saline (1 mL)	7
3	Yes (4 units Kg^{-1})	Beractant (400 $mg \cdot Kg^{-1}$)	7
4	Yes (4 units Kg^{-1})	Beractant-PE (400 $mg \cdot Kg^{-1}$) *	6
5	Yes (4 units Kg^{-1})	PE (400 $mg \cdot Kg^{-1}$)	5

* Beractant-PE: Beractant enriched with PE sonicated in normal saline at a final concentration of 0.75 mg/mL.

As illustrated in Figure 10A, morphological analysis of the lungs showed that surfactant enriched with PE and PE decreased lung inflammation and fibrosis induced by bleomycin. The effect on the fibrotic response was corroborated by the semi quantitative analysis of the fibrotic index (Figure 10B). This finding was consistent with the quantification of collagen. As shown in Figure 11, bleomycin mice almost doubled the concentration of collagen in the lungs when compared to the control group challenged with normal saline solution (99.2 ± 8.1 versus 49.1 ± 16.9 μg/mL; $p < 0.05$). Treatment with Beractant tended to decrease the concentration of collagen but the difference was not statistically significant (73.7 ± 18.8; $p < 0.05$). Treatment of the bleomycin-injured mice with PE (56.6 ± 24.2 μg/mL; $p < 0.05$) or, even more, with surfactant enriched with PE (51.2 ± 29.7 μg/mL; $p < 0.05$) showed a significant decrease in lung collagen accumulation. All comparisons were made with ANOVA and *post-hoc* Dunnett test against the bleomycin-injured mice treated with normal saline.

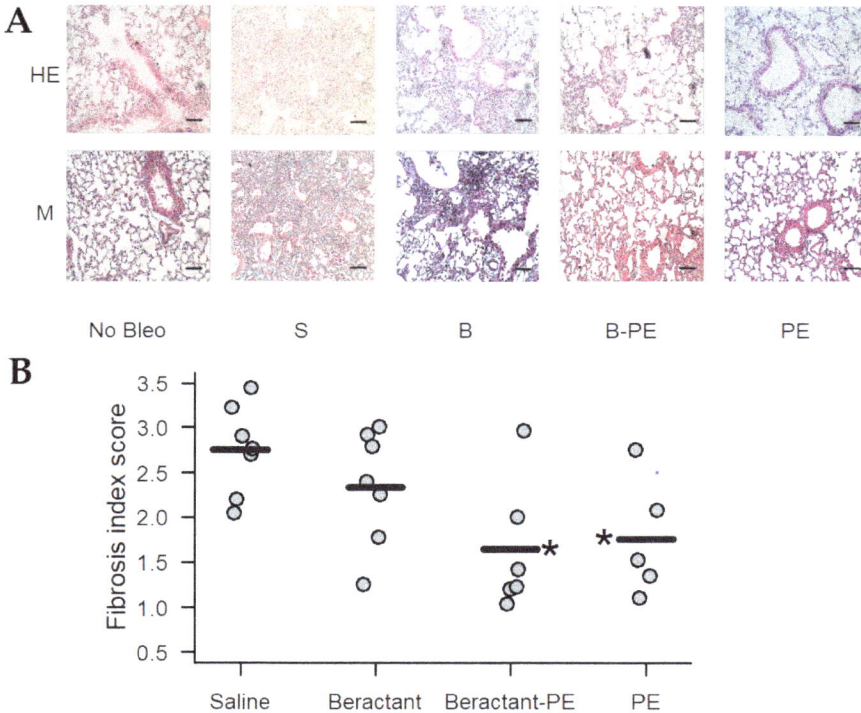

Figure 10. Effect of 1.5 mL of inhaled Beractant, phosphatidylethanolamine (PE)-enriched Beractant and PE on pulmonary fibrosis in bleomycin treated C57/BL mice. The first dose was given the same day of bleomycin administration. **Saline**: after intratracheal administration of 3 U/Kg bleomycin, 1.5 mL of normal saline was nebulized every 48 h for 6 doses. **Beractant**: after the administration of bleomycin, Beractant 400 mg/Kg were nebulized every 48 h for 6 doses. **Beractant-PE**: the treatment consisted of 400 mg·mL^{-1} of Beractant enriched with PE at a final concentration of 0.75 mg mL^{-1}. **PE**: treatment consisted of PE 0.75 mg·mL^{-1}. (**A**) Representative photomicrographs from experimental groups (scale bar, 150 μm). Lung tissue samples were stained with hematoxylin-eosin (HE) and Masson (M). No bleo: control with no bleomycin and saline treatment; S: control with inhaled saline; B: inhaled Beractant; B-PE: inhaled PE-enriched Beractant; PE: inhaled PE; (**B**) Semi quantitative evaluation of lung lesions. Asterisks indicate a significant difference ($p < 0.05$, ANOVA with *post-hoc* Dunnett test against the control (Saline).

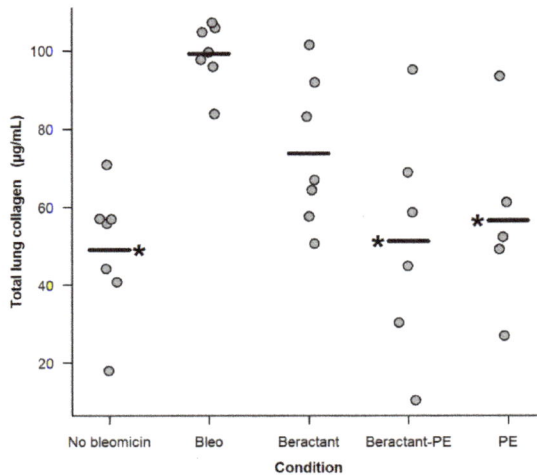

Figure 11. Effect of 1.5 mL of inhaled Beractant, phosphatidylethanolamine (PE)-enriched Beractant and PE on total lung collagen content in bleomycin treated C57/BL mice. The first dose was given the same day of bleomycin administration. **Bleo**: after intratracheal administration of 4 U/Kg bleomycin, 1.5 mL of normal saline was nebulized every 48 h for 6 doses. **Beractant**: after the administration of bleomycin, Beractant 400 mg/Kg were nebulized every 48 h for 6 doses. **Beractant-PE**: the treatment consisted of 400 mg mL^{-1} of Beractant enriched with PE at a final concentration of 0.75 mg mL^{-1}. **PE**: treatment consisted of phosphatidlethanolamine 0.75 mg mL^{-1}. Lines represent the mean of the groups. Total collagen content was measured with the Sircoll assay (Biocolor Ltd., Northern Ireland, UK). (* $p < 0.05$, ANOVA with *post-hoc* Dunnett test against the saline-treated group).

3. Discussion

Lung surfactant is a complex mixture of phospholipids and specific proteins that has many essential functions. Its role in lung fibrosis has not been clearly established but there is some evidence in experimental models that modifications in the concentration and/or composition of surfactant are present when lung fibrosis occurs [12].

In the clinical context, some interstitial lung diseases, such as sarcoidosis, idiopathic pulmonary fibrosis and hypersensitivity pneumonitis, are associated with decreased phospholipid content of surfactant [13]. Likewise, increased levels of SP-A have been described in untreated patients with idiopathic pulmonary fibrosis (IPF) and hypersensitivity pneumonitis [14]. Even though the mechanism has not been clearly established, it has been shown that SP-A interacts differentially with surfactant phospholipids, both in the polar group and the fatty acid chains in order to form aggregates [15] suggesting that an increase in SP-A could favor alterations in the relative composition of the lipid fraction.

We have previously shown that Beractant decreases collagen expression in normal human lung fibroblasts through an increase in [Ca^{2+}]$_i$ [7,8]. Type I collagen represents the major fibrous collagen synthesized by wound fibroblasts during the repair process and is exceedingly deposited in response to a traumatic injury, thereby leading to lung fibrosis [16,17]. In the present study, we found that when DPCC, PG and PE were used individually at physiological concentrations, only PE was able to diminish collagen expression and this effect was time and dose-dependent. The exact mechanism of action remains to be elucidated. In neutrophils, there is evidence that pulmonary surfactant could act through the incorporation of ionic channels in the membrane, producing depolarization and activation of G coupled receptor which produce intracellular Ca^{2+} release [18]. Furthermore, we previously showed that Beractant caused an increase in [Ca^{2+}]$_i$ that was patterned by the interplay between InsP$_3$-dependent ER [Ca^{2+}]$_i$ release and SOCE [8]. In that study, we found that neither DPCC nor PG were able to elevate [Ca^{2+}]$_i$. Herein, we demonstrate that PE elicits an intracellular Ca^{2+} signal in lung fibroblasts that mimics the Ca^{2+} response to Beractant. Accordingly, PE induced different patterns of intracellular Ca^{2+} signals, including a transient Ca^{2+} spike, a biphasic Ca^{2+} elevation and intracellular Ca^{2+} oscillations. Moreover, the onset of PE-induced intracellular Ca^{2+} signals was prevented by inhibiting the PLC/InsP$_3$ signaling pathway with U73122 or 2-APB and was sustained by SOCE. Therefore, these findings lend credit to the notion that PE is the major phospholipid component that underlies the Ca^{2+} response to Beractant. Furthermore, the effect of PE on collagen expression was attenuated by U73122. This is in accordance with our findings reported previously with Beractant and confirms that PE regulates gene expression in NHLF through an increase in [Ca^{2+}]$_i$ [8]. Accordingly, intracellular Ca^{2+} signaling represents an established mechanism to control gene expression in mammalian cells [19–21]. Moreover, an increase in [Ca^{2+}]$_i$ has been shown to regulate collagen expression also in cultured human cardiac [22,23] and pulmonary [24] fibroblasts as well as in human glomerular mesangial cells [25]. Finally, we found that PE-induced apoptosis was prevented by interfering with the accompanying Ca^{2+} signal by pre-treating the cells with U73122 to block PLC, CPA to deplete the ER Ca^{2+} store and Gd^{3+} to inhibit SOCE. These data confirm the finding that beractant induces apoptosis in a Ca^{2+}-dependent manner in human ling fibroblasts [8].

However, we cannot rule out alternative mechanisms to explain the effect of PE on collagen expression in NHLF. For instance, PE is the main phospholipid that favors the negative curvatures of the cell membrane. It is also a molecule with a net neutral charge due to its amino group (zwitterionic). Previous works have studied the interaction of phospholipids with integral membrane proteins. The first factor that could be modified by phospholipids is the hydrophobic width of the lipid bilayer which is determined by the fatty acid chains of phospholipids. This modification determines how much of the protein is included in the membrane and what part is responsible for interaction. In addition, the polar group can interact with the protein and change its conformation [26,27]. Future work is mandatory to assess whether this mechanism is somehow involved in PLC recruitment by PE. Moreover, PE has been shown to activate the extracellular signal-regulated kinases (Erk) and Stat signaling pathways [28], which may inhibit type 1 collagen expression in human dermal fibroblasts [29,30]. Of note, intracellular Ca^{2+} signaling may recruit both Erk [31] and Stat signaling [32].

The deposition of excess collagen in tissues is the mainstay in the pathogenesis of fibrosis. Drugs that inhibit fibroblast proliferation and collagen expression in vitro potentially have a therapeutic effect. In preclinical studies with pirfenidone and nintedanib, currently in clinical use, a decrease in lung fibroblast proliferation and reduced collagen expression was shown [33]. In view of this, we investigated the effect of aerosolized Beractant, PE-enriched Beractant and PE alone in C57/BL mice with bleomycin-induced lung fibrosis. Our results show that Beractant enriched with PE reduces the fibrotic response induced by bleomycin in vivo. More importantly, PE alone displayed a similar effect to that of PE-enriched Beractant, thereby supporting the notion that this molecule plays a pivotal role in the antifibrotic properties of surfactant. Future work will have to confirm that intracellular Ca^{2+} signaling plays a crucial role in reducing fibroblast proliferation and collagen expression also in vivo. In addition to the molecular mechanisms previously proposed, PE could also act in vivo by modifying the biophysical properties of surfactant. Rapid adsorption of lipids to the air-liquid interface is an essential property of lung surfactant and the hydrophobic surfactant proteins SP-B and SP-C play a crucial role in this effect but recently it has been reported that PE also induces acceleration in the adsorption rate [34,35]. In other experiments, aerosolized nanovesicles prepared with DPPC and dioleylphosphatidylethanolamine improved the resistance of pulmonary surfactants to inhibition in mice with acid-induced lung injury [36]. In our model, these effects could be acting in concert to avoid the alveolar rupture and epithelial damage caused by bleomycin.

However, it is worth noting that the antifibrotic effects are limited to the early administration of exogenous surfactant in the bleomycin model. We did not see such effect when administered at 48 h after the insult (data not shown). Proving different doses of enrichment and using a different animal model would be required in order to establish whether PE enriched surfactant may be a possible treatment for fibrosing diseases of the lung. This work, in turn, calls for future investigations that would elucidate the molecular mechanisms responsible for the effect of PE.

4. Materials and Methods

4.1. Isolation and Purification of Normal Human Lung Fibroblasts

NHLF were obtained from kidney donors with brain death upon previous consent of the family, clearly stating that the lung tissue will be used to extract cells for research purposes and data will remain anonymous. The protocol was accepted by the ethical boards of the Hospital General de Puebla and the Benemérita Universidad Autónoma de Puebla College of Medicine. After clamping the aorta, a lung sample was obtained from the left lower lobe and placed in HAM F-12 medium. The tissue was processed with Trypsin-EDTA solution 1× (Sigma, St. Louis, MO, USA) and serum-free F-12 medium. The filtrate was centrifuged at 200 g for 10 min and the pelleted cells were re-suspended in F-12 (10% fetal bovine serum) and layered in T-25 flasks. Cells were grown to a 75% confluence in F-12 medium, at 37 °C on an atmosphere of 95% of air and 5% of CO_2. For the experiments, cells from passages 5 to 10 were placed in 6 well plates and when confluence was reached, culture medium was substituted with serum-free F-12 for 24 h. DPPC, PG, PE (all from Sigma-Aldrich) or Beractant (Abbot) were used at different concentrations. Three cell lines were used for all the experiments, obtained from a thirty-year-old male, a sixty-one-year-old male and a fifteen-year-old female.

4.2. RNA Isolation and RT-PCR Analysis

Total RNA extraction was performed by the TRIzol reagent (Invitrogen Life Technologies, Grand Island, NY, USA) and one µg of RNA was reverse transcribed into complementary DNA (cDNA) (Advantage RT-for-PCR Kit, Clontech, Palo Alto, CA, USA) according to the manufacturer's instructions. Step One Real-Time PCR System (Applied Biosystems, Carlsbad, CA, USA) was used for amplification by real-time PCR, using FAM-labeled Taqman probes (Applied Biosystems, Thermo Fisher Scientific, Waltham, MA, USA), included Hs00164004_m1 (Collagen Ia1) and Hs00172187_m1 POLR2A (reference gene). All PCRs were carried out in a mixture of 25 µL, which contained four µL of cDNA (100 ng) and 12.5 µL of 2× PCR Master Mix (Applied Biosystems). The PCR conditions were 2 min at 94 °C followed by 40 cycles of 15 s at 95 °C and 60 °C 1 min concluding with an infinite loop of refrigeration. Results from two different experiments performed in triplicate are expressed as the mean ± SD of $2^{-\Delta\Delta Ct}$ of target gene normalized against POLR2A.

4.3. Detection of Apoptosis

Cell apoptosis was measured by using either the APOPercentage apoptosis assay kit (Biocolor Ltd., Northern Ireland, UK), or the Caspase-Glo 3/7 assay (Promega, Madison, IL, USA). The first one indirectly measures the phosphatidylserine transmembrane movement. Forty-eight well plates were seeded with 40,000 cells per well and incubated for the desired times in free serum medium. After the incubation period, three control wells were incubated with 5 mM H_2O_2 for 30 min (positive control). Cells were stained with the APOPercentage dye (100 µL) for 30 min at 37 °C in humidified air with 5% CO_2. Cells were washed twice with PBS 500 µL and trypsinized (50 µL) for 10 min. Finally, 200 µL of dye release reagent were added to each well and the plate was shaken for 10 min and 100 µL aliquots of each well were transferred to a 96 well plate and absorbance was measured at 550 nm. Caspase-Glo 3/7 assay measures caspase-3 and -7 activities. Ninety-six well plates were seeded with 20,000 cells and incubated for the desired times in free serum medium. After the incubation period, conditioned medium was substituted with the Caspase-Glo reagent following the manufacturer recommendations. Luminescence was measured with the Filter Max Pro 5 Multi-Mode Microplate Reader (Molecular Devices, San Jose, CA, USA), results are reported as relative light units (RLU).

4.4. Cytosolic Ca²⁺ Measurements

The technique for intracellular Ca^{2+} signal measurement has been described elsewhere [8]. Briefly, cultured NHLF were loaded with 3 μM Fura-2/AM during 45 min at room temperature and then visualized by an upright epifluorescence microscope (Axiolab, Carl Zeiss, Oberkochen, Germany) equipped with a Zeiss 63× Achroplan objective. $[Ca^{2+}]_i$ was measured by calculating the ratio of the average fluorescence emitted at 510 nm when the cells were excited alternatively at 340 and 380 nm (ratio F340/F380). An elevation in $[Ca^{2+}]_i$ leads to an elevation in the Ratio F340/F380. Ratio measurements were carried out and plotted every 3 s by custom-made software. All recordings were carried out at room temperature (21–23 °C). In some experiments, U73122 (a widely employed PLC inhibitor) at 10 μM in dimethyl sulfoxide (DMSO) was used. Cells were pre-incubated for 15 min with this substance before adding the phospholipids and the corresponding controls had the same amount of DMSO as the experiments. The analysis of the Ca^{2+} data was performed by using ImageJ software (National Institutes of Health, Bethesda, MD, USA, http://rsbweb.nih.gov/ij/).

4.5. In Vivo Murine Model

C57/BL mice (8 weeks age, 20 g average weight) were used. All the experiments were performed according to protocols approved by the Animal Care and Use Committee of Animals of the Benemerita Universidad Autonoma de Puebla College of Medicine. Mice were anesthetized with Ketamine/Xylazine (100 mg·Kg⁻¹/10 mg·Kg⁻¹) and a single intratracheal instillation of saline (0.9%, controls) or saline containing bleomycin sulfate (4 U/Kg, Sigma-Aldrich) in a volume of 30 μL was administered.

All treatments were administered via aerosol as described elsewhere [37]. Briefly, mice were anesthetized and placed in 50 mL plastic tube with an orifice at the bottom where the nose was directed. The tube was connected to a micronebulizer (PARI GmbH, Starnberg, Germany) driven by an air line with a flow rate of 4 L/min. The final volume in the micronebulizer was adjusted to 1.5 mL in all cases. The tube was assembled in such a way that only the nose of the animal was in contact with the aerosol. All treatments were given every 48 h for a total of 6 doses starting on day one after bleomycin instillation.

4.6. Soluble Collagen Assay

Lungs were weighted, minced and agitated for 24 h in acetic acid 0.5 M 1:10 (*w/v*). Samples were filtered and soluble collagen content was measured using Sircol assay (Biocolor Ltd.), according to manufacturer specifications. Briefly, the contents of all tubes were adjusted to 100 μL with distilled water; a collagen standard was prepared with 100 mL aliquots containing 5, 10, 25 and 50 μg. 1 mL of Sircol Dye reagent was added and the content was mixed by inverting. Tubes were shaken mechanically for 30 min; during this time, collagen-dye complex precipitate out of solution. Samples were centrifuged 10,000× *g* for a 10 min period. The unbound dye solution was removed by carefully inverting and draining the tubes. One mL of Alkali reagent was added and vortexed. Samples were read at 540 nm in a spectrophotometer. Results are reported as μg of collagen/μg of lung tissue.

4.7. Histologic Examination

Lung sections from bleomycin-injured and control mice were stained, coded and scored blindly for percent of fibrosis as described elsewhere [38]. The presence of interstitial fibrosis was scored as follows: 0 = no fibrosis, 1 = up to 25% of the field, 2 = 25–50% of field, 3 = 50–75% of field and 4 = 75–100% of field. The assessment was done on the slide scanned completely in zigzag fashion, at 20× magnification. At least 20 fields were assessed in each slide. The mean and median of the scores for each animal was computed and the average of for each group was calculated for statistical comparison. The same pathologist evaluated the percentage of fibrosis twice on the same slides with a 6-month difference (intraclass correlation coefficient of 0.82, $p < 0.05$).

4.8. Preparation of Phospholipids

1,2-Dipalmitoyl-*sn*-glycero-3-phosphorylcholine (DPPC), L-α-Phosphatidyl-DL-glycerol (PG), 1,2-Dioleoyl-sn-glycero-3-phosphoethanolamine (PE) were purchased from Sigma-Aldrich (St. Louis, MO, USA). DPPC and PG were sonicated (CPX-400, Cole-Parmer, EUA, Vernon Hills, IL, USA) in Ham F12 medium at 4° C for 2 min. PE was re-suspended in chloroform at a concentration of 20 mg mL^{-1} and stored at $-70°$ C. Prior to use, the desired amount of PE was evaporated with nitrogen, Ham F12 medium with 1% albumin was added to the pellet and sonicated as specified above.

4.9. Statistical Analysis

Unless otherwise stated, results are expressed as the mean ± SD of triplicates of two or three pooled experiments. For statistical analysis ANOVA or *t*-test was used; a $p < 0.05$ was considered significant. The software used was R version 3.3.3 with the multcomp package version 1.4-6 (R Foundation for Statistical Computing, Vienna, Austria). The data for all the experiments presented in this work, can be found at Supplementary Materials.

Supplementary Materials: Supplementary materials can be found at http://www.mdpi.com/1422-0067/19/9/2758/s1.

Author Contributions: Conceptualization, L.G.V.-d.-L.; Validation, L.G.V.-d.-L., R.B.-R., F.M., N.F.-T.; Formal Analysis, L.G.V.-d.-L., R.B.-R., F.M.; Investigation, B.T.-R., Y.R., N.F.-T., F.V.-d.-L., J.G.C.-L., C.M.-M.; Resources, R.d.-l.-R.P., J.M.J.-J., M.G.-C.; Writing-Original Draft Preparation, L.G.V.-d.-L.; Writing-Review & Editing, L.G.V.-d.-L., C.M.-M., R.B.-R., F.M., F.V.-d.-L.; Visualization, L.G.V.-d.-L., R.B.-R., F.M.; Supervision, L.G.V.-d.-L.; Project Administration, L.G.V.-d.-L.

Funding: This research received no external funding.

Abbreviations

0Ca^{2+}	Removal of extracellular Ca^{2+}
2-APB	2-Aminoethoxydiphenyl borate
CPA	Cyclopiazonic acid
DMSO	Dimethyl sulfoxide
DPPC	Dipalmitoilphosphatidilcoline
ER	Endoplasmic reticulum
InsP$_3$	Inositol-1,4,5-trisphosphate
IPF	Idiopathic pulmonary fibrosis
NHLF	Normal human lung fibroblasts
PC	Phosphatidylcholine
PE	Phosphatidylethanolamine
PG	Phosphatidylglycerol
PL	Phospholipids
PLC	Phospholipase C
RLU	Relative Light Units
SERCA	Sarco-endoplasmic reticulum Ca^{2+}-ATPase
SOCE	Store-operated Ca^{2+} entry

References

1. Lederer, D.J.; Martinez, F.J. Idiopathic Pulmonary Fibrosis. *N. Engl. J. Med.* **2018**, *378*, 1811–1823. [CrossRef] [PubMed]
2. King, T.E., Jr.; Pardo, A.; Selman, M. Idiopathic pulmonary fibrosis. *Lancet* **2011**, *378*, 1949–1961. [CrossRef]
3. Harari, S.; Caminati, A. IPF: New insight on pathogenesis and treatment. *Allergy* **2010**, *65*, 537–553. [CrossRef] [PubMed]

4. Strieter, R.M. What differentiates normal lung repair and fibrosis? Inflammation, resolution of repair and fibrosis. *Proc. Am. Thorac. Soc.* **2008**, *5*, 305–310. [CrossRef] [PubMed]

5. Zuo, Y.Y.; Veldhuizen, R.A.; Neumann, A.W.; Petersen, N.O.; Possmayer, F. Current perspectives in pulmonary surfactant—Inhibition, enhancement and evaluation. *Biochim. Biophys. Acta* **2008**, *1778*, 1947–1977. [CrossRef] [PubMed]

6. Poets, C.F.; Lorenz, L. Prevention of bronchopulmonary dysplasia in extremely low gestational age neonates: Current evidence. *Arch. Dis. Child. Fetal Neonatal Ed.* **2018**, *103*, F285–F291. [CrossRef] [PubMed]

7. Vazquez de Lara, L.G.; Becerril, C.; Montano, M.; Ramos, C.; Maldonado, V.; Melendez, J.; Phelps, D.S.; Pardo, A.; Selman, M. Surfactant components modulate fibroblast apoptosis and type I collagen and collagenase-1 expression. *Am. J. Physiol. Lung Cell. Mol. Physiol.* **2000**, *279*, L950–L957. [CrossRef] [PubMed]

8. Guzman-Silva, A.; Vazquez de Lara, L.G.; Torres-Jacome, J.; Vargaz-Guadarrama, A.; Flores-Flores, M.; Pezzat, S.E.; Lagunas-Martinez, A.; Mendoza-Milla, C.; Tanzi, F.; Moccia, F.; et al. Lung beractant increases free cytosolic levels of Ca^{2+} in human lung fibroblasts. *PLoS ONE* **2015**, *10*, e0134564. [CrossRef]

9. Darby, I.A.; Hewitson, T.D. Fibroblast differentiation in wound healing and fibrosis. *Int. Rev. Cytol.* **2007**, *257*, 143–179. [CrossRef] [PubMed]

10. Dragoni, S.; Laforenza, U.; Bonetti, E.; Lodola, F.; Bottino, C.; Berra-Romani, R.; Carlo, B.G.; Cinelli, M.P.; Guerra, G.; Pedrazzoli, P.; et al. Vascular endothelial growth factor stimulates endothelial colony forming cells proliferation and tubulogenesis by inducing oscillations in intracellular Ca^{2+} concentration. *Stem Cells* **2011**, *29*, 1898–1907. [CrossRef] [PubMed]

11. Moccia, F.; Dragoni, S.; Poletto, V.; Rosti, V.; Tanzi, F.; Ganini, C.; Porta, C. Orai1 and transient receptor potential channels as novel molecular targets to impair tumor neovascularization in renal cell carcinoma and other malignancies. *Anticancer Agents Med. Chem.* **2014**, *14*, 296–312. [CrossRef] [PubMed]

12. Schmidt, R.; Ruppert, C.; Markart, P.; Lubke, N.; Ermert, L.; Weissmann, N.; Breithecker, A.; Ermert, M.; Seeger, W.; Gunther, A. Changes in pulmonary surfactant function and composition in bleomycin-induced pneumonitis and fibrosis. *Toxicol. Appl. Pharmacol.* **2004**, *195*, 218–231. [CrossRef] [PubMed]

13. Akella, A.; Deshpande, S.B. Pulmonary surfactants and their role in pathophysiology of lung disorders. *Indian J. Exp. Biol.* **2013**, *51*, 5–22. [PubMed]

14. Phelps, D.S.; Umstead, T.M.; Mejia, M.; Carrillo, G.; Pardo, A.; Selman, M. Increased surfactant protein-A levels in patients with newly diagnosed idiopathic pulmonary fibrosis. *Chest* **2004**, *125*, 617–625. [CrossRef] [PubMed]

15. Casals, C. Role of surfactant protein A (SP-A)/lipid interactions for SP-A functions in the lung. *Pediatr. Pathol. Mol. Med.* **2001**, *20*, 249–268. [CrossRef] [PubMed]

16. Cutroneo, K.R.; White, S.L.; Phan, S.H.; Ehrlich, H.P. Therapies for bleomycin induced lung fibrosis through regulation of TGF-β1 induced collagen gene expression. *J. Cell. Physiol.* **2007**, *211*, 585–589. [CrossRef] [PubMed]

17. Calabresi, C.; Arosio, B.; Galimberti, L.; Scanziani, E.; Bergottini, R.; Annoni, G.; Vergani, C. Natural aging, expression of fibrosis-related genes and collagen deposition in rat lung. *Exp. Gerontol.* **2007**, *42*, 1003–1011. [CrossRef] [PubMed]

18. Boston, M.E.; Frech, G.C.; Chacon-Cruz, E.; Buescher, E.S.; Oelberg, D.G. Surfactant releases internal calcium stores in neutrophils by G protein-activated pathway. *Exp. Biol. Med.* **2004**, *229*, 99–107. [CrossRef]

19. Lewis, R.S. Calcium oscillations in T-cells: Mechanisms and consequences for gene expression. *Biochem. Soc. Trans.* **2003**, *31*, 925–929. [CrossRef] [PubMed]

20. Parekh, A.B. Decoding cytosolic Ca^{2+} oscillations. *Trends Biochem. Sci.* **2011**, *36*, 78–87. [CrossRef] [PubMed]

21. Gomez, A.M.; Ruiz-Hurtado, G.F.; Benitah, J.P.; Dominguez-Rodriguez, A. Ca^{2+} fluxes involvement in gene expression during cardiac hypertrophy. *Curr. Vasc. Pharmacol.* **2013**, *11*, 497–506. [CrossRef] [PubMed]

22. Janssen, L.J.; Farkas, L.; Rahman, T.; Kolb, M.R. ATP stimulates Ca^{2+}-waves and gene expression in cultured human pulmonary fibroblasts. *Int. J. Biochem. Cell Biol.* **2009**, *41*, 2477–2484. [CrossRef] [PubMed]

23. Mukherjee, S.; Duan, F.; Kolb, M.R.; Janssen, L.J. Platelet derived growth factor-evoked Ca^{2+} wave and matrix gene expression through phospholipase C in human pulmonary fibroblast. *Int. J. Biochem. Cell Biol.* **2013**, *45*, 1516–1524. [CrossRef] [PubMed]

24. Mukherjee, S.; Sheng, W.; Sun, R.; Janssen, L.J. Ca^{2+}/calmodulin-dependent protein kinase IIβ and IIdelta mediate TGFβ-induced transduction of fibronectin and collagen in human pulmonary fibroblasts. *Am. J. Physiol. Lung Cell. Mol. Physiol.* **2017**, *312*, L510–L519. [CrossRef] [PubMed]

25. Wu, P.; Ren, Y.; Ma, Y.; Wang, Y.; Jiang, H.; Chaudhari, S.; Davis, M.E.; Zuckerman, J.E.; Ma, R. Negative regulation of Smad1 pathway and collagen IV expression by store-operated Ca^{2+} entry in glomerular mesangial cells. *Am. J. Physiol. Ren. Physiol.* **2017**, *312*, F1090–F1100. [CrossRef] [PubMed]

26. Klenz, U.; Saleem, M.; Meyer, M.C.; Galla, H.J. Influence of lipid saturation grade and headgroup charge: A refined lung surfactant adsorption model. *Biophys. J.* **2008**, *95*, 699–709. [CrossRef] [PubMed]

27. Lee, A.G. How lipids affect the activities of integral membrane proteins. *Biochim. Biophys. Acta* **2004**, *1666*, 62–87. [CrossRef] [PubMed]

28. Xue, L.; Li, M.; Chen, T.; Sun, H.; Zhu, J.; Li, X.; Wu, F.; Wang, B.; Li, J.; Chen, Y. PE-induced apoptosis in SMMC7721 cells: Involvement of Erk and Stat signaling pathways. *Int. J. Mol. Med.* **2014**, *34*, 119–129. [CrossRef] [PubMed]

29. Shi, J.; Li, J.; Guan, H.; Cai, W.; Bai, X.; Fang, X.; Hu, X.; Wang, Y.; Wang, H.; Zheng, Z.; et al. Anti-fibrotic actions of interleukin-10 against hypertrophic scarring by activation of PI3K/AKT and STAT3 signaling pathways in scar-forming fibroblasts. *PLoS ONE* **2014**, *9*, e98228. [CrossRef] [PubMed]

30. Yang, Y.; Kim, H.J.; Woo, K.J.; Cho, D.; Bang, S.I. Lipo-PGE1 suppresses collagen production in human dermal fibroblasts via the ERK/Ets-1 signaling pathway. *PLoS ONE* **2017**, *12*, e0179614. [CrossRef] [PubMed]

31. Zuccolo, E.; Di, B.C.; Lodola, F.; Orecchioni, S.; Scarpellino, G.; Kheder, D.A.; Poletto, V.; Guerra, G.; Bertolini, F.; Balduini, A.; et al. Stromal cell-derived factor-1α promotes endothelial colony-forming cell migration through the Ca^{2+}-dependent activation of the extracellular signal-regulated kinase 1/2 and phosphoinositide 3-kinase/AKT pathways. *Stem Cells Dev.* **2018**, *27*, 23–34. [CrossRef] [PubMed]

32. Araki, T.; van Egmond, W.N.; van Haastert, P.J.; Williams, J.G. Dual regulation of a Dictyostelium STAT by cGMP and Ca^{2+} signaling. *J. Cell Sci.* **2010**, *123*, 837–841. [CrossRef] [PubMed]

33. Kolb, M.; Bonella, F.; Wollin, L. Therapeutic targets in idiopathic pulmonary fibrosis. *Respir. Med.* **2017**, *131*, 49–57. [CrossRef] [PubMed]

34. Loney, R.W.; Anyan, W.R.; Biswas, S.C.; Rananavare, S.B.; Hall, S.B. The accelerated late adsorption of pulmonary surfactant. *Langmuir* **2011**, *27*, 4857–4866. [CrossRef] [PubMed]

35. Jordanova, A.; Georgiev, G.A.; Alexandrov, S.; Todorov, R.; Lalchev, Z. Influence of surfactant protein C on the interfacial behavior of phosphatidylethanolamine monolayers. *Eur. Biophys. J.* **2009**, *38*, 369–379. [CrossRef] [PubMed]

36. Kaviratna, A.S.; Banerjee, R. Nanovesicle aerosols as surfactant therapy in lung injury. *Nanomedicine* **2012**, *8*, 665–672. [CrossRef] [PubMed]

37. Sun, Y.; Yang, R.; Zhong, J.G.; Fang, F.; Jiang, J.J.; Liu, M.Y.; Lu, J. Aerosolised surfactant generated by a novel noninvasive apparatus reduced acute lung injury in rats. *Crit. Care* **2009**, *13*, R31. [CrossRef] [PubMed]

38. Fattman, C.L.; Gambelli, F.; Hoyle, G.; Pitt, B.R.; Ortiz, L.A. Epithelial expression of TIMP-1 does not alter sensitivity to bleomycin-induced lung injury in C57BL/6 mice. *Am. J. Physiol. Lung Cell. Mol. Physiol.* **2008**, *294*, L572–L581. [CrossRef] [PubMed]

International Journal of
Molecular Sciences

MDPI

Article

Automated Intracellular Calcium Profiles Extraction from Endothelial Cells Using Digital Fluorescence Images

Marcial Sanchez-Tecuatl [1,*], **Ajelet Vargaz-Guadarrama** [2], **Juan Manuel Ramirez-Cortes** [1,*],
Pilar Gomez-Gil [3], **Francesco Moccia** [4] and **Roberto Berra-Romani** [2,*]

1   Electronics Department, National Institute of Astrophysics, Optics and Electronics, 72840 Puebla, Mexico
2   Biomedicine School, Benemérita Universidad Autónoma de Puebla, 72410 Puebla, Mexico;
    ajeletvargaz7@hotmail.com
3   Computer Science Department, National Institute of Astrophysics, Optics and Electronics,
    72840 Puebla, Mexico; pgomez@inaoep.mx
4   Department of Biology and Biotechnology "Lazzaro Spallanzani", University of Pavia, 27100 Pavia, Italy;
    francesco.moccia@unipv.it
*   Correspondence: marcial_st@hotmail.com (M.S.-T.); jmram@inaoep.mx (J.M.R.-C.);
    rberra001@hotmail.com (R.B.-R.)

Received: 1 September 2018; Accepted: 24 October 2018; Published: 2 November 2018

Abstract: Endothelial cells perform a wide variety of fundamental functions for the cardiovascular system, their proliferation and migration being strongly regulated by their intracellular calcium concentration. Hence it is extremely important to carefully measure endothelial calcium signals under different stimuli. A proposal to automate the intracellular calcium profiles extraction from fluorescence image sequences is presented. Digital image processing techniques were combined with a multi-target tracking approach supported by Kalman estimation. The system was tested with image sequences from two different stimuli. The first one was a chemical stimulus, that is, ATP, which caused small movements in the cells trajectories, thereby suggesting that the bath application of the agonist does not generate significant artifacts. The second one was a mechanical stimulus delivered by a glass microelectrode, which caused major changes in cell trajectories. The importance of the tracking block is evidenced since more accurate profiles were extracted, mainly for cells closest to the stimulated area. Two important contributions of this work are the automatic relocation of the region of interest assigned to the cells and the possibility of data extraction from big image sets in efficient and expedite way. The system may adapt to different kind of cell images and may allow the extraction of other useful features.

Keywords: intracellular calcium; endothelium; cell tracking; multi-target tracking; Kalman

---

1. Introduction

Vascular endothelium has long been considered as a homogeneous population of inert cells that forms a permeable barrier between flowing blood and surrounding tissues, regulating the exchange of nutrients, oxygen and catabolic waste at capillary level. This rather conventional view has recently given the way to a more realistic interpretation of the multifaceted role played by endothelial cells (ECs) within the cardiovascular system. The endothelium constitutes a multifunctional signal transduction surface that is now regarded as the largest endocrine organ of the body by virtue of its extension (about 350 m^2) and ability to release a broad spectrum of paracrine mediators [1]. For instance, ECs control the vascular tone, platelet activation and aggregation and angiogenesis. Endothelial dysfunction ultimately results in severe cardiovascular disorders, such as hypertension, atherosclerosis, coronary artery disease and stroke, which may cause patient's death [2,3].

Int. J. Mol. Sci. **2018**, *19*, 3440; doi:10.3390/ijms19113440   217   www.mdpi.com/journal/ijms

An increase in intracellular Ca^{2+} concentration ($[Ca^{2+}]_i$) represents one of the most important signaling pathways whereby ECs exert their sophisticated control of cardiovascular homeostasis [4–6]. For instance, extracellular autacoids, such as adenosine trisphosphate (ATP) and acetylcholine, cause a robust increase in endothelial $[Ca^{2+}]_i$ to recruit the Ca^{2+}/Calmodulin-dependent endothelial nitric oxide (NO) synthase (eNOS), thereby inducing NO release and vasodilation [7]. Likewise, endothelial Ca^{2+} signals drive the production of additional vasorelaxing mediators, such as prostacyclin I2 and hydrogen sulphide [8] and engage in Ca^{2+}-dependent intermediate and small-conductance K^+ channels to activate endothelial-dependent hyperpolarization (EDH) [9].

The intact endothelium of excised rat aorta provides an ideal preparation to investigate endothelial Ca^{2+} handling as the intracellular Ca^{2+} toolkit of native endothelium may differ from that described in cultured ECs [10–12]. This preparation, therefore, is suitable for both studying the Ca^{2+} response to physiological agonists, including ATP and acetylcholine [11,13,14] and investigating the Ca^{2+} response to mechanical injury that ultimately results in the process of endothelial wound healing [6,12,15]. Unraveling the signal transduction pathways engaged by physiological agonists and mechanical injury is essential to design novel therapeutic treatments aimed at restoring endothelial integrity and limiting cardiovascular diseases. Nevertheless, tracking endothelial Ca^{2+} signaling in their native microenvironment is a rather challenging task.

The ratiometric Ca^{2+}-sensitive fluorophore Fura-2 is the most suitable fluorescent dye to obtain quantitative data on $[Ca^{2+}]_i$ and to carry out long-lasting recording due to its little photobleaching [15,16]. It has been widely used to measure intracellular Ca^{2+} signals in intact endothelium of excised rat aorta [6,11–15] as well as of other vascular segments [17,18].

The term calcium profile (CP) indicates a time-dependent change in the mean fluorescence emitted by a Ca^{2+} indicator, such as Fura-2, within a user-defined area, also termed region of interest (ROI), manually drawn around a single cell. The CP extraction procedure consists in collecting the mean intensity within each ROI for every frame throughout a ratio image sequence. This procedure provides a meaningful profile of the Ca^{2+} signal as a function of time when the ROI surrounds only one cell nucleus, which appears brighter as it is thicker cell area. For a conventional experiment, the CP of multiple cells is extracted from ratio image sequences acquired for hundreds or even thousands of frames (acquisition frequency 0.3 Hz): manual ROI analysis is time consuming and labor intensive and may introduce artificial signals as the position and plane of focus of the cells may change radically during an experiment. This leads the users to manually relocate the ROIs frame per frame and one by one if needed. Consequently, the process becomes tedious, impractical and even more time consuming.

The Ca^{2+} response to ATP or acetylcholine in the intact endothelium of excised rat aorta consists in a biphasic increase in $[Ca^{2+}]_i$, which comprises a rapid Ca^{2+} transient followed by a plateau phase of intermediate amplitude between the resting Ca^{2+} levels and the initial Ca^{2+} response [11,13,14]. Similarly, the Ca^{2+} signal induced by mechanical injury consists in a rapid Ca^{2+} peak followed by a long-lasting decline towards the baseline in the majority ECs adjacent to the lesion site. The remaining fraction of cells in the first row also displayed periodic Ca^{2+} oscillations which overlapped the decay phase and were initiated after the initial Ca^{2+} spike. Furthermore, repetitive Ca^{2+} spikes and heterogeneous Ca^{2+} signals were detected also in ECs further back from the wound edge (see Supplementary Material in Reference [15]). As discussed elsewhere [6], injury-induced intra- and intercellular oscillations in $[Ca^{2+}]_i$ might play a key role in healing damaged endothelium by inducing surviving ECs to proliferate and migrate in order to cover the denuded membrane basement.

Despite the obvious importance of recording the Ca^{2+} signal in all the cells that lie beyond the injury edge, several technical difficulties prevented us from successfully accomplishing this task. The main technical difficulties were: (1) cell selection was performed by manual ROI drawing around the cells: in the typical microscope visual field of intact endothelium at 40× magnification, around 100–150 cells can be visualized, this resulted in a time consuming cell selection; (2) as endothelial borders on in situ endothelium were not clearly identifiable, a ROI might include part of the cytoplasm of adjacent cells, so a special attention should be taken in order to draw ROI very close to the nucleus area, increasing the ROI time selection and error possibilities; (3) as ECs of intact vessels are located over a thick layer of vascular smooth muscle that contract or relax in response to vasoactive substances released by the chemical or mechanical stimulated endothelium, this results in the movement of ECs during the experiment protocol in such a way that ROI must be repositioned manually throughout the experiment and according to the specific trajectory of each endothelial cell; (4) under the mechanical injury procedure, cells that are damaged from the glass micro-pipette tip (see Section 4. Materials and Methods), immediately loose Fura-2 fluorescence and disappear from the microscope field (here defined as OFF-LINE cells, in contrast to ON-LINE which are defined as the cells present); in addition, as shown in Reference [15], they are able to incorporate ethidium bromide, a fluorescent molecule unable to cross an intact plasma membrane and therefore indicative of damaged/dead cells. A wrong interpretation of data will occur if the CPs from ROIs that enclose those damaged/dead cells are mistakenly taken in to account. It should also point out that the simple response to Ca^{2+}-releasing agonists, such as ATP and acetylcholine, is hampered by the fact that bath application of these agonists stimulates the underlying layer of vascular smooth muscle cells (VMSCs). The following contraction of the muscular component of the artery wall causes many ECs to transiently disappear during some frames because of movement artefacts, thereby emitting less fluorescence and introducing a false negative in the Ca^{2+} signal.

In this work, a proposal to automate the CP extraction process is presented. Taking advantage from the slow cell movement, a tracking scheme based on intersections was implemented, which proved to be a simple and fast solution to avoid the movement artefact of the endothelial monolayer in situ caused by either physiological (due to VSMC contraction) of mechanical (due to the direct damage of the micro-pipette tip) stimulation. Although overlap-based tracking is a straightforward solution in many cases [19,20], problems related to partially missing data require the use of complementary approaches for robustness improvement. In this work, a Kalman filter is incorporated into the proposed system as a computational tool to keep track of cells in movement even when they momentarily vanish. Kalman estimation has been successfully used for object tracking in several applications [21–23]. The position, velocity and acceleration of the cells are estimated in every iteration to be able to predict the position of the current cell in the case it has disappeared.

2. Results

The intracellular CPs are extracted with the aim of modeling the endothelium behavior under different stimuli. With that purpose in mind, the application was tested with several image sequences obtained from two principal types of experiments: Endothelium under chemical (ATP) stimulus and Endothelium under mechanical stimulus. First, we will illustrate the procedure that we exploited to automate CP extraction in in situ ECs and then get into the details of this novel algorithm.

Online Supplementary Video 1 shows the typical experimental procedure to automate CP extraction in in situ ECs that undergo a mechanical injury imposed by the tip of a glass micro-pipette, as described in Materials and Methods. The upper left movie shows the inner wall of the aortic rings loaded with the Ca^{2+}-sensitive fluorophore, Fura-2, which was fully covered by fluorescent ECs. At frame 0, an example of the adaptive segmentation outcome is shown, bordering all ECs visualized in the microscope field. For illustrative proposes, a single EC was selected and a orange ROI was automatically drawn around it in order to extract its CP and tracking its frame to frame position during the experiment (see Video 1 at frame 3). The CP was simultaneously measured and illustrated in the bottom part of the movie. Scraping the lumen of the aortic ring with the tip of a glass micro-pipette (approximately at frame 63) caused a rapid Ca^{2+} peak following by a long-lasting decline to the baseline in the ECs adjacent to the injury site, such as the EC enclosed by the orange ROI. As it can be observed on the "Position graph (upper right movie)" when the injury occurs, a drastic position change takes places for the cell surrounded by the tracking ROI. However, despite the clear movement of the EC, the automated data processing procedure enables the ROI to remain positioned over the cell of interest, which leads to a careful measurement of the Ca^{2+} signal. In Figure 1A, a typical CP obtained by the automated tracking system is shown, where the arrow indicates when mechanical injury was performed.

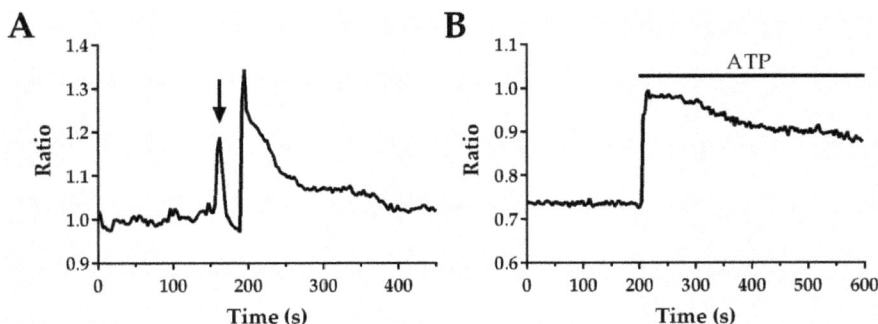

Figure 1. Ca^{2+} profiles from in situ endothelial cells obtained under two different stimuli (**A**) Damaging rat aorta inner wall by using the tip of a glass micro-pipette evokes a long-lasting Ca^{2+} signal in endothelial cells (ECs) located at the injury site. The arrow indicates the time when injury was performed; (**B**) Intracellular Ca^{2+} transient evoked by ATP 300 (μM) in in situ ECs from rat aorta. Calcium profile (CPs) were obtained from one single cell using the automated CP extraction method.

Online Supplementary Video 2 shows a similar procedure used to track an EC in its native microenvironment stimulated by the application of 300 μM of ATP at frame 63. Under this experimental condition, a smallest cell movement was observed. The CP obtained by using the automated CP extraction algorithm is showed in Figure 1B. In the following paragraphs, we will discuss how our Automatic Tracking procedure enables to record the true Ca^{2+} signals induced by the two distinct stimuli without missing any relevant information.

The blocks that will be presented in the Materials and Methods section were merged to conform a functional, easy and user-friendly software application. It is worth pointing out that the proposed blocks are the fundamental components of this proposal to automate the methodology for $[Ca^{2+}]_i$ profiles extraction. To achieve the functionality of the software application, these components were connected with control flows as shown in the high-level flowchart, provided in Flowchart file as Supplementary Material. Programming provides certain degree of freedom, therefore, the implementation of the blocks and their integration into a system may be enhanced as required. The application was built using the Matlab® GUIDE.

2.1. ATP Stimulus

This is a chemical stimulus that has been delivered as described in Materials and Methods. The dimensions of the images are 640 × 480 pixels (px) and they have *. TIFF format. Application of 300 μM ATP to in situ endothelium from rat aorta is able to evoke a biphasic increase in $[Ca^{2+}]_i$, as shown in Figure 2 right (blue line, data obtained with Automatic Tracking approach). However, ECs undergo a little displacement and intensity variations when exposed to bath application of the agonist. For comparison purposes, fifteen cells were selected randomly and their intracellular CPs were obtained using the proposed system and ImageJ program, manually repositioning the ROI every time the cell moves outside of it (ImageJ-Supervised Tracking). The mean square error (MSE) between both obtained data sets is shown in Table 1. Please note that MSE is insignificant compared with the mean of its respective CP extracted with our automated approach ($\overline{CP}$). Although small movements were registered in the cells trajectory, the kinetics of CPs are very similar and the MSE is small, as was expected. This observation demonstrates that this approach, which consists in the simple bath application of the agonist, does not generate significant artifacts and does not induce significant VSMC contraction. It should, however, be pointed out that the magnitude of the initial peak, which is due to Ca^{2+} release from the endoplasmic release, is significantly larger than the amplitude measured without tracking (red line). Cyan line, represent the CP obtained by using ImageJ-Supervised Tracking. In this particular experiment, the single ROI was repositioned at least 3 times along the experiment. As expected, the CP obtained by ImageJ-Supervised Tracking method, overlaps with the CP obtained with our automated approach. This observation reinforces the notion that automatic tracking is recommendable to speed off-line analysis and to gather all the information content of the Ca^{2+} signal.

Table 1. Mean of CPs obtained using the proposed approach ($\overline{CP}$) and the mean square error (MSE) between CPs extracted using ImageJ-Supervised Tracking and our automatic tracking proposal. Adenosine triphosphate (ATP) stimulus experiment.

Cell	$\overline{CP}$	MSE	Cell	$\overline{CP}$	MSE
1	0.9098	1.6242×10^{-4}	9	0.8851	2.9395×10^{-5}
2	1.0111	4.4606×10^{-4}	10	0.8607	5.5402×10^{-5}
3	0.9515	6.0467×10^{-5}	11	0.8728	7.9154×10^{-5}
4	0.9490	4.7925×10^{-5}	12	0.8504	1.1613×10^{-4}
5	0.8713	1.2525×10^{-5}	13	0.8498	1.3611×10^{-5}
6	0.9013	1.7247×10^{-5}	14	0.8648	8.0574×10^{-5}
7	0.8699	2.6627×10^{-5}	15	0.8851	2.2170×10^{-5}
8	0.8481	3.4207×10^{-5}	-	-	-

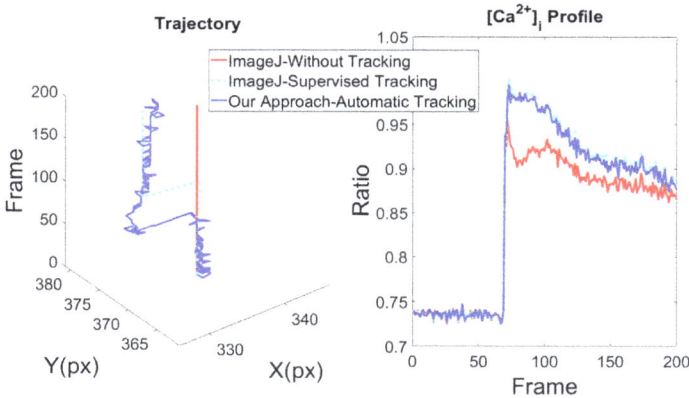

Figure 2. (**Left**), shows the cell trajectory, cell position in X and Y axes respectively, measured in pixels (px) and compared between ImageJ without tracking (red line), ImageJ with manually tracking (cyan line) and trajectory obtained with our automated tracking proposed system (blue line); (**Right**), shows an example of CPs calculated in each condition previously described.

In order to carry on performance evaluation of the system, the density of tracked cells per frame ρ was proposed as figure of merit. The indicator represents an instantaneous measurement of the performance at a given frame i and it is defined in (1).

$$\rho_i = 100 \times \frac{\text{Number of ONLINE cells at frame } i}{\text{Number of binary objects at frame } i} \tag{1}$$

The fluorescence of ECs may decrease or be lost, temporarily or definitively, along a given experiment. A parameter defined as reliability limit (RL) was introduced to provide to the user the possibility to control how many times a cell is allowed to disappear consecutively from the microscope field before it is considered as unreliable to be measured and marked with OFF-LINE status. Health science experts are able to set RL based on their expertise. Usage model of RL is detailed in Section 4.4.

Image sequences of two different ATP experiments with 200 images per excitation wavelength were analyzed. All the detected cells were selected to test the performance of this approach with several reliability limits (RL). The average results are shown in Table 2 and the average number of detected and selected cells was 190.

Table 2. Average performance results with three different reliability limit (RL) values.

RL	ρ_{max} (%)	ρ_{min} (%)	ρ_{mean} (%)
3	104.3956	67.1378	84.8415
5	104.3956	76.4045	89.4309
7	104.3956	79.7753	92.3110

It can be seen that ρmax exceeds the 100%, which is due to the fact that a few cells disappear in some frames thereby causing a mismatch between a given number of cells and a higher number of ROIs than cells. On the other hand, the big amount of noise in some frames produces more binary blobs than ROIs.

2.2. Mechanical Stimulus

In this experiment, a region of the tissue is damaged with the tip of a glass micro-pipette. Detailed experimental procedure is described in Section 4. Materials and Methods. The image sequences are in *.TIFF format with a dimension of 768 × 576 pixels.

Three zones were considered in the image as shown in the Figure 3a and three cells were selected per zone. Zone 2 represents the area onto which the mechanical lesion is imposed, whereas zones 1 and 3 represent the most proximal regions to the damaged site. The fluorescence and position of cells in zones 1 and 3 display a more stable behavior than cells in the zone 2. Some cells of the inner zone (red rectangle on Figure 3a,b) die and then disappear, please see the transition from Figure 3a to Figure 3b; this is reflected as an OFF-LINE status (i.e., blue A001 and A002 ROIs).

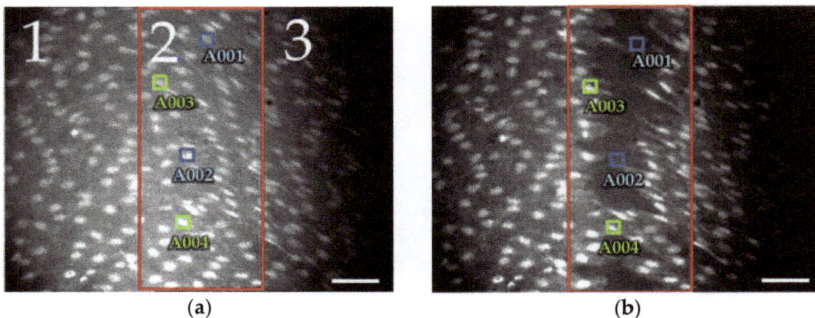

(a) (b)

Figure 3. Images from mechanical stimulus experiment. (**a**) Before and (**b**) after the tissue injury. Suggested zones for the analysis (1, 2 and 3). Blue ROIs, A001 and A002 in (**a**) shown cells in the injured zone that were damaged by the tip of glass micro-pipette and lost Fura-2 fluorescence and then die and disappeared ((**b**) blue ROIs, A001 and A002); these ROIs were marked as OFF-LINE cells. Green ROIs, A003 and A004, enclosed cells adjacent to the injured area, which have partially lost the contact with neighboring ECs, but preserved Fura-2 fluorescence after injury ((**b**) green ROIs, A003 and A004); considered as ON-LINE cells. Bar length is 40 μm.

As the cells moves slightly in the outer zones (1 and 3), it can be seen that the intracellular CPs obtained with ImageJ and with the proposed approach are very similar (Figure 4a). The results obtained from zone 2 show a more notable difference in the structure of the CPs and in the trajectory of the cells (Figure 4b).

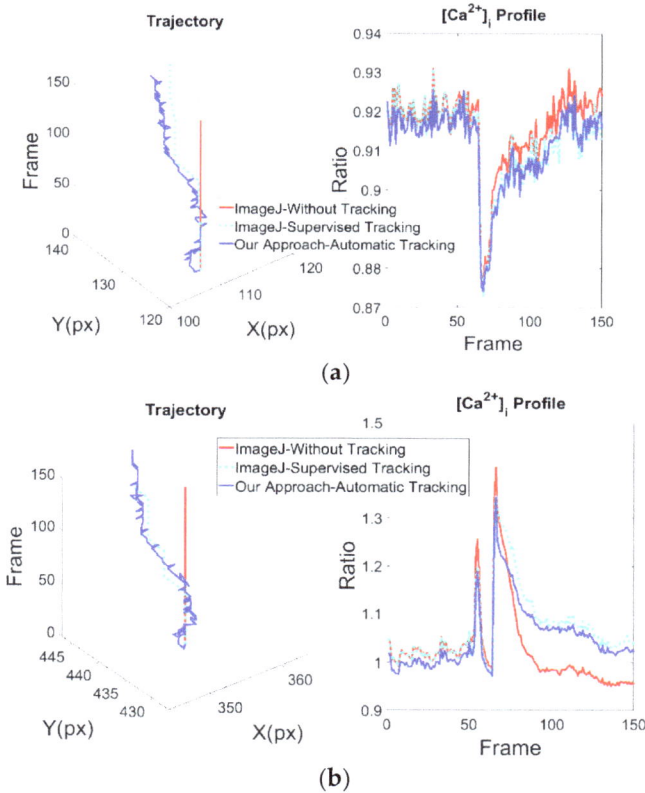

(a)

(b)

Figure 4. Results per defined zones due to mechanical stimuli: (a) Results obtained from zones 1 and 3; (b) Results obtained from zone 2. (**Left**), shows trajectories of cells, i.e., calculated position on X and Y axes respectively, measured in pixels (px). (**Right**), shows typical recordings of CP analyzed by: ImageJ without tracking (red line); with manually tracking (cyan line) and automated tracking proposed system (blue line).

Based on the mean squared difference of the results obtained from zones 1 and 3, indicated in Table 3, the proposed scheme delivers consistent results as shown in the previous subsection. While the results of the zone 2 highlights the importance of tracking block since the structure of CPs changes dramatically, it is noteworthy that there are patterns of intracellular CPs associated to different physiological functions. It turns out that a better profile extraction may help a better understanding of cellular behavior [15]. Accordingly, the CP of zone 2 provided by the proposed approach shows a clearly discernible plateau following the initial peak that is missed by ImageJ without tracking. This signal, therefore, reflects the true dynamics of the Ca^{2+} response induced by mechanical stimulation in ECs directly exposed to mechanical injury. The difference in the waveform of the Ca^{2+} signal is rather important, as the Ca^{2+} plateau has long been known to recruit a number of

Ca^{2+}-sensitive decoders involved in cell proliferation and migration, which are the crucial steps for the regeneration process.

As for ATP experiments, two mechanical stimulus experiments with 200 frames per wavelength were used to test the approach. The average results are in Table 4 where the average number of selected cells was 283. As expected, compared to the results of ATP experiments, a lower performance can be which is caused by the more random and more noticeable movement of the cells.

Table 3. Mean of CPs obtained using the proposed approach ($\overline{CP}$) and the mean square error (MSE) between CPs extracted using ImageJ-Supervised Tracking and our automatic tracking proposal. Mechanical stimulus experiment.

Zone	Cell	$\overline{CP}$	MSE
1	1	0.8991	3.5821×10^{-6}
	2	0.8765	4.5569×10^{-6}
	3	0.8937	3.2133×10^{-5}
2	4	1.0551	1.6675×10^{-3}
	5	1.0483	4.2171×10^{-4}
	6	1.0488	9.2146×10^{-4}
3	7	0.9941	3.9861×10^{-6}
	8	0.9676	2.2033×10^{-5}
	9	1.0074	8.7559×10^{-5}

Table 4. Average performance results with different RLs.

RL	ρ_{max} (%)	ρ_{min} (%)	ρ_{mean} (%)
3	100.7117	58.0756	69.9652
5	101.0714	59.7938	73.1079
7	101.0714	61.1684	75.1659

3. Discussion

A multi-target tracking system to automate the intracellular CP extraction from fluorescence image sequences was designed and implemented. The approach was based on logical intersections supported by Kalman estimation which was fed with a constant acceleration model.

A figure of merit was proposed to measure the performance and the system was tested using several image sequences from two different types of experiments, one of them (i.e., ATP experiment) showed the extracted data consistency and the other one (i.e., mechanical lesion) showed the actual need of the cell tracking.

The results allow us to conclude that an important contribution of this work is the automatic relocation of the regions of interest based on statistical measurements which favors massive data extraction and time saving during the analysis. Some appropriate technological improvements of the acquisition systems, for example, faster sample rate, combined with this approach may ensure a meaningful performance improvement, that is, more frames equals more intersections. Finally, the simplicity and efficiency of the proposed approach would allow a real-time application.

In conclusion, this novel algorithm provides an additional tool to accelerate the analysis of the hundreds of endothelial Ca^{2+} signals arising within the intact wall of excised vessels in response to physiological stimuli and to uncover the real dynamics of Ca^{2+} waves at the edge of the lesion site. This approach will be exploited to get more insights into injury-induced intra- and intercellular Ca^{2+} waves, that could in turn lead to the identification of alternative targets to prevent endothelial dysfunction induced by the implantation of medical devices, such as stents and angioplasty, or turbulent blood flow.

4. Materials and Methods

4.1. Dissection of the Aorta

Male Wistar rats aged 2–3 months were provided by the Benemérita Universidad Autónoma de Puebla animal core facilities, where they were kept under conditions of constant environmental temperature and exposed to dark light cycles of 12 h, with water and food *ad libitum*. All the experimental procedures on animals were performed according to protocols approved by the University Animal Care and Use Committee, identification code: BERRSAL71, 18-05-2017. The procedure for obtaining in situ ECs from rat aorta, was developed by Berra-Romani and Cols., 2004 [24] and it is briefly described below. The day of the experiment rats were sacrificed with an overdose of intraperitoneal ketamine-xylazine solution, 0.2 mL per 100 g of weight; subsequently they underwent to an anterior thoracotomy to expose the aortic arch and the heart. The heart was extracted, to cannulate and perfuse the aorta with a physiological salt solution (PSS). Thoracic and abdominal aorta were dissected out and placed in a Petri dish with PSS at room temperature. Using a stereomicroscope (Nikon SMZ-2T), the connective and fatty tissues surrounding the aorta were removed. Subsequently the aorta was cut into ~5 mm wide rings, stored in PSS at room temperature (22–24 °C) and used within 5 h. Using a microdissection scissors, the aortic rings were carefully cut to open and care was taken to avoid any damage to the endothelium and obtain aortic strips with intact endothelium.

4.2. [Ca²⁺]ᵢ Measurements

The aortic strips with intact endothelium were bathed in PSS with 16 µmol Fura-2/AM for 60 min at room temperature, washed for 30 min to allow intracellular de-esterification of Fura-2/AM and fixed (with the luminal face up) to the bottom of a Petri dish covered by inert silicone (Silgard® 184 Silicone Elastomer, Down Corning, MI, USA) by using four 0.4 mm diameter stainless steel pins. In situ ECs were visualized by an upright epifluorescence Axiolab microscope (Carl Zeiss, Oberkochen, Germany), equipped with a Zeiss 40× Achroplan objective (water-immersion, 2.05 mm working distance, 1.0 numerical aperture). To monitor the $[Ca^{2+}]_i$, ECs were excited alternately at 340 and 380 nm and the emitted light was detected at 510 nm. A neutral density filter (optical density = 1.0) was coupled to the 380 nm filter to approach the intensity of the 340 nm light. A round diaphragm was used to increase the contrast. A filter wheel (Lambda 10, Sutter Instrument, Novato, CA, USA) commanded by a computer positioned alternately along the optical path the two filters that allowed the passage of light respectively at 340 and 380 nm. Custom software, working in the LINUX environment, was used to drive the camera (Extended-ISIS Camera, Photonic Science, Millham, UK) and the filter wheel and to measure and plot in real time (on-line) the fluorescence from ~80–100 drawn manually rectangular "regions of interest" (ROI) enclosing one single cell. $[Ca^{2+}]_i$ was monitored by measuring, for each ROI, the ratio of the mean fluorescence emitted at 510 nm when exciting alternatively at 340 and 380 nm (shortly termed "ratio"). An increase in $[Ca^{2+}]_i$ causes an increase in the ratio. Ratio measurements were performed and plotted on-line every 3 s. The on-line procedure allowed monitoring of $[Ca^{2+}]_i$ at the same time as micro-pipette injury. The fluorescence images obtained at 340 nm and 380 nm were stored on the hard disk and data analysis was carried out later as described in data analysis section. The experiments were performed at room temperature (20–22 °C) to limit time-dependent decreases in the intensity of the fluorescence signal.

4.3. ATP Stimulation of in Situ Endothelial Cells

Medium exchange and administration of ATP (300 µM) was carried out by first removing the bathing medium (2 mL) by a suction pump and then adding the desired solution. The medium could be substituted quickly without producing artifacts in the fluorescence signal because a small meniscus of liquid remained between the tip of the objective and the facing surface of the aorta.

4.4. Mechanical Disruption of in Situ Endothelial Cells

As shown in Reference [15], aortic endothelium was injured under microscopic control by means of a glass micro-pipette with a tip of about 30 μm diameter, driven by an XYZ hydraulic micromanipulator (Narishige Scientific Instrument Lab., Tokyo, Japan). The electrodes were fabricated using a flaming-brown micropipette puller P-1000 (Sutter Instruments, Novato, CA, USA). Images of Fura-2-loaded ECs, together with numbered ROI, were taken before the lesion, to identify the cells facing the injury site. The microelectrode was first positioned almost parallel and very near to the endothelium surface. Then it was moved downward, along the z-axis, until the electrode tip gently touched the endothelium and moved horizontally across the visual field to scrape 1–3 consecutive rows of ECs.

4.4.1. Chemicals

Fura-2/AM was obtained from Molecular Probes (Molecular Probes Europe BV, Leiden, The Netherlands). All other chemicals were purchased from Sigma.

4.4.2. Solution

The composition of the PSS expressed in mmol/L was: NaCl (140), KCl (4.7), MgCl2 (1.2), Glucose (6), CaCl2 (2.5) and HEPES (5). Solution was titrated to pH 7.4 with NaOH.

4.4.3. Data Analysis

Endothelial Ca^{2+} imaging data were processed using an automated data processing procedure. The data processing procedure approach is represented in the block diagram shown in Figure 5. As the first step, the fluorescence images obtained at 340 and 380 nm are separately loaded; thereafter, a calibration stage is inserted to determine the parameters used by the image segmentation.

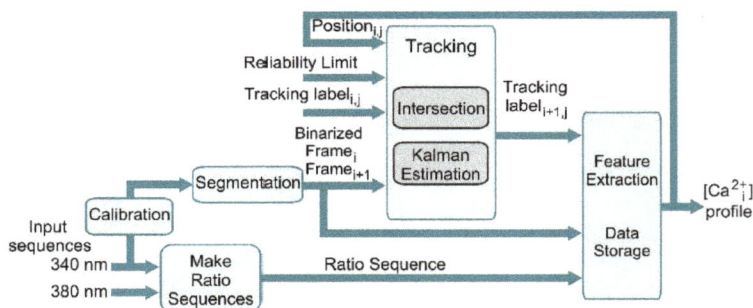

Figure 5. Block diagram of the proposed solution.

The sequence of ratio images is concurrently generated (Make Ratio Sequences). A noise reduction and an adaptive thresholding processes are performed over an image sequence. The resulting binary images are taken as the basis for the tracking scheme based on logical intersections; the tracking scheme is, in turn, enhanced with state estimation by using a Kalman filter. Feature extraction of the ROI and the storage of useful data are performed in the last sub-block.

4.5. Ratio Sequence Block

Fura-2 indicator is excited with $\lambda = 340$ nm and $\lambda = 380$ nm wavelengths and its fluorescent emissions are both monitored at $\lambda = 510$ nm; therefore, two image sequences are generated as experimental data, that is, an image sequence per excitation wavelength.

In this module the pixel to pixel ratio, $\lambda = 340$ nm over $\lambda = 380$ nm excitation wavelength images, is carried out for each frame to generate the Ratio image sequence. At this point, there are three image sequences available: two 8-bit resolution (uint8) *.TIFF and one float type image sequences. The used arrays have $m \times n \times k$ dimensions, where $m = 576$ and $n = 768$ are the images height and width, respectively and k is the number of images per sequence. It is noteworthy to remember that CP are intensity measurements extracted from the ratio images.

As stated above, Fura-2 has a maximum fluorescence when it is excited with 340 nm wavelength, so the contrast of these images is usually better than the contrast of images obtained with 380 nm excitation. Therefore, the 340 nm image sequence was taken as the basis for the calibration, segmentation, tracking and feature extraction blocks. To guarantee the integrity of Ca^{2+} measurements, the 340 nm image sequence is modified only right after the ratio sequence was generated. The 380 nm image sequence was used only to generate the ratiometric image sequence.

4.6. Calibration Block

In this stage, the user manually selects several sample cells, ROIs, from a $\lambda = 340$ nm frame. Then the average of the selected ROIs areas, defined as μ_{Areas}, is used to calculate the window size w to be applied on the next stages, that is, filtering and binarization.

As this is based on area statistics, the more ROIs are selected the better w approximation can be computed. This stage allows the system to be adapted to different cell sizes. Based on the assumption that selected cells are bounded by $w \times w$ squares, the proposed window size is defined as $w = [a \times \sqrt{\mu_{Areas}}]$, where a is scalar.

A set of images was manually segmented by health science experts, aka ground truth segmentation. The same images were also segmented and conditioned automatically with algorithm presented in Section 4.7 and different a values. Then the mean square error between corresponding couples of images was calculated. To simplify the calculation, the binarized images were converted to double data type to use the mmse Matlab built in function. As shown in Figure 6, the minimum mean square error was obtained when $a = 2$.

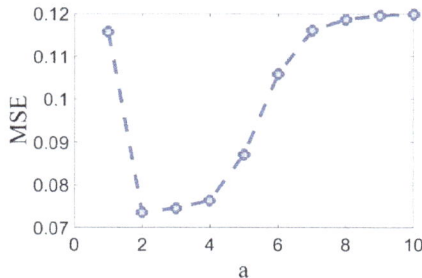

Figure 6. Mean square error (MSE) of binarization and ground truth segmentation for different a value.

4.7. Segmentation

In order to reduce noise, preserve the sharp high frequency detail and enhance the contrast of the images, a median filter in cascade with linear histogram stretching was used before the thresholding process. The median filter square window size was set to $w_{median} = w/8$.

Adaptive binarization based on local statistics was taken as a solution to work with non-uniform illumination images. Since the efficiency of integral image representation has been proven for mean and variance calculation [21], it was used to implement the Niblack technique [22], whose threshold is defined in (2).

$$T(x,y) = m(x,y) + k_T s(x,y) \tag{2}$$

where m and s stand for local mean and local standard deviation, respectively; the scalar gain was experimentally set to $k_T = 1$.

The last conditioning sub stage was composed of binary image opening with disk of radius 3 as structural element; then, the holes in binary blobs were filled up and finally a dilation with radius 1 disk was applied.

4.8. Tracking Block

The purpose of this block is to follow the selected cells through the image sequences, this task is performed using labels associated to the cells, termed tracking labels. A cell tracking scheme based on intersections was implemented, which proved to be a simple and fast solution, with the drawback of missing cells in subsequent images due to several conditions. In consequence, cell tracking was enhanced using Kalman estimation. The goal of health science experts is to understand the behavior of vascular endothelium, through its Ca^{2+} signals and to discover its regeneration mechanisms under some conditions and stimuli. When endothelium is stimulated, many cells transiently disappear during some frames because of movement artefacts, thereby emitting less fluorescence. In this situation, it is very probable to detect no intersection. Kalman filtering has been successfully used for object tracking in several applications [21–23]. In this work, a Kalman filter is incorporated into the proposed system as a computational tool to keep track of cells in movement even when they momentarily vanish. The position, velocity and acceleration of the cells are estimated in every iteration to be able to predict the position of the current cell in the case it has disappeared.

A counter, per cell, called *reliability counter* is compared with RL to determine the cell status, that is, ONLINE or OFF-LINE. When a cell is selected by the user, it is surrounded by a ROI and the software application sets ONLINE status as an initial condition. Only ONLINE cells are tracked to avoid loss of performance. As of this point, when the *position* of an object is mentioned, it is referring to its *centroid*.

4.8.1. Logical Intersections Approach

As tracking block is based on logical intersections, that is and operations, it works iteratively with two frames of binarized and labeled images, f_{n-1} and f_n. The main goal is to identify the selected cells from f_{n-1} to f_n through the correspondence of the labels, which is termed track_label. A kernel is obtained for each selected cell with the intersection of the current cell with the f_n. When intersection detects only one object, the centroid of that object is obtained in order to get the tracking label in f_n. When more than one or no object is detected, the prediction equations of Kalman filter are used to estimate the position and to find the new label for the current cell in f_n. This process is illustrated in Figure 7.

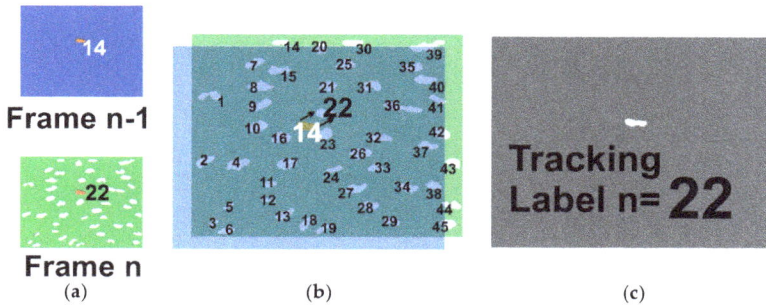

Figure 7. The intersection process to track cells, from frame n-1 to frame n, is illustrated. (**a**) The kernel of a selected cell with track_label = 14 in frame $n - 1$ (blue background and orange highlighted) and the binarized and labeled frame n are the input data, notice that the corresponding label for selected cell will be 22; (**b**) Kernel and binary image are overlapped to show the intersection process, which is a logical and operation between them; (**c**) The outcome is a new kernel, in this case only one object was generated by the intersection; based on its position in frame n the new track_label value is updated to 22. This iterative process is performed per each selected cell (ROI) and each frame.

Once the new tracking label is found, a block termed *store data* is responsible of storing the position and intensity measurement for each cell and resets the reliability counter. If no label is detected with intersections neither with Kalman estimation, it is assumed that the cell disappeared in that frame and the reliability counter value is increased by one. When this counter is equal to the user defined RL, the tracking is considered as unreliable and the status of the current cell is changed to OFFLINE and its new tracking label is no longer searched. The block is initialized with $f_{n-1} = 1$ and $f_n = 2$ and finishes with $f_{n-1} = n - 1$ to $f_n = k$.

4.8.2. Kalman Estimator

Although the cell motion is relatively slow, its randomness entails that the tracking based on intersection may detect in the kernel more than one or none object due to the uncertainty in the target trajectory or acceleration at any given time. Since it is extremely complex to find an exact mathematical model to describe the motion of cells, for tracking purposes, it was assumed that cells have a constant acceleration motion. The deterministic description for two dimensions in state space model is shown in (3).

$$
\begin{bmatrix} x \\ \dot{x} \\ \ddot{x} \\ y \\ \dot{y} \\ \ddot{y} \end{bmatrix}_{n+1} =
\begin{bmatrix}
1 & T & \frac{T^2}{2} & 0 & 0 & 0 \\
0 & 1 & T & 0 & 0 & 0 \\
0 & 0 & 1 & 0 & 0 & 0 \\
0 & 0 & 0 & 1 & T & \frac{T^2}{2} \\
0 & 0 & 0 & 0 & 1 & T \\
0 & 0 & 0 & 0 & 0 & 1
\end{bmatrix}
\begin{bmatrix} x \\ \dot{x} \\ \ddot{x} \\ y \\ \dot{y} \\ \ddot{y} \end{bmatrix}_n +
\begin{bmatrix} \frac{T^3 J_x}{3} \\ \frac{T^2 J_x}{2} \\ T J_x \\ \frac{T^3 J_y}{3} \\ \frac{T^2 J_y}{2} \\ T J_y \end{bmatrix}_n
$$

$$
z = \begin{bmatrix}
1 & 0 & 0 & 0 & 0 & 0 \\
0 & 0 & 0 & 1 & 0 & 0
\end{bmatrix}
\begin{bmatrix} x \\ \dot{x} \\ \ddot{x} \\ y \\ \dot{y} \\ \ddot{y} \end{bmatrix}_n
\tag{3}
$$

where J_n is a random change in acceleration, a.k.a. jerk, occurring between time n and $n + 1$. The random jerk J_n has the auto correlation function given by (4) and it is characterized as white noise.

$$J_n \bar{J}_m = \begin{cases} \sigma_J^2 & \text{for } n = m \\ 0 & \text{for } n \neq m \end{cases} \tag{4}$$

With the sampling period of the experimental data, $T = 3$, the observability matrix has full rank, that is, rank = 6, therefore the proposed system model is observable and the Kalman filter can work correctly. Assuming statistical independence between the coordinates x and y, the system covariance matrix is shown in (5) and the observation error covariance matrix in (6).

$$Q = q^T q$$

$$q = \begin{bmatrix} \frac{T^3 \sigma_{Jx}}{3} & \frac{T^2 \sigma_{Jx}}{2} & T\sigma_{Jx} & \frac{T^3 \sigma_{Jy}}{3} & \frac{T^2 \sigma_{Jy}}{2} & T\sigma_{Jy} \end{bmatrix} \tag{5}$$

$$R = \begin{bmatrix} \sigma_x^2 & 0 \\ 0 & \sigma_y^2 \end{bmatrix} \tag{6}$$

4.8.3. Statistical Parameters

The parameters σ_{Jx} and σ_{Jy} belongs to the random nature of the system, for this reason, they were extracted from experimental data. A number $c = 15$ of cells were randomly selected and manually tracked, through $p = 100$ frames, to register their vector positions, x and y. Three numerical derivatives were performed in order to get the jerk vector for each dimension, then the standard deviation of each jerk vector was computed. Finally, σ_{Jx} is the average of those σ_{Jxc}, see (7) and (8). The same process was done for σ_{Jy}.

$$\sigma_{Jx_c}^2 = \frac{1}{p-1} \sum_{i=1}^{p} \left(J x_i - \bar{J}_x \right)^2 \tag{7}$$

$$\sigma_{Jx} = \frac{1}{c} \sum_{n=1}^{c} \sigma_{Jx_n} \tag{8}$$

The state estimation with prediction and correction is performed in every iteration once the current cell has been found with the aim of updating the second and third order derivatives because a single prediction is used only when intersection scheme fails to find the selected cell. For the observation error covariance matrix, the standard deviations were set in such a way that the estimates were very close to measurements and that the influence of Q was incorporated to avoid singularities in the calculus. The parameters are shown in Table 5.

Table 5. Statistical parameters for Kalman filter.

Parameter	σ_{Jx}	σ_{Jy}	σ_x	σ_y
Value	0.0269	0.0993	0.001	0.001

A set of artificial images was generated in order to test the Kalman filter estimation. The image sequence was designed to simulate the trajectory of a cell, the coordinates of the cell were determined mathematically as $x = c$ and $y = A\cos(2\pi f t)$, where c is a constant or a bias, $f = 4$, A was adjusted to produce a 36 pixels peak to peak amplitude and t is a vector with the same number of elements than x. The area of the object in the images was the average cells area obtained from the experimental data.

The estimation was obtained with Kalman filter configured with the statistical parameters described above. In Figure 8 it can be seen the trajectory estimation using the whole scheme predictor–corrector. As some cells disappear during some frames, statistical measurements for Kalman correction are not available along the entire image sequence. Thus, several missed points were used as a focused test of the Kalman prediction, that is, with no evaluation of correction equations.

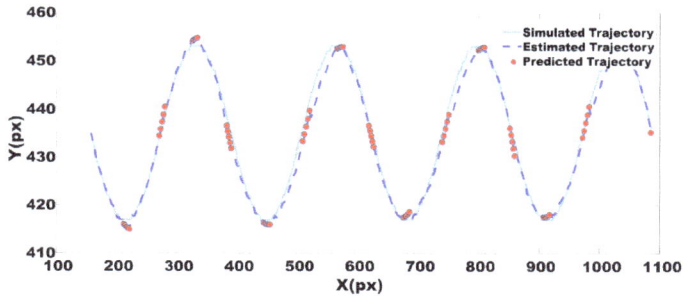

Figure 8. Kalman filter testing. Simulated, estimated and predicted trajectories in solid, dashed and dotted lines, respectively. Position in X and Y axes respectively, measured in pixels (px).

The mean square error between trajectories in both cases is reported in Table 6. By using the centroid as the position of the cell, the reported error is tolerable since it falls inside the average ROI area and it is possible to find again the selected cell. Thus, the Kalman filter has a good performance.

Table 6. Mean square error between trajectories.

Estimator Configuration	MSE_x	MSE_y
Prediction-Correction	9.6537	0.8958
Only Prediction	10.5347	0.8929

Tracking results with RL = 3 are shown in Figure 9. It can be seen that the value of reliability counter of cell "A007" increased to three, from frame 1 to 4, then its status changed to OFF-LINE and the cell "A007" is not tracked anymore as of frame 5.

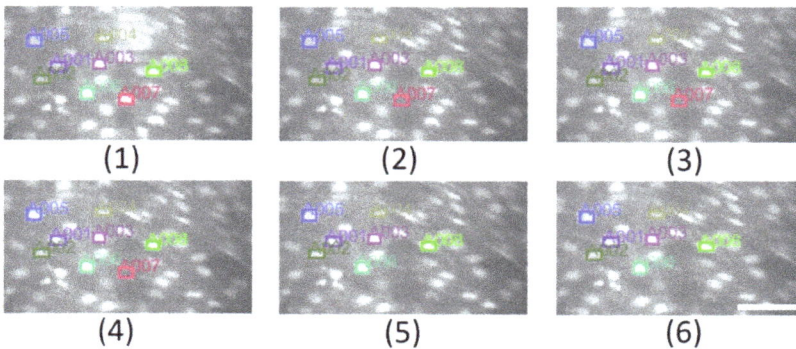

Figure 9. Tracking results through six consecutive frames. Eight cells were selected by the user, only ONLINE cells are labeled, ROI surrounded and tracked. OFFLINE cells are not tracked as shown with cell A007; the cell is visible at frame (**1**), then it vanishes as of frame (**2**) and its reliability counter increases by one between frames (**2–4**). At frame (**4**) its reliability counter is equal to RL = 3, then it is tagged as OFFLINE and is not tracked as of frame (**5**). Frame (**6**) shows the tracking of remaining ONLINE cells. Bar length is 40 μm.

4.9. Feature Extraction and Data Storage

This is the last block of the proposed solution. The block receives as input the *new tracking label*. The cell under investigation is identified with that label, the ROI is relocated and the intensity average is computed from the respective ratio frame, that is, the CP is extracted. In this stage, more features can be extracted from the binarized image. Online Supplementary Video 3 shows the analysis complete flow on the software application developed with the approach described throughout this manuscript. This system has, therefore, the potential to help health experts to expand their studies by greatly limiting the movement artefacts due to the manipulation of the extracellular solution or due the delivery of a mechanical stimulus.

Supplementary Materials: The supplementary materials are available online at www.mdpi.com/1422-0067/19/11/3440/s1.

Author Contributions: Conceptualization, M.S.-T., J.M.R.-C. and P.G.-G.; Data curation, A.V.-G., F.M. and R.B.-R.; Investigation, M.S.-T., J.M.R.-C., F.M. and R.B.-R.; Methodology, M.S.-T., J.M.R.-C. and R.B.-R.; Software, M.S.-T.; Supervision, J.M.R.-C., P.G.-G. and R.B.-R.; Validation, M.S.-T., A.V.-G., J.M.R.-C. and R.B.-R.; Writing—original draft, M.S.-T.; Writing—review & editing, J.M.R.-C., F.M. and R.B.-R.

Funding: This research received no external funding.

Conflicts of Interest: The authors declare no conflict of interest.

Abbreviations

ATP	Adenosine trisphosphate
CP	Calcium profile
Ca^{2+}	Divalent calcium ion
$[Ca^{2+}]_i$	Intracellular Ca^{2+} concentration
ECs	Endothelial cells
EDH	Endothelial-dependent hyperpolarization
NO	Nitric oxide
eNOS	Nitric oxide synthase
PSS	Physiological salt solution
ROI	Region of interest
RL	Reliability limit
VSMC	Vascular smooth muscle cell
MSE	Mean square error

References

1. Pries, A.R.; Secomb, T.W.; Gaehtgens, P. The endothelial surface layer. *Pflug. Arch.* **2000**, *440*, 653–666. [CrossRef] [PubMed]
2. Ayaori, M.; Iwakami, N.; Uto-Kondo, H.; Sato, H.; Sasaki, M.; Komatsu, T.; Iizuka, M.; Takiguchi, S.; Yakushiji, E.; Nakaya, K.; et al. Dipeptidyl peptidase-4 inhibitors attenuate endothelial function as evaluated by flow-mediated vasodilatation in type 2 diabetic patients. *J. Am. Heart Assoc.* **2013**, *2*, 7–10. [CrossRef] [PubMed]
3. Balakumar, P.; Sharma, R.; Kalia, A.N.; Singh, M. Hyperuricemia: Is it a risk factor for vascular endothelial dysfunction and associated cardiovascular disorders? *Curr. Hypertens. Rev.* **2009**, *5*, 1–6. [CrossRef]
4. Dupont, G.; Croisier, H. Spatiotemporal organization of Ca^{2+} dynamics: A modeling-based approach. *HFSP J.* **2010**, *4*, 43–51. [CrossRef] [PubMed]
5. Moccia, F.; Berra-Romani, R.; Tanzi, F. Update on vascular endothelial Ca^{2+} signalling: A tale of ion channels, pumps and transporters. *World J. Biol. Chem.* **2012**, *3*, 127. [CrossRef] [PubMed]
6. Moccia, F.; Tanzi, F.; Munaron, L. Endothelial Remodelling and Intracellular Calcium Machinery. *Curr. Mol. Med.* **2014**, *14*, 457–480. [CrossRef] [PubMed]
7. Berra-Romani, R.; Avelino-Cruz, J.E.; Raqeeb, A.; Della Corte, A.; Cinelli, M.; Montagnani, S.; Guerra, G.; Moccia, F.; Tanzi, F. Ca^{2+}-dependent nitric oxide release in the injured endothelium of excised rat aorta: A

promising mechanism applying in vascular prosthetic devices in aging patients. *BMC Surg.* **2013**, *13*, S40. [CrossRef] [PubMed]

8. Potenza, D.M.; Guerra, G.; Avanzato, D.; Poletto, V.; Pareek, S.; Guido, D.; Gallanti, A.; Rosti, V.; Munaron, L.; Tanzi, F.; et al. Hydrogen sulphide triggers VEGF-induced intracellular Ca^{2+} signals in human endothelial cells but not in their immature progenitors. *Cell Calcium* **2014**, *56*, 225–234. [CrossRef] [PubMed]

9. Behringer, E.J. Calcium and electrical signaling in arterial endothelial tubes: New insights into cellular physiology and cardiovascular function. *Microcirculation* **2017**, *24*, e12328. [CrossRef] [PubMed]

10. Bondarenko, A. Sodium-calcium exchanger contributes to membrane hyperpolarization of intact endothelial cells from rat aorta during acetylcholine stimulation. *Br. J. Pharmacol.* **2004**, *143*, 9–18. [CrossRef] [PubMed]

11. Usachev, Y.M.; Marchenko, S.M.; Sage, S.O. Cytosolic calcium concentration in resting and stimulated endothelium of excised intact rat aorta. *J. Physiol.* **1995**, *489*, 309–317. [CrossRef] [PubMed]

12. Berra-Romani, R.; Raqeeb, A.; Avelino-Cruz, J.E.; Moccia, F.; Oldani, A.; Speroni, F.; Taglietti, V.; Tanzi, F. Ca^{2+} signaling in injured in situ endothelium of rat aorta. *Cell Calcium* **2008**, *44*, 298–309. [CrossRef] [PubMed]

13. Berra-Romani, R.; Raqeeb, A.; Guzman-Silva, A.; Torres-Jácome, J.; Tanzi, F.; Moccia, F. Na^{+}-Ca^{2+} exchanger contributes to Ca^{2+} extrusion in ATP-stimulated endothelium of intact rat aorta. *Biochem. Biophys. Res. Commun.* **2010**, *395*, 126–130. [CrossRef] [PubMed]

14. Huang, T.-Y.; Chu, T.-F.; Chen, H.-I.; Jen, C.J. Heterogeneity of $[Ca^{2+}]_i$ signaling in intact rat aortic endothelium. *FASEB J.* **2000**, *14*, 797–804. [CrossRef] [PubMed]

15. Berra-Romani, R.; Raqeeb, A.; Torres-Jácome, J.; Guzman-Silva, A.; Guerra, G.; Tanzi, F.; Moccia, F. The mechanism of injury-induced intracellular calcium concentration oscillations in the endothelium of excised rat aorta. *J. Vasc. Res.* **2011**, *49*, 65–76. [CrossRef] [PubMed]

16. Bootman, M.D.; Rietdorf, K.; Collins, T.; Walker, S.; Sanderson, M. Ca^{2+}-Sensitive Fluorescent Dyes and Intracellular Ca^{2+} Imaging. *Cold Spring Harb. Protoc.* **2013**, *2013*. [CrossRef] [PubMed]

17. Yeon, S.I.; Kim, J.Y.; Yeon, D.S.; Abramowitz, J.; Birnbaumer, L.; Muallem, S.; Lee, Y.H. Transient receptor potential canonical type 3 channels control the vascular contractility of mouse mesenteric arteries. *PLoS ONE* **2014**, *9*, 17–19. [CrossRef] [PubMed]

18. Laskey, R.E.; Adams, D.J.; Van Breemen, C. Cytosolic $[Ca^{2+}]$ measurements in endothelium of rabbit cardiac valves using imaging fluorescence microscopy. *Am. J. Physiol. Heart Circ. Physiol.* **1994**, *266*, H2130–H2135. [CrossRef] [PubMed]

19. Čehovin, L.; Leonardis, A.; Kristan, M. Visual object tracking performance measures revisited. *IEEE Trans. Image Process.* **2015**, *25*, 1261–1274. [CrossRef]

20. Huang, C.H.; Boyer, E.; Navab, N.; Ilic, S. Human shape and pose tracking using keyframes. In Proceedings of the IEEE Conference on Computer Vision and Pattern Recognition (CVPR), Columbus, OH, USA, 24–27 June 2014; pp. 3446–3453. [CrossRef]

21. Shantaiya, S.; Verma, K.; Mehta, K. Multiple Object Tracking using Kalman Filter and Optical Flow. *Eur. J. Adv. Eng. Technol.* **2015**, *2*, 34–39.

22. Qiang, Y.; Lee, J.Y.; Bartenschlager, R.; Rohr, K.; Feld, I.N. Colocalization analysis and particle tracking on multi-channel fluorescence microscopy images. In Proceedings of the 2017 IEEE 14th International Symposium on Biomedical Imaging (ISBI 2017), Melbourne, VIC, Australia, 18–21 April 2017; pp. 646–649.

23. Ojha, S.; Sakhare, S. Image processing techniques for object tracking in video surveillance—A survey. In Proceedings of the 2015 International Conference on Pervasive Computing (ICPC), Pune, India, 8–10 January 2015; pp. 1–6. [CrossRef]

24. Berra-Romani, R.; Rinaldi, C.; Raqeeb, A.; Castelli, L.; Magistretti, J.; Taglietti, V.; Tanzi, F. The duration and amplitude of the plateau phase of ATP- and ADP-Evoked Ca^{2+} signals are modulated by ectonucleotidases in in situ endothelial cells of rat aorta. *J. Vasc. Res.* **2004**, *41*, 166–173. [CrossRef] [PubMed]

International Journal of
Molecular Sciences

Review

The Different Facets of Extracellular Calcium Sensors: Old and New Concepts in Calcium-Sensing Receptor Signalling and Pharmacology

Andrea Gerbino * and Matilde Colella *

Department of Biosciences, Biotechnology and Biopharmaceutics, University of Bari, 70121 Bari, Italy;
andrea.gerbino@uniba.it (A.G.); matilde.colella@uniba.it (M.C.);
Tel.: +39-080-544-3334 (A.G.); +39-080-544-3028 (M.C.)

Received: 2 March 2018; Accepted: 25 March 2018; Published: 27 March 2018

Abstract: The current interest of the scientific community for research in the field of calcium sensing in general and on the calcium-sensing Receptor (CaR) in particular is demonstrated by the still increasing number of papers published on this topic. The extracellular calcium-sensing receptor is the best-known G-protein-coupled receptor (GPCR) able to sense external Ca^{2+} changes. Widely recognized as a fundamental player in systemic Ca^{2+} homeostasis, the CaR is ubiquitously expressed in the human body where it activates multiple signalling pathways. In this review, old and new notions regarding the mechanisms by which extracellular Ca^{2+} microdomains are created and the tools available to measure them are analyzed. After a survey of the main signalling pathways triggered by the CaR, a special attention is reserved for the emerging concepts regarding CaR function in the heart, CaR trafficking and pharmacology. Finally, an overview on other Ca^{2+} sensors is provided.

Keywords: Ca^{2+}-signalling; microelectrode; fluorophore; exocytosis; cAMP signalling; parathyroid extracellular-Ca^{2+}-sensing receptor (CaR); cancer; apoptosis; ischemia/reperfusion; hypertrophy; heart; cardiomyocytes

1. Introduction

Over the years, the intracellular Ca^{2+} signal has emerged as the most intensely studied second messenger pathway [1]. Ca^{2+} is the most versatile intracellular messenger, not only in the cytoplasm but also in diverse subcellular compartments, such as mitochondria [2], nuclei [3], lysosomes [4], and the endoplasmic reticulum [5]. The cloning of the extracellular Calcium-sensing receptor (CaR) 25 years ago by Brown et al. [6] paved the way for a heightened understanding of the mechanisms used by Ca^{2+} to signal in the microenvironments outside the cell. Thus, Ca^{2+} acts as a multipurpose messenger, exerting effects at the cellular, subcellular, and extracellular levels. A very interesting and comprehensive list of reviews on the CaR has been recently published by leading experts in the field of extracellular Calcium-sensing and signalling [7]. Other key topics, such as the structure-function relationships [8] and role of the CaR in cancer [9], have also been very recently covered elsewhere and will be named here in the corresponding paragraphs. The readers are thus referred to these fine reviews for a complete understanding of the CaR's function in cell physiology and pathology [10].

The present review will examine the potential conditions under which extracellular Ca^{2+} concentration ($[Ca^{2+}]$) may change in the local microenvironment outside the cell, with a particular emphasis on the experimental approaches utilized to measure ex vivo and in vitro such Ca^{2+} fluctuations. These extracellular Ca^{2+} changes are sensed by the CaR (and other Ca^{2+} sensors) and transduced by means of different signalling pathways that modulate organ-specific physiological processes. Over the last few years, new concepts have emerged regarding the CaR's role in cancer and cardiac physiology and pathology. Also, exciting news have come into view in the fields of CaR signalling, trafficking, and pharmacology. In addition, it is now clear that a number of different Ca^{2+} sensors share important tasks with the CaR. Therefore, the aim of this review is to summarize for the reader of *International Journal of Molecular Science* these new and exciting notions.

2. Extracellular Ca^{2+} Microdomains

Serum-ionized Ca^{2+} concentration (~1.4 mM) is usually precisely regulated by a number of cooperative organs, including parathyroid glands, bones, kidney, and intestine [11]. However, this tight modulation does not prevent local changes in extracellular Ca^{2+}, as those have been identified in the microscopic volume of interstitial fluid surrounding cells of many tissues [12]. This scenario has become increasingly important because the amplitude and the shape of these extracellular Ca^{2+} fluctuations represent an autocrine/paracrine cell-to-cell communication form based on the activity of extracellular Ca^{2+} sensors.

Many factors have been proposed in the onset of such significant extracellular Ca^{2+} changes. At first, we will evaluate how the activity of the cells, in terms of intracellular Ca^{2+} signalling events, can be rapidly translated in asymmetrical Ca^{2+} changes. Then, we will analyze how the exocytosis of Ca^{2+}-loaded vesicles and the synchronous opening of voltage-gated Ca^{2+} conductances perturb Ca^{2+} levels in the close proximity of a cell.

When closely packed to form a functional tissue, cells delimit small diffusion spaces. Within this context, cells themselves represent major sources and/or sinks for Ca^{2+} during intracellular Ca^{2+} signalling events as a result of the activation of Ca^{2+} efflux (e.g., by the plasma membrane Ca^{2+}-ATPase and/or Na+/Ca^{2+} exchanger) and Ca^{2+} influx (e.g., by store-operated channels, SOCs) pathways at the plasma membrane, respectively. Differential dynamics or a polarized distribution of these Ca^{2+} influx/efflux mechanisms [13–16] are required for the onset of significant extracellular Ca^{2+} microdomains. Our group showed that stimulation with carbachol, a Ca^{2+}-mediated agonist, resulted in asymmetrical fluctuations in local extracellular Ca^{2+} levels in a polarized epithelium.

Using Ca^{2+}-selective microelectrodes, we measured a substantial plasma membrane Ca^{2+}-ATPase (PMCA)-dependent local increase in the extracellular Ca^{2+} concentration at the apical face and a similar store-operated channel (SOC)-dependent depletion at the basolateral side of amphibian acid-secreting cells (Figure 1). Thus, in intact epithelia, asymmetrical changes in extracellular Ca^{2+} reflect the polarized distribution of Ca^{2+}-handling mechanisms in the plasma membrane [17]. Of note, PMCA is highly expressed at the apical membrane of these cells where it shares a strategic localization with the extracellular CaR [17]. In fact, these carbachol-induced changes in extracellular Ca^{2+} act as paracrine/autocrine messages sufficient and essential to strictly regulate tissue-specific functions (e.g., alkaline secretion, pepsinogen secretion, and water flux) independently of intracellular $[Ca^{2+}]$ changes [18,19].

Secretory granules are characterized by very high intraluminal Ca^{2+} concentrations [20–22]. Among others, insulin granules of rat insulinoma contain ionized Ca^{2+} in a range between 60 and 120 mM [23]. Thus, it does not come as a surprise that exocytotic events can significantly change extracellular Ca^{2+} in the microenvironment immediately surrounding the cells. Recently, we showed using Ca^{2+}-microelectrodes that stimulation of insulin secretion by high glucose and other secretagogues (glibenclamide, forskolin/IBMX, and ATP) induced a delayed elevation of extracellular Ca^{2+} within rat insulinoma (INS-1E) β-cell pseudoislets [24]. The exocytotic extrusion of Ca^{2+} through Ca^{2+}-loaded vesicles released in the extracellular environment has also been demonstrated in different cell types, such as salivary gland cells [25], bovine adrenal medullary cells [26], neurohypophyseal nerve endings [27], and sea urchin eggs [28].

Int. J. Mol. Sci. **2018**, *19*, 999

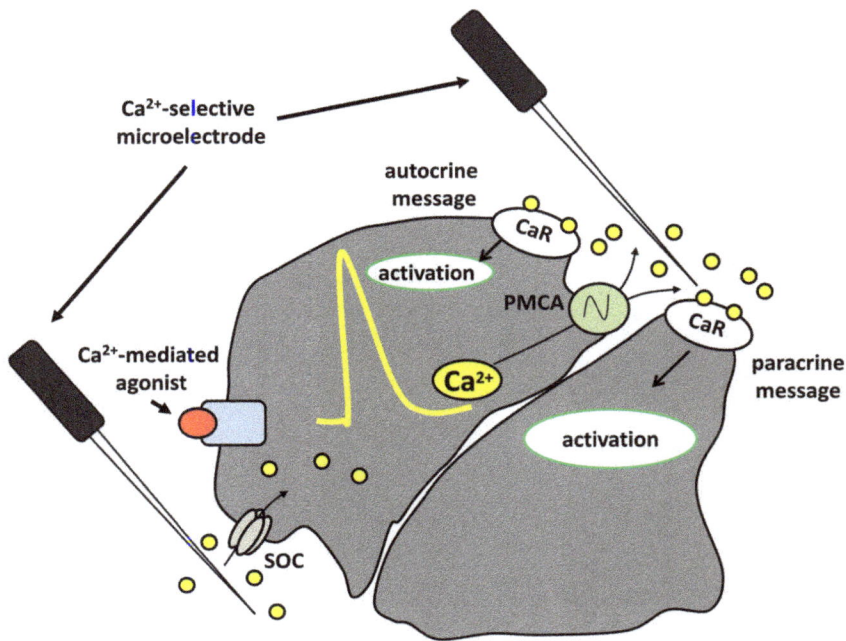

Figure 1. Schematic representation of intercellular communication mediated by the calcium-sensing Receptor (CaR) in polarized epithelial cells. When a Ca^{2+}-mediated agonist (red circle) binds a specific receptor, Ca^{2+} concentration (yellow line) increases within the cytosol. Extrusion of Ca^{2+} via the plasma membrane Ca^{2+}-ATPase (PMCA), localized at the apical membrane, might activate the CaR expressed on cells in close proximity (paracrine message) or on the same cell (autocrine message). The asymmetrical changes in extracellular Ca^{2+} are the result of the polarized localization of PMCA and store-operated channels (SOCs) at the apical and basolateral membranes, respectively. These fluctuations can be recorded with Ca^{2+}-sensitive microelectrodes. Modified from [29].

In addition to the above-mentioned mechanisms, excitable cells express a different kind of voltage-gated Ca^{2+} conductances that, during their physiological activity, might influence extracellular Ca^{2+} levels [30–32]. It has been demonstrated that, in the central nervous system, the synchronized opening of voltage-gated Ca^{2+} channels (VGCCs) elicited significant reductions in extracellular Ca^{2+} levels [33,34]. We and others have shown that in mouse islets of Langerhans and INS-1E pseudoislets, glucose stimulation depleted the concentration of Ca^{2+} in the extracellular milieu (of about 500 μM) as a consequence of Ca^{2+} influx through a VGCC at the plasma membrane [24,35,36]. Transient reductions were also recorded in rat neocortex [33] and in the cardiac muscle [37].

3. Measuring Fluctuations in Extracellular Ca^{2+} Levels

Real-time measurement of extracellular Ca^{2+} changes in the proximity of the plasma membrane (e.g., intercellular spaces) is a technically challenging experimental maneuver. In fact, researchers must deal on the one side with the lack of proper tools to directly access these restricted compartments, and on the other side with the quantification of small Ca^{2+} changes (few hundreds micromolar) over a high extracellular Ca^{2+} background (physiologically around 1.1–1.4 mM). Over the years, many different experimental approaches to measure extracellular Ca^{2+} fluctuations have been proposed in a number of different tissue preparations, although each one of them has its own drawbacks in terms of either sensitivity or spatial resolution.

Microelectrode-based electrophysiological approaches have been applied to diverse cell and tissue types to monitor extracellular Ca^{2+} changes. Early studies directly measured changes in extracellular Ca^{2+} concentration in the rat cerebellum using double-barreled Ca^{2+}-selective micropipettes. During repetitive stimulation of the central nervous system, extracellular Ca^{2+} was significantly reduced, indicating that Ca^{2+} modulation in brain microenvironments may be a significant parameter in the behavior of neuronal ensembles [38].

The vibrating probe technique uses a low-resistance Ca^{2+}-sensitive electrode vibrating in proximity of the surface of cells or tissues in order to measure small extracellular Ca^{2+} gradients (for more technical details refer to [39,40]). Self-referencing, low-resistance vibrating Ca^{2+} microelectrodes were shown to non-invasively quantify, in real time, local extracellular ion gradients with high sensitivity and square micron spatial resolution. Ca^{2+}-selective self-referencing electrodes have been used extensively to show Ca^{2+} flux at different regions of hair cells, cells of the zona pellucida, and the central nervous system [41–44].

As described in the previous section, we extensively used Ca^{2+}-selective microelectrodes to directly record the profile of agonist-induced changes in extracellular $[Ca^{2+}]$ in the restricted domain of both the intact amphibian gastric mucosa [17] and in the microenvironment immediately surrounding INS-1E cells organized as pancreatic pseudoislets [24].

Finally, Mupanomunda and colleagues used in situ microdialysis techniques in anesthetized Wistar rats to monitor extracellular $[Ca^{2+}]$ in the sub-epithelial spaces of the intestine [45] and kidney [46]. In these studies, an increased load of $[Ca^{2+}]$ in the gut lumen or perturbations in whole-animal Ca^{2+} homeostasis were shown to cause significant changes in the external $[Ca^{2+}]$ underlying the epithelium of these tissues.

Ca^{2+}-sensitive dyes have shown extensive utility in the study of the spatio-temporal dynamics of Ca^{2+} fluctuations. These tools, compared with electrophysiological techniques, are more sensitive and provide for a better time resolution and more reliable access to limited spaces. However, these approaches are performed in free extracellular Ca^{2+} (thus, non-physiological conditions) in order to avoid saturation of the fluorophores.

Ca^{2+}-sensitive metallochromic indicators (antipyrylazo III and tetramethylmurexide) were used to estimate extracellular Ca^{2+} dynamics in intact beating hearts. These indicators revealed significant depletion of external Ca^{2+} following the synchronous opening of a Ca^{2+} influx pathway during contraction [37,47,48]. Later on, Tepikin and Petersen introduced the droplet technique [49–51], an elegant approach to measure extracellular Ca^{2+} changes elicited by the activity of the PMCA. This method employs Fluo-3 under Ca^{2+}-free media conditions to quantify extracellular Ca^{2+} changes in small clusters of exocrine gland cells contained in a small droplet of solution (20–50 times the volume of the cells). Extracellular Ca^{2+} was also measured in cell clusters digested from exocrine glands (pancreas, submandibular gland) and placed in a small experimental chamber containing Ca^{2+} green-1 bound to high molecular weight dextran beads in Ca^{2+}-free solution [25,52].

Ca^{2+}-sensitive probes conjugated to a hydrophobic tail have been created to cross the external leaflet of the cell membrane and directly measure extracellular Ca^{2+} levels. A classic example is Ca^{2+}-Green-C-18, the fluorophore conjugated with a lipid derivative that has been successfully used to record Ca^{2+} extrusion and fluctuations $^{2+}$inside the T-tubules [53]. In smooth muscle cells, the Ca^{2+}-sensitive indicator Fura-C18 was used to measure extracellular Ca^{2+} changes near the plasma membrane. The same dye was used to compare the extracellular Ca^{2+} changes recorded with signals measured from the bulk cytosol [54,55]. In addition, Fura-C18 has been used to directly measure near-membrane extracellular $[Ca^{2+}]$ changes in HEK-293 cells, showing that Ca^{2+} extruded to the extracellular space by PMCA, as a consequence of intracellular Ca^{2+} events, activates a positive feedback through the CaR [56].

4. The Extracellular Calcium-Sensing Receptor (CaR)

4.1. Discovery and Cloning of the Extracellular Calcium-Sensing Receptor

The first hints about the existence of an extracellular Ca^{2+} sensor come from a very early study (1966) correlating serum Ca^{2+} concentration and parathyroid hormone (PTH) secretion in whole animals [57,58]. About 10 years later, further evidences corroborated such a correlation in isolated glands and dispersed parathyroid cells [11,59–64].

The direct link between extracellular $[Ca^{2+}]$ changes, PTH secretion, and intracellular Ca^{2+} signals was finally demonstrated by Nemeth and Scarpa [65], while the actual molecular identification of the Ca^{2+} sensor had to wait until 1993, when Brown and colleagues cloned a peculiar G protein coupled receptor (GPCR) from the bovine parathyroid glands [6] and named it "extracellular Calcium-sensing receptor". This receptor, here named CaR, is also referred to as CaSR, after its gene name, or more recently CaS, as recommended by NC-IUPHAR [66].

4.2. Structural Features of the CaR

The extracellular calcium-sensing receptor belongs to class C (or 3) of the GPCR superfamily, which consists of eight metabotropic glutamate receptors (mGlu1-8), two type B metabotropic-aminobutyric acid receptors (GABA$_{B1}$ and GABA$_{B2}$), three taste receptors (T1R1-3), a promiscuous L-α-amino acid (GPRC6A) receptor [67], and seven orphan receptors [68]. The primary structure of human CaR consists of four main parts: a large N-terminal extracellular domain (ECD) (612 amino acids) [69], a cysteine-rich domain, a seven-transmembrane domain (TMD, 250 amino acids), and a 216-residue carboxyterminal (C)-tail [70].

The ECD contains a large bilobed, nutrient-binding Venus Flytrap (VFT) domain related to bacterial periplasmic binding proteins which has been crystallized in several members of GPCR class C (mGlu1 [71], mGlu3 and mGlu7 [72], and the GABA$_{B2}$ subunit [73]). It seems that upon Ca^{2+} (or other agonists) binding, the open cleft of the VFT closes inducing, in turn, conformational changes in both the TMD and the intracellular domains that are believed to initiate signal transduction [74,75].

The functional, highly glycosilated CaR proteins reside on the cell membrane as homodimers [76,77] or hetero-dimers with other class C receptors (mGluR [78] or GABA$_B$ [79]). While non-covalent interactions across the dimer interface [80] are essential for dimer formation, the intermolecular disulfide between two conserved cysteines of the VFTs (C129 and C131 [81–83]) seems not to influence the formation of dimer but to stabilize it [84]. Critical for CaR expression and activity seem to be other three intra-molecular disulfides formed by cysteines in the ECD of the CaR, including C60–C101, C358–C395, and C437–C449 [85]. More recently, Brauner-Osborn showed that one allosteric site per CaR dimer is sufficient for achieving the modulatory effect of positive allosteric modulators, while prevention of activation in both 7TM domains is needed in order to obtain complete inhibition by negative allosteric modulators [86].

The nine-cysteine structure domain (found in all members of the GPCR family C but the GABA B receptor) serves as a linker between the ECD and TMD and is essential for the CaR signalling capacity as demonstrated by the consequences of its removal [87]. The intracellular domain of the CaR shows a low degree of conservation among species [88] except for residues located at the beginning of the C-tail (863–925) that are crucial for surface expression [89] and those (960–984) involved in the interaction with downstream proteins [8,90,91]. As for other GPCRs, CaR activity can be regulated via phosphorylation by different protein kinases, such as PKC [76], which plays a significant role in the negative regulation of the CaR's activity and the generation of CaR-induced intracellular Ca^{2+} oscillations. On the other hand, PKA [92] is involved in the intracellular retention of the receptor [93]. For detailed reviews on the structure of the CaR and new insights on ligand binding pockets, see [8,70,94]. The structural requirements for processes such as bias and allosteric modulation of the CaR have been analyzed by recent works [95]. In addition, a number of papers have highlighted the importance of the potential Ca^{2+} binding sites and their relevance for associated diseases [8,96–102].

Recently, the first high-resolution crystal structure of the ECD of the human CaR bound with Mg^{2+} has been proposed [102]. In this work, the authors found a high-affinity tryptophan derivative in the crystal structure of the CaR, which seems to play a role in potentiating the function of the receptor. The proposed CaR co-activation operated by extracellular Ca^{2+} and amino acids might have a wide impact for the regulation of the receptor in districts of our body that are constantly exposed to both CaR agonists, such as the intestine. An exhaustive review about the recent insights regarding the structure and functional cooperativity of the CaR (and other members of cGPCRs) in health and disease has been published by the group of Jenny Yang [103].

4.3. CaR Promiscuity: Many Orthosteric Agonists and Allosteric Modulators Concur in Modulating CaR Function at Multiple Sites

Although, as the name states, the main ligands of the CaR are extracellular Ca^{2+} ions, which bind to the receptor with a high positive cooperativity, the CaR is also able to respond to other extracellular stimuli (see Table 1), such as inorganic di- and tri-valent cations, including Sr^{2+}, Ba^{2+}, Gd^{3+}, Al^{3+} [11,104], various organic polycations, such as the polyamines spermine and spermidine [105], and basic polypeptides, including polylysine, polyarginine [11,106], and amyloid β-peptides [107]. The CaR has also been shown to be activated by some aminoglycoside antibiotics, such as neomycin and gentamycin [108,109]. These ligands, referred to as type I calcimimetics or orthosteric agonists, can stimulate the receptor also in the absence of Ca^{2+} or any other ligands.

Table 1. Orthosteric agonists and allosteric modulators of the calcium-sensing receptor (modified from [110]).

Orthosteric Agonists (Type I Calcimimetics)	Class Members
A. Cations	High potency: Gd^{3+}; Eu^{3+}; Tb^{3+} Intermediate potency: Zn^{2+}; Ni^{2+}; Cd^{2+}; Pb^{2+}; Co^{2+}; Fe^{2+} Low potency: Ca^{2+}; Mg^{2+}; Ba^{2+}; Sr^{2+}; Mn^{2+}
B. Polyamines	spermine, spermidine, putrescine
C. Aminoglycoside antibiotics	neomycin, gentamycin, tobramycin, poromomycin, kanamycin, ribostamycin
D. Basic polypeptides	Poly-L-arginine, poly-L-lysine, protamine, amyloid β-peptides
Allosteric Modulators (Type II Calcimimetics)	
A. L-amino acids	phenylalanine, tryptophan, tyrosine, histidine
B. Glutathione analogs	γ-glutamyl-tripeptides: glutathione, S-methylglutathione, S-propylglutathione, γ-glutamyl-tripeptides (γ-Glu-Ala, γ-Glu-Cys)
C. Small molecule calcimimetics	The first generation: NPS R-568, NPS R-467, AMG 073, AMG 416 The second generation: cinacalcet The third generation: dibenzylamine calcimimetics, R,R-calcimimetic B, AC-265347
D. Small molecule calcilytics	NPS 2143, Calhex 231, ATF396, AXT914, ronacaleret, NPSP795, SB-423557, SB-423562

The other class of CaR agonists is represented by type II calcimimetics (or allosteric modulators), composed by ligands which bind to a site different from that of the orthosteric agonists, thus altering the receptor conformation and, as a consequence, the affinity and/or the signaling capacity of the orthosteric agonist. This action can be exerted in a positive (calcimimetics) or a negative direction (calcilytics).

Belonging to this class of CaR agonists are natural L-amino acids, especially aromatic [111–113] and glutathione analogs [114,115], which synergize with extracellular Ca^{2+}. The receptor's activity is also negatively modulated by low extracellular pH [116] and high ionic strength [117].

A detailed review on the binding sites of orthosteric and allosteric agonists of the CaR has been recently published by Zhang and colleagues [8].

Among the allosteric modulators of the CaR, there are drugs with a high therapeutical potential: first, second, and third generation calcimimetics [118–121].

After the first generation of calcimimetics, which is represented by phenylalkylamine derivatives, such as NPS R-568 and NPS R-467 [122], the second generation reached the clinic: cinacalcet is the only CaR modulator currently used in clinical practice to reduce parathyroid glands activity in patients under dialysis for end-stage renal disease [123–125]. The third generation calcimimetics, which comprise the dibenzylamine calcimimetic, *R,R*-calcimimetic B, and the structurally distinct calcimimetic AC-265347, appear to have enhanced tissue-selective effects on PTH and calcitonin release in rats [126,127] most probably via biased allosteric modulation of the CaR's activity [128] (see below).

Allosteric antagonists or calcilytics, such as NPS-2143 and Calhex 231, antagonize the stimulatory effects of extracellular Ca^{2+} and are currently in clinical trials for treating osteoporosis [129].

In addition to affecting the signaling capacity of the CaR, allosteric CaR modulators have been shown more recently to act as pharmacochaperones that rescue cell surface expression of mutant CaRs that are retained intracellularly (see the paragraphs below).

4.4. Trimeric G Proteins and Signalling Pathways Activated by the CaR

Besides its ability to sense changes in multiple physiological parameters via interaction with different ligands, the CaR has also been shown to activate a wide array of different intracellular pathways (Figure 2) (for comprehensive reviews see [130,131]).

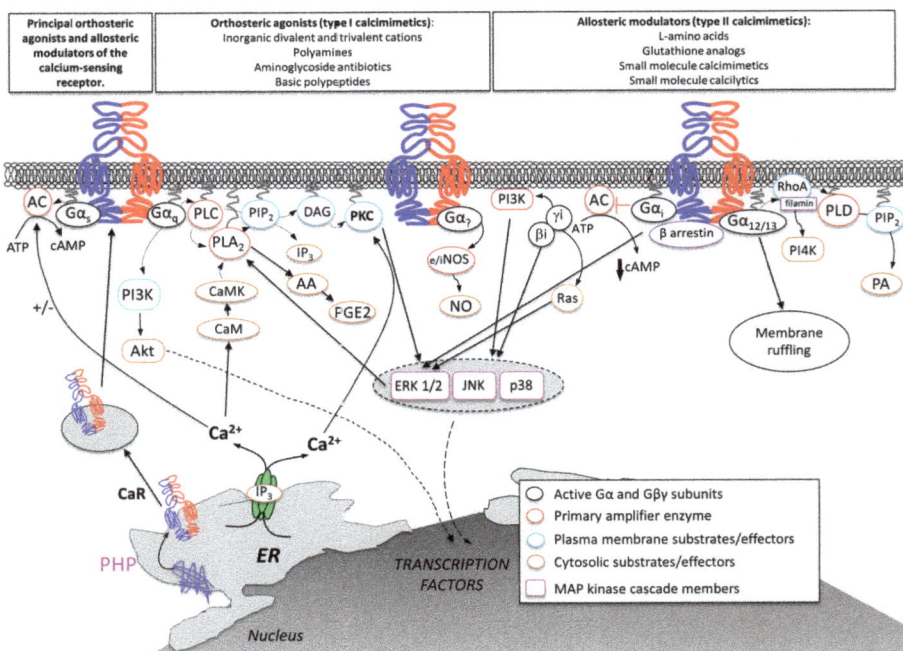

Figure 2. Schematic illustration of intracellular transduction cascades activated by orthosteric agonists or allosteric modulators on the CaR. Modified from [110].

4.4.1. Gq/G11

Most of the studies available in the literature on CaR physiopathological functions in diverse tissues of the body have demonstrated that this receptor, as shown in parathyroids [61,132], mainly interacts with the Gq/11 heterotrimeric G protein, inducing in turn the activation of phospholipase C (PLC)/inositol-1,4,5-trisphosphate (InsP3)/Ca^{2+} mobilization and activation of

PKC and mitogen-activated protein kinase (MAPK) pathways [133]. These intracellular events cause a decrease in parathyroid hormone (PTH) secretion and reduction in renal tubular Ca^{2+} reabsorption [133].

PKC has been described to phosphorylate specific CaR residues and attenuate signalling and/or activate the mitogen-activated protein kinase (MAPK) cascade (ERK1/2, p38 MAPK, JNK), ultimately causing phosphorylation and activation of ERK1/2 [134]. CaR-induced and Gq-mediated activation of phospholipase A2 and phospholipase D have also been reported [135]. Also, activation of the Akt pathway by Gq (via phosphatidylinositol 3-kinase, PI3K) has been described as a determinant in CaR-mediated cell migration [136] and metalloproteases production [137]. The activation of the intracellular Ca^{2+} signalling pathway is also responsible for a Gi/o-independent inhibition of cyclic AMP synthesis mediated by Ca^{2+}-induced inhibition of adenylyl cyclase isoforms 5, 6, and/or 9 [138–140]. For a recent review see also [141].

4.4.2. Gi/o

Early studies have led to the suggestion that the CaR can interact with the pertussis-toxin-sensitive inhibitory G protein, $G\alpha i/o$, causing inhibitory control of PTH secretion [142,143] although the matter is still debated [144]. More recent work has clearly demonstrated CaR-mediated inhibition of adenylyl cyclase (AC) by a pertussis toxin (PTX)-sensitive mechanism. In gastric mucosa, CaR activation led to PTX-dependent alkaline and pepsinogen secretion [18]. In CaR-expressing HEK-293 cells, CaR stimulation quickly reversed or significantly prevented the agonist-induced rise in cytosolic cAMP via a dual mechanism involving pertussis-toxin-sensitive Gi and the CaR-stimulated increase in intracellular Ca^{2+} level [139]. The beta gamma subunits of the Gi/o protein have also been described as mediators of CaR -induced ERK1/2 phosphorylation via Ras and MAPK activation [134]. More recently, the role of beta-arrestin in Gi/o protein-dependent activation of (ERK) and c-Jun N-terminal kinases (JNK) has been directly demonstrated [145]. Gi activation by the CaR has also been reported to induce activation of phosphatidylinositol 4-kinase (PI4K) [146,147].

4.4.3. G12/13

G12/13 pathways have been described in different cell types to differentially modulate CaR-mediated intracellular Ca^{2+} kinetics.

Both early and recent studies have described the capacity of aromatic amino acids to induce intracellular Ca^{2+} oscillations via a G12/13 /Rho/ filamin-A pathway, which in turn induce Ca^{2+} influx from the extracellular fluid via the TRPC1 channel opening [148–150]. On the contrary, in pre-osteoblasts, a G12-dependent dephosphorylation of CaR residue T888 by phosphatase PP2A [151,152] seems to be involved in the appearance of the sustained Ca^{2+} increases necessary to activate chemotaxis [153]. In bone, CaR-mediated G12/13 activation positively modulates osteoblast differentiation [154] and negatively modulates osteoclastogenesis [155]. Importantly, CaR-dependent G12/13 signalling seems to be also implicated in the spreading of breast and prostate cancer cells [156].

4.4.4. Gs

In some cell contexts, such as human breast cancer [157] and pituitary cells [158], the CaR has been shown to modulate hormone secretion by activating Gs. Interestingly, a CaR-induced Gs-independent mechanism for the increase of cyclic AMP synthesis has been recently identified in human fetal lung [159].

Finally, a role for nitric oxide (NO) in mediating CaR action on ion channels has been recently demonstrated in the vasculature [160–164].

4.5. Physiopathological Significance of CaR: Old and New Aspects

As the name suggests, the main physiological function of the CaR is to sense subtle changes in serum Ca^{2+} and maintain systemic Ca^{2+} homeostasis within a narrow range via the oppositely directed changes in secretion of PTH, which in turn modulate the equilibrium between Ca^{2+} uptake and conservation via a direct action on kidney, bone, and thyroid C cells [165] and indirect actions on intestine. Also, the CaR controls calcitonin secretion from thyroidal C-cells upon detection of variations in $[Ca^{2+}]$ext (for a review, see [11]).

When the extracellular Ca^{2+} concentration is reduced, CaR-induced parathyroid hormone (PTH) secretion is elicited by the parathyroid glands. Consequences of PTH release are: (a) reduced excretion of Ca^{2+} by the kidney; (b) increased intestinal absorption of Ca^{2+}, which is mediated by PTH-induced synthesis of the active vitamin D metabolite 1,25-dihydroxyvitamin D; and (c) net release of skeletal Ca^{2+} under sustained PTH secretion. All these tasks cooperate to increase serum Ca^{2+} back to physiological levels. On the other hand, when serum Ca^{2+} levels are increased over the physiological range, the receptor is activated inhibiting PTH synthesis and secretion [166].

The physiological significance of the CaR in tissues involved in systemic Ca^{2+} homeostasis is well-highlighted by the identification of mutations of the CaR gene that cause several inherited disorders of extracellular calcium sensing (for a review, see [167]).

Familial hypocalciuric hypercalcemia (FHH) and neonatal severe hyperparathyroidism (NSHPT) [168] are caused by loss-of-function mutations of the CaR gene, whereas autosomal dominant hypocalcemia (ADH) and Bartter Syndrome type V [169] occur from gain-of-function mutations of the CaR [170–172].

Besides its expression in tissues involved in Ca^{2+} homeostasis, such as thyroid glands, kidneys, and bone, the CaR has been identified in an astonishing number of diverse cell types (reviewed in [11] and [173]), i.e., brain [174], pancreas [175], stomach [89], liver [176], and heart [177,178]. In these tissues, the CaR has been shown to regulate a variety of cellular processes, such as secretion, differentiation, and gene expression. For recent reviews on this topic see also [74,179].

4.5.1. CaR Involvement in Cancer

The massive impact of the CaR in such a wide array of cellular processes, including proliferation, differentiation, migration, adhesion, and apoptosis foresees a role for this receptor in many aspects (incidence, progression, recurrence, and lethality) of various cancers (e.g., parathyroid, pancreatic, prostate, breast, colorectal, ovarian, gastric, skin, and neuroblastoma) [9,180]. However, mutations of the CaR are not early events in tumorigenesis. Still, an increasing amount of literature is concordant on CaR participation in various aspects of the neoplastic progression. In this line, a number of papers evaluating the role of CaR polymorphisms in the risk, incidence, recurrence, or lethality of different kinds of cancer has been published [9].

The CaR is associated with both anti- and pro-proliferative signals. Thus, it can either prevent or promote tumorigenesis depending on the cellular context and the type of cancer [180]. Tissue-specific alterations of CaR expression level, functional activity, or coupling to signaling pathways might explain such opposing behavior.

In tissues such as parathyroid or colon, CaR inhibits cell proliferation and induces terminal differentiation and apoptosis. In tumors of these organs (and more recently also of the gastric mucosa), the observed and progressive reduction in CaR expression (e.g., from normal tissue to carcinomas) grants malignant potential, indicating for the receptor and the physiological significance of its activation a putative role as tumor suppressor [181–185]. Accordingly, when taking also CaR functional activity into consideration, inactivating mutations of the CaR or calcilytics might induce hyperplasia, while calcimimetics have the opposite outcome [186,187]. This scenario is extremely important assuming that the CaR is a novel target for the prevention/treatment of these tumors. However, as reported throughout this review, the incredible number of CaR interactors (both extracellular and intracellular) increases the complexity associated with this kind of pharmacological approach.

Indeed, some of the first evidence connecting extracellular Ca^{2+} and tumors was the inverse correlation between dietary Ca^{2+} level and risk of colorectal cancer [188]. This effect seemed to be mediated by the CaR and potentiated by the active vitamin D metabolite calcitriol [189]. Since CaR transcription is strongly upregulated by vitamin D [190], a Ca^{2+}-containing diet could protect against colon cancer partly via upregulation of CaR expression. The work by the group of Enikö Kallay provided clear evidence about the role of dietary Ca^{2+} and dissected the reciprocal interplay between extracellular Ca^{2+}/CaR and the Vitamin D system (for an exhaustive review refer to [189]). Significant reduction in CaR expression level has also been observed in unfavorable neuroblastomas when compared with benign, differentiated neuroblastic tumors. However, information about the mechanisms underlying the loss of CaR expression is still scattered and thus under intense investigation. Yet, in colon cancer and neuroblastomas, genetic or epigenetic mechanisms, such as hypermethylation and histone deacetylation of CaR promoter or monosomy of the chromosome 3, have been proposed [9,191].

On the other hand, in prostate and breast tumors, the expression level of the CaR is increased [192,193]. This experimental evidence is in line with an oncogene-like behavior of the CaR (e.g., increasing proliferation and inhibiting cell death), which seems strictly related to the $G\alpha s$/cAMP-dependent production of the parathyroid hormone-related protein (PTHrP). Strikingly, in normal mammary epithelial cells, during lactation, the CaR inhibits PTHrP synthesis and secretion through a $G\alpha i$, clearly indicating the ability of the receptor to bind and activate opposing G proteins in normal and breast cancer cells [157]. In addition, in breast cancer cells, CaR signalling through $G\alpha 12/13$ affects cell migration via rho-dependent actin filament formation facilitating the metastatic spread of the tumor [194]. These tumors often lead to bone metastasis, since bone provides homing signals (such as PTHrP-induced high extracellular Ca^{2+}) to several cancers, such as breast, prostate, lung, and kidney cancer.

4.5.2. The CaR in the Heart

Another field of emerging interest for its implications on human health concerns CaR function in cardiovascular physiology and pathology.

The CaR has been demonstrated to exert a fundamental role in the vasculature and to affect blood pressure in different experimental models [160,195–198]. Nice reviews on this topic have been recently published by experts in the field [199–202].

Hence, the focus of this section is on the specific role of the CaR in the heart, a topic which is still quite controversial and, in our opinion, needs further study in the near future.

The first paper describing the functional expression of the CaR in the heart was published 15 years ago [177]. In that first report, the authors identified CaR mRNA and protein both in atrial and ventricular tissues from adult rats. Also, a tentative imaging approach was performed by using high extracellular Ca^{2+} concentrations, spermine (1–10 mM), and gadolinium on isolated adult ventricular cardiomyocytes. Although the interpretation of data collected with high extracellular Ca^{2+} on Fura-2-loaded cells could be hampered by a protocol in which the cardiomyocytes were first bathed in 0 Ca^{2+}, thus activating a serious Ca^{2+} overload upon Ca^{2+} re-addition, the authors provided evidence of effects of extracellular calcium, spermine, and gadolinium on changes of Inositol Phosphate (IP) levels, thus implicating the PLC/InsP3 signalling pathway in CaR activation.

The existence of a functional CaR was next clearly demonstrated in neonatal rat ventricular cardiomyocytes by Tfelt-Hansen and collaborators [178]. Importantly, in this study the calcimimetic AMG 073 was used to activate the CaR and proven to be effective both in IPs accumulation and ERK1/2 activation. Also, the involvement of CaR in IPs modulation was straightforwardly demonstrated by the significant inhibition of Ca^{2+}-induced IP accumulation upon infection of neonatal cardiac myocytes with adeno-associated viruses containing the dominant negative CaR R185Q.

Starting from 2006, a conspicuous number of papers has been published by researchers from the Harbin University, where the first paper on the CaR in cardiac tissues was published [177]. The results collected during the last decade by this group depict a general scenario in which CaR activation is mostly causative of cardiac diseases.

On the contrary, other groups have provided evidence of a cardioprotective role of the CaR. Various experimental models and approaches have been used to inquire into the CaR function in cardiac apoptosis induced by different drugs or by ischemia/reperfusion, hypertrophy and fibrosis, and diabetic cardiomyopathy. The role of the CaR in pre- and post-conditioning has also been investigated.

Role of the CaR in Cardiac Apoptosis

A pro-apototic effect of GdCl3, used as the only CaR agonist (in the presence of NiCl2 and CdCl2 to block L-type Ca^{2+} channels and Na+/Ca^{2+} exchangers, respectively) was reported by Sun and colleagues in neonatal cardiac myocytes [203]. CaR pro-apoptotic action was suggested to be mediated by intracellular Ca^{2+}-induced activation of extracellular signal-regulated protein kinase (ERK), c-Jun NH2-terminal protein kinases (JNK), and p38. GdCl3 was also shown to activate caspase 9 under these experimental conditions.

In parallel, another work was published on isolated rat hearts subjected to an ischemia/reperfusion protocol [204]. By using GdCl3 as CaR agonist, the authors suggested that an increased CaR expression is involved in the induction of cardiomyocytes apoptosis via Ca^{2+} overload-induced cytochrome C/caspase 3 axis activation. Later, Sun and colleagues suggested, by using the same experimental protocol, that one of the molecular players involved in CaR-mediated apoptosis in neonatal cardiomyocytes is the Ca^{2+} channel TRPC6 [205], whose expression was found, like the CaR, to be increased after Gd^{3+} stimulation. Similarly, an increase of TRPC3 expression was described in neonatal rat cardiomyocytes in a following paper, although no evaluation of apoptosis was performed in this study [206].

The effects of known proapoptotic substances were also investigated. Ciclosporin A (CsA) was described to exert its adverse effects (also) by increasing CaR expression both in vitro [207] and in vivo [208].

In neonatal rat cardiomyocytes [207], the proapoptotic role of the CaR in CsA-induced apoptosis was investigated by using Gd^{3+} and NPS-2390 as CaR agonist and inhibitor, respectively. While in the neonatal cell model the CsA-induced CaR-mediated cell death was attributed to Ca^{2+} overload [207], in a similar work performed two years later on H9c2 cardiomyoblasts the involvement of the MAP kinases pathway was suggested [209]. In the same year, the CaR was also implicated in lipopolysaccaride (LPS)-induced apoptosis via its increased expression and consequent Ca^{2+} overload and TNF alpha and IL-6 release [210].

Role of the CaR in Apoptotic Pathways Activated by Ischemia/Reperfusion

During the years 2006–2011, some of the same researchers from Harbin University made a considerable effort to disclose the involvement of the CaR in the apoptotic pathway activated by ischemia/reperfusion (I/R) or hypoxia/reoxigenation protocols both on in vivo and in vitro models, i.e., neonatal rat cardiomyocytes [211–214], isolated adult rat cardiomyocytes [204,215], and adult rat hearts [212,215]. Altogether, the collected results suggest that I/R induces a significant increase of CaR expression. Subsequent CaR-mediated signalling pathways activate apoptosis. The involvement of mitochondria in such an effect was suggested to be mediated by Ca^{2+} overload via inter-organelle signalling [213], ER stress [212], and PKC delta [216]. Activation of the C-Jun NH2 terminal protein kinase signalling pathway was also described as one of the mechanisms relevant for CaR-mediated injury during ischemia/reperfusion in adult rats [215].

Later, another group at the Nanjing University confirmed the work of Jiang et al. [211,217] and implicated a downregulation of CaR expression and signalling in the anti-apoptotic effect of hepatocyte growth factor in neonatal cardiomyocytes subjected to simulated ischemia/reperfusion [218].

Role of CaR Inhibition in Post-Conditioning

Almost in the same years, Zhang et al. [219] suggested on isolated rat hearts subjected to an ischemia/reperfusion protocol followed by post-conditioning (PC) that PKCε is involved in the protective effect of the PC maneuver via its negative feedback on the CaR and the subsequent reduction of intracellular Ca^{2+}. Also, in these studies $GdCl_3$ + $NiCl_2$ + $CdCl_2$ was the only experimental maneuver used to activate the CaR.

Similar results were obtained some years later on the neonatal rat cardiomyocytes model, this time using also the calcilytic NPS 2390 and implicating the attenuation of CaR-induced sarcoplasmic reticulum–mitochondria crosstalk in the protective effects of PKCε during post-conditioning [220].

In 2012, in a work performed in vivo and on isolated adult rat cardiomyocytes subjected to a hypoxia post-conditioning protocol, the authors suggested that the pro-apoptotic action of CaR exerted via ER stress activation was counteracted during post-conditioning by a reduction of CaR protein expression [221]. Very recently, Zhang and colleagues confirmed that ischemic post-conditioning and KATP channel agonists diazoxide and pinacidil exert a protective effect against ischemia/reperfusion injury through a mitigation of Ca^{2+} overload induced by downregulation of the CaR [222].

Given the relevance of CaR expression levels for the rationale of these studies, it is worth mentioning a recent short communication from the group of Schluter [223] focused on the assessment of CaR expression levels after ischemia reperfusion and cardiac pre- and post-conditioning. This report pointed out that different CaR regulation patterns seem to exist between left and right ventricular tissues and that also the size of infarcts could be a determinant in regulating CaR expression [223].

Role of the CaR in Pre-Conditioning

In opposition to the theory of a deleterious action of the CaR in the heart, there was already in 2010 the work by Sun and Murphy in which the authors suggested a cardioprotective role for the CaR, strategically localized in caveolae, during ischemic pre-conditioning (IPC) [224].

In this study, the perfusion of adult rat hearts with the CaR antagonist NPS 2143 during the application of the ischemic pre-conditioning protocol was indeed able to attenuate cardioprotection of IPC, as assessed by measuring the levels of the cardioprotective phospho ERK1/2, AKT, and GSK3β kinases, infarct size, and functional recovery. The suggestive speculation of the authors depicts a scenario in which the changes in pH and ionic strength imaginable at the particular microdomains of caveolae may activate the CaR, which could thus "sense" and "mediate" IPC.

Another controversial topic is the role of CaR activation in the progress of diabetic cardiomyopathy [225,226].

On the one hand, Bai et al. suggested that CaR expression is reduced in diabetic rat hearts and could contribute to the progress of diabetic cardiomyopathy via impaired intracellular Ca^{2+} signalling in adult cardiac cells [225]. On the other hand, Qi and colleagues showed that CaR expression is rather increased in diabetic rat hearts and by using Gd^{3+}, NPS 2390, and a siRNA to silence CaR protein, suggested that CaR signalling contributes to apoptosis of neonatal cardiomyocytes exposed to high concentrations of glucose in vitro via Ca^{2+} overload, reduction of the Bcl2/Bax ratio protein expression, and modulation of the MAP kinase pathway [226]. The apparent discrepancy of results on CaR expression was explained by invoking a difference in the experimental models which, according to Qi et al., were representative of type 2 diabetes in the first work [225] and type 1 diabetes in the latter [226].

Role of the CaR in Hypertrophy and Heart Failure

A role for the CaR in cardiac hypertrophy has been postulated since 2006, when Tfelt-Hansen and collaborators [178] observed a CaR-induced decrease in DNA synthesis in neonatal cardiomyocytes. Considering that this parameter increases in cells undergoing hypertrophy [227], the authors supposed a protective role for the CaR against cardiac hypertrophy.

In opposition to this theory, in a study performed by Wang and colleagues [228] on an in vitro model of hypertrophy, an increased expression of the CaR was found and exacerbating effects of Gd^{3+} on Ca^{2+} signalling and hypertrophic markers were argued to be mediated by the CaR.

Later, Zhong et al. suggested a role for the CaR in the Gd^{3+}-induced increase of nuclear Ca^{2+} spikes' frequency and consequent activation of the Calcineurin/NFAT pathway [229], a well-known player in cardiac hypertrophy [230].

The role of the CaR in cardiac hypertrophy and heart failure was further investigated in a successive work [231], in which Calindol and Calhex 231, as agonist and inhibitor of the CaR, respectively, were subcutaneously injected in isoproterenol-treated Wistar rats to assess the role of the CaR in ER stress in vivo. CaR expression was found to significantly increase during heart failure induced by isoproterenol and Chalex 231 significantly reduced the cross-sectional diameter of hypertrophic cardiomyocytes. In the same study, the CaR was suggested to play a role in cardiac hypertrophy and heart failure also in mice subjected to thoracic aorta constriction (TAC) [231]. In this model, Calindol was found to induce significant cardiac hypertrophy.

The involvement of the CaR in ER stress-mediated apoptosis in failing hearts was also suggested by the observation that Calindol and Chalex 231 were able to increase and decrease, respectively, ER stress protein expression and apoptosis in both the rat and mouse models. Finally, a CaR-induced IP3-dependent Ca+ overload of mitochondria and subsequent initiation of apoptosis were suggested in rat hearts and neonatal rat cardiomyocytes incubated for 48 h with isoproterenol. On these bases, the authors proposed that the CaR could represent a putative target for pharmacological intervention in heart failure.

Very recently, Liu et al. suggested the possibility to ameliorate isoproterenol-induced cardiac hypertrophy through suppression of CaR signalling with the CaR inhibitor Chalex 231 both in vivo and in vitro [232]. In this paper, the authors suggested that the anti-hypertrophic effect and amelioration of cardiac function by Chalex 231 may be mediated through inhibition of autophagy and the CaMKK-AMPK-mTOR signalling pathway.

A recent in vitro and in vivo study also suggested a role for the CaR in promoting cardiac fibrosis [233].

In opposition to a maladaptive pro-hypertrophic effect of the CaR, which is the common denominator of the above-described studies, there are recent contributions from the group of Schluter [223] which sustain the theory of a cardioprotective effect of the CaR in a pharmacological model of hypertension in adult rats. Interestingly, these authors showed that upon NO-synthase inhibition an increased expression of endothelin A and/or B1 receptors and CaR was measurable. This upregulation of CaR, most probably mediated by a paracrine action of endothelin 1, was suggestive of a CaR-induced adaptive hypertrophy. Indeed, CaR inhibition or destabilization by NPS 2390 and siRNA against RAMP, respectively, decreased load-free cell shortening, which was used as a read-out of cardiac function.

Role of the CaR on the Electrical Properties of Cardiac Cells

Notwithstanding the abrupt proliferation of literature on the role of the CaR in the heart in the last 15 years, very few data were available, until recently, on the effect of CaR agonists and modulators on the electrical properties of cardiac cells.

In very recent times, the group of Schluter [234] and Shieh [235] have published interesting data on this topic on adult rat hearts and guinea pig cardiomyocytes, respectively, by using different physiologically relevant experimental approaches.

First, Schreckenberg and colleagues examined the effects of CaR acute stimulation on the cardiac physiology of adult male rats [234]. The experimental approaches were diverse: the contractile response of ventricular cardiomyocytes was recorded with a cell-edge-detection system in parallel with Fura-2 measurements of cytosolic Ca^{2+}, while cardiac performance in response to acute CaR stimulation or inhibition was assessed on left ventricle muscle stripes and whole hearts by the Langendorff system.

Putrescine and Gd^{3+} were used as CaR agonists, while NPS-2390 (10/100 μM) and downregulation of CaR by siRNA were used to assess the specificity of the responses.

The data collected, while confirming the expression of CaR by adult rat ventricular cardiomyocytes, demonstrated that CaR action is relevant for basal cell shortening at physiological extracellular Ca^{2+} concentrations and that additional activation of the CaR speeds up cardiomyocytes relaxation and augments cell shortening, supposedly via increased Ca^{2+} transients induced by activation of the Gq/PLC/IP3 pathway.

Very interestingly, Liu et al. clearly demonstrated that activation of the CaR in guinea pig cardiomyocytes by the allosteric agonist NPS R 568 activates both the PLC and phsphatidylinositol-4 kinase (PI4) Kinase pathways and that the PI4 Kinase pathway predominates [235]. The net effect is an increase in PIP2 at the plasma membrane, which in turn determines a significant increase in currents through inward rectifier K+ channels. This action could thus exert a stabilization of the resting membrane potential of cardiomyocytes and contribute to a re-evaluation of old data on the stabilizing and cardioprotective effects of polyamines [236].

Finally, interesting hints on the CaR's role in the heart come from a paper focused on the function of the CaR in regulating blood vessel tone and blood pressure [237].

Since the experimental model of CaR KO mice (SM22αCaSRΔflox/ΔfloxKO mice) used in this work also display a deletion of the CaR in cardiac cells, the authors tested the hearts of KO mice by perfusion in the Langendorff mode. Given the significantly reduced basal heart rate found in KO mice versus wild-type (WT) animals, the authors suggested that CaR deletion from heart directly affects chronotropy. Also, ex vivo lung slice preparations from KO mice showed a reduced rate of spontaneous activity, suggesting that the decreased chronotropy in KO mice could be due to decreased pacemaker activity.

In summary, notwithstanding the numerous results collected on physio-pathological roles of the CaR in the heart, the current findings are far from providing a coherent picture.

Considering CaR promiscuity and pleiotropicity and the emerging concept of biased signalling at the CaR (see below), an interesting field of future investigation would be the accurate pharmacological assessment of the effects exerted by panels of CaR agonists and modulators on different signalling pathways in physiologically relevant experimental models.

Given the actual and potential use of calcimimetics and calcilytics in the clinic, these studies would be highly significant for their implication for human health.

4.6. Novel Trafficking/Signaling Modes of the CaR

Although GPCR have often been seen as cell surface proteins able to go through a simple on-off switch, research of the past few years has drawn a much more complex scenario, in which receptors exist in multiple conformations and, in dependence of the different ligands, can be stabilized in different conformational states, in turn activating multiple, yet specifically selected, downstream pathways whose identity depends on the combination of ligands and the specific signalling toolkit of the cell where they are expressed [238–242].

The development of new technologies, such as microscopy techniques and probes to follow receptor trafficking and subcellular signal dynamics (such as FRET and single molecule microscopy) as well as BRET-based high throughput assays for biased signalling [243], have been essential to demonstrate that these proteins are capable of trafficking among a variety of subcellular compartments, initiating specific signalling pathways at different locations: nuclear [244], mitochondrial [245], and endosomal–Golgi membrane [246,247].

As stated in the paragraph regarding the structural features of the CaR, functional receptors localized at the plasma membrane are dimers of the glycosilated CaR which are formed intracellularly [248] and are stabilized by ligand binding at the cell surface [249]. As for other GPCRs, the expression level of CaR is the result of a fine regulation of the dynamic equilibrium between mechanisms that precede its biosynthesis (transcription/mRNA level, translation efficiency, epigenetic control [250,251]) and regulate its anterograde trafficking and those controlling its retrograde trafficking, including degradation. In fact, it has been known since a long time that the fine balance between maturation, internalization, recycling, and degradation can influence the net level of cell surface receptor and thus represent a mechanism for the cell to regulate receptor desensitization and modulate strength of signal transduction [252]. Thus, the strength of signaling depends upon the level of GPCRs expression at the plasma membrane and the consequent receptor availability for agonists and modulators. This is also the case for CaR signalling as recently established by Conigrave's group [253].

A significant contribution to the field of CaR trafficking has come in the last ten years from different groups who have focused on the elucidation of the key steps involved in CaR biosynthesis and trafficking from the ER membrane to the Golgi membrane and then to the plasma membrane (see [254] for a recent review). A fundamental role in this field is also played by research focused on the CaR-interacting proteins, which finely modulate anterograde and retrograde trafficking of the CaR (for comprehensive reviews see [10,153,254–256]).

The Agonist-Driven Insertional Signalling (ADIS)

Both early and recent studies have revealed two peculiar characteristics of the CaR: a minimal functional desensitization (both by phosphorylation and/or β-arrestins) and a significant "reservoir" of CaR proteins in intracellular membrane compartments. Western blotting or immunohistochemistry analyses [11,76,257] have clearly showed, since early studies, that CaR immunoreactivity is predominantly represented by an intracellular, core-glycosilated form. Recent data are making increasingly clear that such an observation is not an artifact but is strictly related to the complex interaction between CaR trafficking and signalling. In fact, both the hallmarks of the CaR, minimal desensitization and high levels of intracellular proteins, can be explained by a novel signalling mode described by the group of Dr. Breitwieser and termed Agonist-driven insertional signalling (ADIS) [250,258].

In this model, CaR signalling is regulated by an agonist-dependent modulation of the net amount of receptor proteins at the plasma membrane. In particular, the level of CaR cell-surface expression results from the balance between the rate of trafficking from a large, pre-plasma membrane pool located at the Golgi/trans-Golgi/post-Golgi vesicles (which depends upon the concentration of CaR agonists and/or allosteric modulators) and the process of constitutive endocytosis of the receptor already at the plasma membrane. The sensitivity of parathyroid and renal tubular cells to extracellular Ca^{2+}, for example, is likely influenced by the balance between ADIS, which enhances anterograde trafficking of newly synthesized CaR proteins to the plasma membrane [258] and clathrin-mediated endocytosis and retrograde trafficking of cell-surface CaR receptors [10].

4.7. Synthetic Allosteric Modulators of the CaR Can Function as Pharmacoperones

Strictly related to the research in the field of CaR trafficking is the observation that CaR allosteric modulators (see Paragraph 4.3) can work as pharmacoperones. Pharmacoperones are permeant substances (which could be agonists, antagonists, or allosteric modulators) that can reach a misfolded protein at its intracellular location (most commonly the endoplasmic reticulum), stabilize it, and rescue the receptor to the cell membrane surface.

A relevant effort on the capacity of CaR allosteric modulators to work as pharmacoperones comes from Breitwieser's group [259–261]. These authors firstly demonstrated that the positive allosteric modulator NPS-R-568 was able to rescue almost fifty percent of the loss-of-function mutant CaRs analyzed [259,260]. In parallel, another group demonstrated the same ability of NPS-R-568 to rescue intracellular signalling in four mutants (out of seven) although the plasma membrane localization of the receptors was not assessed [262].

In the last five years, relevant findings in the field of pharmacoperone action of type II calcimimetics and calcilytics have been provided by studies from the group of Dr. Leach. These authors, while confirming data on NPS-R-568, showed the ability of cinacalcet, the only CaR modulator approved in the clinic, to effectively rescue trafficking and signalling of CaR mutants with loss of cell surface expression. [128,263]. They also showed that NPS-2143, a negative allosteric modulator, is able to positively modulate CaR trafficking to the cell membrane while negatively modulating CaR signalling in the presence of the orthosteric substrate [144,263], underlining a striking example of positive bias (see also above) toward trafficking by a CaR allosteric modulator.

Nonetheless, both early and recent studies suggest that the pharmacoperone action of NPS R-568 [264] and NPS-2143 [259,264] is mutant-specific.

Importantly, the effectiveness of old and new calcilytics on gain-of-function mutations of the CaR was further explored in vitro [265,266], demonstrating that a novel treatment option for patients with activating CaR mutations could be represented by the new quinazolinone-derived calcilytics, which have been shown to be effective in attenuating enhanced cytosolic Ca^{2+} signalling in four known mutations causing Bartter syndrome (BS) type 5 and autosomal dominant hypocalcemia (ADH). Very interestingly, recent in vivo works involving a mouse model harboring an ADH gain-of-function CaR mutation demonstrated the effectiveness of both old (i.e., NPS 2143) [267] and new (i.e., JTT-305/MK-5442) [268] calcilytics in normalizing the defect and improving the associated hypocalcemia.

4.8. Biased Signalling

Biased signalling (also termed biased agonism or ligand-directed signalling or stimulus bias) represents a recently described, yet quite general, modality of the signalling characteristic of diverse GPCRs [240].

It concerns the capability of different ligands to stabilize diverse conformations of the receptor and thus preferentially head GPCR signalling towards a specific transduction cascade.

This new concept, which substitutes the idea of one receptor/one GPCR conformation/one-uniform activation of signaling pathway(s), greatly complicates the scenario of GPCRs signalling, but opens new perspectives in the design of smart and tissue-specific drugs that avoid off-target actions or on-target effects on multiple tissues [269].

Given the promiscuity and pleiotropicity of the CaR and the wide expression of this receptor in tissues involved or not in the regulation of systemic Ca^{2+}, it is not surprising that fundamental insights into this new ligand- and tissue-specific pharmacology come from the study of biased signalling at the CaR.

In fact, the existence of ligand- and tissue-specific effects in the signalling pathways activated by the CaR, although not precisely quantified/measured, are traceable in a vast number of papers published throughout the years.

It could also be hypothesized that the experimental approaches used in the recent past could have contributed to an underestimation of the phenomenon of biased signalling at the CaR. Most of the work on CaR signalling is in fact performed by testing only a small number of CaR agonists and modulators on a limited array of signal transduction cascades (most frequently cytosolic Ca^{2+} dynamics). A striking example of biased signalling at the CaR comes from an acquired hypocalciuric hypocalcemia autoantibody, which was shown to increase the effects of external Ca^{2+} on Gq signalling to the detriment of the Gi-induced signalling pathway [270]. As far as the differential effects of CaR agonists and modulators are concerned, both early studies by Riccardi's group on rat pancreas [271] and more recent work by Ziegelstein and colleagues on human aortic endothelial cells showed that only some CaR ligands were able to induce intracellular Ca^{2+} release, while others were ineffective [160]. Also, at the level of physiological functions, Tfelt-Hansen's group demonstrated that while cinacalcet was able to induce vasodilatation of rat aorta, the neomycin and Gd^{3+} were ineffective [161].

In the last years, a large number of findings has highlighted the physiopathological significance of biased signalling.

A major contribution in this area has come from Bräuner-Osborne's group [145,272,273]. Specifically, these authors explored the action of a panel of well-known orthosteric CaR agonists on the Gq/11 protein, Gi/o protein, and extracellular signal-regulated kinases 1 and 2 (ERK1/2) pathways. Thomsen and colleagues [145] revealed that in HEK-293 cells stably transfected with rat

CaR, the natural agonist Ca^{2+} is biased toward cAMP inhibition and IP3 accumulation over ERK1/2 phosphorylation. Differently, a significant bias toward ERK1/2 phosphorylation was shown for the polyamine spermine.

Interestingly, while two different groups showed that strontium ranelate, a drug used as treatment for osteoporosis, is a biased agonist of the CaR toward the ERK1/2 pathway and another cascade independent of Gq/11 signalling [273,274], different results were reported by others [275]. In another study, while both Sr^{2+} and Ca^{2+} were shown to stimulate PLC and NF-kB, only strontium ranelate induced apoptosis via PKC activation [276].

In recent years, the possibility of exploiting and "inducing" stimulus bias at the CaR has been deeply explored also by the group of Christopoulos and Leach [95,263,277,278] (reviewed in [144,269]). The first quantitative evidence that an allosteric modulator used in clinical practice exhibits stimulus bias was provided a few years ago by Davey and colleagues [277], which straightforwardly proved that allosteric modulators are biased toward intracellular Ca^{2+} signalling in a HEK293-TREx c-myc-CaR cell line.

Next, the same group analyzed in detail the effect of CaR allosteric modulators on naturally occurring CaR mutations [278], demonstrating that mutations can alter signalling bias by stabilizing receptor conformations that differentially couple to downstream signalling pathways. Importantly, and as stated above, in a subsequent paper, both Cinacalcet and the calcylitic NPS-2143 were demonstrated to have a bias toward Ca^{2+} signalling and rescue CaR mutants to the cell membrane [263]. The therapeutic implications of such data are enormous.

Just as an example, the calcimimetic cinacalcet has been used in clinic for the treatment of patients carrying loss-of-function CaR mutations only in the cases of end-stage renal disease. This is because of its hypocalcemic properties, most probably due to the Ca^{2+}-mediated increase of calcitonin secretion [272]. It is thus clear that designing a novel drug able to act on parathyroid glands without affecting calcitonin secretion from parafollicular cells would be the best therapeutic option.

Remarkably, in a very recent paper, the search for a calcimimetic able to suppress PTH levels at concentrations with low/no effect on serum calcitonin seems to have found good candidates among the third generation of calcimimetics [128]. Altogether, these findings highlight the importance of a deeper comprehension of the mechanisms of action of different orthosteric agonists and modulators in the different tissues for the better design of drugs that are devoid of side effects and are able to activate specific pathways in a cell-selective fashion [144,269].

It is worthwhile noting that an even higher level of complexity in the scenario of biased agonism and pharmacoperones action of CaR ligands is the observation that GPCRs have been recently shown to sustain G (Gs or Gi) protein signalling at different subcellular sites even after internalization of ligand–receptor complexes in endosomes (for a recent review see [242]).

5. Old and New Extracellular Ca^{2+} Sensors

Even though the CaR is so far the best-studied extracellular Ca^{2+} sensor, other receptors (either membrane proteins or channels) able to change their function following significant fluctuations in extracellular Ca^{2+} level have been identified [279]. Of note, those of the same family of the CaR have been the focus of intense studies based on a high degree of structural homologies, conservation of specific domains, and a common functional role as nutrient/salinity sensor. This family includes eight metabotropic glutamate receptors (mGluR1–8), two gamma-aminobutyric acid (GABA) receptors, three taste receptors (T1-T3), the recently "deorphanized" GPRC6A, and seven orphan receptors [280].

5.1. Metabotropic Glutamate Receptors

Metabotropic glutamate receptors (mGluRs) are neurotransmitters expressed mainly in the central nervous system, where they modulate key functions in a variety of different neurological processes (e.g., memory, learning, pain, and synaptic plasticity) [281]. mGluRs (especially mGluR1 and -R5) possess significant sensitivity to extracellular Ca^{2+} [282,283]; however, it is still unclear whether they exist in nature as proper Ca^{2+} sensors [282] or whether extracellular Ca^{2+} just sensitizes the receptor to its principal ligand, the excitatory amino acid L-glutamate [284,285]. For example, Ca^{2+} depletion in the synaptic cleft during burst firing induces significant inhibition of postsynaptic mGluR1 function [286]. In addition, Gerber et al. showed that increasing concentrations of extracellular Ca^{2+} might enhance the potency/efficacy of glutamate action on mGluRs1 [287]. Recently, a novel potential extracellular Ca^{2+}-binding site was identified within the extracellular domain of the mGluR1 adjacent to the known L-Glu-binding site [288,289]. When both L-Glu and Ca^{2+} are bound to their partially overlapping binding sites, they synergistically modulate mGluR1 activation. Strikingly, a significant reduction in the sensitivity of mGluR1 to extracellular Ca^{2+} and, in some cases, to L-Glu is induced by a mutation in the predicted Ca^{2+}-binding residue [290]. Furthermore, this Ca^{2+}-binding site is also adjacent to the interaction sites of orthosteric agonists and antagonists; thus, extracellular Ca^{2+} might modulate the sensitivity of mGluR1 not only to orthosteric agonists and antagonists but also to allosteric modulators [291].

5.2. The GABA_B Receptor

The gamma-aminobutyric acid subtype B ($GABA_B$) receptor plays a key role in regulating membrane excitability and synaptic transmission in the brain. The functional receptor consists of a heterodimer of $GABA_{B-R1}$ and $GABA_{B-R2}$ [292]. As for the CaR and mGluRs, the $GABA_B$ receptor has the ability to sense extracellular Ca^{2+} (0.001–1 mM) but not other divalent and trivalent cations. Wise et al. showed that an addition of 1 mM extracellular Ca^{2+} potentiated GABA stimulation (EC50 was reduced from 72 to 7.7 μM) in membrane preparations from both CHO cells (that stably express the $GABA_{B-R1/R2}$ heterodimer) and rat brain cortex. Of note, even though Ca^{2+} does not act as a direct agonist of the receptor, extracellular Ca^{2+} may allosterically enhance the GABA-induced responses of the $GABA_B$ receptor [293].

5.3. Taste Receptors

T1R1 and T1R3 belong to the T1R family and form an heterodimer that is considered a broadly tuned L-amino acid sensor. Silve et al. reported a Ca^{2+}-binding pocket for T1R3. Of note, this receptor subunit is characterized by Ca^{2+}-binding residues that are conserved in the Venus fly trap region of the CaR [294]. Recently, in vivo taste studies have suggested the involvement of T1R3 in Ca^{2+}-Mg^{2+} taste in mice and humans [295,296]. These studies have also supposed that T1R3 might form a dimer with another receptor able to sense extracellular Ca^{2+}, such as the CaR.

5.4. The G Protein-Coupled Receptor Family C Group 6 Member A (GPRC6A)

The G protein-coupled receptor family C group 6 member A (GPRC6A) displays the closest mammalian homology to the CaR. This receptor, cloned in 2004, is extensively expressed at the tissue level [297] while limited information is available at the sub-tissue level. So far, experiments with KO mice for the receptor have shown GPCR6 involvement in the regulation of biological processes such as inflammation, metabolism, and hormonal functions. GPCR6 is activated by a large number of basic and small aliphatic L-α-amino acids (being L-arginine, L-lysine, and L-ornithine the most potent compounds). In addition, high extracellular Ca^{2+} levels (5–10 mM) either directly activate or positively modulate the receptor. This range of extracellular Ca^{2+} concentration can be physiological only in a tissue, such as the bone, where the receptor seems to be involved in bone remodelling processes [298]. However, it remains unclear which of the proposed agonists are the most important physiologically.

Furthermore, CaR allosteric modulators, such as NPS568, Calindol, and NPS2143 (IC50~10 µM), were shown to be effective also on GPRC6A [67,298,299].

5.5. Notch Receptor

The epidermal growth factor (EGF)-like domain contains a Ca^{2+}-binding domain with a wide range of Ca^{2+} affinities [300]. This type of module is probably the most prevalent extracellular Ca^{2+}-binding site. Among the proteins containing Ca^{2+}-binding EGF-like domains, there are proteins involved in blood coagulation, fibrinolysis and its complement system, matrix proteins, and the developmentally important cell surface receptors Notch [300]. Left–right asymmetry in all vertebrates is established during embryo development by several different molecular events. Raya et al., using a combination of modelling and experimental procedures, showed that, in chick embryos, asymmetric patterns of gene expression were the result of an extracellular Ca^{2+} gradient toward the left margin of the developing embryo. Here, extracellular Ca^{2+}-induced activation of Notch results in the asymmetric secretion of Nodal, a specific signalling molecule [301]. Of note is that the perturbed establishment of the left–right patterning in embryo maturation was induced by the disruption of the extracellular Ca^{2+} gradient.

5.6. Cadherins

Cadherins are a family of Ca^{2+}-dependent adherent receptors involved in development, morphogenesis, synaptogenesis, differentiation, and carcinogenesis. Most of the members are single membrane-spanning proteins possessing Ca^{2+}-binding extracellular domains with a wide range of Ca^{2+} affinities (Kd from µM to mM) [302]. The cadherin extracellular domains act as a Ca^{2+}-activated "mechanical switch" through which extracellular Ca^{2+} fluctuations are detected and integrated to control their mechanical integrity and mechano-transduction capability.

5.7. Stromal Interaction Molecule 1 (STIM1)

Ca^{2+}-binding proteins containing EF-hand domains are present in the extracellular environment or matrix. These proteins do also localize at the plasma membrane, with the EF-hand domain facing the extracellular medium [303,304]. An important example of this class of proteins is the single transmembrane-spanning Ca^{2+}-binding protein, STIM1, generally known as the endoplasmic reticulum Ca^{2+} sensor, which provides the trigger for store-operated channels (SOCs) activation. STIM1 at the cell surface may represent a fundamental pharmacological target to finely tune Ca^{2+}-regulated functions that are mediated by SOCs.

5.8. Hemichannels

Gap junctions are constituted by the juxtaposition of two membrane-spanning connexons (hemichannels) positioned end-to-end. However, functionally active hemichannels exist at the plasma membrane where they allow the rapid bi-directional movement of ions, second messengers, and metabolites between the cytosol and the extracellular space and vice-versa [305]. Opening of most hemichannels depends critically on decreases in extracellular Ca^{2+} levels (of about ~200 µM) around the physiological Ca^{2+} levels [306,307]. The vestibule of Connexin 32 has a Ca^{2+}-binding site involved in the regulatory effect of external Ca^{2+}. Of note is that it has been shown that the complete Ca^{2+} deregulation of connexin 32 hemichannel activity is induced by a mutation causing an inherited peripheral neuropathy. This evidence led to the assumption that the pathogenesis of the neuropathy may be strictly related to this dysfunction [308]. The pore of connexin 43 hemichannels might be in an open or closed conformation, with the open probability significantly dependent on external Ca^{2+} levels (from 1.8 to 1.0 mM) [309].

5.9. Ion Channels

Dynamic changes in extracellular Ca^{2+} are known to impact on a number of ion channels, especially those located in the brain.

Ca^{2+} homeostasis modulator 1 (CALHM1) is an ion channel expressed at the cell membrane of cortical neurons and type II taste bud cells of the tongue [310]. This peculiar channel, closed at resting membrane potential and modulated by significant membrane depolarizations, is also able to sense a decrease in extracellular Ca^{2+} that thus works, in synergy with voltage, to allosterically increase the CALHM channel gating open probability.

Interestingly, Ca^{2+} and other cations, anions, and ATP are able to permeate through the non-specific pore of CALHM1 [310].

Acid-sensing ion channels (ASICs) are trimeric cationic channels activated by extracellular protons and primarily expressed in the nervous system. These channels are permeable to Na+ and Ca^{2+} and "sense" extracellular Ca^{2+}, which modulates their open probability by competing with protons [311–313]. In addition, two negatively charged residues near the entrance of the channel pore have been shown to be critical for the direct Ca^{2+} blockade of ASIC1a currents [313]. Of note is that these channels seem to be also modulated by small fluctuations in extracellular Zn^{2+}, Mg^{2+}, and spermine [311,314].

Another class of ion channels modulated by dynamic changes of extracellular $[Ca^{2+}]$ are the voltage-gated Human ether-à-go-go-related gene (HERG) encoded K+ channels, which have been shown to be implicated in a variety of physiopathological mechanisms, such as neuronal spike frequency adaptation and tumor cell growth [315]. Johnson et al. showed that small changes in extracellular Ca^{2+} under physiological conditions shift the voltage dependence of channel activation to more positive membrane potentials, allowing for fine-tuned cell repolarization [316]. Under this scenario, HERG channels seem to be responsible for a delayed rectifier K+ current functional to cellular repolarization of the cardiac contractile cycle in the heart [317].

Finally, the Transient receptor potential cation channel, subfamily M, member 7 (TRPM7) was shown to sense a reduction of extracellular Ca^{2+} in hippocampal and pyramidal neurons and aid an inward cation current [318–320]. These 36-pS channels (originally named Ca^{2+}-sensing nonselective cation channels, IcsNSC), which are almost silent at the 2.0 mM extracellular Ca^{2+} level, become robustly activated by small decreases in extracellular Ca^{2+} [321].

The history of research on the CaR's role in physiology and pathological states is a perfect example of the relevance of basic research for ameliorating human health. It is in fact clear that none of the clinically relevant discoveries in the pharmacological field today could have ever taken place without a deep knowledge of the specific mechanisms underlying the functionality of the tissues and organs on whom CaR mutations have their more striking effects.

It is also clear that the novel methodological tools available to researchers in the field of GPCRs and signal transduction will be of paramount importance to further improve the knowledge of basic signalling modes, a step that is indispensable for the design of smart drugs able to rescue GPCR defects in human diseases.

Acknowledgments: The authors gratefully acknowledge the contribution and support of all the colleagues and students who participated in the studies described in the present article.

Author Contributions: Matilde Colella and Andrea Gerbino contributed equally to this work.

Conflicts of Interest: The authors declare no conflict of interest.

Abbreviations

AA	arachidonic acid
AC	adenylate cyclase
ATP	adenosine triphosphate
CaM	Calmodulin
CaMK	Ca^{2+}/calmodulin-dependent protein kinase
cAMP	cyclic AMP
DAG	Diacylglycerol
eNOS	endothelial nitric oxide synthase
ER	endoplasmic reticulum
ERK 1/2	extracellular-signal regulated kinase
iNOS	inducible nitric oxide synthase
InsP3	inositol-1,4,5-trisphosphate
JNK	Jun amino-terminal kinase
MAPK	mitogen-activated protein kinase
MEK	MAPK kinase
NO	nitric oxide
p38	p38 mitogen-activated protein kinase
PA	phosphatidic acid
PHP	Pharmacoperones
PI3K	phosphatidylinositol 3-kinase
PI4K	phosphatidylinositol 4-kinase
PIP2	phosphatidylinositol 4,5-bisphosphate
PKC	protein kinase C
PLA2	phospholipase A2
PLC	phospholipase C
PLD	phospholipase D
RhoA	Ras homolog gene family, member A
SOC	store-operated Ca^{2+} channel

References

1. Berridge, M.J.; Bootman, M.D.; Roderick, H.L. Calcium signalling: Dynamics, homeostasis and remodelling. *Nat. Rev. Mol. Cell Biol.* **2003**, *4*, 517–529. [CrossRef] [PubMed]
2. Rizzuto, R.; Bernardi, P.; Pozzan, T. Mitochondria as all-round players of the calcium game. *J. Physiol.* **2000**, *529 Pt 1*, 37–47. [CrossRef] [PubMed]
3. Bootman, M.D.; Fearnley, C.; Smyrnias, I.; MacDonald, F.; Roderick, H.L. An update on nuclear calcium signalling. *J. Cell Sci.* **2009**, *122*, 2337–2350. [CrossRef] [PubMed]
4. Medina, D.L.; Di Paola, S.; Peluso, I.; Armani, A.; De Stefani, D.; Venditti, R.; Montefusco, S.; Scotto-Rosato, A.; Prezioso, C.; Forrester, A.; et al. Lysosomal calcium signalling regulates autophagy through calcineurin and tfeb. *Nat. Cell Biol.* **2015**, *17*, 288–299 [CrossRef] [PubMed]
5. Corbett, E.F.; Michalak, M. Calcium, a signaling molecule in the endoplasmic reticulum? *Trends Biochem. Sci.* **2000**, *25*, 307–311. [CrossRef]
6. Brown, E.M.; Gamba, G.; Riccardi, D.; Lombardi, M.; Butters, R.; Kifor, O.; Sun, A.; Hediger, M.A.; Lytton, J.; Hebert, S.C. Cloning and characterization of an extracellular Ca^{2+}-sensing receptor from bovine parathyroid. *Nature* **1993**, *366*, 575–580. [CrossRef] [PubMed]
7. Brown, E.M.; Conigrave, A.D. Preface. *Best Pract. Res. Clin. Endocrinol. Metab.* **2013**, *27*, 283–284. [CrossRef] [PubMed]
8. Zhang, C.; Miller, C.L.; Brown, E.M.; Yang, J.J. The calcium sensing receptor: From calcium sensing to signaling. *Sci. China Life Sci.* **2015**, *58*, 14–27. [CrossRef] [PubMed]
9. Tennakoon, S.; Aggarwal, A.; Kallay, E. The calcium-sensing receptor and the hallmarks of cancer. *Biochim. Biophys. Acta* **2016**, *1863*, 1398–1407. [CrossRef] [PubMed]
10. Hannan, F.M.; Olesen, M.K.; Thakker, R.V. Calcimimetic and calcilytic therapies for inherited disorders of the calcium-sensing receptor signalling pathway. *Br. J. Pharmacol.* **2017**. [CrossRef] [PubMed]

11. Brown, E.M.; MacLeod, R.J. Extracellular calcium sensing and extracellular calcium signaling. *Physiol. Rev.* **2001**, *81*, 239–297. [CrossRef] [PubMed]

12. Hofer, A.M. Another dimension to calcium signaling: A look at extracellular calcium. *J. Cell Sci.* **2005**, *118*, 855–862. [CrossRef] [PubMed]

13. Ashby, M.C.; Tepikin, A.V. Polarized calcium and calmodulin signaling in secretory epithelia. *Physiol. Rev.* **2002**, *82*, 701–734. [CrossRef] [PubMed]

14. Belan, P.; Gerasimenko, O.; Petersen, O.H.; Tepikin, A.V. Distribution of Ca^{2+} extrusion sites on the mouse pancreatic acinar cell surface. *Cell Calcium* **1997**, *22*, 5–10. [CrossRef]

15. Peng, J.B.; Brown, E.M.; Hediger, M.A. Apical entry channels in calcium-transporting epithelia. *News Physiol. Sci.* **2003**, *18*, 158–163. [CrossRef] [PubMed]

16. Petersen, O.H. Localization and regulation of Ca^{2+} entry and exit pathways in exocrine gland cells. *Cell Calcium* **2003**, *33*, 337–344. [CrossRef]

17. Caroppo, R.; Gerbino, A.; Debellis, L.; Kifor, O.; Soybel, D.I.; Brown, E.M.; Hofer, A.M.; Curci, S. Asymmetrical, agonist-induced fluctuations in local extracellular $[Ca^{2+}]$ in intact polarized epithelia. *EMBO J.* **2001**, *20*, 6316–6326. [CrossRef] [PubMed]

18. Caroppo, R.; Gerbino, A.; Fistetto, G.; Colella, M.; Debellis, L.; Hofer, A.M.; Curci, S. Extracellular calcium acts as a "Third messenger" To regulate enzyme and alkaline secretion. *J. Cell Biol.* **2004**, *166*, 111–119. [CrossRef] [PubMed]

19. Gerbino, A.; Fistetto, G.; Colella, M.; Hofer, A.M.; Debellis, L.; Caroppo, R.; Curci, S. Real time measurements of water flow in amphibian gastric glands: Modulation via the extracellular Ca^{2+}-sensing receptor. *J. Biol. Chem.* **2007**, *282*, 13477–13486. [CrossRef] [PubMed]

20. Andersson, T.; Berggren, P.O.; Gylfe, E.; Hellman, B. Amounts and distribution of intracellular magnesium and calcium in pancreatic beta-cells. *Acta Physiol. Scand.* **1982**, *114*, 235–241. [CrossRef] [PubMed]

21. Nicaise, G.; Maggio, K.; Thirion, S.; Horoyan, M.; Keicher, E. The calcium loading of secretory granules. A possible key event in stimulus-secretion coupling. *Biol. Cell* **1992**, *75*, 89–99. [CrossRef]

22. Gillot, I.; Ciapa, B.; Payan, P.; Sardet, C. The calcium content of cortical granules and the loss of calcium from sea urchin eggs at fertilization. *Dev. Biol.* **1991**, *146*, 396–405. [CrossRef]

23. Hutton, J.C.; Penn, E.J.; Peshavaria, M. Low-molecular-weight constituents of isolated insulin-secretory granules. Bivalent cations, adenine nucleotides and inorganic phosphate. *Biochem. J.* **1983**, *210*, 297–305. [CrossRef] [PubMed]

24. Gerbino, A.; Maiellaro, I.; Carmone, C.; Caroppo, R.; Debellis, L.; Barile, M.; Busco, G.; Colella, M. Glucose increases extracellular $[Ca^{2+}]$ in rat insulinoma (ins-1e) pseudoislets as measured with Ca^{2+}-sensitive microelectrodes. *Cell Calcium* **2012**, *51*, 393–401. [CrossRef] [PubMed]

25. Belan, P.; Gardner, J.; Gerasimenko, O.; Gerasimenko, J.; Mills, C.L.; Petersen, O.H.; Tepikin, A.V. Isoproterenol evokes extracellular Ca^{2+} spikes due to secretory events in salivary gland cells. *J. Biol. Chem.* **1998**, *273*, 4106–4111. [CrossRef] [PubMed]

26. Von Grafenstein, H.R.; Powis, D.A. Calcium is released by exocytosis together with catecholamines from bovine adrenal medullary cells. *J. Neurochem.* **1989**, *53*, 428–435. [CrossRef] [PubMed]

27. Thirion, S.; Stuenkel, E.L.; Nicaise, G. Calcium loading of secretory granules in stimulated neurohypophysial nerve endings. *Neuroscience* **1995**, *64*, 125–137. [CrossRef]

28. Kuhtreiber, W.M.; Gillot, I.; Sardet, C.; Jaffe, L.F. Net calcium and acid release at fertilization in eggs of sea urchins and ascidians. *Cell Calcium* **1993**, *14*, 73–86. [CrossRef]

29. Hofer, A.M.; Gerbino, A.; Caroppo, R.; Curci, S. The extracellular calcium-sensing receptor and cell-cell signaling in epithelia. *Cell Calcium* **2004**, *35*, 297–306. [CrossRef] [PubMed]

30. Cohen, J.E.; Fields, R.D. Extracellular calcium depletion in synaptic transmission. *Neuroscientist* **2004**, *10*, 12–17. [CrossRef] [PubMed]

31. Egelman, D.M.; Montague, P.R. Calcium dynamics in the extracellular space of mammalian neural tissue. *Biophys. J.* **1999**, *76*, 1856–1867. [CrossRef]

32. Vassilev, P.M.; Mitchel, J.; Vassilev, M.; Kanazirska, M.; Brown, E.M. Assessment of frequency-dependent alterations in the level of extracellular Ca^{2+} in the synaptic cleft. *Biophys. J.* **1997**, *72*, 2103–2116. [CrossRef]

33. Pumain, R.; Heinemann, U. Stimulus- and amino acid-induced calcium and potassium changes in rat neocortex. *J. Neurophysiol.* **1985**, *53*, 1–16. [CrossRef] [PubMed]

34. Rusakov, D.A.; Fine, A. Extracellular Ca²⁺ depletion contributes to fast activity-dependent modulation of synaptic transmission in the brain. *Neuron* **2003**, *37*, 287–297. [CrossRef]

35. Moura, A.S. Membrane potential and intercellular calcium during glucose challenge in mouse islet of langerhans. *Biochem. Biophys. Res. Commun.* **1995**, *214*, 798–802. [CrossRef] [PubMed]

36. Perez-Armendariz, E.; Atwater, I. Glucose-evoked changes in [K⁺] and [Ca²⁺] in the intercellular spaces of the mouse islet of langerhans. *Adv. Exp. Med. Biol.* **1986**, *211*, 31–51. [PubMed]

37. Cleemann, L.; Pizarro, G.; Morad, M. Optical measurements of extracellular calcium depletion during a single heartbeat. *Science* **1984**, *226*, 174–177. [CrossRef] [PubMed]

38. Nicholson, C.; Bruggencate, G.T.; Steinberg, R.; Stöckle, H. Calcium modulation in brain extracellular microenvironment demonstrated with ion-selective micropipette. *Proc. Natl. Acad. Sci. USA* **1977**, *74*, 1287–1290. [CrossRef] [PubMed]

39. Jaffe, L.F.; Nuccitelli, R. An ultrasensitive vibrating probe for measuring steady extracellular currents. *J. Cell Biol.* **1974**, *63*, 614–628. [CrossRef] [PubMed]

40. Kuhtreiber, W.M.; Jaffe, L.F. Detection of extracellular calcium gradients with a calcium-specific vibrating electrode. *J. Cell Biol.* **1990**, *110*, 1565–1573. [CrossRef] [PubMed]

41. Yamoah, E.N.; Lumpkin, E.A.; Dumont, R.A.; Smith, P.J.; Hudspeth, A.J.; Gillespie, P.G. Plasma membrane Ca²⁺-atpase extrudes Ca²⁺ from hair cell stereocilia. *J. Neurosci.* **1998**, *18*, 610–624. [PubMed]

42. Pepperell, J.R.; Kommineni, K.; Buradagunta, S.; Smith, P.J.; Keefe, D.L. Transmembrane regulation of intracellular calcium by a plasma membrane sodium/calcium exchanger in mouse ova. *Biol. Reprod.* **1999**, *60*, 1137–1143. [CrossRef] [PubMed]

43. Knox, R.J.; Jonas, E.A.; Kao, L.S.; Smith, P.J.; Connor, J.A.; Kaczmarek, L.K. Ca²⁺ influx and activation of a cation current are coupled to intracellular Ca²⁺ release in peptidergic neurons of aplysia californica. *J. Physiol.* **1996**, *494 Pt 3*, 627–639. [CrossRef] [PubMed]

44. Smith, P.J.; Hammar, K.; Porterfield, D.M.; Sanger, R.H.; Trimarchi, J.R. Self-referencing, non-invasive, ion selective electrode for single cell detection of trans-plasma membrane calcium flux. *Microsc. Res. Tech.* **1999**, *46*, 398–417. [CrossRef]

45. Mupanomunda, M.M.; Ishioka, N.; Bukoski, R.D. Interstitial Ca²⁺ undergoes dynamic changes sufficient to stimulate nerve-dependent Ca²⁺-induced relaxation. *Am. J. Physiol.* **1999**, *276*, H1035–H1042. [CrossRef] [PubMed]

46. Mupanomunda, M.M.; Tian, B.; Ishioka, N.; Bukoski, R.D. Renal interstitial Ca²⁺. *Am. J. Physiol. Renal Physiol.* **2000**, *278*, F644–F649. [CrossRef] [PubMed]

47. Hilgemann, D.W. Extracellular calcium transients at single excitations in rabbit atrium measured with tetramethylmurexide. *J. Gen. Physiol.* **1986**, *87*, 707–735. [CrossRef] [PubMed]

48. Hilgemann, D.W.; Langer, G.A. Transsarcolemmal calcium movements in arterially perfused rabbit right ventricle measured with extracellular calcium-sensitive dyes. *Circ. Res.* **1984**, *54*, 461–467. [CrossRef] [PubMed]

49. Tepikin, A.V.; Llopis, J.; Snitsarev, V.A.; Gallacher, D.V.; Petersen, O.H. The droplet technique: Measurement of calcium extrusion from single isolated mammalian cells. *Pflugers Arch.* **1994**, *428*, 664–670. [CrossRef] [PubMed]

50. Tepikin, A.V.; Voronina, S.G.; Gallacher, D.V.; Petersen, O.H. Pulsatile Ca²⁺ extrusion from single pancreatic acinar cells during receptor-activated cytosolic Ca²⁺ spiking. *J. Biol. Chem.* **1992**, *267*, 14073–14076. [PubMed]

51. Tepikin, A.V.; Voronina, S.G.; Gallacher, D.V.; Petersen, O.H. Acetylcholine-evoked increase in the cytoplasmic Ca²⁺ concentration and Ca²⁺ extrusion measured simultaneously in single mouse pancreatic acinar cells. *J. Biol. Chem.* **1992**, *267*, 3569–3572. [PubMed]

52. Belan, P.V.; Gerasimenko, O.V.; Tepikin, A.V.; Petersen, O.H. Localization of Ca²⁺ extrusion sites in pancreatic acinar cells. *J. Biol. Chem.* **1996**, *271*, 7615–7619. [CrossRef] [PubMed]

53. Blatter, L.A.; Niggli, E. Confocal near-membrane detection of calcium in cardiac myocytes. *Cell Calcium* **1998**, *23*, 269–279. [CrossRef]

54. Etter, E.F.; Kuhn, M.A.; Fay, F.S. Detection of changes in near-membrane Ca²⁺ concentration using a novel membrane-associated Ca²⁺ indicator. *J. Biol. Chem.* **1994**, *269*, 10141–10149. [PubMed]

55. Etter, E.F.; Minta, A.; Poenie, M.; Fay, F.S. Near-membrane [Ca²⁺] transients resolved using the Ca²⁺ indicator ffp18. *Proc. Natl. Acad. Sci. USA* **1996**, *93*, 5368–5373. [CrossRef]

56. De Luisi, A.; Hofer, A.M. Evidence that Ca^{2+} cycling by the plasma membrane Ca^{2+}-atpase increases the 'excitability' of the extracellular Ca^{2+}-sensing receptor. *J. Cell Sci.* **2003**, *116*, 1527–1538. [CrossRef] [PubMed]

57. Care, A.D.; Sherwood, L.M.; Potts, J.T., Jr.; Aurbach, G.D. Perfusion of the isolated parathyroid gland of the goat and sheep. *Nature* **1966**, *209*, 55–57. [CrossRef] [PubMed]

58. Sherwood, L.M.; Potts, J.T., Jr.; Care, A.D.; Mayer, G.P.; Aurbach, G.D. Evaluation by radioimmunoassay of factors controlling the secretion of parathyroid hormone. *Nature* **1966**, *209*, 52–55. [CrossRef] [PubMed]

59. Lopez-Barneo, J.; Armstrong, C.M. Depolarizing response of rat parathyroid cells to divalent cations. *J. Gen. Physiol.* **1983**, *82*, 269–294. [CrossRef] [PubMed]

60. Shoback, D.; Thatcher, J.; Leombruno, R.; Brown, E. Effects of extracellular Ca^{2+} and Mg^{2+} on cytosolic Ca^{2+} and pth release in dispersed bovine parathyroid cells. *Endocrinology* **1983**, *113*, 424–426. [CrossRef] [PubMed]

61. Brown, E.; Enyedi, P.; LeBoff, M.; Rotberg, J.; Preston, J.; Chen, C. High extracellular Ca^{2+} and Mg^{2+} stimulate accumulation of inositol phosphates in bovine parathyroid cells. *FEBS Lett.* **1987**, *218*, 113–118. [CrossRef]

62. Kifor, O.; Brown, E.M. Relationship between diacylglycerol levels and extracellular Ca^{2+} in dispersed bovine parathyroid cells. *Endocrinology* **1988**, *123*, 2723–2729. [CrossRef] [PubMed]

63. Shoback, D.M.; McGhee, J.M. Fluoride stimulates the accumulation of inositol phosphates, increases intracellular free calcium, and inhibits parathyroid hormone release in dispersed bovine parathyroid cells. *Endocrinology* **1988**, *122*, 2833–2839. [CrossRef] [PubMed]

64. Shoback, D.M.; Membreno, L.A.; McGhee, J.G. High calcium and other divalent cations increase inositol trisphosphate in bovine parathyroid cells. *Endocrinology* **1988**, *123*, 382–389. [CrossRef] [PubMed]

65. Nemeth, E.F.; Scarpa, A. Are changes in intracellular free calcium necessary for regulating secretion in parathyroid cells? *Ann. N. Y. Acad. Sci.* **1987**, *493*, 542–551. [CrossRef] [PubMed]

66. Foord, S.M.; Bonner, T.I.; Neubig, R.R.; Rosser, E.M.; Pin, J.P.; Davenport, A.P.; Spedding, M.; Harmar, A.J. International union of pharmacology. Xlvi. G protein-coupled receptor list. *Pharmacol. Rev.* **2005**, *57*, 279–288. [CrossRef] [PubMed]

67. Clemmensen, C.; Smajilovic, S.; Wellendorph, P.; Brauner-Osborne, H. The GPCR, class c, group 6, subtype a (GPRC6A) receptor: From cloning to physiological function. *Br. J. Pharmacol.* **2014**, *171*, 1129–1141. [CrossRef] [PubMed]

68. Norskov-Lauritsen, L.; Brauner-Osborne, H. Role of post-translational modifications on structure, function and pharmacology of class c g protein-coupled receptors. *Eur. J. Pharmacol.* **2016**, *763*, 233–240. [CrossRef] [PubMed]

69. Garrett, J.E.; Capuano, I.V.; Hammerland, L.G.; Hung, B.C.; Brown, E.M.; Hebert, S.C.; Nemeth, E.F.; Fuller, F. Molecular cloning and functional expression of human parathyroid calcium receptor cdnas. *J. Biol. Chem.* **1995**, *270*, 12919–12925. [CrossRef] [PubMed]

70. Hu, J.; Spiegel, A.M. Structure and function of the human calcium-sensing receptor: Insights from natural and engineered mutations and allosteric modulators. *J. Cell. Mol. Med.* **2007**, *11*, 908–922. [CrossRef] [PubMed]

71. Kunishima, N.; Shimada, Y.; Tsuji, Y.; Sato, T.; Yamamoto, M.; Kumasaka, T.; Nakanishi, S.; Jingami, H.; Morikawa, K. Structural basis of glutamate recognition by a dimeric metabotropic glutamate receptor. *Nature* **2000**, *407*, 971–977. [PubMed]

72. Muto, T.; Tsuchiya, D.; Morikawa, K.; Jingami, H. Structures of the extracellular regions of the group ii/iii metabotropic glutamate receptors. *Proc. Natl. Acad. Sci. USA* **2007**, *104*, 3759–3764. [CrossRef] [PubMed]

73. Geng, Y.; Xiong, D.; Mosyak, L.; Malito, D.L.; Kniazeff, J.; Chen, Y.; Burmakina, S.; Quick, M.; Bush, M.; Javitch, J.A.; et al. Structure and functional interaction of the extracellular domain of human GABA$_B$ receptor GBR2. *Nat. Neurosci.* **2012**, *15*, 970–978. [CrossRef] [PubMed]

74. Chakravarti, B.; Chattopadhyay, N.; Brown, E.M. Signaling through the extracellular calcium-sensing receptor (CASR). *Adv. Exp. Med. Biol.* **2012**, *740*, 103–142. [PubMed]

75. Hu, J.; Reyes-Cruz, G.; Goldsmith, P.K.; Gantt, N.M.; Miller, J.L.; Spiegel, A.M. Functional effects of monoclonal antibodies to the purified amino-terminal extracellular domain of the human Ca^{2+} receptor. *J. Bone Miner Res.* **2007**, *22*, 601–608. [CrossRef] [PubMed]

76. Bai, M.; Trivedi, S.; Brown, E.M. Dimerization of the extracellular calcium-sensing receptor (CAR) on the cell surface of car-transfected hek293 cells. *J. Biol. Chem.* **1998**, *273*, 23605–23610. [CrossRef] [PubMed]

77. Bai, M.; Trivedi, S.; Kifor, O.; Quinn, S.J.; Brown, E.M. Intermolecular interactions between dimeric calcium-sensing receptor monomers are important for its normal function. *Proc. Natl. Acad. Sci. USA* **1999**, *96*, 2834–2839. [CrossRef] [PubMed]

78. Gama, L.; Wilt, S.G.; Breitwieser, G.E. Heterodimerization of calcium sensing receptors with metabotropic glutamate receptors in neurons. *J. Biol. Chem.* **2001**, *276*, 39053–39059. [CrossRef] [PubMed]

79. Chang, W.; Tu, C.; Cheng, Z.; Rodriguez, L.; Chen, T.H.; Gassmann, M.; Bettler, B.; Margeta, M.; Jan, L.Y.; Shoback, D. Complex formation with the type b gamma-aminobutyric acid receptor affects the expression and signal transduction of the extracellular calcium-sensing receptor. Studies with HEK-293 cells and neurons. *J. Biol. Chem.* **2007**, *282*, 25030–25040. [CrossRef] [PubMed]

80. Jiang, Y.; Minet, E.; Zhang, Z.; Silver, P.A.; Bai, M. Modulation of interprotomer relationships is important for activation of dimeric calcium-sensing receptor. *J. Biol. Chem.* **2004**, *279*, 14147–14156. [CrossRef] [PubMed]

81. Ray, K.; Hauschild, B.C.; Steinbach, P.J.; Goldsmith, P.K.; Hauache, O.; Spiegel, A.M. Identification of the cysteine residues in the amino-terminal extracellular domain of the human Ca^{2+} receptor critical for dimerization. Implications for function of monomeric Ca^{2+} receptor. *J. Biol. Chem.* **1999**, *274*, 27642–27650. [CrossRef] [PubMed]

82. Reyes-Cruz, G.; Hu, J.; Goldsmith, P.K.; Steinbach, P.J.; Spiegel, A.M. Human Ca^{2+} receptor extracellular domain. Analysis of function of lobe i loop deletion mutants. *J. Biol. Chem.* **2001**, *276*, 32145–32151. [CrossRef] [PubMed]

83. Pace, A.J.; Gama, L.; Breitwieser, G.E. Dimerization of the calcium-sensing receptor occurs within the extracellular domain and is eliminated by cys → ser mutations at cys101 and cys236. *J. Biol. Chem.* **1999**, *274*, 11629–11634. [CrossRef] [PubMed]

84. Zhang, Z.; Sun, S.; Quinn, S.J.; Brown, E.M.; Bai, M. The extracellular calcium-sensing receptor dimerizes through multiple types of intermolecular interactions. *J. Biol. Chem.* **2001**, *276*, 5316–5322. [CrossRef] [PubMed]

85. Fan, G.F.; Ray, K.; Zhao, X.M.; Goldsmith, P.K.; Spiegel, A.M. Mutational analysis of the cysteines in the extracellular domain of the human Ca^{2+} receptor: Effects on cell surface expression, dimerization and signal transduction. *FEBS Lett.* **1998**, *436*, 353–356. [CrossRef]

86. Jacobsen, S.E.; Gether, U.; Brauner-Osborne, H. Investigating the molecular mechanism of positive and negative allosteric modulators in the calcium-sensing receptor dimer. *Sci. Rep.* **2018**, *7*, 46355. [CrossRef] [PubMed]

87. Hu, J.; Hauache, O.; Spiegel, A.M. Human Ca^{2+} receptor cysteine-rich domain. Analysis of function of mutant and chimeric receptors. *J. Biol. Chem.* **2000**, *275*, 16382–16389. [CrossRef] [PubMed]

88. Pin, J.P.; Galvez, T.; Prezeau, L. Evolution, structure, and activation mechanism of family 3/c G-protein-coupled receptors. *Pharmacol. Ther.* **2003**, *98*, 325–354. [CrossRef]

89. Ray, J.M.; Squires, P.E.; Curtis, S.B.; Meloche, M.R.; Buchan, A.M. Expression of the calcium-sensing receptor on human antral gastrin cells in culture. *J. Clin. Invest.* **1997**, *99*, 2328–2333. [CrossRef] [PubMed]

90. Awata, H.; Huang, C.; Handlogten, M.E.; Miller, R.T. Interaction of the calcium-sensing receptor and filamin, a potential scaffolding protein. *J. Biol. Chem.* **2001**, *276*, 34871–34879. [CrossRef] [PubMed]

91. Hjalm, G.; MacLeod, R.J.; Kifor, O.; Chattopadhyay, N.; Brown, E.M. Filamin-a binds to the carboxyl-terminal tail of the calcium-sensing receptor, an interaction that participates in car-mediated activation of mitogen-activated protein kinase. *J. Biol. Chem.* **2001**, *276*, 34880–34887. [CrossRef] [PubMed]

92. Bosel, J.; John, M.; Freichel, M.; Blind, E. Signaling of the human calcium-sensing receptor expressed in HEK293-cells is modulated by protein kinases a and c. *Exp. Clin. Endocrinol. Diabetes* **2003**, *111*, 21–26. [CrossRef] [PubMed]

93. Stepanchick, A.; McKenna, J.; McGovern, O.; Huang, Y.; Breitwieser, G.E. Calcium sensing receptor mutations implicated in pancreatitis and idiopathic epilepsy syndrome disrupt an arginine-rich retention motif. *Cell. Physiol. Biochem.* **2010**, *26*, 363–374. [CrossRef] [PubMed]

94. Hendy, G.N.; Canaff, L.; Cole, D.E. The casr gene: Alternative splicing and transcriptional control, and calcium-sensing receptor (CASR) protein: Structure and ligand binding sites. *Best Pract. Res. Clin. Endocrinol. Metab.* **2013**, *27*, 285–301. [CrossRef] [PubMed]

95. Leach, K.; Gregory, K.J.; Kufareva, I.; Khajehali, E.; Cook, A.E.; Abagyan, R.; Conigrave, A.D.; Sexton, P.M.; Christopoulos, A. Towards a structural understanding of allosteric drugs at the human calcium-sensing receptor. *Cell Res.* **2016**, *26*, 574–592. [CrossRef] [PubMed]

96. Huang, Y.; Zhou, Y.; Yang, W.; Butters, R.; Lee, H.W.; Li, S.; Castiblanco, A.; Brown, E.M.; Yang, J.J. Identification and dissection of Ca^{2+}-binding sites in the extracellular domain of Ca^{2+}-sensing receptor. *J. Biol. Chem.* **2007**, *282*, 19000–19010. [CrossRef] [PubMed]

97. Huang, Y.; Zhou, Y.; Castiblanco, A.; Yang, W.; Brown, E.M.; Yang, J.J. Multiple Ca^{2+}-binding sites in the extracellular domain of the Ca^{2+}-sensing receptor corresponding to cooperative Ca^{2+} response. *Biochemistry* **2009**, *48*, 388–398. [CrossRef] [PubMed]

98. Huang, Y.; Zhou, Y.; Wong, H.C.; Castiblanco, A.; Chen, Y.; Brown, E.M.; Yang, J.J. Calmodulin regulates Ca^{2+}-sensing receptor-mediated Ca^{2+} signaling and its cell surface expression. *J. Biol. Chem.* **2010**, *285*, 35919–35931. [CrossRef] [PubMed]

99. Zhang, C.; Zhuo, Y.; Moniz, H.A.; Wang, S.; Moremen, K.W.; Prestegard, J.H.; Brown, E.M.; Yang, J.J. Direct determination of multiple ligand interactions with the extracellular domain of the calcium-sensing receptor. *J. Biol. Chem.* **2014**, *289*, 33529–33542. [CrossRef] [PubMed]

100. Zhang, C.; Huang, Y.; Jiang, Y.; Mulpuri, N.; Wei, L.; Hamelberg, D.; Brown, E.M.; Yang, J.J. Identification of an L-phenylalanine binding site enhancing the cooperative responses of the calcium-sensing receptor to calcium. *J. Biol. Chem.* **2014**, *289*, 5296–5309. [CrossRef] [PubMed]

101. Zhang, C.; Mulpuri, N.; Hannan, F.M.; Nesbit, M.A.; Thakker, R.V.; Hamelberg, D.; Brown, E.M.; Yang, J.J. Role of Ca^{2+} and l-phe in regulating functional cooperativity of disease-associated "Toggle" Calcium-sensing receptor mutations. *PLoS ONE* **2014**, *9*, e113622. [CrossRef] [PubMed]

102. Zhang, C.; Zhang, T.; Zou, J.; Miller, C.L.; Gorkhali, R.; Yang, J.Y.; Schilmiller, A.; Wang, S.; Huang, K.; Brown, E.M.; et al. Structural basis for regulation of human calcium-sensing receptor by magnesium ions and an unexpected tryptophan derivative co-agonist. *Sci. Adv.* **2016**, *2*, e1600241. [CrossRef] [PubMed]

103. Zhang, C.; Miller, C.L.; Gorkhali, R.; Zou, J.; Huang, K.; Brown, E.M.; Yang, J.J. Molecular basis of the extracellular ligands mediated signaling by the calcium sensing receptor. *Front. Physiol.* **2016**, *7*, 441. [CrossRef] [PubMed]

104. Brown, E.M.; Fuleihan, G.E.-H.; Chen, C.J.; Kifor, O. A comparison of the effects of divalent and trivalent cations on parathyroid hormone release, 3′,5′-cyclic-adenosine monophosphate accumulation, and the levels of inositol phosphates in bovine parathyroid cells. *Endocrinology* **1990**, *127*, 1064–1071. [CrossRef] [PubMed]

105. Quinn, S.J.; Ye, C.P.; Diaz, R.; Kifor, O.; Bai, M.; Vassilev, P.; Brown, E. The Ca^{2+}-sensing receptor: A target for polyamines. *Am. J. Physiol.* **1997**, *273*, C1315–C1323. [CrossRef] [PubMed]

106. Brown, E.M.; Katz, C.; Butters, R.; Kifor, O. Polyarginine, polylysine, and protamine mimic the effects of high extracellular calcium concentrations on dispersed bovine parathyroid cells. *J. Bone Miner Res.* **1991**, *6*, 1217–1225. [CrossRef] [PubMed]

107. Ye, C.; Ho-Pao, C.L.; Kanazirska, M.; Quinn, S.; Rogers, K.; Seidman, C.E.; Seidman, J.G.; Brown, E.M.; Vassilev, P.M. Amyloid-beta proteins activate Ca^{2+}-permeable channels through calcium-sensing receptors. *J. Neurosci. Res.* **1997**, *47*, 547–554. [CrossRef]

108. Brown, E.M.; Butters, R.; Katz, C.; Kifor, O. Neomycin mimics the effects of high extracellular calcium concentrations on parathyroid function in dispersed bovine parathyroid cells. *Endocrinology* **1991**, *128*, 3047–3054. [CrossRef] [PubMed]

109. Nemeth, E.F. Regulation of cytosolic calcium by extracellular divalent cations in c-cells and parathyroid cells. *Cell Calcium* **1990**, *11*, 323–327. [CrossRef]

110. Colella, M.; Gerbino, A.; Hofer, A.M.; Curci, S. Recent advances in understanding the extracellular calcium-sensing receptor. *F1000Res* **2016**, *5*. [CrossRef] [PubMed]

111. Conigrave, A.D.; Quinn, S.J.; Brown, E.M. L-amino acid sensing by the extracellular Ca^{2+}-sensing receptor. *Proc. Natl. Acad. Sci. USA* **2000**, *97*, 4814–4819. [CrossRef] [PubMed]

112. Conigrave, A.D.; Mun, H.C.; Lok, H.C. Aromatic l-amino acids activate the calcium-sensing receptor. *J. Nutr.* **2007**, *137*, 1524S–1527S; discussion 1548S. [CrossRef] [PubMed]

113. Conigrave, A.D.; Hampson, D.R. Broad-spectrum amino acid-sensing class c G-protein coupled receptors: Molecular mechanisms, physiological significance and options for drug development. *Pharmacol. Ther.* **2010**, *127*, 252–260. [CrossRef] [PubMed]

114. Wang, M.; Yao, Y.; Kuang, D.; Hampson, D.R. Activation of family c G-protein-coupled receptors by the tripeptide glutathione. *J. Biol. Chem.* **2006**, *281*, 8864–8870. [CrossRef] [PubMed]

115. Broadhead, G.K.; Mun, H.C.; Avlani, V.A.; Jourdon, O.; Church, W.B.; Christopoulos, A.; Delbridge, L.; Conigrave, A.D. Allosteric modulation of the calcium-sensing receptor by gamma-glutamyl peptides: Inhibition of pth secretion, suppression of intracellular camp levels, and a common mechanism of action with l-amino acids. *J. Biol. Chem.* **2011**, *286*, 8786–8797. [CrossRef] [PubMed]

116. Quinn, S.J.; Bai, M.; Brown, E.M. Ph sensing by the calcium-sensing receptor. *J. Biol. Chem.* **2004**, *279*, 37241–37249. [CrossRef] [PubMed]

117. Quinn, S.J.; Kifor, O.; Trivedi, S.; Diaz, R.; Vassilev, P.; Brown, E. Sodium and ionic strength sensing by the calcium receptor. *J. Biol. Chem.* **1998**, *273*, 19579–19586. [CrossRef] [PubMed]

118. Nemeth, E.F.; Steffey, M.E.; Hammerland, L.G.; Hung, B.C.; van Wagenen, B.C.; DelMar, E.G.; Balandrin, M.F. Calcimimetics with potent and selective activity on the parathyroid calcium receptor. *Proc. Natl. Acad. Sci. USA* **1998**, *95*, 4040–4045. [CrossRef] [PubMed]

119. Kiefer, L.; Leiris, S.; Dodd, R.H. Novel calcium sensing receptor ligands: A patent survey. *Expert Opin. Ther. Pat.* **2011**, *21*, 681–698. [CrossRef] [PubMed]

120. Kiefer, L.; Beaumard, F.; Gorojankina, T.; Faure, H.; Ruat, M.; Dodd, R.H. Design and synthesis of calindol derivatives as potent and selective calcium sensing receptor agonists. *Bioorg. Med. Chem.* **2015**, *24*, 554–569. [CrossRef] [PubMed]

121. Nemeth, E.F.; Goodman, W.G. Calcimimetic and calcilytic drugs: Feats, flops, and futures. *Calcif. Tissue Int.* **2016**, *98*, 341–358. [CrossRef] [PubMed]

122. Nemeth, E.F. Calcimimetic and calcilytic drugs: Just for parathyroid cells? *Cell Calcium* **2004**, *35*, 283–289. [CrossRef] [PubMed]

123. Lindberg, J.S.; Culleton, B.; Wong, G.; Borah, M.F.; Clark, R.V.; Shapiro, W.B.; Roger, S.D.; Husserl, F.E.; Klassen, P.S.; Guo, M.D.; et al. Cinacalcet hcl, an oral calcimimetic agent for the treatment of secondary hyperparathyroidism in hemodialysis and peritoneal dialysis: A randomized, double-blind, multicenter study. *J. Am. Soc. Nephrol.* **2005**, *16*, 800–807. [CrossRef] [PubMed]

124. Hebert, S.C. Therapeutic use of calcimimetics. *Annu. Rev. Med.* **2006**, *57*, 349–364. [CrossRef] [PubMed]

125. Brown, E.M. Clinical utility of calcimimetics targeting the extracellular calcium-sensing receptor (casr). *Biochem. Pharmacol.* **2010**, *80*, 297–307. [CrossRef] [PubMed]

126. Henley, C., 3rd; Yang, Y.; Davis, J.; Lu, J.Y.; Morony, S.; Fan, W.; Florio, M.; Sun, B.; Shatzen, E.; Pretorius, J.K.; et al. Discovery of a calcimimetic with differential effects on parathyroid hormone and calcitonin secretion. *J. Pharmacol. Exp. Ther.* **2011**, *337*, 681–691. [CrossRef] [PubMed]

127. Ma, J.N.; Owens, M.; Gustafsson, M.; Jensen, J.; Tabatabaei, A.; Schmelzer, K.; Olsson, R.; Burstein, E.S. Characterization of highly efficacious allosteric agonists of the human calcium-sensing receptor. *J. Pharmacol. Exp. Ther.* **2011**, *337*, 275–284. [CrossRef] [PubMed]

128. Cook, A.E.; Mistry, S.N.; Gregory, K.J.; Furness, S.G.; Sexton, P.M.; Scammells, P.J.; Conigrave, A.D.; Christopoulos, A.; Leach, K. Biased allosteric modulation at the cas receptor engendered by structurally diverse calcimimetics. *Br. J. Pharmacol.* **2015**, *172*, 185–200. [CrossRef] [PubMed]

129. Nemeth, E.F. The search for calcium receptor antagonists (calcilytics). *J. Mol. Endocrinol.* **2002**, *29*, 15–21. [CrossRef] [PubMed]

130. Magno, A.L.; Ward, B.K.; Ratajczak, T. The calcium-sensing receptor: A molecular perspective. *Endocr. Rev.* **2011**, *32*, 3–30. [CrossRef] [PubMed]

131. Conigrave, A.D.; Ward, D.T. Calcium-sensing receptor (CASR): Pharmacological properties and signaling pathways. *Best Pract. Res. Clin. Endocrinol. Metab.* **2013**, *27*, 315–331. [CrossRef] [PubMed]

132. Nemeth, E.F.; Scarpa, A. Rapid mobilization of cellular Ca^{2+} in bovine parathyroid cells evoked by extracellular divalent cations. Evidence for a cell surface calcium receptor. *J. Biol. Chem.* **1987**, *262*, 5188–5196. [PubMed]

133. Hannan, F.M.; Babinsky, V.N.; Thakker, R.V. Disorders of the calcium-sensing receptor and partner proteins: Insights into the molecular basis of calcium homeostasis. *J. Mol. Endocrinol.* **2016**, *57*, R127–R142. [CrossRef] [PubMed]

134. Kifor, O.; MacLeod, R.J.; Diaz, R.; Bai, M.; Yamaguchi, T.; Yao, T.; Kifor, I.; Brown, E.M. Regulation of map kinase by calcium-sensing receptor in bovine parathyroid and car-transfected hek293 cells. *Am. J. Physiol. Renal Physiol.* **2001**, *280*, F291–F302. [CrossRef] [PubMed]

135. Kifor, O.; Diaz, R.; Butters, R.; Brown, E.M. The Ca^{2+}-sensing receptor (CAR) activates phospholipases c, a2, and d in bovine parathyroid and car-transfected, human embryonic kidney (HEK293) cells. *J. Bone Miner. Res.* **1997**, *12*, 715–725. [CrossRef] [PubMed]

136. Boudot, C.; Saidak, Z.; Boulanouar, A.K.; Petit, L.; Gouilleux, F.; Massy, Z.; Brazier, M.; Mentaverri, R.; Kamel, S. Implication of the calcium sensing receptor and the phosphoinositide 3-kinase/AKT pathway in the extracellular calcium-mediated migration of raw 264.7 osteoclast precursor cells. *Bone* **2010**, *46*, 1416–1423. [CrossRef] [PubMed]

137. Li, H.X.; Kong, F.J.; Bai, S.Z.; He, W.; Xing, W.J.; Xi, Y.H.; Li, G.W.; Guo, J.; Li, H.Z.; Wu, L.Y.; et al. Involvement of calcium-sensing receptor in oxldl-induced MMP-2 production in vascular smooth muscle cells via PI3K/AKT pathway. *Mol. Cell. Biochem.* **2012**, *362*, 115–122. [CrossRef] [PubMed]

138. de Jesus Ferreira, M.C.; Helies-Toussaint, C.; Imbert-Teboul, M.; Bailly, C.; Verbavatz, J.M.; Bellanger, A.C.; Chabardes, D. Co-expression of a Ca^{2+}-inhibitable adenylyl cyclase and of a Ca^{2+}-sensing receptor in the cortical thick ascending limb cell of the rat kidney. Inhibition of hormone-dependent camp accumulation by extracellular Ca^{2+}. *J. Biol. Chem.* **1998**, *273*, 15192–15202. [CrossRef] [PubMed]

139. Gerbino, A.; Ruder, W.C.; Curci, S.; Pozzan, T.; Zaccolo, M.; Hofer, A.M. Termination of camp signals by Ca^{2+} and Gαi via extracellular Ca^{2+} sensors: A link to intracellular Ca^{2+} oscillations. *J. Cell Biol.* **2005**, *171*, 303–312. [CrossRef] [PubMed]

140. Beierwaltes, W.H. The role of calcium in the regulation of renin secretion. *Am. J. Physiol. Renal Physiol.* **2010**, *298*, F1–F11. [CrossRef] [PubMed]

141. Ward, B.K.; Magno, A.L.; Walsh, J.P.; Ratajczak, T. The role of the calcium-sensing receptor in human disease. *Clin. Biochem.* **2012**, *45*, 943–953. [CrossRef] [PubMed]

142. Fitzpatrick, L.A.; Brandi, M.L.; Aurbach, G.D. Prostaglandin F2 alpha and alpha-adrenergic agonists regulate parathyroid cell function via the inhibitory guanine nucleotide regulatory protein. *Endocrinology* **1986**, *118*, 2115–2119. [CrossRef] [PubMed]

143. Chen, C.J.; Barnett, J.V.; Congo, D.A.; Brown, E.M. Divalent cations suppress 3′,5′-adenosine monophosphate accumulation by stimulating a pertussis toxin-sensitive guanine nucleotide-binding protein in cultured bovine parathyroid cells. *Endocrinology* **1989**, *124*, 233–239. [CrossRef] [PubMed]

144. Leach, K.; Sexton, P.M.; Christopoulos, A.; Conigrave, A.D. Engendering biased signalling from the calcium-sensing receptor for the pharmacotherapy of diverse disorders. *Br. J. Pharmacol.* **2014**, *171*, 1142–1155. [CrossRef] [PubMed]

145. Thomsen, A.R.B.; Hvidtfeldt, M.; Bräuner-Osborne, H. Biased agonism of the calcium-sensing receptor. *Cell Calcium* **2012**, *51*, 107–116. [CrossRef] [PubMed]

146. Huang, C.; Handlogten, M.E.; Miller, R.T. Parallel activation of phosphatidylinositol 4-kinase and phospholipase c by the extracellular calcium-sensing receptor. *J. Biol. Chem.* **2002**, *277*, 20293–20300. [CrossRef] [PubMed]

147. Avlani, V.A.; Ma, W.; Mun, H.C.; Leach, K.; Delbridge, L.; Christopoulos, A.; Conigrave, A.D. Calcium-sensing receptor-dependent activation of CREB phosphorylation in HEK293 cells and human parathyroid cells. *Am. J. Physiol. Endocrinol. Metab.* **2013**, *304*, E1097–E1104. [CrossRef] [PubMed]

148. Rey, O.; Young, S.H.; Yuan, J.; Slice, L.; Rozengurt, E. Amino acid-stimulated Ca^{2+} oscillations produced by the Ca^{2+}-sensing receptor are mediated by a phospholipase c/inositol 1,4,5-trisphosphate-independent pathway that requires g12, rho, filamin-a, and the actin cytoskeleton. *J. Biol. Chem.* **2005**, *280*, 22875–22882. [CrossRef] [PubMed]

149. Rey, O.; Young, S.H.; Papazyan, R.; Shapiro, M.S.; Rozengurt, E. Requirement of the trpc1 cation channel in the generation of transient Ca^{2+} oscillations by the calcium-sensing receptor. *J. Biol. Chem.* **2006**, *281*, 38730–38737. [CrossRef] [PubMed]

150. Rey, O.; Young, S.H.; Jacamo, R.; Moyer, M.P.; Rozengurt, E. Extracellular calcium sensing receptor stimulation in human colonic epithelial cells induces intracellular calcium oscillations and proliferation inhibition. *J. Cell. Physiol.* **2010**, *225*, 73–83. [CrossRef] [PubMed]

151. Zhu, D.; Kosik, K.S.; Meigs, T.E.; Yanamadala, V.; Denker, B.M. Galpha12 directly interacts with pp2a: Evidence for Gα12-stimulated PP2A phosphatase activity and dephosphorylation of microtubule-associated protein, tau. *J. Biol. Chem.* **2004**, *279*, 54983–54986. [CrossRef] [PubMed]

152. Zhu, D.; Tate, R.I.; Ruediger, R.; Meigs, T.E.; Denker, B.M. Domains necessary for galpha12 binding and stimulation of protein phosphatase-2a (PP2A): Is galpha12 a novel regulatory subunit of PP2A? *Mol. Pharmacol.* **2007**, *71*, 1268–1276. [CrossRef] [PubMed]

153. McCormick, W.D.; Atkinson-Dell, R.; Campion, K.L.; Mun, H.C.; Conigrave, A.D.; Ward, D.T. Increased receptor stimulation elicits differential calcium-sensing receptor(T888) dephosphorylation. *J. Biol. Chem.* **2010**, *285*, 14170–14177. [CrossRef] [PubMed]

154. Rybchyn, M.S.; Slater, M.; Conigrave, A.D.; Mason, R.S. An AKT-dependent increase in canonical wnt signaling and a decrease in sclerostin protein levels are involved in strontium ranelate-induced osteogenic effects in human osteoblasts. *J. Biol. Chem.* **2011**, *286*, 23771–23779. [CrossRef] [PubMed]

155. Brennan, T.C.; Rybchyn, M.S.; Green, W.; Atwa, S.; Conigrave, A.D.; Mason, R.S. Osteoblasts play key roles in the mechanisms of action of strontium ranelate. *Br. J. Pharmacol.* **2009**, *157*, 1291–1300. [CrossRef] [PubMed]

156. Brennan, S.C.; Thiem, U.; Roth, S.; Aggarwal, A.; Fetahu, I.; Tennakoon, S.; Gomes, A.R.; Brandi, M.L.; Bruggeman, F.; Mentaverri, R.; et al. Calcium sensing receptor signalling in physiology and cancer. *Biochim. Biophys. Acta* **2013**, *1833*, 1732–1744. [CrossRef] [PubMed]

157. Mamillapalli, R.; VanHouten, J.; Zawalich, W.; Wysolmerski, J. Switching of G-protein usage by the calcium-sensing receptor reverses its effect on parathyroid hormone-related protein secretion in normal versus malignant breast cells. *J. Biol. Chem.* **2008**, *283*, 24435–24447. [CrossRef] [PubMed]

158. Mamillapalli, R.; Wysolmerski, J. The calcium-sensing receptor couples to Galpha(s) and regulates pthrp and acth secretion in pituitary cells. *J. Endocrinol.* **2010**, *204*, 287–297. [CrossRef] [PubMed]

159. Brennan, S.C.; Wilkinson, W.J.; Tseng, H.E.; Finney, B.; Monk, B.; Dibble, H.; Quilliam, S.; Warburton, D.; Galietta, L.J.; Kemp, P.J.; et al. The extracellular calcium-sensing receptor regulates human fetal lung development via cftr. *Sci. Rep.* **2016**, *6*, 21975. [CrossRef] [PubMed]

160. Ziegelstein, R.C.; Xiong, Y.; He, C.; Hu, Q. Expression of a functional extracellular calcium-sensing receptor in human aortic endothelial cells. *Biochem. Biophys. Res. Commun.* **2006**, *342*, 153–163. [CrossRef] [PubMed]

161. Smajilovic, S.; Sheykhzade, M.; Holmegard, H.N.; Haunso, S.; Tfelt-Hansen, J. Calcimimetic, AMG 073, induces relaxation on isolated rat aorta. *Vascul. Pharmacol.* **2007**, *47*, 222–228. [CrossRef] [PubMed]

162. Greenberg, H.Z.E.; Shi, J.; Jahan, K.S.; Martinucci, M.C.; Gilbert, S.J.; Vanessa Ho, W.S.; Albert, A.P. Stimulation of calcium-sensing receptors induces endothelium-dependent vasorelaxations via nitric oxide production and activation of ikca channels. *Vascul. Pharmacol.* **2016**, *80*, 75–84. [CrossRef] [PubMed]

163. Qu, Y.Y.; Wang, L.M.; Zhong, H.; Liu, Y.M.; Tang, N.; Zhu, L.P.; He, F.; Hu, Q.H. TRPC1 stimulates calciumsensing receptorinduced storeoperated Ca^{2+} entry and nitric oxide production in endothelial cells. *Mol. Med. Rep.* **2017**, *16*, 4613–4619. [CrossRef] [PubMed]

164. Greenberg, H.Z.E.; Carlton-Carew, S.R.E.; Khan, D.M.; Zargaran, A.K.; Jahan, K.S.; Vanessa Ho, W.S.; Albert, A.P. Heteromeric TRPV4/TRPC1 channels mediate calcium-sensing receptor-induced nitric oxide production and vasorelaxation in rabbit mesenteric arteries. *Vascul. Pharmacol.* **2017**, *96–98*, 53–62. [CrossRef] [PubMed]

165. Garrett, J.E.; Tamir, H.; Kifor, O.; Simin, R.T.; Rogers, K.V.; Mithal, A.; Gagel, R.F.; Brown, E.M. Calcitonin-secreting cells of the thyroid express an extracellular calcium receptor gene. *Endocrinology* **1995**, *136*, 5202–5211. [CrossRef] [PubMed]

166. Hofer, A.M.; Brown, E.M. Extracellular calcium sensing and signalling. *Nat. Rev. Mol. Cell Biol.* **2003**, *4*, 530–538. [CrossRef] [PubMed]

167. Brown, E.M. Mutations in the calcium-sensing receptor and their clinical implications. *Horm. Res.* **1997**, *48*, 199–208. [CrossRef] [PubMed]

168. Pollak, M.R.; Brown, E.M.; Chou, Y.H.; Hebert, S.C.; Marx, S.J.; Steinmann, B.; Levi, T.; Seidman, C.E.; Seidman, J.G. Mutations in the human Ca^{2+}-sensing receptor gene cause familial hypocalciuric hypercalcemia and neonatal severe hyperparathyrcidism. *Cell* **1993**, *75*, 1297–1303. [CrossRef]

169. Watanabe, S.; Fukumoto, S.; Chang, H.; Takeuchi, Y.; Hasegawa, Y.; Okazaki, R.; Chikatsu, N.; Fujita, T. Association between activating mutations of calcium-sensing receptor and Bartter's syndrome. *Lancet* **2002**, *360*, 692–694. [CrossRef]

170. Pollak, M.R.; Brown, E.M.; Estep, H.L.; McLaine, P.N.; Kifor, O.; Park, J.; Hebert, S.C.; Seidman, C.E.; Seidman, J.G. Autosomal dominant hypocalcaemia caused by a Ca^{2+}-sensing receptor gene mutation. *Nat. Genet.* **1994**, *8*, 303–307. [CrossRef] [PubMed]

171. Zhao, X.M.; Hauache, O.; Goldsmith, P.K.; Collins, R.; Spiegel, A.M. A missense mutation in the seventh transmembrane domain constitutively activates the human Ca^{2+} receptor. *FEBS Lett.* **1999**, *448*, 180–184. [CrossRef]

172. Carmosino, M.; Gerbino, A.; Hendy, G.N.; Torretta, S.; Rizzo, F.; Debellis, L.; Procino, G.; Svelto, M. NKCC2 activity is inhibited by the Bartter's syndrome type 5 gain-of-function CAR-A843E mutant in renal cells. *Biol. Cell* **2015**, *107*, 98–110. [CrossRef] [PubMed]

173. Tfelt-Hansen, J.; Brown, E.M. The calcium-sensing receptor in normal physiology and pathophysiology: A review. *Crit. Rev. Clin. Lab. Sci.* **2005**, *42*, 35–70. [CrossRef] [PubMed]

174. Ruat, M.; Molliver, M.E.; Snowman, A.M.; Snyder, S.H. Calcium sensing receptor: Molecular cloning in rat and localization to nerve terminals. *Proc. Natl. Acad. Sci. USA* **1995**, *92*, 3161–3165. [CrossRef] [PubMed]

175. Squires, P.E. Non-Ca^{2+}-homeostatic functions of the extracellular Ca^{2+}-sensing receptor (CAR) in endocrine tissues. *J. Endocrinol.* **2000**, *165*, 173–177. [CrossRef] [PubMed]

176. Canaff, L.; Petit, J.L.; Kisiel, M.; Watson, P.H.; Gascon-Barre, M.; Hendy, G.N. Extracellular calcium-sensing receptor is expressed in rat hepatocytes. Coupling to intracellular calcium mobilization and stimulation of bile flow. *J. Biol. Chem.* **2001**, *276*, 4070–4079. [CrossRef] [PubMed]

177. Wang, R.; Xu, C.; Zhao, W.; Zhang, J.; Cao, K.; Yang, B.; Wu, L. Calcium and polyamine regulated calcium-sensing receptors in cardiac tissues. *Eur. J. Biochem.* **2003**, *270*, 2680–2688. [CrossRef] [PubMed]

178. Tfelt-Hansen, J.; Hansen, J.L.; Smajilovic, S.; Terwilliger, E.F.; Haunso, S.; Sheikh, S.P. Calcium receptor is functionally expressed in rat neonatal ventricular cardiomyocytes. *Am. J. Physiol. Heart Circ. Physiol.* **2006**, *290*, H1165–H1171. [CrossRef] [PubMed]

179. Alfadda, T.I.; Saleh, A.M.; Houillier, P.; Geibel, J.P. Calcium-sensing receptor 20 years later. *Am. J. Physiol. Cell Physiol.* **2014**, *307*, C221–C231. [CrossRef] [PubMed]

180. Saidak, Z.; Mentaverri, R.; Brown, E.M. The role of the calcium-sensing receptor in the development and progression of cancer. *Endocr. Rev.* **2009**, *30*, 178–195. [CrossRef] [PubMed]

181. Corbetta, S.; Mantovani, G.; Lania, A.; Borgato, S.; Vicentini, L.; Beretta, E.; Faglia, G.; di Blasio, A.M.; Spada, A. Calcium-sensing receptor expression and signalling in human parathyroid adenomas and primary hyperplasia. *Clin. Endocrinol.* **2000**, *52*, 339–348. [CrossRef]

182. Gogusev, J.; Duchambon, P.; Hory, B.; Giovannini, M.; Goureau, Y.; Sarfati, E.; Drueke, T.B. Depressed expression of calcium receptor in parathyroid gland tissue of patients with hyperparathyroidism. *Kidney Int.* **1997**, *51*, 328–336. [CrossRef] [PubMed]

183. Farnebo, F.; Enberg, U.; Grimelius, L.; Backdahl, M.; Schalling, M.; Larsson, C.; Farnebo, L.O. Tumor-specific decreased expression of calcium sensing receptor messenger ribonucleic acid in sporadic primary hyperparathyroidism. *J. Clin. Endocrinol. Metab.* **1997**, *82*, 3481–3486. [CrossRef] [PubMed]

184. Sheinin, Y.; Kallay, E.; Wrba, F.; Kriwanek, S.; Peterlik, M.; Cross, H.S. Immunocytochemical localization of the extracellular calcium-sensing receptor in normal and malignant human large intestinal mucosa. *J. Histochem. Cytochem.* **2000**, *48*, 595–602. [CrossRef] [PubMed]

185. Chakrabarty, S.; Wang, H.; Canaff, L.; Hendy, G.N.; Appelman, H.; Varani, J. Calcium sensing receptor in human colon carcinoma: Interaction with Ca^{2+} and 1,25-dihydroxyvitamin D_3. *Cancer Res.* **2005**, *65*, 493–498. [PubMed]

186. Miller, G.; Davis, J.; Shatzen, E.; Colloton, M.; Martin, D.; Henley, C.M. Cinacalcet hcl prevents development of parathyroid gland hyperplasia and reverses established parathyroid gland hyperplasia in a rodent model of ckd. *Nephrol. Dial. Transplant.* **2012**, *27*, 2198–2205. [CrossRef] [PubMed]

187. Mizobuchi, M.; Hatamura, I.; Ogata, H.; Saji, F.; Uda, S.; Shiizaki, K.; Sakaguchi, T.; Negi, S.; Kinugasa, E.; Koshikawa, S.; et al. Calcimimetic compound upregulates decreased calcium-sensing receptor expression level in parathyroid glands of rats with chronic renal insufficiency. *J. Am. Soc. Nephrol.* **2004**, *15*, 2579–2587. [CrossRef] [PubMed]

188. Garland, C.; Shekelle, R.B.; Barrett-Connor, E.; Criqui, M.H.; Rossof, A.H.; Paul, O. Dietary vitamin d and calcium and risk of colorectal cancer: A 19-year prospective study in men. *Lancet* **1985**, *1*, 307–309. [CrossRef]

189. Aggarwal, A.; Hobaus, J.; Tennakoon, S.; Prinz-Wohlgenannt, M.; Graca, J.; Price, S.A.; Heffeter, P.; Berger, W.; Baumgartner-Parzer, S.; Kallay, E. Active vitamin d potentiates the anti-neoplastic effects of calcium in the colon: A cross talk through the calcium-sensing receptor. *J. Steroid Biochem. Mol. Biol.* **2016**, *155*, 231–238. [CrossRef] [PubMed]

190. Canaff, L.; Hendy, G.N. Human calcium-sensing receptor gene. Vitamin d response elements in promoters P1 and P2 confer transcriptional responsiveness to 1,25-dihydroxyvitamin D. *J. Biol. Chem.* **2002**, *277*, 30337–30350. [CrossRef] [PubMed]

191. Hendy, G.N.; Canaff, L. Calcium-sensing receptor gene: Regulation of expression. *Front. Physiol.* **2016**, *7*, 394. [CrossRef] [PubMed]

192. Mihai, R.; Stevens, J.; McKinney, C.; Ibrahim, N.B. Expression of the calcium receptor in human breast cancer—A potential new marker predicting the risk of bone metastases. *Eur. J. Surg. Oncol.* **2006**, *32*, 511–515. [CrossRef] [PubMed]

193. Sanders, J.L.; Chattopadhyay, N.; Kifor, O.; Yamaguchi, T.; Brown, E.M. Ca^{2+}-sensing receptor expression and pthrp secretion in pc-3 human prostate cancer cells. *Am. J. Physiol. Endocrinol. Metab.* **2001**, *281*, E1267–E1274. [CrossRef] [PubMed]

194. Ward, D.T.; Riccardi, D. New concepts in calcium-sensing receptor pharmacology and signalling. *Br. J. Pharmacol.* **2012**, *165*, 35–48. [CrossRef] [PubMed]

195. Weston, A.H.; Absi, M.; Ward, D.T.; Ohanian, J.; Dodd, R.H.; Dauban, P.; Petrel, C.; Ruat, M.; Edwards, G. Evidence in favor of a calcium-sensing receptor in arterial endothelial cells: Studies with calindol and calhex 231. *Circ. Res.* **2005**, *97*, 391–398. [CrossRef] [PubMed]

196. Smajilovic, S.; Hansen, J.L.; Christoffersen, T.E.; Lewin, E.; Sheikh, S.P.; Terwilliger, E.F.; Brown, E.M.; Haunso, S.; Tfelt-Hansen, J. Extracellular calcium sensing in rat aortic vascular smooth muscle cells. *Biochem. Biophys. Res. Commun.* **2006**, *348*, 1215–1223. [CrossRef] [PubMed]

197. Berra Romani, R.; Raqeeb, A.; Laforenza, U.; Scaffino, M.F.; Moccia, F.; Avelino-Cruz, J.E.; Oldani, A.; Coltrini, D.; Milesi, V.; Taglietti, V.; et al. Cardiac microvascular endothelial cells express a functional Ca^{2+}-sensing receptor. *J. Vascul. Res.* **2009**, *46*, 73–82. [CrossRef] [PubMed]

198. Weston, A.H.; Absi, M.; Harno, E.; Geraghty, A.R.; Ward, D.T.; Ruat, M.; Dodd, R.H.; Dauban, P.; Edwards, G. The expression and function of Ca^{2+}-sensing receptors in rat mesenteric artery; comparative studies using a model of type ii diabetes. *Br. J. Pharmacol.* **2008**, *154*, 652–662. [CrossRef] [PubMed]

199. Smajilovic, S.; Tfelt-Hansen, J. Calcium acts as a first messenger through the calcium-sensing receptor in the cardiovascular system. *Cardiovasc. Res.* **2007**, *75*, 457–467. [CrossRef] [PubMed]

200. Weston, A.H.; Geraghty, A.; Egner, I.; Edwards, G. The vascular extracellular calcium-sensing receptor: An update. *Acta Physiol.* **2011**, *203*, 127–137. [CrossRef] [PubMed]

201. Smajilovic, S.; Yano, S.; Jabbari, R.; Tfelt-Hansen, J. The calcium-sensing receptor and calcimimetics in blood pressure modulation. *Br. J. Pharmacol.* **2011**, *164*, 884–893. [CrossRef] [PubMed]

202. Schreckenberg, R.; Schluter, K.D. Calcium sensing receptor expression and signalling in cardiovascular physiology and disease. *Vascul. Pharmacol.* **2018**. [CrossRef] [PubMed]

203. Sun, Y.H.; Liu, M.N.; Li, H.; Shi, S.; Zhao, Y.J.; Wang, R.; Xu, C.Q. Calcium-sensing receptor induces rat neonatal ventricular cardiomyocyte apoptosis. *Biochem. Biophys. Res. Commun.* **2006**, *350*, 942–948. [CrossRef] [PubMed]

204. Zhang, W.H.; Fu, S.B.; Lu, F.H.; Wu, B.; Gong, D.M.; Pan, Z.W.; Lv, Y.J.; Zhao, Y.J.; Li, Q.F.; Wang, R.; et al. Involvement of calcium-sensing receptor in ischemia/reperfusion-induced apoptosis in rat cardiomyocytes. *Biochem. Biophys. Res. Commun.* **2006**, *347*, 872–881. [CrossRef] [PubMed]

205. Sun, Y.H.; Li, Y.Q.; Feng, S.L.; Li, B.X.; Pan, Z.W.; Xu, C.Q.; Li, T.T.; Yang, B.F. Calcium-sensing receptor activation contributed to apoptosis stimulates trpc6 channel in rat neonatal ventricular myocytes. *Biochem. Biophys. Res. Commun.* **2010**, *394*, 955–961. [CrossRef] [PubMed]

206. Feng, S.L.; Sun, M.R.; Li, T.T.; Yin, X.; Xu, C.Q.; Sun, Y.H. Activation of calcium-sensing receptor increases TRPC3 expression in rat cardiomyocytes. *Biochem. Biophys. Res. Commun.* **2011**, *406*, 278–284. [CrossRef] [PubMed]

207. Tang, J.; Wang, G.; Liu, Y.; Fu, Y.; Chi, J.; Zhu, Y.; Zhao, Y.; Yin, X. Cyclosporin a induces cardiomyocyte injury through calcium-sensing receptor-mediated calcium overload. *Pharmazie* **2011**, *66*, 52–57. [PubMed]

208. Zhao, Y.; Hou, G.; Zhang, Y.; Chi, J.; Zhang, L.; Zou, X.; Tang, J.; Liu, Y.; Fu, Y.; Yin, X. Involvement of the calcium-sensing receptor in cyclosporin a-induced cardiomyocyte apoptosis in rats. *Pharmazie* **2011**, *66*, 968–974. [PubMed]

209. Chi, J.; Zhu, Y.; Fu, Y.; Liu, Y.; Zhang, X.; Han, L.; Yin, X.; Zhao, D. Cyclosporin a induces apoptosis in H9C2 cardiomyoblast cells through calcium-sensing receptor-mediated activation of the erk MAPK and p38 MAPK pathways. *Mol. Cell. Biochem.* **2013**, *367*, 227–236. [CrossRef] [PubMed]

210. Wang, H.-Y.; Liu, X.-Y.; Han, G.; Wang, Z.-Y.; Li, X.-X.; Jiang, Z.-M.; Jiang, C.-M. LPS induces cardiomyocyte injury through calcium-sensing receptor. *Mol. Cell. Biochem.* **2013**, *379*, 153–159. [CrossRef] [PubMed]

211. Jiang, C.M.; Han, L.P.; Li, H.Z.; Qu, Y.B.; Zhang, Z.R.; Wang, R.; Xu, C.Q.; Li, W.M. Calcium-sensing receptors induce apoptosis in cultured neonatal rat ventricular cardiomyocytes during simulated ischemia/reperfusion. *Cell Biol. Int.* **2008**, *32*, 792–800. [CrossRef] [PubMed]

212. Lu, F.; Tian, Z.; Zhang, W.; Zhao, Y.; Bai, S.; Ren, H.; Chen, H.; Yu, X.; Wang, J.; Wang, L.; et al. Calcium-sensing receptors induce apoptosis in rat cardiomyocytes via the Endo (sarco) plasmic reticulum pathway during hypoxia/reoxygenation. *Basic Clin. Pharmacol. Toxicol.* **2010**, *106*, 396–405. [CrossRef] [PubMed]

213. Lu, F.H.; Tian, Z.; Zhang, W.H.; Zhao, Y.J.; Li, H.L.; Ren, H.; Zheng, H.S.; Liu, C.; Hu, G.X.; Tian, Y.; et al. Calcium-sensing receptors regulate cardiomyocyte Ca^{2+} signaling via the sarcoplasmic reticulum-mitochondrion interface during hypoxia/reoxygenation. *J. Biomed. Sci.* **2010**, *17*, 50. [CrossRef] [PubMed]

214. Guo, J.; Li, H.Z.; Zhang, W.H.; Wang, L.C.; Wang, L.N.; Zhang, L.; Li, G.W.; Li, H.X.; Yang, B.F.; Wu, L.; et al. Increased expression of calcium-sensing receptors induced by ox-LDL amplifies apoptosis of cardiomyocytes during simulated ischaemia-reperfusion. *Clin. Exp. Pharmacol. Physiol.* **2010**, *37*, e128–e135. [CrossRef] [PubMed]

215. Jiang, C.M.; Xu, C.Q.; Mi, Y.; Li, H.Z.; Wang, R.; Li., W.-M. Calcium-sensing receptor induced myocardial ischemia/reperfusion injury via the c-jun nh2-terminal protein kinase pathway. *Acta Cardiol. Sin.* **2010**, *26*, 102–110.

216. Zheng, H.S.; Liu, J.; Liu, C.; Lu, F.H.; Zhao, Y.J.; Jin, Z.F.; Ren, H.; Leng, X.N.; Jia, J.; Hu, G.X.; et al. Calcium-sensing receptor activating phosphorylation of PKC delta translocation on mitochondria to induce cardiomyocyte apoptosis during ischemia/reperfusion. *Mol. Cell. Biochem.* **2011**, *358*, 335–343. [CrossRef] [PubMed]

217. Yan, L.; Zhu, T.; Sun, T.; Wang, L.; Pan, S.; Tao, Z.; Yang, Z.; Cao, K. Activation of calcium-sensing receptors is associated with apoptosis in a model of simulated cardiomyocytes ischemia/reperfusion. *J. Biomed. Res.* **2010**, *24*, 301–307. [CrossRef]

218. Yan, L.; Zhu, T.B.; Wang, L.S.; Pan, S.Y.; Tao, Z.X.; Yang, Z.; Cao, K.; Huang, J. Inhibitory effect of hepatocyte growth factor on cardiomyocytes apoptosis is partly related to reduced calcium sensing receptor expression during a model of simulated ischemia/reperfusion. *Mol. Biol. Rep.* **2011**, *38*, 2695–2701. [CrossRef] [PubMed]

219. Zhang, W.H.; Lu, F.H.; Zhao, Y.J.; Wang, L.N.; Tian, Y.; Pan, Z.W.; Lv, Y.J.; Wang, Y.L.; Du, L.J.; Sun, Z.R.; et al. Post-conditioning protects rat cardiomyocytes via pkcepsilon-mediated calcium-sensing receptors. *Biochem. Biophys. Res. Commun.* **2007**, *361*, 659–664. [CrossRef] [PubMed]

220. Dong, S.; Teng, Z.; Lu, F.H.; Zhao, Y.J.; Li, H.; Ren, H.; Chen, H.; Pan, Z.W.; Lv, Y.J.; Yang, B.F.; et al. Post-conditioning protects cardiomyocytes from apoptosis via PKC (epsilon)-interacting with calcium-sensing receptors to inhibit Endo (sarco) plasmic reticulum-mitochondria crosstalk. *Mol. Cell. Biochem.* **2010**, *341*, 195–206. [CrossRef] [PubMed]

221. Gan, R.T.; Hu, G.X.; Zhao, Y.J.; Li, H.L.; Jin, Z.F.; Ren, H.; Dong, S.Y.; Zhong, X.; Li, H.Z.; Yang, B.F.; et al. Post-conditioning protecting rat cardiomyocytes from apoptosis via attenuating calcium-sensing receptor-induced Endo (sarco) plasmic reticulum stress. *Mol. Cell. Biochem.* **2012**, *361*, 123–134. [CrossRef] [PubMed]

222. Zhang, L.; Cao, S.; Deng, S.; Yao, G.; Yu, T. Ischemic postconditioning and pinacidil suppress calcium overload in anoxia-reoxygenation cardiomyocytes via down-regulation of the calcium-sensing receptor. *PeerJ* **2016**, *4*, e2612. [CrossRef] [PubMed]

223. Dyukova, E.; Schreckenberg, R.; Sitdikova, G.; Schlüter, K.-D. Influence of ischemic pre- and post-conditioning on cardiac expression of calcium-sensing receptor. *BioNanoScience* **2017**, *7*, 112–114. [CrossRef]

224. Sun, J.; Murphy, E. Calcium sensing receptor: A sensor and mediator of ischemic preconditioning in the hearts. *Am. J. Physiol. Heart Circ. Physiol.* **2010**, *299*. [CrossRef] [PubMed]

225. Bai, S.Z.; Sun, J.; Wu, H.; Zhang, N.; Li, H.X.; Li, G.W.; Li, H.Z.; He, W.; Zhang, W.H.; Zhao, Y.J.; et al. Decrease in calcium-sensing receptor in the progress of diabetic cardiomyopathy. *Diabetes Res. Clin. Pract.* **2012**, *95*, 378–385. [CrossRef] [PubMed]

226. Qi, H.; Cao, Y.; Huang, W.; Liu, Y.; Wang, Y.; Li, L.; Liu, L.; Ji, Z.; Sun, H. Crucial role of calcium-sensing receptor activation in cardiac injury of diabetic rats. *PLoS ONE* **2013**, *8*, e65147. [CrossRef] [PubMed]

227. Matturri, L.; Milei, J.; Grana, D.R.; Lavezzi, A.M. Characterization of myocardial hypertrophy by DNA content, pcna expression and apoptotic index. *Int. J. Cardiol.* **2002**, *82*, 33–39. [CrossRef]

228. Wang, L.N.; Wang, C.; Lin, Y.; Xi, Y.H.; Zhang, W.H.; Zhao, Y.J.; Li, H.Z.; Tian, Y.; Lv, Y.J.; Yang, B.F.; et al. Involvement of calcium-sensing receptor in cardiac hypertrophy-induced by angiotensinii through calcineurin pathway in cultured neonatal rat cardiomyocytes. *Biochem. Biophys. Res. Commun.* **2008**, *369*, 584–589. [CrossRef] [PubMed]

229. Zhong, X.L.J.; Lu, F.; Wang, Y.; Zhao, Y.; Dong, S.; Leng, X.; Jia, J.; Ren, H.; Xu, C.; Zhang, W. Calcium sensing receptor regulates cardiomyocyte function through nuclear calcium. *Cell Biol. Int.* **2012**, *36*, 937–943. [CrossRef] [PubMed]

230. Colella, M.; Pozzan, T. Cardiac cell hypertrophy in vitro: Role of calcineurin/nfat as Ca^{2+} signal integrators. *Ann. N. Y. Acad. Sci.* **2008**, *1123*, 64–68. [CrossRef] [PubMed]

231. Lu, F.H.; Fu, S.B.; Leng, X.; Zhang, X.; Dong, S.; Zhao, Y.J.; Ren, H.; Li, H.; Zhong, X.; Xu, C.Q.; et al. Role of the calcium-sensing receptor in cardiomyocyte apoptosis via the sarcoplasmic reticulum and mitochondrial death pathway in cardiac hypertrophy and heart failure. *Cell. Physiol. Biochem.* **2013**, *31*, 728–743. [CrossRef] [PubMed]

232. Liu, L.; Wang, C.; Lin, Y.; Xi, Y.; Li, H.; Shi, S.; Li, H.; Zhang, W.; Zhao, Y.; Tian, Y.; et al. Suppression of calcium-sensing receptor ameliorates cardiac hypertrophy through inhibition of autophagy. *Mol. Med. Rep.* **2016**, *14*, 111–120. [CrossRef] [PubMed]

233. Zhang, X.; Zhang, T.; Wu, J.; Yu, X.; Zheng, D.; Yang, F.; Li, T.; Wang, L.; Zhao, Y.; Dong, S.; et al. Calcium sensing receptor promotes cardiac fibroblast proliferation and extracellular matrix secretion. *Cell. Physiol. Biochem.* **2014**, *33*, 557–568. [CrossRef] [PubMed]

234. Schreckenberg, R.; Dyukova, E.; Sitdikova, G.; Abdallah, Y.; Schluter, K.D. Mechanisms by which calcium receptor stimulation modifies electromechanical coupling in isolated ventricular cardiomyocytes. *Pflugers Arch.* **2015**, *467*, 379–388. [CrossRef] [PubMed]

235. Liu, C.-H.; Chang, H.-K.; Lee, S.-P.; Shieh, R.-C. Activation of the Ca^{2+}-sensing receptors increases currents through inward rectifier k+ channels via activation of phosphatidylinositol 4-kinase. *Pflugers Arch.* **2017**, *468*, 1931–1943. [CrossRef] [PubMed]

236. Tagliavini, S.; Genedani, S.; Bertolini, A.; Bazzani, C. Ischemia- and reperfusion-induced arrhythmias are prevented by putrescine. *Eur. J. Pharmacol.* **1991**, *194*, 7–10. [CrossRef]

237. Schepelmann, M.; Yarova, P.L.; Lopez-Fernandez, I.; Davies, T.S.; Brennan, S.C.; Edwards, P.J.; Aggarwal, A.; Graca, J.; Rietdorf, K.; Matchkov, V.; et al. The vascular Ca^{2+}-sensing receptor regulates blood vessel tone and blood pressure. *Am. J. Physiol. Cell Physiol.* **2016**, *310*, C193–C204. [CrossRef] [PubMed]

238. Wess, J. G-protein-coupled receptors: Molecular mechanisms involved in receptor activation and selectivity of g-protein recognition. *FASEB J.* **1997**, *11*, 346–354. [CrossRef] [PubMed]

239. Lefkowitz, R.J. Historical review: A brief history and personal retrospective of seven-transmembrane receptors. *Trends Pharmacol. Sci.* **2004**, *25*, 413–422. [CrossRef] [PubMed]

240. Rajagopal, S.; Ahn, S.; Rominger, D.H.; Gowen-MacDonald, W.; Lam, C.M.; Dewire, S.M.; Violin, J.D.; Lefkowitz, R.J. Quantifying ligand bias at seven-transmembrane receptors. *Mol. Pharmacol.* **2010**, *80*, 367–377. [CrossRef] [PubMed]

241. Lohse, M.J.; Hofmann, K.P. Spatial and temporal aspects of signaling by G-protein-coupled receptors. *Mol. Pharmacol.* **2015**, *88*, 572–578. [CrossRef] [PubMed]

242. Pupo, A.S.; Duarte, D.A.; Lima, V.; Teixeira, L.B.; Parreiras, E.S.L.T.; Costa-Neto, C.M. Recent updates on gpcr biased agonism. *Pharmacol. Res.* **2016**, *112*, 49–57. [CrossRef] [PubMed]

243. Calebiro, D.; Sungkaworn, T.; Maiellaro, I. Real-time monitoring of GPCR/CAMP signalling by fret and single-molecule microscopy. *Horm. Metab. Res.* **2014**, *46*, 827–832. [CrossRef] [PubMed]

244. Ibarra, C.; Vicencio, J.M.; Estrada, M.; Lin, Y.; Rocco, P.; Rebellato, P.; Munoz, J.P.; Garcia-Prieto, J.; Quest, A.F.; Chiong, M.; et al. Local control of nuclear calcium signaling in cardiac myocytes by perinuclear microdomains of sarcolemmal insulin-like growth factor 1 receptors. *Circ. Res.* **2013**, *112*, 236–245. [CrossRef] [PubMed]

245. Benard, G.; Massa, F.; Puente, N.; Lourenco, J.; Bellocchio, L.; Soria-Gomez, E.; Matias, I.; Delamarre, A.; Metna-Laurent, M.; Cannich, A.; et al. Mitochondrial cb1 receptors regulate neuronal energy metabolism. *Nat. Neurosci.* **2012**, *15*, 558–564. [CrossRef] [PubMed]

246. Calebiro, D.; Nikolaev, V.O.; Gagliani, M.C.; de Filippis, T.; Dees, C.; Tacchetti, C.; Persani, L.; Lohse, M.J. Persistent camp-signals triggered by internalized g-protein coupled receptors. *PLoS Biol.* **2009**, *7*, e1000172. [CrossRef] [PubMed]

247. Irannejad, R.; Tomshine, J.C.; Tomshine, J.R.; Chevalier, M.; Mahoney, J.P.; Steyaert, J.; Rasmussen, S.G.; Sunahara, R.K.; El-Samad, H.; Huang, B.; et al. Conformational biosensors reveal GPCR signalling from endosomes. *Nature* **2013**, *495*, 534–538. [CrossRef] [PubMed]

248. Pidasheva, S.; Grant, M.; Canaff, L.; Ercan, O.; Kumar, U.; Hendy, G.N. Calcium-sensing receptor dimerizes in the endoplasmic reticulum: Biochemical and biophysical characterization of casr mutants retained intracellularly. *Hum. Mol. Genet.* **2006**, *15*, 2200–2209. [CrossRef] [PubMed]

249. Ward, D.T.; Brown, E.M.; Harris, H.W. Disulfide bonds in the extracellular calcium-polyvalent cation-sensing receptor correlate with dimer formation and its response to divalent cations in vitro. *J. Biol. Chem.* **1998**, *273*, 14476–14483. [CrossRef] [PubMed]

250. Breitwieser, G.E. Minireview: The intimate link between calcium sensing receptor trafficking and signaling: Implications for disorders of calcium homeostasis. *Mol. Endocrinol.* **2012**, *26*, 1482–1495. [CrossRef] [PubMed]

251. Breitwieser, G.E. Pharmacoperones and the calcium sensing receptor: Exogenous and endogenous regulators. *Pharmacol. Res.* **2014**, *83*, 30–37. [CrossRef] [PubMed]

252. Holtback, U.; Brismar, H.; DiBona, G.F.; Fu, M.; Greengard, P.; Aperia, A. Receptor recruitment: A mechanism for interactions between g protein-coupled receptors. *Proc. Natl. Acad. Sci. USA* **1999**, *96*, 7271–7275. [CrossRef] [PubMed]

253. Brennan, S.C.; Mun, H.C.; Leach, K.; Kuchel, P.W.; Christopoulos, A.; Conigrave, A.D. Receptor expression modulates calcium-sensing receptor mediated intracellular Ca^{2+} mobilization. *Endocrinology* **2015**, *156*, 1330–1342. [CrossRef] [PubMed]

254. Breitwieser, G.E. The calcium sensing receptor life cycle: Trafficking, cell surface expression, and degradation. *Best Pract. Res. Clin. Endocrinol. Metab.* **2013**, *27*, 303–313. [CrossRef] [PubMed]

255. Ray, K. Calcium-sensing receptor: Trafficking, endocytosis, recycling, and importance of interacting proteins. *Prog. Mol. Biol. Transl. Sci.* **2015**, *132*, 127–150. [PubMed]

256. Huang, C.; Miller, R.T. The calcium-sensing receptor and its interacting proteins. *J. Cell Mol. Med.* **2007**, *11*, 923–934. [CrossRef] [PubMed]

257. Bai, M.; Quinn, S.; Trivedi, S.; Kifor, O.; Pearce, S.H.; Pollak, M.R.; Krapcho, K.; Hebert, S.C.; Brown, E.M. Expression and characterization of inactivating and activating mutations in the human Ca^{2+} o-sensing receptor. *J. Biol. Chem.* **1996**, *271*, 19537–19545. [CrossRef] [PubMed]

258. Grant, M.P.; Stepanchick, A.; Cavanaugh, A.; Breitwieser, G.E. Agonist-driven maturation and plasma membrane insertion of calcium-sensing receptors dynamically control signal amplitude. *Sci. Signal.* **2011**, *4*, ra78. [CrossRef] [PubMed]

259. Huang, Y.; Breitwieser, G.E. Rescue of calcium-sensing receptor mutants by allosteric modulators reveals a conformational checkpoint in receptor biogenesis. *J. Biol. Chem.* **2007**, *282*, 9517–9525. [CrossRef] [PubMed]

260. White, E.; McKenna, J.; Cavanaugh, A.; Breitwieser, G.E. Pharmacochaperone-mediated rescue of calcium-sensing receptor loss-of-function mutants. *Mol. Endocrinol.* **2009**, *23*, 1115–1123. [CrossRef] [PubMed]

261. Cavanaugh, A.; McKenna, J.; Stepanchick, A.; Breitwieser, G.E. Calcium-sensing receptor biosynthesis includes a cotranslational conformational checkpoint and endoplasmic reticulum retention. *J. Biol. Chem.* **2010**, *285*, 19854–19864. [CrossRef] [PubMed]

262. Rus, R.; Haag, C.; Bumke-Vogt, C.; Bahr, V.; Mayr, B.; Mohlig, M.; Schulze, E.; Frank-Raue, K.; Raue, F.; Schofl, C. Novel inactivating mutations of the calcium-sensing receptor: The calcimimetic nps r-568 improves signal transduction of mutant receptors. *J. Clin. Endocrinol. Metab.* **2008**, *93*, 4797–4803. [CrossRef] [PubMed]

263. Leach, K.; Wen, A.; Cook, A.E.; Sexton, P.M.; Conigrave, A.D.; Christopoulos, A. Impact of clinically relevant mutations on the pharmacoregulation and signaling bias of the calcium-sensing receptor by positive and negative allosteric modulators. *Endocrinology* **2013**, *154*, 1105–1116. [CrossRef] [PubMed]

264. Nakamura, A.; Hotsubo, T.; Kobayashi, K.; Mochizuki, H.; Ishizu, K.; Tajima, T. Loss-of-function and gain-of-function mutations of calcium-sensing receptor: Functional analysis and the effect of allosteric modulators NPS R-568 and NPS 2143. *J. Clin. Endocrinol. Metab.* **2013**, *98*, E1692–E1701. [CrossRef] [PubMed]

265. Letz, S.; Rus, R.; Haag, C.; Dorr, H.G.; Schnabel, D.; Mohlig, M.; Schulze, E.; Frank-Raue, K.; Raue, F.; Mayr, B.; et al. Novel activating mutations of the calcium-sensing receptor: The calcilytic NPS-2143 mitigates excessive signal transduction of mutant receptors. *J. Clin. Endocrinol. Metab.* **2010**, *95*, E229–E233. [CrossRef] [PubMed]

266. Letz, S.; Haag, C.; Schulze, E.; Frank-Raue, K.; Raue, F.; Hofner, B.; Mayr, B.; Schofl, C. Amino alcohol-(NPS-2143) and quinazolinone-derived calcilytics (ATF936 and AXT914) differentially mitigate excessive signalling of calcium-sensing receptor mutants causing bartter syndrome type 5 and autosomal dominant hypocalcemia. *PLoS ONE* **2014**, *9*, e115178. [CrossRef] [PubMed]

267. Hannan, F.M.; Walls, G.V.; Babinsky, V.N.; Nesbit, M.A.; Kallay, E.; Hough, T.A.; Fraser, W.D.; Cox, R.D.; Hu, J.; Spiegel, A.M.; et al. The calcilytic agent NPS 2143 rectifies hypocalcemia in a mouse model with an activating calcium-sensing receptor (CASR) mutation: Relevance to autosomal dominant hypocalcemia type 1 (ADH1). *Endocrinology* **2015**, *156*, 3114–3121. [CrossRef] [PubMed]

268. Dong, B.; Endo, I.; Ohnishi, Y.; Kondo, T.; Hasegawa, T.; Amizuka, N.; Kiyonari, H.; Shioi, G.; Abe, M.; Fukumoto, S.; et al. Calcilytic ameliorates abnormalities of mutant calcium-sensing receptor (CASR) knock-in mice mimicking autosomal dominant hypocalcemia (ADH). *J. Bone Miner. Res.* **2015**, *30*, 1980–1993. [CrossRef] [PubMed]

269. Leach, K.; Conigrave, A.D.; Sexton, P.M.; Christopoulos, A. Towards tissue-specific pharmacology: Insights from the calcium-sensing receptor as a paradigm for GPCR (patho)physiological bias. *Trends Pharmacol. Sci.* **2015**, *36*, 215–225. [CrossRef] [PubMed]

270. Makita, N.; Sato, J.; Manaka, K.; Shoji, Y.; Oishi, A.; Hashimoto, M.; Fujita, T.; Iiri, T. An acquired hypocalciuric hypercalcemia autoantibody induces allosteric transition among active human ca-sensing receptor conformations. *Proc. Natl. Acad. Sci. USA* **2007**, *104*, 5443–5448. [CrossRef] [PubMed]

271. Bruce, J.I.; Yang, X.; Ferguson, C.J.; Elliott, A.C.; Steward, M.C.; Case, R.M.; Riccardi, D. Molecular and functional identification of a Ca^{2+} (polyvalent cation)-sensing receptor in rat pancreas. *J. Biol. Chem.* **1999**, *274*, 20561–20568. [CrossRef] [PubMed]

272. Thomsen, A.R.; Smajilovic, S.; Brauner-Osborn, H. Novel strategies in drug discovery of the calcium-sensing receptor based on biased signaling. *Curr. Drug. Targets* **2012**, *13*, 1324–1335. [CrossRef] [PubMed]

273. Thomsen, A.R.; Worm, J.; Jacobsen, S.E.; Stahlhut, M.; Latta, M.; Brauner-Osborne, H. Strontium is a biased agonist of the calcium-sensing receptor in rat medullary thyroid carcinoma 6-23 cells. *J. Pharmacol. Exp. Ther.* **2012**, *343*, 638–649. [CrossRef] [PubMed]

274. Chattopadhyay, N.; Quinn, S.J.; Kifor, O.; Ye, C.; Brown, E.M. The calcium-sensing receptor (CAR) is involved in strontium ranelate-induced osteoblast proliferation. *Biochem. Pharmacol.* **2007**, *74*, 438–447. [CrossRef] [PubMed]

275. Coulombe, J.; Faure, H.; Robin, B.; Ruat, M. In vitro effects of strontium ranelate on the extracellular calcium-sensing receptor. *Biochem. Biophys. Res. Commun.* **2004**, *323*, 1184–1190. [CrossRef] [PubMed]

276. Hurtel-Lemaire, A.S.; Mentaverri, R.; Caudrillier, A.; Cournarie, F.; Wattel, A.; Kamel, S.; Terwilliger, E.F.; Brown, E.M.; Brazier, M. The calcium-sensing receptor is involved in strontium ranelate-induced osteoclast apoptosis: New insights into the associated signaling pathways. *J. Biol. Chem.* **2009**, *284*, 575–584. [CrossRef] [PubMed]

277. Davey, A.E.; Leach, K.; Valant, C.; Conigrave, A.D.; Sexton, P.M.; Christopoulos, A. Positive and negative allosteric modulators promote biased signaling at the calcium-sensing receptor. *Endocrinology* **2012**, *153*, 1232–1241. [CrossRef] [PubMed]

278. Leach, K.; Wen, A.; Davey, A.E.; Sexton, P.M.; Conigrave, A.D.; Christopoulos, A. Identification of molecular phenotypes and biased signaling induced by naturally occurring mutations of the human calcium-sensing receptor. *Endocrinology* **2012**, *153*, 4304–4316. [CrossRef] [PubMed]

279. Hofer, A.M.; Lefkimmiatis, K. Extracellular calcium and cAMP: Second messengers as "Third messengers"? *Physiology* **2007**, *22*, 320–327. [CrossRef] [PubMed]

280. Wellendorph, P.; Johansen, L.D.; Brauner-Osborne, H. Molecular pharmacology of promiscuous seven transmembrane receptors sensing organic nutrients. *Mol. Pharmacol.* **2009**, *76*, 453–465. [CrossRef] [PubMed]

281. Kano, M.; Kato, M. Quisqualate receptors are specifically involved in cerebellar synaptic plasticity. *Nature* **1987**, *325*, 276–279. [CrossRef] [PubMed]

282. Francesconi, A.; Duvoisin, R.M. Divalent cations modulate the activity of metabotropic glutamate receptors. *J. Neurosci. Res.* **2004**, *75*, 472–479. [CrossRef] [PubMed]

283. Kubo, Y.; Miyashita, T.; Murata, Y. Structural basis for a Ca^{2+}-sensing function of the metabotropic glutamate receptors. *Science* **1998**, *279*, 1722–1725. [CrossRef] [PubMed]

284. Nash, M.S.; Saunders, R.; Young, K.W.; Challiss, R.A.; Nahorski, S.R. Reassessment of the Ca^{2+} sensing property of a type i metabotropic glutamate receptor by simultaneous measurement of inositol 1,4,5-trisphosphate and Ca^{2+} in single cells. *J. Biol. Chem.* **2001**, *276*, 19286–19293. [CrossRef] [PubMed]

285. Tabata, T.; Kano, M. Calcium dependence of native metabotropic glutamate receptor signaling in central neurons. *Mol. Neurobiol.* **2004**, *29*, 261–270. [CrossRef]

286. Hardingham, N.R.; Bannister, N.J.; Read, J.C.; Fox, K.D.; Hardingham, G.E.; Jack, J.J. Extracellular calcium regulates postsynaptic efficacy through group 1 metabotropic glutamate receptors. *J. Neurosci.* **2006**, *26*, 6337–6345. [CrossRef] [PubMed]

287. Gerber, U.; Gee, C.E.; Benquet, P. Metabotropic glutamate receptors: Intracellular signaling pathways. *Curr. Opin. Pharmacol.* **2007**, *7*, 56–61. [CrossRef] [PubMed]

288. Wang, X.; Kirberger, M.; Qiu, F.; Chen, G.; Yang, J.J. Towards predicting Ca^{2+}-binding sites with different coordination numbers in proteins with atomic resolution. *Proteins* **2009**, *75*, 787–798. [CrossRef] [PubMed]

289. Wang, X.; Zhao, K.; Kirberger, M.; Wong, H.; Chen, G.; Yang, J.J. Analysis and prediction of calcium-binding pockets from apo-protein structures exhibiting calcium-induced localized conformational changes. *Protein Sci.* **2010**, *19*, 1180–1190. [CrossRef] [PubMed]

290. Jiang, Y.; Huang, Y.; Wong, H.C.; Zhou, Y.; Wang, X.; Yang, J.; Hall, R.A.; Brown, E.M.; Yang, J.J. Elucidation of a novel extracellular calcium-binding site on metabotropic glutamate receptor 1α (mGluR1α) that controls receptor activation. *J. Biol. Chem.* **2010**, *285*, 33463–33474. [CrossRef] [PubMed]

291. Jiang, J.Y.; Nagaraju, M.; Meyer, R.C.; Zhang, L.; Hamelberg, D.; Hall, R.A.; Brown, E.M.; Conn, P.J.; Yang, J.J. Extracellular calcium modulates actions of orthosteric and allosteric ligands on metabotropic glutamate receptor 1α. *J. Biol. Chem.* **2014**, *289*, 1649–1661. [CrossRef] [PubMed]

292. Jones, K.A.; Borowsky, B.; Tamm, J.A.; Craig, D.A.; Durkin, M.M.; Dai, M.; Yao, W.J.; Johnson, M.; Gunwaldsen, C.; Huang, L.Y.; et al. Gaba(b) receptors function as a heteromeric assembly of the subunits $GABA_bR1$ and $GABA_bR2$. *Nature* **1998**, *396*, 674–679. [CrossRef] [PubMed]

293. Wise, A.; Green, A.; Main, M.J.; Wilson, R.; Fraser, N.; Marshall, F.H. Calcium sensing properties of the $GABA_b$ receptor. *Neuropharmacology* **1999**, *38*, 1647–1656. [CrossRef]

294. Silve, C.; Petrel, C.; Leroy, C.; Bruel, H.; Mallet, E.; Rognan, D.; Ruat, M. Delineating a Ca^{2+} binding pocket within the venus flytrap module of the human calcium-sensing receptor. *J. Biol. Chem.* **2005**, *280*, 37917–37923. [CrossRef] [PubMed]

295. Tordoff, M.G.; Shao, H.; Alarcón, L.K.; Margolskee, R.F.; Mosinger, B.; Bachmanov, A.A.; Reed, D.R.; McCaughey, S. Involvement of t1r3 in calcium-magnesium taste. *Physiol. Genomics* **2008**, *34*, 338–348. [CrossRef] [PubMed]

296. Tordoff, M.G.; Alarcón, L.K.; Valmeki, S.; Jiang, P. T1R3: A human calcium taste receptor. *Sci. Rep.* **2012**, *2*, 496. [CrossRef] [PubMed]

297. Wellendorph, P.; Bräuner-Osborne, H. Molecular cloning, expression, and sequence analysis of gprc6a, a novel family c G-protein-coupled receptor. *Gene* **2004**, *335*, 37–46. [CrossRef] [PubMed]

298. Pi, M.; Quarles, L.D. Osteoblast calcium-sensing receptor has characteristics of ANF/7TM receptors. *J. Cell Biochem.* **2005**, *95*, 1081–1092. [CrossRef] [PubMed]

299. Faure, H.; Gorojankina, T.; Rice, N.; Dauban, P.; Dodd, R.H.; Bräuner-Osborne, H.; Rognan, D.; Ruat, M. Molecular determinants of non-competitive antagonist binding to the mouse gprc6a receptor. *Cell Calcium* **2009**, *46*, 323–332. [CrossRef] [PubMed]

300. Stenflo, J.; Stenberg, Y.; Muranyi, A. Calcium-binding EGF-like modules in coagulation proteinases: Function of the calcium ion in module interactions. *Biochim. Biophys. Acta* **2000**, *1477*, 51–63. [CrossRef]

301. Raya, A.; Kawakami, Y.; Rodríguez-Esteban, C.; Ibañes, M.; Rasskin-Gutman, D.; Rodríguez-León, J.; Büscher, D.; Feijó, J.A.; Izpisúa Belmonte, J.C. Notch activity acts as a sensor for extracellular calcium during vertebrate left-right determination. *Nature* **2004**, *427*, 121–128. [CrossRef] [PubMed]

302. Halbleib, J.M.; Nelson, W.J. Cadherins in development: Cell adhesion, sorting, and tissue morphogenesis. *Genes Dev.* **2006**, *20*, 3199–3214. [CrossRef] [PubMed]

303. Spassova, M.A.; Soboloff, J.; He, L.P.; Xu, W.; Dziadek, M.A.; Gill, D.L. Stim1 has a plasma membrane role in the activation of store-operated Ca^{2+} channels. *Proc. Natl. Acad. Sci. USA* **2006**, *103*, 4040–4045. [CrossRef] [PubMed]

304. Jardin, I.; López, J.J.; Redondo, P.C.; Salido, G.M.; Rosado, J.A. Store-operated Ca^{2+} entry is sensitive to the extracellular Ca^{2+} concentration through plasma membrane stim1. *Biochim. Biophys. Acta* **2009**, *1793*, 1614–1622. [CrossRef] [PubMed]

305. Fasciani, I.; Temperán, A.; Pérez-Atencio, L.F.; Escudero, A.; Martínez-Montero, P.; Molano, J.; Gómez-Hernández, J.M.; Paino, C.L.; González-Nieto, D.; Barrio, L.C. Regulation of connexin hemichannel activity by membrane potential and the extracellular calcium in health and disease. *Neuropharmacology* **2013**, *75*, 479–490. [CrossRef] [PubMed]

306. Ebihara, L.; Liu, X.; Pal, J.D. Effect of external magnesium and calcium on human connexin46 hemichannels. *Biophys. J.* **2003**, *84*, 277–286. [CrossRef]

307. Quist, A.P.; Rhee, S.K.; Lin, H.; Lal, R. Physiological role of Gap-junctional hemichannels. Extracellular calcium-dependent isosmotic volume regulation. *J. Cell Biol.* **2000**, *148*, 1063–1074. [CrossRef] [PubMed]

308. Gomez-Hernandez, J.M.; de Miguel, M.; Larrosa, B.; Gonzalez, D.; Barrio, L.C. Molecular basis of calcium regulation in connexin-32 hemichannels. *Proc. Natl. Acad. Sci. USA* **2003**, *100*, 16030–16035. [CrossRef] [PubMed]

309. Thimm, J.; Mechler, A.; Lin, H.; Rhee, S.; Lal, R. Calcium-dependent open/closed conformations and interfacial energy maps of reconstituted hemichannels. *J. Biol. Chem.* **2005**, *280*, 10646–10654. [CrossRef] [PubMed]

310. Ma, Z.; Tanis, J.E.; Taruno, A.; Foskett, J.K. Calcium homeostasis modulator (CALHM) ion channels. *Pflugers Arch.* **2016**, *468*, 395–403. [CrossRef] [PubMed]

311. Babini, E.; Paukert, M.; Geisler, H.S.; Grunder, S. Alternative splicing and interaction with di- and polyvalent cations control the dynamic range of acid-sensing ion channel 1 (ASIC1). *J. Biol. Chem.* **2002**, *277*, 41597–41603. [CrossRef] [PubMed]

312. Immke, D.C.; McCleskey, E.W. Protons open acid-sensing ion channels by catalyzing relief of Ca^{2+} blockade. *Neuron* **2003**, *37*, 75–84. [CrossRef]

313. Paukert, M.; Babini, E.; Pusch, M.; Gründer, S. Identification of the Ca^{2+} blocking site of acid-sensing ion channel (ASIC) 1: Implications for channel gating. *J. Gen. Physiol.* **2004**, *124*, 383–394. [CrossRef] [PubMed]

314. Jiang, Q.; Inoue, K.; Wu, X.; Papasian, C.J.; Wang, J.Q.; Xiong, Z.G.; Chu, X.P. Cysteine 149 in the extracellular finger domain of acid-sensing ion channel 1b subunit is critical for zinc-mediated inhibition. *Neuroscience* **2011**, *193*, 89–99. [CrossRef] [PubMed]

315. Johnson, J.P.; Mullins, F.M.; Bennett, P.B. Human ether-à-go-go-related gene k+ channel gating probed with extracellular Ca^{2+}. Evidence for two distinct voltage sensors. *J. Gen. Physiol.* **1999**, *113*, 565–580. [CrossRef] [PubMed]

316. Johnson, J.P.; Balser, J.R.; Bennett, P.B. A novel extracellular calcium sensing mechanism in voltage-gated potassium ion channels. *J. Neurosci.* **2001**, *21*, 4143–4153. [PubMed]

317. Sanguinetti, M.C.; Jiang, C.; Curran. M.E.; Keating, M.T. A mechanistic link between an inherited and an acquired cardiac arrhythmia: Herg encodes the ikr potassium channel. *Cell* **1995**, *81*, 299–307. [CrossRef]

318. Xiong, Z.G.; Chu, X.P.; MacDonald, J.F. Effect of lamotrigine on the Ca^{2+}-sensing cation current in cultured hippocampal neurons. *J. Neurophysiol.* **2001**, *86*, 2520–2526. [CrossRef] [PubMed]

319. Xiong, Z.; Lu, W.; MacDonald, J.F. Extracellular calcium sensed by a novel cation channel in hippocampal neurons. *Proc. Natl. Acad. Sci. USA* **1997**, *94*, 7012–7017. [CrossRef] [PubMed]

320. Burgo, A.; Carmignoto, G.; Pizzo, P.; Pozzan, T.; Fasolato, C. Paradoxical Ca^{2+} rises induced by low external Ca^{2+} in rat hippocampal neurones. *J. Physiol.* **2003**, *549*, 537–552. [CrossRef] [PubMed]

321. Wei, W.L.; Sun, H.S.; Olah, M.E.; Sun, X.; Czerwinska, E.; Czerwinski, W.; Mori, Y.; Orser, B.A.; Xiong, Z.G.; Jackson, M.F.; et al. Trpm7 channels in hippocampal neurons detect levels of extracellular divalent cations. *Proc. Natl. Acad. Sci. USA* **2007**, *104*, 16323–16328. [CrossRef] [PubMed]

International Journal of
Molecular Sciences

MDPI

Review

Calcium Signaling in Vertebrate Development and Its Role in Disease

Sudip Paudel [†], Regan Sindelar [†] and Margaret Saha *

College of William and Mary, Williamsburg, VA 23187, USA; spaudel@email.wm.edu (S.P.);
rssindelar@email.wm.edu (R.S.)
* Correspondence: mssaha@wm.edu; Tel.: +1-757-221-2407
† These authors contributed equally to this work.

Received: 27 September 2018; Accepted: 22 October 2018; Published: 30 October 2018

Abstract: Accumulating evidence over the past three decades suggests that altered calcium signaling during development may be a major driving force for adult pathophysiological events. Well over a hundred human genes encode proteins that are specifically dedicated to calcium homeostasis and calcium signaling, and the majority of these are expressed during embryonic development. Recent advances in molecular techniques have identified impaired calcium signaling during development due to either mutations or dysregulation of these proteins. This impaired signaling has been implicated in various human diseases ranging from cardiac malformations to epilepsy. Although the molecular basis of these and other diseases have been well studied in adult systems, the potential developmental origins of such diseases are less well characterized. In this review, we will discuss the recent evidence that examines different patterns of calcium activity during early development, as well as potential medical conditions associated with its dysregulation. Studies performed using various model organisms, including zebrafish, *Xenopus*, and mouse, have underscored the critical role of calcium activity in infertility, abortive pregnancy, developmental defects, and a range of diseases which manifest later in life. Understanding the underlying mechanisms by which calcium regulates these diverse developmental processes remains a challenge; however, this knowledge will potentially enable calcium signaling to be used as a therapeutic target in regenerative and personalized medicine.

Keywords: calcium; development; embryo; human disease; animal model; *Xenopus*; zebrafish; mouse

1. Introduction

Calcium is an ancient and ubiquitous signaling ion involved in a wide array of physiological processes throughout the entire lifespan of the individual, from fertilization through senescence. Given its central position in virtually every cellular process and tissue type, it is not surprising that its dysregulation leads to a wide spectrum of pathophysiological conditions. These conditions include, but are not limited to, infertility, miscarriage, developmental defects, neuropathic pain, epilepsy and seizures [1].

Under normal conditions, the basal intracellular concentration of calcium ion (Ca^{2+}) is maintained at very low levels compared to the extracellular spaces by a number of proteins, including pumps and transporters that have Ca^{2+} binding ability. Calcium pumps, using ATP, remove Ca^{2+} from the cytosol and help cells maintain roughly hundred to thousand-fold concentration gradients between intracellular and extracellular environments. Ca^{2+}-ATPases are an example of these pumps, which are localized in various cellular organelles; for example, P2A ATPase is localized in the endoplasmic reticulum, P2B ATPase is localized in the plasma membrane, and P2C ATPase is localized in the Golgi apparatus [2–4]. In addition, Ca^{2+} enters the cytosol passively from the extracellular matrix, neighboring cells, and cytoplasmic Ca^{2+} stores; the opening and closing of these channels is driven by both chemical and electrical gradients. Cells create these gradients in order to use calcium as a signaling molecule [3]. During calcium-mediated signaling events, Ca^{2+} permeant channels open transiently allowing spontaneous influx of Ca^{2+} into the cytosol. This influx of Ca^{2+} increases intracellular concentration of Ca^{2+}, and is often referred to generically as a calcium transient [5–7]. The short transient fluctuation in cytoplasmic Ca^{2+} within a given (typically isolated) cell due to these influxes are often defined as calcium spikes [8]. When a calcium spike originates from one or a few cells and then propagates along several other neighboring cells via gap junctions or paracrine signaling, it is called an intercellular wave [5,8–10]. The change in concentration of cytoplasmic Ca^{2+} is sensed by a number of calcium sensing proteins with calcium binding affinities ranging from hundreds to million-fold [11]. These calcium sensing proteins transduce calcium signals, decoding spatial and temporal changes in amplitude, duration and localization of calcium activity [7,11,12].

In turn, these signaling cascades drive a wide array of physiological responses throughout the life of the organism from fertilization onwards. While the role of calcium and its associated pathologies have been intensively studied in the adult organism, less attention has been focused on the role of calcium and its relationship to disease during embryonic development. Given the importance of calcium activity as a regulator of embryonic development from fertilization through organogenesis, dysregulation of these calcium dynamics, due to either mutations or epigenetic developmental dysregulation of some element or component of the calcium signaling process, results in various diseases and disorders (Supplementary Table S1). For example, Timothy syndrome is caused by constitutive opening of voltage gated calcium channel CaV1.2 due to mutations in the *cacna1c* gene. This gene is expressed embryonically and the syndrome is characterized by syndactyly, intellectual disability, congenital heart defects, distinctive facial features and developmental delay. Similarly, mutations in the gene ATP2A1 (sarco(endo)plasmic reticulum calcium-ATPase 1 (SERCA1)) on chromosome 16p11 result in Brody myopathy, which is characterized by a decrease or loss of sarcoplasmic reticulum Ca^{2+}-ATPase activity and problems with muscle contraction [13]. While some of these mutations and dysregulated processes are embryonic lethal, many manifest their effects at birth, and others may not show symptoms until later in life due to their indirect effects within a complex genetic network [14].

Calcium activity during development is varied and complex with embryos exhibiting different patterns of spikes and waves. Animal model studies during early stages of development have provided a broad understanding of human developmental defects and diseases related to the dysregulation of calcium activity. In this review, we will provide an overview of the current state of knowledge regarding the role of calcium activity in embryonic and fetal development and disease. Given the obvious challenges of studying calcium activity in human embryonic development, much of the information we will discuss derives from model systems, particularly frogs, fish, and mice. We will discuss each stage of development from fertilization through organogenesis chronologically. Each section will begin with a brief overview of the key developmental events that occur during that particular stage and then proceed to analyze the role of calcium in those processes, including how dysregulation of calcium dynamics can, and does, lead to disease.

2. Calcium Activity during Development and Its Role in Disease

2.1. Fertilization and Egg Activation

Fertilization is the process by which DNA of the sperm and egg unite to give rise to a new diploid organism. Sperm entry then triggers the oocyte to transition into a developing embryo in a process known as egg activation. Egg activation is characterized by the occurrence of a number of sequential events in the oocyte during fertilization: recruitment of maternal mRNA and formation of polysomes, completion of meiosis, modification of the plasma membrane and zona pellucida in order to prevent polyspermy, cortical granule exocytosis, formation of male and female proneuclei, and syngamy, the fusion of two genomes [15,16]. While species differences exist, the process of egg activation is a relatively conserved mechanism that is mediated and coordinated by calcium; failure in any step of this process typically results in infertility.

The importance of calcium activity in the process of fertilization and egg activation cannot be underestimated. Fertilization initiates elevations of intracellular Ca^{2+} concentration in all vertebrate oocytes studied to date [17]. These elevations are initiated from the site of sperm-egg fusion, and are caused by transient influxes of Ca^{2+} from both the extracellular milieu and intracellular calcium stores. The patterns of these influxes do vary somewhat across species. For example, the oocytes from some lower vertebrates such as *Xenopus* and zebrafish achieve this elevation via a single calcium transient, while mammalian oocytes exhibit an initial transient increase within a few minutes of the sperm binding to the egg surface, followed by subsequent oscillations in cytoplasmic Ca^{2+} concentration at 20 to 30 min intervals [17–19]. This calcium activity was visualized for the first time in a mammalian egg by imaging zona-free mouse oocytes using aequorin during in vitro fertilization [20]. Similar calcium behavior was observed in mouse and human oocytes during in vitro fertilization and intracytoplasmic sperm injection using aequorin or various other calcium sensitive dyes [21,22]. Inhibition of this calcium activity results in fertilization failure. For example, when extracellular Ca^{2+} was restricted from entering the oocyte cytoplasm using bivalent cation chelators such as BAPTA or EGTA, calcium insulators such as gadolinium, or a Ca^{2+} free culture medium, *Xenopus* oocytes failed to develop normal cleavage furrows, and mouse oocytes failed to resume meiosis during the egg activation process [18,23,24]. Similarly, wild type sperm was unable to activate zona-free mouse eggs lacking a gene (cacna1h$^{-/-}$) that codes for a T-type voltage gated calcium channel, which mediates extracellular calcium entry [25]. Pharmacological inhibition of this channel, using mibefradil or pimozide, also resulted in a phenotype similar to cacna1h$^{-/-}$ following in vitro fertilization [25]. These observations suggest that impaired entry of calcium from extracellular matrix to the oocyte cytoplasm may result in fertilization failure and lead to infertility.

Sperm specific phospholipase C, such as phospholipase C isozyme zeta, on its own is sufficient to trigger calcium release from intracellular calcium stores via channels, including STIM1, in order to accomplish egg activation in different species [17]. Mouse eggs injected with sperm-supernatants treated with an anti-phospholipase C zeta antibody failed to show Ca^{2+} elevation and subsequent egg activation events. However, when the eggs were subsequently microinjected with only sperm specific phospholipase C zeta, normal elevation of Ca^{2+} was observed and egg activation was rescued [26, 27]. These observations have been further strengthened by recent gene editing technology. For example, intracytoplasmic sperm injection or in vivo fertilization with phospholipase C zeta knockout sperm failed to trigger egg activation in wild type mouse eggs, but subsequent injection of wild type phospholipase C zeta-RNA successfully rescued egg activation and zygotic development [28]. Various measures, including histological analysis, sperm viability, sperm motility and hyperactivity assays, showed that phospholipase C zeta gene knockout using CRISPR/Cas9 results in no defects in spermatogenesis [28]. Similarly, the knock down of STIM1 from pig oocytes using STIM1 specific siRNA resulted in failure to generate Ca^{2+} oscillations, decreased oocyte survival rate, and decreased efficiency of cleavage and blastocyst development [29]. Taken together, these observations suggest that impaired function of sperm-specific phospholipase C zeta also results in impaired egg activation leading to mammalian infertility. Inhibition of these Ca^{2+} elevations results in failure to initiate and complete functional fertilization in humans and other model organisms studied to date, including mice, *Xenopus* and zebrafish [16,30,31].

Recent molecular advances, including genomics, transcriptomics, proteomics, gene editing technology, and imaging technology, are opening new avenues of exploration for the potential therapeutic use of calcium in infertility treatments. For instance, assisted oocyte activation using mechanical, chemical, or electrical artificial triggers that cause calcium influx into the oocyte cytoplasm have been shown to trigger egg activation [30]. Therefore, manipulation of calcium activity could serve as a potential therapy for severe infertility associated with egg activation failure.

2.2. Cleavage and Blastula Stages

Following egg activation, vertebrate embryos undergo a rapid series of cell divisions leading to a relatively unstructured and undifferentiated mass of fluid-filled cells termed the blastula. It is during this stage that the maternal to zygotic transition occurs in which the zygotic genome is activated, an event that happens very early in mammals. It is also during this stage of mammalian development that the first important developmental lineage decisions occur, namely the determination of the inner cell mass that will give rise to the embryo proper and the trophoblast that will eventually give rise to the extraembryonic tissues. The blastula stage occurs during the first few weeks of embryonic development in humans; developmental defects during this stage of development affect germ layer formation and morphogenetic movements and will typically result in a non-viable embryo [32].

Calcium plays a key role in the regulation of the cell divisions during this early stage of development. During the initial cleavage stages, the embryo exhibits slowly moving intracellular waves of calcium. When imaging transgenic zebrafish that express genetically encoded calcium indicators, including cameleon YC2.60 and GCaMP6 driven by promoters of constitutive genes such as hspa8 and βactin, these waves were localized exclusively to the cleavage furrow [19,33]. Immunocytochemistry assays in zebrafish embryos have shown that this furrow-associated calcium activity is mediated by store-operated Ca^{2+} entry channels, including STIM1 and ORAI1 [34,35]. Furrow-associated waves have been shown to remodel microtubules into a furrow microtubule array that is required for cytokinesis in zebrafish. The furrow microtubule array is analogous to the mammalian midbody during the final stage of cytokinesis. The furrow microtubule array remodeling process is mediated by *nebel*, a cytokinesis regulator gene. When *nebel* mutant embryos were imaged using Oregon-Green BAPTA dextran, they not only showed reductions in both amplitude and frequency of slowly moving intracellular waves of calcium, but also failed to undergo cleavage. However, when calcium activity was enhanced by injecting IP3, NAADP, and $CaCl_2$ solution into *nebel* mutant eggs, furrow microtubule array formation was increased and the eggs progressed towards pseudocleavage [36]. These observations underscore the importance of calcium during the first divisions of zygotic development.

As embryos develop, furrow-associated Ca^{2+} activity accompanies successive cleavages until the 64-cell stage and then becomes undetectable, at least in zebrafish [37]. These results were consistent with findings from previous studies in wild type zebrafish [38]. During later blastula stages, namely the 128 cell-stage in zebrafish, both forms of calcium activity (spikes and waves) are observed. During a calcium spike, cytoplasmic calcium activity spreads over a whole cell, whereas in a calcium wave, it propagates across a group of contiguous or nearly contiguous cells in the blastoderm [37,39]. In *Xenopus*, this activity has been shown to be enriched in anterior ectoderm at late blastula stages [40]. Dysregulation of this calcium activity results in misalignment of microtubules, which leads ultimately to gastrulation failure [41]. These stages are very challenging to image in mammalian development, so it remains unknown if similar mechanisms are at play in human embryogenesis.

2.3. Gastrulation

During gastrulation, embryos exhibit characteristic calcium activity patterns which regulate well-coordinated cell movements that result in the formation of the characteristic vertebrate body plan and the induction of the nervous system. The three germ layers began to form during blastula stages, but become well established and exhibit regional differences during gastrulation [42]. The massive degree of cell movement leads to significant tissue rearrangements and inductive interactions that will ultimately give rise to specific organ systems [43,44]. The ectoderm gives rise to the nervous system and epidermis, while the mesoderm predominantly gives rise to the circulatory system, skeletal-muscular system, connective tissue and urinary system. The digestive, respiratory and glandular systems including organs such as the liver and the pancreas are derived from the endoderm. Typically, failure of gastrulation is embryonic lethal; however, some rare birth defects such as caudal dysgenesis and conjoined twinning are considered to be associated with impaired gastrulation [45].

Not surprisingly, calcium activity is essential for virtually every aspect of gastrulation. Because of the size and accessibility of the embryos, the majority of our knowledge of the critical role of calcium derives from studying amphibian and fish embryos. As embryos progress from the mid-blastula transition to early gastrula stage of development, calcium activity drastically decreases in its frequency of occurrence and its amplitude in zebrafish and *Xenopus* [37,39]. This period of decreased activity, in *Xenopus* and zebrafish embryos, is referred to as a "quiet period" [37,40,46,47]. After mid-blastula transition, however, calcium activity is more complex and tissue specific, that is, different tissue domains of the embryo exhibit differential propagating multicellular waves and non-propagating spikes [39]. For example, the dorsal quadrant of embryos, in comparison with ventral and lateral quadrants, exhibits higher calcium activity in terms of both frequency of occurrence and amplitude in *Xenopus* and zebrafish [37,40,48]. Similarly, calcium activity is enriched in the frequency of occurrence and amplitude in Blastula Chordin and Noggin-expressing region which is located in dorsal animal cells and presumptive neural tissue in *Xenopus* and Zebrafish [40,47,49–51]. Additionally, a gradual elevation of Ca^{2+} in the dorsal quadrant is superimposed at early gastrula stages and continues until late neurula in zebrafish and *Xenopus* [40,52]. Immunohistochemical assays have shown that newt embryos express various calcium channels, including L-type voltage-gated calcium channels and ryanodine receptors, throughout the presumptive ectoderm during gastrulation. Inhibition of this calcium activity, using low concentrations of potent inhibitors of the sarco/endoplasmic reticulum Ca^{2+} ATPase pumps, including thapsigargin and cyclopiazonic acid, during late blastula/early gastrula stages of development resulted in severe developmental defects including cyclopia and tail defects in zebrafish [53].

Calcium activity has been shown to regulate the extensive tissue movements that occur during gastrulation; disruption of calcium activity severely impedes cell and tissue movements. During massive tissue rearrangement processes, intercellular calcium waves are generated, which are considered an extension of the spikes and waves observed in the blastula stage [54]. The leading cells of this rearrangement initiate gastrulation by sensing and translating the signal from the ectoderm. Interestingly, these leader cells exhibit remarkably high calcium activity compared to other cells from second and third rows throughout gastrulation in *Xenopus* explants [44,55]. Disruption of calcium activity during gastrulation using calcium chelators such as BAPTA-AM, calcium channel blockers such as R(+)BayK and nicardipine, and morpholino antisense constructs which inhibit translation of calcium channels such as L-type calcium channels and purinergic receptor P2Y11, significantly reduces leader cells' protrusion activity and migration speed, thus impairing gastrulation and convergent extension [43,44]. When calcium activity was increased by ionomycin treatment, higher migratory activity of the leading edge mesoderm was induced during gastrulation in *Xenopus* [44]. In humans, some extremely rare congenital defects such as sirenomelia and caudal dysgenesis are thought to occur due to impaired gastrulation, however, the relationship to calcium activity remains unclear. In general, defects in gastrulation will result in embryonic lethality [45,56].

2.4. Neural Induction

The tissue rearrangements that occur during gastrulation form the basis for neural induction, as the overlying dorsal ectoderm comes into contact with the dorsal mesendoderm. This process commences in the dorsal ectoderm near the onset of gastrulation, from embryonic day 6.0–8.5 in mice, due to a complex interaction between the blastula-chordin and noggin-expressing center and Nieuwkoop center in *Xenopus* [42,57]. This interaction triggers the secretion of neural inducing factors, such as noggin, chordin, follistatin, Xnr3 and Cerberus. These factors inhibit a potent epidermal inducer, bone morphogenetic protein (BMP), and also coordinate regulation of fibroblast growth factor (FGF), Ca^{2+} and Wnt signaling. Therefore, neural induction has been considered the default pathway during development as it occurs in the absence of the suppressive influence of BMP [40,51]. Calcium activity is strongly implicated in the process of neural induction.

Barth and Barth, using amphibian embryos, first introduced the concept that Ca^{2+} treatment can activate neural induction [58]. Stableford (1967) supported this idea by showing an eight-fold increase in Ca^{2+} concentration in the internal fluid of amphibian neurula when compared with the blastula [59]. Recently, further detailed analysis of calcium activity has been achieved with both spatial and temporal resolution during early neural development. These studies have shown that the increase in cytoplasmic Ca^{2+} concentration in the dorsal ectoderm is critical in enabling cross-talk between different signaling pathways, including BMP and FGF/Erk, required for neural induction in *Xenopus* and zebrafish [51,60]. Similar results in mice are consistent with the findings in lower vertebrates. For example, pharmacological inhibition of calcineurin, a calcium-calmodulin-dependent serine/threonine phosphatase, using cyclosporin A and FK506 on day 3 to day 8 of gestation in mice results in reduced expression of neuroectodermal markers *FezF1*, *FezF2*, and *Six3*. This inhibition also increased expression of mesodermal marker Brachyury and BMP responsive genes including BMP4-Smad1/5 target genes, and induced phosphorylation of Smad1/5. However, FGF- or ionophore elevated Ca^{2+} dependent activation of calcineurin rescued these effects [51]. Similarly, when elevation of cytoplasmic Ca^{2+} from neuroectoderm is inhibited using L-type voltage gated calcium channel antagonists, such as R(+)BayK8644, and calcium chelators, such as BAPTA, neuralization of the dorsal ectoderm is prevented. This inhibition blocks the expression of neural markers, including geminin, *Zic3*, and *NCAM* and also results in developmental defects, including head and CNS deformation, absence of eyes, abnormal notochord formation, and spina bifida in *Xenopus* [40]. Similar results have been obtained by inhibiting calcium influx from both the extracellular matrix via other channels including TRP and intracellular stores via mediators such as IP3R and RyR [40,49,51,61]. Taken together, these experiments indicate that calcium signaling is essential for neural induction in a wide range of vertebrate animals, including mammals, and that deregulated calcium activity may result in severe developmental defects.

2.5. Organogenesis

Gastrulation establishes the basic vertebrate body plan with recognizable anterior-posterior, dorsal-ventral, and left-right axes. During organogenesis, the regional differences in the three germ layers that developed during the blastula and gastrula stages give rise to different organ systems. While calcium activity is critical for the development of all organ systems, in this review, we will focus on the importance of calcium activity during nervous system, heart, kidney, muscle and immune system development, as well as diseases associated with abnormal calcium activity.

2.5.1. Nervous System Development

Neural Tube Closure

Following gastrulation and neural induction, the neural tube forms the rudimentary basis of the central nervous system. By the end of gastrulation, undifferentiated neural progenitor cells form the neural plate on the dorsal surface of the embryo as a flat neuro-epithelial sheet. As embryonic development progresses, this plate lengthens along the anterior-posterior axis in a process called rostro-caudal extension, the lateral borders narrow via a process called mediolateral convergence, elevate, and bend to form ridged/grooved neural folds. The edges of the neural fold meet and fuse to form the hollow neuroepithelial tube. This process is characterized by apical constriction, cell shape change, and morphogenetic movements. Apical constriction is believed to drive convergent extension, movement, and transformation of two-dimensional planar neural tissue to a three-dimensional structure with dorsoventral, mediolateral and rostrocaudal axes in the neural tube between weeks three and four of human gestation [62,63]. This seemingly simple process of neural tube formation is orchestrated by a complex gene expression network, which is tightly regulated by spatiotemporal interactions between different tissue types as well as a number of molecular processes involving ~300 genes, epigenetic modifications, and multiple signaling pathways [63,64].

Calcium activity has been shown to regulate biomechanical processes and epithelial re-modeling during neural tube closure. Smedley and Stanisstreet (1986) cultured rat embryos at 10.4 days in papaverine and D-600, which are pharmacological inhibitors of calcium channels, and found that this inhibition results in severe neural tube defects [65]. This experiment provided the first evidence that calcium is necessary for neural tube closure. Recent studies using vertebrate model organisms including zebrafish, *Xenopus*, and mice have supported the idea that calcium is a major regulator of neural tube closure [4,66–68]. Visualization of calcium influxes during neural tube closure using genetically encoded calcium indicator showed distinct spikes and waves. Spikes were short, only lasting less than 40 s, and limited to single cells, while waves were long-lasting transients that originate from a single cell or a group of a few cells and propagate across several to hundreds of cells in *Xenopus* [4,66]. As neural tube closure progresses, intracellular calcium activity increases in terms of both incidence and frequency of occurrence, becoming extremely frequent during late stages of closure. Interestingly, the frequency of calcium activity correlates with a number of biomechanical processes that are necessary for neural tube closure: apical contraction, cell polarization and intercalation, morphogenetic movement, and cell shape change [4,66,68]. The apical regions of neuroepithelial cells are highly enriched with motor proteins that mediate contractions of actin filaments in the apical regions of these cells. When this actin activity is imaged using F-actin marker Lifeact-EGFP at the same time as calcium activity, F-actin mesh-like structures develop in the center of the cells following the calcium transients during neural tube closure in *Xenopus* [4]. In addition, calcium transients trigger non-muscle myosin II activation within a minute of their occurrence in *Xenopus* and *Drosophila* [4,64,67].

Pharmacologically disrupting calcium transients by blocking IP3R and other membrane-localized Ca^{2+} channels, using 2ABP and nifedipine, abolished both apical constriction and movements of the neural plate towards closing. Also, this disruption reduced the width of the neural gene *Sox2* expression zone and ultimately resulted in neural tube closure defects; these defects included impaired forebrain, hindbrain and midbrain in *Xenopus* and *Ciona* [4,66,68–71]. These results suggest that dysregulation of calcium during convergent extension and neural tube closure results in severe central nervous system defects. Central nervous system defects, including neural tube closure defects, accounted for about 8% of all birth defects in the United States from 2004–2006; examples include anencephaly, spina bifida without anencephaly, encephalocele, and anophthalmia/microphthalmia [72]. The causes of these defects are not completely understood; however, impaired calcium activity is likely to be a potential factor.

Neurotransmitter Phenotype Specification

A functional brain is composed of heterogeneous populations of cells that express different neurotransmitters and neurotransmitter receptors. The process of neural differentiation and acquisition of functional identity, also known as neurotransmitter phenotype specification, is achieved by complex interactions between transcription factors, morphogenetic proteins such as BMP and Shh, and other signaling molecules [73]. Common examples of neurotransmitter phenotypes include cholinergic and GABAergic. Interestingly, the neurotransmitter phenotype for a single neuron is plastic and can change depending upon its extrinsic and intrinsic signaling environment [74].

Calcium signaling has been established as a key regulator of neural differentiation and phenotype specification. In 1993, Spitzer and colleagues first introduced the concept of calcium activity-mediated neurotransmitter specification. When presumptive spinal cord neurons of *Xenopus* were cultured in calcium free medium, the total number of GABA immunoreactive GABAergic neurons decreased by 30% [75]. Similarly, when calcium activity of primary neural cells isolated from 3 different stages of development, namely neural plate, neural fold and neural tube, in *Xenopus* was enhanced via exposure to (−)BayK 8644, a pharmacological activator of voltage gated calcium channels, the total number of excitatory neurons was found to be decreased [76]. Furthermore, these results were strengthened by evidence that enhancing calcium spike frequency, either by overexpression of the voltage-gated rat brain sodium channels $rNa_v2a\alpha$ and $rNav_2a\beta$ or application of the sodium channel antagonist veratridine, resulted in increased GABA and Glycine immunoreactive inhibitory neurons in *Xenopus* embryonic spinal cord in vitro as well as in vivo. Conversely, chronic induction of hyperpolarization, either by overexpression of mRNA encoding the human inward rectifier potassium channel hKir2.1 or pharmacological blockers, including EGTA or a mixture of GVIA w-conotoxin, calcicludine, flunarizine and tetrodotoxin, has been shown to suppress calcium activity. This suppression of calcium activity increased the number of glutamatergic and cholinergic immunoreactive excitatory neurons in *Xenopus* spinal cord, in vivo as well as in tissue culture [77,78]. These changes in the ratio between excitatory neurons and inhibitory neurons lead to neurodevelopmental diseases, including autism spectrum disorder, Rett syndrome, and fragile X syndrome [79,80]. For example, reduction in the number of inhibitory synapses and over-activation of class I metabotropic glutamate receptor in the cerebral cortex and hippocampus due to lack of a modulator of voltage gated calcium channel Cav2.2 called Fragile X Mental Retardation 1(*fmr1*) gene results in fragile X syndrome in mice. Strikingly, when these mice were treated with the GABA agonist gaboxadol, the phenotype was rescued [80].

Furthermore, different cell types have been shown to exhibit their signature frequencies and incidence patterns of calcium activity during a 10-h period of *Xenopus* spinal cord development after neural tube closure. Dorsal sensory Rohon–Beard neurons exhibit spontaneous calcium activity characterized by a low and constant frequency, dorsolateral interneurons exhibit monotonically increasing frequency, ventral motoneurons exhibit low to high stepping frequency, and ventral interneurons maintain a high frequency throughout the larval period [77]. Differential proportions of these different cell types have been implicated in developmental diseases: for example, decreased GABAergic, glutamate decarboxylase 65 KDa isoform immunoreactive, neurons in the cerebellar dentate nuclei and increased GABAergic interneurons in the hippocampus have been implicated in idiopathic autism. Similarly, in vivo studies have associated reduced GABA$_A$ receptor density in the fronto-temporal cortex in Rett Syndrome [81]. Involvement of calcium channels in a wide array of disorders associated with impaired neurotransmitter phenotype specification suggests that dysregulation of calcium homeostasis might play a role in the etiology of various pathophysiological conditions, including chronic pain, cerebellar ataxia, autism schizophrenia and migraine [82]. Therefore, dysregulation of calcium signaling could be a potential cause of neurodevelopmental diseases.

2.5.2. Muscle Development

Muscle tissue, both skeletal and cardiac, is the derivative of mesoderm, the middle of the three germ layers. Paraxial mesoderm gives rise to subcutaneous tissue and to somitomeres, which develop into head mesenchyme, skeletal muscles and the axial skeleton, intermediate mesoderm gives rise to urogenital structures, and lateral plate mesoderm gives rise to components of limbs and the circulatory system, including the heart. It is apparent that derivation of these structures from a primitive germ layer is the product of a complex spatio-temporally regulated interaction network which is affected by multiple factors, calcium being one of them. Disruption of calcium activity during muscle development results in severe developmental defects, including but not limited to reduced and amorphous skeletal muscle [83,84]. Differentiating muscle cells, both in culture and in situ, exhibit spontaneous calcium transients as a first step of muscle development (myogenesis) in vertebrates. Chernoff and Hilfer for the first time cultured chick embryo trunk in media supplemented with and without calcium, in addition to various other calcium transport agonists and antagonists, including caffeine, ionophore A23187, papaverine, and verapamil, and determined that calcium activity was required for somitogenesis [85]. In addition, somite maturation and calcium activity have been shown to correlate in *Xenopus*. When somites of early tailbud stage *Xenopus* embryos, namely anterior somites, maturing somites, segmenting somites and unsegmented paraxial mesoderm, were exposed and imaged using a calcium sensitive dye (fluo-3) along the AP-axis, calcium activity, measured by incidence, frequency and duration of transients, was inversely correlated with the maturity of somites. For instance, anterior somites were more mature than unsegmented paraxial mesoderm as reflected in the expression of sarcomeric myosin, and anterior somites exhibited no calcium activity while unsegmented paraxial mesoderm exhibited the highest activity [86]. Interestingly, two distinct calcium active periods with a frequency ranging from 0.02 Hz to 0.12 Hz and a quiet period lasting about 3.5 h at the 28-somite stage have been shown to occur during slow muscle cell development after the 17-somite stage in transgenic zebrafish. This transgenic zebrafish constitutively expressed apoaequorin under an actin promoter only in developing muscle cells [87–89]. All of these observations underscore the importance of calcium during early muscle development.

Calcium activity observed during myogenesis, as opposed to neurogenesis, mainly relies upon store-operated calcium entry, which is mediated by RyR, TPC2, IP3R, ORAI/STIM, VGCC, and TRPC, among others [90–92]. Dysregulation of cytoplasmic Ca^{2+} homeostasis due to mutations or pharmacological manipulation of these mediators during myogenesis results in severe fetal neuromuscular disorders [82,93,94]. For instance, mice lacking STIM1 die perinatally [95]. Loss of function mutations in *STIM1* and *Orai* have also been identified in human patients suffering from hypotonia [96,97]. Additionally, blocking the the RyR channel using ryanodine resulted in inhibited differentiation of fetal myoblasts in E9.5 mice [98], hydrocephaly and scoliokyphosis in *Xenopus* [86]; and disruption of the vertical myoseptae and smooth muscle cell spanning in zebrafish [87,99]. Also, a homozygous Ile4898 to I4898T mutation in RYR1 in E9.5 mice causes severe asphyxia characterized by reduced and amorphous skeletal muscle, disorganized myofibrils, delayed cardiovascular development and perinatal death [92]. Similarly, when TPC2 expression was blocked using a TPC2 specific morpholino or a CRISPR/Cas system, decreased myotome width, U-shaped somites in the anterior trunk instead of the usual chevron-shaped, and smooth muscle cell myofibrils that were not aligned into bundles were observed in zebrafish [88,89]. These phenotypic effects were rescued when additional calcium release from intracellular stores was triggered using caffeine, IP3 or expression of mRNA construct of TPC2 [88,89]. These observations demonstrate the crucial role of calcium signaling in diseases of the muscle.

2.5.3. Heart Development

The heart and other components of the circulatory system derive from lateral plate mesoderm during the third week of pregnancy in humans. A plate of promyocardial cells intermixed with endothelial strands on either side of the neural fold represents the developing heart at the neural fold stage of development. As embryos progress, two massive growth events, a massive growth of the anterior portion of the neural tube and an endodermal invagination to form the foregut, drive the folding of the promyocardial plate to form endocardial tubes in either side of the neural tube. At this stage, the heart is bilaterally symmetrical, with an inverted Y shape. Although establishment of the pulmonary system happens later in development, the endothelial strand ensures the presence of the circulatory system. These two tubes elongate and converge towards each other to form the primitive heart tube, which quickly forms different cardiac structures including the truncus arteriosus and sinus venosus [100–102]. It is indisputable that a genetic blueprint is required to regulate all the developmental processes and necessary cell differentiations in the developing heart, including formation of cardiomyocytes and valvular interstitial cells. For example, NKX2.5 is expressed in early heart progenitor cells, and is considered a master controller of cardiac development. In addition, various members of the TGF-β superfamily including nodal and activin, BMPs, and other signaling molecules including calcium have been shown to play an indispensable role during cardiogenesis [100].

Dysregulation of calcium activity results in failure in every step of cardiac development, including differentiation of cardiac progenitor cells and cardiac tube formation. Two decades ago, using aequorin, Creton et al. found a 10-fold increase in cytoplasmic Ca^{2+} in the cardiac region during cardiogenesis at the segmentation period of zebrafish development. This increase was achieved by a large number of calcium spikes that occurred every 10–20 min for several hours [103]. When this calcium activity was blocked by injecting low concentrations of BAPTA buffer into the zygote (higher concentrations stalled fertilized embryos in the one cell stage), the heart defects were apparent.

The heart was smaller and stretched longitudinally, and it was not able to pump the blood [103]. Recent findings using a diverse group of organisms, both in vivo and in vitro, have strengthened our understanding of calcium's role in cardiac development. For example, elevated levels of calcium were observed while imaging the heart region in both transgenic zebrafish and mice at comparable stages of development. These transgenic animals expressed genetically encoded calcium markers GCaMP6 and GCaMP2 respectively, which express under the promoter of ubiquitously expressing genes (Tg[βactin2:GCaMP6s]stl351, Tg[ubi:GCaMP6s]stl352 in zebrafish and Tet-Off αMHC-CaMP2 in mice) in order to visualize cytosolic Ca^{2+} ions [37,104]. When these calcium spikes during cardiogenesis are blocked by culturing mouse embryos at E7.5 to E8.5 in media supplemented with the L-type voltage gated calcium channel blockers, nifedipine and verapamil, various heart defects, including lack of a right ventricle and a large left ventricle, were apparent [105]. In addition, these blockers reduced DNA synthesis, (assayed using tritiated [3H]-thymidine incorporation), cell division (via reduction of the mRNA expression of a positive cell cycle regulator), cyclin B1 (detected using qPCR), and differentiation of cardiac progenitors (detected using immunolabeling of differentiation marker sarcomeric myosin), in cardiac cells from E11.5 mice in vitro [106]. Similarly, mouse embryos treated with bivalent cation chelators such as BAPTA or EGTA, or embryos that lack calreticulin, which is a Ca^{2+}-binding chaperone of the endoplasmic reticulum, die in utero due to defective heart development. However, the calcium ionophore ionomycin restored myofibrillogenesis in cardiomyocytes [107,108]. These observations underscore the importance of calcium activity during cardiogenesis and associated developmental heart defects.

2.5.4. Kidney Development

Disruptions to calcium signaling pathways have also been implicated in a number of developmental dysfunctions and diseases in the kidney. Calcium activity mediates the initiation and formation of the kidney field; the kidney field is defined as the mesodermal territory where expression of *pax8* and *lhx1* overlap during organogenesis. Overexpression of *pax8* and/or *lhx1* leads to the formation of enlarged and ectopic pronephroi [52,109]. In addition, calcium regulates the transcription of genes that are involved in the formation of renal structures, including *pax8*, *lhx1*, *osr1*, and *osr2*. Loss of function of any of these genes leads to impaired pronephros development in *Drosophila* and *Xenopus* [109–111]. Using aequorin imaging, it was shown that calcium transients occur in intact *Xenopus* embryos in the lateral mesoderm during kidney tubule formation. When these transients were inhibited with a calcium chelator during the late gastrula or mid neurula stage, kidney tubule development was disrupted. Interestingly, incubating *Xenopus* ectoderm with activin A, which is a IP3 modulator/mesodermal inducer, followed by retinoic acid, which is a voltage-gated calcium channels modulator, rescued pronephric tubule formation [112]. Note that retinoic acid stimulates calcium transients, while activin A alone does not. However, treating *Xenopus* ectoderm with activin A followed by caffeine or ionomycin not only increased cytoplasmic Ca^{2+}, but also triggered normal tubule differentiation [112]. This evidence suggests that calcium signaling is necessary and sufficient to regulate the tubule differentiation process.

Additionally, several calcium-dependent proteins have been linked to pronephros development and function. For example, the proteins polycystin-1 and polycystin-2, encoded by the genes PKD1 and PKD2 respectively, are expressed in renal tissue of both humans and mice during renal development and have been shown to function together to create a calcium-permeable channel [107]. Mutations in PKD1 and PKD2 result in autosomal dominant polycystic kidney disease in humans [113]. The protein products of these mutant forms of PKD1 and PKD2 do not form their normal calcium permeable channel during embryogenesis, which results in the development of large fluid-filled cysts within the kidney tubules and collecting ducts. Collectively these observations suggest that calcium activity is instrumental to multiple aspects of kidney development and associated diseases.

2.5.5. Immune System

Although the immune system does not appear until later stages of development, for example until 12 weeks of development in humans, immune cell differentiation and migration, which are mediated by calcium signaling, are hallmarks of development. Boucek and Snyderman first demonstrated the requirement for calcium activity during neutrophil functioning using lanthanum chloride, which is a calcium influx inhibitor [114]. More recent studies have strengthened the idea that a transient increase in cytoplasmic Ca^{2+} is required for immune cell differentiation and migration. *Stim1* and *Stim2* double-knockout mice at 5–6 weeks showed reduced regulatory T cells in the thymus, spleen, and lymph nodes, and these T-cells showed no Ca^{2+} influx when imaged using calcium indicator Fura-2-AM even after thapsigargin treatment [115]. These knockout mice exhibited a decrease in regulatory T-cells in all immune organs, leading to autoimmune disease symptoms [115,116]. Similarly, imaging transgenic zebrafish embryos that expressed GCaMP3 specifically in neutrophils under the promoter *LysC* showed elevated cytoplasmic Ca^{2+} as well as enhanced calcium spiking behavior during migration and phagocytosis at wound sites in vivo. Inhibition of this calcium activity using the calcium channel antagonist SKF 96365 resulted in impaired recruitment of these neutrophils due to their undirected movement [117]. Likewise, when fura-2 loaded human neutrophils were cultured in media with and without calcium supplement, cells cultured in calcium free medium were unable to migrate on poly-D-lysine-coated glass [118].

In addition, cytoplasmic Ca^{2+} elevation is also facilitated by store-operated calcium entry through calcium release activated channels in immune cells, including lymphocytes and mast cells [115,119]. Dysregulation of calcium release activated channels results in a severely compromised immune system. For example, mice lacking STIM1 or with a gene-trap insertion in the *Orai1* gene show defective mast cell degranulation and hereditary immunodeficiencies [120,121]. Patients with these mutations displayed symptoms such as severe T cell immunodeficiency, impaired T cell activation and proliferation leading to recurrent viral, bacterial, and fungal infections, and muscular hypotonia [119]. Similarly, alterations in *STIM1* expression affected the sensitivity of immunoglobin E-mediated immediate-phase anaphylactic responses in vivo in mice [120]. These observations underscore the importance of calcium homeostasis and signaling behaviors in immune system development and function.

3. Wound Healing and Regeneration

3.1. Wound Healing before Formation of Immune System

In order to reach maturity, embryos must be able to respond to a wide variety of perturbations that are likely to occur during development. Embryos in the early stages of development exhibit inflammatory-response-free and scar-free wound closure with a mechanism that closely parallels neural tube closure during morphogenesis [122]. This healing is often called regenerative healing, and does not occur in most adult organisms [123]. The mechanisms of these abilities rely mostly on calcium signaling. Imaging the optic tectum of *Xenopus* larva using Oregon Green BAPTA-1 showed that fine micropipette mediated mechanical insults and targeted induction of cell death by a high-voltage electrical stimulus triggered a rapid influx of Ca^{2+} that generated further spontaneous Ca^{2+} influxes, which propagated through multiple rows of cells surrounding the injury site within seconds. These rapid calcium waves were accompanied by contractions of the neuroepithelium and expulsion of potentially damaged cells [124]. When this calcium activity was blocked and/or inhibited using BAPTA, thapsigargin, 2-aminoethoxydiphenyl borate, which is an IP3R blocker, and an array of purinergic receptor blockers, tissue contractions and potentially damaged cell expulsions were hindered. This blockage also negatively impacted wound healing [124]. In addition, damaged cells also release ATP into the extracellular milieu, which is recognized by the metabotropic purinergic receptors P2Ys on adjacent undamaged cells [125]. Activation of P2Y receptors further increases the cytoplasmic Ca^{2+} and activated metalloproteinase. This leads to heparin-binding EGF-like growth factor activation and subsequent cell proliferation, as well as formation of actomyosin cables and actin-rich protrusions, which guide tissue contraction, cell elongation, and migration in order to close the wound [125]. Therefore, calcium activity plays an indispensable role in embryonic scar-free regenerative wound healing.

3.2. Wound Healing Following the Formation of Immune System

At later stages of development, following the formation of functional immune cells, wound healing differs mechanistically from early embryonic wound healing, however, calcium activity continues to play a critical role. During this later stage of development, immune cells are also involved in wound healing and healing is no longer inflammation-response-free. Interestingly, this mechanism of healing is also regulated by calcium as a first responder. In this mechanism, calcium regulates both modulation of actomyosin and recruitment of immune cells. In *Drosophila* embryos, elevated calcium induced by laser wounding was shown to activate hydrogen peroxide synthase, resulting in a rapid accumulation of H_2O_2 around wound sites as indicated by the fluorigenic reporter Amplex Ultrared. A reduction in H_2O_2 signals at the wound site was observed in TRPM2 channel and innexin 2 mutant embryos, in which calcium elevation did not occur, indicating that calcium was necessary to this inflammatory response [126]. Likewise, using a genetically encoded H_2O_2 ratiometric sensor HyPer, Niethammer et al. for the first time showed that H_2O_2 is required for rapid recruitment of leukocytes to the wound in zebrafish larva [127]. Inhibition of this calcium activity using thapsigargin or EGTA not only inhibits H_2O_2 release healing efficiency, but also reduces the average hemocyte response during laser-induced epithelial wound healing in *Drosophila* embryos [126]. Even after the immune system develops and matures, calcium continues to play a critical role in embryonic wound healing.

3.3. Regeneration

Equally importantly, embryos also show calcium-mediated tissue regenerative capabilities during the early stages of development. Spontaneous calcium activity has been observed in the regenerating tail in *Xenopus* during the first hours of recovery. Inhibiting this calcium activity using ryanodine reduced the number of activated muscle progenitor cells and skeletal muscle stem cells, also known as muscle satellite cells, in the regenerating tissue. This, in turn, inhibited the regeneration process [128]. During regeneration, calcium activity increases secretion and mobilization of growth factors including insulin-like growth factor, interleukins, including interleukin-1, interleukin-6 and interleukin-8, parathyroid hormone, transforming growth factor-β and platelet-derived growth factor. Platelets are widely recognized as inducers of cell proliferation, and are also involved in stem cell differentiation, cell migration, and revascularization of damaged tissue via increased release of growth factors during tissue regeneration. Therefore, platelet-rich plasma derivatives have been used in regenerative medicine [129]. This increase in growth factor release has been proven- to be mediated by calcium activity. This calcium activity has been shown to be critical for platelet functioning using double knockouts of various calcium channels, including TRPM7, TRPC6, Orai1 and SERCA, and pharmacological agonists and antagonists in mice [130–132]. These factors facilitate changes in the expression of specific genes that are responsible for proper cell proliferation and migration during regeneration [122,133]. Using microarrays, Patterson et al. reported 624 upregulated and 826 downregulated genes associated with epidermal regeneration in *Drosophila* embryos, including genes related to cellular component organization and stress response [134]. Similarly, using an RNA-Seq approach, Tsujioka et al. reported 25 candidate genes involved in tail regeneration in *Xenopus*, including genes related to cell proliferation, for example, CDK1 and cyclin B2 [135]. These results indicate that calcium activates expression of a number of gene cascades that are responsible for tissue regeneration. Understanding the underlying mechanisms of embryonic regeneration could lead to new developments in the fields of regenerative clinical medicine and tissue engineering.

4. Conclusions and Future Directions

Over the last three decades, studies have shown that dysregulation of calcium signaling results in severe medical conditions, including neural tube and other developmental defects. Given the versatility of calcium as a signaling molecule and its near ubiquitous involvement with virtually every aspect of embryonic development (Figure 1), this is scarcely surprising. Yet despite its importance in development and the genesis of human disease, there is much that we do not understand about calcium and many avenues for future research. For example, currently, a comprehensive analysis of the regulation of spatiotemporal expression patterns of calcium-associated genes and their interactions is lacking. Such an analysis would suggest specific tissues and developmental timepoints where calcium-regulated processes can go awry. New gene editing and tissue engineering techniques as well as novel next generation sequencing approaches may aid in understanding the etiology and mechanisms of medical conditions associated with dysregulation of calcium, as well as potential therapeutic solutions for these conditions. In addition, determination of the embryonic expression patterns, both at the mRNA and protein level, of genes known to be involved in calcium activity and screens for novel genes implicated in calcium activity are warranted.

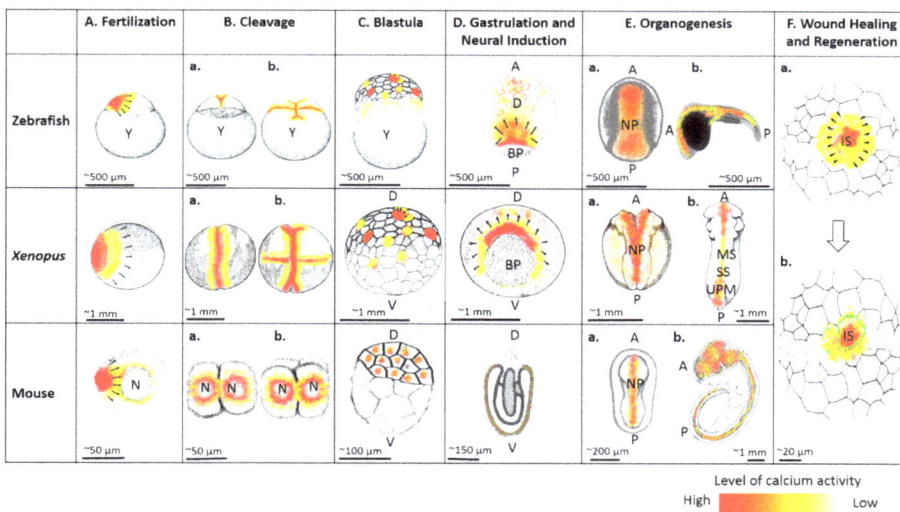

Figure 1. Comparative calcium activity among zebrafish, frogs, and mice during egg activation and fertilization (**A**), first cleavage (**Ba**), second cleavage (**Bb**), blastula (**C**), gastrulation and neural induction (**D**) and organogenesis (**E**), including neural tube closure (**Ea**) and muscle development (**Eb**) in zebrafish, *Xenopus*, and mouse; also, depiction of calcium dynamics immediately after tissue damage (**Fa**) and calcium mediated actomyosin filament (green lining) during wound healing and regeneration (**Fb**). Black arrows show direction of propagation of calcium waves. Y, yolk; A, anterior; P, posterior; D, dorsal; V, ventral; BP, blastopore; NP, neural plate; UPM, unsegmented paraxial mesoderm; MS, matured somites; SS, segmenting somites; IS, injury site; N, nucleus. Images were adapted and re-drawn from [136–138].

In addition to many avenues for further research, there is also a huge potential to utilize this versatile signaling molecule therapeutically and in regenerative medicine. It is well known that developing embryos up to the blastocyst stage of development and embryonic stem cells exhibit the ability to either maintain stem cell identity or differentiate into the lineages of all three germ layers. This opens new avenues for the exploration of these cells' potential uses in regenerative medicine [139–141]. The ultimate goal of regenerative medicine is to restore or substitute damaged tissue so the tissue can maintain its function(s); this may require boosting natural defense systems, which could harness and accelerate innate healing processes in order to cure previously unmanageable medical problems [142]. Towards this end, Jergova et al. engineered neural progenitor cells to generate transgenic GABAergic cells that release an analgesic peptide serine-histogranin, which antagonizes NMDA receptors. These cells were then transplanted onto an injury site in order to assess their effect on peripheral neuropathic pain after 1 week of induced injury in rat spinal cord. This transplant resulted in improved pain sensitivity based upon a pain-related behavioral assessment with four pain parameters: mechanical, heat, cold, and tactile [143]. Similarly, Mery et al. engineered pacemaker cells from embryonic stem cells that induce InsP3 mediated oscillatory calcium release from the endoplasmic reticulum; these cells had a potential to serve as pacemaker cells prior to the development of pacemaker ionic channels in early embryos. This oscillatory calcium release was required for pacemaker activity. When InsP3 receptor antisense cDNA was expressed constitutively in these cells, expression of type I InsP3 receptors and spontaneous calcium activity, both frequency of occurrence and amplitude, were reduced, and beating activity was also impaired [144]. These observations underscore the potential use of calcium signaling toolkits in medical cell and tissue engineering.

Current approaches used to understand calcium-related defects include mapping and cloning causative gene(s), recapitulating clinical features, and using pharmacological reagents in order to ameliorate symptoms. Gene editing is also beginning to be explored as a way to understand and possibly treat these defects. For example, Limpitikul et al. reprogrammed dermal fibroblasts into iPSC-derived cardiomyocytes, then they used these cardiomyocytes to model calm2-associated Long-QT syndrome, which is a congenital birth defect associated with a mutation in the calcium binding protein calmodulin2, *calm2* gene. Using CRISPRi technology, they were able to selectively correct the mutated allele and rescue the long QT syndrome phenotype [145]. Molecular gene editing tools like CRISPR have tremendous capabilities for therapeutic use, and these capabilities will continue to expand as molecular technology improves even further. However, off-target effects of these tools and complex etiologies of some birth defects continue to create critical therapeutic challenges. These etiologies can be influenced by a wide array of factors, including multiple genes, impaired gene-gene interactions, combinatorial effects of polymorphisms, changes in protein levels in time and space, epigenetic modifications, environmental effects, and mutations in epigenetic regulators. For example, ~300 genes have been shown to be critically required for neural tube closure [63], and RNAseq data identified at least 30 causative genes for congenital fetal akinesia deformation [83]. Addressing all of these factors during the process of therapeutic development remains a challenge in the scientific community.

Taken together, these results indicate that calcium is involved in a diverse array of physiological processes, and dysregulation of calcium activity results in wide array of developmental defects. There is a tremendous demand for scar-free healing after major surgical procedures, an ability the embryo possesses following unique patterns of calcium activity. The capacity of the embryo to either maintain stem cell identity or differentiate into specific lineages such as neural cells could also be used to treat neurodegenerative diseases. While great progress has been made in recent decades, translation of these fetal abilities into adult systems still remains a challenge. Understanding how an ion as simple as Ca^{2+} plays diverse physiological roles may help solve these medical problems. Some aspects of calcium signaling are understood in a variety of processes already, and recent advances in genomics, transcriptomics, proteomics, and gene editing technology may help expand our knowledge about the underlying mechanisms of calcium function and enable its potential therapeutic use.

Supplementary Materials: Supplementary materials can be found at http://www.mdpi.com/1422-0067/19/11/3390/s1.

Author Contributions: All authors read the primary literature and participated in writing and reviewing the manuscript.

Acknowledgments: This work was funded by the National Institutes of Health (1R15HD077624-01) and the National Science Foundation (IOS-1257895) to MSS.

Int. J. Mol. Sci. **2018**, *19*, 3390

Conflicts of Interest: The authors declare no conflict of interest.

References

1. Database GeneCards. GeneCards—Human Genes | Gene Database | Gene Search. Available online: https://www.genecards.org/ (accessed on 23 September 2018).
2. Lew, V.L.; Tsien, R.Y.; Miner, C.; Bookchin, R.M. Physiological $[Ca^{2+}]_i$level and Pump-Leak Turnover in Intact Red Cells Measured Using an Incorporated Ca Chelator. *Nature* **1982**, *298*, 478–481. [CrossRef] [PubMed]
3. Kramer, I. Intracellular Calcium. In *Signal Transduction*; Elsevier: Amsterdam, The Netherlands, 2016; pp. 381–439. [CrossRef]
4. Suzuki, M.; Sato, M.; Koyama, H.; Hara, Y.; Hayashi, K.; Yasue, N.; Imamura, H.; Fujimori, T.; Nagai, T.; Campbell, R.E.; et al. Distinct Intracellular Ca^{2+} Dynamics Regulate Apical Constriction and Differentially Contribute to Neural Tube Closure. *Development* **2017**, *144*, 1307–1316. [CrossRef] [PubMed]
5. Balaji, R.; Bielmeier, C.; Harz, H.; Bates, J.; Stadler, C.; Hildebrand, A.; Classen, A.K. Calcium Spikes, Waves and Oscillations in a Large, Patterned Epithelial Tissue. *Sci. Rep.* **2017**, *7*, 42786. [CrossRef] [PubMed]
6. Suzuki, J.; Kanemaru, K.; Ishii, K.; Ohkura, M.; Okubo, Y.; Iino, M. Imaging Intraorganellar Ca^{2+} at Subcellular Resolution Using CEPIA. *Nat. Commun.* **2014**, *5*, 4153. [CrossRef] [PubMed]
7. Berridge, M.J.; Lipp, P.; Bootman, M.D. The Versatility and Universality of Calcium Signalling. *Nat. Rev. Mol. Cell Biol.* **2000**, *1*, 11–21. [CrossRef] [PubMed]
8. Dupont, G.; Combettes, L.; Leybaert, L. Calcium Dynamics: Spatio-Temporal Organization from the Subcellular to the Organ Level. *Int. Rev. Cytol.* **2007**, *261*, 193–245. [PubMed]
9. Markova, O.; Sénatore, S.; Chardès, C.; Lenne, P.F. Calcium Spikes in Epithelium: Study on Drosophila Early Embryos. *Sci. Rep.* **2015**, *5*, 11379. [CrossRef] [PubMed]
10. Gu, X.; Olson, E.C.; Spitzer, N.C. Spontaneous Neuronal Calcium Spikes and Waves during Early Differentiation. *J. Neurosci. Off. J. Soc. Neurosci.* **1994**, *14*, 6325–6335. [CrossRef]
11. Clapham, D.E. Calcium Signaling. *Cell* **2007**, *131*, 1047–1058. [CrossRef] [PubMed]
12. Berridge, M.J.; Bootman, M.D.; Roderick, H.L. Calcium Signalling: Dynamics, Homeostasis and Remodelling. *Nat. Rev. Mol. Cell Biol.* **2003**, *4*, 517–529. [CrossRef] [PubMed]
13. National Institutes of Health. National Center for Advancing Translational Sciences (NCATS). Available online: https://ncats.nih.gov/ (accessed on 23 September 2018).
14. Zhaurova, K. Genetic Causes of Adult-Onset Disorders. *Nat. Educ.* **2008**, *1*, 49.
15. Robertson, S.; Lin, R. Oocyte-to-Zygote Transition. *Semin. Cell Dev. Biol.* **2018**. [CrossRef] [PubMed]
16. Kuroda, K.; Brosens, J.J.; Quenby, S.; Takeda, S. (Eds.) *Treatment Strategy for Unexplained Infertility and Recurrent Miscarriage*; Springer: Midtown Manhattan, NY, USA, 2018.
17. Kashir, J.; Deguchi, R.; Jones, C.; Coward, K.; Stricker, S.A. Comparative Biology of Sperm Factors and Fertilization-Induced Calcium Signals across the Animal Kingdom. *Mol. Reprod. Dev.* **2013**. [CrossRef] [PubMed]
18. Wozniak, K.L.; Mayfield, B.L.; Duray, A.M.; Tembo, M.; Beleny, D.O.; Napolitano, M.A.; Sauer, M.L.; Wisner, B.W.; Carlson, A.E. Extracellular Ca^{2+} Is Required for Fertilization in the African Clawed Frog, Xenopus Laevis. *PLoS ONE* **2017**, *12*, e0170405. [CrossRef] [PubMed]
19. Mizuno, H.; Sassa, T.; Higashijima, S.; Okamoto, H.; Miyawaki, A. Transgenic Zebrafish for Ratiometric Imaging of Cytosolic and Mitochondrial Ca^{2+} response in Teleost Embryo. *Cell Calcium* **2013**, *54*, 236–245. [CrossRef] [PubMed]
20. Cuthbertson, K.S.; Whittingham, D.G.; Cobbold, P.H. Free Ca^{2+} Increases in Exponential Phases during Mouse Oocyte Activation. *Nature* **1981**, *294*, 754–757. [CrossRef] [PubMed]
21. Deguchi, R.; Shirakawa, H.; Oda, S.; Mohri, T.; Miyazaki, S. Spatiotemporal Analysis of Ca(2+) Waves in Relation to the Sperm Entry Site and Animal-Vegetal Axis during Ca(2+) Oscillations in Fertilized Mouse Eggs. *Dev. Biol.* **2000**, *218*, 299–313. [CrossRef] [PubMed]
22. Ferrer-Buitrago, M.; Bonte, D.; De Sutter, P.; Leybaert, L.; Heindryckx, B. Single Ca^{2+} Transients vs. Oscillatory Ca^{2+} Signaling for Assisted Oocyte Activation: Limitations and Benefits. *Reproduction* **2018**, *155*, R105–R119. [CrossRef] [PubMed]

23. Miao, Y.-L.; Stein, P.; Jefferson, W.N.; Padilla-Banks, E.; Williams, C.J. Calcium Influx-Mediated Signaling Is Required for Complete Mouse Egg Activation. *Proc. Natl. Acad. Sci. USA* **2012**, *109*, 4169–4174. [CrossRef] [PubMed]

24. Miao, Y.L.; Williams, C.J. Calcium Signaling in Mammalian Egg Activation and Embryo Development: The Influence of Subcellular Localization. *Mol. Reprod. Dev.* **2012**, *79*, 742–756. [CrossRef] [PubMed]

25. Bernhardt, M.L.; Zhang, Y.; Erxleben, C.F.; Padilla-Banks, E.; McDonough, C.E.; Miao, Y.-L.; Armstrong, D.L.; Williams, C.J. CaV3.2 T-Type Channels Mediate Ca^{2+} Entry during Oocyte Maturation and Following Fertilization. *J. Cell Sci.* **2015**, *128*, 4442–4452. [CrossRef] [PubMed]

26. Tang, T.S.; Dong, J.B.; Huang, X.Y.; Sun, F.Z.; Royse, J.; Blayney, L.M.; Swann, K.; Lai, F.A. Ca(2+) Oscillations Induced by a Cytosolic Sperm Protein Factor Are Mediated by a Maternal Machinery That Functions only Once in Mammalian Eggs. *Development* **2000**, *127*, 1141–1150. [PubMed]

27. Saunders, C.M. Ca^{2+} Oscillations Triggered by Sperm PLCz. 2002.

28. Hachem, A.; Godwin, J.; Ruas, M.; Lee, H.C.; Ferrer Buitrago, M.; Ardestani, G.; Bassett, A.; Fox, S.; Navarrete, F.; de Sutter, P.; et al. PLCζ Is the Physiological Trigger of the Ca^{2+} Oscillations That Induce Embryogenesis in Mammals but Conception Can Occur in Its Absence. *Development* **2017**, *144*, 2914–2924. [CrossRef] [PubMed]

29. Lee, K.; Wang, C.; Machaty, Z. STIM1 Is Required for Ca^{2+} Signaling during Mammalian Fertilization. *Dev. Biol.* **2012**, *367*, 154–162. [CrossRef] [PubMed]

30. Yamaguchi, T.; Kuroda, K.; Tanaka, A.; Watanabe, S. Fertilization Failure. In *Treatment Strategy for Unexplained Infertility and Recurrent Miscarriage*; Springer: Singapore, 2018; pp. 7–17.

31. Ferrer-Buitrago, M.; Dhaenens, L.; Lu, Y.; Bonte, D.; Vanden Meerschaut, F.; De Sutter, P.; Leybaert, L.; Heindryckx, B. Human Oocyte Calcium Analysis Predicts the Response to Assisted Oocyte Activation in Patients Experiencing Fertilization Failure after ICSI. *Hum. Reprod.* **2018**, *33*, 416–425. [CrossRef] [PubMed]

32. Halliday, J.L.; Ukoumunne, O.C.; Baker, H.W.G.; Breheny, S.; Jaques, A.M.; Garrett, C.; Healy, D.; Amor, D. Increased Risk of Blastogenesis Birth Defects, Arising in the First 4 Weeks of Pregnancy, after Assisted Reproductive Technologies. *Hum. Reprod.* **2010**, *25*, 59–65. [CrossRef] [PubMed]

33. Chen, C.; Jin, J.; Lee, G.A.; Silva, E.; Donoghue, M. Cross-Species Functional Analyses Reveal Shared and Separate Roles for Sox11 in Frog Primary Neurogenesis and Mouse Cortical Neuronal Differentiation. *Biol. Open* **2016**, *5*, 409–417. [CrossRef] [PubMed]

34. Chan, C.M.; Chen, Y.; Hung, T.S.; Miller, A.L.; Shipley, A.M.; Webb, S.E. Inhibition of SOCE Disrupts Cytokinesis in Zebrafish Embryos via Inhibition of Cleavage Furrow Deepening. *Int. J. Dev. Biol.* **2015**, *59*, 289–301. [CrossRef] [PubMed]

35. Chan, C.M.; Aw, J.T.M.; Webb, S.E.; Miller, A.L. SOCE Proteins, STIM1 and Orai1, Are Localized to the Cleavage Furrow during Cytokinesis of the First and Second Cell Division Cycles in Zebrafish Embryos. *Zygote* **2016**, *24*, 880–889. [CrossRef] [PubMed]

36. Eno, C.; Gomez, T.; Slusarski, D.C.; Pelegri, F. Slow Calcium Waves Mediate Furrow Microtubule Reorganization and Germ Plasm Compaction in the Early Zebrafish Embryo. *Development* **2018**, *145*, dev156604. [CrossRef] [PubMed]

37. Chen, J.; Xia, L.; Bruchas, M.R.; Solnica-Krezel, L. Imaging Early Embryonic Calcium Activity with GCaMP6s Transgenic Zebrafish. *Dev. Biol.* **2017**, *430*, 385–396. [CrossRef] [PubMed]

38. Webb, S.E.; Miller, A.L. Ca^{2+} Signaling during Vertebrate Somitogenesis. *Acta Pharmacol. Sin.* **2006**, *27*, 781–790. [CrossRef] [PubMed]

39. Ma, L.H.; Webb, S.E.; Chan, C.M.; Zhang, J.; Miller, A.L. Establishment of a Transitory Dorsal-Biased Window of Localized Ca^{2+} Signaling in the Superficial Epithelium Following the Mid-Blastula Transition in Zebrafish Embryos. *Dev. Biol.* **2009**, *327*, 143–157. [CrossRef] [PubMed]

40. Leclerc, C.; Webb, S.E.; Daguzan, C.; Moreau, M.; Miller, A.L. Imaging Patterns of Calcium Transients during Neural Induction in Xenopus Laevis Embryos. *J. Cell Sci.* **2000**, *113*, 3519–3529. [PubMed]

41. Hara, Y.; Nagayama, K.; Yamamoto, T.S.; Matsumoto, T.; Suzuki, M.; Ueno, N. Directional Migration of Leading-Edge Mesoderm Generates Physical Forces: Implication in Xenopus Notochord Formation during Gastrulation. *Dev. Biol.* **2013**, *382*, 482–495. [CrossRef] [PubMed]

42. Levine, A.J.; Brivanlou, A.H. Proposal of a Model of Mammalian Neural Induction. *Dev. Biol.* **2007**, *308*, 247–256. [CrossRef] [PubMed]

43. Shindo, A.; Hara, Y.; Yamamoto, T.S.; Ohkura, M.; Nakai, J.; Ueno, N. Tissue-Tissue Interaction-Triggered Calcium Elevation Is Required for Cell Polarization during Xenopus Gastrulation. *PLoS ONE* **2010**, *5*, e8897. [CrossRef] [PubMed]

44. Hayashi, K.; Yamamoto, T.S.; Ueno, N. Intracellular Calcium Signal at the Leading Edge Regulates Mesodermal Sheet Migration during Xenopus Gastrulation. *Sci. Rep.* **2018**, *8*, 2433. [CrossRef] [PubMed]

45. Ferrer-Vaquer, A.; Hadjantonakis, A.K. Birth Defects Associated with Perturbations in Preimplantation, Gastrulation, and Axis Extension: From Conjoined Twinning to Caudal Dysgenesis. *Wiley Interdiscip. Rev. Dev. Biol.* **2013**, *2*, 427–442. [CrossRef] [PubMed]

46. Gilland, E.; Miller, A.L.; Karplus, E.; Baker, R.; Webb, S.E. Imaging of Multicellular Large-Scale Rhythmic Calcium Waves during Zebrafish Gastrulation. *Proc. Natl. Acad. Sci. USA* **1999**, *96*, 157–161. [CrossRef] [PubMed]

47. Webb, S.E.; Moreau, M.; Leclerc, C.; Miller, A.L. Calcium Transients and Neural Induction in Vertebrates. *Cell Calcium* **2005**, *37*, 375–385. [CrossRef] [PubMed]

48. Yuen, M.Y.F.; Webb, S.E.; Chan, C.M.; Thisse, B.; Thisse, C.; Miller, A.L. Characterization of Ca^{2+} signaling in the External Yolk Syncytial Layer during the Late Blastula and Early Gastrula Periods of Zebrafish Development. *Biochim. Biophys. Acta* **2013**, *1833*, 1641–1656. [CrossRef] [PubMed]

49. Moreau, M.; Neant, I.; Webb, S.E.; Miller, A.L.; Leclerc, C. Calcium Signalling during Neural Induction in Xenopus Laevis Embryos. *Philos. Trans. R. Soc. B Biol. Sci.* **2008**, *363*, 1371–1375. [CrossRef] [PubMed]

50. Leclerc, C.; Néant, I.; Moreau, M. Early Neural Development in Vertebrates Is also a Matter of Calcium. *Biochimie* **2011**, *93*, 2102–2111. [CrossRef] [PubMed]

51. Cho, A.; Tang, Y.; Davila, J.; Deng, S.; Chen, L.; Miller, E.; Wernig, M.; Graef, I.A. Calcineurin Signaling Regulates Neural Induction through Antagonizing the BMP Pathway. *Neuron* **2014**, *82*, 109–124. [CrossRef] [PubMed]

52. Moreau, M.; Néant, I.; Webb, S.E.; Miller, A.L.; Riou, J.F.; Leclerc, C. Ca^{2+} coding and Decoding Strategies for the Specification of Neural and Renal Precursor Cells during Development. *Cell Calcium* **2016**, *59*, 75–83. [CrossRef] [PubMed]

53. Creton, R. The Calcium Pump of the Endoplasmic Reticulum Plays a Role in Midline Signaling during Early Zebrafish Development. *Dev. Brain Res.* **2004**, *151*, 33–41. [CrossRef] [PubMed]

54. Webb, S.E.; Miller, A.L. Ca^{2+} signaling and Early Embryonic Patterning during the Blastula and Gastrula Periods of Zebrafish and Xenopus Development. *Biochim. Biophys. Acta* **2006**, *1763*, 1192–1208. [CrossRef] [PubMed]

55. Wallingford, J.B.; Ewald, A.J.; Harland, R.M.; Fraser, S.E. Calcium Signaling during Convergent Extension in Xenopus. *Curr. Biol.* **2001**, *11*, 652–661. [CrossRef]

56. Lange, L.; Marks, M.; Liu, J.; Wittler, L.; Bauer, H.; Piehl, S.; Bläß, G.; Timmermann, B.; Herrmann, B.G. Patterning and Gastrulation Defects Caused by the Tw18 Lethal Are Due to Loss of Ppp2r1a. *Biol. Open* **2017**, *6*, 752–764. [CrossRef] [PubMed]

57. Yasuoka, Y.; Taira, M. The Molecular Basis of the Gastrula Organizer in Amphibians and Cnidarians. In *Reproductive and Developmental Strategies. Diversity and Commonality in Animals*; Springer: Tokyo, Japan, 2018; pp. 667–708.

58. Barth, L.G.; Barth, L.J. Sequential Induction of the Presumptive Epidermis of the Rana Pipiens Gastrula. *Biol. Bull.* **1964**, *127*, 413–427. [CrossRef]

59. Stableford, L.T. A Study of Calcium in the Early Development of the Amphibian Embryo. *Dev. Biol.* **1967**, *16*, 303–314. [CrossRef]

60. Leclerc, C.; Néant, I.; Moreau, M. The Calcium: An Early Signal That Initiates the Formation of the Nervous System during Embryogenesis. *Front. Mol. Neurosci.* **2012**, *5*, 1–12. [CrossRef] [PubMed]

61. Lee, H.-K.; Lee, H.-S.; Moody, S.A. Neural Transcription Factors: From Embryos to Neural Stem Cells. *Mol. Cells* **2014**, *37*, 705–712. [CrossRef] [PubMed]

62. Flanagan, M.; Sonnen, J.A.; Keene, C.D.; Hevner, R.F.; Montine, T.J. *Molecular Basis of Diseases of the Nervous System*, 2nd ed.; Elsevier Inc.: Amsterdam, The Netherlands, 2018.

63. Wilde, J.J.; Petersen, J.R.; Niswander, L. Genetic, Epigenetic, and Environmental Contributions to Neural Tube Closure. *Annu. Rev. Genet.* **2014**, *48*, 583–611. [CrossRef] [PubMed]

64. Nikolopoulou, E.; Galea, G.L.; Rolo, A.; Greene, N.D.E.; Copp, A.J. Neural Tube Closure: Cellular, Molecular and Biomechanical Mechanisms. *Development* **2017**, *144*, 552–566. [CrossRef] [PubMed]

65. Smedley, M.J.; Stanisstreet, M. Calcium and Neurulation in Mammalian Embryos II. Effects of Cytoskeletal Inhibitors and Calcium Antagonists on the Neural Folds of Rat Embryos. *Development* **1986**, *93*, 167–178.

66. Christodoulou, N.; Skourides, P.A.A. Cell-Autonomous Ca^{2+} Flashes Elicit Pulsed Contractions of an Apical Actin Network to Drive Apical Constriction during Neural Tube Closure. *Cell Rep.* **2015**, *13*, 2189–2202. [CrossRef] [PubMed]

67. Kong, D.; Wolf, F.; Großhans, J. In Vivo Optochemical Control of Cell Contractility at Single Cell Resolution by Ca^{2+} Induced Myosin Activation. *bioRxiv* **2018**. [CrossRef]

68. Sahu, S.U.; Visetsouk, M.R.; Garde, R.J.; Hennes, L.; Kwas, C.; Gutzman, J.H. Calcium Signals Drive Cell Shape Changes during Zebrafish Midbrain–hindbrain Boundary Formation. *Mol. Biol. Cell* **2017**, *28*, 875–882. [CrossRef] [PubMed]

69. Abdul-Wajid, S.; Morales-Diaz, H.; Khairallah, S.M.; Smith, W.C. T-Type Calcium Channel Regulation of Neural Tube Closure and EphrinA/EPHA Expression. *Cell Rep.* **2015**, *13*, 829–839. [CrossRef] [PubMed]

70. Sequerra, E.B.; Goyal, R.; Castro, P.A.; Levin, J.B.; Borodinsky, L.N. NMDA Receptor Signaling Is Important for Neural Tube Formation and for Preventing Antiepileptic Drug-Induced Neural Tube Defects NMDA Receptor Signaling Is Important for Neural Tube Formation and for Preventing Antiepileptic Drug-Induced Neural Tube Defec. *J. Neurosci.* **2018**. [CrossRef] [PubMed]

71. Borodinsky, L.N. Xenopus Laevis as a Model Organism for the Study of Spinal Cord Formation, Development, Function and Regeneration. *Front. Neural Circuits* **2017**, *11*, 90. [CrossRef] [PubMed]

72. Parker, S.E.; Mai, C.T.; Canfield, M.A.; Rickard, R.; Wang, Y.; Meyer, R.E.; Anderson, P.; Mason, C.A.; Collins, J.S.; Kirby, R.S.; et al. Updated National Birth Prevalence Estimates for Selected Birth Defects in the United States, 2004–2006. *Birth Defects Res. Part A—Clin. Mol. Teratol.* **2010**, *88*, 1008–1016. [CrossRef] [PubMed]

73. Borodinsky, L.N.; Belgacem, Y.H. Crosstalk among Electrical Activity, Trophic Factors and Morphogenetic Proteins in the Regulation of Neurotransmitter Phenotype Specification. *J. Chem. Neuroanat.* **2016**, *73*, 3–8. [CrossRef] [PubMed]

74. Spitzer, N.C. Neurotransmitter Switching? No Surprise. *Neuron* **2015**, *86*, 1131–1144. [CrossRef] [PubMed]

75. Spitzer, N.C.; Debaca, R.C.; Allen, K.A.; Holliday, J. Calcium Dependence of Differentiation of GABA Immunoreactivity in Spinal Neurons. *J. Comp. Neurol.* **1993**, *337*, 168–175. [CrossRef] [PubMed]

76. Lewis, B.B.; Miller, L.E.; Herbst, W.A.; Saha, M.S. The Role of Voltage-Gated Calcium Channels in Neurotransmitter Phenotype Specification: Coexpression and Functional Analysis in Xenopus Laevis. *J. Comp. Neurol.* **2014**, *522*, 2518–2531. [CrossRef] [PubMed]

77. Borodinsky, L.N.; Root, C.M.; Cronin, J.A.; Sann, S.B.; Gu, X.; Spitzer, N.C. Activity-Dependent Homeostatic Specification of Transmitter Expression in Embryonic Neurons. *Nature* **2004**, *429*, 523–530. [CrossRef] [PubMed]

78. Marek, K.W.; Kurtz, L.M.; Spitzer, N.C. CJun Integrates Calcium Activity and Tlx3 Expression to Regulate Neurotransmitter Specification. *Nat. Neurosci.* **2010**, *13*, 944–950. [CrossRef] [PubMed]

79. Nelson, S.B.; Valakh, V. Excitatory/Inhibitory Balance and Circuit Homeostasis in Autism Spectrum Disorders. *Neuron* **2015**, *87*, 684–698. [CrossRef] [PubMed]

80. Olmos-Serrano, J.L.; Paluszkiewicz, S.M.; Martin, B.S.; Kaufmann, W.E.; Corbin, J.G.; Huntsman, M.M. Defective GABAergic Neurotransmission and Pharmacological Rescue of Neuronal Hyperexcitability in the Amygdala in a Mouse Model of Fragile X Syndrome. *J. Neurosci.* **2010**, *30*, 9929–9938. [CrossRef] [PubMed]

81. Cellot, G.; Cherubini, E. GABAergic Signaling as Therapeutic Target for Autism Spectrum Disorders. *Front. Pediatr.* **2014**, *2*. [CrossRef] [PubMed]

82. Nanou, E.; Catterall, W.A. Calcium Channels, Synaptic Plasticity, and Neuropsychiatric Disease. *Neuron* **2018**. [CrossRef] [PubMed]

83. Jungbluth, H.; Treves, S.; Zorzato, F.; Sarkozy, A.; Ochala, J.; Sewry, C.; Phadke, R.; Gautel, M.; Muntoni, F. Congenital Myopathies: Disorders of Excitation-Contraction Coupling and Muscle Contraction. *Nat. Rev. Neurol.* **2018**, *14*, 151–167. [CrossRef] [PubMed]

84. Agrawal, A.; Suryakumar, G.; Rathor, R. Role of Defective Ca^{2+} signaling in Skeletal Muscle Weakness: Pharmacological Implications. *J. Cell Commun. Signal.* **2018**, 1–15. [CrossRef] [PubMed]

85. Chernoff, E.A.G.; Hilfer, S.R. Calcium Dependence and Contraction in Somite Formation. *Tissue Cell* **1982**, *14*, 435–449. [CrossRef]

86. Ferrari, M.B.; Spitzer, N.C. Calcium Signaling in the Developing *Xenopus* Myotome. *Dev. Biol.* **1999**, *213*, 269–282. [CrossRef] [PubMed]

87. Cheung, C.Y.; Webb, S.E.; Love, D.R.; Miller, A.L. Visualization, Characterization and Modulation of Calcium Signaling during the Development of Slow Muscle Cells in Intact Zebrafish Embryos. *Int. J. Dev. Biol.* **2011**, *55*, 153–174. [CrossRef] [PubMed]

88. Kelu, J.J.; Webb, S.E.; Parrington, J.; Galione, A.; Miller, A.L. Ca^{2+} release via Two-Pore Channel Type 2 (TPC2) Is Required for Slow Muscle Cell Myofibrillogenesis and Myotomal Patterning in Intact Zebrafish Embryos. *Dev. Biol.* **2017**, *425*, 109–129. [CrossRef] [PubMed]

89. Kelu, J.J.; Chan, H.L.H.; Webb, S.E.; Cheng, A.H.H.; Ruas, M.; Parrington, J.; Galione, A.; Miller, A.L. Two-Pore Channel 2 Activity Is Required for Slow Muscle Cell-Generated Ca^{2+} Signaling during Myogenesis in Intact Zebrafish. *Int. J. Dev. Biol.* **2015**. [CrossRef] [PubMed]

90. Ferrari, M.B.; Rohrbough, J.; Spitzer, N.C. Spontaneous Calcium Transients Regulate Myofibrillogenesis in Embryonic Xenopus Myocytes. *Dev. Biol.* **1996**, *178*, 484–497. [CrossRef] [PubMed]

91. Campbell, N.R.; Podugu, S.P.; Ferrari, M.B. Spatiotemporal Characterization of Short versus Long Duration Calcium Transients in Embryonic Muscle and Their Role in Myofibrillogenesis. *Dev. Biol.* **2006**, *292*, 253–264. [CrossRef] [PubMed]

92. Zvaritch, E.; Depreux, F.; Kraeva, N.; Loy, R.E.; Goonasekera, S.A.; Boncompagni, S.; Kraev, A.; Gramolini, A.O.; Dirksen, R.T.; Franzini-Armstrong, C.; et al. An Ryr1I4895T Mutation Abolishes Ca^{2+} Release Channel Function and Delays Development in Homozygous Offspring of a Mutant Mouse Line. *Proc. Natl. Acad. Sci. USA* **2007**, *104*, 18537–18542. [CrossRef] [PubMed]

93. Farini, A.; Sitzia, C.; Cassinelli, L.; Colleoni, F.; Parolini, D.; Giovanella, U.; Maciotta, S.; Colombo, A.; Meregalli, M.; Torrente, Y. Inositol 1,4,5-Trisphosphate (IP3)-Dependent Ca^{2+} Signaling Mediates Delayed Myogenesis in Duchenne Muscular Dystrophy Fetal Muscle. *Development* **2016**, *143*, 658–669. [CrossRef] [PubMed]

94. Allard, B. From Excitation to Intracellular Ca^{2+} movements in Skeletal Muscle: Basic Aspects and Related Clinical Disorders. *Neuromuscul. Disord.* **2018**. [CrossRef] [PubMed]

95. Stiber, J.A.; Rosenberg, P.B. The Role of Store-Operated Calcium Influx in Skeletal Muscle Signaling. *Cell Calcium* **2011**, *49*, 341–349. [CrossRef] [PubMed]

96. McCarl, C.A.; Picard, C.; Khalil, S.; Kawasaki, T.; Röther, J.; Papolos, A.; Kutok, J.; Hivroz, C.; LeDeist, F.; Plogmann, K.; et al. ORAI1 Deficiency and Lack of Store-Operated Ca^{2+} entry Cause Immunodeficiency, Myopathy, and Ectodermal Dysplasia. *J. Allergy Clin. Immunol.* **2009**, *124*, 1311–1318. [CrossRef] [PubMed]

97. Picard, C.; McCarl, C.-A.; Papolos, A.; Khalil, S.; Lüthy, K.; Hivroz, C.; LeDeist, F.; Rieux-Laucat, F.; Rechavi, G.; Rao, A.; et al. STIM1 Mutation Associated with a Syndrome of Immunodeficiency and Autoimmunity. *N. Engl. J. Med.* **2009**, *360*, 1971–1980. [CrossRef] [PubMed]

98. Pisaniello, A. The Block of Ryanodine Receptors Selectively Inhibits Fetal Myoblast Differentiation. *J. Cell Sci.* **2003**. [CrossRef]

99. Webb, S.E.; Cheung, C.C.Y.; Chan, C.M.; Love, D.R.; Miller, A.L. Application of Complementary Luminescent and Fluorescent Imaging Techniques to Visualize Nuclear and Cytoplasmic Ca^{2+} Signalling during the In Vivo Differentiation of Slow Muscle Cells in Zebrafish Embryos under Normal and Dystrophic Conditions. *Clin. Exp. Pharmacol. Physiol.* **2012**, *39*, 78–86. [CrossRef] [PubMed]

100. Kloesel, B.; Dinardo, J.A.; Body, S.C. Cardiac Embryology and Molecular Mechanisms of Congenital Heart Disease: A Primer for Anesthesiologists. *Anesthesia Analgesia* **2016**. [CrossRef] [PubMed]

101. Schleich, J.M.; Abdulla, T.; Summers, R.; Houyel, L. An Overview of Cardiac Morphogenesis. *Arch. Cardiovasc. Dis.* **2013**. [CrossRef] [PubMed]

102. Moorman, A.; Webb, S.; Brown, N.; Lamers, W.; Anderson, R. Development of the Heart: Formation of the Cardiac Chambers and Arterial Trunks. *Heart* **2003**, *89*, 806–814. [CrossRef] [PubMed]

103. Créton, R.; Speksnijder, J.E.; Jaffe, L.F. Patterns of Free Calcium in Zebrafish Embryos. *J. Cell Sci.* **1998**, *111*, 1613–1622. [CrossRef] [PubMed]

104. Tallini, Y.N.; Ohkura, M.; Choi, B.-R.; Ji, G.; Imoto, K.; Doran, R.; Lee, J.; Plan, P.; Wilson, J.; Xin, H.-B.; et al. Imaging Cellular Signals in the Heart in Vivo: Cardiac Expression of the High-Signal Ca^{2+} Indicator GCaMP2. *Proc. Natl. Acad. Sci. USA* **2006**, *103*, 4753–4758. [CrossRef] [PubMed]

105. Porter, G.A.; Makuck, R.F.; Rivkees, S.A. Intracellular Calcium Plays an Essential Role in Cardiac Development. *Dev. Dyn.* **2003**, *227*, 280–290. [CrossRef] [PubMed]

106. Hotchkiss, A.; Feridooni, T.; Zhang, F.; Pasumarthi, K.B.S. The Effects of Calcium Channel Blockade on Proliferation and Differentiation of Cardiac Progenitor Cells. *Cell Calcium* **2014**, *55*, 238–251. [CrossRef] [PubMed]

107. Webb, S.E.; Miller, A.L. Calcium Signalling during Embryonic Development. *Nat. Rev. Mol. Cell Biol.* **2003**, *4*, 539–551. [CrossRef] [PubMed]

108. Harvey, R.P. Patterning the Vertebrate Heart. *Nat. Rev. Genet.* **2002**, *3*, 544–556. [CrossRef] [PubMed]

109. Carroll, T.J.; Vize, P.D. Synergism between Pax-8 and Lim-1 in Embryonic Kidney Development. *Dev. Biol.* **1999**, *214*, 46–59. [CrossRef] [PubMed]

110. Buisson, I.; Le Bouffant, R.; Futel, M.; Riou, J.F.; Umbhauer, M. Pax8 and Pax2 Are Specifically Required at Different Steps of Xenopus Pronephros Development. *Dev. Biol.* **2015**, *397*, 175–190. [CrossRef] [PubMed]

111. Tena, J.J.; Neto, A.; de la Calle-Mustienes, E.; Bras-Pereira, C.; Casares, F.; Gómez-Skarmeta, J.L. Odd-Skipped Genes Encode Repressors That Control Kidney Development. *Dev. Biol.* **2007**, *301*, 518–531. [CrossRef] [PubMed]

112. Leclerc, C.; Webb, S.E.; Miller, A.L.; Moreau, M. An Increase in Intracellular Ca^{2+} is Involved in Pronephric Tubule Differentiation in the Amphibian Xenopus Laevis. *Dev. Biol.* **2008**, *321*, 357–367. [CrossRef] [PubMed]

113. Gallagher, A.R.; Hidaka, S.; Gretz, N.; Witzgall, R. Molecular Basis of Autosomal-Dominant Polycystic Kidney Disease. *Cell. Mol. Life Sci.* **2002**, *59*, 682–693. [CrossRef] [PubMed]

114. Boucek, M.M.; Snyderman, R. Calcium Influx Requirement for Human Neutrophil Chemotaxis: Inhibition by Lanthanum Chloride. *Science* **1976**, *193*, 905–907. [CrossRef] [PubMed]

115. Oh-hora, M.; Rao, A. Calcium Signaling in Lymphocytes. *Curr. Opin. Immunol.* **2008**, *20*, 250–258. [CrossRef] [PubMed]

116. Oh-Hora, M. Calcium Signaling in the Development and Function of T-Lineage Cells. *Immunol. Rev.* **2009**. [CrossRef] [PubMed]

117. Beerman, R.W.W.; Matty, M.A.A.; Au, G.G.G.; Looger, L.L.L.; Choudhury, K.R.R.; Keller, P.J.J.; Tobin, D.M.M. Direct In Vivo Manipulation and Imaging of Calcium Transients in Neutrophils Identify a Critical Role for Leading-Edge Calcium Flux. *Cell Rep* **2015**, *13*, 2107–2117. [CrossRef] [PubMed]

118. Marks, P.W.; Maxfield, F.R. Transient Increases in Cytosolic Free Calcium Appear to Be Required for the Migration of Adherent Human Neutrophils. *J. Cell Biol.* **1990**, *110*, 43–52. [CrossRef] [PubMed]

119. Hogan, P.G.; Lewis, R.S.; Rao, A. Molecular Basis of Calcium Signaling in Lymphocytes: STIM and ORAI. *Annu. Rev. Immunol.* **2010**, *28*, 491–533. [CrossRef] [PubMed]

120. Baba, Y.; Nishida, K.; Fujii, Y.; Hirano, T.; Hikida, M.; Kurosaki, T. Essential Function for the Calcium Sensor STIM1 in Mast Cell Activation and Anaphylactic Responses. *Nat. Immunol.* **2008**, *9*, 81–88. [CrossRef] [PubMed]

121. Vig, M.; DeHaven, W.I.; Bird, G.S.; Billingsley, J.M.; Wang, H.; Rao, P.E.; Hutchings, A.B.; Jouvin, M.-H.; Putney, J.W.; Kinet, J.-P. Defective Mast Cell Effector Functions in Mice Lacking the CRACM1 Pore Subunit of Store-Operated Calcium Release–activated Calcium Channels. *Nat. Immunol.* **2008**, *9*, 89–96. [CrossRef] [PubMed]

122. Degen, K.E.; Gourdie, R.G. Embryonic Wound Healing: A Primer for Engineering Novel Therapies for Tissue Repair. *Birth Defects Res. Part C—Embryo Today Rev.* **2012**, *96*, 258–270. [CrossRef] [PubMed]

123. Ud-Din, S.; Volk, S.W.; Bayat, A. Regenerative Healing, Scar-Free Healing and Scar Formation across the Species: Current Concepts and Future Perspectives. *Exp. Dermatol.* **2014**, *23*, 615–619. [CrossRef] [PubMed]

124. Herrgen, L.; Voss, O.P.; Akerman, C.J. Calcium-Dependent Neuroepithelial Contractions Expel Damaged Cells from the Developing Brain. *Dev. Cell* **2014**, *31*, 599–613. [CrossRef] [PubMed]

125. Cordeiro, J.V.; Jacinto, A. The Role of Transcription-Independent Damage Signals in the Initiation of Epithelial Wound Healing. *Nat. Rev. Mol. Cell Biol.* **2013**, *14*, 249–262. [CrossRef] [PubMed]

126. Razzell, W.; Evans, I.R.; Martin, P.; Wood, W. Calcium Flashes Orchestrate the Wound Inflammatory Response through Duox Activation and Hydrogen Peroxide Release. *Curr. Biol.* **2013**, *23*, 424–429. [CrossRef] [PubMed]

127. Niethammer, P.; Grabher, C.; Look, A.T.; Mitchison, T.J. A Tissue-Scale Gradient of Hydrogen Peroxide Mediates Rapid Wound Detection in Zebrafish. *Nature* **2009**, *459*, 996–999. [CrossRef] [PubMed]

128. Tu, M.K.; Borodinsky, L.N. Spontaneous Calcium Transients Manifest in the Regenerating Muscle and Are Necessary for Skeletal Muscle Replenishment. *Cell Calcium* **2014**, *56*, 34–41. [CrossRef] [PubMed]

129. Etulain, J. Platelets in Wound Healing and Regenerative Medicine. *Platelets* **2018**, *15*, 1–3. [CrossRef] [PubMed]

130. Wolf, K.; Braun, A.; Haining, E.J.; Tseng, Y.-L.; Kraft, P.; Schuhmann, M.K.; Gotru, S.K.; Chen, W.; Hermanns, H.M.; Stoll, G.; et al. Partially Defective Store Operated Calcium Entry and Hem(ITAM) Signaling in Platelets of Serotonin Transporter Deficient Mice. *PLoS ONE* **2016**, *11*, e0147664. [CrossRef] [PubMed]

131. Chen, W.; Thielmann, I.; Gupta, S.; Subramanian, H.; Stegner, D.; van Kruchten, R.; Dietrich, A.; Gambaryan, S.; Heemskerk, J.W.M.; Hermanns, H.M.; et al. Orai1-Induced Store-Operated Ca $^{2+}$ Entry Enhances Phospholipase Activity and Modulates Canonical Transient Receptor Potential Channel 6 Function in Murine Platelets. *J. Thromb. Haemost.* **2014**, *12*, 528–539. [CrossRef] [PubMed]

132. Gotru, S.K.; Chen, W.; Kraft, P.; Becker, I.C.; Wolf, K.; Stritt, S.; Zierler, S.; Hermanns, H.M.; Rao, D.; Perraud, A.L.; et al. TRPM7 Kinase Controls Calcium Responses in Arterial Thrombosis and Stroke in Mice. *Arterioscler. Thromb. Vasc. Biol.* **2018**, *38*, 344–352. [CrossRef] [PubMed]

133. Lysaght, M.J.; Jaklenec, A.; Deweerd, E. Great Expectations: Private Sector Activity in Tissue Engineering, Regenerative Medicine, and Stem Cell Therapeutics. *Tissue Eng. Part A* **2008**, *14*, 305–315. [CrossRef] [PubMed]

134. Patterson, R.A.; Juarez, M.T.; Hermann, A.; Sasik, R.; Hardiman, G.; McGinnis, W. Serine Proteolytic Pathway Activation Reveals an Expanded Ensemble of Wound Response Genes in Drosophila. *PLoS ONE* **2013**, *8*. [CrossRef] [PubMed]

135. Tsujioka, H.; Kunieda, T.; Katou, Y.; Shirahige, K.; Kubo, T. Unique Gene Expression Profile of the Proliferating Xenopus Tadpole Tail Blastema Cells Deciphered by RNA-Sequencing Analysis. *PLoS ONE* **2015**. [CrossRef] [PubMed]

136. Mouse Gene Expression Data Search. Available online: http://www.informatics.jax.org/gxd (accessed on 25 September 2018).

137. ZFIN Expression Search. Available online: https://zfin.org/action/expression/search (accessed on 25 September 2018).

138. Gene Expression Search. Available online: http://www.xenbase.org/geneExpression/geneExpressionSearch.do?method=display (accessed on 25 September 2018).

139. Kolios, G.; Moodley, Y. Introduction to Stem Cells and Regenerative Medicine. *Respiration* **2013**, *85*, 3–10. [CrossRef] [PubMed]

140. Ermakov, A.; Daks, A.; Fedorova, O.; Shuvalov, O.; Barlev, N.A. Ca^{2+}-Depended Signaling Pathways Regulate Self-Renewal and Pluripotency of Stem Cells. *Cell Biol. Int.* **2018**, *42*, 1086–1096. [CrossRef] [PubMed]

141. Rahaman, M.N.; Mao, J.J. Stem Cell-Based Composite Tissue Constructs for Regenerative Medicine. *Biotechnol. Bioeng.* **2005**, *91*, 261–284. [CrossRef] [PubMed]

142. Hasan, A. *Tissue Engineering for Artificial Organs: Regenerative Medicine, Smart Diagnostics and Personalized Medicine*; John Wiley & Sons: Hoboken, NJ, USA, 2016; Volume 1–2.

143. Jergova, S.; Gajavelli, S.; Varghese, M.S.; Shekane, P.; Sagen, J. Analgesic Effect of Recombinant GABAergic Cells in a Model of Peripheral Neuropathic Pain. *Cell Transplant.* **2016**, *25*, 629–643. [CrossRef] [PubMed]

144. Méry, A.; Aimond, F.; Ménard, C.; Mikoshiba, K.; Michalak, M.; Pucéat, M. Initiation of Embryonic Cardiac Pacemaker Activity by Inositol 1,4,5-Trisphosphate-Dependent Calcium Signaling. *Mol. Biol. Cell* **2005**, *16*, 2414–2423. [CrossRef] [PubMed]

145. Limpitikul, W.B.; Dick, I.E.; Tester, D.J.; Boczek, N.J.; Limphong, P.; Yang, W.; Choi, M.H.; Babich, J.; Disilvestre, D.; Kanter, R.J.; et al. A Precision Medicine Approach to the Rescue of Function on Malignant Calmodulinopathic Long-QT Syndrome. *Circ. Res.* **2017**, *120*, 39–48. [CrossRef] [PubMed]

International Journal of
Molecular Sciences

MDPI

Review

T Cell Calcium Signaling Regulation by the Co-Receptor CD5

Claudia M. Tellez Freitas, Deborah K. Johnson and K. Scott Weber *

Department of Microbiology and Molecular Biology, Brigham Young University, Provo, UT 84604, USA;
claudiamsmicrobiology@gmail.com (C.M.T.F.); deborahkj@gmail.com (D.K.J.)
* Correspondence: scott_weber@byu.edu; Tel.: +1-801-422-6259

Received: 27 February 2018; Accepted: 24 April 2018; Published: 26 April 2018

Abstract: Calcium influx is critical for T cell effector function and fate. T cells are activated when T cell receptors (TCRs) engage peptides presented by antigen-presenting cells (APC), causing an increase of intracellular calcium (Ca^{2+}) concentration. Co-receptors stabilize interactions between the TCR and its ligand, the peptide-major histocompatibility complex (pMHC), and enhance Ca^{2+} signaling and T cell activation. Conversely, some co-receptors can dampen Ca^{2+} signaling and inhibit T cell activation. Immune checkpoint therapies block inhibitory co-receptors, such as cytotoxic T-lymphocyte associated antigen 4 (CTLA-4) and programmed death 1 (PD-1), to increase T cell Ca^{2+} signaling and promote T cell survival. Similar to CTLA-4 and PD-1, the co-receptor CD5 has been known to act as a negative regulator of T cell activation and to alter Ca^{2+} signaling and T cell function. Though much is known about the role of CD5 in B cells, recent research has expanded our understanding of CD5 function in T cells. Here we review these recent findings and discuss how our improved understanding of CD5 Ca^{2+} signaling regulation could be useful for basic and clinical research.

Keywords: calcium signaling; T cell receptor (TCR); co-receptors; CD-5; PD-1; CTL-4

1. Introduction

T cells are a critical component of the adaptive immune system. T cell responses are influenced by signals that modulate the effects of the T cell receptor (TCR) and peptide-major histocompatibility complex (pMHC) interaction and initiate the transcription of genes involved in cytokine production, proliferation, and differentiation [1–3]. T cell activation requires multiple signals. First, the TCR engages the pMHC leading to tyrosine phosphorylation of CD3 and initiation of the Ca^{2+}/Calcineurin/Nuclear factor of activated T cells (NFAT) or Protein kinase C-theta (PKCθ)/Nuclear factor-κ-light chain enhancer of activated B cells (NF-κB) or Mitogen-activated protein kinase (MAP kinase)/AP-1 pathways [4–6]. Second, cell surface costimulatory molecules, such as co-receptor CD28, amplify TCR-pMHC complex signals and promote stronger intracellular interactions to prevent T cell anergy [7,8]. Finally, cytokines such as interleukin-12 (IL-12), interferon α (INFα), and interleukin-1 (IL-1) promote T cell proliferation, differentiation, and effector functions [6].

Co-receptors such as CD4 and CD8 interact with MHC molecules and additional co-receptors interact with surface ligands present on antigen-presenting cells (APCs) to regulate T cell homeostasis, survival, and effector functions with stimulatory or inhibitory signals [9]. Altering co-receptor levels, balance, or function dramatically affects immune responses and their dysfunction is implicated in autoimmune diseases [10]. Stimulatory co-receptors such as CD28, inducible T cell co-stimulator (ICOS), Tumor necrosis factor receptor superfamily member 9 (TNFRSF9 or 4-1BB), member of the TNR-superfamily receptor (CD134 or OX40), glucocorticoid-induced tumor necrosis factor (TNF) receptor (GITR), CD137, and CD77 promote T cell activation and protective responses [11]. Co-receptor signaling is initiated by the phosphorylation of tyrosine residues located in immunoreceptor tyrosine-based activation motifs (ITAMs) or immunoreceptor tyrosine-based inhibitory motifs (ITIMs) [7,12]. The phosphorylated tyrosines serve as docking sites for spleen tyrosine kinase (Syk) family members such as zeta-chain-associated protein kinase 10 (ZAP-70) and Syk which activate the phospholipase C γ (PLCγ), RAS, and extracellular signal-regulated kinase (ERK) pathways in addition to mobilizing intracellular Ca^{2+} stores [13].

One of the best described T cell co-receptors, CD28, is a stimulatory T cell surface receptor from the Ig superfamily with a single Ig variable-like domain which binds to B7-1 (CD80) and B7-2 (CD86) [2]. Ligand binding phosphorylates CD28 cytoplasmic domain tyrosine motifs such as YMNM and PYAP and initiates binding and activation of phosphatidylinositide 3 kinase (PI3K) which interacts with protein kinase B (Akt) and promotes T cell proliferation and survival [1]. CD28 also activates the NFAT pathway and mobilizes intracellular Ca^{2+} stores through association with growth factor receptor-bound protein 2 (GRB2) and the production of phosphatidylinositol 4,5-bisphosphate (PIP2), the substrate of PLCγ1, respectively [2,14]. Blocking stimulatory co-receptors suppresses T cell effector function. For example, blocking stimulatory CD28 with anti-CD28 antibodies promotes regulatory T cell function and represses activation of auto- and allo-reactive T effector cells after organ transplantation [8,15].

T cells also have inhibitory co-receptors which regulate T cell responses [8]. The best characterized are immunoglobulin (Ig) superfamily members cytotoxic T-lymphocyte-associated protein 4 (CTLA-4) and programmed cell death protein 1 (PD-1) [8,16]. CTLA-4 binds CD80 and CD86 with greater avidity than CD28, and its inhibitory role refines early phase activation signals for proliferation and cytokine production [16–19]. PD-1, another CD28/B7 family member, regulates late phase effector and memory response [20]. Inhibitory co-receptors such as CTLA-4 and PD-1, known as "immune checkpoints", block the interaction between CD28 and its ligands altering downstream secondary T cell activation signals [19]. Therefore, blocking CTLA-4 or PD-1 promotes effector T cell function in immunosuppressive environments [19,21].

There are also a number of co-receptors that have differential modulatory properties. For example, CD5, a lymphocyte glycoprotein expressed on thymocytes and all mature T cells, has contradictory roles at different time points. CD5 expression is set during thymocyte development and decreases the perceived strength of TCR-pMHC signaling in naïve T cells by clustering at the TCR-pMHC complex and reducing TCR downstream signals such as the Ca^{2+} response when its cytoplasmic pseudo-ITAM domain is phosphorylated [22–25]. The CD5 cytoplasmic domain has four tyrosine residues (Y378, Y429, Y411, and Y463), and residues Y429 and Y441 are found in a YSQP-(x8)-YPAL pseudo ITAM motif while other tyrosine residues make up a pseudo-ITIM domain [23]. Phosphorylated tyrosines recruit several effector molecules and may sequester activation kinases away from the TCR complex, effectively reducing activation signaling strength [23]. Recruited proteins include Src homology-2 protein phosphatase-1 (SHP-1), Ras GTPase protein (rasGAP), CBL, casein kinase II (CK2), zeta-chain-associated protein kinase 70 (ZAP70), and PI3K which are involved in regulating both positive and negative TCR-induced responses [26–28]. For example, ZAP-70 phosphorylates other substrates and eventually recruits effector molecules such as PLC gamma and promotes Ca^{2+} signaling and Ras activation which stimulates the ERK pathway and leads to cellular activation [13,29]. Conversely, SHP1 inhibits Ca^{2+} signaling and PKC activation via decreased tyrosine phosphorylation of PLCγ [13,26,30,31]. Further, Y463 serves as a docking site for c-Cb1, a ubiquitin ligase, which is phosphorylated upon CD3–CD5 ligation and leads to increased ubiquitylation and lysosomal/proteasomal degradation of TCR downstream signaling effectors and CD5 itself [32]. Thus, CD5 has a mix of downstream effects that both promote and inhibit T cell activation. Curiously, recent work suggests that in contrast to its initial inhibitory nature, CD5 also co-stimulates resting and mature T cells by augmenting CD3-mediated signaling [25,33–35].

Ca^{2+} is an important second messenger in many cells types, including lymphocytes, and plays a key role in shaping immune responses. In naïve T cells, intracellular Ca^{2+} is maintained at low levels, but when TCR-pMHC complexes are formed, inositol triphosphate (IP3) initiates Ca^{2+} release from intracellular stores of the endoplasmic reticulum (ER) which opens the Ca^{2+} release-activated Ca^{2+} channels (CRAC) and initiates influx of extracellular Ca^{2+} through store-operated Ca^{2+} entry (SOCE) [36–41]. The resulting elevation of intracellular Ca^{2+} levels activates transcription factors involved in T cell proliferation, differentiation, and cytokine production (e.g., nuclear factor of activated cells (NFAT)) [36,37]. Thus, impaired Ca^{2+} mobilization affects T cell development, activation, differentiation, and function [42,43]. Examples of diseases with impaired Ca^{2+} signaling in T cells include systemic lupus erythematosus, type 1 diabetes mellitus, and others [44,45].

In this review, we will focus on CD5 co-receptor signaling and its functional effects on T cell activation. First, we will discuss how the inhibitory co-receptors CTLA-4 and PD-1 modulate T cell function. Then we will compare CTLA-4 and PD-1 function to CD5 function, examine recent findings that expand our understanding of the role of CD5, and assess how these findings apply to T cell Ca^{2+} signaling. Finally, we will consider CD5 Ca^{2+} signaling regulation in T cells and its potential physiological impact on immunometabolism, cell differentiation, homeostasis, and behavior.

2. Roles of Negative Regulatory T Cell Co-Receptors

2.1. Cytotoxic T-Lymphocyte Antigen-4 (CTLA-4)

Cytotoxic T-lymphocyte antigen-4 (CTLA-4, CD152) inhibits early stages of T cell activation by recruiting inhibitory proteins such as SHP-2 and type II serine/threonine phosphatase PP2A that interfere with T cell synapse signaling [21,46–48]. CTLA-4 binds B7, a protein on activated APCs, with higher affinity than the stimulatory co-receptor CD28; the resulting balance between inhibitory and stimulatory signals controls T cell activation or anergy [19,49]. In naïve T cells, CTLA-4 is located in intracellular vesicles which localize at TCR binding sites following antigen recognition and intracellular Ca^{2+} mobilization [19,50]. Like CD28, CTLA-4 aggregates to the central supramolecular activation complex (cSMAC) where it then extrinsically controls activation by decreasing immunological synapse contact time [51–53]. This suppresses proactivation signals by activating ligands (B7-1 and B7-2) and induces the enzyme Inoleamine 2,3-dioxygenase (IDO) which impairs Ca^{2+} mobilization and suppresses T cell activation, ultimately altering IL-2 production and other effector functions in T cells [51,54,55]. CTLA-4 also stimulates production of regulatory cytokines, such as transforming growth factor beta (TGF-β), which inhibit APC presentation and T cell effector function [47,52,53]. Compared to effector T cells (T_{eff}), CTLA-4 is highly expressed in regulatory T cells (T_{reg}) and plays a role in maintaining T_{reg} homeostasis, proliferation, and immune responses [16,56,57]. Total or partial CTLA-4 deficiency inhibits T_{reg}'s ability to control cytokine production and can cause immune dysregulation [58–61]. Thus, CTLA-4 has an important role in the T_{reg} suppressive response [60]. Additionally, CTLA-4 mutations are associated with autoimmune diseases as thoroughly reviewed by Kristiansen et al. [62].

The loss of CTLA-4 results in removal of CTLA-4 competition with CD28 for B7-1 and B7-2 and is implicated in autoimmunity and cancer [15,63]. Because CTLA-4 inhibits TCR signaling, CTLA-4 deficiency leads to T cell overactivation as measured by increased CD3ζ phosphorylation and Ca^{2+} mobilization [64]. Thus, modulating CTLA-4 signaling is an attractive target for immunotherapies that seek to boost or impair early TCR signaling for cancer and autoinflammatory diseases [65,66]. For example, Ipilimunab, an IgG1 antibody-based melanoma treatment, is a T cell potentiator that blocks CTLA-4 to stimulate T cell proliferation and stem malignant disease progression by delaying tumor progression and has been shown to significantly increase life expectancy [19,67,68]. Additionally, Tremelimumab, a noncomplement fixing IgG2 antibody, has been tested alone or in combination with other antibodies such as Durvalumab (a PD-1 inhibitor) and improves antitumor activity in patients with non-small cell lung cancer (NSCLC), melanoma, colon cancer, gastric cancer, and mesothelioma treatment [69–74].

2.2. Programmed Death 1 (PD-1)

Programmed cell death protein-1 (PD-1, CD279) is a 288-amino acid (50–55 KDa) type I transmembrane protein and a member of the B7/CD28 immunoglobulin superfamily expressed on activated T cells, B cells, and myeloid cells [19,75,76]. PD-1 has two known ligands, PD-L1 and PD-L2, which inhibit T cell activation signals [77]. Like CTLA-4, PD-1 also inhibits T cell proliferation and cytokine production (INF-γ, TNF and IL-2) but is expressed at a later phase of T cell activation [19]. PD-1 has an extracellular single immunoglobulin (Ig) superfamily domain and a cytoplasmic domain containing an ITIM and an immunoreceptor tyrosine-based switch motif (ITSM) subunit critical for PD-1 inhibitory function [78]. Upon T cell activation, PD-1 is upregulated and initiates ITIM and ITSM tyrosine interaction with SHP-2 which mediates TCR signaling inhibition by decreasing ERK phosphorylation and intracellular Ca^{2+} mobilization [79,80]. PD-1 can block the activation signaling pathways PI3K-Akt and Ras-Mek-ERK, which inhibit or regulate T cell activation [79,81]. Thus, engagement of PD-1 by its ligand affects intracellular Ca^{2+} mobilization, IL-2 and TNF-α production, supporting PD-1's inhibitory role in TCR strength-mediated signals [82].

PD-1 signaling also affects regulatory T cell (T_{reg}) homeostasis, expansion, and function [83]. T_{reg} activation and proliferation are impacted by PD-1 expression which enhances their development and function while inhibiting T effector cells [75,84]. PD-1, PD-L, and T_{regs} help terminate immune responses [85]. Thus, PD-1 deficiency results not only in increased T cell activation, but in the breakdown of tolerance and the development of autoimmunity in diseases such as multiple sclerosis and systemic lupus erythematosus [85–89]. PD-1 and its ligands protect tissues from autoimmune attacks by regulating T cell activation and inducing and maintaining peripheral tolerance [90,91]. Studies done in PD-1-deficient mice observed the development of lupus-like glomerulonephritis and arthritis, cardiomyopathy, autoimmune hydronephrosis, and Type I diabetes, among other ailments [92–94]. PD-1 protects against autoimmunity and promotes T_{reg} function. [85]. Enhancing T_{reg} response with a PD-L1 agonist shows therapeutic potential for asthma and other autoimmune disorders [85,95]. Because PD-1 specifically modulates lymphocyte function, effective FDA-approved monoclonal antibodies targeting PD-1 are clinically available (i.e., Pembrolizumab and Nivolumab) to treat advanced malignancies [20]. Not only does blocking PD-1 decrease immunotolerance of tumor cells, it also increases cytotoxic T lymphocyte antitumor activity [20].

3. CD5: A Contradictory Co-Receptor

3.1. Overview of CD5 Signaling and Ca^{2+} Mobilization in T Cells

CD5, known as Ly-1 antigen in mice or as Leu-1 in humans, is a type I transmembrane glycoprotein (67 kDa) expressed on the surface of thymocytes, mature T cells, and a subset of B cells (B-1a) [96,97]. Although CD5 was discovered over 30 years ago, it was only in the last decade that CD5 gained attention as a key T cell activation regulator [98,99]. CD5 expression is set in the thymus during positive selection and correlates with how tightly the thymocyte TCR binds to self-peptide-MHC (self-pMHC); greater TCR affinity for self-peptide leads to increased CD5 expression in double positive (DP) thymocytes [100]. In other words, DP thymocytes that receive strong activation signals through their TCR express more CD5 than those DP thymocytes that receive weak TCR signals [100]. CD5 knockout mice ($CD5^{-/-}$) have a defective negative and positive selection process, and therefore their thymocytes are hyper-responsive to TCR stimulation with increased Ca^{2+} mobilization, proliferation, and cytokine production [23,98]. On the other hand, because of the increased TCR avidity for self-pMHC, mature T cells with high CD5 expression ($CD5^{hi}$) (peripheral or postpositive selection T cells) respond to foreign peptide with increased survival and activation compared to mature T cells with low CD5 expression ($CD5^{lo}$) [34,101]. Therefore, CD5 is a negative regulator of TCR signaling in the thymus and modulates mature T cell response in the periphery [23,34,100,102].

While CTLA-4 and PD-1 belong to the immunoglobulin (Ig) family, CD5 belongs to group B of the scavenger receptor cysteine-rich (SRCR) superfamily and contains three extracellular SRCR domains [30,96,103]. The cytoplasmic tail of CD5 contains several tyrosine residues which mediate the negative regulatory activity independent of extracellular engagement [100,104,105]. As CD5 physically associates with TCRζ/CD3 complex upon TCR and pMHC interaction, the tyrosine residues in both

TCRζ and CD5 are phosphorylated by tyrosine kinases associated with the complex [30,106–110]. This interaction is so intrinsic to T cell signaling that CD5 expression levels are proportional to the degree of TCRζ phosphorylation, IL-2 production capacity, and ERK phosphorylation which are critical for CD3-mediated signaling [33,111]. It is unknown whether posttranslational modifications, such as conserved domain 1 and domain 2 glycosylations, impact CD5 signaling [112,113]. CD5 is present in membrane lipids rafts of mature T cells where, upon activation, it helps augment TCR signaling, increases Ca^{2+} mobilization, and upregulates ZAP-70/LAT (linker for activation of T cells) activation [114–116]. This suggests that CD5 is not only a negative regulator in thymocytes, but also appears to positively influence T cell immune response to foreign antigens [117,118]. See Figure 1.

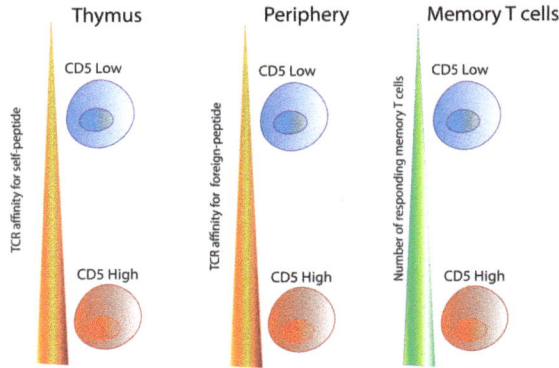

Figure 1. Effects of CD5 on different stages of T cell development. CD5 expression on thymocytes is directly proportional to the signaling intensity of the TCR:self-pMHC interaction. In the periphery, T cells with higher CD5 levels (CD5[hi]) are better responders to foreign-peptide. Long-lived memory cells populations are enriched for CD5[hi] T cells [34,102,119].

CD5 has three known ligands: CD72, a glycoprotein expressed by B cells, CD5 ligand or CD5L, an activation antigen expressed on splenocytes, and CD5 itself [120–122]. Crosslinking CD5L to CD5 increases intracellular Ca^{2+} concentrations [30,120,121,123,124]. Early studies with anti-CD5 monoclonal antibodies also demonstrated enhanced Ca^{2+} mobilization and proliferation, suggesting that CD5 co-stimulates and increases the T cell activation signal [125,126]. Following TCR:pMHC interaction, CD5 cytoplasmic ITAM and ITIM like-domains are phosphorylated by p56lck and bound by Src homology 2 (SH2) domain-containing protein tyrosine phosphatase (SHP-1) [108,127,128]. However, while SHP-1 affects Ca^{2+} mobilization and is a purported down-regulator of thymocyte activation, recent findings suggest that SHP-1 is not necessary for CD5 signaling as T cells deficient in SHP-1 have normal CD5 expression and continue to signal normally [26,129]. Thus, while CD5 is not a SHP-1 substrate and SHP-1 is likely unnecessary for CD5 signaling, CD5 signaling results in increased Ca^{2+} mobilization. It has yet to be resolved how CD5 can act as an inhibiting co-receptor in the thymus and as an activating co-receptor in the periphery.

3.2. CD5 as a Ca^{2+} Signaling Modulator

As previously mentioned, CD5 expression levels are set in the thymus during T cell development and are maintained on peripheral lymphocytes [117]. CD5 expression in T cells plays an important role during development and primes naïve T cells for responsiveness in the periphery [35,111,130]. CD5[hi] T cells have the highest affinity for self-peptides and respond with increased cytokine production and proliferation to infection [101,131,132].

Our laboratory works with two TCR transgenic mouse lines with different levels of CD5 expression: LLO56 (CD5[hi]) and LLO118 (CD5[lo]) [111,117,130]. While LLO56 (CD5[hi]) and LLO118

(CD5^{lo}) have similar affinity for the same immunodominant epitope (listeriolysin O amino acids 190–205 or LLO$_{190-205}$) from *Listeria monocytogenes*, on day 7 of primary response, LLO118 (CD5^{lo}) has approximately three times the number of responding cells compared to LLO56 (CD5^{hi}), and conversely, on day 4 during secondary infection, LLO56 (CD5^{hi}) has approximately fifteen times more cells than LLO118 (CD5^{lo}) [130]. This difference is not due to differential proliferative capacity, rather LLO56 (CD5^{hi}) has higher levels of apoptosis during the primary response [130]. Thus, LLO56 CD5^{hi} and LLO118 CD5^{lo}'s capacity to respond to infection appears to be regulated by their CD5 expression levels [117]. LLO56 (CD5^{hi}) thymocytes have greater affinity for self-peptide, which primes them to be highly apoptotic [130].

Recently we reported that in response to foreign peptide, LLO56 (CD5^{hi}) naïve T cells have higher intracellular Ca^{2+} mobilization than LLO118 (CD5^{lo}), which correlates with increased rate of apoptosis of LLO56 (CD5^{hi}), as Ca^{2+} overloaded mitochondria release cytochrome c which activates caspase and nuclease enzymes, thus initiating the apoptotic pathways [35,133,134]. LLO56 (CD5^{hi}) naïve T cell increased Ca^{2+} mobilization also provides additional support to the idea that CD5^{hi} T cells have an enhanced response to foreign peptide [35,134]. This supports previous research that found that upon T cell activation, increased CD5 expression is correlated with greater basal TCRζ phosphorylation, increased ERK phosphorylation, and more IL-2 production [101,111].

Thus, unlike CTLA-4 and PD-1 which are expressed only on activated T cells in the periphery during early and late phases of immune response, respectively, CD5 is set during T cell development, and influences T cells both during thymic development and during postthymic immune responses [19, 101,111] (see Figure 2). CD5 not only has an important inhibitory role in the thymus, but also appears to positively influence the T cell population response; for example, more CD5^{hi} T cells populate the memory T cell repertoire because CD5^{hi} naïve T cells have a stronger primary response [34,135]. CD5 finetunes the sensitivity of TCR signaling to pMHC, altering intracellular Ca^{2+} mobilization and NFAT transcription, key players in T cell effector function [19,64,126]. As Ca^{2+} signaling plays a key role in T cell activation and function, controlling Ca^{2+} mobilization in T cells through CD5 expression could influence diverse areas of clinical research including metabolism, cancer treatments, and even cognitive behavior.

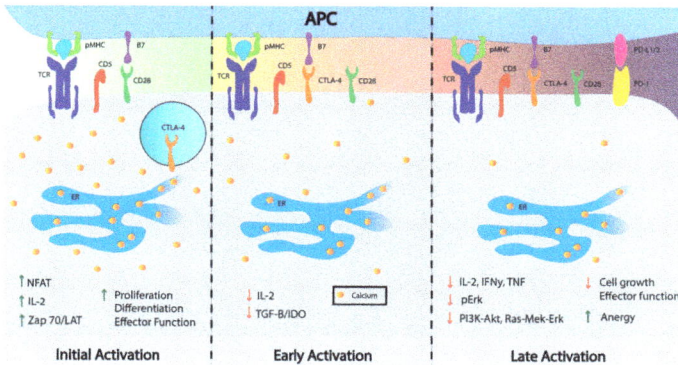

Figure 2. Inhibiting co-receptors modulate T cell activation by increasing (green arrows) or decreasing activity (red arrows). CD5 is present in naïve T cells and localizes to the TCR:pMHC complex during activation. Initial activation cascades signal for the release of CTLA-4 from vesicles to the cell surface while the transcription factor NFAT transcribes PD-1. CTLA-4 provides inhibitory signals during early activation while PD-1 is expressed later and inhibits later stages of T cell activation. The initial Ca^{2+} mobilization is decreased by CTLA-4 and PD-1 downstream signals. A more detailed illustration of the calcium signaling pathway (i.e., IP3, STIM 1/2, CRAC channel, calmodulin, etc.) is outlined in Figure 3.

4. Physiological Impact of CD5 Expression in T Cells

4.1. Metabolism

Naive T cells are in a quiescent state and rely on oxidative phosphorylation (OXPHOS) to generate ATP for survival [136,137]. Upon TCR-pMHC interaction, T cells undergo metabolic reprograming to meet energetic demands by switching from OXPHOS to glycolysis [138]. Glycolysis is a rapid source of ATP and regulates posttranscriptional production of INF-γ, a critical effector cytokine [139]. Following the immune response, most effector T cells undergo apoptosis while a subset become quiescent memory T cells. Memory T cells have lower energetic requirements and rely on OXPHOS and Fatty Acid Oxidation (FAO) to enhance mitochondrial capacity for maintenance and survival [140].

Ca^{2+} signaling is a key second messenger in T cell activation and Ca^{2+} ions also modulate T cell metabolism through CRAC channel activity and NFAT activation [3,141]. During TCR-pMHC binding Ca^{2+} is released from the endoplasmic reticulum (ER) where it is absorbed by the mitochondria and initiates an influx of extracellular Ca^{2+} [3]. First, the rise of cytoplasmic Ca^{2+} activates stromal interaction molecule 1 (STIM1) located on the ER membrane to interact with the CRAC channel located on the cell membrane [142]. The release of the ER store and resulting extracellular Ca^{2+} influx increases the intracellular Ca^{2+} concentration and promotes AMPK (adenosine monophosphates activated protein kinase) expression and CaMKK (calmodulin-dependent protein kinase kinase) activity [3,142,143]. AMPK senses cellular energy levels through the ratio of AMP to ATP and generates ATP by inhibiting ATP-dependent pathways and stimulating catabolic pathways [144]. This indirectly controls T cell fate as AMPK indirectly inhibits mTOR (mammalian target of rapamycin complex) [145]. Because mTOR coordinates the metabolic cues that control T cell homeostasis, it plays a critical role in T cell fate [146]. T cells that are TSC1 (Tuberous sclerosis complex 1)-deficient show metabolic alterations through increased glucose uptake and glycolytic flux [147].

The rise of cytoplasmic Ca^{2+} also encourages mitochondria to uptake cytoplasmic Ca^{2+} through the mitochondrial Ca^{2+} uniporter (MCU) [148]. This MCU uptake increases Ca^{2+} influx by depleting Ca^{2+} near the ER which further activates the CRAC channels and promotes STIM1 oligomerization [3,149–151]. Ca^{2+} uptake in the mitochondria also enhances the function of the tricarboxylic acid cycle (TAC), which generates more ATP through OXFHOS [152,153]. OXPHOS is maintained by a glycolysis product, phosphoenolpyruvate (PEP), which sustains TCR-mediated Ca^{2+}-NFAT signaling by inhibiting the sarcoendoplasmic reticulum (SR) calcium transport ATPase (SERCA) pump, thus promoting T cell effector function [154,155]. Downregulation of calmodulin kinase, CaMKK2, which controls NFAT signaling, decreases glycolytic flux, glucose uptake, and lactate and citrate metabolic processes [156]. Ca^{2+} may also orchestrate the metabolic reprogramming of naïve T cells by promoting glycolysis and OXPHOS through the SOCE/calcineurin pathway which controls the expression of glucose transporters GLUT1/GLUT3 and transcriptional cc-regulator proteins important for the expression of electron transport chain complexes required for mitochondria respiration [141].

Co-receptor stimulation plays a pivotal role in T cell metabolism and function. A decrease in T cell Ca^{2+} signaling represses glycolysis and affects T cell effector function [152]. PD-1 and CTLA-4 depress Ca^{2+} signaling and glycolysis while promoting FAO and antibodies against CTLA-4 and PD-1 increase Ca^{2+} mobilization and glycolysis during T cell activation [157,158]. Like CTLA-4 and PD-1, CD5 modulatory function has the potential to influence T cell metabolism. Analysis of gene families modulated by CD5 in B cells found that CD5 upregulates metabolic-related genes including VEFG, Wnt signaling pathways genes, MAPK cascade genes, I-kB/NF-kB cascade genes, TGF β signaling genes, and adipogenesis process genes [159]. Therefore, proliferation differences correlated with CD5 expression in T cells may be caused by improved metabolic function as CD5^{lo} T cells seem to be more quiescent than CD5^{hi} T cells [160]. Although not much is known about how CD5 alters metabolic function in T cells, signaling strength differences of CD5^{hi} and CD5^{lo} T cell populations correlate with intracellular Ca^{2+} mobilization during activation and influence their immune response [35,111,130]. This implies that different metabolic processes may be initiated which would influence proliferation, memory cell generation, and cytokine production. Figure 3 summarizes how Ca^{2+} may be mobilized in $CD5^{hi}$ and $CD5^{lo}$ naïve T cells and the role Ca^{2+} may play on metabolism.

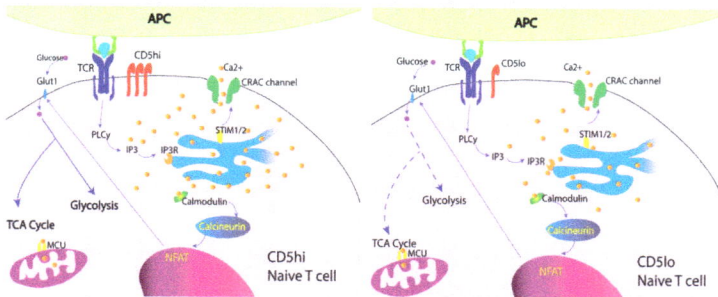

Figure 3. CD5 expression levels in naïve T cells may influence T cell metabolism and function. Differential levels of CD5 result in differences in Ca^{2+} mobilization in naïve T cells. $CD5^{hi}$ naïve T cells have higher Ca^{2+} influx than $CD5^{lo}$ naïve T cells upon TCR:pMHC interaction [35]. Ca^{2+} signaling plays a significant role in T cell activation and influences metabolism and T cell function. Differential Ca^{2+} mobilization and expression of calcineurin and NFAT affect glycolysis and mitochondrial respiration (hypothetical levels of metabolic activation are shown with dashed (low) or solid (high) arrows), suggesting CD5 expression may affect metabolic reprograming during T cell activation [141].

4.2. Neuroimmunology

The field of neuroimmunology examines the interplay between the immune system and the central nervous system (CNS) [161]. The adaptive immune system does influence the CNS as cognition is impaired by the absence of mature T cells [162]. In wild type mice, there is an increase in the number of T cells present in the meninges during the learning process, in stark contrast to mice with T helper 2 cytokine deficiencies (such as IL-4 and IL-13) who have decreased T cell recruitment and impaired learning [163]. Furthermore, regulation of T cell activation and cytokine production critically assists neuronal function and behavior, suggesting that manipulation of T cells could be a potential therapeutic target in treating neuroimmunological diseases [164,165].

T cells go through several microenvironments before reaching the CNS [166]. Many of the signal interactions present in these microenvironments affect T cell function and involve changes in intracellular Ca^{2+} levels [166,167]. In experimental autoimmune encephalitis (EAE), a model for human multiple sclerosis, autoreactive T cells have Ca^{2+} fluctuations throughout their journey to the CNS [166]. Prior to reaching the CNS, T cells interact with splenic stroma cells that do not display the cognate auto-antigen and this interaction produces short-lived low Ca^{2+} mobilization spikes [166]. Following entrance into the CNS, T cells encounter autoantigen-presenting cells and have sustained Ca^{2+} mobilization which results in NFAT translocation and T cell activation [166,168]. EAE mice display reduced social interaction and cognition demonstrating that autoimmune response impairs neuronal function and organismal behavior [169].

Inhibitory T cell co-receptors are implicated in CNS dysregulation and disease. Varicella zoster virus (VZV) infection is characterized by lifelong persistence in neurons. VZV increases the expression of CTLA-4 and PD-1 in infected T cells which reduces IL-2 production and increases T cell anergy [170,171]. PD-1-deficient mice ($Pdcd1^{-/-}$) have increased T cell activation, leading to greater intracellular Ca^{2+} mobilization, and as previously discussed, increased glycolysis [86]. PD-1 deficiency causes elevated concentration of aromatic amino acids in the serum, specifically tryptophan and tyrosine, which decreases their availability in the brain where they are important for the synthesis of neurotransmitters such as dopamine and serotonin; consequently, there is an increase in anxiety-like behavior and fear in Pdcd1$^{-/-}$ mice [86]. Therefore, increased T cell activation caused by PD-1 deficiency can affect brain function and thus, affects cognitive behavior [86].

4.3. Cancer

T cells are critical components of the immune response to cancer. Helper T cells directly activate killer T cells to eradicate tumors and are essential in generating a strong antitumor response alone or in concert with killer T cells by promoting killer T cell activation, infiltration, persistence, and memory formation [172–177]. Tumor-specific T cells may not mount a robust response towards cancerous cells because the tumor microenvironment has numerous immunosuppressive factors; cancerous cells also downregulate cell surface co-stimulatory and MHC proteins which suppresses T cell activation [178–182]. Potent antitumor immune checkpoint blockade therapies using CTLA-4 and PD-1 monoclonal antibodies augment T cell response by suppressing the co-receptors' inhibitory signals, thereby promoting increased Ca^{2+} mobilization, glycolysis, and activation [183,184]. CTLA-4 monoclonal antibodies such as ipilimumab (Yervoy) and tremelimumab block B7-interaction and have been used to treat melanoma [47,185,186]. The monoclonal antibody pembrolizumab is highly selective for PD-1 and prevents PD-1 from engaging PD-L1 and PD-L2, thus enhancing T cell immune response [19,187,188]. Further research will address whether combining anti-CTLA-4 and anti-PD-1 antibodies will improve cancer treatments [19].

As previously mentioned, Ca^{2+} is critical for T cell activation and immune response. Manipulating Ca^{2+} signaling to enhance T cell-directed immune response against cancer is an intriguing notion, yet the means to target the Ca^{2+} response of specific cells without tampering with the metabolic processes of other cells remains elusive [189]. Antitumor activity of tumor-infiltrating lymphocytes (TIL) is inversely related to CD5 expression [99]. CD5 levels in naïve T cells are constantly tuned in the periphery by interactions with self pMHC complexes to maintain homeostasis; therefore, CD5 expression on TILs can be downregulated in response to low affinity for cancer antigens [190–192]. Thus, the majority of TILs are CD5$^{.0}$ which increase their reactivity while CD5^{hi} TILs do not elicit a Ca^{2+} response and become anergic and are unable to eliminate malignant cells [99,192]. While downregulation of CD5 on TILs enhances antitumor T cell activity, CD5^{lo} T cells are also more likely to experience activation-induced cell death (AICD) as CD5 protects T cells from overstimulation [23]. To maximize TIL effectiveness, the inhibitory effects of CD5 could be blocked by neutralizing monoclonal antibodies or soluble CD5-Fc molecules combined with soluble FAS-Fc molecules to reduce the inherent AICD [23,193,194]. Soluble human CD5 (shCD5) may have a similar effect but avoids targeting issues by blocking CD5-mediated interaction via a "decoy receptor" effect. Mice constitutively expressing shCD5 had reduced melanoma and thyoma tumor cell growth and increased numbers of CD4$^+$ and CD8$^+$ T cells [195]. Wild type mice treated with an injection of recombinant shCD5 also had reduced tumor growth [195]. Finally, CD5-deficient mice engrafted with B16-F10 melanoma cells had slower tumor growth compared to wild type C57BL/6 mice [196]. This evidence suggests that CD5, along with PD-1 and CTLA-4, may be a potential target to specifically modulate T cell Ca^{2+} mobilization in an immunosuppressive tumor setting.

4.4. Microbiome

The gut microbiome, including the bacteria and their products, forms a dynamic beneficial symbiosis with the immune system influencing host genes and cellular response. The gut microbiome shapes and directs immune responses while the immune system dictates the bacterial composition of the gut microbiome [197]. As the gut is the major symbiotic system intersecting the immune system and microbiota, understanding their connection has implications for immune system development and function as the gut microbiome is involved in protecting against pathogens, influencing states of inflammation, and even affecting cancer patient outcomes [198,199].

The gut microbiome primes immune responses [200]. Alteration in the microbial composition can induce changes in T cell function in infectious disease, autoimmunity, and cancer [201]. For example, mice treated with antibiotics which restrict or reduce the microbial environment exhibit impaired immune response because their T cells have altered TCR signaling and compromised intracellular Ca^{2+} mobilization in infectious disease and cystic fibrosis models [202–204]. In contrast, administering oral antibiotics to mice with EAE increases the frequency of $CD5^+$ B cell subpopulations in distal lymphoid sites and confers disease protection [205]. In cancer, the microbiome also influences patient response to immune checkpoint inhibitors such as CTLA-4 and PD-1 [206,207]. Mice and melanoma patients immunized or populated with *Bacteriodes fragilis* respond better to treatment with Ipilimumab, a monoclonal antibody against CTLA-4 [198]. Similarly, tumor-specific immunity improved when anti-PD-1/PD-L1 monoclonal antibodies where used in the presence of *Bifidobacterium* [208].

Though little is known about how CD5 influences T cell interaction with the microbiome, some tantalizing details are available. As specific bacterium promotes cancer regression during CTLA-4 and PD-1 checkpoint blockades, a CD5 blockade in conjunction with bacterial selection may also improve immune response. Such studies would lead to novel immunotherapeutic treatments for cancer and autoimmune diseases.

5. Conclusions

CD5, widely known as an inhibitory co-receptor in the thymus, appears to modulate the signaling intensity of peripheral T cells by increasing Ca^{2+} signaling activity and efficacy of $CD5^{hi}$ T cells. CD5 expression levels in the periphery correlates with intracellular Ca^{2+} mobilization, suggesting that CD5 promotes peripheral T cell activation and immune response. As such, CD5 may be a novel checkpoint therapy to regulate T cell activation and metabolism through altering Ca^{2+} mobilization, and could be used to affect neurological behavior, alter microbiome interactions, and treat cancer and autoinflammatory diseases. While this paper focuses on the role of co-receptor CD5 effects on calcium signaling and activation of T cells, CD5 itself may be regulated through posttranslational modifications, such as *N*-glycosylation, which may affect Ca^{2+} mobilization, T cell metabolism, activation, and function. In the future it would be interesting to determine the role of other posttranslational modifications (e.g., *N*-glycosylation, *S*-glutathionylation, lipidation) in CD5 signaling.

Author Contributions: C.M.T.F. is the first author and wrote the manuscript, D.K.J. contributed additional material and editing help, K.S.W. helped with the plan for the manuscript and editing and is the corresponding author.

Funding: This work was supported by a National Institute of Allergy and Infectious Diseases grant (R0102063) to K.S.W. The funder had no role in preparation of the manuscript.

Acknowledgments: We thank Kiara Vaden Whitley, Jeralyn Jones Fransen, Tyler Cox and Josie Tueller for their critical reviews of this manuscript.

Conflicts of Interest: The authors declare no conflict of interest.

Abbreviations

CTLA-4	Cytotoxic T-lymphocyte antigen 4
CD	Cluster of differenciation
PD-1	Programmed cell death protein 1
AMP	Adenosine monophosphate
ATP	Adenosine triphosphate
CaMKK	Calmodulin-dependent protein kinase kinase
AMPK	AMP-activated protein kinase
SOCE	Store-operated calcium channels
CRAC	$Calcium^+$-release-activated channel
STIM	Stromal interaction molecule
SERCA	Sarcoendoplasmic reticulum calcium transport ATPase
ER	Endoplasmic reticulum
NFAT	Nuclear factor of activated T cells
INF-γ	Interferon gamma

Abbreviations

TNF	Tumor necrosis factor
IL-2	Interleukin 2
GLUT1	Glucose transporter 1
GLUT3	Glucose transporter 3
TIL	Tumor infiltrating lymphocytes
ERK	Extracellular signal-regulated kinases

References

1. Chen, L.; Flies, D.B. Molecular mechanisms of T cell co-stimulation and co-inhibition. *Nat. Rev. Immunol.* **2013**, *13*, 227–242. [CrossRef] [PubMed]
2. Beyersdorf, N.; Kerkau, T.; Hünig, T. CD28 co-stimulation in T cell homeostasis: A recent perspective. *Immunotargets Ther.* **2015**, *4*, 111–122. [PubMed]
3. Fracchia, K.M.; Pai, C.Y.; Walsh, C.M. Modulation of T cell metabolism and function through calcium signaling. *Front. Immunol.* **2013**, *4*, 324. [CrossRef] [PubMed]
4. Cunningham, A.J.; Lafferty, K.J. Letter: Cellular proliferation can be an unreliable index of immune competence. *J. Immunol.* **1974**, *112*, 436–437. [CrossRef] [PubMed]
5. Nakayama, T.; Yamashita, M. The TCR-mediated signaling pathways that control the direction of helper T cell differentiation. *Semin. Immunol.* **2010**, *22*, 303–309. [CrossRef] [PubMed]
6. Goral, S. The three-signal hypothesis of lymphocyte activation/targets for immunosuppression. *Dial. Transplant.* **2011**, *40*, 14–16. [CrossRef]
7. Pennock, N.D.; White, J.T.; Cross, E.W.; Cheney, E.E.; Tamburini, B.A.; Kedl, R.M. T cell responses: Naïve to memory and everything in between. *Adv. Physiol. Educ.* **2013**, *37*, 273–283. [CrossRef] [PubMed]
8. Sharpe, A.H.; Abbas, A.K. T cell costimulation—Biology, therapeutic potential, and challenges. *N. Engl. J. Med.* **2006**, *355*, 973–975. [CrossRef] [PubMed]
9. Artyomov, M.N.; Lis, M.; Devadas, S.; Davis, M.M.; Chakraborty, A.K. CD4 and CD8 binding to MHC molecules primarily acts to enhance LCK delivery. *Proc. Natl. Acad. Sci. USA* **2010**, *107*, 16916–16921. [CrossRef] [PubMed]
10. Ravetch, J.V.; Lanier, L.L. Immune inhibitory receptors. *Science* **2000**, *290*, 84–89. [CrossRef] [PubMed]
11. Mellman, I.; Coukos, G.; Dranoff, G. Cancer immunotherapy comes of age. *Nature* **2011**, *480*, 480. [CrossRef] [PubMed]
12. Fuertes Marraco, S.A.; Neubert, N.J.; Verdeil, G.; Speiser, D.E. Inhibitory receptors beyond T cell exhaustion. *Front. Immunol.* **2015**, *6*, 310. [CrossRef] [PubMed]
13. Barrow, A.D.; Trowsdale, J. You say ITAM and I say ITIM, let's call the whole thing off: The ambiguity of immunoreceptor signalling. *Eur. J. Immunol.* **2006**, *36*, 1646–1653. [CrossRef] [PubMed]
14. Esensten, J.H.; Helou, Y.A.; Chopra, G.; Weiss, A.; Bluestone, J.A. CD28 costimulation: From mechanism to therapy. *Immunity* **2016**, *44*, 973–988. [CrossRef] [PubMed]
15. Dilek, N.; Poirier, N.; Hulin, P.; Coulon, F.; Mary, C.; Ville, S.; Vie, H.; Clémenceau, B.; Blancho, G.; Vanhove, B. Targeting CD28, CTLA-4 and PD-L1 costimulation differentially controls immune synapses and function of human regulatory and conventional t cells. *PLoS ONE* **2013**, *8*, e83139. [CrossRef] [PubMed]
16. Chambers, C.A.; Sullivan, T.J.; Allison, J.P. Lymphoproliferation in CTLA-4-deficient mice is mediated by costimulation-dependent activation of CD4+ T cells. *Immunity* **1997**, *7*, 885–895. [CrossRef]
17. Lindsten, T.; Lee, K.P.; Harris, E.S.; Petryniak, B.; Craighead, N.; Reynolds, P.J.; Lombard, D.B.; Freeman, G.J.; Nadler, L.M.; Gray, G.S.; et al. Characterization of CTLA-4 structure and expression on human T cells. *J. Immunol.* **1993**, *151*, 3489–3499. [PubMed]
18. Boise, L.H.; Minn, A.J.; Noel, P.J.; June, C.H.; Accavitti, M.A.; Lindsten, T.; Thompson, C.B. CD28 costimulation can promote T cell survival by enhancing the expression of Bcl-XL. *Immunity* **1995**, *3*, 87–98. [CrossRef]
19. Buchbinder, E.I.; Desai, A. CTLA-4 and PD-1 pathways: Similarities, differences, and implications of their inhibition. *Am. J. Clin. Oncol.* **2016**, *39*, 98–106. [CrossRef] [PubMed]
20. Iwai, Y.; Hamanishi, J.; Chamoto, K.; Honjo, T. Cancer immunotherapies targeting the PD-1 signaling pathway. *J. Biomed. Sci.* **2017**, *24*, 26. [CrossRef] [PubMed]

21. Chambers, C.A.; Kuhns, M.S.; Egen, J.G.; Allison, J.P. CTLA-4-mediated inhibition in regulation of T cell responses: Mechanisms and manipulation in tumor immunotherapy. *Annu. Rev. Immunol.* **2001**, *19*, 565–594. [CrossRef] [PubMed]

22. Brossard, C.; Semichon, M.; Trautmann, A.; Bismuth, G. CD5 inhibits signaling at the immunological synapse without impairing its formation. *J. Immunol.* **2003**, *170*, 4623–4629. [CrossRef] [PubMed]

23. Tabbekh, M.; Mokrani-Hammani, M.B.; Bismuth, G.; Mami-Chouaib, F. T cell modulatory properties of CD5 and its role in antitumor immune responses. *Oncoimmunology* **2013**, *2*, e22841. [CrossRef] [PubMed]

24. Mahoney, K.M.; Freeman, G.J.; McDermott, D.F. The next immune-checkpoint inhibitors: PD-1/PD-L1 blockade in melanoma. *Clin. Ther.* **2015**, *37*, 764–782. [CrossRef] [PubMed]

25. De Wit, J.; Souwer, Y.; van Beelen, A.J.; de Groot, R.; Muller, F.J.; Klaasse Bos, H.; Jorritsma, T.; Kapsenberg, M.L.; de Jong, E.C.; van Ham, S.M. CD5 costimulation induces stable Th17 development by promoting IL-23R expression and sustained STAT3 activation. *Blood* **2011**, *118*, 6107–6114. [CrossRef] [PubMed]

26. Perez-Villar, J.J.; Whitney, G.S.; Bowen, M.A.; Hewgill, D.H.; Aruffo, A.A.; Kanner, S.B. CD5 negatively regulates the T cell antigen receptor signal transduction pathway: Involvement of SH2-containing phosphotyrosine phosphatase SHP-1. *Mol. Cell. Biol.* **1999**, *19*, 2903–2912. [CrossRef] [PubMed]

27. Gary-Gouy, H.; Harriague, J.; Dalloul, A.; Donnadieu, E.; Bismuth, G. CD5-negative regulation of B cell receptor signaling pathways originates from tyrosine residue Y429 outside an immunoreceptor tyrosine-based inhibitory motif. *J. Immunol.* **2002**, *168*, 232–239. [CrossRef] [PubMed]

28. Dennehy, K.M.; Broszeit, R.; Garnett, D.; Durrheim, G.A.; Spruyt, L.L.; Beyers, A.D. Thymocyte activation induces the association of phosphatidylinositol 3-kinase and pp120 with CD5. *Eur. J. Immunol.* **1997**, *27*, 679–686. [CrossRef] [PubMed]

29. Samelson, L.E. Signal transduction mediated by the T cell antigen receptor: The role of adapter proteins. *Annu. Rev. Immunol.* **2002**, *20*, 371–394. [CrossRef] [PubMed]

30. Burgess, K.E.; Yamamoto, M.; Prasad, K.V.S.; Rudd, C.E. CD5 acts as a tyrosine kinase substrate within a receptor complex comprising T cell receptor ζ-chain CD3 and protein-tyrosine kinases P56LCK and P59FYN. *Proc. Natl. Acad. Sci. USA* **1992**, *89*, 9311–9315. [CrossRef] [PubMed]

31. Consuegra-Fernandez, M.; Aranda, F.; Simoes, I.; Orta, M.; Sarukhan, A.; Lozano, F. CD5 as a Target for Immune-Based Therapies. *Crit. Rev. Immunol.* **2015**, *35*, 85–115. [CrossRef] [PubMed]

32. Roa, N.S.; Ordonez-Rueda, D.; Chavez-Rios, J.R.; Raman, C.; Garcia-Zepeda, E.A.; Lozano, F.; Soldevila, G. The carboxy-terminal region of CD5 is required for c-CBL mediated TCR signaling downmodulation in thymocytes. *Biochem. Biophys. Res. Commun.* **2013**, *432*, 52–59. [CrossRef] [PubMed]

33. Berney, S.M.; Schaan, T.; Wolf, R.E.; Kimpel, D.L.; van der Heyde, H.; Atkinson, T.P. CD5 (OKT1) augments CD3-mediated intracellular signaling events in human T lymphocytes. *Inflammation* **2001**, *25*, 215–221. [CrossRef] [PubMed]

34. Azzam, H.S.; DeJarnette, J.B.; Huang, K.; Emmons, R.; Park, C.-S.; Sommers, C.L.; El-Khoury, D.; Shores, E.W.; Love, P.E. Fine tuning of TCR signaling by CD5. *J. Immunol.* **2001**, *166*, 5464–5472. [CrossRef] [PubMed]

35. Freitas, C.M.T.; Hamblin, G.J.; Raymond, C.M.; Weber, K.S. Naive helper T cells with high CD5 expression have increased calcium signaling. *PLoS ONE* **2017**, *12*, e0178799. [CrossRef] [PubMed]

36. Feske, S. Calcium signalling in lymphocyte activation and disease. *Nat. Rev. Immunol.* **2007**, *7*, 690–702. [CrossRef] [PubMed]

37. Joseph, N.; Reicher, B.; Barda-Saad, M. The calcium feedback loop and T cell activation: How cytoskeleton networks control intracellular calcium flux. *Biochim. Biophys. Acta Biomembr.* **2014**, *1838*, 557–568. [CrossRef] [PubMed]

38. Vig, M.; Kinet, J.-P. Calcium signaling in immune cells. *Nat. Immunol.* **2009**, *10*, 21–27. [CrossRef] [PubMed]

39. Wolf, I.M.A.; Guse, A.H. Ca^{2+} microdomains in T-lymphocytes. *Front. Oncol.* **2017**, *7*, 73. [CrossRef] [PubMed]

40. Hogan, P.G.; Lewis, R.S.; Rao, A. Molecular basis of calcium signaling in lymphocytes: STIM and ORAI. *Annu. Rev. Immunol.* **2010**, *28*, 491–533. [CrossRef] [PubMed]

41. Oh-hora, M.; Rao, A. Calcium signaling in lymphocytes. *Curr. Opin. Immunol.* **2008**, *20*, 250–258. [CrossRef] [PubMed]

42. Janeway, C.A., Jr. The co-receptor function of CD4. *Semin. Immunol.* **1991**, *3*, 153–160. [PubMed]

43. Moran, A.E.; Hogquist, K.A. T cell receptor affinity in thymic development. *Immunology* **2012**, *135*, 261–267. [CrossRef] [PubMed]

44. Kyttaris, V.C.; Zhang, Z.; Kampagianni, O.; Tsokos, G.C. Calcium signaling in systemic lupus erythematosus T cells: A treatment target. *Arthritis Rheum.* **2011**, *63*, 2058–2066. [CrossRef] [PubMed]

45. Demkow, U.; Winklewski, P.; Ciepiela, O.; Popko, K.; Lipinska, A.; Kucharska, A.; Michalska, B.; Wasik, M. Modulatory effect of insulin on T cell receptor mediated calcium signaling is blunted in long lasting type 1 diabetes mellitus. *Pharmacol. Rep.* **2012**, *64*, 150–156. [CrossRef]

46. Parry, R.V.; Chemnitz, J.M.; Frauwirth, K.A.; Lanfranco, A.R.; Braunstein, I.; Kobayashi, S.V.; Linsley, P.S.; Thompson, C.B.; Riley, J.L. CTLA-4 and PD-1 receptors inhibit T cell activation by distinct mechanisms. *Mol. Cell. Biol.* **2005**, *25*, 9543–9553. [CrossRef] [PubMed]

47. Grosso, J.F.; Jure-Kunkel, M.N. CTLA-4 blockade in tumor models: An overview of preclinical and translational research. *Cancer Immun.* **2013**, *13*, 5. [PubMed]

48. Rudd, C.E.; Taylor, A.; Schneider, H. CD28 and CTLA-4 coreceptor expression and signal transduction. *Immunol. Rev.* **2009**, *229*, 12–26. [CrossRef] [PubMed]

49. Jago, C.B.; Yates, J.; Olsen Saraiva CÂMara, N.; Lechler, R.I.; Lombardi, G. Differential expression of CTLA-4 among T cell subsets. *Clin. Exp. Immunol.* **2004**, *136*, 463–471. [CrossRef] [PubMed]

50. Linsley, P.S.; Bradshaw, J.; Greene, J.; Peach, R.; Bennett, K.L.; Mittler, R.S. Intracellular trafficking of CTLA-4 and focal localization towards sites of TCR engagement. *Immunity* **1996**, *4*, 535–543. [CrossRef]

51. Schneider, H.; Smith, X.; Liu, H.; Bismuth, G.; Rudd, C.E. CTLA-4 disrupts ZAP70 microcluster formation with reduced T cell/APC dwell times and calcium mobilization. *Eur. J. Immunol.* **2008**, *38*, 40–47. [CrossRef] [PubMed]

52. Grohmann, U.; Orabona, C.; Fallarino, F.; Vacca, C.; Calcinaro, F.; Falorni, A.; Candeloro, P.; Belladonna, M.L.; Bianchi, R.; Fioretti, M.C.; et al. CTLA-4-Ig regulates tryptophan catabolism in vivo. *Nat. Immunol.* **2002**, *3*, 1097–1101. [CrossRef] [PubMed]

53. Chen, W.; Jin, W.; Wahl, S.M. Engagement of cytotoxic T lymphocyte-associated antigen 4 (CTLA-4) induces transforming growth factor beta (TGF-β) production by murine CD4$^+$ T cells. *J. Exp. Med.* **1998**, *188*, 1849–1857. [CrossRef] [PubMed]

54. Hryniewicz, A.; Boasso, A.; Edghill-Smith, Y.; Vaccari, M.; Fuchs, D.; Venzon, D.; Nacsa, J.; Betts, M.R.; Tsai, W.-P.; Heraud, J.-M.; et al. CTLA-4 blockade decreases TGF-β, IDO, and viral RNA expression in tissues of SIVmac251-infected macaques. *Blood* **2006**, *108*, 3834–3842. [CrossRef] [PubMed]

55. Iken, K.; Liu, K.; Liu, H.; Bizargity, P.; Wang, L.; Hancock, W.W.; Visner, G.A. Indoleamine 2,3-dioxygenase and metabolites protect murine lung allografts and impair the calcium mobilization of T cells. *Am. J. Respir. Cell Mol. Biol.* **2012**, *47*, 405–416. [CrossRef] [PubMed]

56. Walker, L.S.K.; Sansom, D.M. Confusing signals: Recent progress in CTLA-4 biology. *Trends Immunol.* **2015**, *36*, 63–70. [CrossRef] [PubMed]

57. Cederbom, L.; Hall, H.; Ivars, F. CD4$^+$CD25$^+$ regulatory T cells down-regulate co-stimulatory molecules on antigen-presenting cells. *Eur. J. Immunol.* **2000**, *30*, 1538–1543. [CrossRef]

58. Burnett, D.L.; Parish, I.A.; Masle-Farquhar, E.; Brink, R.; Goodnow, C.C. Murine LRBA deficiency causes CTLA-4 deficiency in Tregs without progression to immune dysregulation. *Immunol. Cell Biol.* **2017**, *95*, 775–778. [CrossRef] [PubMed]

59. Verma, N.; Burns, S.O.; Walker, L.S.K.; Sansom, D.M. Immune deficiency and autoimmunity in patients with CTLA-4 (CD152) mutations. *Clin. Exp. Immunol.* **2017**, *190*, 1–7. [CrossRef] [PubMed]

60. Wing, K.; Onishi, Y.; Prieto-Martin, P.; Yamaguchi, T.; Miyara, M.; Fehervari, Z.; Nomura, T.; Sakaguchi, S. CTLA-4 control over Foxp3$^+$ regulatory T cell function. *Science* **2008**, *322*, 271–275. [CrossRef] [PubMed]

61. Sojka, D.K.; Hughson, A.; Fowell, D.J. CTLA-4 is Required by CD4$^+$CD25$^+$ treg to control CD4$^+$ T cell lymphopenia-induced proliferation. *Eur. J. Immunol.* **2009**, *39*, 1544–1551. [CrossRef] [PubMed]

62. Kristiansen, O.P.; Larsen, Z.M.; Pociot, F. CTLA-4 in autoimmune diseases–a general susceptibility gene to autoimmunity? *Genes Immun.* **2000**, *1*, 170–184. [CrossRef] [PubMed]

63. Chikuma, S. CTLA-4, an essential immune-checkpoint for T cell activation. *Curr. Top. Microbiol. Immunol.* **2017**, *410*, 99–126. [PubMed]

64. Tai, X.; Van Laethem, F.; Pobezinsky, L.; Guinter, T.; Sharrow, S.O.; Adams, A.; Granger, L.; Kruhlak, M.; Lindsten, T.; Thompson, C.B.; et al. Basis of CTLA-4 function in regulatory and conventional CD4$^+$ T cells. *Blood* **2012**, *119*, 5155–5163. [CrossRef] [PubMed]

65. Lo, B.; Abdel-Motal, U.M. Lessons from CTLA-4 deficiency and checkpoint inhibition. *Curr. Opin. Immunol.* **2017**, *49*, 14–19. [CrossRef] [PubMed]

66. Avogadri, F.; Yuan, J.; Yang, A.; Schaer, D.; Wolchok, J.D. Modulation of CTLA-4 and GITR for cancer immunotherapy. *Curr. Top. Microbiol. Immunol.* **2011**, *344*, 211–244. [PubMed]

67. Royal, R.E.; Levy, C.; Turner, K.; Mathur, A.; Hughes, M.; Kammula, U.S.; Sherry, R.M.; Topalian, S.L.; Yang, J.C.; Lowy, I.; et al. Phase 2 trial of single agent Ipilimumab (anti-CTLA-4) for locally advanced or metastatic pancreatic adenocarcinoma. *J. Immunother.* **2010**, *33*, 828–833. [CrossRef] [PubMed]

68. Le, D.T.; Lutz, E.; Uram, J.N.; Sugar, E.A.; Onners, B.; Solt, S.; Zheng, L.; Diaz, L.A., Jr.; Donehower, R.C.; Jaffee, E.M.; et al. Evaluation of ipilimumab in combination with allogeneic pancreatic tumor cells transfected with a GM-CSF gene in previously treated pancreatic cancer. *J. Immunother.* **2013**, *36*, 382–389. [CrossRef] [PubMed]

69. Chung, K.Y.; Gore, I.; Fong, L.; Venook, A.; Beck, S.B.; Dorazio, P.; Criscitiello, P.J.; Healey, D.I.; Huang, B.; Gomez-Navarro, J.; et al. Phase II study of the anti-cytotoxic T-lymphocyte-associated antigen 4 monoclonal antibody, tremelimumab, in patients with refractory metastatic colorectal cancer. *J. Clin. Oncol.* **2010**, *28*, 3485–3490. [CrossRef] [PubMed]

70. Ribas, A.; Camacho, L.H.; Lopez-Berestein, G.; Pavlov, D.; Bulanhagui, C.A.; Millham, R.; Comin-Anduix, B.; Reuben, J.M.; Seja, E.; Parker, C.A.; et al. Antitumor activity in melanoma and anti-self responses in a phase I trial with the anti-cytotoxic T lymphocyte-associated antigen 4 monoclonal antibody CP-675,206. *J. Clin. Oncol.* **2005**, *23*, 8968–8977. [CrossRef] [PubMed]

71. Calabro, L.; Morra, A.; Fonsatti, E.; Cutaia, O.; Fazio, C.; Annesi, D.; Lenoci, M.; Amato, G.; Danielli, R.; Altomonte, M.; et al. Efficacy and safety of an intensified schedule of tremelimumab for chemotherapy-resistant malignant mesothelioma: An open-label, single-arm, phase 2 study. *Lancet Respir. Med.* **2015**, *3*, 301–309. [CrossRef]

72. Comin-Anduix, B.; Escuin-Ordinas, H.; Ibarrondo, F.J. Tremelimumab: Research and clinical development. *OncoTargets Ther.* **2016**, *9*, 1767–1776.

73. Ribas, A.; Comin-Anduix, B.; Chmielowski, B.; Jalil, J.; de la Rocha, P.; McCannel, T.A.; Ochoa, M.T.; Seja, E.; Villanueva, A.; Oseguera, D.K.; et al. Dendritic cell vaccination combined with CTLA4 blockade in patients with metastatic melanoma. *Clin. Cancer Res.* **2009**, *15*, 6267–6276. [CrossRef] [PubMed]

74. Antonia, S.; Goldberg, S.B.; Balmanoukian, A.; Chaft, J.E.; Sanborn, R.E.; Gupta, A.; Narwal, R.; Steele, K.; Gu, Y.; Karakunnel, J.J.; et al. Safety and antitumour activity of durvalumab plus tremelimumab in non-small cell lung cancer: A multicentre, phase 1b study. *Lancet Oncol.* **2016**, *17*, 299–308. [CrossRef]

75. Dong, Y.; Sun, Q.; Zhang, X. PD-1 and its ligands are important immune checkpoints in cancer. *Oncotarget* **2017**, *8*, 2171–2186. [CrossRef] [PubMed]

76. Shi, L.; Chen, S.; Yang, L.; Li, Y. The role of PD-1 and PD-L1 in T cell immune suppression in patients with hematological malignancies. *J. Hematol. Oncol.* **2013**, *6*, 74. [CrossRef] [PubMed]

77. Keir, M.E.; Butte, M.J.; Freeman, G.J.; Sharpe, A.H. PD-1 and its ligands in tolerance and immunity. *Annu. Rev. Immunol.* **2008**, *26*, 677–704. [CrossRef] [PubMed]

78. Okazaki, T.; Wang, J. PD-1/PD-L pathway and autoimmunity. *Autoimmunity* **2005**, *38*, 353–357. [CrossRef] [PubMed]

79. Boussiotis, V.A. Molecular and biochemical aspects of the PD-1 checkpoint pathway. *N. Engl. J. Med.* **2016**, *375*, 1767–1778. [CrossRef] [PubMed]

80. Wang, S.-F.; Fouquet, S.; Chapon, M.; Salmon, H.; Regnier, F.; Labroquère, K.; Badoual, C.; Damotte, D.; Validire, P.; Maubec, E.; et al. Early T cell signalling is reversibly altered in PD-1+ T lymphocytes infiltrating human tumors. *PLoS ONE* **2011**, *6*, e17621. [CrossRef] [PubMed]

81. Gorentla, B.K.; Zhong, X.-P. T cell receptor signal transduction in T lymphocytes. *J. Clin. Cell. Immunol.* **2012**, *2012*, 005.

82. Wei, F.; Zhong, S.; Ma, Z.; Kong, H.; Medvec, A.; Ahmed, R.; Freeman, G.J.; Krogsgaard, M.; Riley, J.L. Strength of PD-1 signaling differentially affects T cell effector functions. *Proc. Natl. Acad. Sci. USA* **2013**, *110*, E2480–E2489. [CrossRef] [PubMed]

83. Cochain, C.; Chaudhari, S.M.; Koch, M.; Wiendl, H.; Eckstein, H.-H.; Zernecke, A. Programmed cell death-1 deficiency exacerbates T cell activation and atherogenesis despite expansion of regulatory T cells in atherosclerosis-prone mice. *PLoS ONE* **2014**, *9*, e93280. [CrossRef] [PubMed]

84. Asano, T.; Kishi, Y.; Meguri, Y.; Yoshioka, T.; Iwamoto, M.; Maeda, Y.; Yagita, H.; Tanimoto, M.; Koreth, J.; Ritz, J.; et al. PD-1 signaling has a critical role in maintaining regulatory T cell homeostasis; implication for treg depletion therapy by PD-1 blockade. *Blood* **2015**, *126*, 848.

85. Francisco, L.M.; Sage, P.T.; Sharpe, A.H. The PD-1 pathway in tolerance and autoimmunity. *Immunol. Rev.* **2010**, *236*, 219–242. [CrossRef] [PubMed]

86. Miyajima, M.; Zhang, B.; Sugiura, Y.; Sonomura, K.; Guerrini, M.M.; Tsutsui, Y.; Maruya, M.; Vogelzang, A.; Chamoto, K.; Honda, K.; et al. Metabolic shift induced by systemic activation of T cells in PD-1-deficient mice perturbs brain monoamines and emotional behavior. *Nat. Immunol.* **2017**, *18*, 1342–1352. [CrossRef] [PubMed]

87. Riella, L.V.; Paterson, A.M.; Sharpe, A.H.; Chandraker, A. Role of the PD-1 pathway in the immune response. *Am. J. Transplant.* **2012**, *12*, 2575–2587. [CrossRef] [PubMed]

88. Kroner, A.; Mehling, M.; Hemmer, B.; Rieckmann, P.; Toyka, K.V.; Maurer, M.; Wiendl, H. A PD-1 polymorphism is associated with disease progression in multiple sclerosis. *Ann. Neurol* **2005**, *58*, 50–57. [CrossRef] [PubMed]

89. Pawlak-Adamska, E.; Nowak, O.; Karabon, L.; Pokryszko-Dragan, A.; Partyka, A.; Tomkiewicz, A.; Ptaszkowski, J.; Frydecka, I.; Podemski, R.; Dybko, J.; et al. *PD-1* gene polymorphic variation is linked with first symptom of disease and severity of relapsing-remitting form of MS. *J. Neuroimmunol.* **2017**, *305*, 115–127. [CrossRef] [PubMed]

90. Dai, S.; Jia, R.; Zhang, X.; Fang, Q.; Huang, L. The PD-1/PD-Ls pathway and autoimmune diseases. *Cell. Immunol.* **2014**, *290*, 72–79. [CrossRef] [PubMed]

91. Gianchecchi, E.; Delfino, D.V.; Fierabracci, A. Recent insights into the role of the PD-1/PD-L1 pathway in immunological tolerance and autoimmunity. *Autoimmun. Rev.* **2013**, *12*, 1091–1100. [CrossRef] [PubMed]

92. Wang, J.; Yoshida, T.; Nakaki, F.; Hiai, H.; Okazaki, T.; Honjo, T. Establishment of NOD-Pdcd1$^{-/-}$ mice as an efficient animal model of type I diabetes. *Proc. Natl. Acad. Sci. USA* **2005**, *102*, 11823–11828. [CrossRef] [PubMed]

93. Okazaki, T.; Otaka, Y.; Wang, J.; Hiai, H.; Takai, T.; Ravetch, J.V.; Honjo, T. Hydronephrosis associated with antiurothelial and antinuclear autoantibodies in BALB/c-Fcgr2b$^{-/-}$Pdcd1$^{-/-}$ mice. *J. Exp. Med.* **2005**, *202*, 1643–1648. [CrossRef] [PubMed]

94. Nishimura, H.; Okazaki, T.; Tanaka, Y.; Nakatani, K.; Hara, M.; Matsumori, A.; Sasayama, S.; Mizoguchi, A.; Hiai, H.; Minato, N.; et al. Autoimmune dilated cardiomyopathy in PD-1 receptor-deficient mice. *Science* **2001**, *291*, 319–322. [CrossRef] [PubMed]

95. Xiao, Y.; Yu, S.; Zhu, B.; Bedoret, D.; Bu, X.; Francisco, L.M.; Hua, P.; Duke-Cohan, J.S.; Umetsu, D.T.; Sharpe, A.H.; et al. RGMb is a novel binding partner for PD-L2 and its engagement with PD-L2 promotes respiratory tolerance. *J. Exp. Med.* **2014**, *211*, 943–959. [CrossRef] [PubMed]

96. Masuda, K.; Kishimoto, T. CD5: A new partner for IL-6. *Immunity* **2016**, *44*, 720–722. [CrossRef] [PubMed]

97. Huang, H.J.; Jones, N.H.; Strominger, J.L.; Herzenberg, L.A. Molecular cloning of Ly-1, a membrane glycoprotein of mouse T lymphocytes and a subset of B cells: Molecular homology to its human counterpart Leu-1/T1 (CD5). *Proc. Natl. Acad. Sci. USA* **1987**, *84*, 204–208. [CrossRef] [PubMed]

98. Tarakhovsky, A.; Kanner, S.B.; Hombach, J.; Ledbetter, J.A.; Muller, W.; Killeen, N.; Rajewsky, K. A role for CD5 in TCR-mediated signal transduction and thymocyte selection. *Science* **1995**, *269*, 535–537. [CrossRef] [PubMed]

99. Dalloul, A. CD5: A safeguard against autoimmunity and a shield for cancer cells. *Autoimmun. Rev.* **2009**, *8*, 349–353. [CrossRef] [PubMed]

100. Bhandoola, A.; Bosselut, R.; Yu, Q.; Cowan, M.L.; Feigenbaum, L.; Love, P.E.; Singer, A. CD5-mediated inhibition of TCR signaling during intrathymic selection and development does not require the CD5 extracellular domain. *Eur. J. Immunol.* **2002**, *32*, 1811–1817. [CrossRef]

101. Mandl, J.N.; Monteiro, J.P.; Vrisekoop, N.; Germain, R.N. T cell positive selection uses self-ligand binding strength to optimize repertoire recognition of foreign antigens. *Immunity* **2013**, *38*, 263–274. [CrossRef] [PubMed]

102. Henderson, J.G.; Opejin, A.; Jones, A.; Gross, C.; Hawiger, D. CD5 instructs extrathymic regulatory T cell development in response to self and tolerizing antigens. *Immunity* **2015**, *42*, 471–483. [CrossRef] [PubMed]

103. Gringhuis, S.I.; de Leij, L.F.; Wayman, G.A.; Tokumitsu, H.; Vellenga, E. The Ca^{2+}/calmodulin-dependent kinase type IV is involved in the CD5-mediated signaling pathway in human T lymphocytes. *J. Biol. Chem.* **1997**, *272*, 31809–31820. [CrossRef] [PubMed]

104. Hassan, N.J.; Simmonds, S.J.; Clarkson, N.G.; Hanrahan, S.; Puklavec, M.J.; Bomb, M.; Barclay, A.N.; Brown, M.H. CD6 regulates T cell responses through activation-dependent recruitment of the positive regulator SLP-76. *Mol. Cell. Biol.* **2006**, *26*, 6727–6738. [CrossRef] [PubMed]

105. Pena-Rossi, C.; Zuckerman, L.A.; Strong, J.; Kwan, J.; Ferris, W.; Chan, S.; Tarakhovsky, A.; Beyers, A.D.; Killeen, N. Negative regulation of CD4 lineage development and responses by CD5. *J. Immunol.* **1999**, *163*, 6494–6501. [PubMed]

106. Davies, A.A.; Ley, S.C.; Crumpton, M.J. CD5 is phosphorylated on tyrosine after stimulation of the T cell antigen receptor complex. *Proc. Natl. Acad. Sci. USA* **1992**, *89*, 6368–6372. [CrossRef] [PubMed]

107. Samelson, L.E.; Phillips, A.F.; Luong, E.T.; Klausner, R.D. Association of the fyn protein-tyrosine kinase with the T cell antigen receptor. *Proc. Natl. Acad. Sci. USA* **1990**, *87*, 4358–4362. [CrossRef] [PubMed]

108. Raab, M.; Yamamoto, M.; Rudd, C.E. The T cell antigen CD5 acts as a receptor and substrate for the protein-tyrosine kinase p56lck. *Mol. Cell. Biol.* **1994**, *14*, 2862–2870. [CrossRef] [PubMed]

109. Beyers, A.D.; Spruyt, L.L.; Williams, A.F. Molecular associations between the T-lymphocyte antigen receptor complex and the surface antigens CD2, CD4, or CD8 and CD5. *Proc. Natl. Acad. Sci. USA* **1992**, *89*, 2945–2949. [CrossRef] [PubMed]

110. Spertini, F.; Stohl, W.; Ramesh, N.; Moody, C.; Geha, R.S. Induction of human T cell proliferation by a monoclonal antibody to CD5. *J. Immunol.* **1991**, *146*, 47–52. [PubMed]

111. Persaud, S.P.; Parker, C.R.; Lo, W.-L.; Weber, K.S.; Allen, P.M. Intrinsic CD4$^+$ T cell sensitivity and response to pathogen are set and sustained by avidity for thymic and peripheral self-pMHC. *Nat. Immunol.* **2014**, *15*, 266–274. [CrossRef] [PubMed]

112. Calvo, J.; Padilla, O.; Places, L.; Vigorito, E.; Vila, J.M.; Vilella, R.; Mila, J.; Vives, J.; Bowen, M.A.; Lozano, F. Relevance of individual CD5 extracellular domains on antibody recognition, glycosylation and co-mitogenic signalling. *Tissue Antigen.* **1999**, *54*, 16–26. [CrossRef]

113. McAlister, M.S.; Brown, M.H.; Willis, A.C.; Rudd, P.M.; Harvey, D.J.; Aplin, R.; Shotton, D.M.; Dwek, R.A.; Barclay, A.N.; Driscoll, P.C. Structural analysis of the CD5 antigen—Expression, disulphide bond analysis and physical characterisation of CD5 scavenger receptor superfamily domain 1. *Eur J. Biochem.* **1998**, *257*, 131–141. [CrossRef] [PubMed]

114. Cho, J.-H.; Kim, H.-O.; Surh, C.D.; Sprent, J. T cell receptor-dependent regulation of lipid rafts controls naive CD8$^+$ T cell homeostasis. *Immunity* **2010**, *32*, 214–226. [CrossRef] [PubMed]

115. Yashiro-Ohtani, Y.; Zhou, X.-Y.; Toyo-oka, K.; Tai, X.-G.; Park, C.-S.; Hamaoka, T.; Abe, R.; Miyake, K.; Fujiwara, H. Non-CD28 costimulatory molecules present in T cell rafts induce T cell costimulation by enhancing the association of TCR with rafts. *J. Immunol.* **2000**, *164*, 1251–1259. [CrossRef] [PubMed]

116. König, R. Chapter 315—Signal Transduction in T Lymphocytes A2—Bradshaw, Ralph A. In *Handbook of Cell Signaling*, 2nd ed.; Dennis, E.A., Ed.; Academic Press: San Diego, CA, USA, 2010; pp. 2679–2688.

117. Milam, A.V.; Allen, P.M. Functional heterogeneity in CD4$^+$ T cell responses against a bacterial pathogen. *Front. Immunol* **2015**, *6*, 621. [CrossRef] [PubMed]

118. Lozano, F.; Simarro, M.; Calvo, J.; Vila, J.M.; Padilla, O.; Bowen, M.A.; Campbell, K.S. CD5 signal transduction: Positive or negative modulation of antigen receptor signaling. *Crit. Rev. Immunol.* **2000**, *20*, 347–358. [CrossRef] [PubMed]

119. Hogquist, K.A.; Jameson, S.C. The self-obsession of T cells: How TCR signaling thresholds affect fate decisions in the thymus and effector function in the periphery. *Nat. Immunol.* **2014**, *15*, 815–823. [CrossRef] [PubMed]

120. Van de Velde, H.; von Hoegen, I.; Luo, W.; Parnes, J.R.; Thielemans, K. The B-cell surface protein CD72/Lyb-2 is the ligand for CD5. *Nature* **1991**, *351*, 662–665. [CrossRef] [PubMed]

121. Biancone, L.; Bowen, M.A.; Lim, A.; Aruffo, A.; Andres, G.; Stamenkovic, I. Identification of a novel inducible cell-surface ligand of CD5 on activated lymphocytes. *J. Exp. Med.* **1996**, *184*, 811–819. [CrossRef] [PubMed]

122. Brown, M.H.; Lacey, E. A ligand for CD5 is CD5. *J. Immunol.* **2010**, *185*, 6068–6074. [CrossRef] [PubMed]

123. Luo, W.; Van de Velde, H.; von Hoegen, I.; Parnes, J.R.; Thielemans, K. Ly-1 (CD5), a membrane glycoprotein of mouse T lymphocytes and a subset of B cells, is a natural ligand of the B cell surface protein Lyb-2 (CD72). *J. Immunol.* **1992**, *148*, 1630–1634. [PubMed]

124. Vandenberghe, P.; Verwilghen, J.; Van Vaeck, F.; Ceuppens, J.L. Ligation of the CD5 or CD28 molecules on resting human T cells induces expression of the early activation antigen CD69 by a calcium- and tyrosine kinase-dependent mechanism. *Immunology* **1993**, *78*, 210–217. [PubMed]

125. Ceuppens, J.L.; Baroja, M.L. Monoclonal antibodies to the CD5 antigen can provide the necessary second signal for activation of isolated resting T cells by solid-phase-bound OKT3. *J. Immunol.* **1986**, *137*, 1816–1821. [PubMed]

126. June, C.H.; Rabinovitch, P.S.; Ledbetter, J.A. CD5 antibodies increase intracellular ionized calcium concentration in T cells. *J. Immunol.* **1987**, *138*, 2782–2792. [PubMed]

127. Reth, M. Antigen receptor tail clue. *Nature* **1989**, *338*, 383–384. [CrossRef] [PubMed]

128. Unkeless, J.C.; Jin, J. Inhibitory receptors, ITIM sequences and phosphatases. *Curr. Opin. Immunol* **1997**, *9*, 338–343. [CrossRef]

129. Dong, B.; Somani, A.K.; Love, P.E.; Zheng, X.; Chen, X.; Zhang, J. CD5-mediated inhibition of TCR signaling proceeds normally in the absence of SHP-1. *Int. J. Mol. Med.* **2016**, *38*, 45–56. [CrossRef] [PubMed]

130. Weber, K.S.; Li, Q.J.; Persaud, S.P.; Campbell, J.D.; Davis, M.M.; Allen, P.M. Distinct CD4+ helper T cells involved in primary and secondary responses to infection. *Proc. Natl. Acad. Sci. USA* **2012**, *109*, 9511–9516. [CrossRef] [PubMed]

131. Fulton, R.B.; Hamilton, S.E.; Xing, Y.; Best, J.A.; Goldrath, A.W.; Hogquist, K.A.; Jameson, S.C. The TCR's sensitivity to self peptide–MHC dictates the ability of naive CD8+ T cells to respond to foreign antigens. *Nat. Immunol.* **2014**, *16*, 107–117. [CrossRef] [PubMed]

132. Palin, A.C.; Love, P.E. CD5 helps aspiring regulatory T cells ward off unwelcome cytokine advances. *Immunity* **2015**, *42*, 395–396. [CrossRef] [PubMed]

133. Mattson, M.P.; Chan, S.L. Calcium orchestrates apoptosis. *Nat. Cell Biol.* **2003**, *5*, 1041–1043. [CrossRef] [PubMed]

134. Orrenius, S.; Nicotera, P. The calcium ion and cell death. *J. Neural Transm. Suppl.* **1994**, *43*, 1–11. [PubMed]

135. Zhao, C.; Davies, J.D. A peripheral CD4+ T cell precursor for naive, memory, and regulatory T cells. *J. Exp. Med.* **2010**, *207*, 2883–2894. [CrossRef] [PubMed]

136. Wahl, D.R.; Byersdorfer, C.A.; Ferrara, J.L.M.; Opipari, A.W.; Glick, G.D. Distinct metabolic programs in activated T cells: Opportunities for selective immunomodulation. *Immunol. Rev.* **2012**, *249*, 104–115. [CrossRef] [PubMed]

137. Pearce, E.L.; Pearce, E.J. Metabolic pathways in immune cell activation and quiescence. *Immunity* **2013**, *38*, 633–643. [CrossRef] [PubMed]

138. Van der Windt, G.J.; Pearce, E.L. Metabolic switching and fuel choice during T cell differentiation and memory development. *Immunol. Rev.* **2012**, *249*, 27–42. [CrossRef] [PubMed]

139. Chang, C.-H.; Curtis, J.D.; Maggi, L.B., Jr.; Faubert, B.; Villarino, A.V.; O'Sullivan, D.; Huang, S.C.-C.; van der Windt, G.J.W.; Blagih, J.; Qiu, J.; et al. Posttranscriptional control of T cell effector function by aerobic glycolysis. *Cell* **2013**, *153*, 1239–1251. [CrossRef] [PubMed]

140. Almeida, L.; Lochner, M.; Berod, L.; Sparwasser, T. Metabolic pathways in T cell activation and lineage differentiation. *Semin. Immunol.* **2016**, *28*, 514–524. [CrossRef] [PubMed]

141. Vaeth, M.; Maus, M.; Klein-Hessling, S.; Freinkman, E.; Yang, J.; Eckstein, M.; Cameron, S.; Turvey, S.E.; Serfling, E.; Berberich-Siebelt, F.; et al. Store-operated Ca2+ entry controls clonal expansion of T cells through metabolic reprogramming. *Immunity* **2017**, *47*, 664–679. [CrossRef] [PubMed]

142. Feske, S.; Skolnik, E.Y.; Prakriya, M. Ion channels and transporters in lymphocyte function and immunity. *Nat. Rev. Immunol.* **2012**, *12*, 532–547. [CrossRef] [PubMed]

143. Tamás, P.; Hawley, S.A.; Clarke, R.G.; Mustard, K.J.; Green, K.; Hardie, D.G.; Cantrell, D.A. Regulation of the energy sensor AMP-activated protein kinase by antigen receptor and Ca2+ in T lymphocytes. *J. Exp. Med.* **2006**, *203*, 1665–1670. [CrossRef] [PubMed]

144. Ma, E.H.; Poffenberger, M.C.; Wong, A.H.; Jones, R.G. The role of AMPK in T cell metabolism and function. *Curr. Opin. Immunol.* **2017**, *46*, 45–52. [CrossRef] [PubMed]

145. Huang, J.; Manning, B.D. The TSC1–TSC2 complex: A molecular switchboard controlling cell growth. *Biochem. J.* **2008**, *412*, 179–190. [CrossRef] [PubMed]

146. Chi, H. Regulation and function of mTOR signalling in T cell fate decision. *Nat. Rev. Immunol.* **2012**, *12*, 325–338. [CrossRef] [PubMed]

147. MacIver, N.J.; Blagih, J.; Saucillo, D.C.; Tonelli, L.; Griss, T.; Rathmell, J.C.; Jones, R.G. The liver kinase B1 is a central regulator of T cell development, activation, and metabolism. *J. Immunol.* **2011**, *187*, 4187–4198. [CrossRef] [PubMed]

148. Kirichok, Y.; Krapivinsky, G.; Clapham, D.E. The mitochondrial calcium uniporter is a highly selective ion channel. *Nature* **2004**, *427*, 360–364. [CrossRef] [PubMed]

149. Gilabert, J.A.; Bakowski, D.; Parekh, A.B. Energized mitochondria increase the dynamic range over which inositol 1,4,5-trisphosphate activates store-operated calcium influx. *EMBO J.* **2001**, *20*, 2672–2679. [CrossRef] [PubMed]

150. Gilabert, J.A.; Parekh, A.B. Respiring mitochondria determine the pattern of activation and inactivation of the store-operated Ca^{2+} current I (CRAC). *EMBO J.* **2000**, *19*, 6401–6407. [CrossRef] [PubMed]

151. Singaravelu, K.; Nelson, C.; Bakowski, D.; de Brito, O.M.; Ng, S.W.; di Capite, J.; Powell, T.; Scorrano, L.; Parekh, A.B. Mitofusin 2 regulates STIM1 migration from the Ca^{2+} store to the plasma membrane in cells with depolarized mitochondria. *J. Biol. Chem.* **2011**, *286*, 12189–12201. [CrossRef] [PubMed]

152. Dimeloe, S.; Burgener, A.V.; Grahlert, J.; Hess, C. T cell metabolism governing activation, proliferation and differentiation; a modular view. *Immunology* **2017**, *150*, 35–44. [CrossRef] [PubMed]

153. Jouaville, L.S.; Pinton, P.; Bastianutto, C.; Rutter, G.A.; Rizzuto, R. Regulation of mitochondrial ATP synthesis by calcium: Evidence for a long-term metabolic priming. *Proc. Natl. Acad. Sci. USA* **1999**, *96*, 13807–13812. [CrossRef] [PubMed]

154. Ho, P.C.; Bihuniak, J.D.; Macintyre, A.N.; Staron, M.; Liu, X.; Amezquita, R.; Tsui, Y.C.; Cui, G.; Micevic, G.; Perales, J.C.; et al. Phosphoenolpyruvate is a metabolic checkpoint of anti-tumor T cell responses. *Cell* **2015**, *162*, 1217–1228. [CrossRef] [PubMed]

155. Rumi-Masante, J.; Rusinga, F.I.; Lester, T.E.; Dunlap, T.B.; Williams, T.D.; Dunker, A.K.; Weis, D.D.; Creamer, T.P. Structural basis for activation of calcineurin by calmodulin. *J. Mol. Biol.* **2012**, *415*, 307–317. [CrossRef] [PubMed]

156. Racioppi, L.; Means, A.R. Calcium/calmodulin-dependent protein kinase kinase 2: Roles in signaling and pathophysiology. *J. Biol. Chem.* **2012**, *287*, 31658–31665. [CrossRef] [PubMed]

157. Chang, C.-H.; Qiu, J.; O'Sullivan, D.; Buck, M.D.; Noguchi, T.; Curtis, J.D.; Chen, Q.; Gindin, M.; Gubin, M.M.; van der Windt, G.J.W.; et al. Metabolic competition in the tumor microenvironment is a driver of cancer progression. *Cell* **2015**, *162*, 1229–1241. [CrossRef] [PubMed]

158. Patsoukis, N.; Bardhan, K.; Chatterjee, P.; Sari, D.; Liu, B.; Bell, L.N.; Karoly, E.D.; Freeman, G.J.; Petkova, V.; Seth, P.; et al. PD-1 alters T cell metabolic reprogramming by inhibiting glycolysis and promoting lipolysis and fatty acid oxidation. *Nat. Commun.* **2015**, *6*, 6692. [CrossRef] [PubMed]

159. Gary-Gouy, H.; Sainz-Perez, A.; Marteau, J.-B.; Marfaing-Koka, A.; Delic, J.; Merle-Beral, H.; Galanaud, P.; Dalloul, A. Natural phosphorylation of CD5 in chronic lymphocytic leukemia B cells and analysis of CD5-regulated genes in a B cell line suggest a role for CD5 in malignant phenotype. *J. Immunol.* **2007**, *179*, 4335–4344. [CrossRef] [PubMed]

160. Palmer, M.J.; Mahajan, V.S.; Chen, J.; Irvine, D.J.; Lauffenburger, D.A. Signaling thresholds govern heterogeneity in IL-7-receptor-mediated responses of naive CD8+ T cells. *Immunol Cell. Biol.* **2011**, *89*, 581–594. [CrossRef] [PubMed]

161. Kipnis, J.; Gadani, S.; Derecki, N.C. Pro-cognitive properties of T cells. *Nat. Rev. Immunol.* **2012**, *12*, 663–669. [CrossRef] [PubMed]

162. Kipnis, J.; Cohen, H.; Cardon, M.; Ziv, Y.; Schwartz, M. T cell deficiency leads to cognitive dysfunction: Implications for therapeutic vaccination for schizophrenia and other psychiatric conditions. *Proc. Natl. Acad. Sci. USA* **2004**, *101*, 8180–8185. [CrossRef] [PubMed]

163. Brombacher, T.M.; Nono, J.K.; De Gouveia, K.S.; Makena, N.; Darby, M.; Womersley, J.; Tamgue, O.; Brombacher, F. IL-13–mediated regulation of learning and memory. *J. Immunol.* **2017**, *198*, 2681–2688. [CrossRef] [PubMed]

164. Oliveira-dos-Santos, A.J.; Matsumoto, G.; Snow, B.E.; Bai, D.; Houston, F.P.; Whishaw, I.Q.; Mariathasan, S.; Sasaki, T.; Wakeham, A.; Ohashi, P.S.; et al. Regulation of T cell activation, anxiety, and male aggression by RGS2. *Proc. Natl. Acad. Sci. USA* **2000**, *97*, 12272–12277. [CrossRef] [PubMed]

165. Filiano, A.J.; Gadani, S.P.; Kipnis, J. How and why do T cells and their derived cytokines affect the injured and healthy brain? *Nat. Rev. Neurosci.* **2017**, *18*, 375. [CrossRef] [PubMed]

166. Kyratsous, N.I.; Bauer, I.J.; Zhang, G.; Pesic, M.; Bartholomäus, I.; Mues, M.; Fang, P.; Wörner, M.; Everts, S.; Ellwart, J.W.; et al. Visualizing context-dependent calcium signaling in encephalitogenic T cells in vivo by two-photon microscopy. *Proc. Natl. Acad. Sci. USA* **2017**, *114*, E6381–E6389. [CrossRef] [PubMed]

167. Smedler, E.; Uhlén, P. Frequency decoding of calcium oscillations. *Biochim. Biophys. Acta Gen. Subj.* **2014**, *1840*, 964–969. [CrossRef] [PubMed]

168. Pesic, M.; Bartholomaus, I.; Kyratsous, N.I.; Heissmeyer, V.; Wekerle, H.; Kawakami, N. 2-photon imaging of phagocyte-mediated T cell activation in the CNS. *J. Clin. Investig.* **2013**, *123*, 1192–1201. [CrossRef] [PubMed]

169. de Bruin, N.M.W.J.; Schmitz, K.; Schiffmann, S.; Tafferner, N.; Schmidt, M.; Jordan, H.; Häußler, A.; Tegeder, I.; Geisslinger, G.; Parnham, M.J. Multiple rodent models and behavioral measures reveal unexpected responses to FTY720 and DMF in experimental autoimmune encephalomyelitis. *Behav. Brain Res.* **2016**, *300*, 160–174. [CrossRef] [PubMed]

170. Schub, D.; Janssen, E.; Leyking, S.; Sester, U.; Assmann, G.; Hennes, P.; Smola, S.; Vogt, T.; Rohrer, T.; Sester, M.; et al. Altered phenotype and functionality of varicella zoster virus–specific cellular immunity in individuals with active infection. *J. Infect. Dis.* **2015**, *211*, 600–612. [CrossRef] [PubMed]

171. Schub, D.; Fousse, M.; Faßbender, K.; Gärtner, B.C.; Sester, U.; Sester, M.; Schmidt, T. CTLA-4-expression on VZV-specific T cells in CSF and blood is specifically increased in patients with VZV related central nervous system infections. *Eur. J. Immunol.* **2018**, *48*, 151–160. [CrossRef] [PubMed]

172. Koebel, C.M.; Vermi, W.; Swann, J.B.; Zerafa, N.; Rodig, S.J.; Old, L.J.; Smyth, M.J.; Schreiber, R.D. Adaptive immunity maintains occult cancer in an equilibrium state. *Nature* **2007**, *450*, 903–907. [CrossRef] [PubMed]

173. Mattes, J.; Hulett, M.; Xie, W.; Hogan, S.; Rothenberg, M.E.; Foster, P.; Parish, C. Immunotherapy of cytotoxic T cell-resistant tumors by T helper 2 cells: An eotaxin and STAT6-dependent process. *J. Exp. Med.* **2003**, *197*, 387–393. [CrossRef] [PubMed]

174. Hung, K.; Hayashi, R.; Lafond-Walker, A.; Lowenstein, C.; Pardoll, D.; Levitsky, H. The central role of CD4+ T cells in the antitumor immune response. *J. Exp. Med.* **1998**, *188*, 2357–2368. [CrossRef] [PubMed]

175. Scholler, J.; Brady, T.L.; Binder-Scholl, G.; Hwang, W.T.; Plesa, G.; Hege, K.M.; Vogel, A.N.; Kalos, M.; Riley, J.L.; Deeks, S.G.; et al. Decade-long safety and function of retroviral-modified chimeric antigen receptor T cells. *Sci. Transl. Med.* **2012**, *4*, 132ra153. [CrossRef] [PubMed]

176. Ho, Y.C.; Shan, L.; Hosmane, N.N.; Wang, J.; Laskey, S.B.; Rosenbloom, D.I.S.; Lai, J.; Blankson, J.N.; Siliciano, J.D.; Siliciano, R.F. Replication-competent noninduced proviruses in the latent reservoir increase barrier to HIV-1 cure. *Cell* **2013**, *155*, 540–551. [CrossRef] [PubMed]

177. Huetter, G.; Nowak, D.; Mossner, M.; Ganepola, S.; Muessig, A.; Allers, K.; Schneider, T.; Hofmann, J.; Kuecherer, C.; Blau, O.; et al. Long-Term Control of HIV by CCR5 Δ32/Δ32 Stem-Cell Transplantaion. *N. Engl. J. Med.* **2009**, *360*, 692–698. [CrossRef] [PubMed]

178. Ahmadzadeh, M.; Johnson, L.A.; Heemskerk, B.; Wunderlich, J.R.; Dudley, M.E.; White, D.E.; Rosenberg, S.A. Tumor antigen-specific CD8 T cells infiltrating the tumor express high levels of PD-1 and are functionally impaired. *Blood* **2009**, *114*, 1537–1544. [CrossRef] [PubMed]

179. Baitsch, L.; Baumgaertner, P.; Devevre, E.; Raghav, S.K.; Legat, A.; Barba, L.; Wieckowski, S.; Bouzourene, H.; Deplancke, B.; Romero, P.; et al. Exhaustion of tumor-specific CD8+ T cells in metastases from melanoma patients. *J. Clin. Investig.* **2011**, *121*, 2350–2360. [CrossRef] [PubMed]

180. Staveley-O'Carroll, K.; Sotomayor, E.; Montgomery, J.; Borrello, I.; Hwang, L.; Fein, S.; Pardoll, D.; Levitsky, H. Induction of antigen-specific T cell anergy: An early event in the course of tumor progression. *Proc. Natl. Acad. Sci. USA* **1998**, *95*, 1178–1183. [CrossRef] [PubMed]

181. Rosenberg, S.A.; Yang, J.C.; Sherry, R.M.; Kammula, U.S.; Hughes, M.S.; Phan, G.Q.; Citrin, D.E.; Restifo, N.P.; Robbins, P.F.; Wunderlich, J.R.; et al. Durable complete responses in heavily pretreated patients with metastatic melanoma using T cell transfer immunotherapy. *Clin. Cancer Res.* **2011**, *17*, 4550–4557. [CrossRef] [PubMed]

182. Dudley, M.E.; Wunderlich, J.R.; Robbins, P.F.; Yang, J.C.; Hwu, P.; Schwartzentruber, D.J.; Topalian, S.L.; Sherry, R.; Restifo, N.P.; Hubicki, A.M; et al. Cancer regression and autoimmunity in patients after clonal repopulation with antitumor lymphocytes. *Science* **2002**, *298*, 850–854. [CrossRef] [PubMed]

183. Postow, M.A.; Callahan, M.K.; Wolchok, J.D. Immune checkpoint blockade in cancer therapy. *J. Clin. Oncol.* **2015**, *33*, 1974–1982. [CrossRef] [PubMed]

184. Wei, S.C.; Levine, J.H.; Cogdill, A.P.; Zhao, Y.; Anang, N.-A.A.S.; Andrews, M.C.; Sharma, P.; Wang, J.; Wargo, J.A.; Pe'er, D.; et al. Distinct cellular mechanisms underlie anti-CTLA-4 and anti-PD-1 checkpoint blockade. *Cell* **2017**, *170*, 1120–1133.e17. [CrossRef] [PubMed]

185. Barbee, M.S.; Ogunniyi, A.; Horvat, T.Z.; Dang, T.O. Current status and future directions of the immune checkpoint inhibitors ipilimumab, pembrolizumab, and nivolumab in oncology. *Ann. Pharmacother.* **2015**, *49*, 907–937. [CrossRef] [PubMed]

186. Sangro, B.; Gomez-Martin, C.; de la Mata, M.; Inarrairaegui, M.; Garralda, E.; Barrera, P.; Riezu-Boj, J.I.; Larrea, E.; Alfaro, C.; Sarobe, P.; et al. A clinical trial of CTLA-4 blockade with tremelimumab in patients with hepatocellular carcinoma and chronic hepatitis C. *J. Hepatol.* **2013**, *59*, 81–88. [CrossRef] [PubMed]

187. Reck, M.; Rodríguez-Abreu, D.; Robinson, A.G.; Hui, R.; Csőszi, T.; Fülöp, A.; Gottfried, M.; Peled, N.; Tafreshi, A.; Cuffe, S.; et al. Pembrolizumab versus chemotherapy for PD-L1–positive non–small-cell lung cancer. *N. Engl. J. Med.* **2016**, *375*, 1823–1833. [CrossRef] [PubMed]

188. Hersey, P.; Gowrishankar, K. Pembrolizumab joins the anti-PD-1 armamentarium in the treatment of melanoma. *Future Oncol.* **2015**, *11*, 133–140. [CrossRef] [PubMed]

189. Rooke, R. Can calcium signaling be harnessed for cancer immunotherapy? *Biochim. Biophys. Acta Mol. Cell Res.* **2014**, *1843*, 2334–2340. [CrossRef] [PubMed]

190. Ernst, B.; Lee, D.S.; Chang, J.M.; Sprent, J.; Surh, C.D. The peptide ligands mediating positive selection in the thymus control T cell survival and homeostatic proliferation in the periphery. *Immunity* **1999**, *11*, 173–181. [CrossRef]

191. Smith, K.; Seddon, B.; Purbhoo, M.A.; Zamoyska, R.; Fisher, A.G.; Merkenschlager, M. Sensory adaptation in naive peripheral CD4 T cells. *J. Exp. Med.* **2001**, *194*, 1253–1261. [CrossRef] [PubMed]

192. Dorothée, G.; Vergnon, I.; El Hage, F.; Chansac, B.L.M.; Ferrand, V.; Lécluse, Y.; Opolon, P.; Chouaib, S.; Bismuth, G.; Mami-Chouaib, F. In situ sensory adaptation of tumor-infiltrating T lymphocytes to peptide-MHC levels elicits strong antitumor reactivity. *J. Immunol.* **2005**, *174*, 6888–6897. [CrossRef] [PubMed]

193. Friedlein, G.; El Hage, F.; Vergnon, I.; Richon, C.; Saulnier, P.; Lecluse, Y.; Caignard, A.; Boumsell, L.; Bismuth, G.; Chouaib, S.; et al. Human CD5 protects circulating tumor antigen-specific CTL from tumor-mediated activation-induced cell death. *J. Immunol.* **2007**, *178*, 6821–6827. [CrossRef] [PubMed]

194. Axtell, R.C.; Webb, M.S.; Barnum, S.R.; Raman, C. Cutting edge: Critical role for CD5 in experimental autoimmune encephalomyelitis: Inhibition of engagement reverses disease in mice. *J. Immunol.* **2004**, *173*, 2928–2932. [CrossRef] [PubMed]

195. Simoes, I.T.; Aranda, F.; Carreras, E.; Andres, M.V.; Casado-Llombart, S.; Martinez, V.G.; Lozano, F. Immunomodulatory effects of soluble CD5 on experimental tumor models. *Oncotarget* **2017**, *8*, 108156–108169. [CrossRef] [PubMed]

196. Tabbekh, M.; Franciszkiewicz, K.; Haouas, H.; Lecluse, Y.; Benihoud, K.; Raman, C.; Mami-Chouaib, F. Rescue of tumor-infiltrating lymphocytes from activation-induced cell death enhances the antitumor CTL response in CD5-deficient mice. *J. Immunol.* **2011**, *187*, 102–109. [CrossRef] [PubMed]

197. Round, J.L.; Mazmanian, S.K. The gut microbiome shapes intestinal immune responses during health and disease. *Nat. Rev. Immunol.* **2009**, *9*, 313–323. [CrossRef] [PubMed]

198. Vétizou, M.; Pitt, J.M.; Daillère, R.; Lepage, P.; Waldschmitt, N.; Flament, C.; Rusakiewicz, S.; Routy, B.; Roberti, M.P.; Duong, C.P.M.; et al. Anticancer immunotherapy by CTLA-4 blockade relies on the gut microbiota. *Science* **2015**, *350*, 1079–1084. [CrossRef] [PubMed]

199. Botticelli, A.; Zizzari, I.; Mazzuca, F.; Ascierto, P.A.; Putignani, L.; Marchetti, L.; Napoletano, C.; Nuti, M.; Marchetti, P. Cross-talk between microbiota and immune fitness to steer and control response to anti PD-1/PDL-1 treatment. *Oncotarget* **2017**, *8*, 8890–8899. [CrossRef] [PubMed]

200. Kosiewicz, M.M.; Dryden, G.W.; Chhabra, A.; Alard, P. Relationship between gut microbiota and development of T cell associated disease. *FEBS Lett.* **2014**, *588*, 4195–4206. [CrossRef] [PubMed]

201. Lathrop, S.K.; Bloom, S.M.; Rao, S.M.; Nutsch, K.; Lio, C.-W.; Santacruz, N.; Peterson, D.A.; Stappenbeck, T.S.; Hsieh, C.-S. Peripheral education of the immune system by colonic commensal microbiota. *Nature* **2011**, *478*, 250. [CrossRef] [PubMed]

202. Gonzalez-Perez, G.; Lamousé-Smith, E.S.N. Gastrointestinal microbiome dysbiosis in infant mice alters peripheral CD8+ T cell receptor signaling. *Front. Immunol.* **2017**, *8*, 265. [CrossRef] [PubMed]

203. Huang, T.; Wei, B.; Velazquez, P.; Borneman, J.; Braun, J. Commensal microbiota alter the abundance and TCR responsiveness of splenic naïve CD4+ T lymphocytes. *Clin. Immunol.* **2005**, *117*, 221–230. [CrossRef] [PubMed]

204. Bazett, M.; Bergeron, M.-E.; Haston, C.K. Streptomycin treatment alters the intestinal microbiome, pulmonary T cell profile and airway hyperresponsiveness in a cystic fibrosis mouse model. *Sci. Rep.* **2016**, *6*, 19189. [CrossRef] [PubMed]

205. Ochoa-Repáraz, J.; Mielcarz, D.W.; Haque-Begum, S.; Kasper, L.H. Induction of a regulatory B cell population in experimental allergic encephalomyelitis by alteration of the gut commensal microflora. *Gut Microbes* **2010**, *1*, 103–108. [CrossRef] [PubMed]

206. Allison, J.P.; Krummel, M.F. The Yin and Yang of T cell costimulation. *Science* **1995**, *270*, 932–933. [CrossRef] [PubMed]

207. Allison, J.P. Checkpoints. *Cell* **2015**, *152*, 1202–1205. [CrossRef] [PubMed]

208. Sivan, A.; Corrales, L.; Hubert, N.; Williams, J.B.; Aquino-Michaels, K.; Earley, Z.M.; Benyamin, F.W.; Lei, Y.M.; Jabri, B.; Alegre, M.-L.; et al. Commensal Bifidobacterium promotes antitumor immunity and facilitates anti–PD-L1 efficacy. *Science* **2015**, *350*, 1084–1089. [CrossRef] [PubMed]

International Journal of
Molecular Sciences

MDPI

Review

The Role of Endothelial Ca^{2+} Signaling in Neurovascular Coupling: A View from the Lumen

Germano Guerra [1], Angela Lucariello [2], Angelica Perna [1], Laura Botta [3], Antonio De Luca [2] and Francesco Moccia [3],*

[1] Department of Medicine and Health Sciences "Vincenzo Tiberio", University of Molise, via F. De Santis, 86100 Campobasso, Italy; germano.guerra@unimol.it (G.G.); angelicaperna@gmail.com (A.P.)

[2] Department of Mental Health and Preventive Medicine, Section of Human Anatomy, University of Campania "L. Vanvitelli", 81100 Naples, Italy; angela.lucariello@gmail.com (A.L.); antonio.deluca@unicampania.it (A.D.L.)

[3] Laboratory of General Physiology, Department of Biology and Biotechnology "L. Spallanzani", University of Pavia, via Forlanini 6, 27100 Pavia, Italy; laura.botta@unipv.it

* Correspondence: francesco.moccia@unipv.it; Tel.: +39-382-987-619

Received: 23 February 2018; Accepted: 17 March 2018; Published: 21 March 2018

Abstract: Background: Neurovascular coupling (NVC) is the mechanism whereby an increase in neuronal activity (NA) leads to local elevation in cerebral blood flow (CBF) to match the metabolic requirements of firing neurons. Following synaptic activity, an increase in neuronal and/or astrocyte Ca^{2+} concentration leads to the synthesis of multiple vasoactive messengers. Curiously, the role of endothelial Ca^{2+} signaling in NVC has been rather neglected, although endothelial cells are known to control the vascular tone in a Ca^{2+}-dependent manner throughout peripheral vasculature. Methods: We analyzed the literature in search of the most recent updates on the potential role of endothelial Ca^{2+} signaling in NVC. Results: We found that several neurotransmitters (i.e., glutamate and acetylcholine) and neuromodulators (e.g., ATP) can induce dilation of cerebral vessels by inducing an increase in endothelial Ca^{2+} concentration. This, in turn, results in nitric oxide or prostaglandin E2 release or activate intermediate and small-conductance Ca^{2+}-activated K$^+$ channels, which are responsible for endothelial-dependent hyperpolarization (EDH). In addition, brain endothelial cells express multiple transient receptor potential (TRP) channels (i.e., TRPC3, TRPV3, TRPV4, TRPA1), which induce vasodilation by activating EDH. Conclusions: It is possible to conclude that endothelial Ca^{2+} signaling is an emerging pathway in the control of NVC.

Keywords: neurovascular coupling; neuronal activity; brain endothelial cells; Ca^{2+} signaling; glutamate; acetylcholine; ATP; nitric oxide; endothelial-dependent hyperpolarization; TRP channels

1. Introduction

The brain comprises only 2% of the total body mass, yet it accounts for 20% of the overall energy metabolism [1]. As the brain has a limited capacity to store energy and lacks survival mechanisms that render other organs, such as heart and liver, more tolerant to short periods of anoxia or ischemia, the continuous supply of oxygen (O$_2$) and nutrients and removal of catabolic waste are critical to maintain neuronal integrity and overall brain function. Accordingly, the brain receives up to 20% of cardiac output and consumes ≈20% and ≈25% of total body's O$_2$ and glucose [2–4]. Brain functions cease within seconds after the interruption of cerebral blood flow (CBF), while irreversible neuronal injury occurs within minutes [2,3]. Cerebral autoregulation is the mechanism whereby CBF remains relatively stable in spite of physiological fluctuations in arterial pressure, at least within a certain range. Thus, cerebral arteries constrict in response to an increase in arterial pressure and relax upon a decrease in blood pressure [4,5]. A subtler mechanism, known as functional hyperemia or neurovascular coupling (NVC), intervenes to locally increase the rate of CBF to active brain areas, thereby ensuring adequate matching between the enhanced metabolic needs of neural cells and blood supply [2,4,6]. Through NVC, a local elevation in neuronal activity (NA) causes a significant vasodilation of neighboring microvessels, which increases CBF and generates the blood oxygenation level-dependent (BOLD) signals that are used to monitor brain function through functional magnetic resonance imaging [7,8]. NVC is finely orchestrated by an intercellular signaling network comprised of neurons, astrocytes and vascular cells (endothelial cells, smooth muscle cells and pericytes), which altogether form the neurovascular unit (NVU) [4,6,9]. Synaptically-activated neurons may signal to adjacent vessels either directly or through the interposition of glial cells. Recent evidence has shown that astrocytes modulate NVC at arteriole levels, whereas they mediate neuronal-to-vascular communication at capillaries [2,4,10–14]. NA controls cerebrovascular tone through a number of Ca^{2+}-dependent vasoactive mediators, which regulate the contractile state of either vascular smooth muscle cells (VSMCs) (arteries and arterioles) or pericytes (capillaries). These include nitric oxide (NO) and the arachidonic acid (AA) derivatives, prostaglandin E2 (PGE2), epoxyeicosatrienoic acids (EETs) and (20-HETE) [2,4,9]. While the role played by astrocytes and mural cells, i.e., VSMCs and pericytes, in CBF regulation has been extensively investigated, the contribution of microvascular endothelial cells to NVC has been largely underestimated [7]. This is quite surprising as the endothelium regulates the vascular tone in both systemic and pulmonary circulation by lining the innermost layer of all blood vessels [15–17]. Endothelial cells decode a multitude of chemical (e.g., transmitters and autacoids) and physical (e.g., shear stress pulsatile stretch) signals by generating spatio-temporally-patterned intracellular Ca^{2+} signals, which control VSMC contractility by inducing the synthesis and release of the vasorelaxing factors, NO, prostacyclin (or PGI2), carbon monoxide and hydrogen sulfide, as well as the vasoconstricting prostaglandin H2 and thromboxane A2 [15,18–21]. Moreover, an increase in sub-membranal Ca^{2+} concentration stimulates intermediate- and small-conductance Ca^{2+}-activated K^+ channels (IK_{Ca} and SK_{Ca}, respectively), thereby causing an abrupt hyperpolarization in endothelial membrane potential, which is electronically transmitted to adjacent VSMCs and causes vessel dilation [16,22,23]. Herein, we aim at surveying the role of endothelial Ca^{2+} signaling in NVC by examining the physiological stimuli, e.g., neurotransmitters, neuromodulators and dietary molecules, that modulate CBF through an increase in intracellular Ca^{2+} concentration ($[Ca^{2+}]_i$) in brain microvascular endothelial cells.

2. Neurovascular Coupling

2.1. Cerebral Circulation and the Neurovascular Unit

The concept of NVU was first introduced during the first Stroke Progress Review Group meeting of the National Institute of Neurological Disorders and Stroke of the NIH (July 2001) to highlight the relevance to CBF regulation of the intimate association between neurons and cerebral vessels [4]. The NVU represents a functional unit where neurons, interneurons and astrocytes are in close proximity and are functionally coupled to smooth muscle cells, pericytes, endothelial cells and extracellular matrix (Figure 1). Within the NVU, neurons, glial and vascular cells establish a mutual influence on each other to ensure a highly efficient system to match blood and nutrient supply to the local needs of brain cells [2,4,6]. The interaction between neurons, glial cells and blood vessels may, however, vary along the cerebrovascular tree. Large cerebral arteries arise from Circle of Willis at the base of the brain and give rise to a heavily-interconnected network of pial arteries and arterioles, which run along the cortical surface. Cerebral vessels are lined by a single layer of endothelial cells (*tunica intima*), which is separated from an intermediate layer of smooth muscle cells (*tunica media*) by the internal elastic lamina. An outermost layer mainly comprised of collagen fibers and fibroblasts and enriched in perivascular nerves, known *as tunica adventitia*, is separated from the brain by the Virchow–Robin space, which is an extension of the sub-arachnoid space [4,6]. Penetrating arterioles branch off smaller pial arteries and dive into the brain tissue, thereby giving rise to parenchymal arterioles. The muscular component of the vascular wall is reduced to a single layer of VSMCs in intracerebral arterioles: of note, endothelial cells may project through the internal elastic lamina by establishing direct connections, known as myoendothelial projections, with the adjoining smooth muscle cells. Myoendothelial projections are enriched with gap junctions and allow the bidirectional transfer of information between endothelial cells and VSMCs [24]. While penetrating arterioles are still separated from the substance of the brain by the Virchow–Robin space, perivascular astrocytic processes (i.e., end feet) enter in contact and fuse together with the basal lamina of parenchymal arterioles, so that the perivascular space is obliterated [4,6]. Relevant to local CBF regulation, pyramidal cells and γ-aminobutyric acid (GABA) interneurons provide extensive innervation to parenchymal arterioles, often with the interposition of glial cells [4,10,25–28]. In addition, sub-cortical neurons from locus coeruleus, raphe nucleus and basal forebrain may also send projection fibers, containing respectively acetylcholine, norepinephrine and 5-hydroxytryptamine (5-HT), to intracortical microvessels and surrounding astrocytes [29]. Finally, parenchymal arterioles supply the cerebral circulation by giving rise to a dense network of intercommunicating capillary vessels, which are composed only of specialized endothelial cells and lack VSMCs. However, ≈30% of the brain capillary surface is covered by spatially-isolated contractile cells, i.e., pericytes, while the remaining endothelial capillary tubes are almost entirely wrapped by astrocytic end feet [30,31]. Astrocytes and pericytes located outside of brain capillaries may be innervated by local neurons, as described for astrocytes and VSMCs situated in the upper districts of the vascular tree [2]. The average capillary density of human brain amounts to ≈400 capillaries mm^{-2}, which is enough to ensure that each neuron is endowed with its own capillary and to reduce on average to less than 15 μM the diffusion distance for O_2, nutrients and catabolic waste [32,33]. Therefore, brain microvascular endothelial cells are ideally positioned to sense neurotransmitters released by axonal terminals during local synaptic activity and by sub-cortical projections provided that they express their specific membrane receptors. The following increase in $[Ca^{2+}]_i$ could, in turn, drive the synthesis of Ca^{2+}-dependent vasoactive mediators. At the same time, brain microvascular endothelial cells have the potential to regulate local CBF independently of NA by detecting changes in blood flow variations and in the concentration of blood-borne agonists through distinct patterns of intracellular Ca^{2+} elevation.

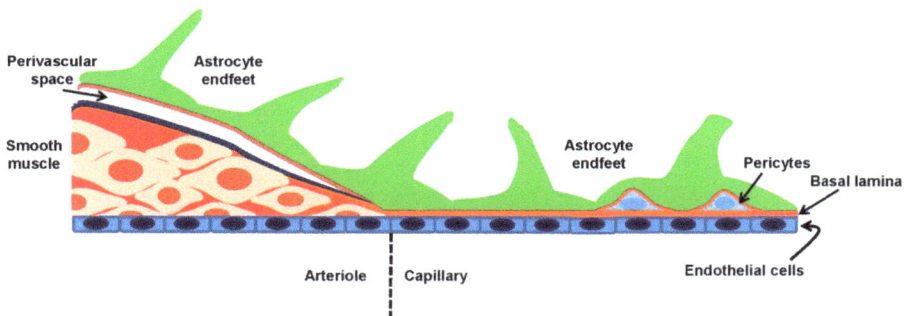

Figure 1. Cellular composition of the neurovascular unit. The vascular wall presents a different structure in arterioles and capillaries, which control the local supply of cerebral blood. Smooth muscle cells form one or more continuous layers around arterioles and change in their contractile state determine vessel diameter and regulate blood perfusion. Capillary diameter is regulated by contractile pericytes, which extend longitudinally and circumferentially along the capillary wall. Astrocyte end feet envelope arterioles and capillaries and are able to release vasoactive mediators, which regulate the contractile state of smooth muscle cells (arterioles) and pericytes (capillary) in response to neuronal activity.

2.2. Cellular and Biochemical Pathways of Neurovascular Coupling: The Role of Neurons and Astrocytes in Arterioles and Capillaries

The mechanisms whereby synaptic activity controls the microvascular tone vary depending on the brain structure and may differ along the vascular tree (Figure 2) [2,4,9]. An increase in intracellular Ca^{2+} concentration ($[Ca^{2+}]_i$) within the dendritic tree is the crucial signal that triggers the synthesis and release of vasoactive messengers in response to excitatory synaptic inputs [2,4,9,34]. For instance, glutamate stimulates post-synaptic N-methyl-D-aspartate (NMDA) and a-amino-3-hydroxy-5-methyl-4-isoxazol epropionic acid (AMPA) receptors to induce extracellular Ca^{2+} entry and recruit the Ca^{2+}/calmodulin (Ca^{2+}/CaM)-dependent neuronal nitric oxide (NO) synthase (nNOS) in hippocampal and cerebellar GABA interneurons [10,25,28,33,35]. The following NO release elicits arteriole vasodilation by inducing VSMC hyperpolarization and relaxation through a soluble guanylate cyclase/protein kinase G (PKG)-dependent mechanism (see below) [13,26,35–37]. Conversely, NMDA receptor (NMDAR)-mediated Ca^{2+} entry in synaptically-activated pyramidal neurons of the somatosensory cortex engages cyclooxygenase 2, which catalyzes the synthesis of the powerful vasodilator, prostaglandin E2 (PGE2), which acts through EP2 and EP4 receptors on VSMCs [27,28]. Intriguingly, NO plays a permissive role in PGE2-dependent vasodilation by maintaining the hemodynamic response during sustained NA [25,38]. Long-lasting synaptic activity could lead to an increase in $[Ca^{2+}]_i$ within perisynaptic astrocytic processes (Figure 2), which lags behind the onset of CBF, but is able to activate phospholipase A2 (PLA2) and cleave AA from the plasma membrane. AA, in turn, diffuses to adjacent VSMCs and is converted into the vasoconstricting messenger, 20-hydroxyeicosatetraenoic (20-HETE), by cytochrome P450 4A (CYP4A) [33,38]. However, neuronal-derived NO inhibits CYP4A, thereby preventing 20-HETE formation and maintaining PGE2-dependent vasodilation [9,33,38]. The mechanism(s) whereby NA induces astrocytic Ca^{2+} signals is still unclear. Glutamate has been predicted to increase astrocyte $[Ca^{2+}]_i$ by binding to metabotropic glutamate receptors (mGluRs) 1 and 5 (mGluR1 and mGluR5, respectively), which stimulate phospholipase Cβ (PLCβ) through Gqα monomer and induce inositol-1,4,5-trisphosphate (InsP$_3$)-dependent Ca^{2+} release from the endoplasmic reticulum (ER) [9,39]. However, mGluR1 and mGluR5 are lacking in adult astrocytes, and the genetic deletion of type 2 InsP$_3$ receptor (InsP$_3$R2), which represents the primary InsP$_3$R isoform in glial cells, does not prevent NVC [2,30,39]. Nevertheless, there is indisputable evidence that astrocytes require mGluRs to drive the hemodynamic response to sensory stimulation also in adult mice [24,40,41]. Furthermore, alternative mechanisms may drive astrocyte Ca^{2+} signaling, including NMDARs, purinergic receptors and multiple transient receptor potential (TRP) channels [14,39]. We refer the reader to a number of exhaustive and recent reviews about the controversial role of astrocyte $[Ca^{2+}]_i$ in NVC [14,30,39,42]. Intriguingly, astrocytes are able to sense transmural pressure across the vascular wall through vanilloid TRP 4 (TRPV4) channels [43,44], which are located on their perivascular end feet. It has, therefore, been proposed that the initial hemodynamic response to NA activates these mechanosensitive Ca^{2+}-permeable channels, thereby causing an increase in astrocyte $[Ca^{2+}]_i$ and recruiting the Ca^{2+}-dependent PLA2 [14]. In addition, synaptically-released ATP could mobilize ER Ca^{2+} by activating P2Y2 and P2Y4 receptors, which are located on astrocytic end feet wrapped around cerebral vessels [45].

Although it has long been thought that CBF regulation occurs at the arteriole level [6], recent studies have convincingly shown that most of the hydraulic resistance that must be decreased in order to increase cortical perfusion is located in the capillary bed, which are wrapped by contractile pericytes [30,46,47]. This model makes physiological sense as, on average, firing neurons are remarkably closer to capillaries than to arterioles (8–23 μm away versus 70–160 μm) [47]. Therefore, capillaries are located in a much more suitable position to rapidly detect NA and initiate the hemodynamic response that ultimately generates BOLD signals. In addition, the regulation of CBF at the capillary level could selectively increase CBF only in active areas, thereby finely matching the local tissue O_2 supply to local cerebral demand [30,46,47]. The cerebellar cortex is composed of three layers: molecular layer, Purkinje cells and granular layer, respectively, from outermost to innermost [48]. Recent studies demonstrated that, at the molecular layer, synaptic activity caused an increase in astrocytic end feet Ca^{2+} concentration by inducing Ca^{2+} entry through P2X1 channels. This spatially-restricted Ca^{2+} signal, in turn, recruited phospholipase D2 (PLD2) and diacylglycerol lipase to synthesize AA, which was then metabolized by cyclooxygenase 1 into PGE2. Finally, PGE2 evoked capillary dilation by binding to EP4 receptors, which were presumably located on pericytes [13]. Conversely, in the granular layer, synaptic activity induced robust NO release by promoting NMDARs-mediated Ca^{2+} entry in granule cells without astrocyte involvement [11].

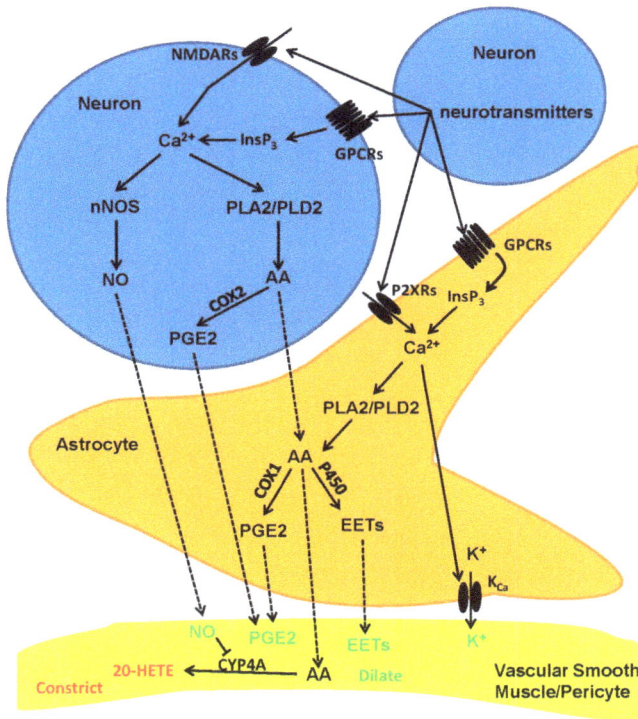

Figure 2. The mechanisms by which neurons and astrocytes stimulate arteriole and capillary dilation in response to synaptic activity. Synaptic activity increases intracellular Ca^{2+} levels within the postsynaptic neuron by stimulating metabotropic G-protein-coupled receptors (GPCRs), ionotropic receptors (e.g., NMDARs) or L-type voltage-operated Ca^{2+} channels (VOCs). This increase in $[Ca^{2+}]_i$ leads to the synthesis of NO and PGE2, which may relax both smooth muscle cells (arterioles) and pericytes (capillaries). Synaptically-released neurotransmitters may also increase $[Ca^{2+}]_i$ in perisynaptic astrocytes, thereby triggering NO release and PGE2/EET production. AA, which may be synthesized by PLA2 in both neurons and astrocytes, may be converted in the vasoconstricting factor, 20-HETE, in perivascular cells. Abbreviations: 20-HETE: 20-hydroxyeicosatetraenoic acid; AA: arachidonic acid; COX1: cyclooxygenase 1; COX2: cyclooxygenase 2; EETs: epoxyeicosatrienoic acids; GPCRs: G-protein-coupled receptors; InsP$_3$: inositol-1,4,5-trisphosphate; K$_{Ca}$: Ca^{2+}-activated intermediate and small conductance K$^+$ channels; NO: nitric oxide; nNOS: neuronal NO synthase; P450: cytochrome P450; PGE2: prostaglandin E2; PLA2: phospholipase A2; PLD2: phospholipase D2; VOCs: L-type voltage-operated Ca^{2+} channels.

2.3. Is There a Role for Endothelial Ca^{2+} Signaling in Neurovascular Coupling?

Vascular endothelial cells control vascular tone by releasing a myriad of vasoactive mediators in response to an increase in $[Ca^{2+}]_i$ evoked by either chemical or mechanical inputs [16,18–20,49]. For instance, endothelial Ca^{2+} signals trigger the selective increase in blood flow to skeletal, respiratory and cardiac muscles induced by physical exercise/training [15]. Curiously, it is still unclear whether and how endothelial Ca^{2+} signals play any role in translating NA into vasoactive signals in the brain [7,24]. Luminal perfusion of neurotransmitters and neuromodulators, such as acetylcholine, ATP, ADP and bradykinin, results in the dilation of cerebral arterioles by inducing the activation of specific Gq-protein-coupled receptors (GPCRs), which are located on the endothelial membrane [29,50–53]. GPCRs stimulate PLCβ to synthesize the ER Ca^{2+}-releasing messenger, InsP3, thereby triggering the Ca^{2+}-mediated signaling cascade that leads to the synthesis and release of most endothelial-derived vasoactive messengers [16,18–20,49,54]. Nevertheless, only scarce information is available about the role of endothelial Ca^{2+} signaling in the hemodynamic response to NA. As described elsewhere [50], synaptically-activated neurons and/or astrocytes could directly stimulate brain microvessels by releasing mediators that traverse the VSMC or pericyte layers and bind to receptors located on the abluminal side of endothelium. Interestingly, brain microvascular endothelial cells may also express NMDARs [55,56] and mGluR1 [57–60], although their potential contribution to NVC has been barely appreciated [4]. Accordingly, recent studies demonstrated that endothelial NMDARs participate in glutamate-induced vasodilation of intraparenchymal arterioles by mediating Ca^{2+} entry and subsequent eNOS activation [55,56]. Finally, brain endothelium could actively mediate the retrograde propagation of the vasodilation signal from the capillaries feeding the sites of NA to the upstream pial vessels (arteries and arterioles) that supply the activated area [6,7,61]. Retrograde vasodilation into the proximal arterial supply is required to achieve an optimally-localized increase in blood flow to active neurons and to avoid a "flow steal" from interconnected vascular networks [4,62]. Retrograde propagation in peripheral vessels is accomplished by endothelial Ca^{2+} signals that impinge on two distinct components to drive the conducted vasomotor response: (1) the Ca^{2+}-dependent rapid activation of EDH, which is restricted to stimulated endothelial cells, but is rapidly transmitted to more remote sites along the vessel and (2) a slower intercellular Ca^{2+} wave that spreads vasodilation via endothelial release of NO and PGE2 [7,63,64]. In the following section, we will discuss the evidence supporting the contribution of endothelial Ca^{2+} signals in NVC and the possibility that interendothelial Ca^{2+} waves are involved in the control of cerebrovessel diameters and microhemodynamics during intense synaptic activity.

3. The Role of Endothelial Ca^{2+} Signaling in Neurovascular Coupling

3.1. Endothelial Ca^{2+} Signaling in Brief

It has long been known that an increase in endothelial $[Ca^{2+}]_i$ delivers the crucial signal to induce the synthesis of multiple vasoactive mediators [16–20]. For instance, intracellular Ca^{2+} signals recruit the Ca^{2+}/CaM-dependent endothelial NOS (eNOS) and the Ca^{2+}-dependent phospholipase A2 (PLA2) to generate NO and prostaglandin I2 (PGI2 or prostacyclin), respectively [16,19]. The Ca^{2+} response to extracellular autacoids typically consists of an initial Ca^{2+} peak, which is due to InsP3-dependent ER Ca^{2+} release, followed by sustained Ca^{2+} entry through store-operated Ca^{2+} channels (Figure 3) [65–67]. Store-operated Ca^{2+} entry (SOCE) is a major Ca^{2+} entry pathway in endothelial cells, being activated by any stimulus leading to the depletion of the ER Ca^{2+} pool [68–70]. The dynamic interplay between InsP3-dependent Ca^{2+} release, which could be amplified by adjoining ryanodine receptors (RyRs) through the process of Ca^{2+}-induced Ca^{2+} release (CICR) (Figure 3), and SOCE results in biphasic Ca^{2+} signals [71,72] or repetitive $[Ca^{2+}]_i$ oscillations in brain vascular endothelium [73–75].

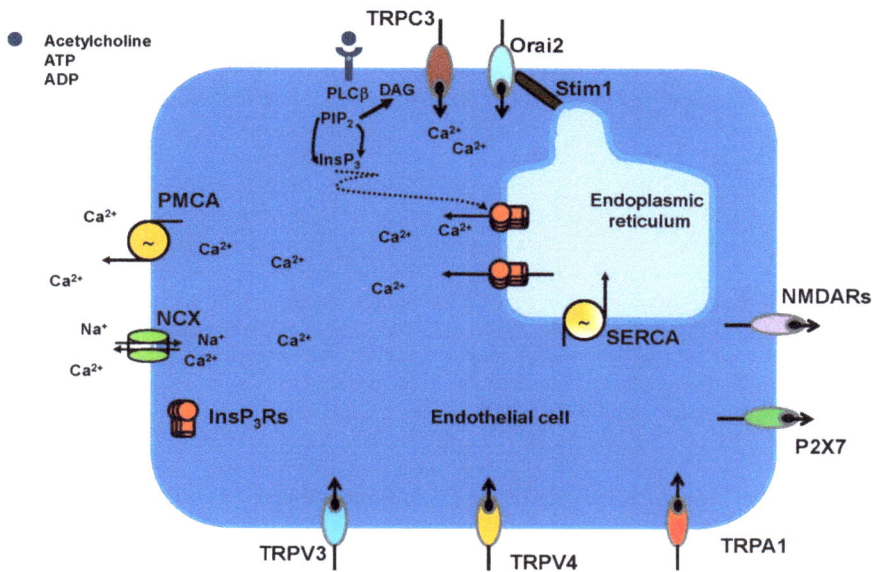

Figure 3. The Ca^{2+} signaling toolkit in brain microvascular endothelial cells. There is scarce information available regarding the molecular components of the Ca^{2+} signaling toolkit in brain microvascular endothelial cells. A recent investigation, however, provided a thorough characterization of the Ca^{2+} machinery in bEND5 cells [76], which represent an established mouse brain microvascular endothelial cell line. Further information was obtained by the analysis of endothelial Ca^{2+} signals in rodent parenchymal arterioles and in the human hCMEC/D3 cell line. Extracellular autacoids bind to specific G-protein-coupled receptors, such as M-AchRs and P2Y1 receptors, thereby activating PLCβ, which in turn cleaves PIP_2 into $InsP_3$ and DAG. $InsP_3$ triggers ER-dependent Ca^{2+} releasing by gating $InsP_3Rs$, while DAG could activate TRPC3. The $InsP_3$-dependent drop in ER Ca^{2+} levels induces SOCE, which is mediated by the interaction between Stim1 and Orai2 in bEND5 cells. Moreover, extracellular Ca^{2+} entry may occur through TRPV3, TRPV4 and TRPA1, which are coupled to either eNOS or EDH [77,78]. Finally, brain microvascular endothelial cells may express Ca^{2+}-permeable ionotropic receptors, such as NMDARs and P2X7 receptors. The elevation in $[Ca^{2+}]_i$ decays to the baseline via the concerted interaction between SERCA and PMCA pumps, as well as through NCX [72,79,80]. Abbreviations: $InsP_3$, inositol-1,4,5-trisphosphate; DAG, diacylglycerol; $InsP_3Rs$, $InsP_3$ receptors; NCX, Na^+–Ca^{2+} exchanger; PMCA, plasma membrane Ca^{2+} ATPase; PIP_2, phosphatidylinositol-4,5-bisphosphate; PLCβ, phospholipase Cβ; SERCA, sarco-endoplasmic reticulum Ca^{2+}-ATPase. The thicker line connecting PIP_2 to $InsP_3$ indicates a high amount of second messenger produced upon PLCβ activation.

3.1.1. The Endothelial Ca^{2+} Toolkit in Brain Microvascular Endothelial Cells: Endogenous Ca^{2+} Release

Scarce information is available regarding the composition of the Ca^{2+} toolkit in brain endothelial cells [74]. For instance, human brain microvascular endothelial cells expressed $InsP_3R1$ and displayed ER-dependent Ca^{2+} release in response to an increase in $InsP_3$ levels [81]. Conversely, only $InsP_3R2$ was expressed in capillary endothelium of rat hippocampus, while RyRs could not be detected in the same study [82]. Nevertheless, RyRs sustained $InsP_3$-dependent Ca^{2+} release in rat brain microvascular endothelial cells in vitro [71], which suggests that the brain endothelial Ca^{2+} toolkit could undergo substantial remodeling in cell cultures [19]. A recent investigation carried out a thorough investigation of the Ca^{2+} toolkit in bEND5 cells [76], a widely-employed mouse brain microvascular endothelial cell line [83]. This analysis revealed that bEND5 cells expressed both

InsP$_3$R1 and InsP$_3$R2, while they lacked InsP$_3$R2 and RyRs [76]. ER-mobilized Ca^{2+} may lead to an increase in mitochondrial Ca^{2+} concentration due to the physical interaction between InsP$_3$Rs and the Ca^{2+}-permeable voltage-dependent anion channels (VDACs), which are embedded in the outer mitochondrial membrane. InsP$_3$Rs-released Ca^{2+} is transferred by VDAC1 into the intermembrane space, from which it is routed towards the mitochondrial matrix though the mitochondrial Ca^{2+} uniporter (MCU) [84,85]. This ER-to-mitochondria Ca^{2+} transfer boosts cellular bioenergetics by stimulating intramitochondrial Ca^{2+}-dependent dehydrogenases, such as oxoglutarate dehydrogenase, NAD-isocitrate dehydrogenase and pyruvate dehydrogenase [86–88]. Subsequently, endothelial mitochondria may contribute to silently (i.e., without a global increase in [Ca^{2+}]$_i$) refill the ER in a sarco-endoplasmic reticulum Ca^{2+}-ATPase (SERCA)-mediated fashion by releasing Ca^{2+} though the mitochondrial Na$^+$/Ca^{2+} exchanger [89,90]. Ca^{2+} entry through the MCU is driven by the negative (i.e., -180 mV) membrane potential ($\Delta\Psi$) that exists across the inner mitochondrial membrane [86]. Mitochondrial content in cerebrovascular endothelium is significantly higher as compared to other vascular districts [91], and InsP$_3$Rs-mediated mitochondrial Ca^{2+} signals arise in rat [92,93] and human [94] brain capillary endothelial cells. Moreover, mitochondrial depolarization hampers the ER-to-mitochondrial Ca^{2+} shuttle, thereby causing a remarkable increase in [Ca^{2+}]$_i$ in rat brain microvascular endothelial cells [95]. An additional mode of Ca^{2+}-mediated cross-talk in endothelial cells may be established between ER and the acidic vesicles of the endolysosomal (EL) Ca^{2+} store [96,97]. The Ca^{2+}-releasing messenger, nicotinic acid adenine dinucleotide phosphate (NAADP), mobilizes EL Ca^{2+} by gating two-pore channels 1 (TPC1) or TPC2 in several types of endothelial cells [54,98–100]. Lysosomal Ca^2 release may be amplified by adjacent ER-embedded InsP$_3$Rs through CICR [97]. The role of lysosomal Ca^{2+} signaling in brain microvascular endothelial cells remains, however, elusive.

3.1.2. The Endothelial Ca^{2+} Toolkit in Brain Microvascular Endothelial Cells: Endogenous Ca^{2+} Release

Likewise, the molecular structure of SOCE in brain vascular endothelium remains to be fully elucidated. Typically, endothelial SOCE is comprised of Stromal interaction molecule 1 and 2 (Stim1 and Stim2, respectively), which sense the drop in ER Ca^{2+} concentration ([Ca^{2+}]$_{ER}$) and Orai1-2 channels, which provide the Ca^{2+}-permeable channel-forming subunits on the plasma membrane [65, 76,101–104]. More specifically, Stim2 is activated by small fluctuations in [Ca^{2+}]$_{ER}$, drives resting Ca^{2+} entry and maintains basal Ca^{2+} levels in endothelial cells [103], while Stim1 is engaged by massive ER Ca^{2+} depletion and sustains agonists-induced extracellular Ca^{2+} entry [65,76,101,102, 104]. However, exceptions to this widespread model may exist. For instance, Stim2 expression is rather modest, and Orai2 represents the only Orai isoform endowed to mouse brain microvascular cells [76]. Of note, Orai2 constitutes the prominent pore-forming subunit of store-operated channels also in mouse neurons [105], which is consistent with the notion that endothelial cells are sensitive to both environmental cues and epigenetic modifications [106]. As widely illustrated in [106,107], the endothelial phenotype is unique in its plasticity and depends on both site-specific signal inputs delivered by the surrounding milieu (which may be diluted in cell culture) and by site-specific epigenetic modifications (i.e., DNA methylation, histone methylation and histone acetylation), which persist under in vitro culture conditions. Similar to other endothelial cells types [108,109], SOCE is constitutively activated to refill the InsP$_3$-dependent ER Ca^{2+} pool and maintains basal Ca^{2+} levels in mouse brain microvascular endothelial cells [76]. AA may induce extracellular Ca^{2+} entry in vascular endothelial cells by prompting Orai1 to interact with Orai3 independently on Stim1 [110]. Nevertheless, Orai3 is not expressed in bEND5 cells, and therefore, this mechanism is unlikely to work in NVC [76]. The endothelial SOCE machinery could involve additional components, such as members of the canonical TRP (TRPC) sub-family of non-selective cation channels, of which seven isoforms exist (TRPC1-7) [68,111]. For instance, Stim1 could also recruit TRPC1 and TRPC4 to assemble into a ternary complex [67,112], whose Na$^+$/Ca^{2+} permeability is determined by Orai1 [113,114]. TRPC1 was expressed in bEND5 cells, while TRPC4 was absent [76]. However, TRPC1 requires Orai1 to be recruited into the SOCE complex upon ER Ca^{2+} depletion [112,115]. It is, therefore, unlikely that it contributes to SOCE in mouse brain microvascular endothelial cells, which lack Orai1 [76], while it may assemble with polycystic TRP2 (TRPP2) to form a stretch-sensitive Ca^{2+}-permeable channel [116]. In addition to SOCE, PLCβ-dependent signaling may lead to the activation of diacylglycerol (DAG)-sensitive Ca^{2+} channels, such as TRPC3 [117,118] and TRPC6 [119,120]. The TRP superfamily of non-selective cation channels consists of 28 members that are classified into six sub-families based on their amino acid sequence homology and structural homology [121,122]. These subfamilies are designated as canonical (TRPC1-7), vanilloid (TRPV1-6), melastatin (TRPM1-8), ankyrin (TRPA1), mucolipin (TRPML1-3) and polycystin (TRPP; TRPP2, TRPP3 and TRPP5) [78,121]. Endothelial cells from different vascular beds may dispose of distinct complements of TRP channels to respond to a myriad of different chemical, mechanical and thermal stimuli [19,78,123,124]. Vascular tone in the brain is specifically regulated by a restricted number of TRP channels, including TRPC3, TRPV3, TRPV4 and TRPA1 (Figure 3), which may stimulate NO release and/or activate IK$_{Ca}$ and SK$_{Ca}$ channels to trigger endothelial-dependent hyperpolarization (EDH) [19,77,78,124]. Endothelial TRPV4 has a potentially relevant role in NVC as it can be activated by AA and its cytochrome P450 epoxygenases-metabolites, i.e., epoxyeicosatrienoic acids (EETs), which evoke vasodilation in intraparenchymal arterioles [9,125]. Finally, human [126], mouse [55,56,127] and rat [128] brain endothelial cells express functional NMDARs (Figure 3) [55], which may mediate glutamate-induced extracellular Ca^{2+} entry. Likewise, P2X7 receptors were recently found in both hCMEC/D3 cells (Figure 3) [129], an immortalized human brain endothelial cell line, and in rat brain endothelial cells in situ [130]. Therefore, the multifaceted endothelial Ca^{2+} toolkit, being located at the interface between neuronal projections and flowing blood, is in an ideal position to trigger and/or modulate NVC by sensing synaptically-released neurotransmitters and blood-borne autacoids.

3.2. Endothelial NMDA Receptors Trigger Glutamate-Induced Nitric Oxide-Mediated Vasodilation

NO represents the major mediator whereby endothelial cells control the vascular tone in large conduit arteries (up to 100%), while its contribution to agonists and/or flow-induced vasodilation decreases (up to 20–50%) as the vascular tree branches into the network of arterioles and capillaries that locally control blood flow [131,132]. NO release is mainly sustained by SOCE rather than by ER-dependent Ca^{2+} release (Figure 4) [133–135], and an oscillatory increase in [Ca^{2+}]$_i$ is the typical waveform that leads to extracellular autacoids-induced NO liberation from vascular endothelial cells [136,137]. In addition to SOCE, TRPC3 [138] and TRPV4 [139] may also evoke extracellular Ca^{2+} entry-mediated eNOS activation, NO release and endothelium-dependent vasodilation. Once released, NO diffuses to adjoining VSMCs to stimulate soluble guanylate cyclase and induce cyclic guanosine-3′,5′-monophosphate (cGMP) production (Figure 4). Then, cGMP activates protein kinase (PKG), which phosphorylates multiple targets to prevent the Ca^{2+}-dependent recruitment of myosin light chain kinase and induce VSMC relaxation. For instance, PKG-dependent phosphorylation inhibits the increase in VSMC [Ca^{2+}]$_i$ induced by L-type voltage-operated Ca^{2+} channels and InsP$_3$Rs; in addition, PKG phosphorylates SERCA, thereby boosting cytosolic Ca^{2+} sequestration into the ER lumen [16,140]. In addition, PKG-dependent phosphorylation stimulates large-conductance Ca^{2+}-activated K$^+$ channels (BK$_{Ca}$), thereby inducing VSMC hyperpolarization and vessel dilation [16,140]. Finally, NO accelerates SERCA-mediated Ca^{2+} reuptake through S-nitrosylation of its cysteine thiols [140]. It is generally assumed that neuronal-derived NO induces vasodilation in the hippocampus and cerebellum [10,25,26,28,33,35–37] and plays a permissive role in PGE2-induced vasorelaxation [9,33,38]. However, alternative sources, e.g., eNOS, have been implicated in glutamate-induced NVC [141,142]. Moreover, glutamate evoked endothelium-dependent vasodilation in the presence of D-serine, a NMDAR co-agonist, in isolated middle cerebral arteries and in brain slice parenchymal arterioles [55,143,144]. NMDARs are heterotetramers comprising seven distinct subunits (GluN1, GluN2A–D and GluN3A and B); GluN1 is strictly required for the assembly of a functional channel and for correct trafficking of the other subunits [144–146].

NMDAR activation requires binding of glutamate to GluN1 and of a co-agonist, D-serine or glycine, to GluN2 [145]. Two GluN1 subunits associate with GluN2A and GluN2B to form neuronal NMDARs, which are therefore sensitive to extracellular Mg^{2+} block [145]. As a consequence, NMDAR activation requires simultaneous binding of synaptically-released glutamate and release of Mg^{2+} inhibition by AMPARs-dependent membrane depolarization [145]. However, in non-neuronal cells, GluN1 subunits assemble with GluN2C and GluN2D, which confer a lower sensitivity to Mg^{2+}, while incorporation of GluN3 further decreases the inhibitory effect of extracellular Mg^{2+} and limits Ca^{2+} permeability [144,145]. NMDAR subunits (i.e., GluN1 and GluN2A-D) have been detected in brain endothelial cells in vitro [126–128,143,147,148] and in cerebral cortex in situ [127] (Figure 3). Intriguingly, NMDARs are more abundant on the basolateral endothelial membrane, which place them in the most suitable position to mediate direct neuronal-to-vascular communication [127]. Recently, Anderson's group demonstrated that glutamate and NMDA induced vasodilation in isolated middle cerebral arteries and in brain slices penetrating arterioles only in the presence of the NMDAR co-agonist, D-serine [55]. This feature could explain why previous investigations, which omitted D-serine or glycine from the bathing solution, failed to observe NMDA-induced dilation of cerebral vessels [149]. Subsequently, the same group showed that astrocytic Ca^{2+} signals induced D-serine release, which was in turn able to activate endothelial NMDARs, thereby activating eNOS in a Ca^{2+}-dependent manner [56,127]. Interestingly, NO evoked dilation of cortical penetrating arterioles by suppressing 20-hydroxyeicosatetraenoic acid (20-HETE) synthesis and boosting PGE2-induced vasorelaxation [56,127]. Future work will have to assess whether eNOS is also activated in response to somatosensory stimulation in vivo and to elucidate the physiological transmitter that induces the Ca^{2+} response in astrocytes. However, these findings clearly show that endothelial NMDARs may control CBF by engaging eNOS at the arteriole level.

3.3. Intracellular Ca^{2+} Signals Drive Acetylcholine-Induced Nitric Oxide Release from Brain Microvascular Endothelial Cells

The cortex receives a widespread acetylcholine innervation mainly arising from the basal forebrain nucleus [28,150]. Basal forebrain acetylcholine neurons broadly projects on intraparenchymal arterioles and capillaries and are, therefore, optimally suited to directly control CBF [28,29,150]. Accordingly, electrical stimulation of basal forebrain neurons results in dilation of intracortical arterioles and increases local CBF [53,151,152]. Acetylcholine-induced cerebral vasodilation is mediated by Gq-coupled muscarinic M5 receptors (M5-AchRs) [53,153], which activate eNOS and induce NO release from cerebrovascular endothelium, in mouse and pig [53,150]. Acetylcholine evokes NO release by triggering repetitive $[Ca^{2+}]_i$ oscillations in several types of endothelial cells [137,154,155]. A recent study focused on bEND5 cells to unravel how acetylcholine induces Ca^{2+}-dependent NO release from cerebrovascular endothelium [76,156,157]. Acetylcholine triggered intracellular oscillations in $[Ca^{2+}]_i$, which were driven by rhythmical InsP$_3$-dependent ER Ca^{2+} release and maintained by SOCE (Figure 3) [76]. The Ca^{2+} response to acetylcholine was mediated by M3-AchRs, which represent another PLCβ-coupled M-AchR isoform. Conversely, nicotine did not cause any detectable increase in $[Ca^{2+}]_i$ [76], as also observed in other endothelial cell types [158]. Acetylcholine-induced Ca^{2+} oscillations led to robust NO release, with eNOS requiring both intracellular Ca^{2+} release and extracellular Ca^{2+} entry to be fully activated [76]. These data were partially confirmed in hCMEC/D3 cells, which displayed a biphasic increase in $[Ca^{2+}]_i$ in response to acetylcholine [76]; of note, human brain microvascular endothelial cells express M5-AchRs [52]. Our preliminary data suggest that the distinct waveforms of acetylcholine-induced Ca^{2+} signals in human vs. mouse brain endothelial cells reflect crucial differences in their Ca^{2+} toolkit. Accordingly, hCMEC/D3 only express InsP$_3$R3, while they lack InsP$_3$R1 and InsP$_3$R2 [159]). InsP$_3$R2, which shows the sharpest dependence on ambient Ca^{2+} and is the most sensitive InsP$_3$R isoform to InsP$_3$, has long been known as the main oscillatory Ca^{2+} unit [160,161]. Conversely, InsP$_3$R3, which is not inhibited by surrounding Ca^{2+}, tends to suppress intracellular Ca^{2+} oscillations [160,161]. Therefore, endothelial Ca^{2+} signaling can be truly regarded as a crucial determinant for acetylcholine-induced NVC.

3.4. Endothelial Ca^{2+} Signals Could Mediate ATP-Induced Vasodilation

Following synaptic activity, neurons and astrocytes release ATP, which serves as a modulator of cellular excitability, synaptic strength and plasticity [162]. ATP is rapidly (200 ms) hydrolyzed by extracellular ectonucleotidases into ADP and adenosine [162], all these mediators being able to physiologically increase CBF [52,163]. Luminal application of ATP and ADP has long been known to induce endothelium-dependent vasodilation in isolated cerebral vessels. For instance, ATP and ADP promoted vasodilation in rat middle cerebral arteries by stimulating NO release, although ATP could also act through cytochrome P450-metabolites [51]. Of note, extraluminal administration of ATP evoked a biphasic vasomotor response in rat penetrating arterioles, consisting of an initial transient vasoconstriction followed by local vasodilation, which was subsequently propagated to upstream locations ($\approx$500 µm) along the vascular wall [164,165]. ATP-induced local vasoconstriction was mediated by ionotropic P2X receptors on VSMCs, while endothelial P2Y1 receptors triggered local vasodilation [164]. ATP induced vasodilation by stimulating NO release and stimulating EET production to activate EDH (see below); EETs were also responsible for the upstream propagation of the vasomotor response [164]. These findings were supported by a recent study, showing that the hemodynamic response to whisker stimulation in the mouse somatosensory cortex required P2Y1 receptor-dependent eNOS activation. Previous administration of fluoroacetate, a glial-specific metabolic toxin, prevented eNOS-dependent functional hyperemia, thereby suggesting that ATP was mainly released by perivascular astrocytes [166]. P2Y1 receptors are GPCRs, which bind to both ATP and ADP and control the vascular tone by inducing intracellular Ca^{2+} signals in endothelial cells through distinct signaling pathways [167–169]. For instance, P2Y1 receptors induced InsP$_3$-dependent ER Ca^{2+} release in rat cardiac microvascular endothelial cells [167], whereas they stimulated cyclic nucleotide-gated channels in the H5V endothelial cell line, in primary cultures of bovine aortic endothelial cells and in mouse aorta endothelial cells in situ [169]. Moreover, ATP induced local and conducted vasodilation in hamster cheek pouch arterioles by triggering an increase in endothelial $[Ca^{2+}]_i$ both locally and $\approx$1200 µm upstream along the same vessel (see also below) [170]. Therefore, it is conceivable that endothelial Ca^{2+} signals mediate P2Y1 receptor-dependent NO release and functional hyperemia in the somatosensory cortex.

3.5. TRP Channels Trigger Endothelial-Dependent Hyperpolarization in Cerebrovascular Endothelial Cells

EDH provides the largest contribution to endothelial vasorelaxing mechanisms in resistance-sized arteries and arterioles, as shown in coronary, renal and mesenteric circulation [22,23,131,171]. EDH is initiated by an increase in endothelial $[Ca^{2+}]_i$, which stimulates IK$_{Ca}$ (KCa$_{3.1}$) and SK$_{Ca}$ (KCa$_{2.3}$) channels to hyperpolarize the endothelial cell membrane (Figure 4). Hyperpolarizing current spreads from vascular endothelium to overlying smooth muscle cells to trigger VSMC relaxation and vessel dilation by inhibiting voltage-dependent Ca^{2+} entry (Figure 4) [22,23,77]. This vasorelaxing mechanism requires a tightly-regulated disposition of the Ca^{2+} sources and the Ca^{2+}-sensitive decoders, i.e., SK$_{Ca}$ and IK$_{Ca}$ channels, which effect membrane hyperpolarization. To achieve such a precise spatial arrangement, endothelial cells extend cellular protrusions through the internal elastic lamina, which establish a heterocellular coupling with adjacent smooth muscle cells through connexin-based myo-endothelial gap-junctions (MEGJs) [22,23,77]. The endothelial ER also protrudes into these myo-endothelial microdomain sites and forms spatially-discrete InsP$_3$-sensitive Ca^{2+} pools, which are juxtaposed with IK$_{Ca}$ channels, which are located on the endothelial protrusions traversing the holes in the vascular wall [23,172]. Conversely, SK$_{Ca}$ channels are distributed throughout the endothelial cell membrane, although they are enriched at MEGJs and may, therefore, sense ER-released Ca^{2+} [173]. InsP$_3$Rs are constitutively activated to produce repetitive, spatially-restricted Ca^{2+} release events, termed Ca^{2+} pulsars, whose frequency can be increased by extracellular vasoactive agonists, such as acetylcholine. Spontaneous InsP$_3$-driven Ca^{2+} pulsars are selectively coupled to IK$_{Ca}$ and SK$_{Ca}$ channels, thereby hyperpolarizing the endothelial membrane potential at myo-endothelial projections and inducing dilation in adjoining VSMCs [172]. K$^+$ signaling in the myo-endothelial space could then be boosted by the stimulation of endothelial inward rectifying K$^+$ (K$_{ir}$) channels or of Na$^+$/K$^+$ ATPase in VSMCs. In addition to InsP$_3$-induced ER-dependent Ca^{2+} release, IK$_{Ca}$ and SK$_{Ca}$ channels can be activated by extracellular Ca^{2+} entry through TRPV4 channels, which is also largely expressed at MEGJs [174]. EDH does not only evoke local vasodilation at the site of endothelial stimulation; the hyperpolarizing current spreads along vascular endothelium to upstream arteries (up to ≈2 mm) and drives the retrograde vasodilation, which is ultimately responsible for the drop in vascular resistance that increases blood supply to active regions [175–177]. Preliminary data revealed that the same clustered architecture of TRPV4 and IK$_{Ca}$ channels is maintained at MEGJs of rodent cerebral arteries [24,78,178–180], which suggests that discrete InsP$_3$Rs-mediated Ca^{2+} pulsars arise also in parenchymal vessels [24]. Moreover, cerebral MEGJs are enriched with TRPA1 channels, which may also support EDH in cortical circulation [178,181]. EDH mediates ATP-, UTP- and SLIGR (a selective agonist of protease-activated receptor 2)-induced vasodilation in rat middle cerebral arteries [164,180,182–185] and acetylcholine-induced vasodilation in mouse posterior cerebral arteries [186].

3.5.1. TRPV4

TRPV4 is emerging as a major regulator of CBF [24,39,163] and represents one of the most important Ca^{2+}-entry pathways in vascular endothelial cells [111,187–191]. TRPV4 channels are polymodal non-selective cation channels that mediate Ca^{2+} entry in response to distinct chemical, thermal and mechanical stimuli [54,125,187,191–194], thereby integrating the diverse surrounding cues acting on the vascular wall. For instance, TRPV4 may be activated by EETs, which are produced via AA epoxygenation by cytochrome P450 epoxygenase enzymes. Briefly, an increase in endothelial [Ca^{2+}]$_i$ induced by either mechanical or chemical stimuli may stimulate the Ca^{2+}-dependent phospholipase A2 (PLA2), which cleaves AA from membrane phospholipids [77,125]. AA is, turn, is converted into EETs, such as 5,6-EET, 8,9-EET, 11,12-EET and 14,15-EET, by cytochrome P450 2C9 or cytochrome P450 2J2 [125]. In particular, 5,6-EET, 8,9-EET and 11,12-EET were shown to stimulate TRPV4-mediated non-selective cation channels and intracellular Ca^{2+} signals in endothelial cells from several vascular beds [191,195–198]. Luminal application of UTP, which is a selective agonist of P2Y2 receptors, in rat middle cerebral arteries induced TRPV4-mediated Ca^{2+} entry across the luminal and abluminal face of the endothelial monolayer. TRPV4-mediated Ca^{2+} entry, in turn, activated PLA2, whose activation was polarized to the abluminal side, to activate EDH and induce vasodilation [199,200]. Moreover, TRPV4-mediated Ca^{2+} entry was necessary to activate IK$_{Ca}$ and SK$_{Ca}$ channels and induce the vasodilatory response to acetylcholine in mouse posterior cerebral arteries [186]. TRPV4 is, therefore, the most likely candidate to trigger ATP-induced retrograde vasodilation in rat parenchymal arterioles, which is initiated by EETs and involves IK$_{Ca}$, but not SK$_{Ca}$, channels [164]. Intriguingly, EETs mediate vasodilation in rodent cortical arterioles [27,42,201,202]. It has been proposed that NA-induced elevation in astrocyte [Ca^{2+}]$_i$ engages PLA2 to liberate AA, thereby resulting in EET synthesis, either within astrocytes or adjacent smooth muscle cells [33,42]. Future work is necessary to assess whether: (1) astrocyte-derived EETs also activate endothelial TRPV4 within the NVU; and/or (2) neural activity directly stimulates brain endothelial cells to produce EET synthesis and activate TRPV4. A recent investigation suggested that EDH was responsible for propagating the hemodynamic response to somatosensory stimulation from the capillary bed to upstream penetrating arterioles and pial arteries [7,203]. However, IK$_{Ca}$ and IK$_{Ca}$ currents could not be recorded in mouse brain capillary endothelial cells [204]. These findings strongly suggest that: (1) EDH is restricted to pial arteries and arterioles, but it cannot be activated in the capillary bed; and (2) an alternative mechanism spreads the vasomotor signal from brain capillaries to upstream vessels (see below) [204].

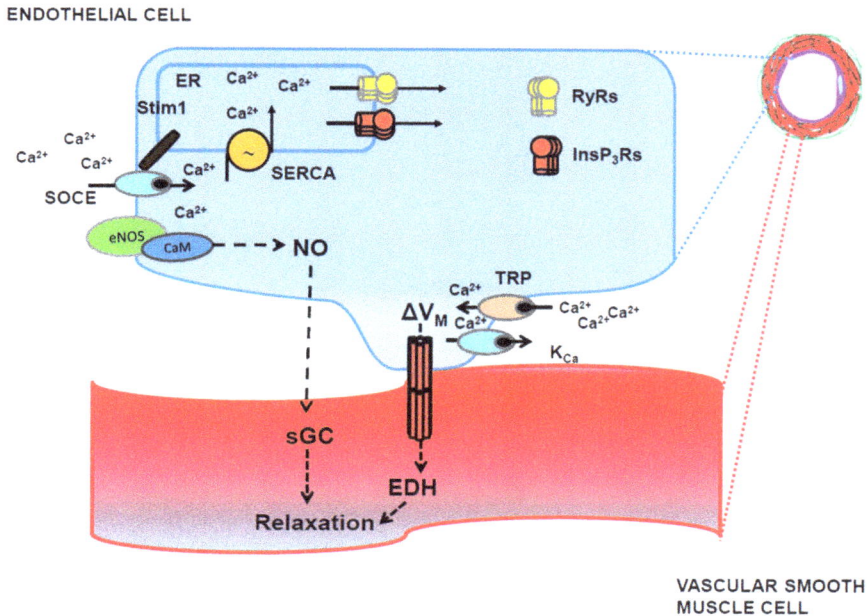

Figure 4. Ca^{2+}-regulated endothelium-dependent vasodilation. Agonists-induced $InsP_3$-dependent ER Ca^{2+} depletion in vascular endothelial cells leads to store-operated Ca^{2+} entry (SOCE). SOCE is tightly coupled to eNOS, thereby triggering robust NO release. NO, in turn, diffuses towards adjacent vascular smooth muscle cells (VSMCs) at myo-endothelial projections and activates soluble guanylyl cyclase (sGC) to induce vasorelaxation. Extracellular agonists may also activate TRP channels (e.g., TRPC3, TRPV3, TRPV4 and TRPA1), which is preferentially coupled to Ca^{2+}-activated K^+ channels (K_{Ca}), such as IK_{Ca} and SK_{Ca}. Endothelial hyperpolarization spreads through myo-endothelial gap junctions to adjoining VSMCs to induce vasorelaxation according to a mechanism known endothelial-dependent hyperpolarization (EDH).

3.5.2. TRPA1, TRPC3 and TRPV3

In addition to TRPV4, TRPA1 channels are also enriched at MEGJs in rat cerebral arteries and may induce EDH in isolated vessels [78]. Accordingly, allyl isothiocyanate (AITC) (mustard oil), a selective TRPA1 agonist, induced local Ca^{2+} entry events, IK_{Ca} channel activation and vasodilation in rat cerebral arteries [205,206]. TRPA1 was physiologically activated by superoxide anions generated by the reactive oxygen species-generating enzyme NADPH oxidase isoform 2 (NOX2), which colocalizes with TRPA1 in cerebral endothelium, but not in other vascular districts [178]. Furthermore, EDH in cerebral arteries may be elicited by Ca^{2+} influx through TRPC3 and TRPV3 channels [78]. TRPC3 is gated by DAG to conduct extracellular Ca^{2+} into vascular endothelial cells upon PLC activation [117,118,138,207]. A recent investigation revealed that TRPC3 contributed to ATP-induced vasodilation in mouse middle cerebral arteries and posterior cerebral arteries by delivering the Ca^{2+} necessary for IK_{Ca} and SK_{Ca} channel activation. IK_{Ca} channels mediated the initial phase of ATP-induced hyperpolarization, whereas SK_{Ca} channels sustained the delayed phase of endothelial hyperpolarization [208]. IK_{Ca} and SK_{Ca} channels in rat brain vessels may also be activated by TRPV3. Accordingly, carvacrol, a monoterpenoid phenol compound highly concentrated in the essential oil of oregano, caused vasodilation by selectively activating TRPV3 in rat isolated posterior cerebral and superior cerebellar arteries. TRPV3-mediated extracellular Ca^{2+} entry, in turn, engaged IK_{Ca} and SK_{Ca} channels to trigger EDH. Unlike TRPV4 and TRPA1, TRPV3 was not specifically localized to MEGJs, but was distributed throughout the endothelial membrane [209]. In addition to carvacrol, TRPV3, as well as TRPA1 are sensitive to several dietary agonists. For instance, TRPV3 may also be activated by eugenol (clove oil) and thymol (found in thyme) [210], whereas allicin (garlic) and cinnamaldehyde induce TRPA1-mediated Ca^{2+} entry [211]. These observations, therefore, suggest that dietary manipulation of TRP channels could be exploited to rescue CBF in cerebrovascular pathologies [6,212].

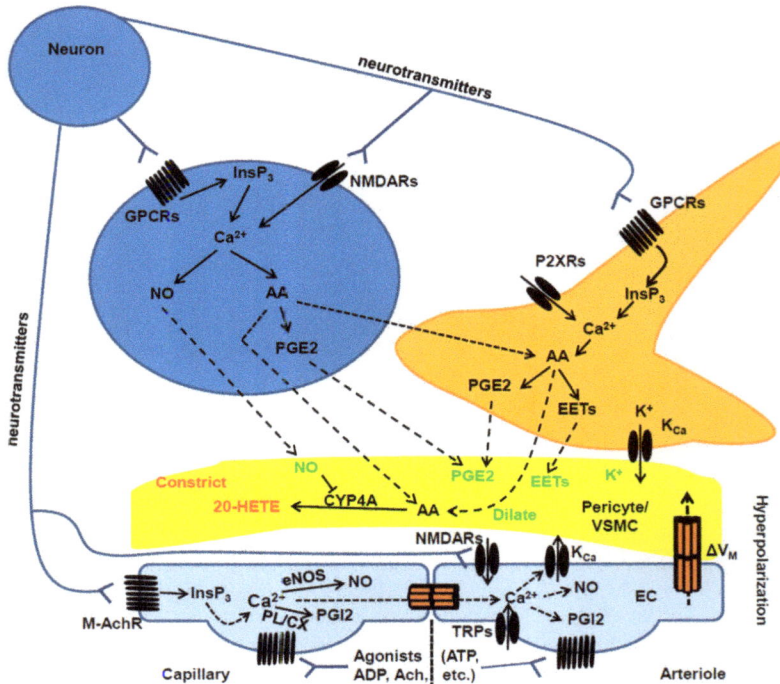

Figure 5. The putative role of endothelial Ca^{2+} signaling in neurovascular coupling. Synaptic activity leads to arteriole and capillary vasodilation by inducing an increase in $[Ca^{2+}]_i$ in postsynaptic neurons and perisynaptic astrocytes, as shown in Figure 2. Recent evidence indicated that synaptically-released glutamate may activate endothelial NMDARs, thereby eliciting the Ca^{2+}-dependent activation of eNOS, in intraparenchymal arterioles. Moreover, acetylcholine may induce Ca^{2+}-dependent NO release from brain endothelial cells by initiating the concerted interplay between InsP$_3$Rs and SOCE [76]. This local vasodilation may be spread to more remote sites ($\approx$ 500 μm) through the initiation of an interendothelial Ca^{2+} wave, which ignites NO release and PGI2 production as long as it travels along the endothelial monolayer. Moreover, this propagating Ca^{2+} sweep could induce vasodilation by also stimulating K$_{Ca}$ channels and evoking EDH in arterioles. Finally, blood-borne autacoids and dietary agonists could induce vasodilation by, respectively, binding to their specific GPCRs and stimulating multiple TRP channels (TRPC3, TRPV3, TRPV4, TRPA1) to initiate EDH. Abbreviations: 20-HETE: 20-hydroxyeicosatetraenoic acid; AA: arachidonic acid; CX: cyclooxygenases 1 and 2; EETs: epoxyeicosatrienoic acids; GPCRs; G-protein-coupled receptors; InsP$_3$: inositol-1,4,5-trisphosphate; K$_{Ca}$: Ca^{2+}-activated intermediate and small conductance K$^+$ channels; NO: nitric oxide; nNOS: neuronal NO synthase; P450: cytochrome P450; PGE2: prostaglandin E2; PL: phospholipase A2; PLD2: phospholipase D2; TRPs: TRP channels; VSMC: vascular smooth muscle cell.

4. How to Integrate Endothelial Ca²⁺ Signals in the Current Models of Neurovascular Coupling and Propagated Vasodilation

The role of endothelial Ca^{2+} signaling in NVC has largely been neglected as most authors focused on other cellular components of the NVU, such as neurons, interneurons, astrocytes, smooth muscle cells and, more recently, pericytes [2,4,6,9,30]. However, the literature discussed in the previous paragraphs suggests that the endothelial Ca^{2+} toolkit could play a crucial role in detecting and translating NA into a local vasoactive signal that is then propagated to upstream vessels [7,24]. In order to interpret the precise function of the ion channel network that controls CBF, it is mandatory to understand when and where the hemodynamic response starts. As mentioned earlier, synaptic activity-evoked NVC could be initiated at the capillary level and conducted upstream [12,47,203,213–217]. This hypothesis makes physiological sense as capillaries are closer to active neurons compared to arterioles and could represent the earliest vascular component to detect NA. In addition, endothelial signaling underpins long-range and almost unattenuated propagation (up to 2 mm) of vasomotor responses along peripheral vessels [22,62,63,176,177,218]. Endothelial engagement during NVC could, therefore, fulfill the same function observed in peripheral circulation, i.e., initiating (or contributing to initiate) vasodilation in proximity of the most metabolically active areas and conducting the vasomotor signal to upstream feeding arteries and arterioles to ensure an adequate increase in local blood supply (Figure 5) [62,177].

We believe that there is wide evidence to conclude that acetylcholine, which is liberated by cholinergic afferents emanated from the basal forebrain neurons during alert wakefulness, arousal, learning and attentional effort [28,150], increases CBF by inducing an increase in endothelial $[Ca^{2+}]_i$, which results in robust NO release [53,76,150,151,153,157]. This model does not rule out the possibility that the sub-population of acetylcholine and NO-synthesizing basal forebrain neurons that send projections onto cortical microvessels could directly control CBF though NO liberation [150,219]; in addition, acetylcholine and NO-synthesizing fibers may also contact NO cortical interneurons, which could contribute to NO-dependent vasodilation [150,220]. Similar to acetylcholine, glutamate could induce NMDARs-mediated NO release from brain endothelium [55,56,127]. Endothelial-derived NO could represent the alternative, i.e., non-neuronal, source of NO supporting vasodilation in cerebellar [141] and cortical [142] parenchymal arterioles. Future work will have to assess whether capillary brain endothelial cells express functional NMDARs and generate NO in response to glutamate stimulation. Intriguingly, the role of glial Ca^{2+} signals [13], which lead to the indispensable release of D-serine in arterioles, and of NO in glutamate-induced capillary dilation have been recently reported in cerebellum [11]. Alternatively, glutamate-induced endothelial-dependent vasodilation could contribute to pre-dilate upstream pial arteries and parenchymal arterioles prior to the retrograde propagation of the vasomotor signal from the capillary bed. Retrograde propagation of the initial vasodilation to upstream cortical vessels still represents a matter of hot debate [4,7,221]. Hillman's group demonstrated that light-dye disruption of endothelial lining dampens propagation of stimulus (electrical hind paw stimulation)-induced vasodilation in pial arteries in vivo, thereby blunting the increase in CBF [203]. In Section 2.3, we anticipated that endothelial Ca^{2+} signals mediate retrograde propagation of the vasomotor response by engaging two distinct mechanisms: (1) IK_{Ca} and SK_{Ca} channels, which effect the fast component of conducted vasodilation to upstream feeding vessels (i.e., pial arteries and arterioles); and (2) an intercellular Ca^{2+} wave that mediates the slower component of conducted vasodilation by inducing the release of NO and PGI2 from vascular endothelial cells [7,63,64]. EDH cannot, however, be initiated by brain capillary endothelial cells, which lack functional IK_{Ca} and SK_{Ca} channels [204]. Nelson's group recently revealed that a modest increase in the extracellular K^+ concentration (≈ 10 mM), which is likely to truly reflect NA, activated brain capillary cell endothelial inward rectifier K^+ ($K_{IR}2.1$) channels to generate a hyperpolarizing signal that rapidly propagated to upstream arterioles (monitored up to ≈ 500 μm) to cause vasodilation and increase CBF in vivo [204]. Therefore, it appears that $K_{IR}2.1$, rather than IK_{Ca} and SK_{Ca}, channels mediate the fast component of conducted vasodilation in cerebral circulation. Of note, the hyperpolarization conduction velocity

Int. J. Mol. Sci. **2018**, *19*, 938

was in the same range, $\approx$2 mm/s, as the propagation speed of retrograde pial artery dilation [204,213]. The secondary slow component of conducted vasodilation is sustained by interendothelial Ca^{2+} waves, as discussed elsewhere [4,7,62]. Earlier work carried out by monitoring endothelial Ca^{2+} with a Ca^{2+}-sensitive dye, i.e., Fura-2, showed that local delivery of acetylcholine triggered a Ca^{2+} wave travelling bidirectionally along the endothelium of hamster feed arteries for no less than 1 mm [222]. This intercellular Ca^{2+} wave, in turn, sustained the slow component of conducted vasodilation by promoting NO and PGI2 release [223]. The use of a genetic Ca^{2+} indicator, GCaMP2, confirmed these data in cremaster muscle arterioles in vivo. Accordingly, acetylcholine induced a local increase in endothelial $[Ca^{2+}]_i$ that activated K_{Ca} channels to induce rapidly-conducting vasodilation at distances >1 mm and travelled along the endothelium as an intercellular Ca^{2+} wave for 300–400 μm. As shown in previous studies [222,223], the intercellular Ca^{2+} wave preceded a secondary vasodilation that was mediated by NO and PGE2 [63]. These findings led to the suggestion that the initial hyperpolarization conducts vasodilation far away (> several millimeters) from the local site of stimulation (including daughter and parent branches), whereas the slower Ca^{2+} wave encompasses only the vessel segments closer to the active region, thereby finely tuning the magnitude and duration of the vasodilatory response [62,63]. A recent study showed that removal of extracellular Ca^{2+} may induce intercellular Ca^{2+} waves in immortalized (RBE4) and primary (from bovine origin) brain microvascular endothelial cells [224]. Although this finding is not sufficient to confirm that interendothelial Ca^{2+} waves may propagate NA-induced local vasodilation to upstream arteries and arterioles, it confirms that brain endothelial cells are able to generate this type of propagating Ca^{2+} signal, which is likely to impinge on InsP3Rs [225–227]. Future work will have to assess whether: (1) an intercellular Ca^{2+} wave contributes to propagating the hemodynamic response from the capillary bed to feeding vessels; and (2) if so, whether EDH is stimulated by the incoming Ca^{2+} wave in parenchymal arterioles to sustain the vasomotor response.

5. Conclusions

Neurovascular coupling is the crucial process to adjust local CBF to the metabolic requirements of active neurons and maintain brain function. Although functional magnetic resonance imaging is routinely employed to monitor the changes in neuronal spiking activity, BOLD signals actually reflect the increases in CBF induced by NA. Therefore, understanding the cellular and molecular bases of NVC is mandatory to interpret the complex relationship between neuronal firing, metabolism and blood flow in both physiological and pathological conditions [6–8]. Intracellular Ca^{2+} signaling has long been known to drive NVC by coupling synaptic activity with the production of vasoactive messengers. Emerging evidence, however, suggests that the endothelial Ca^{2+} toolkit could also be recruited by neurotransmitters (i.e., glutamate and acetylcholine) and neuromodulators (e.g., ATP) to induce NO and PGE2 release or activate EDH. Future work will benefit from the availability of transgenic mice selectively expressing a genetic Ca^{2+} indicator, e.g., GCaMP2, in vascular endothelial cells [63,174]. The combination of endothelial GCaMP2-expressing mice with the novel advances in imaging techniques (i.e., two- or three-photon microscopy) will permit assessing whether synaptic activity increases endothelial $[Ca^{2+}]_i$ in vivo. It will also be important to decipher the role played by IK_{Ca} and SK_{Ca} channels in the hemodynamic response to NA. Earlier work showed that EDH is involved in local and retrograde vasodilation in parenchymal arterioles and middle cerebral arteries, but is absent in capillaries, where functional hyperemia is likely to initiate in response to sensory stimulation. The local vasodilation induced by NA increases the strength of laminar shear stress acting on vascular endothelium, a mechanism that leads to TRPV4-dependent NO release and EDH in peripheral circulation [191,228]. There is, however, conflicting evidence regarding flow-induced vasodilation in pial arteries and parenchymal arteries in the brain [229–231]. Nevertheless, EDH could be exploited by dietary manipulation to treat the vascular dysfunctions associated with aging and neurodegenerative disorders, which cause cognitive impairment by halting CBF [2,3,6,212]. For instance, NVC is severely impaired in Alzheimer's disease due to a defect in NO release [217,232,233]. A recent study revealed that the intracellular Ca^{2+} toolkit is severely compromised in rat brain microvascular endothelial cells exposed to amyloid-beta ($A\beta$) peptide [93], whose accumulation in brain parenchyma and in the cerebrovasculature represents a major pathogenic factor in AD [4]. Future work will have to assess whether dietary and/or pharmacological manipulation is able to rescue or halt functional hyperemia in these subjects by activating the TRP channels (i.e., TRPC3, TRPV3, TRPV4 and TRPA1), which are coupled to EDH in the brain. An emerging area of research for NVC is represented by the vasodilating role of mitochondrial Ca^{2+} in brain microvascular endothelial cells. A recent series of studies revealed that mitochondrial depolarization, induced by BMS-291095 (BMS), an opener of mitochondrial ATP-dependent K^+ (K_{ATP}) channels, prevented Ca^{2+} accumulation within the mitochondrial matrix, thereby resulting in a large increase in $[Ca^{2+}]_i$ in rat cerebral arteries endothelial cells. This mitochondrial-derived Ca^{2+} signal, in turn, induced eNOS activation, NO release and endothelium-dependent vasodilation [95]. Intriguingly, subsequent work demonstrated that mitochondrial-dependent NO release and vasodilation were impaired in cerebral arteries of obese Zucker rats [234]. The physiological stimulus responsible for mitochondrial-induced NO release in cerebral endothelium is yet to be elucidated, but it could be implicated in many other cerebrovascular disorders. Finally, astrocyte-released AA is the precursor of multiple vasoactive messengers, such as PGE2, EETs and 20-HETE, which are all involved in NVC, although their contributions depend on the vascular segment and/or the brain area. However, AA was shown to induce Ca^{2+}-dependent NO release by directly activating TRPV4 in several types of endothelial cells [54,235]. Some evidence suggested that AA was able to induce Ca^{2+} signals in human brain microvascular endothelial cells [94]. Given the key role of this lipid mediator within the NVU, we predict that future work will unveil that AA-dependent endothelial Ca^{2+} signals may also contribute to NVC.

Conflicts of Interest: The authors declare no conflict of interest.

Abbreviations

20-HETE	20-hydroxyeicosatetraenoic acid
AA	arachidonic acid
AMPA	a-amino-3-hydroxy-5-methyl-4-isoxazol epropionic acid
BK_{Ca}	large-conductance Ca^{2+}-activated K^+ channels
BOLD	blood oxygenation level dependent
CaM	Calmodulin
CBF	cerebral blood flow
cGMP	cyclic guanosine-$3',5'$-monophosphate
CICR	Ca^{2+}-induced Ca^{2+} release
CYP4A	cytochrome P450 4A
DAG	Diacylglycerol
EDH	endothelial-dependent hyperpolarization
EETs	epoxyeicosatrienoic acids
EL	Endolysosomal
eNOS	neuronal NO synthase
ER	endoplasmic reticulum
GPCRs	G-protein-coupled receptors
IK_{Ca}	intermediate-conductance Ca^{2+}-activated K^+ channels
$InsP_3$	inositol-1,4,5-trisphosphate
$InsP_3Rs$	$InsP_3$ receptors
K_{Ca}	Ca^{2+}-activated K^+ channels
K_{ir}	inward rectifying K^+ channels
M-AchRs	muscarinic acetylcholine receptors
MCU	mitochondrial Ca^{2+} uniporter

MEGJs	myo-endothelial gap-junctions
mGluRs	metabotropic glutamate receptors
NA	neuronal activity
NAADP	nicotinic acid adenine dinucleotide phosphate
NMDA	N-methyl-D-aspartate
NMDARs	NMDA receptors
NO	nitric oxide
nNOS	neuronal NO synthase
NOX2	NADPH oxidase isoform 2
NVC	neurovascular coupling
NVU	neurovascular unit
O_2	Oxygen
P450	cytochrome P450
PGE2	prostaglandin E2
PIP_2	phosphatidylinositol-4,5-bisphosphate
PLA2	phospholipase A2
PLD2	phospholipase D2
RyRs	ryanodine receptors
SERCA	sarco-endoplasmic reticulum Ca^{2+}-ATPase
SK_{Ca}	small-conductance Ca^{2+}-activated K^+ channels
SOCE	store-operated Ca^{2+} entry
Stim	stromal interaction molecule
TPCs	two-pore channels
TRPs	TRP channels
TRPA1	ankyrin TRP 1
TRPC	canonical TRP
TRPV4	vanilloid TRP 4
VDACs	voltage-dependent anion channels
VSMC	vascular smooth muscle cells

References

1. Attwell, D.; Laughlin, S.B. An energy budget for signaling in the grey matter of the brain. *J. Cereb. Blood Flow Metab.* **2001**, *21*, 1133–1145. [CrossRef] [PubMed]
2. Kisler, K.; Nelson, A.R.; Montagne, A.; Zlokovic, B.V. Cerebral blood flow regulation and neurovascular dysfunction in Alzheimer disease. *Nat. Rev. Neurosci.* **2017**, *18*, 419–434. [CrossRef] [PubMed]
3. Zlokovic, B.V. Neurovascular pathways to neurodegeneration in Alzheimer's disease and other disorders. *Nat. Rev. Neurosci.* **2011**, *12*, 723–738. [CrossRef] [PubMed]
4. Iadecola, C. The Neurovascular Unit Coming of Age: A Journey through Neurovascular Coupling in Health and Disease. *Neuron* **2017**, *96*, 17–42. [CrossRef] [PubMed]
5. Koller, A.; Toth, P. Contribution of flow-dependent vasomotor mechanisms to the autoregulation of cerebral blood flow. *J. Vasc. Res.* **2012**, *49*, 375–389. [CrossRef] [PubMed]
6. Iadecola, C. Neurovascular regulation in the normal brain and in Alzheimer's disease. *Nat. Rev. Neurosci.* **2004**, *5*, 347–360. [CrossRef] [PubMed]
7. Hillman, E.M. Coupling mechanism and significance of the BOLD signal: A status report. *Annu. Rev. Neurosci.* **2014**, *37*, 161–181. [CrossRef] [PubMed]
8. Attwell, D.; Iadecola, C. The neural basis of functional brain imaging signals. *Trends Neurosci.* **2002**, *25*, 621–625. [CrossRef]
9. Attwell, D.; Buchan, A.M.; Charpak, S.; Lauritzen, M.; Macvicar, B.A.; Newman, E.A. Glial and neuronal control of brain blood flow. *Nature* **2010**, *468*, 232–243. [CrossRef] [PubMed]
10. Rancillac, A.; Rossier, J.; Guille, M.; Tong, X.K.; Geoffroy, H.; Amatore, C.; Arbault, S.; Hamel, E.; Cauli, B. Glutamatergic Control of Microvascular Tone by Distinct GABA Neurons in the Cerebellum. *J. Neurosci.* **2006**, *26*, 6997–7006. [CrossRef] [PubMed]

11. Mapelli, L.; Gagliano, G.; Soda, T.; Laforenza, U.; Moccia, F.; D'Angelo, E.U. Granular Layer Neurons Control Cerebellar Neurovascular Coupling Through an NMDA Receptor/NO-Dependent System. *J. Neurosci.* **2017**, *37*, 1340–1351. [CrossRef] [PubMed]

12. Hall, C.N.; Reynell, C.; Gesslein, B.; Hamilton, N.B.; Mishra, A.; Sutherland, B.A.; O'Farrell, F.M.; Buchan, A.M.; Lauritzen, M.; Attwell, D. Capillary pericytes regulate cerebral blood flow in health and disease. *Nature* **2014**, *508*, 55–60. [CrossRef] [PubMed]

13. Mishra, A.; Reynolds, J.P.; Chen, Y.; Gourine, A.V.; Rusakov, D.A.; Attwell, D. Astrocytes mediate neurovascular signaling to capillary pericytes but not to arterioles. *Nat. Neurosci.* **2016**, *19*, 1619–1627. [CrossRef] [PubMed]

14. Rosenegger, D.G.; Gordon, G.R. A slow or modulatory role of astrocytes in neurovascular coupling. *Microcirculation* **2015**, *22*, 197–203. [CrossRef] [PubMed]

15. Laughlin, M.H.; Davis, M.J.; Secher, N.H.; van Lieshout, J.J.; Arce-Esquivel, A.A.; Simmons, G.H.; Bender, S.B.; Padilla, J.; Bache, R.J.; Merkus, D.; et al. Peripheral circulation. *Compr. Physiol.* **2012**, *2*, 321–447. [PubMed]

16. Khaddaj Mallat, R.; Mathew John, C.; Kendrick, D.J.; Braun, A.P. The vascular endothelium: A regulator of arterial tone and interface for the immune system. *Crit. Rev. Clin. Lab. Sci.* **2017**, *54*, 458–470. [CrossRef] [PubMed]

17. Mancardi, D.; Pla, A.F.; Moccia, F.; Tanzi, F.; Munaron, L. Old and new gasotransmitters in the cardiovascular system: Focus on the role of nitric oxide and hydrogen sulfide in endothelial cells and cardiomyocytes. *Curr. Pharm. Biotechnol.* **2011**, *12*, 1406–1415. [CrossRef] [PubMed]

18. Vanhoutte, P.M.; Tang, E.H. Endothelium-dependent contractions: When a good guy turns bad! *J. Physiol.* **2008**, *586*, 5295–5304. [CrossRef] [PubMed]

19. Moccia, F.; Berra-Romani, R.; Tanzi, F. Update on vascular endothelial Ca^{2+} signalling: A tale of ion channels, pumps and transporters. *World J. Biol. Chem.* **2012**, *3*, 127–158. [CrossRef] [PubMed]

20. Moccia, F.; Tanzi, F.; Munaron, L. Endothelial remodelling and intracellular calcium machinery. *Curr. Mol. Med.* **2014**, *14*, 457–480. [CrossRef] [PubMed]

21. Altaany, Z.; Moccia, F.; Munaron, L.; Mancardi, D.; Wang, R. Hydrogen sulfide and endothelial dysfunction: Relationship with nitric oxide. *Curr. Med. Chem.* **2014**, *21*, 3646–3661. [CrossRef] [PubMed]

22. Behringer, E.J. Calcium and electrical signaling in arterial endothelial tubes: New insights into cellular physiology and cardiovascular function. *Microcirculation* **2017**, *24*. [CrossRef] [PubMed]

23. Garland, C.J.; Dora, K.A. EDH: Endothelium-dependent hyperpolarization and microvascular signalling. *Acta Physiologica* **2017**, *219*, 152–161. [CrossRef] [PubMed]

24. Longden, T.A.; Hill-Eubanks, D.C.; Nelson, M.T. Ion channel networks in the control of cerebral blood flow. *J. Cereb. Blood Flow Metab.* **2016**, *36*, 492–512. [CrossRef] [PubMed]

25. Cauli, B.; Hamel, E. Revisiting the role of neurons in neurovascular coupling. *Front. Neuroenerg.* **2010**, *2*, 9. [CrossRef] [PubMed]

26. Cauli, B.; Tong, X.K.; Rancillac, A.; Serluca, N.; Lambolez, B.; Rossier, J.; Hamel, E. Cortical GABA interneurons in neurovascular coupling: Relays for subcortical vasoactive pathways. *J. Neurosci.* **2004**, *24*, 8940–8949. [CrossRef] [PubMed]

27. Lecrux, C.; Toussay, X.; Kocharyan, A.; Fernandes, P.; Neupane, S.; Levesque, M.; Plaisier, F.; Shmuel, A.; Cauli, B.; Hamel, E. Pyramidal neurons are "neurogenic hubs" in the neurovascular coupling response to whisker stimulation. *J. Neurosci.* **2011**, *31*, 9836–9847. [CrossRef] [PubMed]

28. Lecrux, C.; Hamel, E. Neuronal networks and mediators of cortical neurovascular coupling responses in normal and altered brain states. *Philos. Trans. R. Soc. Lond. B Biol. Sci.* **2016**, *371*. pii: 20150350. [CrossRef] [PubMed]

29. Hamel, E. Perivascular nerves and the regulation of cerebrovascular tone. *J. Appl. Physiol. (1985)* **2006**, *100*, 1059–1064. [CrossRef] [PubMed]

30. Nortley, R.; Attwell, D. Control of brain energy supply by astrocytes. *Curr. Opin. Neurobiol.* **2017**, *47*, 80–85. [CrossRef] [PubMed]

31. Mathiisen, T.M.; Lehre, K.P.; Danbolt, N.C.; Ottersen, O.P. The perivascular astroglial sheath provides a complete covering of the brain microvessels: An electron microscopic 3D reconstruction. *Glia* **2010**, *58*, 1094–1103. [CrossRef] [PubMed]

32. Zlokovic, B.V. Neurovascular mechanisms of Alzheimer's neurodegeneration. *Trends Neurosci.* **2005**, *28*, 202–208. [CrossRef] [PubMed]

33. Dalkara, T.; Alarcon-Martinez, L. Cerebral microvascular pericytes and neurogliovascular signaling in health and disease. *Brain Res.* **2015**, *1623*, 3–17. [CrossRef] [PubMed]

34. Lauritzen, M. Reading vascular changes in brain imaging: Is dendritic calcium the key? *Nat. Rev. Neurosci.* **2005**, *6*, 77–85. [CrossRef] [PubMed]

35. Lourenco, C.F.; Ledo, A.; Barbosa, R.M ; Laranjinha, J. Neurovascular-neuroenergetic coupling axis in the brain: Master regulation by nitric oxide and consequences in aging and neurodegeneration. *Free Radic. Biol. Med.* **2017**, *108*, 668–682. [CrossRef] [PubMed]

36. Yang, G.; Chen, G.; Ebner, T.J.; Iadecola, C. Nitric oxide is the predominant mediator of cerebellar hyperemia during somatosensory activation in rats. *Am. J. Physiol.* **1999**, *277 (Pt 2)*, R1760–R1770. [CrossRef] [PubMed]

37. Lovick, T.A.; Brown, L.A.; Key, B.J. Neurovascular relationships in hippocampal slices: Physiological and anatomical studies of mechanisms underlying flow-metabolism coupling in intraparenchymal microvessels. *Neuroscience* **1999**, *92*, 47–60. [CrossRef]

38. Duchemin, S.; Boily, M.; Sadekova, N.; Girouard, H. The complex contribution of NOS interneurons in the physiology of cerebrovascular regulation. *Front. Neural Circ.* **2012**, *6*, 51. [CrossRef] [PubMed]

39. Filosa, J.A.; Morrison, H.W.; Iddings, J.A.; Du, W.; Kim, K.J. Beyond neurovascular coupling, role of astrocytes in the regulation of vascular tone. *Neuroscience* **2016**, *323*, 96–109. [CrossRef] [PubMed]

40. Wang, X.; Lou, N.; Xu, Q.; Tian, G.F.; Peng, W.G.; Han, X.; Kang, J.; Takano, T.; Nedergaard, M. Astrocytic Ca^{2+} signaling evoked by sensory stimulation in vivo. *Nat. Neurosci.* **2006**, *9*, 816–823. [CrossRef] [PubMed]

41. Takano, T.; Tian, G.F.; Peng, W.; Lou, N.; Libionka, W.; Han, X.; Nedergaard, M. Astrocyte-mediated control of cerebral blood flow. *Nat. Neurosci.* **2006**, *9*, 260–267. [CrossRef] [PubMed]

42. Mishra, A. Binaural blood flow control by astrocytes: Listening to synapses and the vasculature. *J. Physiol.* **2017**, *595*, 1885–1902. [CrossRef] [PubMed]

43. Kim, K.J.; Iddings, J.A.; Stern, J.E.; Blanco, V.M.; Croom, D.; Kirov, S.A.; Filosa, J.A. Astrocyte contributions to flow/pressure-evoked parenchymal arteriole vasoconstriction. *J. Neurosci.* **2015**, *35*, 8245–8257. [CrossRef] [PubMed]

44. Kim, K.J.; Ramiro Diaz, J.; Iddings, J.A.; Filosa, J.A. Vasculo-Neuronal Coupling: Retrograde Vascular Communication to Brain Neurons. *J. Neurosci.* **2016**, *36*, 12624–12639. [CrossRef] [PubMed]

45. Simard, M.; Arcuino, G.; Takano, T.; Liu, Q.S.; Nedergaard, M. Signaling at the gliovascular interface. *J. Neurosci.* **2003**, *23*, 9254–9262. [PubMed]

46. Attwell, D.; Mishra, A.; Hall, C.N.; O'Farrell, F.M.; Dalkara, T. What is a pericyte? *J. Cereb. Blood Flow Metab.* **2016**, *36*, 451–455. [CrossRef] [PubMed]

47. Hamilton, N.B.; Attwell, D.; Hall, C.N. Pericyte-mediated regulation of capillary diameter: A component of neurovascular coupling in health and disease. *Front. Neuroenerg.* **2010**, *2*. pii: 5. [CrossRef] [PubMed]

48. D'Angelo, E. The organization of plasticity in the cerebellar cortex: From synapses to control. *Prog. Brain Res.* **2014**, *210*, 31–58. [PubMed]

49. Berra-Romani, R.; Avelino-Cruz, J.E.; Raqeeb, A.; Della Corte, A.; Cinelli, M.; Montagnani, S.; Guerra, G.; Moccia, F.; Tanzi, F. Ca(2)(+)-dependent nitric oxide release in the injured endothelium of excised rat aorta: A promising mechanism applying in vascular prosthetic devices in aging patients. *BMC Surg.* **2013**, *13* (Suppl. 2), S40. [CrossRef] [PubMed]

50. Andresen, J.; Shafi, N.I.; Bryan, R.M., Jr. Endothelial influences on cerebrovascular tone. *J. Appl. Physiol. (1985)* **2006**, *100*, 318–327. [CrossRef] [PubMed]

51. You, J.; Johnson, T.D.; Childres, W.F.; Bryan, R.M., Jr. Endothelial-mediated dilations of rat middle cerebral arteries by ATP and ADP. *Am. J. Physiol.* **1997**, *273 (Pt 2)*, H1472–H1477. [CrossRef] [PubMed]

52. Elhusseiny, A.; Cohen, Z.; Olivier, A.; Stanimirovic, D.B.; Hamel, E. Functional acetylcholine muscarinic receptor subtypes in human brain microcirculation: Identification and cellular localization. *J. Cereb. Blood Flow Metab.* **1999**, *19*, 794–802. [CrossRef] [PubMed]

53. Elhusseiny, A.; Hamel, E. Muscarinic–but not nicotinic–acetylcholine receptors mediate a nitric oxide-dependent dilation in brain cortical arterioles: A possible role for the M5 receptor subtype. *J. Cereb. Blood Flow Metab.* **2000**, *20*, 298–305. [CrossRef] [PubMed]

54. Zuccolo, E.; Dragoni, S.; Poletto, V.; Catarsi, P.; Guido, D.; Rappa, A.; Reforgiato, M.; Lodola, F.; Lim, D.; Rosti, V.; et al. Arachidonic acid-evoked Ca^{2+} signals promote nitric oxide release and proliferation in human endothelial colony forming cells. *Vascul. Pharmacol.* **2016**, *87*, 159–171. [CrossRef] [PubMed]

55. LeMaistre, J.L.; Sanders, S.A.; Stobart, M.J.; Lu, L.; Knox, J.D.; Anderson, H.D.; Anderson, C.M. Coactivation of NMDA receptors by glutamate and D-serine induces dilation of isolated middle cerebral arteries. *J. Cereb. Blood Flow Metab.* **2012**, *32*, 537–547. [CrossRef] [PubMed]

56. Stobart, J.L.; Lu, L.; Anderson, H.D.; Mori, H.; Anderson, C.M. Astrocyte-induced cortical vasodilation is mediated by D-serine and endothelial nitric oxide synthase. *Proc. Natl. Acad. Sci. USA* **2013**, *110*, 3149–3154. [CrossRef] [PubMed]

57. Beard, R.S., Jr.; Reynolds, J.J.; Bearden, S.E. Metabotropic glutamate receptor 5 mediates phosphorylation of vascular endothelial cadherin and nuclear localization of beta-catenin in response to homocysteine. *Vascul. Pharmacol.* **2012**, *56*, 159–167. [CrossRef] [PubMed]

58. Lv, J.M.; Guo, X.M.; Chen, B.; Lei, Q.; Pan, Y.J.; Yang, Q. The Noncompetitive AMPAR Antagonist Perampanel Abrogates Brain Endothelial Cell Permeability in Response to Ischemia: Involvement of Claudin-5. *Cell. Mol. Neurobiol.* **2016**, *36*, 745–753. [CrossRef] [PubMed]

59. Collard, C.D.; Park, K.A.; Montalto, M.C.; Alapati, S.; Buras, J.A.; Stahl, G.L.; Colgan, S.P. Neutrophil-derived glutamate regulates vascular endothelial barrier function. *J. Biol. Chem.* **2002**, *277*, 14801–14811. [CrossRef] [PubMed]

60. Gillard, S.E.; Tzaferis, J.; Tsui, H.C.; Kingston, A.E. Expression of metabotropic glutamate receptors in rat meningeal and brain microvasculature and choroid plexus. *J. Comp. Neurol.* **2003**, *461*, 317–332. [CrossRef] [PubMed]

61. Itoh, Y.; Suzuki, N. Control of brain capillary blood flow. *J. Cereb. Blood Flow Metab.* **2012**, *32*, 1167–1176. [CrossRef] [PubMed]

62. Segal, S.S. Integration and Modulation of Intercellular Signaling Underlying Blood Flow Control. *J. Vasc. Res.* **2015**, *52*, 136–157. [CrossRef] [PubMed]

63. Tallini, Y.N.; Brekke, J.F.; Shui, B.; Doran, R.; Hwang, S.M.; Nakai, J.; Salama, G.; Segal, S.S.; Kotlikoff, M.I. Propagated endothelial Ca^{2+} waves and arteriolar dilation in vivo: Measurements in Cx40BAC GCaMP2 transgenic mice. *Circ. Res.* **2007**, *101*, 1300–1309. [CrossRef] [PubMed]

64. Wolfle, S.E.; Chaston, D.J.; Goto, K.; Sandow, S.L.; Edwards, F.R.; Hill, C.E. Non-linear relationship between hyperpolarisation and relaxation enables long distance propagation of vasodilatation. *J. Physiol.* **2011**, *589 (Pt 10)*, 2607–2623. [CrossRef] [PubMed]

65. Abdullaev, I.F.; Bisaillon, J.M.; Potier, M.; Gonzalez, J.C.; Motiani, R.K.; Trebak, M. STIM1 and ORAI1 mediate CRAC currents and store-operated calcium entry important for endothelial cell proliferation. *Circ. Res.* **2008**, *103*, 1289–1299. [CrossRef] [PubMed]

66. Zuccolo, E.; Di Buduo, C.; Lodola, F.; Orecchioni, S.; Scarpellino, G.; Kheder, D.A.; Poletto, V.; Guerra, G.; Bertolini, F.; Balduini, A.; et al. Stromal Cell-Derived Factor-1alpha Promotes Endothelial Colony-Forming Cell Migration Through the Ca(2+)-Dependent Activation of the Extracellular Signal-Regulated Kinase 1/2 and Phosphoinositide 3-Kinase/AKT Pathways. *Stem Cells Dev.* **2018**, *27*, 23–34. [CrossRef] [PubMed]

67. Freichel, M.; Suh, S.H.; Pfeifer, A.; Schweig, U.; Trost, C.; Weissgerber, P.; Biel, M.; Philipp, S.; Freise, D.; Droogmans, G.; et al. Lack of an endothelial store-operated Ca^{2+} current impairs agonist-dependent vasorelaxation in TRP4-/- mice. *Nat. Cell Biol.* **2001**, *3*, 121–127. [CrossRef] [PubMed]

68. Blatter, L.A. Tissue Specificity: SOCE: Implications for Ca^{2+} Handling in Endothelial Cells. *Adv. Exp. Med. Biol.* **2017**, *993*, 343–361. [PubMed]

69. Moccia, F.; Lodola, F.; Dragoni, S.; Bonetti, E.; Bottino, C.; Guerra, G.; Laforenza, U.; Rosti, V.; Tanzi, F. Ca^{2+} signalling in endothelial progenitor cells: A novel means to improve cell-based therapy and impair tumour vascularisation. *Curr. Vasc. Pharmacol.* **2014**, *12*, 87–105. [CrossRef] [PubMed]

70. Moccia, F.; Dragoni, S.; Lodola, F.; Bonetti, E.; Bottino, C.; Guerra, G.; Laforenza, U.; Rosti, V.; Tanzi, F. Store-dependent Ca(2+) entry in endothelial progenitor cells as a perspective tool to enhance cell-based therapy and adverse tumour vascularization. *Curr. Med. Chem.* **2012**, *19*, 5802–5018. [CrossRef] [PubMed]

71. Brailoiu, E.; Shipsky, M.M.; Yan, G.; Abood, M.E.; Brailoiu, G.C. Mechanisms of modulation of brain microvascular endothelial cells function by thrombin. *Brain Res.* **2017**, *1657*, 167–175. [CrossRef] [PubMed]

72. Domotor, E.; Abbott, N.J.; Adam-Vizi, V. Na^+-Ca^{2+} exchange and its implications for calcium homeostasis in primary cultured rat brain microvascular endothelial cells. *J. Physiol.* **1999**, *515 (Pt 1)*, 147–155. [CrossRef] [PubMed]

73. De Bock, M.; Culot, M.; Wang, N.; Bol, M.; Decrock, E.; De Vuyst, E.; da Costa, A.; Dauwe, I.; Vinken, M.; Simon, A.M.; et al. Connexin channels provide a target to manipulate brain endothelial calcium dynamics and blood-brain barrier permeability. *J. Cereb. Blood Flow Metab.* **2011**, *31*, 1942–1957. [CrossRef] [PubMed]

74. De Bock, M.; Wang, N.; Decrock, E.; Bol, M.; Gadicherla, A.K.; Culot, M.; Cecchelli, R.; Bultynck, G.; Leybaert, L. Endothelial calcium dynamics, connexin channels and blood-brain barrier function. *Prog. Neurobiol.* **2013**, *108*, 1–20. [CrossRef] [PubMed]

75. Scharbrodt, W.; Abdallah, Y.; Kasseckert, S.A.; Gligorievski, D.; Piper, H.M.; Boker, D.K.; Deinsberger, W.; Oertel, M.F. Cytosolic Ca^{2+} oscillations in human cerebrovascular endothelial cells after subarachnoid hemorrhage. *J. Cereb. Blood Flow Metab.* **2009**, *29*, 57–65. [CrossRef] [PubMed]

76. Zuccolo, E.; Lim, D.; Kheder, D.A.; Perna, A.; Catarsi, P.; Botta, L.; Rosti, V.; Riboni, L.; Sancini, G.; Tanzi, F.; et al. Acetylcholine induces intracellular Ca$_{2+}$ oscillations and nitric oxide release in mouse brain endothelial cells. *Cell Calcium* **2017**, *66*, 33–47. [CrossRef] [PubMed]

77. Earley, S. Endothelium-dependent cerebral artery dilation mediated by transient receptor potential and Ca^{2+}-activated K^{+} channels. *J. Cardiovasc. Pharmacol.* **2011**, *57*, 148–153. [CrossRef] [PubMed]

78. Earley, S.; Brayden, J.E. Transient receptor potential channels in the vasculature. *Physiol. Rev.* **2015**, *95*, 645–690. [CrossRef] [PubMed]

79. Manoonkitiwongsa, P.S.; Whitter, E.F.; Wareesangtip, W.; McMillan, P.J.; Nava, P.B.; Schultz, R.L. Calcium-dependent ATPase unlike ecto-ATPase is located primarily on the luminal surface of brain endothelial cells. *Histochem. J.* **2000**, *32*, 313–324. [CrossRef] [PubMed]

80. Moccia, F.; Berra-Romani, R.; Baruffi, S.; Spaggiari, S.; Signorelli, S.; Castelli, L.; Magistretti, J.; Taglietti, V.; Tanzi, F. Ca^{2+} uptake by the endoplasmic reticulum Ca^{2+}-ATPase in rat microvascular endothelial cells. *Biochem. J.* **2002**, *364 (Pt 1)*, 235–244. [CrossRef] [PubMed]

81. Haorah, J.; Knipe, B.; Gorantla, S.; Zheng, J.; Persidsky, Y. Alcohol-induced blood-brain barrier dysfunction is mediated via inositol 1,4,5-triphosphate receptor (IP3R)-gated intracellular calcium release. *J. Neurochem.* **2007**, *100*, 324–336. [CrossRef] [PubMed]

82. Hertle, D.N.; Yeckel, M.F. Distribution of inositol-1,4,5-trisphosphate receptor isotypes and ryanodine receptor isotypes during maturation of the rat hippocampus. *Neuroscience* **2007**, *150*, 625–638. [CrossRef] [PubMed]

83. Czupalla, C.J.; Liebner, S.; Devraj, K. In vitro models of the blood-brain barrier. *Methods Mol. Biol.* **2014**, *1135*, 415–437. [PubMed]

84. Pedriali, G.; Rimessi, A.; Sbano, L.; Giorgi, C.; Wieckowski, M.R.; Previati, M.; Pinton, P. Regulation of Endoplasmic Reticulum-Mitochondria Ca^{2+} Transfer and Its Importance for Anti-Cancer Therapies. *Front. Oncol.* **2017**, *7*, 180. [CrossRef] [PubMed]

85. Dong, Z.; Shanmughapriya, S.; Tomar, D.; Siddiqui, N.; Lynch, S.; Nemani, N.; Breves, S.L.; Zhang, X.; Tripathi, A.; Palaniappan, P.; et al. Mitochondrial Ca^{2+} Uniporter Is a Mitochondrial Luminal Redox Sensor that Augments MCU Channel Activity. *Mol. Cell* **2017**, *65*, 1014–1028.e7. [CrossRef] [PubMed]

86. De Stefani, D.; Rizzuto, R.; Pozzan, T. Enjoy the Trip: Calcium in Mitochondria Back and Forth. *Annu. Rev. Biochem.* **2016**, *85*, 161–192. [CrossRef] [PubMed]

87. Moccia, F. Remodelling of the Ca^{2+} Toolkit in Tumor Endothelium as a Crucial Responsible for the Resistance to Anticancer Therapies. *Curr. Signal Transduct. Ther.* **2017**, *12*. [CrossRef]

88. Kluge, M.A.; Fetterman, J.L.; Vita, J.A. Mitochondria and endothelial function. *Circ. Res.* **2013**, *112*, 1171–1188. [CrossRef] [PubMed]

89. Malli, R.; Frieden, M.; Hunkova, M.; Trenker, M.; Graier, W.F. Ca^{2+} refilling of the endoplasmic reticulum is largely preserved albeit reduced Ca^{2+} entry in endothelial cells. *Cell Calcium* **2007**, *41*, 63–76. [CrossRef] [PubMed]

90. Malli, R.; Frieden, M.; Osibow, K.; Zoratti, C.; Mayer, M.; Demaurex, N.; Graier, W.F. Sustained Ca^{2+} transfer across mitochondria is Essential for mitochondrial Ca^{2+} buffering, sore-operated Ca^{2+} entry, and Ca^{2+} store refilling. *J. Biol. Chem.* **2003**, *278*, 44769–44779. [CrossRef] [PubMed]

91. Busija, D.W.; Rutkai, I.; Dutta, S.; Katakam, P.V. Role of Mitochondria in Cerebral Vascular Function: Energy Production, Cellular Protection, and Regulation of Vascular Tone. *Compr. Physiol.* **2016**, *6*, 1529–1548. [PubMed]

92. Gerencser, A.A.; Adam-Vizi, V. Selective, high-resolution fluorescence imaging of mitochondrial Ca^{2+} concentration. *Cell Calcium* **2001**, *30*, 311–321. [CrossRef] [PubMed]

93. Fonseca, A.C.; Moreira, P.I.; Oliveira, C.R.; Cardoso, S.M.; Pinton, P.; Pereira, C.F. Amyloid-beta disrupts calcium and redox homeostasis in brain endothelial cells. *Mol. Neurobiol.* **2015**, *51*, 610–622. [CrossRef] [PubMed]

94. Evans, J.; Ko, Y.; Mata, W.; Saquib, M.; Eldridge, J.; Cohen-Gadol, A.; Leaver, H.A.; Wang, S.; Rizzo, M.T. Arachidonic acid induces brain endothelial cell apoptosis via p38-MAPK and intracellular calcium signaling. *Microvasc. Res.* **2015**, *98*, 145–158. [CrossRef] [PubMed]

95. Katakam, P.V.; Wappler, E.A.; Katz, P.S.; Rutkai, I.; Institoris, A.; Domoki, F.; Gaspar, T.; Grovenburg, S.M.; Snipes, J.A.; Busija, D.W. Depolarization of mitochondria in endothelial cells promotes cerebral artery vasodilation by activation of nitric oxide synthase. *Arterioscler. Thromb. Vasc. Biol.* **2013**, *33*, 752–759. [CrossRef] [PubMed]

96. Ronco, V.; Potenza, D.M.; Denti, F.; Vullo, S.; Gagliano, G.; Tognolina, M.; Guerra, G.; Pinton, P.; Genazzani, A.A.; Mapelli, L.; et al. A novel Ca(2)(+)-mediated cross-talk between endoplasmic reticulum and acidic organelles: Implications for NAADP-dependent Ca(2)(+) signalling. *Cell Calcium* **2015**, *57*, 89–100. [CrossRef] [PubMed]

97. Zuccolo, E.; Lim, D.; Poletto, V.; Guerra, G.; Tanzi, F.; Rosti, V.; Moccia, F. Acidic Ca$_{2+}$ stores interact with the endoplasmic reticulum to shape intracellular Ca^{2+} signals in human endothelial progenitor cells. *Vascul. Pharmacol.* **2015**, *75*, 70–71. [CrossRef]

98. Di Nezza, F.; Zuccolo, E.; Poletto, V.; Rosti, V.; De Luca, A.; Moccia, F.; Guerra, G.; Ambrosone, L. Liposomes as a Putative Tool to Investigate NAADP Signaling in Vasculogenesis. *J. Cell. Biochem.* **2017**, *118*, 3722–3729. [CrossRef] [PubMed]

99. Favia, A.; Desideri, M.; Gambara, G.; D'Alessio, A.; Ruas, M.; Esposito, B.; Del Bufalo, D.; Parrington, J.; Ziparo, E.; Palombi, F.; et al. VEGF-induced neoangiogenesis is mediated by NAADP and two-pore channel-2-dependent Ca^{2+} signaling. *Proc. Natl. Acad. Sci. USA* **2014**, *111*, E4706–E4715. [CrossRef] [PubMed]

100. Favia, A.; Pafumi, I.; Desideri, M.; Padula, F.; Montesano, C.; Passeri, D.; Nicoletti, C.; Orlandi, A.; Del Bufalo, D.; Sergi, M.; et al. NAADP-Dependent Ca(2+) Signaling Controls Melanoma Progression, Metastatic Dissemination and Neoangiogenesis. *Sci. Rep.* **2016**, *6*, 18925. [CrossRef] [PubMed]

101. Lodola, F.; Laforenza, U.; Bonetti, E.; Lim, D.; Dragoni, S.; Bottino, C.; Ong, H.L.; Guerra, G.; Ganini, C.; Massa, M.; et al. Store-operated Ca^{2+} entry is remodelled and controls in vitro angiogenesis in endothelial progenitor cells isolated from tumoral patients. *PLoS ONE* **2012**, *7*, e42541. [CrossRef] [PubMed]

102. Li, J.; Cubbon, R.M.; Wilson, L.A.; Amer, M.S.; McKeown, L.; Hou, B.; Majeed, Y.; Tumova, S.; Seymour, V.A.L.; Taylor, H.; et al. ORAI1 and CRAC channel dependence of VEGF-activated Ca^{2+} entry and endothelial tube formation. *Circ. Res.* **2011**, *108*, 1190–1198. [CrossRef] [PubMed]

103. Brandman, O.; Liou, J.; Park, W.S.; Meyer, T. STIM2 is a feedback regulator that stabilizes basal cytosolic and endoplasmic reticulum Ca^{2+} levels. *Cell* **2007**, *131*, 1327–1339. [CrossRef] [PubMed]

104. Antigny, F.; Jousset, H.; Konig, S.; Frieden, M. Thapsigargin activates Ca(2)+ entry both by store-dependent, STIM1/ORAI1-mediated, and store-independent, TRPC3/PLC/PKC-mediated pathways in human endothelial cells. *Cell Calcium* **2011**, *49*, 115–127. [CrossRef] [PubMed]

105. Moccia, F.; Zuccolo, E.; Soda, T.; Tanzi, F.; Guerra, G.; Mapelli, L.; Lodola, F.; D'Angelo, E. Stim and Orai proteins in neuronal Ca(2)+ signaling and excitability. *Front. Cell. Neurosci.* **2015**, *9*, 153. [CrossRef] [PubMed]

106. Aird, W.C. Endothelial cell heterogeneity. *Cold Spring Harb. Perspect. Med.* **2012**, *2*, a006429. [CrossRef] [PubMed]

107. Regan, E.R.; Aird, W.C. Dynamical systems approach to endothelial heterogeneity. *Circ. Res.* **2012**, *111*, 110–130. [CrossRef] [PubMed]

108. Moccia, F.; Berra-Romani, R.; Baruffi, S.; Spaggiari, S.; Adams, D.J.; Taglietti, V.; Tanzi, F. Basal nonselective cation permeability in rat cardiac microvascular endothelial cells. *Microvasc. Res.* **2002**, *64*, 187–197. [CrossRef] [PubMed]

109. Zuccolo, E.; Bottino, C.; Diofano, F.; Poletto, V.; Codazzi, A.C.; Mannarino, S.; Campanelli, R.; Fois, G.; Marseglia, G.L.; Guerra, G.; et al. Constitutive Store-Operated Ca(2+) Entry Leads to Enhanced Nitric Oxide Production and Proliferation in Infantile Hemangioma-Derived Endothelial Colony-Forming Cells. *Stem Cells Dev.* **2016**, *25*, 301–319. [CrossRef] [PubMed]

110. Li, J.; Bruns, A.F.; Hou, B.; Rode, B.; Webster, P.J.; Bailey, M.A.; Appleby, H.L.; Moss, N.K.; Ritchie, J.E.; Yuldasheva, N.Y.; et al. ORAI3 Surface Accumulation and Calcium Entry Evoked by Vascular Endothelial Growth Factor. *Arterioscler. Thromb. Vasc. Biol.* **2015**, *35*, 1987–1994. [CrossRef] [PubMed]

111. Moccia, F. Endothelial Ca(2+) Signaling and the Resistance to Anticancer Treatments: Partners in Crime. *Int. J. Mol. Sci.* **2018**, *19*. pii: E217. [CrossRef] [PubMed]

112. Sundivakkam, P.C.; Freichel, M.; Singh, V.; Yuan, J.P.; Vogel, S.M.; Flockerzi, V.; Malik, A.B.; Tiruppathi, C. The Ca(2+) sensor stromal interaction molecule 1 (STIM1) is necessary and sufficient for the store-operated Ca(2+) entry function of transient receptor potential canonical (TRPC) 1 and 4 channels in endothelial cells. *Mol. Pharmacol.* **2012**, *81*, 510–526. [CrossRef] [PubMed]

113. Cioffi, D.L.; Wu, S.; Chen, H.; Alexeyev, M.; St Croix, C.M.; Pitt, B.R.; Uhlig, S.; Stevens, T. ORAI1 determines calcium selectivity of an endogenous TRPC heterotetramer channel. *Circ. Res.* **2012**, *110*, 1435–1444. [CrossRef] [PubMed]

114. Xu, N.; Cioffi, D.L.; Alexeyev, M.; Rich, T.C.; Stevens, T. Sodium entry through endothelial store-operated calcium entry channels: Regulation by ORAI1. *Am. J. Physiol. Cell Physiol.* **2015**, *308*, C277–C288. [CrossRef] [PubMed]

115. Cioffi, D.L.; Barry, C.; Stevens, T. Store-operated calcium entry channels in pulmonary endothelium: The emerging story of TRPCS and Orai1. *Adv. Exp. Med. Biol.* **2010**, *661*, 137–154. [PubMed]

116. Berrout, J.; Jin, M.; O'Neil, R.G. Critical role of TRPP2 and TRPC1 channels in stretch-induced injury of blood-brain barrier endothelial cells. *Brain Res.* **2012**, *1436*, 1–12. [CrossRef] [PubMed]

117. Dragoni, S.; Laforenza, U.; Bonetti, E ; Lodola, F.; Bottino, C.; Guerra, G.; Borghesi, A.; Stronati, M.; Rosti, V.; Tanzi, F.; et al. Canonical transient receptor potential 3 channel triggers vascular endothelial growth factor-induced intracellular Ca^{2+} oscillations in endothelial progenitor cells isolated from umbilical cord blood. *Stem Cells Dev.* **2013**, *22*, 2561–2580. [CrossRef] [PubMed]

118. Moccia, F.; Lucariello, A.; Guerra, G. TRPC3-mediated Ca(2+) signals as a promising strategy to boost therapeutic angiogenesis in failing hearts: The role of autologous endothelial colony forming cells. *J. Cell. Physiol.* **2017**. [CrossRef]

119. Hamdollah Zadeh, M.A.; Glass, C.A.; Magnussen, A.; Hancox, J.C.; Bates, D.O. VEGF-mediated elevated intracellular calcium and angiogenesis in human microvascular endothelial cells in vitro are inhibited by dominant negative TRPC6. *Microcirculation* **2008**, *15*, 605–614. [CrossRef] [PubMed]

120. Chaudhuri, P.; Rosenbaum, M.A.; Birnbaumer, L.; Graham, L.M. Integration of TRPC6 and NADPH oxidase activation in lysophosphatidylcholine-induced TRPC5 externalization. *Am. J. Physiol. Cell Physiol.* **2017**, *313*, C541–C555. [CrossRef] [PubMed]

121. Gees, M.; Colsoul, B.; Nilius, B. The role of transient receptor potential cation channels in Ca^{2+} signaling. *Cold Spring Harb. Perspect. Biol.* **2010**, *2*, a003962. [CrossRef] [PubMed]

122. Parenti, A.; De Logu, F.; Geppetti, P.; Benemei, S. What is the evidence for the role of TRP channels in inflammatory and immune cells? *Br. J. Pharmacol.* **2016**, *173*, 953–969. [CrossRef] [PubMed]

123. Di, A.; Malik, A.B. TRP channels and the control of vascular function. *Curr. Opin. Pharmacol.* **2010**, *10*, 127–132. [CrossRef] [PubMed]

124. Wong, C.O.; Yao, X. TRP channels in vascular endothelial cells. *Adv. Exp. Med. Biol.* **2011**, *704*, 759–780. [PubMed]

125. Ellinsworth, D.C.; Earley, S.; Murphy, T.V.; Sandow, S.L. Endothelial control of vasodilation: Integration of myoendothelial microdomain signalling and modulation by epoxyeicosatrienoic acids. *Pflugers Arch.* **2014**, *466*, 389–405. [CrossRef] [PubMed]

126. Sharp, C.D.; Hines, I.; Houghton, J.; Warren, A.; Jackson, T.H. t.; Jawahar, A.; Nanda, A.; Elrod, J.W.; Long, A.; Chi, A.; et al. Glutamate causes a loss in human cerebral endothelial barrier integrity through activation of NMDA receptor. *Am. J. Physiol. Heart Circ. Physiol.* **2003**, *285*, H2592–H2598. [CrossRef] [PubMed]

127. Lu, L.; Hogan-Cann, A.D.; Globa, A.K.; Lu, P.; Nagy, J.I.; Bamji, S.X.; Anderson, C.M. Astrocytes drive cortical vasodilatory signaling by activating endothelial NMDA receptors. *J. Cereb. Blood Flow Metab.* **2017**. [CrossRef] [PubMed]

128. Krizbai, I.A.; Deli, M.A.; Pestenacz, A.; Siklos, L.; Szabo, C.A.; Andras, I.; Joo, F. Expression of glutamate receptors on cultured cerebral endothelial cells. *J. Neurosci. Res.* **1998**, *54*, 814–819. [CrossRef]

129. Yang, F.; Zhao, K.; Zhang, X.; Zhang, J.; Xu, B. ATP Induces Disruption of Tight Junction Proteins via IL-1 Beta-Dependent MMP-9 Activation of Human Blood-Brain Barrier In Vitro. *Neural Plast.* **2016**, *2016*, 8928530. [CrossRef] [PubMed]

130. Zhao, H.; Zhang, X.; Dai, Z.; Feng, Y.; Li, Q.; Zhang, J.H.; Liu, X.; Chen, Y.; Feng, H. P2X7 Receptor Suppression Preserves Blood-Brain Barrier through Inhibiting RhoA Activation after Experimental Intracerebral Hemorrhage in Rats. *Sci. Rep.* **2016**, *6*, 23286. [CrossRef] [PubMed]

131. De Wit, C.; Wolfle, S.E. EDHF and gap junctions: Important regulators of vascular tone within the microcirculation. *Curr. Pharm. Biotechnol.* **2007**, *8*, 11–25. [CrossRef] [PubMed]

132. Shu, X.; Keller, T.C. t.; Begandt, D.; Butcher, J.T.; Biwer, L.; Keller, A.S.; Columbus, L.; Isakson, B.E. Endothelial nitric oxide synthase in the microcirculation. *Cell. Mol. Life Sci.* **2015**, *72*, 4561–4575. [CrossRef] [PubMed]

133. Dedkova, E.N.; Blatter, L.A. Nitric oxide inhibits capacitative Ca_{2+} entry and enhances endoplasmic reticulum Ca^{2+} uptake in bovine vascular endothelial cells. *J. Physiol.* **2002**, *539 (Pt 1)*, 77–91. [CrossRef] [PubMed]

134. Parekh, A.B. Functional consequences of activating store-operated CRAC channels. *Cell Calcium* **2007**, *42*, 111–121. [CrossRef] [PubMed]

135. Lin, S.; Fagan, K.A.; Li, K.X.; Shaul, P.W.; Cooper, D.M.; Rodman, D.M. Sustained endothelial nitric-oxide synthase activation requires capacitative Ca^{2+} entry. *J. Biol. Chem.* **2000**, *275*, 17979–17985. [CrossRef] [PubMed]

136. Kimura, C.; Oike, M.; Ohnaka, K.; Nose, Y.; Ito, Y. Constitutive nitric oxide production in bovine aortic and brain microvascular endothelial cells: A comparative study. *J. Physiol.* **2004**, *554 (Pt 3)*, 721–730. [CrossRef] [PubMed]

137. Boittin, F.X.; Alonso, F.; Le Gal, L.; Allagnat, F.; Beny, J.L.; Haefliger, J.A. Connexins and M3 muscarinic receptors contribute to heterogeneous Ca(2+) signaling in mouse aortic endothelium. *Cell. Physiol. Biochem.* **2013**, *31*, 166–178. [CrossRef] [PubMed]

138. Yeon, S.I.; Kim, J.Y.; Yeon, D.S.; Abramowitz, J.; Birnbaumer, L.; Muallem, S.; Lee, Y.H. Transient receptor potential canonical type 3 channels control the vascular contractility of mouse mesenteric arteries. *PLoS ONE* **2014**, *9*, e110413. [CrossRef] [PubMed]

139. Zhang, D.X.; Mendoza, S.A.; Bubolz, A.H.; Mizuno, A.; Ge, Z.D.; Li, R.; Warltier, D.C.; Suzuki, M.; Gutterman, D.D. Transient receptor potential vanilloid type 4-deficient mice exhibit impaired endothelium-dependent relaxation induced by acetylcholine in vitro and in vivo. *Hypertension* **2009**, *53*, 532–538. [CrossRef] [PubMed]

140. Zhao, Y.; Vanhoutte, P.M.; Leung, S.W. Vascular nitric oxide: Beyond eNOS. *J. Pharmacol. Sci.* **2015**, *129*, 83–94. [CrossRef] [PubMed]

141. Iadecola, C.; Zhang, F.; Xu, X. Role of nitric oxide synthase-containing vascular nerves in cerebrovasodilation elicited from cerebellum. *Am. J. Physiol.* **1993**, *264 (Pt 2)*, R738–R346. [CrossRef] [PubMed]

142. De Labra, C.; Rivadulla, C.; Espinosa, N.; Dasilva, M.; Cao, R.; Cudeiro, J. Different sources of nitric oxide mediate neurovascular coupling in the lateral geniculate nucleus of the cat. *Front. Syst. Neurosci.* **2009**, *3*, 9. [CrossRef] [PubMed]

143. Fiumana, E.; Parfenova, H.; Jaggar, J.H.; Leffler, C.W. Carbon monoxide mediates vasodilator effects of glutamate in isolated pressurized cerebral arterioles of newborn pigs. *Am. J. Physiol. Heart Circ. Physiol.* **2003**, *284*, H1073–H1079. [CrossRef] [PubMed]

144. Hogan-Cann, A.D.; Anderson, C.M. Physiological Roles of Non-Neuronal NMDA Receptors. *Trends Pharmacol. Sci.* **2016**, *37*, 750–767. [CrossRef] [PubMed]

145. Paoletti, P.; Bellone, C.; Zhou, Q. NMDA receptor subunit diversity: Impact on receptor properties, synaptic plasticity and disease. *Nat. Rev. Neurosci.* **2013**, *14*, 383–400. [CrossRef] [PubMed]

146. Fukaya, M.; Kato, A.; Lovett, C.; Tonegawa, S.; Watanabe, M. Retention of NMDA receptor NR2 subunits in the lumen of endoplasmic reticulum in targeted NR1 knockout mice. *Proc. Natl. Acad. Sci. USA* **2003**, *100*, 4855–4860. [CrossRef] [PubMed]

147. Neuhaus, W.; Freidl, M.; Szkokan, P.; Berger, M.; Wirth, M.; Winkler, J.; Gabor, F.; Pifl, C.; Noe, C.R. Effects of NMDA receptor modulators on a blood-brain barrier in vitro model. *Brain Res.* **2011**, *1394*, 49–61. [CrossRef] [PubMed]

148. Kamat, P.K.; Kalani, A.; Tyagi, S.C.; Tyagi, N. Hydrogen Sulfide Epigenetically Attenuates Homocysteine-Induced Mitochondrial Toxicity Mediated Through NMDA Receptor in Mouse Brain Endothelial (bEnd3) Cells. *J. Cell. Physiol.* **2015**, *230*, 378–394. [CrossRef] [PubMed]

149. Busija, D.W.; Bari, F.; Domoki, F.; Louis, T. Mechanisms involved in the cerebrovascular dilator effects of N-methyl-D-aspartate in cerebral cortex. *Brain Res. Rev.* **2007**, *56*, 89–100. [CrossRef] [PubMed]

150. Hamel, E. Cholinergic modulation of the cortical microvascular bed. *Prog. Brain Res.* **2004**, *145*, 171–178. [PubMed]

151. Zhang, F.; Xu, S.; Iadecola, C. Role of nitric oxide and acetylcholine in neocortical hyperemia elicited by basal forebrain stimulation: Evidence for an involvement of endothelial nitric oxide. *Neuroscience* **1995**, *69*, 1195–1204. [CrossRef]

152. Iadecola, C.; Zhang, F. Permissive and obligatory roles of NO in cerebrovascular responses to hypercapnia and acetylcholine. *Am. J. Physiol.* **1996**, *271 (Pt 2)*, R990–R1001. [CrossRef] [PubMed]

153. Yamada, M.; Lamping, K.G.; Duttaroy, A.; Zhang, W.; Cui, Y.; Bymaster, F.P.; McKinzie, D.L.; Felder, C.C.; Deng, C.X.; Faraci, F.M.; et al. Cholinergic dilation of cerebral blood vessels is abolished in M(5) muscarinic acetylcholine receptor knockout mice. *Proc. Natl. Acad. Sci. USA* **2001**, *98*, 14096–14101. [CrossRef] [PubMed]

154. Sandow, S.L.; Senadheera, S.; Grayson, T.H.; Welsh, D.G.; Murphy, T.V. Calcium and endothelium-mediated vasodilator signaling. *Adv. Exp. Med. Biol.* **2012**, *740*, 811–831. [PubMed]

155. Kasai, Y.; Yamazawa, T.; Sakurai, T.; Taketani, Y.; Iino, M. Endothelium-dependent frequency modulation of Ca^{2+} signalling in individual vascular smooth muscle cells of the rat. *J. Physiol.* **1997**, *504 (Pt 2)*, 349–357. [CrossRef] [PubMed]

156. Weikert, S.; Freyer, D.; Weih, M.; Isaev, N.; Busch, C.; Schultze, J.; Megow, D.; Dirnagl, U. Rapid Ca^{2+}-dependent NO-production from central nervous system cells in culture measured by NO-nitrite/ozone chemoluminescence. *Brain Res.* **1997**, *748*, 1–11. [CrossRef]

157. Radu, B.M.; Osculati, A.M.M.; Suku, E.; Banciu, A.; Tsenov, G.; Merigo, F.; Di Chio, M.; Banciu, D.D.; Tognoli, C.; Kacer, P.; et al. All muscarinic acetylcholine receptors (M1–M5) are expressed in murine brain microvascular endothelium. *Sci. Rep.* **2017**, *7*, 5083. [CrossRef] [PubMed]

158. Moccia, F.; Frost, C.; Berra-Romani, R.; Tanzi, F.; Adams, D.J. Expression and function of neuronal nicotinic ACh receptors in rat microvascular endothelial cells. *Am. J. Physiol. Heart Circ. Physiol.* **2004**, *286*, H486–H491. [CrossRef] [PubMed]

159. Laforenza, U.; Pellavio, G.; Moccia, F. Muscarinic M5 receptors trigger acetylcholine-induced Ca^{2+} signals and nitric oxide release in human brain microvascular endothelial cells. 2017; manuscript in preparation.

160. Uhlen, P.; Fritz, N. Biochemistry of calcium oscillations. *Biochem. Biophys. Res. Commun.* **2010**, *396*, 28–32. [CrossRef] [PubMed]

161. Mikoshiba, K. IP3 receptor/Ca^{2+} channel: From discovery to new signaling concepts. *J. Neurochem.* **2007**, *102*, 1426–1046. [CrossRef] [PubMed]

162. Guzman, S.J.; Gerevich, Z. P2Y Receptors in Synaptic Transmission and Plasticity: Therapeutic Potential in Cognitive Dysfunction. *Neural Plast.* **2016**, *2016*, 1207393. [CrossRef] [PubMed]

163. Munoz, M.F.; Puebla, M.; Figueroa, X.F. Control of the neurovascular coupling by nitric oxide-dependent regulation of astrocytic Ca(2+) signaling. *Front. Cell. Neurosci.* **2015**, *9*, 59. [CrossRef] [PubMed]

164. Dietrich, H.H.; Horiuchi, T.; Xiang, C.; Hongo, K.; Falck, J.R.; Dacey, R.G., Jr. Mechanism of ATP-induced local and conducted vasomotor responses in isolated rat cerebral penetrating arterioles. *J. Vasc. Res.* **2009**, *46*, 253–264. [CrossRef] [PubMed]

165. Dietrich, H.H.; Kajita, Y.; Dacey, R.G., Jr. Local and conducted vasomotor responses in isolated rat cerebral arterioles. *Am. J. Physiol.* **1996**, *71*, H1109–H1116. [CrossRef] [PubMed]

166. Toth, P.; Tarantini, S.; Davila, A.; Valcarcel-Ares, M.N.; Tucsek, Z.; Varamini, B.; Ballabh, P.; Sonntag, W.E.; Baur, J.A.; Csiszar, A.; et al. Purinergic glio-endothelial coupling during neuronal activity: Role of P2Y1 receptors and eNOS in functional hyperemia in the mouse somatosensory cortex. *Am. J. Physiol. Heart Circ. Physiol.* **2015**, *309*, H1837–H1845. [CrossRef] [PubMed]

167. Moccia, F.; Baruffi, S.; Spaggiari, S.; Coltrini, D.; Berra-Romani, R.; Signorelli, S.; Castelli, L.; Taglietti, V.; Tanzi, F. P2y1 and P2y2 receptor-operated Ca^{2+} signals in primary cultures of cardiac microvascular endothelial cells. *Microvasc. Res.* **2001**, *61*, 240–252. [CrossRef] [PubMed]

168. Marrelli, S.P. Mechanisms of endothelial P2Y(1)- and P2Y(2)-mediated vasodilatation involve differential $[Ca^{2+}]$i responses. *Am. J. Physiol. Heart Circ. Physiol.* **2001**, *281*, H1759–H1766. [CrossRef] [PubMed]

169. Kwan, H.Y.; Cheng, K.T.; Ma, Y.; Huang, Y.; Tang, N.L.; Yu, S.; Yao, X. CNGA2 contributes to ATP-induced noncapacitative Ca^{2+} influx in vascular endothelial cells. *J. Vasc. Res.* **2010**, *47*, 148–156. [CrossRef] [PubMed]

170. Duza, T.; Sarelius, I.H. Conducted dilations initiated by purines in arterioles are endothelium dependent and require endothelial Ca^{2+}. *Am. J. Physiol. Heart Circ. Physiol.* **2003**, *285*, H26–H37. [CrossRef] [PubMed]

171. Kohler, R.; Hoyer, J. The endothelium-derived hyperpolarizing factor: Insights from genetic animal models. *Kidney Int.* **2007**, *72*, 145–150. [CrossRef] [PubMed]

172. Ledoux, J.; Taylor, M.S.; Bonev, A.D.; Hannah, R.M.; Solodushko, V.; Shui, B.; Tallini, Y.; Kotlikoff, M.I.; Nelson, M.T. Functional architecture of inositol 1,4,5-trisphosphate signaling in restricted spaces of myoendothelial projections. *Proc. Natl. Acad. Sci. USA* **2008**, *105*, 9627–9632. [CrossRef] [PubMed]

173. Dora, K.A.; Gallagher, N.T.; McNeish, A.; Garland, C.J. Modulation of endothelial cell KCa3.1 channels during endothelium-derived hyperpolarizing factor signaling in mesenteric resistance arteries. *Circ. Res.* **2008**, *102*, 1247–1255. [CrossRef] [PubMed]

174. Sonkusare, S.K.; Bonev, A.D.; Ledoux, J.; Liedtke, W.; Kotlikoff, M.I.; Heppner, T.J.; Hill-Eubanks, D.C.; Nelson, M.T. Elementary Ca^{2+} signals through endothelial TRPV4 channels regulate vascular function. *Science* **2012**, *336*, 597–601. [CrossRef] [PubMed]

175. Behringer, E.J.; Segal, S.S. Spreading the signal for vasodilatation: Implications for skeletal muscle blood flow control and the effects of ageing. *J. Physiol.* **2012**, *590*, 6277–6284. [CrossRef] [PubMed]

176. Behringer, E.J.; Segal, S.S. Tuning electrical conduction along endothelial tubes of resistance arteries through Ca(2+)-activated K(+) channels. *Circ. Res.* **2012**, *110*, 1311–1321. [CrossRef] [PubMed]

177. Kapela, A.; Behringer, E.J.; Segal, S.S.; Tsoukias, N.M. Biophysical properties of microvascular endothelium: Requirements for initiating and conducting electrical signals. *Microcirculation* **2017**, *25*. [CrossRef] [PubMed]

178. Sullivan, M.N.; Gonzales, A.L.; Pires, P.W.; Bruhl, A.; Leo, M.D.; Li, W.; Oulidi, A.; Boop, F.A.; Feng, Y.; Jaggar, J.H.; et al. Localized TRPA1 channel Ca^{2+} signals stimulated by reactive oxygen species promote cerebral artery dilation. *Science Signal.* **2015**, *8*, ra2. [CrossRef] [PubMed]

179. Hannah, R.M.; Dunn, K.M.; Bonev, A.D.; Nelson, M.T. Endothelial SK(Ca) and IK(Ca) channels regulate brain parenchymal arteriolar diameter and cortical cerebral blood flow. *J. Cereb. Blood Flow Metab.* **2011**, *31*, 1175–1186. [CrossRef] [PubMed]

180. Sokoya, E.M.; Burns, A.R.; Setiawan, C.T.; Coleman, H.A.; Parkington, H.C.; Tare, M. Evidence for the involvement of myoendothelial gap junctions in EDHF-mediated relaxation in the rat middle cerebral artery. *Am. J. Physiol. Heart Circ. Physiol.* **2006**, *291*, H385–H393. [CrossRef] [PubMed]

181. Bagher, P.; Beleznai, T.; Kansui, Y.; Mitchell, R.; Garland, C.J.; Dora, K.A. Low intravascular pressure activates endothelial cell TRPV4 channels, local Ca^{2+} events, and IKCa channels, reducing arteriolar tone. *Proc. Natl. Acad. Sci. USA* **2012**, *109*, 18174–18179. [CrossRef] [PubMed]

182. Golding, E.M.; Marrelli, S.P.; You, J.; Bryan, R.M., Jr. Endothelium-derived hyperpolarizing factor in the brain: A new regulator of cerebral blood flow? *Stroke* **2002**, *33*, 661–663. [PubMed]

183. McNeish, A.J.; Dora, K.A.; Garland, C.J. Possible role for K^+ in endothelium-derived hyperpolarizing factor-linked dilatation in rat middle cerebral artery. *Stroke* **2005**, *36*, 1526–1532. [CrossRef] [PubMed]

184. Marrelli, S.P.; Eckmann, M.S.; Hunte, M.S. Role of endothelial intermediate conductance KCa channels in cerebral EDHF-mediated dilations. *Am. J. Physiol. Heart Circ. Physiol.* **2003**, *285*, H1590–H1599. [CrossRef] [PubMed]

185. McNeish, A.J.; Sandow, S.L.; Neylon, C.B.; Chen, M.X.; Dora, K.A.; Garland, C.J. Evidence for involvement of both IKCa and SKCa channels in hyperpolarizing responses of the rat middle cerebral artery. *Stroke* **2006**, *37*, 1277–1282. [CrossRef] [PubMed]

186. Zhang, L.; Papadopoulos, P.; Hamel, E. Endothelial TRPV4 channels mediate dilation of cerebral arteries: Impairment and recovery in cerebrovascular pathologies related to Alzheimer's disease. *Br. J. Pharmacol.* **2013**, *170*, 661–670. [CrossRef] [PubMed]

187. Dragoni, S.; Guerra, G.; Fiorio Pla, A.; Bertoni, G.; Rappa, A.; Poletto, V.; Bottino, C.; Aronica, A.; Lodola, F.; Cinelli, M.P.; et al. A functional Transient Receptor Potential Vanilloid 4 (TRPV4) channel is expressed in human endothelial progenitor cells. *J. Cell. Physiol.* **2015**, *230*, 95–104. [CrossRef] [PubMed]

188. Thodeti, C.K.; Matthews, B.; Ravi, A.; Mammoto, A.; Ghosh, K.; Bracha, A.L.; Ingber, D.E. TRPV4 channels mediate cyclic strain-induced endothelial cell reorientation through integrin-to-integrin signaling. *Circ. Res.* **2009**, *104*, 1123–1130. [CrossRef] [PubMed]

189. He, D.; Pan, Q.; Chen, Z.; Sun, C.; Zhang, P.; Mao, A.; Zhu, Y.; Li, H.; Lu, C.; Xie, M.; et al. Treatment of hypertension by increasing impaired endothelial TRPV4-KCa2.3 interaction. *EMBO Mol. Med.* **2017**, *9*, 1491–1503. [CrossRef] [PubMed]

190. Ho, W.S.; Zheng, X.; Zhang, D.X. Role of endothelial TRPV4 channels in vascular actions of the endocannabinoid, 2-arachidonoylglycerol. *Br. J. Pharmacol.* **2015**, *172*, 5251–5264. [CrossRef] [PubMed]

191. Mendoza, S.A.; Fang, J.; Gutterman, D.D.; Wilcox, D.A.; Bubolz, A.H.; Li, R.; Suzuki, M.; Zhang, D.X. TRPV4-mediated endothelial Ca^{2+} influx and vasodilation in response to shear stress. *Am. J. Physiol. Heart Circ. Physiol.* **2010**, *298*, H466–H476. [CrossRef] [PubMed]

192. Zheng, X.; Zinkevich, N.S.; Gebremedhin, D.; Gauthier, K.M.; Nishijima, Y.; Fang, J.; Wilcox, D.A.; Campbell, W.B.; Gutterman, D.D.; Zhang, D.X. Arachidonic acid-induced dilation in human coronary arterioles: Convergence of signaling mechanisms on endothelial TRPV4-mediated Ca^{2+} entry. *J. Am. Heart Assoc.* **2013**, *2*, e000080. [CrossRef] [PubMed]

193. Watanabe, H.; Vriens, J.; Suh, S.H.; Benham, C.D.; Droogmans, G.; Nilius, B. Heat-evoked activation of TRPV4 channels in a HEK293 cell expression system and in native mouse aorta endothelial cells. *J. Biol. Chem.* **2002**, *277*, 47044–47051. [CrossRef] [PubMed]

194. White, J.P.; Cibelli, M.; Urban, L.; Nilius, B.; McGeown, J.G.; Nagy, I. TRPV4: Molecular Conductor of a Diverse Orchestra. *Physiol. Rev.* **2016**, *96*, 911–973. [CrossRef] [PubMed]

195. Watanabe, H.; Vriens, J.; Prenen, J.; Droogmans, G.; Voets, T.; Nilius, B. Anandamide and arachidonic acid use epoxyeicosatrienoic acids to activate TRPV4 channels. *Nature* **2003**, *424*, 434–438. [CrossRef] [PubMed]

196. Graier, W.F.; Simecek, S.; Sturek, M. Cytochrome P450 mono-oxygenase-regulated signalling of Ca^{2+} entry in human and bovine endothelial cells. *J. Physiol.* **1995**, *482 (Pt 2)*, 259–274. [CrossRef] [PubMed]

197. Vriens, J.; Owsianik, G.; Fisslthaler, B.; Suzuki, M.; Janssens, A.; Voets, T.; Morisseau, C.; Hammock, B.D.; Fleming, I.; Busse, R.; et al. Modulation of the Ca^2 permeable cation channel TRPV4 by cytochrome P450 epoxygenases in vascular endothelium. *Circ. Res.* **2005**, *97*, 908–915. [CrossRef] [PubMed]

198. Earley, S.; Pauyo, T.; Drapp, R.; Tavares, M.J.; Liedtke, W.; Brayden, J.E. TRPV4-dependent dilation of peripheral resistance arteries influences arterial pressure. *Am. J. Physiol. Heart Circ. Physiol.* **2009**, *297*, H1096–H1102. [CrossRef] [PubMed]

199. Marrelli, S.P.; O'Neil R, G.; Brown, R.C.; Bryan, R.M., Jr. PLA2 and TRPV4 channels regulate endothelial calcium in cerebral arteries. *Am. J. Physiol. Heart Circ. Physiol.* **2007**, *292*, H1390–H1397. [CrossRef] [PubMed]

200. You, J.; Marrelli, S.P.; Bryan, R.M., Jr. Role of cytoplasmic phospholipase A2 in endothelium-derived hyperpolarizing factor dilations of rat middle cerebral arteries. *J. Cereb. Blood Flow Metab.* **2002**, *22*, 1239–1247. [CrossRef] [PubMed]

201. Liu, X.; Li, C.; Gebremedhin, D.; Hwang, S.H.; Hammock, B.D.; Falck, J.R.; Roman, R.J.; Harder, D.R.; Koehler, R.C. Epoxyeicosatrienoic acid-dependent cerebral vasodilation evoked by metabotropic glutamate receptor activation in vivo. *Am. J. Physiol. Heart Circ. Physiol.* **2011**, *301*, H373–H381. [CrossRef] [PubMed]

202. Peng, X.; Zhang, C.; Alkayed, N.J.; Harder, D.R.; Koehler, R.C. Dependency of cortical functional hyperemia to forepaw stimulation on epoxygenase and nitric oxide synthase activities in rats. *J. Cereb. Blood Flow Metab.* **2004**, *24*, 509–517. [CrossRef] [PubMed]

203. Chen, B.R.; Kozberg, M.G.; Bouchard, M.B.; Shaik, M.A.; Hillman, E.M. A critical role for the vascular endothelium in functional neurovascular coupling in the brain. *J. Am. Heart Assoc.* **2014**, *3*, e000787. [CrossRef] [PubMed]

204. Longden, T.A.; Dabertrand, F.; Koide, M.; Gonzales, A.L.; Tykocki, N.R.; Brayden, J.E.; Hill-Eubanks, D.; Nelson, M.T. Capillary K(+)-sensing initiates retrograde hyperpolarization to increase local cerebral blood flow. *Nat. Neurosci.* **2017**, *20*, 717–726. [CrossRef] [PubMed]

205. Earley, S.; Gonzales, A.L.; Crnich, R. Endothelium-dependent cerebral artery dilation mediated by TRPA1 and Ca^{2+}-Activated K^+ channels. *Circ. Res.* **2009**, *104*, 987–994. [CrossRef] [PubMed]

206. Qian, X.; Francis, M.; Solodushko, V.; Earley, S.; Taylor, M.S. Recruitment of dynamic endothelial Ca^{2+} signals by the TRPA1 channel activator AITC in rat cerebral arteries. *Microcirculation* **2013**, *20*, 138–148. [CrossRef] [PubMed]

207. Senadheera, S.; Kim, Y.; Grayson, T.H.; Toemoe, S.; Kochukov, M.Y.; Abramowitz, J.; Housley, G.D.; Bertrand, R.L.; Chadha, P.S.; Bertrand, P.P.; et al. Transient receptor potential canonical type 3 channels facilitate endothelium-derived hyperpolarization-mediated resistance artery vasodilator activity. *Cardiovasc. Res.* **2012**, *95*, 439–447. [CrossRef] [PubMed]

208. Kochukov, M.Y.; Balasubramanian, A.; Abramowitz, J.; Birnbaumer, L.; Marrelli, S.P. Activation of endothelial transient receptor potential C3 channel is required for small conductance calcium-activated potassium

channel activation and sustained endothelial hyperpolarization and vasodilation of cerebral artery. *J. Am. Heart Assoc.* **2014**, *3*, e000913. [CrossRef] [PubMed]

209. Pires, P.W.; Sullivan, M.N.; Pritchard, H.A.; Robinson, J.J.; Earley, S. Unitary TRPV3 channel Ca^{2+} influx events elicit endothelium-dependent dilation of cerebral parenchymal arterioles. *Am. J. Physiol. Heart Circ. Physiol.* **2015**, *309*, H2031–H2041. [CrossRef] [PubMed]

210. Yang, P.; Zhu, M.X. Trpv3. In *Handbook of Experimental Pharmacology*; Springer: Berlin, Germany, 2014; Volume 222, pp. 273–291.

211. Zygmunt, P.M.; Hogestatt, E.D. Trpa1. In *Handbook of Experimental Pharmacology*; Springer: Berlin, Germany, 2014; Volume 222, pp. 583–630.

212. Iadecola, C. The pathobiology of vascular dementia. *Neuron* **2013**, *80*, 844–866. [CrossRef] [PubMed]

213. Chen, B.R.; Bouchard, M.B.; McCaslin, A.F.; Burgess, S.A.; Hillman, E.M. High-speed vascular dynamics of the hemodynamic response. *NeuroImage* **2011**, *54*, 1021–1030. [CrossRef] [PubMed]

214. Hillman, E.M.; Devor, A.; Bouchard, M.B.; Dunn, A.K.; Krauss, G.W.; Skoch, J.; Bacskai, B.J.; Dale, A.M.; Boas, D.A. Depth-resolved optical imaging and microscopy of vascular compartment dynamics during somatosensory stimulation. *NeuroImage* **2007**, *35*, 89–104. [CrossRef] [PubMed]

215. Stefanovic, B.; Hutchinson, E.; Yakovleva, V.; Schram, V.; Russell, J.T.; Belluscio, L.; Koretsky, A.P.; Silva, A.C. Functional reactivity of cerebral capillaries. *J. Cereb. Blood Flow Metab.* **2008**, *28*, 961–972. [CrossRef] [PubMed]

216. O'Herron, P.; Chhatbar, P.Y.; Levy, M.; Shen, Z.; Schramm, A.E.; Lu, Z.; Kara, P. Neural correlates of single-vessel haemodynamic responses in vivo. *Nature* **2016**, *534*, 378–382. [CrossRef] [PubMed]

217. Kisler, K.; Nelson, A.R.; Rege, S.V.; Ramanathan, A.; Wang, Y.; Ahuja, A.; Lazic, D.; Tsai, P.S.; Zhao, Z.; Zhou, Y.; et al. Pericyte degeneration leads to neurovascular uncoupling and limits oxygen supply to brain. *Nat. Neurosci.* **2017**, *20*, 406–416. [CrossRef] [PubMed]

218. Behringer, E.J.; Socha, M.J.; Polo-Parada, L.; Segal, S.S. Electrical conduction along endothelial cell tubes from mouse feed arteries: Confounding actions of glycyrrhetinic acid derivatives. *Br. J. Pharmacol.* **2012**, *166*, 774–787. [CrossRef] [PubMed]

219. Vaucher, E.; Hamel, E. Cholinergic basal forebrain neurons project to cortical microvessels in the rat: Electron microscopic study with anterogradely transported Phaseolus vulgaris leucoagglutinin and choline acetyltransferase immunocytochemistry. *J. Neurosci.* **1995**, *15*, 7427–7441. [PubMed]

220. Vaucher, E.; Linville, D.; Hamel, E. Cholinergic basal forebrain projections to nitric oxide synthase-containing neurons in the rat cerebral cortex. *Neuroscience* **1997**, *79*, 827–836. [CrossRef]

221. Jensen, L.J.; Holstein-Rathlou, N.H. The vascular conducted response in cerebral blood flow regulation. *J. Cereb. Blood Flow Metab.* **2013**, *33*, 649–656. [CrossRef] [PubMed]

222. Uhrenholt, T.R.; Domeier, T.L.; Segal, S.S. Propagation of calcium waves along endothelium of hamster feed arteries. *Am. J. Physiol. Heart Circ. Physiol.* **2007**, *292*, H1634–H1640. [CrossRef] [PubMed]

223. Domeier, T.L.; Segal, S.S. Electromechanical and pharmacomechanical signalling pathways for conducted vasodilatation along endothelium of hamster feed arteries. *J. Physiol.* **2007**, *579*, 175–186. [CrossRef] [PubMed]

224. De Bock, M.; Culot, M.; Wang, N.; da Costa, A.; Decrock, E.; Bol, M.; Bultynck, G.; Cecchelli, R.; Leybaert, L. Low extracellular Ca^{2+} conditions induce an increase in brain endothelial permeability that involves intercellular Ca^{2+} waves. *Brain Res.* **2012**, *1487*, 78–87. [CrossRef] [PubMed]

225. Berra-Romani, R.; Raqeeb, A.; Torres-Jácome, J.; Guzman-Silva, A.; Guerra, G.; Tanzi, F.; Moccia, F. The mechanism of injury-induced intracellular calcium concentration oscillations in the endothelium of excised rat aorta. *J. Vasc. Res.* **2012**, *49*, 65–76. [CrossRef] [PubMed]

226. Francis, M.; Waldrup, J.R.; Qian, X.; Solodushko, V.; Meriwether, J.; Taylor, M.S. Functional Tuning of Intrinsic Endothelial Ca^{2+} Dynamics in Swine Coronary Arteries. *Circ. Res.* **2016**, *118*, 1078–1090. [CrossRef] [PubMed]

227. Wilson, C.; Saunter, C.D.; Girkin, J.M.; McCarron, J.G. Clusters of specialized detector cells provide sensitive and high fidelity receptor signaling in the intact endothelium. *FASEB J.* **2016**, *30*, 2000–2013. [CrossRef] [PubMed]

228. Kohler, R.; Hoyer, J. Role of TRPV4 in the Mechanotransduction of Shear Stress in Endothelial Cells. In *TRP Ion Channel Function in Sensory Transduction and Cellular Signaling Cascades*; Liedtke, W.B., Heller, S., Eds.; CRC Press: Boca Raton, FL, USA, 2007.

Int. J. Mol. Sci. **2018**, *19*, 938

229. Bryan, R.M., Jr.; Marrelli, S.P.; Steenberg, M.L.; Schildmeyer, L.A.; Johnson, T.D. Effects of luminal shear stress on cerebral arteries and arterioles. *Am. J. Physiol. Heart Circ. Physiol.* **2001**, *280*, H2011–H2022. [CrossRef] [PubMed]

230. Bryan, R.M., Jr.; Steenberg, M.L.; Marrelli, S.P. Role of endothelium in shear stress-induced constrictions in rat middle cerebral artery. *Stroke* **2001**, *32*, 1394–1400. [CrossRef] [PubMed]

231. Ngai, A.C.; Winn, H.R. Modulation of cerebral arteriolar diameter by intraluminal flow and pressure. *Circ. Res.* **1995**, *77*, 832–840. [CrossRef] [PubMed]

232. Lourenco, C.F.; Ledo, A.; Barbosa, R.M.; Laranjinha, J. Neurovascular uncoupling in the triple transgenic model of Alzheimer's disease: Impaired cerebral blood flow response to neuronal-derived nitric oxide signaling. *Exp. Neurol.* **2017**, *291*, 36–43. [CrossRef] [PubMed]

233. Lourenco, C.F.; Ledo, A.; Dias, C.; Barbosa, R.M.; Laranjinha, J. Neurovascular and neurometabolic derailment in aging and Alzheimer's disease. *Front. Aging Neurosci.* **2015**, *7*, 103. [CrossRef] [PubMed]

234. Katakam, P.V.; Domoki, F.; Snipes, J.A.; Busija, A.R.; Jarajapu, Y.P.; Busija, D.W. Impaired mitochondria-dependent vasodilation in cerebral arteries of Zucker obese rats with insulin resistance. *Am. J. Physiol. Regul. Integr. Comp. Physiol.* **2009**, *296*, R289–R298. [CrossRef] [PubMed]

235. Mottola, A.; Antoniotti, S.; Lovisolo, D.; Munaron, L. Regulation of noncapacitative calcium entry by arachidonic acid and nitric oxide in endothelial cells. *FASEB J.* **2005**, *19*, 2075–2077. [CrossRef] [PubMed]

International Journal of
Molecular Sciences

MDPI

Review

Correcting Calcium Dysregulation in Chronic Heart Failure Using *SERCA2a* Gene Therapy

T. Jake Samuel [1], Ryan P. Rosenberry [1], Seungyong Lee [1] and Zui Pan [2,*]

[1]   Department of Kinesiology, College of Nursing and Health Innovation,
      The University of Texas at Arlington, Arlington, TX 76019, USA; Thomas.Samuel@uta.edu (T.J.S.);
      ryan.rosenberry@uta.edu (R.P.R.); seungyong.lee@uta.edu (S.L.)
[2]   Department of Graduate Nursing, College of Nursing and Health Innovation,
      The University of Texas at Arlington, Arlington, TX 76019, USA
*   Correspondence: zui.pan@uta.edu; Tel.: +1-817-272-2595

Received: 27 February 2018; Accepted: 3 April 2018; Published: 5 April 2018

Abstract: Chronic heart failure (CHF) is a major contributor to cardiovascular disease and is the leading cause of hospitalization for those over the age of 65, which is estimated to account for close to seventy billion dollars in healthcare costs by 2030 in the US alone. The successful therapies for preventing and reversing CHF progression are urgently required. One strategy under active investigation is to restore dysregulated myocardial calcium (Ca^{2+}), a hallmark of CHF. It is well established that intracellular Ca^{2+} concentrations are tightly regulated to control efficient myocardial systolic contraction and diastolic relaxation. Among the many cell surface proteins and intracellular organelles that act as the warp and woof of the regulatory network controlling intracellular Ca^{2+} signals in cardiomyocytes, sarco/endoplasmic reticulum Ca^{2+} ATPase type 2a (SERCA2a) undoubtedly plays a central role. SERCA2a is responsible for sequestrating cytosolic Ca^{2+} back into the sarcoplasmic reticulum during diastole, allowing for efficient uncoupling of actin-myosin and subsequent ventricular relaxation. Accumulating evidence has demonstrated that the expression of SERCA2a is downregulated in CHF, which subsequently contributes to severe systolic and diastolic dysfunction. Therefore, restoring SERCA2a expression and improving cardiomyocyte Ca^{2+} handling provides an excellent alternative to currently used transplantation and mechanical assist devices in the treatment of CHF. Indeed, advancements in safe and effective gene delivery techniques have led to the emergence of *SERCA2a* gene therapy as a potential therapeutic choice for CHF patients. This mini-review will succinctly detail the progression of *SERCA2a* gene therapy from its inception in plasmid and animal models, to its clinical trials in CHF patients, highlighting potential avenues for future work along the way.

Keywords: calcium handling; SERCA2a; chronic heart failure; clinical trial; adeno-associated virus (AAV); gene delivery

---

1. Introduction

Chronic heart failure (CHF) is a major form of cardiovascular disease and is the leading cause of hospitalization for those over the age of 65 [1]. It is expected to account for almost seventy billion dollars in healthcare costs by 2030 in the US alone [2]. The current pharmacological therapies for CHF merely target symptom management, with limited success in treating the underlying etiology of the disease. The late stage CHF patients eventually need expensive and hard-to-obtain heart transplantation or mechanical assist devices. Therefore, successful therapies for preventing and reversing CHF progression are urgently required.

Understanding the basic mechanisms involved in the development of CHF has been an active field of research in the quest to identify abnormalities that could potentially be targeted by gene transfer.

It has been well established that myocardial calcium (Ca^{2+}) dysregulation is a hallmark of CHF. Tight control of cardiac intracellular Ca^{2+} handling plays an integral role in synchronous actin-myosin cross-bridge cycling and the resulting systolic contraction and diastolic relaxation. Many proteins either at the cell surface or intracellular organelles form the warp and woof of the regulatory network for intracellular Ca^{2+} signals in cardiomyocytes (Figure 1). During systole, Ca^{2+} enters the sarcolemma through L-type Ca^{2+} channels, i.e., dihydropyrodine receptor (DHPR) located on the transverse-tubules (T-tubules), before diffusing across the short distance to the ryanodine receptors (RyRs) located to the sarcoplasmic reticulum (SR). This small amount of Ca^{2+} influx triggers RyRs to generate further Ca^{2+}-induced-Ca^{2+}-release (CICR) from the SR to form magnified Ca^{2+} transient [3]. Ca^{2+} then binds to troponin-C in the sarcomere and stimulates actin-myosin cross-bridge linking, i.e., systolic contraction [4]. During diastole, in order to reduce the likelihood of Ca^{2+} binding to troponin-C and causing prolonged myocardial systolic contraction, Ca^{2+} is actively sequestered back into the SR via a sarco/endoplasmic reticulum Ca^{2+} ATPase, SERCA2a in cardiomyocyte [5,6], or expelled from the sarcolemma by sodium-Ca^{2+} exchanger pumps (NCX) [3]. As such, SERCA2a is undoubtedly essential for complete myocyte Ca^{2+} homeostasis.

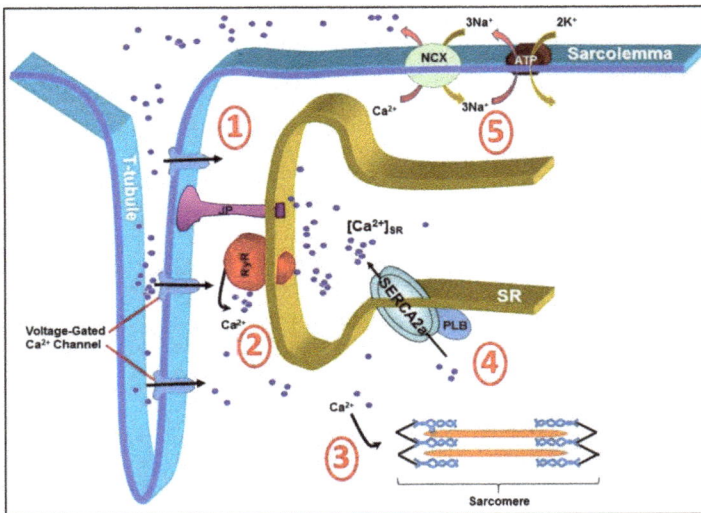

Figure 1. Cardiac intracellular Ca^{2+} is tightly regulated by several proteins. Efficient systolic contraction and diastolic relaxation is reliant on efficient Ca^{2+} handling through 5 main processes. (1) Diffusion of Ca^{2+} in to the cytosol via voltage-gated Ca^{2+} channels (DHPRs) located on the surface of the transverse-tubule (T-tubule); (2) Ca^{2+}-induced-Ca^{2+}-release from the ryanodine receptors (RyRs); (3) Binding of Ca^{2+} to troponin-C in the sarcomere, stimulating actin-myosin cross-linking; (4) Sequestration of Ca^{2+} back in to the sarcoplasmic reticulum via the important Ca^{2+} pump sarco/endoplasmic reticulum Ca^{2+} ATPase (SERCA2a); and (5) Expulsion of Ca^{2+} from the cell via sodium-calcium exchanger pumps (NCX). JP, junctophilins; PLB, phopholamban; ATP, ATP pump; $3Na^+$, sodium; K^+, potassium.

Dysregulation of any of the above Ca^{2+} handling processes will result in impaired ventricular contractility and impaired myocyte relaxation, leading to cardiac dysfunction [4,7,8]. While efforts have been devoted to investigating the therapeutic value of many Ca^{2+} channels and proteins in regulating cytosolic Ca^{2+} concentrations, SERCA2a has received the most interest in recent years.

Reduction in SERCA2a levels in the SR has been reported in failing heart tissues [9,10]. Impaired Ca^{2+}-ATPase pump function and decreased *SERCA2a* gene transcription has recently been attributed to delayed left ventricular relaxation and increased left ventricular filling pressures [11] which may lead to pulmonary edema, a common symptom in CHF (Figure 2). Biotechnological advances over the past two decades have led to gene therapy becoming a viable option for treating many pathologic conditions—including CHF [12–14]. Matter of fact, SERCA2a had been identified as a possible gene therapy target as early as 1978 [15]. This review will focus on *SERCA2a* gene therapy in CHF and attempt to highlight the seminal investigations using early gene delivery techniques in animals that contributed to the early promise for its use in humans. It will also provide a summary of the major human clinical trials which identified several limitations to these methods for treating human CHF. Finally, recent novel developments in gene delivery and targeting will be discussed and future directions for this field of research will be proposed.

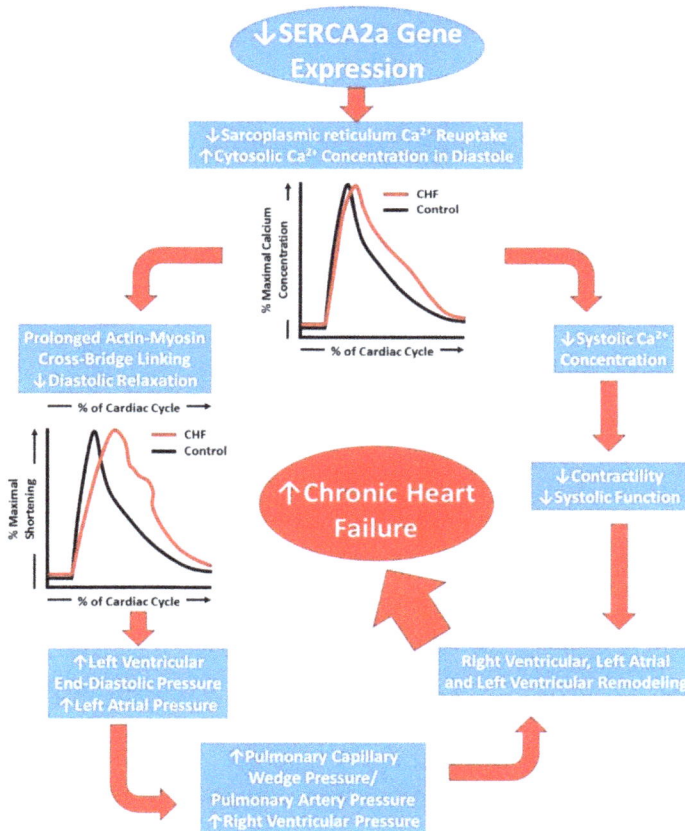

Figure 2. Schematic representation of the progression of chronic heart failure. Initial downregulation of SERCA2a function leads to increased cytosolic Ca^{2+}, which ultimately compromises left ventricular contractility (systolic function) and leads to prolonged left ventricular relaxation. Prolonged relaxation leads to increased filling pressures and a backlog of pressure in to the pulmonary circulation and right heart, the result of which leads to severe right and left heart remodeling and chronic heart failure development.

2. Early Techniques of *SERCA2a* Gene Manipulation in Animals

The role of cytosolic Ca^{2+} in cardiac myocyte contraction has been intensely studied for decades. As far back as 40 years ago, investigators were purifying SERCA proteins to better understand the molecular mechanisms through which Ca^{2+} is transported from the cytosol into the sarcoplasmic reticulum (SR) [15]. Shortly following the identification of Ca^{2+} channels as key mediators of cardiac function, they became therapeutic targets in the treatment of various cardiac ailments [16–18]. With the rise of interest in genomics and gene editing techniques, the direct manipulation of Ca^{2+} channels or pumps, and thus manipulation of Ca^{2+} handling in cardiac myocytes, became possible.

2.1. Plasmid and Direct DNA Injection

The earliest successful approaches of manipulating *SERCA2a* gene expression in animal models were performed using direct insertion of plasmids into developing mouse oocytes. The rat SERCA2a cDNA was cloned into plasmids containing mouse cardiac α-MHC promoter. While certain repetitive and unnecessary exons were removed, an additional human growth hormone polyadenylation site was included to promote both polyadenylation and termination [9]. Once these sequences are successfully cloned into the plasmid, restriction endonucleases are applied to cleave the gene from the plasmid. The linear DNA fragment can then be purified for direct injection into oocytes. The resulting transgenic mice possessed extra copies of the *SERCA2a* gene and exhibited increased expression of the protein [9,19,20].

Initial studies applying this transgenic technique sought to characterize and measure the effects of the increased SERCA2a expression in otherwise unaltered mice. Early investigations found that overexpression of the *SERCA2a* gene in a transgenic mouse model resulted in improved Ca^{2+} handling which translated to augmented myocyte contraction and relaxation [9,10,20]. Of note, the transgenic mice had a significantly shorter relaxation time indicating improved relaxation kinetics, consistent with improved Ca^{2+} sequestration via SERCA2a [20]. With these positive proof-of-concept results, investigators next sought to determine whether SERCA2a could improve cardiac functionality in disease stages. To test this potential treatment, an aortic stenosis-induced CHF model was employed in transgenic mice. The increased afterload from the artificial stenosis caused left ventricular hypertrophy, mimicking the conditions of early CHF. In this investigation, the transgenic expression of SERCA2a provided chronic protective effects against the progression of CHF [19].

Although these early studies demonstrated positive and profound effects from direct injection of DNA fragment containing *SERCA2a* gene, there are substantial limitations inherent to this technique. First, this laboratory procedure requires a significant time commitment and many uncontrollable steps before the success of the transfection can be determined. In the case of He et al. super-ovulated eggs were treated with purified SERCA2a DNA, gestated for 20 days, and the newborn mice were allowed to develop for three weeks before testing tail samples for SERCA2a expression by Southern blotting [20]. This process relies on integration of the gene into the embryonic genome, normal gestation and delivery, and inerrant expression in the developed animal. Even if each step is performed to perfection, the imprecision introduces excessive error and contributes to study failure. In fact, only three mice out of nine integrated the SERCA2a plasmid. Of these three mice, the single male was sterile, leaving only two female transgenic mice that successfully passed the gene on to their offspring [20].

A second limitation of this technique is that for additional copies of SERCA2a to be expressed, all biological machinery in the transcription/translation/post-translational modification pathway must also function in a normal and healthy manner. Direct injection of the DNA into oocytes carries the potential risk for improper integration of the gene into the genome. As evidenced by the low number of successfully reared transgenic mice, the possibility for lethality or sterility is considerable. Furthermore, this approach cannot be applied to developed organisms, hindering its expansion and/or adaptation into other animal models. To further investigate SERCA2a as an intervention, it was necessary to explore more advanced and versatile gene modifying techniques.

Last and foremost, direct manipulation of human oocytes raises serious ethical issues, which disqualify this approach from developing into a useful human gene therapy for CHF. Rather, these transgenic animal studies laid the theoretical understanding that prompted improvements in biotechnological techniques to tackle this problem. Alternative ethical and safer approaches have been called upon.

2.2. Adenoviral (Ad)-Based Vectors

Ad are double-strand DNA vectors that bind to and enter the cell membrane before being transported to the nucleus, allowing transfection of genes into a variety of cells such as cardiomyocytes, skeletal muscle and smooth muscle myocytes [21]. Delivery of Ad-based vectors to cardiomyocytes through intracoronary infusion and direct myocardial injection has showed some success in transduction of the target gene [22]. Although Ad-based vectors attain a high level of cellular transduction, the various Ad serotypes result in varied efficiency [21], leading to mixed success in modulating target gene expression.

Ad mediated *SERCA2a* gene delivery (Ad/SERCA2a) was reported to present improvements in cardiac function in CHF due to the upregulation of *SERCA2a* genes. Ad/*SERCA2a* gene transfer in mice with CHF enhances contractile function [23] and restores phosphocreatine and ATP level in the heart, which represents the recovery of heart energetic state and removal of myocardial ischemia [24]. Furthermore, there is a dose response to Ad/*SERCA2a* transfer as overexpression of the *SERCA2a* gene promotes an additive increase in SERCA2a activity which further influences improved ventricular contractility [25] but has no effect on the atrium [26]. In addition to improvements in left ventricular contractility, Ad/*SERCA2a* gene transfer has also been shown to promote reverse myocardial remodeling by reducing left ventricular anterior wall thickening in Wistar rats with CHF, which reduced arrhythmic events and caused the restoration of left ventricular function similar to the observation in control animals [27].

The benefits of *SERCA2a* gene transfer via Ad-based vectors were not only limited to the myocardial tissue, however, as it also showed benefits on the vascular smooth muscle cells by preventing vascular remodeling and inhibition of neointimal thickening in a human ex vivo model of the coronary artery [28] and augmented coronary blood flow [29]. Although the Ad-based vector technique was successful overall, there were some inherent limitations to this gene delivery method, such as the presence of neutralizing antibodies and the severe immune response against Ad vector and/or the gene-modified cells [21]. Therefore, the development of more sophisticated viral vectors techniques was explored to overcome these limitations.

2.3. Adeno-Associated Virus (AAV)-Based Vectors

The AAV-based vector has significant advantage over Ad-based vector in gene delivery. To date, several generations of AAV systems have been adapted and adopted in the *SERCA2a* gene delivery with each generation improving delivery efficiency and safety over the previous one. AAV-based vector has been shown to safely and successfully alter gene expression in cardiac tissue [21], with much less inflammation than that associated with the Ad-based vector technique [21]. AAVs comprise more than 100 serotypes and each serotype presents distinct transduction efficiency in different tissues [21]. Among these AAVs, the AAV-1 was the best studied and AAV1-based vector is most commonly used. Indeed, several earlier trials were conducted using the AAV1/SERCA2a intra-coronary injection method in animal CHF models with good success. Hadri and colleagues demonstrated that long-term SERCA2a overexpression via in vivo AAV1-mediated gene transfer improves left ventricular ejection fraction, coronary artery blood flow, and expression of endothelial nitric oxide synthase in a swine model of CHF [30]. Similarly, AAV1/*SERCA2a* gene delivery improves echocardiography derived myocardial function and decreases myocardial apoptosis in pigs [31], as well as restores SERCA2a activity and protein expression following atrial fibrillation-related declines in SERCA2a activity and expression [32]. Finally, *SERCA2a* gene transfer not only effects cardiac tissue but also directly influences vascular endothelial and smooth muscle cell function by improving Ca^{2+} signaling [33] which highlights SERCA2a as an important therapeutic target for treating cardiovascular disease—including CHF.

Overall, AAV1 and AAV6 mediated-*SERCA2a* gene delivery methods have provided beneficial improvements in myocardial and coronary artery Ca^{2+} handling, contributing to improved cardiac function and reverse remodeling in animal models of CHF. This initial success in improving cardiac function in animal models led to large-scale clinical trials in humans which will be discussed in the following section.

3. Human Trials

Over the past decade several large-scale clinical trials have been performed using AAV1/SERCA2a in the treatment of human CHF (Table 1) and are discussed below.

3.1. The Calcium Upregulation by Percutaneous Administration of Gene Therapy in Cardiac Disease (CUPID) Trial

In response to the promising preclinical models in animals that demonstrated AAV1/SERCA2a were well tolerated and improved cardiac function, CUPID was the first-in-human clinical trial to evaluate the efficacy of recombinant AAV1/SERCA2a in human CHF [12,13]. The initial CUPID multicenter trial included two phases, wherein phase 1 was an open-label dose escalation protocol and phase 2, a randomized, double-blind, placebo-controlled trial. The overall primary outcome of the CUPID trials was to monitor safety, while secondary outcomes included improvements in physical activity and efficacy.

In CUPID phase 1, 12 patients were administered single intracoronary infusions, comparing 3 dose levels of AAV1/SERCA2a (AAV1 capsid with human SERCA2a cDNA flanked with inverted terminal repeats) with placebo, and primary outcomes were measured at 6 and 12 months post-infusion. Patients were enrolled independent of CHF etiology, were NYHA class III/IV, and had exhausted all current pharmacologic therapy. The results of phase 1 suggested that intracoronary infusion of AAV1/SERCA2a demonstrated an acceptable safety profile given the advanced CHF population. Several of the patients exhibited improvements in CHF symptoms and left ventricular structure and function at 6 months follow-up [12]. Of note, two of the patients who did not respond to therapy already had pre-existing neutralizing antibodies (NAb) for the viral capsid proteins. Pre-existing natural exposure to NAb, which are known to inhibit vector uptake, limit the efficacy of AAV1 treatment in CHF [34–36]. Discussion of the impact of NAb on vector uptake are beyond the scope of the current review and have been comprehensively described elsewhere [37].

The relative success of phase 1 led to initiation of phase 2, including 39 advanced CHF patients randomized to 3 doses of viral vector; low (6×10^{11} DNase resistant particles; DRP), medium (3×10^{12} DRP), and high doses (1×10^{13} DRP). Endpoints included patients' symptomatic (NYHA functional class, Minnesota Living with Heart Failure Questionnaire; MLWHFQ) and functional (maximal oxygen uptake, 6-minute Walk Test) status, blood biomarker levels (N-Terminal-pro Brain Natriuretic Peptide, NT-proBNP), and left ventricular function and remodeling (Ejection fraction and End-systolic volume). One-year follow-up suggested the patients randomized to the "high" treatment dose of AAV1/SERCA2a improved in functional class (Mean reduction in MLWHFQ: -10.3 ± 12.21), blood biomarker (NT-proBNP mean reduction: $12.4 \pm 47.83\%$) and left ventricular function (end-systolic volume mean reduction: -9.6 ± 27.55 mL; $-4 \pm 13.76\%$). Importantly, there was a significant increase in the time to cardiac event in all AAV1/SERCA2a groups in the CUPID trial, suggesting that perhaps even the lower doses of viral vector had beneficial effects on patient outcome, independent of physiological changes that are not appreciable at smaller sample sizes [13].

Table 1. Current and previous clinical trials to test the efficacy of AAV1/SERCA2a on the Heart Failure.

National Clinical Trial Code	Status	Study Title	Conditions	Interventions	Study Results	References
NCT02772068	Recruiting	Hemodynamic Response to Exercise in HFpEF Patients After Upregulation of SERCA2a	Congestive Heart Failure	Drug: Istaroxime Other: Exercise	No Published Results	-
NCT01966887	Terminated	AAV1-CMV-Serca2a GENe Therapy Trial in Heart Failure (AGENT-HF)	Congestive Heart Failure Ischemic and non-ischemic Cardiomyopathies	Genetic: AAV1/SERCA2a (MYDICAR)-single intracoronary infusion Genetic: Placebo; single intracoronary infusion	No Positive or Negative Effects	[N]
NCT00534703	Terminated	Investigation of the Safety and Feasibility of AAV1/SERCA2a Gene Transfer in Patients with Chronic Heart Failure (SERCA-LVAD)	Chronic Heart Failure Left Ventricular Assist Device	Genetic: AAV1/SERCA2a Drug; Placebo	Terminated Early—No Results	-
NCT01643330	Completed	A Study of Genetically Targeted Enzyme Replacement Therapy for Advanced Heart Failure (CUPID-2b)	Ischemic and non-ischemic Cardiomyopathies Heart Failure	Genetic: AAV1/SERCA2a (MYDICAR) Genetic: Placebo	No Positive or Negative Effects	[12–14,37,39–41]
NCT00454818	Completed	Efficacy and Safety Study of Genetically Targeted Enzyme Replacement Therapy for Advanced Heart Failure (CUPID)	Heart Failure, Congestive Dilated Cardiomyopathy	Genetic: MYDICAR Phase 1 (Open-label, Serial Dose-Escalation Study) Procedure: Placebo Infusion Genetic: MYDICAR Phase 2 (Placebo-controlled, Randomized Study)	Positive Results	[12,13,40–42]
NCT02346422	Terminated	A Phase 1/2 Study of High-Dose Genetically Targeted Enzyme Replacement Therapy for Advanced Heart Failure	Heart Failure Cardiomyopathy	Genetic: MYICAR Phase 1/2 (Dose 2.5 × 10¹³ DRP)	Terminated Early—No Results	[12,13,40]

3.2. Post-CUPID

Since the conclusion of CUPID, three clinical trials utilizing AAV1 have targeted SERCA2a in human CHF [14,38,39,43]. CUPID2, a continuation of the CUPID clinical trials, was the first to recruit patients from outside the United States, enrolling a total of 250 patients with CHF. Patients were randomized to 10-minute intracoronary infusion of either placebo or "high" dose of AAV1/SERCA2a. In contrast to the results of the original CUPID trial, CUPID2 did not reduce either recurrent heart failure events, or terminal events in the study population [14,39].

The other two post-CUPID trails began in the United Kingdom, known as SERCA-LVAD and AGENT-HF. SERCA-LVAD aimed to take tissue biopsies at the time of left ventricular assist device implantation (pre-infusion of AAV1/SERCA2a) and at the time of transplant or LVAD removal (post-infusion) in order to correlate changes in clinical outcome with SERCA2a expression levels [43]. AGENT-HF utilized AAV1/SERCA2a with left ventricular structure and function as the primary outcome variables assessed at 6 months post-treatment [38]. Unfortunately, due to the negative results of the CUPID2 trial and lack of preliminary success, both studies were terminated prematurely. The studies showed no improvement in ventricular remodeling but were underpowered to demonstrate any effect of AAV1/*SERCA2a* gene therapy [38]. Taken together, the human trials utilizing AAV1/SERCA2a to treat HF have been widely unsuccessful to date.

The early clinical trial CUPID provided the basis for several more aggressive clinical trials (CUPID2, Agent-HF, SERCA-LVAD), all of which failed to demonstrate improvement in survival and/or were terminated prior to sufficient data collection. The reason for the failure of CUPID2 remains largely unknown. One possible reason is attributable to the insufficient delivery of SERCA2a DNA using intracoronary infusion of AAV1 [44], as intracoronary infusion may result in inadequate uptake of the viral vector into cardiomyocytes. In addition, work by Mingozzi and colleagues suggests that viral loads which include empty viral capsid particles enhance the gene delivery as they act as "decoys" and may block the inhibitory activity of antibodies [45]. Thus, *SERCA2a* gene therapy may not be at a dead end yet and there is some light through the tunnel to show it still as a valuable approach for HF patients. The challenge is rather how to effectively deliver the gene to cardiomyocytes. Are there better, more potent viral vectors and more advanced gene editing techniques for gene delivery and expression in human heart?

4. The Promise of Novel Gene Delivery Techniques in Animals

Among more than 100 wild-type AAV serotypes, some serotypes have been demonstrated to have distinct features on cardiomyocyte transduction [21]. Compared to AAV1, AAV8 showed 20-fold higher gene expression in the myocardium in mice. Systemic venous infusion of AAV9 triggers robust level (>200-fold) of gene expression in the cardiomyocytes when compared to AAV1 [46]. To date, AAV9-based vectors are by far the most efficient methods for delivering genes to cardiac tissue in mice [46–48]. However, AAV6 was also reported to mediate the most efficient transduction in mouse cardiac tissue [49].

Based on these findings, many recent studies have adopted the AAV9 method to transfer *SERCA2a* genes for treating CHF in animal models. Lyon et al. showed that *SERCA2a* gene transfer with AAV9 restored SERCA2a proteins in the rat heart and improved left ventricular function by reducing Ca^{2+} leakage from sarcoplasmic reticulum [50]. In addition, AAV9/SERCA2a therapy recovers the electrical signal constancy by improving ventricular arrhythmias [50,51] and the treatment rectifies ECG abnormalities such as tachycardia, shortened P-R interval, and prolonged Q-T intervals [52]. Similarly, *SERCA2a* gene therapy overexpressed SERCA2a protein and repressed atrial fibrillation induced by rapid pacing of the atrium in rabbit model [53]. Toward clinical application, this line of study has been moved forward to large animals as well. AAV9/*SERCA2a* gene therapies were performed in German pigs and ovine with ischemic CHF models [54]. AAV9 targeting the *S100A1* gene, which augments the activity of SERCA2a proteins so that Ca^{2+} signaling is improved, prevents left ventricular remodeling, and restores SERCA2a protein expression [54]. Likewise, injection of AAV9/SERCA2a into the coronary arteries significantly increased *SERCA2a* gene expression in the walls of the left ventricle in sheep [55]. This successful overexpression of SERCA2a triggered improvements in left ventricular systolic and diastolic function and subsequent left ventricular remodeling [55]. These positive outcomes provide the basis for future clinical trials in humans utilizing AAV9-based vectors. Current investigations in humans using AAV9-based vectors include Pompe disease (NCT02240407) and Batten disease (NCT02725580). Taken together, these data suggest that the AAV9/*SERCA2a* gene delivery is an appealing therapeutic method for targeting CHF due to its effects on the ventricular contractile function and restoration of electrical complications associated with the failing heart.

Besides AAVs, gene delivery method using other viral vectors is emerging. For example, lentiviral vectors have been demonstrated to be effective for SERCA2 gene transfer in rat ischemic HF model [56]. Since lentivirus can integrate the gene of interest into the host chromosome, this approach can achieve long-term effect of gene expression [57–59]. Many research groups have begun to examine safety and efficacy of this gene therapy approach in CHF animal models [56,60–62].

5. Concluding Remarks and Future Directions

The beneficial left ventricular functional and structural adaptations associated with AAV9/*SERCA2a* gene therapy provide solid preclinical evidence for future work to focus on translating these findings from animal work to human clinical applications. Compared to AAV1 which used in failed CUPID2 and other earlier trials, AAV9 presents superior efficiency and safety. Therefore, the improved Ca^{2+} handling in cardiac tissues in response to AAV9/SERCA2a would have significant benefits for patients suffering from CHF—particularly CHF of ischemic origin.

In addition to the focus on *SERCA2a* gene expression, there are a host of other molecules that both indirectly and directly modify SERCA2a function that are worthwhile investigating using modern gene therapy techniques. These molecules include, but are not limited to, small ubiquitin-related modifier-1 (SUMO-1) [63,64]. SUMO-1 is a post-transcriptional protein, which plays an important role in modifying other proteins involved in the preservation and stabilization of SERCA2a proteins. In porcine and human cardiac tissue, AAV1/SUMO-1 improved cardiac function. Concurrent transfection of both SUMO-1 and SERCA2a increased both expression and function of SERCA2a [63,64]. As such, this approach may represent a possible dual therapy in humans. Indeed, combined gene therapy has recently shown significant promise in the treatment of animal models of heart failure, with combined transfer of apelin, fibroblast growth factor-2, and SERCA2a demonstrating increased gene expression in ischemic heart failure model [61]. Therefore, future work should seek to translate the use of combined gene therapy in humans.

Moreover, novel pharmaceutical therapies targeting activation of SERCA2a have been developed. Istaroxime, a drug developed by Italian pharmaceutical company Sigma-Tau (Pomezia, Italy), is a lusitropic-inotropic drug recently approved for use in CHF [65]. In fact, currently a Phase 1 clinical trial at the University of Texas Southwestern is utilizing Istaroxime during exercise (NCT02772068), with primary outcome measure changes in cardiac filling pressures, and secondary outcome measure change in cardiac relaxation time. The study hopes to investigate the effect of SERCA2a activation on exercise-dependent changes in cardiac function (NCT02772068). At the same time, many other altered regulatory proteins involving in cardiomyocyte Ca^{2+} homeostasis could be potential targets for gene therapy in CHF as well, such as phospholamban, junctophilin-2 (JP2), RyRs and NCXs. For example, JP2 downregulation has been reported in human failing heart and restoring JP2 by AAV9-mediated gene therapy could rescue heart failure in mice [66–69].

In conclusion, the development of gene therapy and gene delivery methods have received significant progress over the past few decades, contributing to a novel therapeutic approach for treating CHF. Although the early promising AAV1/SERCA2a method in animal models failed to replicate the same results in humans, recent advancements in gene delivery techniques, especially AAV9/SERCA2a have provided compelling results for beneficial outcome of the upregulation of *SERCA2a* gene expression. Targeting SERCA2a for treatment of CHF in human patients should be still

Int. J. Mol. Sci. **2018**, *19*, 1086

an option deserving further investigation. Future work should focus on translating the recent findings using AAV9/SERCA2a techniques into large-scale clinical trials in humans.

Acknowledgments: The authors gratefully acknowledge the contribution of Benjamin Young in helping prepare this manuscript.

Conflicts of Interest: The authors declare no conflict of interest.

References

1. Azad, N.; Lemay, G. Management of chronic heart failure in the older population. *J. Geriatr. Cardiol.* **2014**, *11*, 329–337. [PubMed]
2. Heidenreich, P.A.; Albert, N.M.; Allen, L.A.; Bluemke, D.A.; Butler, J.; Fonarow, G.C.; Ikonomidis, J.S.; Khavjou, O.; Konstam, M.A.; Maddox, T.M.; et al. Forecasting the impact of heart failure in the United States: A policy statement from the american heart association. *Circ. Heart Fail.* **2013**, *6*, 606–619. [CrossRef] [PubMed]
3. Berridge, M.J.; Bootman, M.D.; Roderick, H.L. Calcium signalling: Dynamics, homeostasis and remodelling. *Nat. Rev. Mol. Cell Biol.* **2003**, *4*, 517–529. [CrossRef] [PubMed]
4. Hasenfuss, G.; Pieske, B.; Holubarsch, C.; Alpert, N.R.; Just, H. Excitation-contraction coupling and contractile protein function in failing and nonfailing human myocardium. *Adv. Exp. Med. Biol.* **1993**, *346*, 91–100. [PubMed]
5. Balderas-Villalobos, J.; Molina-Munoz, T.; Mailloux-Salinas, P.; Bravo, G.; Carvajal, K.; Gomez-Viquez, N.L. Oxidative stress in cardiomyocytes contributes to decreased serca2a activity in rats with metabolic syndrome. *Am. J. Physiol. Heart Circ. Physiol.* **2013**, *305*, H1344–H1353. [CrossRef] [PubMed]
6. Kawase, Y.; Ly, H.Q.; Prunier, F.; Lebeche, D.; Shi, Y.; Jin, H.; Hadri, L.; Yoneyama, R.; Hoshino, K.; Takewa, Y.; et al. Reversal of cardiac dysfunction after long-term expression of SERCA2a by gene transfer in a pre-clinical model of heart failure. *J. Am. Coll. Cardiol.* **2008**, *51*, 1112–1119. [CrossRef] [PubMed]
7. Morgan, J.P. Abnormal intracellular modulation of calcium as a major cause of cardiac contractile dysfunction. *N. Engl. J. Med.* **1991**, *325*, 625–632. [CrossRef]
8. Schwarzl, M.; Ojeda, F.; Zeller, T.; Seiffert, M.; Becher, P.M.; Munzel, T.; Wild, P.S.; Blettner, M.; Lackner, K.J.; Pfeiffer, N.; et al. Risk factors for heart failure are associated with alterations of the LV end-diastolic pressure-volume relationship in non-heart failure individuals: Data from a large-scale, population-based cohort. *Eur. Heart J.* **2016**, *37*, 1807–1314. [CrossRef] [PubMed]
9. Baker, D.L.; Hashimoto, K.; Grupp, I.L.; Ji, Y.; Reed, T.; Loukianov, E.; Grupp, G.; Bhagwhat, A.; Hoit, B.; Walsh, R.; et al. Targeted overexpression of the sarcoplasmic reticulum Ca^{2+}-atpase increases cardiac contractility in transgenic mouse hearts. *Circ. Res.* **1998**, *83*, 1205–1214. [CrossRef] [PubMed]
10. Del Monte, F.; Harding, S.E.; Schmidt, U.; Matsui, T.; Kang, Z.B.; Dec, G.W.; Gwathmey, J.K.; Rosenzweig, A.; Hajjar, R.J. Restoration of contractile function in isolated cardiomyocytes from failing human hearts by gene transfer of SERCA2a. *Circulation* **1999**, *100*, 2308–2311. [CrossRef]
11. Studeli, R.; Jung, S.; Mohacsi, P.; Perruchoud, S.; Castiglioni, P.; Wenaweser, P.; Heimbeck, G.; Feller, M.; Hullin, R. Diastolic dysfunction in human cardiac allografts is related with reduced SERCA2a gene expression. *Am. J. Transplant.* **2006**, *6*, 775–782. [CrossRef] [PubMed]
12. Jaski, B.E.; Jessup, M.L.; Mancini, D.M.; Cappola, T.P.; Pauly, D.F.; Greenberg, B.; Borow, K.; Dittrich, H.; Zsebo, K.M.; Hajjar, R.J.; et al. Calcium upregulation by percutaneous administration of gene therapy in cardiac disease (CUPID Trial), a first-in-human phase 1/2 clinical trial. *J. Card. Fail.* **2009**, *15*, 171–181. [CrossRef] [PubMed]
13. Jessup, M.; Greenberg, B.; Mancini, D.; Cappola, T.; Pauly, D.F.; Jaski, B.; Yaroshinsky, A.; Zsebo, K.M.; Dittrich, H.; Hajjar, R.J.; et al. Calcium upregulation by percutaneous administration of gene therapy in cardiac disease (CUPID): A phase 2 trial of intracoronary gene therapy of sarcoplasmic reticulum Ca^{2+}-atpase in patients with advanced heart failure. *Circulation* **2011**, *124*, 304–313. [CrossRef] [PubMed]
14. Greenberg, B.; Butler, J.; Felker, G.M.; Ponikowski, P.; Voors, A.A.; Desai, A.S.; Barnard, D.; Bouchard, A.; Jaski, B.; Lyon, A.R.; et al. Calcium upregulation by percutaneous administration of gene therapy in patients with cardiac disease (CUPID 2): A randomised, multinational, double-blind, placebo-controlled, phase 2b trial. *Lancet* **2016**, *387*, 1178–1186. [CrossRef]

15. Tada, M.; Yamamoto, T.; Tonomura, Y. Molecular mechanism of active calcium transport by sarcoplasmic reticulum. *Physiol. Rev.* **1978**, *58*, 1–79. [CrossRef] [PubMed]

16. Leon, M.B.; Rosing, D.R.; Bonow, R.O.; Epstein, S.E. Combination therapy with calcium-channel blockers and β-blockers for chronic stable angina-pectoris. *Am. J. Cardiol.* **1985**, *55*, B69–B80. [CrossRef]

17. Cannon, R.O., 3rd; Watson, R.M.; Rosing, D.R.; Epstein, S.E. Efficacy of calcium channel blocker therapy for angina pectoris resulting from small-vessel coronary artery disease and abnormal vasodilator reserve. *Am. J. Cardiol.* **1985**, *56*, 242–246. [CrossRef]

18. Antman, E.M.; Stone, P.H.; Muller, J.E.; Braunwald, E. Calcium channel blocking agents in the treatment of cardiovascular disorders. Part I: Basic and clinical electrophysiologic effects. *Ann. Intern. Med.* **1980**, *93*, 875–885. [CrossRef] [PubMed]

19. Ito, K.; Yan, X.H.; Feng, X.; Manning, W.J.; Dillmann, W.H.; Lorell, B.H. Transgenic expression of sarcoplasmic reticulum Ca^{2+} atpase modifies the transition from hypertrophy to early heart failure. *Circ. Res.* **2001**, *89*, 422–429. [CrossRef] [PubMed]

20. He, H.P.; Giordano, F.J.; HilalDandan, R.; Choi, D.J.; Rockman, H.A.; McDonough, P.M.; Bluhm, W.F.; Meyer, M.; Sayen, M.R.; Swanson, E.; et al. Overexpression of the rat sarcoplasmic reticulum Ca^{2+} atpase gene in the heart of transgenic mice accelerates calcium transients and cardiac relaxation. *J. Clin. Investig.* **1997**, *100*, 380–389. [CrossRef] [PubMed]

21. Rincon, M.Y.; VandenDriessche, T.; Chuah, M.K. Gene therapy for cardiovascular disease: Advances in vector development, targeting, and delivery for clinical translation. *Cardiovasc. Res.* **2015**, *108*, 4–20. [CrossRef] [PubMed]

22. Williams, P.D.; Ranjzad, P.; Kakar, S.J.; Kingston, P.A. Development of viral vectors for use in cardiovascular gene therapy. *Viruses* **2010**, *2*, 334–371. [CrossRef] [PubMed]

23. Del Monte, F.; Lebeche, D.; Guerrero, J.L.; Tsuji, T.; Doye, A.A.; Gwathmey, J.K.; Hajjar, R.J. Abrogation of ventricular arrhythmias in a model of ischemia and reperfusion by targeting myocardial calcium cycling. *Proc. Natl. Acad. Sci. USA* **2004**, *101*, 5622–5627. [CrossRef] [PubMed]

24. Del Monte, F.; Williams, E.; Lebeche, D.; Schmidt, U.; Rosenzweig, A.; Gwathmey, J.K.; Lewandowski, E.D.; Hajjar, R.J. Improvement in survival and cardiac metabolism after gene transfer of sarcoplasmic reticulum Ca^{2+}-atpase in a rat model of heart failure. *Circulation* **2001**, *104*, 1424–1429. [CrossRef] [PubMed]

25. Chaudhri, B.; del Monte, F.; Hajjar, R.J.; Harding, S.E. Contractile effects of adenovirally-mediated increases in SERCA2a activity: A comparison between adult rat and rabbit ventricular myocytes. *Mol. Cell. Biochem.* **2003**, *251*, 103–109. [CrossRef] [PubMed]

26. Nassal, M.M.; Wan, X.; Laurita, K.R.; Cutler, M.J. Atrial SERCA2a overexpression has no affect on cardiac alternans but promotes arrhythmogenic SR Ca^{2+} triggers. *PLoS ONE* **2015**, *10*, e0137359. [CrossRef] [PubMed]

27. Miyamoto, M.I.; del Monte, F.; Schmidt, U.; DiSalvo, T.S.; Kang, Z.B.; Matsui, T.; Guerrero, J.L.; Gwathmey, J.K.; Rosenzweig, A.; Hajjar, R.J. Adenoviral gene transfer of SERCA2a improves left-ventricular function in aortic-banded rats in transition to heart failure. *Proc. Natl. Acad. Sci. USA* **2000**, *97*, 793–798. [CrossRef] [PubMed]

28. Lipskaia, L.; Hadri, L.; Le Prince, P.; Esposito, B.; Atassi, F.; Liang, L.; Glorian, M.; Limon, I.; Lompre, A.M.; Lehoux, S.; et al. SERCA2a gene transfer prevents intimal proliferation in an organ culture of human internal mammary artery. *Gene Ther.* **2013**, *20*, 396–406. [CrossRef] [PubMed]

29. Sakata, S.; Lebeche, D.; Sakata, Y.; Sakata, N.; Chemaly, E.R.; Liang, L.; Nakajima-Takenaka, C.; Tsuji, T.; Konishi, N.; del Monte, F.; et al. Transcoronary gene transfer of SERCA2a increases coronary blood flow and decreases cardiomyocyte size in a type 2 diabetic rat model. *Am. J. Physiol. Heart Circ. Physiol.* **2007**, *292*, H1204–H1207. [CrossRef] [PubMed]

30. Hadri, L.; Bobe, R.; Kawase, Y.; Ladage, D.; Ishikawa, K.; Atassi, F.; Lebeche, D.; Kranias, E.G.; Leopold, J.A.; Lompre, A.M.; et al. SERCA2a gene transfer enhances enos expression and activity in endothelial cells. *Mol. Ther.* **2010**, *18*, 1284–1292. [CrossRef] [PubMed]

31. Xin, W.; Lu, X.; Li, X.; Niu, K.; Cai, J. Attenuation of endoplasmic reticulum stress-related myocardial apoptosis by serca2a gene delivery in ischemic heart disease. *Mol. Med.* **2011**, *17*, 201–210. [CrossRef] [PubMed]

32. Kuken, B.N.; Aikemu, A.N.; Xiang, S.Y.; Wulasihan, M.H. Effect of SERCA2a overexpression in the pericardium mediated by the AAV1 gene transfer on rapid atrial pacing in rabbits. *Genet. Mol. Res.* **2015**, *14*, 13625–13632. [CrossRef] [PubMed]

33. Lipskaia, L.; Hadri, L.; Lopez, J.J; Hajjar, R.J.; Bobe, R. Benefit of SERCA2a gene transfer to vascular endothelial and smooth muscle cells: A new aspect in therapy of cardiovascular diseases. *Curr. Vasc. Pharmacol.* **2013**, *11*, 465–479. [CrossRef] [PubMed]

34. Wobus, C.E.; Hugle-Dorr, B.; Girod, A.; Petersen, G.; Hallek, M.; Kleinschmidt, J.A. Monoclonal antibodies against the adeno-associated virus type 2 (AAV-2) capsid: Epitope mapping and identification of capsid domains involved in AAV-2-cell interaction and neutralization of AAV-2 infection. *J. Virol.* **2000**, *74*, 9281–9293. [CrossRef] [PubMed]

35. Scallan, C.D.; Jiang, H.; Liu, T.; Patarrcyo-White, S.; Sommer, J.M.; Zhou, S.; Couto, L.B.; Pierce, G.F. Human immunoglobulin inhibits liver transduction by AAV vectors at low AAV2 neutralizing titers in scid mice. *Blood* **2006**, *107*, 1810–1817. [CrossRef] [PubMed]

36. Moskalenko, M.; Chen, L.; van Roey, M.; Donahue, B.A.; Snyder, R.O.; McArthur, J.G.; Patel, S.D. Epitope mapping of human anti-adeno-associated virus type 2 neutralizing antibodies: Implications for gene therapy and virus structure. *J. Virol.* **2000**, *74*, 1761–1766. [CrossRef] [PubMed]

37. Greenberg, B.; Butler, J.; Felker, G.M.; Ponikowski, P.; Voors, A.A.; Pogoda, J.M.; Provost, R.; Guerrero, J.; Hajjar, R.J.; Zsebo, K.M. Prevalence of AAV1 neutralizing antibodies and consequences for a clinical trial of gene transfer for advanced heart failure. *Gene Ther.* **2016**, *23*, 313–319. [CrossRef] [PubMed]

38. Hulot, J.S.; Salem, J.E.; Redheuil, A.; Collet, J.P.; Varnous, S.; Jourdain, P.; Logeart, D.; Gandjbakhch, E.; Bernard, C.; Hatem, S.N.; et al. Effect of intracoronary administration of AAV1/SERCA2a on ventricular remodelling in patients with advanced systolic heart failure: Results from the agent-hf randomized phase 2 trial. *Eur. J. Heart Fail.* **2017**, *19*, 1534–1541. [CrossRef] [PubMed]

39. Greenberg, B.; Yaroshinsky, A.; Zsebo, K.M.; Butler, J.; Felker, G.M.; Voors, A.A.; Rudy, J.J.; Wagner, K.; Hajjar, R.J. Design of a phase 2b trial of intracoronary administration of AAV1/SERCA2a in patients with advanced heart failure: The CUPID 2 trial (calcium up-regulation by percutaneous administration of gene therapy in cardiac disease phase 2b). *JACC Heart Fail.* **2014**, *2*, 84–92. [CrossRef] [PubMed]

40. Zsebo, K.; Yaroshinsky, A.; Rudy, J.J.; Wagner, K.; Greenberg, B.; Jessup, M.; Hajjar, R.J. Long-term effects of AAV1/SERCA2a gene transfer in patients with severe heart failure: Analysis of recurrent cardiovascular events and mortality. *Circ. Res.* **2014**, *114*, 101–108. [CrossRef] [PubMed]

41. Horowitz, J.D.; Rosenson, R.S.; McMurray, J.J.; Marx, N.; Remme, W.J. Clinical trials update aha congress 2010. *Cardiovasc. Drugs Ther.* **2011**, *25*, 69–76. [CrossRef] [PubMed]

42. Hajjar, R.J.; Zsebo, K.; Deckelbaum, L.; Thompson, C.; Rudy, J.; Yaroshinsky, A.; Ly, H.; Kawase, Y.; Wagner, K.; Borow, K.; et al. Design of a phase 1/2 trial of intracoronary administration of AAV1/SERCA2a in patients with heart failure. *J. Card. Fail.* **2008**, *14*, 355–367. [CrossRef] [PubMed]

43. Hulot, J.S.; Ishikawa, K.; Hajjar, R.J. Gene therapy for the treatment of heart failure: Promise postponed. *Eur. Heart J.* **2016**, *37*, 1651–1658. [CrossRef] [PubMed]

44. Yla-Herttuala, S. Gene therapy for heart failure: Back to the bench. *J. Am. Soc. Gene Ther.* **2015**, *23*, 1551–1552. [CrossRef] [PubMed]

45. Mingozzi, F.; Anguela, X.M.; Pavani, G.; Chen, Y.; Davidson, R.J.; Hui, D.J.; Yazicioglu, M.; Elkouby, L.; Hinderer, C.J.; Faella, A.; et al. Overcoming preexisting humoral immunity to AAV using capsid decoys. *Sci. Transl. Med.* **2013**, *5*, 194ra192. [CrossRef] [PubMed]

46. Pacak, C.A.; Mah, C.S.; Thattaliyath, B.D.; Conlon, T.J.; Lewis, M.A.; Cloutier, D.E.; Zolotukhin, I.; Tarantal, A.F.; Byrne, B.J. Recombinant adeno-associated virus serotype 9 leads to preferential cardiac transduction in vivo. *Circ. Res.* **2006**, *99*, e3–e9. [CrossRef] [PubMed]

47. Vandendriessche, T.; Thorrez, L.; Acosta-Sanchez, A.; Petrus, I.; Wang, L.; Ma, L.; Waele, D.E.; Iwasaki, Y.; Gillijns, V.; Wilson, J.M.; et al. Efficacy and safety of adeno-associated viral vectors based on serotype 8 and 9 vs. Lentiviral vectors for hemophilia B gene therapy. *J. Thromb. Haemost.* **2007**, *5*, 16–24. [CrossRef] [PubMed]

48. Inagaki, K.; Fuess, S.; Storm, T.A.; Gibson, G.A.; McTiernan, C.F.; Kay, M.A.; Nakai, H. Robust systemic transduction with AAV9 vectors in mice: Efficient global cardiac gene transfer superior to that of AAV8. *Mol. Ther.* **2006**, *14*, 45–53. [CrossRef] [PubMed]

49. Zincarelli, C.; Soltys, S.; Rengo, G.; Koch, W.J.; Rabinowitz, J.E. Comparative cardiac gene delivery of adeno-associated virus serotypes 1–9 reveals that AAV6 mediates the most efficient transduction in mouse heart. *Clin. Transl. Sci.* **2010**, *3*, 81–89. [CrossRef] [PubMed]

50. Lyon, A.R.; Bannister, M.L.; Collins, T.; Pearce, E.; Sepehripour, A.H.; Dubb, S.S.; Garcia, E.; O'Gara, P.; Liang, L.; Kohlbrenner, E.; et al. SERCA2a gene transfer decreases sarcoplasmic reticulum calcium leak and reduces ventricular arrhythmias in a model of chronic heart failure. *Circ. Arrhythm. Electrophysiol.* **2011**, *4*, 362–372. [CrossRef] [PubMed]

51. Cutler, M.J.; Wan, X.; Plummer, B.N.; Liu, H.; Deschenes, I.; Laurita, K.R.; Hajjar, R.J.; Rosenbaum, D.S. Targeted sarcoplasmic reticulum Ca^{2+} ATPase 2a gene delivery to restore electrical stability in the failing heart. *Circulation* **2012**, *126*, 2095–2104. [CrossRef] [PubMed]

52. Shin, J.H.; Bostick, B.; Yue, Y.; Hajjar, R.; Duan, D. SERCA2a gene transfer improves electrocardiographic performance in aged mdx mice. *J. Transl. Med.* **2011**, *9*, 132. [CrossRef] [PubMed]

53. Wang, H.L.; Zhou, X.H.; Li, Z.Q.; Fan, P.; Zhou, Q.N.; Li, Y.D.; Hou, Y.M.; Tang, B.P. Prevention of atrial fibrillation by using sarcoplasmic reticulum calcium atpase pump overexpression in a rabbit model of rapid atrial pacing. *Med. Sci. Monit.* **2017**, *23*, 3952–3960. [CrossRef] [PubMed]

54. Pleger, S.T.; Shan, C.; Ksienzyk, J.; Bekeredjian, R.; Boekstegers, P.; Hinkel, R.; Schinkel, S.; Leuchs, B.; Ludwig, J.; Qiu, G.; et al. Cardiac AAV9-S100A1 gene therapy rescues post-ischemic heart failure in a preclinical large animal model. *Sci. Transl. Med.* **2011**, *3*, 92ra64. [CrossRef] [PubMed]

55. Katz, M.G.; Fargnoli, A.S.; Williams, R.D.; Steuerwald, N.M.; Isidro, A.; Ivanina, A.V.; Sokolova, I.M.; Bridges, C.R. Safety and efficacy of high-dose adeno-associated virus 9 encoding sarcoplasmic reticulum Ca^{2+} adenosine triphosphatase delivered by molecular cardiac surgery with recirculating delivery in ovine ischemic cardiomyopathy. *J. Thorac. Cardiovasc. Surg.* **2014**, *148*, 1065–1072, 1073e1–1073e2; discussion 1072–1073. [CrossRef] [PubMed]

56. Niwano, K.; Arai, M.; Koitabashi, N.; Watanabe, A.; Ikeda, Y.; Miyoshi, H.; Kurabayashi, M. Lentiviral vector-mediated SERCA2 gene transfer protects against heart failure and left ventricular remodeling after myocardial infarction in rats. *J. Am. Soc. Gene Ther.* **2008**, *16*, 1026–1032. [CrossRef] [PubMed]

57. Fleury, S.; Simeoni, E.; Zuppinger, C.; Deglon, N.; Von Segesser, L.K.; Kappenberger, L.; Vassalli, G. Multiply attenuated, self-inactivating lentiviral vectors efficiently deliver and express genes for extended periods of time in adult rat cardiomyocytes in vivo. *Circulation* **2003**, *107*, 2375–2382. [CrossRef] [PubMed]

58. Miyoshi, H.; Takahashi, M.; Gage, F.H.; Verma, I.M. Stable and efficient gene transfer into the retina using an hiv-based lentiviral vector. *Proc. Natl. Acad. Sci. USA* **1997**, *94*, 10319–10323. [CrossRef] [PubMed]

59. Naldini, L.; Blomer, U.; Gallay, P.; Ory, D.; Mulligan, R.; Gage, F.H.; Verma, I.M.; Trono, D. In vivo gene delivery and stable transduction of nondividing cells by a lentiviral vector. *Science* **1996**, *272*, 263–267. [CrossRef] [PubMed]

60. Merentie, M.; Lottonen-Raikaslehto, L.; Parviainen, V.; Huusko, J.; Pikkarainen, S.; Mendel, M.; Laham-Karam, N.; Karja, V.; Rissanen, R.; Hedman, M.; et al. Efficacy and safety of myocardial gene transfer of adenovirus, adeno-associated virus and lentivirus vectors in the mouse heart. *Gene Ther.* **2016**, *23*, 296–305. [CrossRef] [PubMed]

61. Renaud-Gabardos, E.; Tatin, F.; Hantelys, F.; Lebas, B.; Calise, D.; Kunduzova, O.; Masri, B.; Pujol, F.; Sicard, P.; Valet, P.; et al. Therapeutic benefit and gene network regulation by combined gene transfer of apelin, FGF2, and SERCA2a into ischemic heart. *Mol. Ther.* **2018**, *26*, 902–916. [CrossRef] [PubMed]

62. Merentie, M.; Rissanen, R.; Lottonen-Raikaslehto, L.; Huusko, J.; Gurzeler, E.; Turunen, M.P.; Holappa, L.; Makinen, P.; Yla-Herttuala, S. Doxycycline modulates VEGF-A expression: Failure of doxycycline-inducible lentivirus shrna vector to knockdown VEGF-A expression in transgenic mice. *PLoS ONE* **2018**, *13*, e0190981. [CrossRef] [PubMed]

63. Tilemann, L.; Lee, A.; Ishikawa, K.; Aguero, J.; Rapti, K.; Santos-Gallego, C.; Kohlbrenner, E.; Fish, K.M.; Kho, C.; Hajjar, R.J. SUMO-1 gene transfer improves cardiac function in a large-animal model of heart failure. *Sci. Transl. Med.* **2013**, *5*, 211ra159. [CrossRef] [PubMed]

64. Kho, C.; Lee, A.; Jeong, D.; Oh, J.G.; Chaanine, A.H.; Kizana, E.; Park, W.J.; Hajjar, R.J. SUMO1-dependent modulation of SERCA2a in heart failure. *Nature* **2011**, *477*, 601–605. [CrossRef] [PubMed]

65. Ferrandi, M.; Barassi, P.; Tadini-Buoninsegni, F.; Bartolommei, G.; Molinari, I.; Tripodi, M.G.; Reina, C.; Moncelli, M.R.; Bianchi, G.; Ferrari, P. Istaroxime stimulates SERCA2a and accelerates calcium cycling in heart failure by relieving phospholamban inhibition. *Br. J. Pharmacol.* **2013**, *169*, 1849–1861. [CrossRef] [PubMed]

66. Reynolds, J.O.; Quick, A.P.; Wang, Q.; Beavers, D.L.; Philippen, L.E.; Showell, J.; Barreto-Torres, G.; Thuerauf, D.J.; Doroudgar, S.; Glembotski, C.C.; et al. Junctophilin-2 gene therapy rescues heart failure by normalizing RyR2-mediated Ca^{2+} release. *Int. J. Cardiol.* **2016**, *225*, 371–380. [CrossRef] [PubMed]
67. Beavers, D.L.; Landstrom, A.P.; Chiang, D.Y.; Wehrens, X.H. Emerging roles of Junctophilin-2 in the heart and implications for cardiac diseases. *Cardiovasc. Res.* **2014**, *103*, 198–205. [CrossRef] [PubMed]
68. Guo, A.; Zhang, X.; Iyer, V.R.; Chen, B.; Zhang, C.; Kutschke, W.J.; Weiss, R.M.; Franzini-Armstrong, C.; Song, L.S. Overexpression of Junctophilin-2 does not enhance baseline function but attenuates heart failure development after cardiac stress. *Proc. Natl. Acad. Sci. USA* **2014**, *111*, 12240–12245. [CrossRef] [PubMed]
69. Zhang, C.; Chen, B.; Guo, A.; Zhu, Y.; Miller, J.D.; Gao, S.; Yuan, C.; Kutschke, W.; Zimmerman, K.; Weiss, R.M.; et al. Microtubule-mediated defects in Junctophilin-2 trafficking contribute to myocyte transverse-tubule remodeling and Ca^{2+} handling dysfunction in heart failure. *Circulation* **2014**, *129*, 1742–1750. [CrossRef] [PubMed]

International Journal of
Molecular Sciences

MDPI

Review

Calcium and Nuclear Signaling in Prostate Cancer

Ivan V. Maly and Wilma A. Hofmann *

Department of Physiology and Biophysics, Jacobs School of Medicine and Biomedical Sciences,
University at Buffalo, 955 Main Street, Buffalo, NY 14203, USA; ivanmaly@buffalo.edu
* Correspondence: whofmann@buffalo.edu; Tel.: +1-716-829-3290

Received: 16 March 2018; Accepted: 17 April 2018; Published: 19 April 2018

Abstract: Recently, there have been a number of developments in the fields of calcium and nuclear signaling that point to new avenues for a more effective diagnosis and treatment of prostate cancer. An example is the discovery of new classes of molecules involved in calcium-regulated nuclear import and nuclear calcium signaling, from the G protein-coupled receptor (GPCR) and myosin families. This review surveys the new state of the calcium and nuclear signaling fields with the aim of identifying the unifying themes that hold out promise in the context of the problems presented by prostate cancer. Genomic perturbations, kinase cascades, developmental pathways, and channels and transporters are covered, with an emphasis on nuclear transport and functions. Special attention is paid to the molecular mechanisms behind prostate cancer progression to the malignant forms and the unfavorable response to anti-androgen treatment. The survey leads to some new hypotheses that connect heretofore disparate results and may present a translational interest.

Keywords: metastasis; nuclear import; myosin IC; calcium; prostate cancer

1. Introduction

Despite the recent progress in diagnosis and treatment, prostate cancer remains one of the most common and lethal malignancies [1,2], with approximately 3.3 million men living with the condition only in the United States. The disease is characterized by the comparative ease of the initial diagnosis, a long period of indolence relative to other common cancers, and an abrupt and less predictable progression to the lethal stages. When sharpened up, the questions facing the practitioners and researchers have been formulated as concerning the possibility of treatment for the patients that need it, and the actual need of those for whom treatment is possible [3]. In many cases, the critical events in disease progression from the stage that is believed to merit only watchful waiting to one that is lethal are the emergence of castration resistance and metastasis [4]. Mechanistically grounded prognostic biomarkers and potential drug targets associated with these steps toward poor survival are of particular interest as outcomes of the molecular biological research.

From the perspective of molecular sciences, the puzzle of prostate cancer is encapsulated in a complex perturbation of multiple regulatory pathways and mechanisms. In the face of this complexity, the biological signal integration by the calcium ion (Ca^{2+}), in addition to the central importance of gene expression regulation, presents a particular interest. The intracellular propagation of a calcium-mediated signal [5] and the import of downstream molecules into the nucleus [6] can be seen as two common physicochemical events in the dysregulation involving the cell and its environment. Moreover, calcium channels and GPCRs are among the favored "druggable" targets in therapeutics development [7–9], and metastasis is manifested primarily by the molecular-motor driven cell motility [10,11]. The unexpected results of the recent experiments implicate the GPCR and myosin molecular motor molecules in the intranuclear calcium signaling and calcium-regulated nuclear transport [12,13]. Motivated by the emergent role of these mechanisms in multiple aspects of prostate cancer cell regulation, below we review the recent advances in molecular biology of prostate cancer with a special focus on calcium signaling and nuclear transport.

2. Genomic Background

2.1. NKX3.1 Insufficiency

85% of high-grade prostate intraepithelial neoplasia (PIN) and prostate adenocarcinomas display a loss of heterozygosity in the 8p21.2 locus that includes the *NKX3.1* homeobox gene [14]. The incidence correlates with disease grade [15], and there is evidence of an epigenetic downregulation at the locus as well [16]. *NKX3.1* is an early marker of developing prostate epithelium during budding from the urogenital sinus, and its expression throughout the development of the gland is important for ductal branching and expression of secreted proteins [17–19]. It is also important in adulthood, when homo- and heterozygous mutants display hyperplasia and often PIN [20,21]. As shown by Lei et al. [22], NKX3.1 stabilizes the tumor suppressor p53, inhibits the activation of protein kinase B (PKB/AKT), and blocks prostate cancer initiation caused by the phosphatase and tensin homolog deleted on chromosome 10 (PTEN) loss in xenografts. At the same time, *NKX3.1* inactivation in mice leads to a deficiency in the response to oxidative stress [23]. In human prostate cell lines, it has been shown that NKX3.1 promotes the response to DNA damage [24] but is ubiquitinated under the action of inflammatory cytokines [25]. Recent experiments have demonstrated, in addition, that the proteolytic control of NKX3.1 expression levels in prostate cancer cells is regulated by mitogens such as epithelial growth factor (EGF) in an intracellular calcium- and protein kinase C (PKC)-sensitive manner [26]. Thus, on the protein level, the effect of most common genomic abnormality that has been identified in prostate cancer converges with the effect of the posttranslational expression regulation by calcium signaling. Additional experiments are needed to establish the relative impact of these two factors at different stages of the disease development.

2.2. Amplification and Susceptibility in MYC

The *MYC* oncogene is somatically amplified (as part of the 8q24 region) in a subset of advanced prostate tumors [27,28]. Specifically nuclear MYC was found to be upregulated in many PIN lesions and most carcinomas in the absence of gene amplification [29]. Numerous single nucleotide polymorphisms have been identified in the gene-poor region next to *MYC* that contains *MYC*'s long-range regulatory elements [30–33]. It has also been established that overexpression of *MYC* in transgenic mice drives formation of PIN as well as progression to invasive adenocarcinoma [34]. The tumors formed under these conditions are characterized by downregulation of *NKX3.1* and a simultaneous upregulation of *Pim1*, *MYC*'s partner in lymphomas. Coexpression of *MYC* and *Pim1*, on the other hand, leads to formation of carcinomas that exhibit the neuroendocrine phenotype [35]. Similarly to the recent results concerning NKX3.1, it has been known for some time that *MYC* in the prostate cancer cells is under the control of calcium-mediated signaling. Specifically, it has been shown that *MYC* is controlled by Notch, which involves the calcium/calmodulin-dependent kinase II (CAMKII)-regulated production of the nuclear import-competent cleaved form of Notch-1 [36]. The crosstalk of this calcium-regulated nuclear signaling pathway, which connects with one of the key regulators of prostate cancer progression, with other factors influencing the intracellular calcium homeostasis appears to be a promising direction for future research.

2.3. TMPRSS2-ERG Fusion

Most prostate carcinomas exhibit deletions (less commonly translocations) on chromosome 21q that activate ETS-family transcription factors, usually via an N-terminal fusion with the androgen receptor (AR)-activated gene *TMPRSS2* [37–41]. *TMPRSS2-ERG* is found in approximately 15% of high-grade PINs and 50% of localized prostate cancer [42,43]. In agreement with these clinical sample analyses, AR binding in the lymph node cancer of prostate (LNCaP) cell line brings the *TMPRSS2* and *ERG* loci in physical proximity [44,45]. This finding suggests a role for androgen signaling in the induction of fusions under conditions leading to DNA damage. In this connection, it is interesting that double-stranded breaks also occur under the action of topoisomerase II, which is recruited to the AR-responsive elements by androgen signaling [46]. At the same time, whole-genome chromatin immunoprecipitation analysis has shown that ERG binding can silence AR-responsive genes [47]. Conceivably, this effect may reinforce the physiological hormonal changes in the aging prostate, as well as those accompanying therapeutic castration. Furthermore, in cell culture and transgenic mice, activation of *ETS* expression contributes quantitatively to promotion of invasivity and epithelial-mesenchymal transition (EMT) [38,48,49]. By itself, expression of truncated human *ERG* in mice leads only to a weak PIN [48,49]. It however synergizes with a loss of *PTEN*, leading to a high-grade PIN and carcinoma [50,51]. Collectively, these results put the *TMPRSS2-ERG* fusion at the center of a web of mechanisms responsible for the prostate cancer progression.

Recent work has begun to clarify the relationship of this common genomic condition with perturbations of calcium signaling and nuclear import. Epigenomic profiling has identified the genes that are differentially methylated in the *TMPRSS2-ERG* fusion vs. non-fusion prostate tumors [52]. Among the top-ranked genes is the calcium-channel gene *CACNA1D*, a target of ERG. The most recent extension of this work employed retrospective analysis to demonstrate a negative correlation of the calcium channel blocker use by the prostate cancer patients with both high-grade disease and occurrence of fusion tumors [53]. The latter result can be interpreted as a selective disadvantage of a fusion in the absence of a functional reinforcement by means of the perturbation of calcium signaling. Prostate cancer and benign epithelial prostate cells overexpressing ERG display a characteristic invasivity in vitro [48], an effect that was most recently reversed using peptidomimetic inhibitors of ERG that were fused with a nuclear localization sequence (NLS) peptides to colocalize with ERG in the nucleus [54]. It is remarkable that of the two forms of the fusion protein found in prostate cancer cells, one lacks the NLS and does not enter the nucleus [55]. It binds the fully functional form of the fusion and results in its downregulation on the protein level as well as a reduced expression of MYC. The evidence points to the importance of further study of prostate oncoprotein isoforms that may be characterized by different nucleocytoplasmic partitioning.

The *TMPRSS2-ERG* fusion activity is also associated with Y-box binding protein 1 (YB-1), a genomic-instability response protein that enters the nucleus in response to genotoxic stress [56]. YB-1 contributes to the upregulation of AR during the development of castration resistance [57]. The nuclear localization of this protein has been most recently associated statistically with a poor prognosis for prostate cancer patients [58]. Its accumulation was at the same time shown to be linked to the *TMPRSS2-ERG* fusion, as well as *PTEN* deletion. It appears worth investigating whether the conditions for nuclear import, for example, the calcium-mediated depolarization of the nuclear envelope that is reviewed below, impact the YB-1 regulated downstream effects of the *TMPRSS2-ERG* fusion.

Additionally, the fusion drives expression of Toll-like receptors (TLR), including TLR4 [59]. This results in an increased phosphorylation of nuclear factor κB (NF-κB) on the serine residue that prevents binding of inhibitor of κB (IκB) that masks the NLS. The active fusion isoform and the phosphorylated factor both localize exclusively in the nucleus, as seen also in the orthotopic tumor samples. In vitro, the same experiments demonstrate, the cells' proliferation is controlled by the transcriptional activity of NF-κB, and the latter is increased with the expression of the fusion product. These findings have led to the speculation that the activation of TLR4 by bacterial lipopolysaccharides, as well as tumor microenvironment proteins, may impact the progression of prostate cancer. Adding to the hypothetical effects that may be induced by the *TMPRSS2-ERG* fusion, it should be noted that NF-κB is known to be able to drive an aberrant ligand-independent nuclear accumulation of AR by direct binding [60]. Conceivably, this may mean that the fusion can aid the cells in bypassing the critical androgen-regulated step during the establishment of androgen independence in the progressing tumor. One more gene activated downstream from *TMPRSS2-ERG* is the chemokine receptor *CXCR4* [61], whose role in intranuclear calcium signaling will be reviewed in Section 8.1.

2.4. PTEN Loss

The phosphatase gene *PTEN* undergoes homozygous deletion early in prostate carcinogenesis [62]. The loss has been found to be correlated with the development of aggressive and castration-resistant disease [63–65]. At the same time, it has been shown that conditional deletion of *PTEN* in prostate epithelium leads to PIN and adenocarcinoma [66]. Inactivation of *PTEN* in genetically engineered mice was found to cooperate with the loss of function of *NKX3.1* [67], upregulation of *c-MYC* [68], and expression of the human *TMPRSS2-ERG* fusion [50,51]. Among the molecular partners required for tumor formation on the *PTEN* loss background, the studies have identified mammalian target of rapamycin complex 2 (mTORC2) and phosphoinositide 3-kinase (PI3K) isoform p110β [69,70]. It is remarkable that prostate epithelial cells from *NKX3.1; PTEN* mutant mice display androgen independence even before development of the cancerous phenotype [71]. If this finding is translatable to the conditions in the human prostate, it indicates sufficiency of the combination of these common early and advanced genomic abnormalities for the critical step in the disease progression. This possibility raises the importance of further investigation, as suggested in Section 2.1, of the calcium-dependent regulation of NKX3.1, which appears to parallel the background genomic effect on the protein level.

In LNCaP cells, PTEN was found to interact with AR directly, resulting in an inhibition of the latter's nuclear localization and promotion of its degradation [72]. PTEN's own localization in heterozygous (+/−) prostate cancer cells was found to be predominantly nuclear as assessed by immunofluorescence in vitro [73]. The nuclear retention of PTEN was investigated also on model non-prostate-cancer cell lines [74] and found to be regulated by reactive oxygen species (ROS). The functional importance of the nuclear localization has been demonstrated in the same study in mouse xenograft models. Using human prostate cancer cells expressing mutant forms of PTEN, it was possible to demonstrate that nuclear PTEN acts in concert with p53 to suppress tumor formation. Most recently, the nuclear transport receptor importin 11 has been identified as a tumor suppressor that protects PTEN from proteolytic degradation [75]. This suggest the existence of a secondary function of the molecular interaction that presumably underpins the phosphatase's nuclear import. Generally, it appears to be an example of the nuclear import serving also as a regulatory mechanism impacting the protein-level expression of a critical tumor suppressor, akin to the chaperone-mediated sequestration mechanisms that will be reviewed in Section 7.

3. Developmental Pathways

Gene expression studies suggest that developmental signaling pathways may be reactivated during neoplastic processes in the adult prostate [76,77]. In the mouse, the wingless-integration adenomatous-polyposis-coli (WNT-APC)-β-catenin pathway is seen as contributing to carcinogenesis positively [78–80]. Consistent with this is the observation of an enhanced nuclear localization of β-catenin specifically in castration-resistant tumors [81]. In man, on the other hand, the nuclear localization of β-catenin has been seen as negatively correlated with tumor progression [82,83]. Further pointing to an explanation whereby the WNT-β-catenin signaling in prostate cancer development may be stage-specific is the finding that the WNT5A protein is upregulated specifically in the metastatic human samples [84]. The same work also demonstrated the existence of calcium waves that are induced by WNT5A in cultured metastatic prostate cancer cells, as well as this protein's stimulation of the actin cytoskeleton activity and cell motility (the latter being also under the control of CAMKII). Subsequent experiments showed that the calcium wave induced by this and other WNT proteins results in an intranuclear calcium mobilization and import of β-catenin into the nucleus, which is dependent on a thapsigargin-sensitive depolarization of the nuclear envelope [85]. The results support an interpretation whereby WNT signaling has a dual role as an anti-dedifferentiation regulator and a promoter of cell motility. It will be interesting to determine whether the intranuclear calcium mobilization by WNT impacts the functions of nuclear myosin (reviewed in Section 8.2), and what mechanisms or quantitative properties distinguish it from the one that has for some time been viewed as associated with apoptosis [86,87].

Among other developmental pathways, Hedgehog and fibroblast growth factor (FGF) signaling should be mentioned. Hedgehog signaling plays a significant role in prostate cancer progression [88]. Although experiments on tumor-derived primary cultures were seen initially as pointing toward an autocrine mechanism [89], targeted experimentation identified the mechanism as paracrine signaling to the epithelium from the prostate stroma [90], similarly to other epithelial cancers [91]. Nuclear accumulation of the downstream effector of Hedgehog signaling, transcription factor Gli1, as seen in clinical samples, is associated with metastasis and poor survival [92]. A breaking report [93] identifies the mode of the intracellular Hedgehog signal propagation in the prostate cancer cells as a novel non-canonical pathway involving AR binding-driven nuclear entry of the Gli transcription factors. The contribution of FGF signaling is analogous, in that both epithelial activation of FGF receptor 1 and stromal overexpression of FGF10 play a role [94,95], potentially leading downstream to the activation of extracellular signal-regulated kinase (ERK). Kwabi-Addo et al. [96] call attention to the alternative translation initiation isoforms of FGF2, which display an intranuclear accumulation and are expressed in transgenic adenocarcinoma of the mouse prostate (TRAMP) models, and posit that these isoforms may present a special interest in the context of paracrine signaling, because of the possibility of direct translation of the extracellular signal to the gene expression effects in the target cell. Overall, the ongoing work on the developmental pathways' dysregulation in prostate cancer emphasizes the need for the elucidation of the pivotal molecular mechanisms behind the calcium-regulated nuclear import, which may involve such novel biophysical principles as gating by the nuclear envelope depolarization.

4. Kinase Cascades

The *PTEN* loss of function, which may result, in particular, from the reviewed genomic abnormality, has been shown to cause an upregulation of the AKT/mTOR pathway, primarily via activation of the kinase AKT1 [97–99]. An alternative activating mechanism for this signaling pathway is the activating mutation in AKT1 itself, which has been found in human prostate cancer [100]. Experimental expression of the constitutively activated isoform p110β of PI3K is also sufficient for induction of neoplasia in mice [101]. In addition, activation of either AKT or ERK enables androgen-independent growth both in vitro and in vivo [102]. Collectively, these findings establish a broad molecular background among the intracellular signaling kinase classes, whose perturbation may support the development of prostate cancer.

Underscoring the functional importance of the activated kinases' localization in the cell, it has been shown that tumor suppressor PML recruits both phospho-AKT (pAKT) and its phosphatase PP2 to nuclear bodies [103]. Complicating the picture, loss of *PML* leads to an impairment of the PP2 activity toward pAKT and an accumulation of the latter in the nucleus. The kinase acts in the nucleus by inhibiting FOXO transcription factors, in particular the FOXO3A-mediated transcription of Bim and p27 (Kip1), suggesting a two-pronged mechanism that involves both the suppression of apoptosis and an acceleration of the cell cycle. Promisingly for the translation of this insight into the practice of prostate cancer drug development, the AKT inhibitor ML-9 has recently been found effective at inducing prostate cancer cell death in vitro in a calcium-dependent manner [104].

Adding to the in vitro and in vivo evidence, ERK is frequently found to be activated in the clinical prostate cancer, often at advanced stages and in combination with AKT. As reviewed above, the simultaneous activation promotes progression and androgen independence [102,105]. Moreover, simultaneous targeting of the two pathways was sufficient to inhibit hormone-refractory prostate cancer in the mouse model [106]. Mechanistically, the activation of mTOR and IκB kinase (IKK) by AKT leads to stimulation of NF-κB [107]. Supporting the significance of this regulation, the abnormal

NF-κB signaling correlates with androgen sensitivity, metastasis, and outcome in human prostate cancer and the mouse model. The studies implicate both the expression and the nuclear localization of NF-κB [108–111], and identify among the downstream mechanisms its ability to regulate expression of AR [112].

Phosphorylation of mitogen-activated protein kinases (MAPK) by expression of constitutively active RAF or RAS in the mouse prostate epithelium also leads to a nuclear accumulation of pERK, and this action promotes induction of tumorigenesis [79,113]. However, such mutations are rare in human prostate cancer. Notwithstanding, the collective perturbation of the RAS/RAF pathway via small but collectively significant alterations in the expression of the individual pathway members is prevalent in advanced human prostate cancer [114]. Phospho-MAPK, whether induced by EGF in vitro or prevalent in certain subsets of prostate cancer epithelial cells as assessed by immunofluorescence in vivo, preferentially partitions into the nucleus [115], which can be seen as indicative of the predominantly nuclear downstream functions. These functions remain to be investigated. Taken together, the data reviewed in the last sections demonstrate how a large number of fundamental molecular perturbations behind the development of prostate cancer are interlinked by calcium signaling and the calcium-regulated nuclear import mechanisms (Table 1).

Table 1. General regulatory factors involved in the calcium and nuclear signaling in prostate cancer. The molecular factors and their cell-physiological effects are categorized, and the corresponding calcium and nuclear events sketched out in the approximate order of their treatment in the text.

Category	Factor	Physiology	Partners	Calcium Regulation/Nuclear Signal
genomic	*NKX3.1* insufficiency	development	p53, AKT, PTEN	Ca^{2+}-regulated proteolysis
	MYC amplification	cell cycle	NKX3.1, PIM1	Ca^{2+}-regulated Notch nuclear entry
	TMPRSS2-ERG fusion	EMT	PTEN, *CACNA1D*	NF-κB nuclear entry, Ca^{2+} channel expression
	PTEN loss	hormone independence	mTORC2, PI3K	AR nuclear localization
developmental	WNT	cell motility	APC	Ca^{2+} wave, β-catenin nuclear entry
	Hedgehog	signaling from stroma	AR	Gli1 nuclear entry
	FGF	signaling from stroma	ERK	nuclear entry of alternative translation isoforms
kinases and phosphatases	AKT-mTOR	apoptosis	FOXO3A, Bim, p27	pAKT accumulation in nuclear bodies, Ca^{2+}-dependent inhibition
	ERK, MAPK	hormone independence	RAS, RAF, AKT	pERK/pMAPK accumulation in nucleus
	calcineurin	anti-senescence	NFATc1, MYC, IL6, STAT3	NFATc1 nuclear import

5. Anti-Senescence Signaling

A limitless proliferation potential achieved through anti-senescence factors has been proposed to be one of the biological hallmarks of cancer cells [116,117]. The recent characterization of store-operated calcium entry (SOCE) channel components STIM1 and ORAI1 in prostate cancer sheds light on the involvement of calcium in these processes. It was found [118] that the expression of both STIM1 and ORAI1 correlated negatively with the Gleason score (the commonly used histopathological assessment of prostate cancer progression) and was lower in prostate carcinomas relative to the normal tissue. STIM1 was also significantly reduced in hyperplasia. In agreement with these results obtained by immunohistochemistry on patient samples, STIM1 expression was reduced in carcinoma cell lines relative to the benign prostate hyperplasia (BPH) cell line BPH-1, and additionally reduced in the malignant PC3 line relative to the less metastatic DU145. The endogenous expression level differences in these experiments had the expected functional consequences for SOCE activity. Overexpression of the two factors, on the other hand, induced growth inhibition via a cell cycle arrest and senescence, as detected morphologically and by β-galactosidase staining. The apoptosis markers were similarly affected. The complex phenotype was additionally characterized by an increased migration of the cultured cells and a decreased ability of their conditioned medium to induce recruitment of macrophages in vitro. These effects were correlated with a signature of downstream TGF-β signaling activation, manifested in an overexpression of Snail and WNT1 and a rise of p-Smads and nuclear β-catenin, as well as a perturbed secretion of cytokines. On the other hand, inoculation of the STIM1- and ORAI1-overexpressing cells into non-obese diabetic, severe combined immunodeficiency (NOD/SCID) mice showed a comparative reduction of tumor growth and a relative loss of BrdU-positive cells and E-cadherin immunostaining. The latter can be seen as indicative of EMT, in agreement with the complex phenotype observed in vitro. An additional complicating finding both in vivo and in vitro has been an induction, in STIM1- and ORAI1-overexpressing cells and tumors, of a relative overexpression of the apoptosis-inhibiting [119,120] form of the tumor necrosis factor (TNF) receptor, decoy receptor 2 (DcR2), even though the anti-apoptotic Bcl-2 and XIAP in these experiments were reduced. A consistent interpretation may be possible, stemming from the fact that DcR2 is, at the same time, an established senescence marker [121,122].

Additional recent findings linking senescence in prostate cancer to calcium signaling pertain to the nuclear factor of activated T-cells c1 (NFATc1), a protein previously implicated in the regulation of expression of prostate specific membrane antigen in an ionomycin-responsive manner in LNCaP cells [123], as well as in the expression of the osteomimicry markers in PC3 and C4-2B cells [124]. NFATc subfamily proteins (the majority of the NFAT types) are activated via dephosphorylation by the calcium/calmodulin-dependent serine/threonine phosphatase calcineurin, upon which the NLS is exposed and the protein undergoes import into the nucleus, where it performs the functions of a transcription activator [125]. In a general agreement with previous work on NFATc1, Manda et al. [126] find the protein expressed in human adenocarcinoma samples (both in the neoplastic epithelium and in the stroma) as well as in tumorigenic prostate cancer cell lines, but not in the epithelium of the normal prostate or benign RWPE-1 cells. Targeted expression of the activated nuclear form of NFATc1 in a mouse model leads to PIN and subsequently adenocarcinoma marked by an elevated expression of proinflammatory cytokines, apparent proliferative influence of NFATc1+ cells on the neighboring cells, and castration resistance. Revealingly, MYC was upregulated in both NFATc1+ and—apparently via the elevated IL6 and pSTAT3—in the neighboring cells. A double transgenic model with a *PTEN* deletion displayed a synergy of the factors and an accelerated tumorigenesis. Compared with the common *PTEN* deletion model, the prostates were marked by a suppressed expression and nuclear localization of the senescence marker p21 and an insignificant β-galactosidase staining. The mechanism of the p21 suppression may involve the STAT3-SKP2 pathway demonstrated in the gastric and cervical cancer cells [127,128]. Altogether, the data suggest that the calcium-regulated nuclear import step may be critical for the establishment of anti-senescence signaling in advanced prostate cancer.

6. Channels and Transporters

6.1. Transient Receptor Potential (TRP) Channels

Channels and transporters are among the best-studied molecular classes orchestrating the calcium signaling. In the recent years, considerable attention has been given to the role of TRP channels in the development of prostate cancer. The function of TRPV6, a vanilloid subfamily plasma membrane calcium channel, in particular, is now comparatively well characterized. Earlier termed CaT-like, it is a marker of locally advanced, metastatic, and androgen-insensitive prostate cancer [129]. On mRNA level, in situ hybridization uncovered a correlation of its expression with a number of cancer progression characteristics, including the Gleason score [130]. Remarkably, while its expression is detectable in most androgen-insensitive lesions, the level is reduced relative to untreated tumors. In-vitro studies have shown that this channel, as well as the related TRPC6, increases prostate cancer cell proliferation, acting through NFAT [131,132]. A subsequent detailed immunohistochemical analysis established also the correlation of the TRPV6 expression with prostate-specific antigen (PSA) and TRPC1, and with a lack of the apoptosis marker caspase-3 cleaved fragment [133]. Intriguingly, the presence of TRPV6 in the more advanced tumors, where it is found in luminal cells, was correlated with the appearance of ORAI1 in the same cells, which normally localizes to the basal cells. It has been possible to establish the involvement of TRPV6 in SOCE in prostate cancer cells, as well as its physical association with ORAI1 via STIM1 and TRPC1 under the conditions of SOCE. The latter conditions also induced a translocation of TRPV6 to the plasma membrane, as observed using a confocal microscope. The same study showed that overexpression of TPRV6 in prostate cancer cells in vitro increases their resistance to apoptosis induced, for example, by cisplatin, while in heterotopic xenografts in mice its experimentally manipulated level of expression correlates with the resulting tumor mass. Most recently, an inhibitor of this channel was evaluated in a first-in-human phase I clinical trial with promising results [134]. The work on TRPV6, to date, can serve as a model of calcium-mediated mechanism elucidation leading to translationally relevant results in the prostate cancer field.

Among the other TRP channels, as reviewed most recently by Cui et al. [8], the nonselective cationic channel vanilloid 2 (TRPV2) has been associated specifically with metastatic prostate cancer. Experiments in an earlier TRPV2 study [135] demonstrated the channel's involvement in the promotion of prostate cancer cells' migration and invasion both in vitro and in vivo. The effect, accompanied by elevated expression levels of the invasion enzymes MMP2 and MMP9, as well as cathepsin B, appeared to be mediated by an elevated basal cytosolic concentration of Ca^{2+} that was dependent on the channel's expression level. For progress in system-level understanding of the mechanism behind the emergence of prostate cancer invasivity, it will be of interest to characterize the functional interplay of these effects with the myosin-driven secretion of metalloproteases [136], a process that is likely open to calcium regulation in the light of the molecular dynamics investigations that will be reviewed in the last section.

Another TRP channel, melastatin family 8 (TRPM8, initially termed TRP-P8), was first found to be overexpressed in prostate cancer by Tsavaler et al. [137]. In androgen-responsive prostate cancer cells in vitro, it is expressed on the plasma membrane and in the endoplasmic reticulum (ER), and the level of expression is positively regulated by androgen [138]. The channel is capable of initiating an agonist-induced elevation of the intracellular calcium ion concentration. In keeping with the general theme of fine-tuned homeostasis, which is widely accepted to characterize the calcium regulation [139], both the agonist and antagonist action on TRPM8 in in-vitro experiments was able to induce apoptosis. In the same experiments, the channel was found to be expressed at a low and unregulated level in androgen-insensitive cells. This finding agrees with the loss of TRPM8 in transition to androgen independence that was detected in a genome-wide expression profiling study [140]. Subsequent work [141] found that experimental re-expression of this channel in androgen-insensitive cells reduced their migration in vitro. At the same time, in non-prostate-cancer model cell culture, PSA was able to potentiate the TRPM8 agonist-induced calcium current by enhancing trafficking of the channel to the plasma membrane via its phosphorylation downstream of the bradykinin 2 receptor signaling pathway. In resection samples and primary cultures, the plasma membrane localization and activity of this channel was found to be characteristic of the luminal cells of normal tissue, BPH, and especially in situ tumors, whereas the dedifferentiated transit amplifying and intermediate phenotype retained primarily the endoplasmic isoform [142]. As pointed out by the authors of this investigation, one functional consequence of the endoplasmic localization of the channel may be the apoptosis-resistant state characterized by a lowered intraendoplasmic calcium concentration, which had been seen in other studies of advanced prostate cancer [143,144]. Thus, the channel's role appears to be different, depending on the stage of prostate cancer, and change from its expression maintaining the pro-survival calcium homeostasis to its partial loss potentially contributing to the development of metastatic potential in parallel with the acquisition of androgen independence.

A recent study [145] has implicated a second melastatin family TRP channel, TRPM7, in the positive regulation of prostate cancer cells' migration, invasion, and expression of MMPs, as well as in downregulation of E-cadherin that accompanies EMT. Furthermore, the related TRPM4 has a function in the proliferation of prostate cancer cells, according to a breaking report [146]. In this instance, the downstream cascade includes a relative dephosphorylation of β-catenin and an elevation of this regulator's nuclear localization. Completing the present picture of a broad and mechanistically diverse involvement of the melastatin subfamily channels in the prostate cancer development, TRPM2 is one more known member that controls proliferation of prostate cancer cells by inhibiting the nuclear ADP-ribosylation [147]. The intracellular localization of TRPM2, which is associated with the plasmalemma of benign prostate epithelial cells, is altered in tumorigenic prostate cell lines, where it appears in intranuclear clusters. It will be interesting to investigate if the channels of this broad class are similar to the GPCRs to be reviewed in Section 8.1 in exhibiting nuclear translocation and influencing the trans-nuclear envelope potential and intranuclear calcium concentration in prostate cancer cells. Their downstream targets, hypothetically, could include the physical nuclear organization and transcription-related functions of nuclear myosins (see Section 8.2).

6.2. Other Channels and Transporters

The other types of channels and transporters that have been implicated in the development of prostate cancer include molecules responsible for a remarkably diverse set of calcium homeostasis mechanisms, such as voltage-gated and calcium-activated channels, as well as intracellular pumps. In particular, there is pharmacological evidence of a contribution from T-type calcium channels, including, specifically, $Ca_v3.2$, to the proliferative activity of prostate cancer cells in vitro and to cancer progression in xenograft experiments [148,149]. A similar involvement of this type of channels has been documented in other tumor types [150]. The $Ca_v3.2$ isoform may also be involved in neuroendocrine differentiation in prostate cancer, which is associated with castration resistance, distant metastasis, and poor prognosis [151]. Revealingly, induction of neuroendocrine differentiation of LNCaP epithelial prostate cancer cells by androgen deprivation in vitro was found to be accompanied by an upregulation of $Ca_v3.2$ and an increase in the associated calcium current [152,153]. These results suggest that a calcium-mediated reinforcement loop similar to the one already discussed in connection with the *CACNA1D* gene (which encodes $Ca_v1.3$, see Section 2.3) may be contributing to the neuroendocrine differentiation in the androgen-deprived prostate.

An intriguing connection to the cytoskeleton signaling has recently been uncovered that involves a plasma membrane channel that is calcium-activated but conducts potassium ions. The big potassium calcium-sensitive (BKCa) channels are expressed in excitable cells [154] and aberrantly in cancers of various origin [155]. They are overexpressed in a significant subset of advanced prostate cancers, and inhibiting their expression can suppress PC3 cell proliferation in vitro [156]. A recent study [157] has extended these results, using a xenograft model, and additionally demonstrated a stimulating effect of these channels' overexpression on cell migration and invasion. Interestingly, the effects depended only slightly on the channels' conductance, but were exerted instead in an integrin- and phosphorylated focal adhesion kinase (FAK)-dependent manner. The immunofluorescence colocalization and immunoprecipitation showed a formation of a BKCa-αVβ3 integrin complex, accompanied by recruitment and phosphorylation of FAK. The possible interplay of this mechanism with the tumor stroma's influence on the focal adhesion signaling, as well as the potential feedback from the extracellular matrix modification by the invasive epithelial cells themselves, present interest for future research.

The sarcoendoplasmic calcium pump SERCA's expression in epithelial prostate cancer cells is upregulated by the action of EGF, dihydrotestosterone (DHT), or serum, according to the in-vitro studies on LNCaP cells [158]. Stimulation of proliferation caused by these agents is suppressed by the SERCA inhibitor thapsigargin. Thus, the pump appears to be both an intermediary of the response to mitogens and subject to the downstream regulation caused by these factors, which may sensitize the tumor to the subsequent stimulation. Although the SERCA isoform implicated by the cited study was the non-tissue specific isoform (2B), a prostate-selective thapsigargin analog has been designed [159] and, most recently, reached phase II in the clinical trials for prostate cancer [8].

Int. J. Mol. Sci. **2018**, *19*, 1237

In addition to the reviewed roles of calcium-release activated calcium channel protein ORAI1 in senescence and SOCE, the channel appears to be involved also in the capacitative entry mediating apoptosis in androgen-sensitive prostate cancer cells in culture [160]. The androgen dependence of expression of ORAI1 itself, which was demonstrated in this study, raises the possibility of its downregulation being part of the mechanism of the emergence of aggressive cancer following the androgen deprivation therapy. Extending these results, the store-independent entry was recently associated with prostate cancer-specific enhanced expression of ORAI3 and formation of a heteromeric channel with ORAI1 [161]. Interestingly, the channel is gated by arachidonic acid, opening a new link between calcium signaling and the regulation by metabolism and inflammation.

The mitochondrial calcium uniporter, which contributes to the proapoptotic accumulation of calcium in mitochondria, is a comparatively new pump that has been studied in relation to cancer [8]. In-silico screens pointed to a negative regulation of its expression by miR-25, and this prediction was borne out by the experimental measurements [162]. Established prostate cancer cell lines express large quantities of this micro-RNA and are, at the same time, characterized by suppressed levels of the uniporter. Experimental restoration of its expression has been shown to sensitize prostate cancer cells in vitro to apoptosis. This is a promising direction for a further investigation in vivo.

An important apoptotic pathway is mediated by a sustained Ca^{2+} flux from the ER to mitochondria through the endoplasmic inositol 1,4,5-trisphosphate receptors, InsP3R [163]. The regulation of this mechanism has been linked to the deficiency of the *PTEN* gene, which, as reviewed above, is common in prostate cancer. The role of InsP3R has been elucidated in a series of experiments on established prostate cancer cell lines, which similarly exhibit the *PTEN* inactivation [164]. LNCaP cells, in particular, are characterized by a frameshift mutation in *PTEN*, and PKB/ACT in these cells was found to be constitutively phosphorylated and activated in the PI3K-dependent manner [165]. The phosphorylation in these experiments has proved determinative of the cells' resistance to apoptosis upon withdrawal of the serum factors. Subsequent work [166] has traced the mechanism of the resistance to phosphorylation of InsP3R by AKT. Most recently, a similar effect has been described in PC3 cells, where it is mediated by the lack of competition from the deficient *PTEN* product for binding to InsP3R3, which opens the latter to binding the F-box protein FBXL2 [167]. FBXL2 has the function of the receptor subunit of the ubiquitin ligase complexes, and can target InsP3R3 for degradation. The different mechanisms in the lymph node- and bone-metastatic cell lines may be indicative of diverging functions that can be discovered even for the comparatively well-characterized molecular classes at the different stages of prostate cancer progression. While the conditions-dependent multifunctionality may play the role of a complicating factor in drug development, it can also present an opportunity to more selectively target the tumor development at its specific crucial steps.

Finally, in the discussion of the channels involved in calcium signaling in prostate cancer, ryanodine receptors (RyR) have also been found to be expressed in prostate cancer cell lines [168]. The isoforms RyR1-3 vary in quantity depending on the cell line, a finding that should be further investigated, as it may be indicative of modification of the response mediated by these channels, depending on the stage in the tumor development or its environment. Calcium release is observed under the action of the RyR agonist caffeine and has been found to mildly promote apoptosis of the lymph-node metastatic LNCaP cells in vitro [169]. A synthetic agonist 4-chloro-*m*-cresol was also effective at eliciting this calcium response in these experiments, pointing to a possibility of targeting the receptor through a prostate-selective drug design in the future. On the whole, the calcium and calcium-operated channels and transporters remain one of the most systematically studied molecular classes in relation to their roles in prostate cancer (Table 2).

Table 2. Calcium and nuclear signaling in prostate cancer that is associated with specific channels and transporters.

Category	Channel/Transporter	Related Cell Physiology	Partners	Calcium Regulation/Nuclear Signal
TRP channels	TRPM2	proliferation	nuclear ADP-ribosylase	channel accumulation in nuclear clusters
	TRPM4	proliferation	β-catenin	β-catenin nuclear accumulation
	TRPM7	EMT	E-cadherin, MMPs	PM Ca^{2+} channel
	TRPM8	apoptosis	AR, PSA	lowered ER Ca^{2+}
	TRPV2	invasion	MMPs	elevated basal Ca^{2+}
	TRPV6	proliferation	NFAT, ORAI1, STIM1	SOCE
other channels, transporters	BKCa	invasion	integrin	Ca^{2+}-activated phosphorylation of FAK
	SERCA	proliferation	EGF, DHT	ER Ca^{2+}
	ORAI1	apoptosis	ORAI3, arachidonic acid	capacitative entry via the gated heteromer
	mitochondrial uniporter	apoptosis	miR-25	mitochondrial Ca^{2+}
	InsP3R	apoptosis	AKT, PTEN, FBXL2	ER-to-mitochondria current

7. Additional Topics

7.1. Calpain Proteolysis

Upon elevation of the intracellular calcium concentration in prostate cancer cells, as demonstrated in experiments in vitro [170], the calcium-regulated protease calpain cleaves β-catenin, producing a stable, transcriptionally active fragment that enters the nucleus. At the same time, microarray experiments demonstrated an elevation of the calpain expression in metastatic prostate cancer, which may suggest sensitization of this nuclear import pathway to calcium signaling in cells undergoing the transition to malignancy. Alongside the calcineurin-NFAT (Section 5) and calmodulin-myosin (Section 8.2) mechanisms, the calpain-β-catenin mechanism can be seen as part of an emerging paradigmatic triad representative of the mechanistically diverse calcium-regulated nuclear import pathways, many of which probably remain to be elucidated.

Another target for the calcium-induced cleavage by calpain in prostate cancer cells is E-cadherin, whose inactivation is characteristic of adenocarcinomas and is often seen as a marker for EMT [171]. The intracellular calcium-induced calpain proteolysis of FAK [172], at the same time, is a contributing factor to the modulation of cell adhesion in developing prostate cancer. The likely BKCa-mediated crosstalk with the mechanism reviewed in the last section is a promising avenue for future research. On the whole, the data on calpain demonstrate a remarkably multifaceted role of the calcium-regulated proteolysis in prostate cancer development and emphasize the simultaneous challenge and opportunity for the systems-based design of novel therapeutics that could reverse the dysregulation of this process.

7.2. 14-3-3 Mediated Nucleocytoplasmic Redistribution

As reviewed above (Section 4), AKT is one of the key signaling kinases involved in prostate cancer development, which exerts its effects, in particular, by inhibiting the FOXO3A transcription activator of the proapoptotic factor Bim through a mechanism that involves nucleocytoplasmic partitioning. Additional light has been shed on this issue by the work that has implicated the chaperone protein 14-3-3 [173]. The serine phosphorylation of FOXO3A by AKT creates a binding site for 14-3-3 and leads to retention of the complex in the cytoplasm, sequestering the transcription factor. The cytoplasmic localization of FOXO3A and 14-3-3 is correlated with the Gleason score in human samples and with disease progression in TRAMP mice [174]. In the reviewed work by Trotman et al. [103], it was shown, in particular, that pAKT in the mouse model prostate lesions accumulates in the nuclei. The mechanistic significance of this fact has been illuminated by the results that were obtained on non-prostate-cancer cells in vitro [175]. In the latter experiments, the nuclear pAKT was found to phosphorylate FOXO3A/FKHL1 immediately prior to the 14-3-3 binding and nuclear export of this transcription factor. The chaperone-mediated sequestration and nuclear export induced by the intranuclear activated kinase is an intriguing element of the nuclear transport-related mechanistic repertoire of prostate cancer, which merits further study.

7.3. Autonomic-System Regulation

In addition to the *PTEN* inactivation and constitutive activation of AKT, an acetylcholine-induced and calcium-dependent pathway leading to this kinase's activation has been uncovered recently [176]. Following the in-vivo indications that the autonomic, including cholinergic, nerve infiltration promoted prostate cancer development [177], the in-vitro work by Wang et al. [176] revealed the existence of autocrine acetylcholine signaling in epithelial prostate cancer cells, which proceeds through the muscarinic GPCR CHRM3 and a Ca^{2+} influx. Phosphorylation of AKT under these conditions was found to be mediated by calmodulin and CAMK kinase (CAMKK). CHRM3 is known to be upregulated in human prostate cancer samples, and experiments in vivo and in vitro have implicated the autocrine mechanism in the prostate cancer cells' proliferative and migratory potential. An antagonist treatment in vivo also increased the percentage of apoptotic cells and reduced the castration-resistant tumor growth. The most recent extension of this work has implicated the autocrine calcium-dependent mechanism in BPH and maintaining the prostate epithelial progenitor cells in the proliferative state [178]. It has also been demonstrated in LNCaP cells that carbachol-induced CAMKK-dependent blockage of apoptosis is mediated by AKT phosphorylation of the Bcl-2 associated death promoter protein BAD [179]. The effect on BAD is linked to its 14-3-3 binding and sequestration in the cytosol from the mitochondria, as shown in non-prostate-cancer cells [180]. Similar to the cytoplasmic sequestration of FOXO3A from the nucleus (last section), this mechanism adds complexity to the web of the spatially-restricted interactions of the molecules involved in the intracellular transport downstream of calcium signaling.

It is interesting to note that although the sympathetic, adrenergic effects in the Magnon et al. experiments [177] were found to be mediated by the stromal β1 and β2 receptors, a similar adrenergic effect on proliferation of primary prostate epithelial cells was demonstrated in the reviewed work of Thebault et al. [131], where it was mediated by α1 receptors. The downstream adrenergic signaling seen in the epithelial cell culture proceeded through a store-independent calcium entry via the TRPC6 channels, calcineurin, and NFAT-dependent transcription, thereby implicating the reviewed (Section 5) calcium-dependent nuclear import pathway.

8. New Molecular Classes in Calcium Signaling to the Nucleus

8.1. GPCRs

Alongside the adrenergic receptors, whose role was reviewed above, the chemokine cysteine (C)-X-C receptor 4 (CXCR4) is another GPCR that is remarkable for its involvement in the nuclear calcium signaling. It has been implicated in bone tropism of prostate cancer cells [181], in which it participates through its interaction with its bone-associated agonist stromal cell-derived factor 1 (SDF-1). It has also been found to promote migration and invasion of prostate cancer cells under the conditions when PTEN is inactivated by ROS [182]. It resides on the plasma membrane but also localizes specifically to the nuclei in human prostate cancer samples, as revealed by means of immunohistochemistry [13]. It similarly accumulates in the nuclear fractions of cultured prostate cancer cells, possesses an NLS, and associates with the nuclear import receptor transportin-β1. Exposure to SDF-1 results in dissociation of the α subunit from the CXCR4 in the nuclear fraction, leading to intranuclear calcium release. In addition to the identification of a new class of nuclear calcium signaling molecules associated with prostate cancer, this recent study points to a novel mechanism of escape from any therapeutic receptor antagonist action that is limited to the cell surface.

8.2. Myosins

Non-muscle myosins have recently been shown to have multiple functions in prostate cancer. The longest-known functions of the myosin family concern cell motility, and recent work on prostate cancer cells has finely delineated this class of functions among the myosin family members including IB, IXB, X, and XVIIIA [183]. The contributions of these molecules to the shaping of the actin cytoskeleton and motile morphology of prostate cancer cells underpin mechanically the acquisition of a metastatic phenotype by the originally non-motile epithelial cells [184]. Myosin IC too has long been known as a contributor to cell migration and to intracellular transport in various cell types [185]. The most recent results have implicated it in the prostate cancer cell migration and invasion [136]. The three isoforms of this protein, including the recently discovered one (A) that is associated with metastatic prostate cancer [186,187]; and the one (B) originally described as the nuclear myosin [188], differ kinetically, according to the latest work [189].

Following the discovery of the NLS [190] within the IQ domain of myosin IC that mediates the interaction with apo-calmodulin, it was demonstrated that in prostate cancer cells, elevation of the intracellular calcium concentration drives the nuclear import of this myosin [12]. This import depends on importin-β1 and is presumed to be mediated by the unmasking of the NLS when, as established for myosin IC in vitro [191], the calcium-calmodulin dissociates from the IQ domain. The most recent work [192] has demonstrated that binding to phosphoinositides, mediated in part by the NLS region, contributes to the nuclear import of myosin IC. It will be of interest to determine whether similar mechanisms regulate the nucleocytoplasmic partitioning of the other myosins known to localize to the nucleus, such as subfamilies II, V, VI, X, XVI, and XVIII [193].

The nuclear functions have to date been best characterized for myosin IC isoform B. They include promotion of transcription, which is effected through an association with RNA polymerases I and II [194–200]. This myosin isoform also associates with RNA transcripts and pre-ribosomal units, participating in the export of the ribosome subunits from the nucleus [201,202]. The movements of chromosomes inside the nucleus similarly depend on nuclear myosin and actin [203,204]. Myosin IC isoform C—long thought to be cytoplasmic but now recognized as the most responsive to the calcium-regulated nuclear import in prostate cancer cells [12]—displays a capacity for interaction with RNA polymerase II and for maintaining the level of polymerase I activity [205]. Since calmodulin, acting as the light chain, modulates the motor properties of myosin IC [206] and itself undergoes nuclear import via the facilitated pathway [207], it appears promising to investigate next the intranuclear calcium regulation of the myosin functions in prostate cancer.

The tumor-suppressor qualities of myosin IC have recently been characterized in non-prostate-cancer cells [208]. Partially similar results involving suppression of anchorage-independent growth had previously been obtained in relation to myosin XVIII in other cancers, as reviewed by Oudekirk and Krendel [10], and most recently in a clinical association study for myosin VC in colorectal cancer [209]. At the same time, in breast and prostate cancer *MYO18A* (which encodes the non-muscle myosin XVIIIA) appeared to be an oncogene [210]. Future work should be able to shed light on the possible reciprocal regulation by means of the calcium-controlled switch between the cytoplasmic and nuclear functions of the non-muscle myosins, leading to a unified picture of these phenomena.

9. Summary and Outlook

The intracellular calcium signal and the nuclear entry of the activated factors can be seen as dual, linked integrating events that undergird the molecular complexity behind the critical stages of prostate cancer development (Tables 1–3). As these multiple examples in the reviewed literature from the recent years show, the nuclear vs. cytoplasmic localization frequently presents a unique differentiating feature that is associated with a prognosis for progression of the disease. Although localized calcium signals have the capacity to channel the information transmitted by the specific pathways, the influences impinging on the intracellular calcium homeostasis establish a network for a physical integration of the conditions of the tumor microenvironment. Of special interest from the molecular biological viewpoint

Int. J. Mol. Sci. **2018**, *19*, 1237

are the events linking the calcium signaling to the regulation of the nuclear entry, as well as the dynamic steady-state nucleocytoplasmic distribution of the transcription regulators and cytoskeletal effector molecules. The polyfunctionality and novel intracellular localizations of the previously known factors in prostate cancer development that have been uncovered in the most recent work point, on the one hand, at a greater complexity that lies ahead in the molecular exploration of the disease mechanisms. On the other hand, however, they hold out a promise for a mechanistically grounded validation of new prognostic markers and potential targets for drug development, which would be able to disrupt the critical nodes in the prostate cancer network.

Table 3. Additional molecular classes involved in the calcium and nuclear signaling in prostate cancer.

Category	Factor	Physiology	Partners	Calcium Regulation/ Nuclear Signal
proteases	calpain	metastasis, EMT	AR, β-catenin, E-cadherin	nuclear entry of cleaved β-catenin
chaperones	14-3-3	apoptosis	AKT, FOXO3A, Bim	nuclear export of FOXO3A
autonomic regulation	CHRM3	castration resistance	CAMKK, AKT, BAD	acetylcholine-induced Ca^{2+} influx
	adrenergic receptors	proliferation	TPRC6, calcineurin, NFAT	store-independent entry
GPCRs	CXCR4	bone tropism	SDF-1, PTEN, ROS, importin-β1	nuclear entry, intranuclear Ca^{2+} release
molecular motors	myosin IC	migration, invasion, metastasis	calmodulin, importin-β1, RNA polymerase I and II, nuclear actin	Ca^{2+}-induced nuclear entry

Acknowledgments: Supported in part by the NCI of the National Institutes of Health under award number R21CA220155 to Wilma A. Hofmann.

Author Contributions: Ivan V. Maly—conceptualization and writing; Wilma A. Hofmann—conceptualization, review and editing.

Conflicts of Interest: The authors declare no conflict of interest. The funding sponsors had no role in the design of the study; in the collection, analyses, or interpretation of data; in the writing of the manuscript, and in the decision to publish the results.

Abbreviations

AR	Androgen receptor
BKCa	Big potassium calcium-sensitive (channel)
BPH	Benign prostate hyperplasia
CAMK	Calcium/calmodulin-dependent kinase
CAMKK	Calcium/calmodulin-dependent kinase kinase
DcR2	Decoy receptor 2
DHT	Dihydrotestosterone
EGF	Epithelial growth factor
EMT	Epithelial-mesenchymal transition
ER	Endoplasmic reticulum
ERK	Extracellular signal-regulated kinase
FAK	Focal adhesion kinase
FGF	Fibroblast growth factor
GPCR	G protein-coupled receptor
IκB	Inhibitor of κB
IKK	Inhibitor of κB kinase
InsP3R	Inositol 1,4,5-trisphosphate receptor

LNCaP	Lymph node cancer of prostate (cell line)
MAPK	Mitogen-activated protein kinase
mTORC2	Mammalian target of rapamycin complex 2
NFAT	Nuclear factor of activated T-cells
NF-κB	Nuclear factor κB
NLS	Nuclear localization signal (sequence)
PI3K	Phosphoinositide 3-kinase
PIN	Prostate intraepithelial neoplasia
PKB	Protein kinase B
PKC	Protein kinase C
PSA	Prostate-specific antigen
PTEN	Phosphatase and tensin homolog deleted on chromosome 10
ROS	Reactive oxygen species
RyR	Ryanodine receptor
SDF-1	Stromal cell-derived factor 1
SOCE	Store-operated calcium entry
TLR	Toll-like receptor
TRAMP	Transgenic adenocarcinoma of the mouse prostate
TRP	Transient receptor potential (channel)
YB-1	Y-box binding protein 1

References

1. Miller, K.D.; Siegel, R.L.; Lin, C.C.; Mariotto, A.B.; Kramer, J.L.; Rowland, J.H.; Stein, K.D.; Alteri, R.; Jemal, A. Cancer treatment and survivorship statistics, 2016. *CA Cancer J. Clin.* **2016**, *66*, 271–289. [CrossRef] [PubMed]

2. Siegel, R.L.; Miller, K.D.; Jemal, A. Cancer statistics, 2016. *CA Cancer J. Clin.* **2016**, *66*, 7–30. [CrossRef] [PubMed]

3. Carlsson, S.; Vickers, A. Spotlight on prostate cancer: The latest evidence and current controversies. *BMC Med.* **2015**, *13*, 60. [CrossRef] [PubMed]

4. Shen, M.M.; Abate-Shen, C. Molecular genetics of prostate cancer: New prospects for old challenges. *Genes Dev.* **2010**, *24*, 1967–2000. [CrossRef] [PubMed]

5. Bootman, M.D. Calcium signaling. *Cold Spring Harb. Perspect. Biol.* **2012**, *4*, a011171. [CrossRef] [PubMed]

6. Cautain, B.; Hill, R.; de Pedro, N.; Link, W. Components and regulation of nuclear transport processes. *FEBS J.* **2015**, *282*, 445–462. [CrossRef] [PubMed]

7. Godfraind, T. Discovery and Development of Calcium Channel Blockers. *Front. Pharmacol.* **2017**, *8*, 286. [CrossRef] [PubMed]

8. Cui, C.; Merritt, R.; Fu, L.; Pan, Z. Targeting calcium signaling in cancer therapy. *Acta Pharm. Sin. B* **2017**, *7*, 3–17. [CrossRef] [PubMed]

9. Hauser, A.S.; Attwood, M.M.; Rask-Andersen, M.; Schioth, H.B.; Gloriam, D.E. Trends in GPCR drug discovery: New agents, targets and indications. *Nat. Rev. Drug Discov.* **2017**, *16*, 829–842. [CrossRef] [PubMed]

10. Ouderkirk, J.L.; Krendel, M. Non-muscle myosins in tumor progression, cancer cell invasion, and metastasis. *Cytoskeleton* **2014**, *71*, 447–463. [CrossRef] [PubMed]

11. Li, Y.R.; Yang, W.X. Myosins as fundamental components during tumorigenesis: Diverse and indispensable. *Oncotarget* **2016**, *7*, 46785–46812. [CrossRef] [PubMed]

12. Maly, I.V.; Hofmann, W.A. Calcium-regulated import of myosin IC into the nucleus. *Cytoskeleton* **2016**, *73*, 341–350. [CrossRef] [PubMed]

13. Don-Salu-Hewage, A.S.; Chan, S.Y.; McAndrews, K.M.; Chetram, M.A.; Dawson, M.R.; Bethea, D.A.; Hinton, C.V. Cysteine (C)-X-C receptor 4 undergoes transportin 1-dependent nuclear localization and remains functional at the nucleus of metastatic prostate cancer cells. *PLoS ONE* **2013**, *8*, e57194. [CrossRef] [PubMed]

14. Emmert-Buck, M.R.; Vocke, C.D.; Pozzatti, R.O.; Duray, P.H.; Jennings, S.B.; Florence, C.D.; Zhuang, Z.; Bostwick, D.G.; Liotta, L.A.; Linehan, W.M. Allelic loss on chromosome 8p12-21 in microdissected prostatic intraepithelial neoplasia. *Cancer Res.* **1995**, *55*, 2959–2962. [PubMed]

15. Bethel, C.R.; Faith, D.; Li, X.; Guan, B.; Hicks, J.L.; Lan, F.; Jenkins, R.B.; Bieberich, C.J.; De Marzo, A.M. Decreased NKX3.1 protein expression in focal prostatic atrophy, prostatic intraepithelial neoplasia, and adenocarcinoma: Association with gleason score and chromosome 8p deletion. *Cancer Res.* **2006**, *66*, 10683–10690. [CrossRef] [PubMed]

16. Asatiani, E.; Huang, W.X.; Wang, A.; Rodriguez Ortner, E.; Cavalli, L.R.; Haddad, B.R.; Gelmann, E.P. Deletion, methylation, and expression of the NKX3.1 suppressor gene in primary human prostate cancer. *Cancer Res.* **2005**, *65*, 1164–1173. [CrossRef] [PubMed]

17. Bhatia-Gaur, R.; Donjacour, A.A.; Sciavolino, P.J.; Kim, M.; Desai, N.; Young, P.; Norton, C.R.; Gridley, T.; Cardiff, R.D.; Cunha, G.R.; et al. Roles for Nkx3.1 in prostate development and cancer. *Genes Dev.* **1999**, *13*, 966–977. [CrossRef] [PubMed]

18. Schneider, A.; Brand, T.; Zweigerdt, R.; Arnold, H. Targeted disruption of the Nkx3.1 gene in mice results in morphogenetic defects of minor salivary glands: Parallels to glandular duct morphogenesis in prostate. *Mech. Dev.* **2000**, *95*, 163–174. [CrossRef]

19. Tanaka, M.; Komuro, I.; Inagaki, H.; Jenkins, N.A.; Copeland, N.G.; Izumo, S. Nkx3.1, a murine homolog of Ddrosophila bagpipe, regulates epithelial ductal branching and proliferation of the prostate and palatine glands. *Dev. Dyn.* **2000**, *219*, 248–260. [CrossRef]

20. Abdulkadir, S.A.; Magee, J.A.; Peters, T.J.; Kaleem, Z.; Naughton, C.K.; Humphrey, P.A.; Milbrandt, J. Conditional loss of Nkx3.1 in adult mice induces prostatic intraepithelial neoplasia. *Mol. Cell. Biol.* **2002**, *22*, 1495–1503. [CrossRef] [PubMed]

21. Kim, M.J.; Bhatia-Gaur, R.; Banach-Petrosky, W.A.; Desai, N.; Wang, Y.; Hayward, S.W.; Cunha, G.R.; Cardiff, R.D.; Shen, M.M.; Abate-Shen, C. Nkx3.1 mutant mice recapitulate early stages of prostate carcinogenesis. *Cancer Res.* **2002**, *62*, 2999–3004. [PubMed]

22. Lei, Q.; Jiao, J.; Xin, L.; Chang, C.J.; Wang, S.; Gao, J.; Gleave, M.E.; Witte, O.N.; Liu, X.; Wu, H. NKX3.1 stabilizes p53, inhibits AKT activation, and blocks prostate cancer initiation caused by PTEN loss. *Cancer Cell* **2006**, *9*, 367–378. [CrossRef] [PubMed]

23. Ouyang, X.; DeWeese, T.L.; Nelson, W.G.; Abate-Shen, C. Loss-of-function of Nkx3.1 promotes increased oxidative damage in prostate carcinogenesis. *Cancer Res.* **2005**, *65*, 6773–6779. [CrossRef] [PubMed]

24. Bowen, C.; Gelmann, E.P. NKX3.1 activates cellular response to DNA damage. *Cancer Res.* **2010**, *70*, 3089–3097. [CrossRef] [PubMed]

25. Markowski, M.C.; Bowen, C.; Gelmann, E.P. Inflammatory cytokines induce phosphorylation and ubiquitination of prostate suppressor protein NKX3.1. *Cancer Res.* **2008**, *68*, 6896–6901. [CrossRef] [PubMed]

26. Decker, J.; Jain, G.; Kießling, T.; Philip, S.; Rid, M.; Barth, T.T.; Möller, P.; Cronauer, M.V.; Marienfeld, R.B. Loss of the Tumor Suppressor NKX3.1 in Prostate Cancer Cells is Induced by Prostatitis Related Mitogens. *J. Clin. Exp. Oncol.* **2016**, *5*. [CrossRef]

27. Sato, K.; Qian, J.; Slezak, J.M.; Lieber, M.M.; Bostwick, D.G.; Bergstralh, E.J.; Jenkins, R.B. Clinical significance of alterations of chromosome 8 in high-grade, advanced, nonmetastatic prostate carcinoma. *J. Natl. Cancer Inst.* **1999**, *91*, 1574–1580. [CrossRef] [PubMed]

28. Jenkins, R.B.; Qian, J.; Lieber, M.M.; Bostwick, D.G. Detection of c-myc oncogene amplification and chromosomal anomalies in metastatic prostatic carcinoma by fluorescence in situ hybridization. *Cancer Res.* **1997**, *57*, 524–531. [PubMed]

29. Gurel, B.; Iwata, T.; Koh, C.M.; Jenkins, R.B.; Lan, F.; Van Dang, C.; Hicks, J.L.; Morgan, J.; Cornish, T.C.; Sutcliffe, S.; et al. Nuclear MYC protein overexpression is an early alteration in human prostate carcinogenesis. *Mod. Pathol.* **2008**, *21*, 1156–1167. [CrossRef] [PubMed]

30. Sotelo, J.; Esposito, D.; Duhagon, M.A.; Banfield, K.; Mehalko, J.; Liao, H.; Stephens, R.M.; Harris, T.J.; Munroe, D.J.; Wu, X. Long-range enhancers on 8q24 regulate c-Myc. *Proc. Natl. Acad. Sci. USA* **2010**, *107*, 3001–3005. [CrossRef] [PubMed]

31. Jia, L.; Landan, G.; Pomerantz, M.; Jaschek, R.; Herman, P.; Reich, D.; Yan, C.; Khalid, O.; Kantoff, P.; Oh, W.; et al. Functional enhancers at the gene-poor 8q24 cancer-linked locus. *PLoS Genet.* **2009**, *5*, e1000597. [CrossRef] [PubMed]

32. Al Olama, A.A.; Kote-Jarai, Z.; Giles, G.G.; Guy, M.; Morrison, J.; Severi, G.; Leongamornlert, D.A.; Tymrakiewicz, M.; Jhavar, S.; Saunders, E.; et al. Multiple loci on 8q24 associated with prostate cancer susceptibility. *Nat. Genet.* **2009**, *41*, 1058–1060. [CrossRef] [PubMed]

33. Gudmundsson, J.; Sulem, P.; Gudbjartsson, D.F.; Blondal, T.; Gylfason, A.; Agnarsson, B.A.; Benediktsdottir, K.R.; Magnusdottir, D.N.; Orlygsdottir, G.; Jakobsdottir, M.; et al. Genome-wide association and replication studies identify four variants associated with prostate cancer susceptibility. *Nat. Genet.* **2009**, *41*, 1122–1126. [CrossRef] [PubMed]

34. Ellwood-Yen, K.; Graeber, T.G.; Wongvipat, J.; Iruela-Arispe, M.L.; Zhang, J.; Matusik, R.; Thomas, G.V.; Sawyers, C.L. Myc-driven murine prostate cancer shares molecular features with human prostate tumors. *Cancer Cell* **2003**, *4*, 223–238. [CrossRef]

35. Wang, J.; Kim, J.; Roh, M.; Franco, O.E.; Hayward, S.W.; Wills, M.L.; Abdulkadir, S.A. Pim1 kinase synergizes with c-MYC to induce advanced prostate carcinoma. *Oncogene* **2010**, *29*, 2477–2487. [CrossRef] [PubMed]

36. Mamaeva, O.A.; Kim, J.; Feng, G.; McDonald, J.M. Calcium/calmodulin-dependent kinase II regulates notch-1 signaling in prostate cancer cells. *J. Cell. Biochem.* **2009**, *106*, 25–32. [CrossRef] [PubMed]

37. Mehra, R.; Tomlins, S.A.; Shen, R.; Nadeem, O.; Wang, L.; Wei, J.T.; Pienta, K.J.; Ghosh, D.; Rubin, M.A.; Chinnaiyan, A.M.; et al. Comprehensive assessment of TMPRSS2 and ETS family gene aberrations in clinically localized prostate cancer. *Mod. Pathol.* **2007**, *20*, 538–544. [CrossRef] [PubMed]

38. Tomlins, S.A.; Laxman, B.; Dhanasekaran, S.M.; Helgeson, B.E.; Cao, X.; Morris, D.S.; Menon, A.; Jing, X.; Cao, Q.; Han, B.; et al. Distinct classes of chromosomal rearrangements create oncogenic ETS gene fusions in prostate cancer. *Nature* **2007**, *448*, 595–599. [CrossRef] [PubMed]

39. Tomlins, S.A.; Rhodes, D.R.; Perner, S.; Dhanasekaran, S.M.; Mehra, R.; Sun, X.W.; Varambally, S.; Cao, X.; Tchinda, J.; Kuefer, R.; et al. Recurrent fusion of TMPRSS2 and ETS transcription factor genes in prostate cancer. *Science* **2005**, *310*, 644–648. [CrossRef] [PubMed]

40. Iljin, K.; Wolf, M.; Edgren, H.; Gupta, S.; Kilpinen, S.; Skotheim, R.I.; Peltola, M.; Smit, F.; Verhaegh, G.; Schalken, J.; et al. TMPRSS2 fusions with oncogenic ETS factors in prostate cancer involve unbalanced genomic rearrangements and are associated with HDAC1 and epigenetic reprogramming. *Cancer Res.* **2006**, *66*, 10242–10246. [CrossRef] [PubMed]

41. Rouzier, C.; Haudebourg, J.; Carpentier, X.; Valerio, L.; Amiel, J.; Michiels, J.F.; Pedeutour, F. Detection of the TMPRSS2-ETS fusion gene in prostate carcinomas: Retrospective analysis of 55 formalin-fixed and paraffin-embedded samples with clinical data. *Cancer Genet. Cytogenet.* **2008**, *183*, 21–27. [CrossRef] [PubMed]

42. Mosquera, J.M.; Perner, S.; Genega, E.M.; Sanda, M.; Hofer, M.D.; Mertz, K.D.; Paris, P.L.; Simko, J.; Bismar, T.A.; Ayala, G.; et al. Characterization of TMPRSS2-ERG fusion high-grade prostatic intraepithelial neoplasia and potential clinical implications. *Clin. Cancer Res.* **2008**, *14*, 3380–3385. [CrossRef] [PubMed]

43. Albadine, R.; Latour, M.; Toubaji, A.; Haffner, M.; Isaacs, W.B.; E, A.P.; Meeker, A.K.; Demarzo, A.M.; Epstein, J.I.; Netto, G.J. TMPRSS2-ERG gene fusion status in minute (minimal) prostatic adenocarcinoma. *Mod. Pathol.* **2009**, *22*, 1415–1422. [CrossRef] [PubMed]

44. Lin, C.; Yang, L.; Tanasa, B.; Hutt, K.; Ju, B.G.; Ohgi, K.; Zhang, J.; Rose, D.W.; Fu, X.D.; Glass, C.K.; et al. Nuclear receptor-induced chromosomal proximity and DNA breaks underlie specific translocations in cancer. *Cell* **2009**, *139*, 1069–1083. [CrossRef] [PubMed]

45. Mani, R.S.; Tomlins, S.A.; Callahan, K.; Ghosh, A.; Nyati, M.K.; Varambally, S.; Palanisamy, N.; Chinnaiyan, A.M. Induced chromosomal proximity and gene fusions in prostate cancer. *Science* **2009**, *326*, 1230. [CrossRef] [PubMed]

46. Haffner, M.C.; Aryee, M.J.; Toubaji, A.; Esopi, D.M.; Albadine, R.; Gurel, B.; Isaacs, W.B.; Bova, G.S.; Liu, W.; Xu, J.; et al. Androgen-induced TOP2B-mediated double-strand breaks and prostate cancer gene rearrangements. *Nat. Genet.* **2010**, *42*, 668–675. [CrossRef] [PubMed]

47. Yu, J.; Yu, J.; Mani, R.S.; Cao, Q.; Brenner, C.J.; Cao, X.; Wang, X.; Wu, L.; Li, J.; Hu, M.; et al. An integrated network of androgen receptor, polycomb, and TMPRSS2-ERG gene fusions in prostate cancer progression. *Cancer Cell.* **2010**, *17*, 443–454. [CrossRef] [PubMed]

48. Tomlins, S.A.; Laxman, B.; Varambally, S.; Cao, X.; Yu, J.; Helgeson, B.E.; Cao, Q.; Prensner, J.R.; Rubin, M.A.; Shah, R.B.; et al. Role of the TMPRSS2-ERG gene fusion in prostate cancer. *Neoplasia* **2008**, *10*, 177–188. [CrossRef] [PubMed]

49. Klezovitch, O.; Risk, M.; Coleman, I.; Lucas, J.M.; Null, M.; True, L.D.; Nelson, P.S.; Vasioukhin, V. A causal role for ERG in neoplastic transformation of prostate epithelium. *Proc. Natl. Acad. Sci. USA* **2008**, *105*, 2105–2110. [CrossRef] [PubMed]

50. Carver, B.S.; Tran, J.; Gopalan, A.; Chen, Z.; Shaikh, S.; Carracedo, A.; Alimonti, A.; Nardella, C.; Varmeh, S.; Scardino, P.T.; et al. Aberrant ERG expression cooperates with loss of PTEN to promote cancer progression in the prostate. *Nat. Genet.* **2009**, *41*, 619–624. [CrossRef] [PubMed]

51. King, J.C.; Xu, J.; Wongvipat, J.; Hieronymus, H.; Carver, B.S.; Leung, D.H.; Taylor, B.S.; Sander, C.; Cardiff, R.D.; Couto, S.S.; et al. Cooperativity of TMPRSS2-ERG with PI3-kinase pathway activation in prostate oncogenesis. *Nat. Genet.* **2009**, *41*, 524–526. [CrossRef] [PubMed]

52. Geybels, M.S.; Alumkal, J.J.; Luedeke, M.; Rinckleb, A.; Zhao, S.; Shui, I.M.; Bibikova, M.; Klotzle, B.; van den Brandt, P.A.; Ostrander, E.A.; et al. Epigenomic profiling of prostate cancer identifies differentially methylated genes in TMPRSS2:ERG fusion-positive versus fusion-negative tumors. *Clin. Epigenet.* **2015**, *7*, 128. [CrossRef] [PubMed]

53. Geybels, M.S.; McCloskey, K.D.; Mills, I.G.; Stanford, J.L. Calcium Channel Blocker Use and Risk of Prostate Cancer by TMPRSS2:ERG Gene Fusion Status. *Prostate* **2017**, *77*, 282–290. [CrossRef] [PubMed]

54. Wang, X.; Qiao, Y.; Asangani, I.A.; Ateeq, B.; Poliakov, A.; Cieslik, M.; Pitchiaya, S.; Chakravarthi, B.; Cao, X.; Jing, X.; et al. Development of Peptidomimetic Inhibitors of the ERG Gene Fusion Product in Prostate Cancer. *Cancer Cell* **2017**, *31*, 532–548. [CrossRef] [PubMed]

55. Rastogi, A.; Tan, S.H.; Mohamed, A.A.; Chen, Y.; Hu, Y.; Petrovics, G.; Sreenath, T.; Kagan, J.; Srivastava, S.; McLeod, D.G.; et al. Functional antagonism of TMPRSS2-ERG splice variants in prostate cancer. *Genes Cancer* **2014**, *5*, 273–284. [PubMed]

56. Gimenez-Bonafe, P.; Fedoruk, M.N.; Whitmore, T.G.; Akbari, M.; Ralph, J.L.; Ettinger, S.; Gleave, M.E.; Nelson, C.C. YB-1 is upregulated during prostate cancer tumor progression and increases P-glycoprotein activity. *Prostate* **2004**, *59*, 337–349. [CrossRef] [PubMed]

57. Shiota, M.; Takeuchi, A.; Song, Y.; Yokomizo, A.; Kashiwagi, E.; Uchiumi, T.; Kuroiwa, K.; Tatsugami, K.; Fujimoto, N.; Oda, Y.; et al. Y-box binding protein-1 promotes castration-resistant prostate cancer growth via androgen receptor expression. *Endocr. Relat. Cancer* **2011**, *18*, 505–517. [CrossRef] [PubMed]

58. Heumann, A.; Kaya, O.; Burdelski, C.; Hube-Magg, C.; Kluth, M.; Lang, D.S.; Simon, R.; Beyer, B.; Thederan, I.; Sauter, G.; et al. Up regulation and nuclear translocation of Y-box binding protein 1 (YB-1) is linked to poor prognosis in ERG-negative prostate cancer. *Sci. Rep.* **2017**, *7*, 2056. [CrossRef] [PubMed]

59. Wang, J.; Cai, Y.; Shao, L.J.; Siddiqui, J.; Palanisamy, N.; Li, R.; Ren, C.; Ayala, G.; Ittmann, M. Activation of NF-{kappa}B by TMPRSS2/ERG Fusion Isoforms through Toll-Like Receptor-4. *Cancer Res.* **2011**, *71*, 1325–1333. [CrossRef] [PubMed]

60. Nadiminty, N.; Lou, W.; Sun, M.; Chen, J.; Yue, J.; Kung, H.J.; Evans, C.P.; Zhou, Q.; Gao, A.C. Aberrant activation of the androgen receptor by NF-kappaB2/p52 in prostate cancer cells. *Cancer Res.* **2010**, *70*, 3309–3319. [CrossRef] [PubMed]

61. Singareddy, R.; Semaan, L.; Conley-Lacomb, M.K.; St John, J.; Powell, K.; Iyer, M.; Smith, D.; Heilbrun, L.K.; Shi, D.; Sakr, W.; et al. Transcriptional regulation of CXCR4 in prostate cancer: Significance of TMPRSS2-ERG fusions. *Mol. Cancer Res.* **2013**, *11*, 1349–1361. [CrossRef] [PubMed]

62. Wang, S.I.; Parsons, R.; Ittmann, M. Homozygous deletion of the PTEN tumor suppressor gene in a subset of prostate adenocarcinomas. *Clin. Cancer Res.* **1998**, *4*, 811–815. [PubMed]

63. Schmitz, M.; Grignard, G.; Margue, C.; Dippel, W.; Capesius, C.; Mossong, J.; Nathan, M.; Giacchi, S.; Scheiden, R.; Kieffer, N. Complete loss of PTEN expression as a possible early prognostic marker for prostate cancer metastasis. *Int. J. Cancer* **2007**, *120*, 1284–1292. [CrossRef] [PubMed]

64. Sircar, K.; Yoshimoto, M.; Monzon, F.A.; Koumakpayi, I.H.; Katz, R.L.; Khanna, A.; Alvarez, K.; Chen, G.; Darnel, A.D.; Aprikian, A.G.; et al. PTEN genomic deletion is associated with p-Akt and AR signalling in poorer outcome, hormone refractory prostate cancer. *J. Pathol.* **2009**, *218*, 505–513. [CrossRef] [PubMed]

65. McMenamin, M.E.; Soung, P.; Perera, S.; Kaplan, I.; Loda, M.; Sellers, W.R. Loss of PTEN expression in paraffin-embedded primary prostate cancer correlates with high Gleason score and advanced stage. *Cancer Res.* **1999**, *59*, 4291–4296. [PubMed]

66. Wang, S.; Gao, J.; Lei, Q.; Rozengurt, N.; Pritchard, C.; Jiao, J.; Thomas, G.V.; Li, G.; Roy-Burman, P.; Nelson, P.S.; et al. Prostate-specific deletion of the murine Pten tumor suppressor gene leads to metastatic prostate cancer. *Cancer Cell* **2003**, *4*, 209–221. [CrossRef]

67. Kim, M.J.; Cardiff, R.D.; Desai, N.; Banach-Petrosky, W.A.; Parsons, R.; Shen, M.M.; Abate-Shen, C. Cooperativity of Nkx3.1 and Pten loss of function in a mouse model of prostate carcinogenesis. *Proc. Natl. Acad. Sci. USA* **2002**, *99*, 2384–2889. [CrossRef] [PubMed]

68. Kim, J.; Eltoum, I.E.; Roh, M.; Wang, J.; Abdulkadir, S.A. Interactions between cells with distinct mutations in c-MYC and Pten in prostate cancer. *PLoS Genet.* **2009**, *5*, e1000542. [CrossRef] [PubMed]
69. Jia, S.; Liu, Z.; Zhang, S.; Liu, P.; Zhang, L.; Lee, S.H.; Zhang, J.; Signoretti, S.; Loda, M.; Roberts, T.M.; et al. Essential roles of PI(3)K-p110β in cell growth, metabolism and tumorigenesis. *Nature* **2008**, *454*, 776–779. [CrossRef] [PubMed]
70. Guertin, D.A.; Stevens, D.M.; Saitoh, M.; Kinkel, S.; Crosby, K.; Sheen, J.H.; Mullholland, D.J.; Magnuson, M.A.; Wu, H.; Sabatini, D.M. mTOR complex 2 is required for the development of prostate cancer induced by Pten loss in mice. *Cancer Cell* **2009**, *15*, 148–159. [CrossRef] [PubMed]
71. Gao, H.; Ouyang, X.; Banach-Petrosky, W.A.; Shen, M.M.; Abate-Shen, C. Emergence of androgen independence at early stages of prostate cancer progression in Nkx3.1; Pten mice. *Cancer Res.* **2006**, *66*, 7929–7933. [CrossRef] [PubMed]
72. Lin, H.K.; Hu, Y.C.; Lee, D.K.; Chang, C. Regulation of androgen receptor signaling by PTEN (phosphatase and tensin homolog deleted on chromosome 10) tumor suppressor through distinct mechanisms in prostate cancer cells. *Mol. Endocrinol.* **2004**, *18*, 2409–2423. [CrossRef] [PubMed]
73. Trotman, L.C.; Wang, X.; Alimonti, A.; Chen, Z.; Teruya-Feldstein, J.; Yang, H.; Pavletich, N.P.; Carver, B.S.; Cordon-Cardo, C.; Erdjument-Bromage, H.; et al. Ubiquitination regulates PTEN nuclear import and tumor suppression. *Cell* **2007**, *128*, 141–156. [CrossRef] [PubMed]
74. Chang, C.J.; Mulholland, D.J.; Valamehr, B.; Mosessian, S.; Sellers, W.R.; Wu, H. PTEN nuclear localization is regulated by oxidative stress and mediates p53-dependent tumor suppression. *Mol. Cell. Biol.* **2008**, *28*, 3281–3289. [CrossRef] [PubMed]
75. Chen, M.; Nowak, D.G.; Narula, N.; Robinson, B.; Watrud, K.; Ambrico, A.; Herzka, T.M.; Zeeman, M.E.; Minderer, M.; Zheng, W.; et al. The nuclear transport receptor Importin-11 is a tumor suppressor that maintains PTEN protein. *J. Cell Biol.* **2017**, *216*, 641–656. [CrossRef] [PubMed]
76. Schaeffer, E.M.; Marchionni, L.; Huang, Z.; Simons, B.; Blackman, A.; Yu, W.; Parmigiani, G.; Berman, D.M. Androgen-induced programs for prostate epithelial growth and invasion arise in embryogenesis and are reactivated in cancer. *Oncogene* **2008**, *27*, 7180–7191. [CrossRef] [PubMed]
77. Pritchard, C.; Mecham, B.; Dumpit, R.; Coleman, I.; Bhattacharjee, M.; Chen, Q.; Sikes, R.A.; Nelson, P.S. Conserved gene expression programs integrate mammalian prostate development and tumorigenesis. *Cancer Res.* **2009**, *69*, 1739–1747. [CrossRef] [PubMed]
78. Bruxvoort, K.J.; Charbonneau, H.M.; Giambernardi, T.A.; Goolsby, J.C.; Qian, C.N.; Zylstra, C.R.; Robinson, D.R.; Roy-Burman, P.; Shaw, A.K.; Buckner-Berghuis, B.D.; et al. Inactivation of Apc in the mouse prostate causes prostate carcinoma. *Cancer Res.* **2007**, *67*, 2490–2496. [CrossRef] [PubMed]
79. Pearson, H.B.; Phesse, T.J.; Clarke, A.R. K-ras and Wnt signaling synergize to accelerate prostate tumorigenesis in the mouse. *Cancer Res.* **2009**, *69*, 94–101. [CrossRef] [PubMed]
80. Yu, X.; Wang, Y.; Jiang, M.; Bierie, B.; Roy-Burman, P.; Shen, M.M.; Taketo, M.M.; Wills, M.; Matusik, R.J. Activation of β-Catenin in mouse prostate causes HGPIN and continuous prostate growth after castration. *Prostate* **2009**, *69*, 249–262. [CrossRef] [PubMed]
81. Wang, G.; Wang, J.; Sadar, M.D. Crosstalk between the androgen receptor and β-catenin in castrate-resistant prostate cancer. *Cancer Res.* **2008**, *68*, 9918–9927. [CrossRef] [PubMed]
82. Horvath, L.G.; Henshall, S.M.; Lee, C.S.; Kench, J.G.; Golovsky, D.; Brenner, P.C.; O'Neill, G.F.; Kooner, R.; Stricker, P.D.; Grygiel, J.J.; et al. Lower levels of nuclear β-catenin predict for a poorer prognosis in localized prostate cancer. *Int. J. Cancer* **2005**, *113*, 415–422. [CrossRef] [PubMed]
83. Whitaker, H.C.; Girling, J.; Warren, A.Y.; Leung, H.; Mills, I.G.; Neal, D.E. Alterations in β-catenin expression and localization in prostate cancer. *Prostate* **2008**, *68*, 1196–1205. [CrossRef] [PubMed]
84. Wang, Q.; Symes, A.J.; Kane, C.A.; Freeman, A.; Nariculam, J.; Munson, P.; Thrasivoulou, C.; Masters, J.R.; Ahmed, A. A novel role for Wnt/Ca^{2+} signaling in actin cytoskeleton remodeling and cell motility in prostate cancer. *PLoS ONE* **2010**, *5*, e10456. [CrossRef] [PubMed]
85. Thrasivoulou, C.; Millar, M.; Ahmed, A. Activation of intracellular calcium by multiple Wnt ligands and translocation of β-catenin into the nucleus: A convergent model of Wnt/Ca^{2+} and Wnt/β-catenin pathways. *J. Biol. Chem.* **2013**, *288*, 35651–35659. [CrossRef] [PubMed]
86. Nicotera, P.; Zhivotovsky, B.; Orrenius, S. Nuclear calcium transport and the role of calcium in apoptosis. *Cell Calcium* **1994**, *16*, 279–288. [CrossRef]

87. Hsu, J.L.; Pan, S.L.; Ho, Y.F.; Hwang, T.L.; Kung, F.L.; Guh, J.H. Costunolide induces apoptosis through nuclear Ca^{2+} overload and DNA damage response in human prostate cancer. *J. Urol.* **2011**, *185*, 1967–1974. [CrossRef] [PubMed]

88. Karhadkar, S.S.; Bova, G.S.; Abdallah, N.; Dhara, S.; Gardner, D.; Maitra, A.; Isaacs, J.T.; Berman, D.M.; Beachy, P.A. Hedgehog signalling in prostate regeneration, neoplasia and metastasis. *Nature* **2004**, *431*, 707–712. [CrossRef] [PubMed]

89. Sanchez, P.; Hernandez, A.M.; Stecca, B.; Kahler, A.J.; DeGueme, A.M.; Barrett, A.; Beyna, M.; Datta, M.W.; Datta, S.; Ruiz i Altaba, A. Inhibition of prostate cancer proliferation by interference with SONIC HEDGEHOG-GLI1 signaling. *Proc. Natl. Acad. Sci. USA* **2004**, *101*, 12561–12566. [CrossRef] [PubMed]

90. Shaw, A.; Gipp, J.; Bushman, W. The Sonic Hedgehog pathway stimulates prostate tumor growth by paracrine signaling and recapitulates embryonic gene expression in tumor myofibroblasts. *Oncogene* **2009**, *28*, 4480–4490. [CrossRef] [PubMed]

91. Yauch, R.L.; Gould, S.E.; Scales, S.J.; Tang, T.; Tian, H.; Ahn, C.P.; Marshall, D.; Fu, L.; Januario, T.; Kallop, D.; et al. A paracrine requirement for hedgehog signalling in cancer. *Nature* **2008**, *455*, 406–410. [CrossRef] [PubMed]

92. Gonnissen, A.; Isebaert, S.; Haustermans, K. Hedgehog signaling in prostate cancer and its therapeutic implication. *Int. J. Mol. Sci.* **2013**, *14*, 13979–14007. [CrossRef] [PubMed]

93. Li, N.; Truong, S.; Nouri, M.; Moore, J.; Al Nakouzi, N.; Lubik, A.A.; Buttyan, R. Non-canonical activation of hedgehog in prostate cancer cells mediated by the interaction of transcriptionally active androgen receptor proteins with Gli3. *Oncogene* **2018**. [CrossRef] [PubMed]

94. Acevedo, V.D.; Gangula, R.D.; Freeman, K.W.; Li, R.; Zhang, Y.; Wang, F.; Ayala, G.E.; Peterson, L.E.; Ittmann, M.; Spencer, D.M. Inducible FGFR-1 activation leads to irreversible prostate adenocarcinoma and an epithelial-to-mesenchymal transition. *Cancer Cell* **2007**, *12*, 559–571. [CrossRef] [PubMed]

95. Memarzadeh, S.; Xin, L.; Mulholland, D.J.; Mansukhani, A.; Wu, H.; Teitell, M.A.; Witte, O.N. Enhanced paracrine FGF10 expression promotes formation of multifocal prostate adenocarcinoma and an increase in epithelial androgen receptor. *Cancer Cell* **2007**, *12*, 572–585. [CrossRef] [PubMed]

96. Kwabi-Addo, B.; Ozen, M.; Ittmann, M. The role of fibroblast growth factors and their receptors in prostate cancer. *Endocr. Relat. Cancer* **2004**, *11*, 709–724. [CrossRef] [PubMed]

97. Chen, M.L.; Xu, P.Z.; Peng, X.D.; Chen, W.S.; Guzman, G.; Yang, X.; Di Cristofano, A.; Pandolfi, P.P.; Hay, N. The deficiency of *Akt1* is sufficient to suppress tumor development in *Pten+/−* mice. *Genes Dev.* **2006**, *20*, 1569–1574. [CrossRef] [PubMed]

98. Thomas, G.V.; Horvath, S.; Smith, B.L.; Crosby, K.; Lebel, L.A.; Schrage, M.; Said, J.; De Kernion, J.; Reiter, R.E.; Sawyers, C.L. Antibody-based profiling of the phosphoinositide 3-kinase pathway in clinical prostate cancer. *Clin. Cancer Res.* **2004**, *10*, 8351–8356. [CrossRef] [PubMed]

99. Shen, M.M.; Abate-Shen, C. Pten inactivation and the emergence of androgen-independent prostate cancer. *Cancer Res.* **2007**, *67*, 6535–6538. [CrossRef] [PubMed]

100. Boormans, J.L.; Hermans, K.G.; van Leenders, G.J.; Trapman, J.; Verhagen, P.C. An activating mutation in AKT1 in human prostate cancer. *Int. J. Cancer* **2008**, *123*, 2725–2726. [CrossRef] [PubMed]

101. Lee, S.H.; Poulogiannis, G.; Pyne, S.; Jia, S.; Zou, L.; Signoretti, S.; Loda, M.; Cantley, L.C.; Roberts, T.M. A constitutively activated form of the p110β isoform of PI3-kinase induces prostatic intraepithelial neoplasia in mice. *Proc. Natl. Acad. Sci. USA* **2010**, *107*, 11002–11007. [CrossRef] [PubMed]

102. Gao, H.; Ouyang, X.; Banach-Petrosky, W.A.; Gerald, W.L.; Shen, M.M.; Abate-Shen, C. Combinatorial activities of Akt and B-Raf/Erk signaling in a mouse model of androgen-independent prostate cancer. *Proc. Natl. Acad. Sci. USA* **2006**, *103*, 14477–14482. [CrossRef] [PubMed]

103. Trotman, L.C.; Alimonti, A.; Scaglioni, P.P.; Koutcher, J.A.; Cordon-Cardo, C.; Pandolfi, P.P. Identification of a tumour suppressor network opposing nuclear Akt function. *Nature* **2006**, *441*, 523–527. [CrossRef] [PubMed]

104. Kondratskyi, A.; Yassine, M.; Slomianny, C.; Kondratska, K.; Gordienko, D.; Dewailly, E.; Lehen'kyi, V.; Skryma, R.; Prevarskaya, N. Identification of ML-9 as a lysosomotropic agent targeting autophagy and cell death. *Cell Death Dis.* **2014**, *5*, e1193. [CrossRef] [PubMed]

105. Uzgare, A.R.; Isaacs, J.T. Enhanced redundancy in Akt and mitogen-activated protein kinase-induced survival of malignant versus normal prostate epithelial cells. *Cancer Res.* **2004**, *64*, 6190–6199. [CrossRef] [PubMed]

106. Kinkade, C.W.; Castillo-Martin, M.; Puzio-Kuter, A.; Yan, J.; Foster, T.H.; Gao, H.; Sun, Y.; Ouyang, X.; Gerald, W.L.; Cordon-Cardo, C.; et al. Targeting AKT/mTOR and ERK MAPK signaling inhibits hormone-refractory prostate cancer in a preclinical mouse model. *J. Clin. Investig.* **2008**, *118*, 3051–3064. [CrossRef] [PubMed]

107. Dan, H.C.; Cooper, M.J.; Cogswell, P.C.; Duncan, J.A.; Ting, J.P.; Baldwin, A.S. Akt-dependent regulation of NF-{kappa}B is controlled by mTOR and Raptor in association with IKK. *Genes Dev.* **2008**, *22*, 1490–1500. [CrossRef] [PubMed]

108. Fradet, V.; Lessard, L.; Begin, L.R.; Karakiewicz, P.; Masson, A.M.; Saad, F. Nuclear factor-kappaB nuclear localization is predictive of biochemical recurrence in patients with positive margin prostate cancer. *Clin. Cancer Res.* **2004**, *10*, 8460–8464. [CrossRef] [PubMed]

109. Ismail, H.A.; Lessard, L.; Mes-Masson, A.M.; Saad, F. Expression of NF-kappaB in prostate cancer lymph node metastases. *Prostate* **2004**, *58*, 308–313. [CrossRef] [PubMed]

110. Lessard, L.; Karakiewicz, P.I.; Bellon-Gagnon, P.; Alam-Fahmy, M.; Ismail, H.A.; Mes-Masson, A.M.; Saad, F. Nuclear localization of nuclear factor-kappaB p65 in primary prostate tumors is highly predictive of pelvic lymph node metastases. *Clin. Cancer Res.* **2006**, *12*, 5741–5745. [CrossRef] [PubMed]

111. Luo, J.L.; Tan, W.; Ricono, J.M.; Korchynskyi, O.; Zhang, M.; Gonias, S.L.; Cheresh, D.A.; Karin, M. Nuclear cytokine-activated IKKα controls prostate cancer metastasis by repressing Maspin. *Nature* **2007**, *446*, 690–694. [CrossRef] [PubMed]

112. Zhang, L.; Altuwaijri, S.; Deng, F.; Chen, L.; Lal, P.; Bhanot, U.K.; Korets, R.; Wenske, S.; Lilja, H.G.; Chang, C.; et al. NF-kappaB regulates androgen receptor expression and prostate cancer growth. *Am. J. Pathol.* **2009**, *175*, 489–499. [CrossRef] [PubMed]

113. Jeong, J.H.; Wang, Z.; Guimaraes, A.S.; Ouyang, X.; Figueiredo, J.L.; Ding, Z.; Jiang, S.; Guney, I.; Kang, G.H.; Shin, E.; et al. BRAF activation initiates but does not maintain invasive prostate adenocarcinoma. *PLoS ONE* **2008**, *3*, e3949. [CrossRef] [PubMed]

114. Taylor, B.S.; Schultz, N.; Hieronymus, H.; Gopalan, A.; Xiao, Y.; Carver, B.S.; Arora, V.K.; Kaushik, P.; Cerami, E.; Reva, B.; et al. Integrative genomic profiling of human prostate cancer. *Cancer Cell* **2010**, *18*, 11–22. [CrossRef] [PubMed]

115. Gioeli, D.; Mandell, J.W.; Petroni, G.R.; Frierson, H.F., Jr.; Weber, M.J. Activation of mitogen-activated protein kinase associated with prostate cancer progression. *Cancer Res.* **1999**, *59*, 279–284. [PubMed]

116. Roderick, H.L.; Cook, S.J. Ca^{2+} signalling checkpoints in cancer: Remodelling Ca^{2+} for cancer cell proliferation and survival. *Nat. Rev. Cancer* **2008**, *8*, 361–375. [CrossRef] [PubMed]

117. Hanahan, D.; Weinberg, R.A. Hallmarks of cancer: The next generation. *Cell* **2011**, *144*, 646–674. [CrossRef] [PubMed]

118. Xu, Y.; Zhang, S.; Niu, H.; Ye, Y.; Hu, F.; Chen, S.; Li, X.; Luo, X.; Jiang, S.; Liu, Y.; et al. STIM1 accelerates cell senescence in a remodeled microenvironment but enhances the epithelial-to-mesenchymal transition in prostate cancer. *Sci. Rep.* **2015**, *5*, 11754. [CrossRef] [PubMed]

119. van Noesel, M.M.; van Bezouw, S.; Salomons, G.S.; Voute, P.A.; Pieters, R.; Baylin, S.B.; Herman, J.G.; Versteeg, R. Tumor-specific down-regulation of the tumor necrosis factor-related apoptosis-inducing ligand decoy receptors DcR1 and DcR2 is associated with dense promoter hypermethylation. *Cancer Res.* **2002**, *62*, 2157–2161. [PubMed]

120. Sheikh, M.S.; Fornace, A.J., Jr. Death and decoy receptors and p53-mediated apoptosis. *Leukemia* **2000**, *14*, 1509–1513. [CrossRef] [PubMed]

121. Collado, M.; Gil, J.; Efeyan, A.; Guerra, C.; Schuhmacher, A.J.; Barradas, M.; Benguria, A.; Zaballos, A.; Flores, J.M.; Barbacid, M.; et al. Tumour biology: Senescence in premalignant tumours. *Nature* **2005**, *436*, 642. [CrossRef] [PubMed]

122. Althubiti, M.; Lezina, L.; Carrera, S.; Jukes-Jones, R.; Giblett, S.M.; Antonov, A.; Barlev, N.; Saldanha, G.S.; Pritchard, C.A.; Cain, K.; et al. Characterization of novel markers of senescence and their prognostic potential in cancer. *Cell Death Dis.* **2014**, *5*, e1528. [CrossRef] [PubMed]

123. Lee, S.J.; Lee, K.; Yang, X.; Jung, C.; Gardner, T.; Kim, H.S.; Jeng, M.H.; Kao, C. NFATc1 with AP-3 site binding specificity mediates gene expression of prostate-specific-membrane-antigen. *J. Mol. Biol.* **2003**, *330*, 749–760. [CrossRef]

124. Kavitha, C.V.; Deep, G.; Gangar, S.C.; Jain, A.K.; Agarwal, C.; Agarwal, R. Silibinin inhibits prostate cancer cells- and RANKL-induced osteoclastogenesis by targeting NFATc1, NF-kappaB, and AP-1 activation in RAW264.7 cells. *Mol. Carcinog.* **2014**, *53*, 169–180. [CrossRef] [PubMed]

125. Mognol, G.P.; Carneiro, F.R.; Robbs, B.K.; Faget, D.V.; Viola, J.P. Cell cycle and apoptosis regulation by NFAT transcription factors: New roles for an old player. *Cell Death Dis.* **2016**, *7*, e2199. [CrossRef] [PubMed]

126. Manda, K.R.; Tripathi, P.; Hsi, A.C.; Ning, J.; Ruzinova, M.B.; Liapis, H.; Bailey, M.; Zhang, H.; Maher, C.A.; Humphrey, P.A.; et al. NFATc1 promotes prostate tumorigenesis and overcomes PTEN loss-induced senescence. *Oncogene* **2016**, *35*, 3282–3292. [CrossRef] [PubMed]

127. Wei, Z.; Jiang, X.; Qiao, H.; Zhai, B.; Zhang, L.; Zhang, Q.; Wu, Y.; Jiang, H.; Sun, X. STAT3 interacts with Skp2/p27/p21 pathway to regulate the motility and invasion of gastric cancer cells. *Cell. Signal.* **2013**, *25*, 931–938. [CrossRef] [PubMed]

128. Huang, H.; Zhao, W.; Yang, D. Stat3 induces oncogenic Skp2 expression in human cervical carcinoma cells. *Biochem. Biophys. Res. Commun.* **2012**, *418*, 186–190. [CrossRef] [PubMed]

129. Wissenbach, U.; Niemeyer, B.A.; Fixemer, T.; Schneidewind, A.; Trost, C.; Cavalie, A.; Reus, K.; Meese, E.; Bonkhoff, H.; Flockerzi, V. Expression of CaT-like, a novel calcium-selective channel, correlates with the malignancy of prostate cancer. *J. Biol. Chem.* **2001**, *276*, 19461–19468. [CrossRef] [PubMed]

130. Fixemer, T.; Wissenbach, U.; Flockerzi, V.; Bonkhoff, H. Expression of the Ca^{2+}-selective cation channel TRPV6 in human prostate cancer: A novel prognostic marker for tumor progression. *Oncogene* **2003**, *22*, 7858–7861. [CrossRef] [PubMed]

131. Thebault, S.; Flourakis, M.; Vanoverberghe, K.; Vandermoere, F.; Roudbaraki, M.; Lehen'kyi, V.; Slomianny, C.; Beck, B.; Mariot, P.; Bonnal, J.L.; et al. Differential role of transient receptor potential channels in Ca^{2+} entry and proliferation of prostate cancer epithelial cells. *Cancer Res.* **2006**, *66*, 2038–2047. [CrossRef] [PubMed]

132. Lehen'kyi, V.; Flourakis, M.; Skryma, R.; Prevarskaya, N. TRPV6 channel controls prostate cancer cell proliferation via Ca^{2+}/NFAT-dependent pathways. *Oncogene* **2007**, *26*, 7380–7385. [CrossRef] [PubMed]

133. Raphael, M.; Lehen'kyi, V.; Vandenberghe, M.; Beck, B.; Khalimonchyk, S.; Vanden Abeele, F.; Farsetti, L.; Germain, E.; Bokhobza, A.; Mihalache, A.; et al. TRPV6 calcium channel translocates to the plasma membrane via Orai1-mediated mechanism and controls cancer cell survival. *Proc. Natl. Acad. Sci. USA* **2014**, *111*, E3870–E3879. [CrossRef] [PubMed]

134. Fu, S.; Hirte, H.; Welch, S.; Ilenchuk, T.T.; Lutes, T.; Rice, C.; Fields, N.; Nemet, A.; Dugourd, D.; Piha-Paul, S.; et al. First-in-human phase I study of SOR-C13, a TRPV6 calcium channel inhibitor, in patients with advanced solid tumors. *Investig. New Drugs* **2017**, *35*, 324–333. [CrossRef] [PubMed]

135. Monet, M.; Lehen'kyi, V.; Gackiere, F.; Firlej, V.; Vandenberghe, M.; Roudbaraki, M.; Gkika, D.; Pourtier, A.; Bidaux, G.; Slomianny, C.; et al. Role of cationic channel TRPV2 in promoting prostate cancer migration and progression to androgen resistance. *Cancer Res.* **2010**, *70*, 1225–1235. [CrossRef] [PubMed]

136. Maly, I.V.; Domaradzki, T.M.; Gosy, V.A.; Hofmann, W.A. Myosin isoform expressed in metastatic prostate cancer stimulates cell invasion. *Sci. Rep.* **2017**, *7*, 8476. [CrossRef] [PubMed]

137. Tsavaler, L.; Shapero, M.H.; Morkowski, S.; Laus, R. Trp-p8, a novel prostate-specific gene, is up-regulated in prostate cancer and other malignancies and shares high homology with transient receptor potential calcium channel proteins. *Cancer Res.* **2001**, *61*, 3760–3769. [PubMed]

138. Zhang, L.; Barritt, G.J. Evidence that TRPM8 is an androgen-dependent Ca^{2+} channel required for the survival of prostate cancer cells. *Cancer Res.* **2004**, *64*, 8365–8373. [CrossRef] [PubMed]

139. Monteith, G.R.; Davis, F.M.; Roberts-Thomson, S.J. Calcium channels and pumps in cancer: Changes and consequences. *J. Biol. Chem.* **2012**, *287*, 31666–31673. [CrossRef] [PubMed]

140. Henshall, S.M.; Afar, D.E.; Hiller, J.; Horvath, L.G.; Quinn, D.I.; Rasiah, K.K.; Gish, K.; Willhite, D.; Kench, J.G.; Gardiner-Garden, M.; et al. Survival analysis of genome-wide gene expression profiles of prostate cancers identifies new prognostic targets of disease relapse. *Cancer Res.* **2003**, *63*, 4196–4203. [PubMed]

141. Gkika, D.; Flourakis, M.; Lemonnier, L.; Prevarskaya, N. PSA reduces prostate cancer cell motility by stimulating TRPM8 activity and plasma membrane expression. *Oncogene* **2010**, *29*, 4611–4616. [CrossRef] [PubMed]

142. Bidaux, G.; Flourakis, M.; Thebault, S.; Zholos, A.; Beck, B.; Gkika, D.; Roudbaraki, M.; Bonnal, J.L.; Mauroy, B.; Shuba, Y.; et al. Prostate cell differentiation status determines transient receptor potential melastatin member 8 channel subcellular localization and function. *J. Clin. Investig.* **2007**, *117*, 1647–1657. [CrossRef] [PubMed]

143. Vanoverberghe, K.; Vanden Abeele, F.; Mariot, P.; Lepage, G.; Roudbaraki, M.; Bonnal, J.L.; Mauroy, B.; Shuba, Y.; Skryma, R.; Prevarskaya, N. Ca^{2+} homeostasis and apoptotic resistance of neuroendocrine-differentiated prostate cancer cells. *Cell Death Differ.* **2004**, *11*, 321–330. [CrossRef] [PubMed]

144. Vanden Abeele, F.; Skryma, R.; Shuba, Y.; Van Coppenolle, F.; Slomianny, C.; Roudbaraki, M.; Mauroy, B.; Wuytack, F.; Prevarskaya, N. Bcl-2-dependent modulation of Ca^{2+} homeostasis and store-operated channels in prostate cancer cells. *Cancer Cell* **2002**, *1*, 169–179. [CrossRef]

145. Chen, L.; Cao, R.; Wang, G.; Yuan, L.; Qian, G.; Guo, Z.; Wu, C.L.; Wang, X.; Xiao, Y. Downregulation of TRPM7 suppressed migration and invasion by regulating epithelial-mesenchymal transition in prostate cancer cells. *Med. Oncol.* **2017**, *34*, 127. [CrossRef] [PubMed]

146. Sagredo, A.I.; Sagredo, E.A.; Cappelli, C.; Baez, P.; Andaur, R.E.; Blanco, C.; Tapia, J.C.; Echeverria, C.; Cerda, O.; Stutzin, A.; et al. TRPM4 regulates Akt/GSK3-β activity and enhances β-catenin signaling and cell proliferation in prostate cancer cells. *Mol. Oncol.* **2018**, *12*, 151–165. [CrossRef] [PubMed]

147. Zeng, X.; Sikka, S.C.; Huang, L.; Sun, C.; Xu, C.; Jia, D.; Abdel-Mageed, A.B.; Pottle, J.E.; Taylor, J.T.; Li, M. Novel role for the transient receptor potential channel TRPM2 in prostate cancer cell proliferation. *Prostate Cancer Prostatic Dis.* **2010**, *13*, 195–201. [CrossRef] [PubMed]

148. Haverstick, D.M.; Heady, T.N.; Macdonald, T.L.; Gray, L.S. Inhibition of human prostate cancer proliferation in vitro and in a mouse model by a compound synthesized to block Ca^{2+} entry. *Cancer Res.* **2000**, *60*, 1002–1008. [PubMed]

149. Gray, L.S.; Perez-Reyes, E.; Gomora, J.C.; Haverstick, D.M.; Shattock, M.; McLatchie, L.; Harper, J.; Brooks, G.; Heady, T.; Macdonald, T.L. The role of voltage gated T-type Ca^{2+} channel isoforms in mediating "capacitative" Ca^{2+} entry in cancer cells. *Cell Calcium* **2004**, *36*, 489–497. [CrossRef] [PubMed]

150. Panner, A.; Wurster, R.D. T-type calcium channels and tumor proliferation. *Cell Calcium* **2006**, *40*, 253–259. [CrossRef] [PubMed]

151. Parimi, V.; Goyal, R.; Poropatich, K.; Yang, X.J. Neuroendocrine differentiation of prostate cancer: A review. *Am. J. Clin. Exp. Urol.* **2014**, *2*, 273–285. [PubMed]

152. Mariot, P.; Vanoverberghe, K.; Lalevee, N.; Rossier, M.F.; Prevarskaya, N. Overexpression of an α1H (Cav3.2) T-type calcium channel during neuroendocrine differentiation of human prostate cancer cells. *J. Biol. Chem.* **2002**, *277*, 10824–10833. [CrossRef] [PubMed]

153. Rossier, M.F.; Lesouhaitier, O.; Perrier, E.; Bockhorn, L.; Chiappe, A.; Lalevee, N. Aldosterone regulation of T-type calcium channels. *J. Steroid Biochem. Mol. Biol.* **2003**, *85*, 383–388. [CrossRef]

154. Contreras, G.F.; Castillo, K.; Enrique, N.; Carrasquel-Ursulaez, W.; Castillo, J.P.; Milesi, V.; Neely, A.; Alvarez, O.; Ferreira, G.; Gonzalez, C.; et al. A BK (Slo1) channel journey from molecule to physiology. *Channels* **2013**, *7*, 442–458. [CrossRef] [PubMed]

155. Ge, L.; Hoa, N.T.; Wilson, Z.; Arismendi-Morillo, G.; Kong, X.T.; Tajhya, R.B.; Beeton, C.; Jadus, M.R. Big Potassium (BK) ion channels in biology, disease and possible targets for cancer immunotherapy. *Int. Immunopharmacol.* **2014**, *22*, 427–443. [CrossRef] [PubMed]

156. Bloch, M.; Ousingsawat, J.; Simon, R.; Schraml, P.; Gasser, T.C.; Mihatsch, M.J.; Kunzelmann, K.; Bubendorf, L. KCNMA1 gene amplification promotes tumor cell proliferation in human prostate cancer. *Oncogene* **2007**, *26*, 2525–2534. [CrossRef] [PubMed]

157. Du, C.; Zheng, Z.; Li, D.; Chen, L.; Li, N.; Yi, X.; Yang, Y.; Guo, F.; Liu, W.; Xie, X.; et al. BKCa promotes growth and metastasis of prostate cancer through facilitating the coupling between αvβ3 integrin and FAK. *Oncotarget* **2016**, *7*, 40174–40188. [CrossRef] [PubMed]

158. Legrand, G.; Humez, S.; Slomianny, C.; Dewailly, E.; Vanden Abeele, F.; Mariot, P.; Wuytack, F.; Prevarskaya, N. Ca^{2+} pools and cell growth. Evidence for sarcoendoplasmic Ca^{2+}-ATPases 2B involvement in human prostate cancer cell growth control. *J. Biol. Chem.* **2001**, *276*, 47608–47614. [CrossRef] [PubMed]

159. Denmeade, S.R.; Mhaka, A.M.; Rosen, D.M.; Brennen, W.N.; Dalrymple, S.; Dach, I.; Olesen, C.; Gurel, B.; Demarzo, A.M.; Wilding, G.; et al. Engineering a prostate-specific membrane antigen-activated tumor endothelial cell prodrug for cancer therapy. *Sci. Transl. Med.* **2012**, *4*, 140ra186. [CrossRef] [PubMed]

160. Flourakis, M.; Lehen'kyi, V.; Beck, B.; Raphael, M.; Vandenberghe, M.; Abeele, F.V.; Roudbaraki, M.; Lepage, G.; Mauroy, B.; Romanin, C.; et al. Orai1 contributes to the establishment of an apoptosis-resistant phenotype in prostate cancer cells. *Cell Death Dis.* **2010**, *1*, e75. [CrossRef] [PubMed]

161. Dubois, C.; Vanden Abeele, F.; Lehen'kyi, V.; Gkika, D.; Guarmit, B.; Lepage, G.; Slomianny, C.; Borowiec, A.S.; Bidaux, G.; Benahmed, M.; et al. Remodeling of channel-forming ORAI proteins determines an oncogenic switch in prostate cancer. *Cancer Cell* **2014**, *26*, 19–32. [CrossRef] [PubMed]

162. Marchi, S.; Lupini, L.; Patergnani, S.; Rimessi, A.; Missiroli, S.; Bonora, M.; Bononi, A.; Corra, F.; Giorgi, C.; De Marchi, E.; et al. Downregulation of the mitochondrial calcium uniporter by cancer-related miR-25. *Curr. Biol.* **2013**, *23*, 58–63. [CrossRef] [PubMed]

163. Orrenius, S.; Zhivotovsky, B.; Nicotera, P. Regulation of cell death: The calcium-apoptosis link. *Nat. Rev. Mol. Cell Biol.* **2003**, *4*, 552–565. [CrossRef] [PubMed]

164. Vlietstra, R.J.; van Alewijk, D.C.; Hermans, K.G.; van Steenbrugge, G.J.; Trapman, J. Frequent inactivation of PTEN in prostate cancer cell lines and xenografts. *Cancer Res.* **1998**, *58*, 2720–2723. [PubMed]

165. Carson, J.P.; Kulik, G.; Weber, M.J. Antiapoptotic signaling in LNCaP prostate cancer cells: A survival signaling pathway independent of phosphatidylinositol 3′-kinase and Akt/protein kinase B. *Cancer Res.* **1999**, *59*, 1449–1453. [PubMed]

166. Khan, M.T.; Wagner, L., 2nd; Yule, D.I.; Bhanumathy, C.; Joseph, S.K. Akt kinase phosphorylation of inositol 1,4,5-trisphosphate receptors. *J. Biol. Chem.* **2006**, *281*, 3731–3737. [CrossRef] [PubMed]

167. Kuchay, S.; Giorgi, C.; Simoneschi, D.; Pagan, J.; Missiroli, S.; Saraf, A.; Florens, L.; Washburn, M.P.; Collazo-Lorduy, A.; Castillo-Martin, M.; et al. PTEN counteracts FBXL2 to promote IP3R3- and Ca^{2+}-mediated apoptosis limiting tumour growth. *Nature* **2017**, *546*, 554–558. [CrossRef] [PubMed]

168. Kobylewski, S.E.; Henderson, K.A.; Eckhert, C.D. Identification of ryanodine receptor isoforms in prostate DU-145, LNCaP, and PWR-1E cells. *Biochem. Biophys. Res. Commun.* **2012**, *425*, 431–435. [CrossRef] [PubMed]

169. Mariot, P.; Prevarskaya, N.; Roudbaraki, M.M.; Le Bourhis, X.; Van Coppenolle, F.; Vanoverberghe, K.; Skryma, R. Evidence of functional ryanodine receptor involved in apoptosis of prostate cancer (LNCaP) cells. *Prostate* **2000**, *43*, 205–214. [CrossRef]

170. Rios-Doria, J.; Kuefer, R.; Ethier, S.P.; Day, M.L. Cleavage of β-catenin by calpain in prostate and mammary tumor cells. *Cancer Res.* **2004**, *64*, 7237–7240. [CrossRef] [PubMed]

171. Rios-Doria, J.; Day, K.C.; Kuefer, R.; Rashid, M.G.; Chinnaiyan, A.M.; Rubin, M.A.; Day, M.L. The role of calpain in the proteolytic cleavage of E-cadherin in prostate and mammary epithelial cells. *J. Biol. Chem.* **2003**, *278*, 1372–1379. [CrossRef] [PubMed]

172. Park, J.J.; Rubio, M.V.; Zhang, Z.; Um, T.; Xie, Y.; Knoepp, S.M.; Snider, A.J.; Gibbs, T.C.; Meier, K.E. Effects of lysophosphatidic acid on calpain-mediated proteolysis of focal adhesion kinase in human prostate cancer cells. *Prostate* **2012**, *72*, 1595–1610. [CrossRef] [PubMed]

173. Shukla, S.; Shukla, M.; Maclennan, G.T.; Fu, P.; Gupta, S. Deregulation of FOXO3A during prostate cancer progression. *Int. J. Oncol.* **2009**, *34*, 1613–1620. [PubMed]

174. Shukla, S.; Bhaskaran, N.; Maclennan, G.T.; Gupta, S. Deregulation of FoxO3a accelerates prostate cancer progression in TRAMP mice. *Prostate* **2013**, *73*, 1507–1517. [CrossRef] [PubMed]

175. Brunet, A.; Kanai, F.; Stehn, J.; Xu, J.; Sarbassova, D.; Frangioni, J.V.; Dalal, S.N.; DeCaprio, J.A.; Greenberg, M.E.; Yaffe, M.B. 14-3-3 transits to the nucleus and participates in dynamic nucleocytoplasmic transport. *J. Cell Biol.* **2002**, *156*, 817–828. [CrossRef] [PubMed]

176. Wang, N.; Yao, M.; Xu, J.; Quan, Y.; Zhang, K.; Yang, R.; Gao, W.Q. Autocrine Activation of CHRM3 Promotes Prostate Cancer Growth and Castration Resistance via CaM/CaMKK-Mediated Phosphorylation of Akt. *Clin. Cancer Res.* **2015**, *21*, 4676–4685. [CrossRef] [PubMed]

177. Magnon, C.; Hall, S.J.; Lin, J.; Xue, X.; Gerber, L.; Freedland, S.J.; Frenette, P.S. Autonomic nerve development contributes to prostate cancer progression. *Science* **2013**, *341*, 1236361. [CrossRef] [PubMed]

178. Wang, N.; Dong, B.J.; Quan, Y.; Chen, Q.; Chu, M.; Xu, J.; Xue, W.; Huang, Y.R.; Yang, R.; Gao, W.Q. Regulation of Prostate Development and Benign Prostatic Hyperplasia by Autocrine Cholinergic Signaling via Maintaining the Epithelial Progenitor Cells in Proliferating Status. *Stem Cell Rep.* **2016**, *6*, 668–678. [CrossRef] [PubMed]

179. Schmitt, J.M.; Smith, S.; Hart, B.; Fletcher, L. CaM kinase control of AKT and LNCaP cell survival. *J. Cell. Biochem.* **2012**, *113*, 1514–1526. [CrossRef] [PubMed]

180. Hekman, M.; Albert, S.; Galmiche, A.; Rennefahrt, U.E.; Fueller, J.; Fischer, A.; Puehringer, D.; Wiese, S.; Rapp, U.R. Reversible membrane interaction of BAD requires two C-terminal lipid binding domains in conjunction with 14-3-3 protein binding. *J. Biol. Chem.* **2006**, *281*, 17321–17336. [CrossRef] [PubMed]

181. Taichman, R.S.; Cooper, C.; Keller, E.T.; Pienta, K.J.; Taichman, N.S.; McCauley, L.K. Use of the stromal cell-derived factor-1/CXCR4 pathway in prostate cancer metastasis to bone. *Cancer Res.* **2002**, *62*, 1832–1837. [PubMed]

182. Chetram, M.A.; Don-Salu-Hewage, A.S.; Hinton, C.V. ROS enhances CXCR4-mediated functions through inactivation of PTEN in prostate cancer cells. *Biochem. Biophys. Res. Commun.* **2011**, *410*, 195–200. [CrossRef] [PubMed]

183. Makowska, K.A.; Hughes, R.E.; White, K.J.; Wells, C.M.; Peckham, M. Specific Myosins Control Actin Organization, Cell Morphology, and Migration in Prostate Cancer Cells. *Cell Rep.* **2015**, *13*, 2118–2125. [CrossRef] [PubMed]

184. Peckham, M. How myosin organization of the actin cytoskeleton contributes to the cancer phenotype. *Biochem. Soc. Trans.* **2016**, *44*, 1026–1034. [CrossRef] [PubMed]

185. Barylko, B.; Jung, G.; Albanesi, J.P. Structure, function, and regulation of myosin 1C. *Acta Biochim. Pol.* **2005**, *52*, 373–380. [PubMed]

186. Ihnatovych, I.; Sielski, N.L.; Hofmann, W.A. Selective expression of myosin IC Isoform A in mouse and human cell lines and mouse prostate cancer tissues. *PLoS ONE* **2014**, *9*, e108609. [CrossRef] [PubMed]

187. Ihnatovych, I.; Migocka-Patrzalek, M.; Dukh, M.; Hofmann, W.A. Identification and characterization of a novel myosin Ic isoform that localizes to the nucleus. *Cytoskeleton* **2012**, *69*, 555–565. [CrossRef] [PubMed]

188. Nowak, G.; Pestic-Dragovich, L.; Hozak, P.; Philimonenko, A.; Simerly, C.; Schatten, G.; de Lanerolle, P. Evidence for the presence of myosin I in the nucleus. *J. Biol. Chem.* **1997**, *272*, 17176–17181. [CrossRef] [PubMed]

189. Zattelman, L.; Regev, R.; Usaj, M.; Reinke, P.Y.A.; Giese, S.; Samson, A.O.; Taft, M.H.; Manstein, D.J.; Henn, A. N-terminal splicing extensions of the human MYO1C gene fine-tune the kinetics of the three full-length myosin IC isoforms. *J. Biol. Chem.* **2017**, *292*, 17804–17818. [CrossRef] [PubMed]

190. Dzijak, R.; Yildirim, S.; Kahle, M.; Novak, P.; Hnilicova, J.; Venit, T.; Hozak, P. Specific nuclear localizing sequence directs two myosin isoforms to the cell nucleus in calmodulin-sensitive manner. *PLoS ONE* **2012**, *7*, e30529. [CrossRef] [PubMed]

191. Gillespie, P.G.; Cyr, J.L. Calmodulin binding to recombinant myosin-1c and myosin-1c IQ peptides. *BMC Biochem.* **2002**, *3*, 31. [CrossRef]

192. Nevzorov, I.; Sidorenko, E.; Wang, W.; Zhao, H.; Vartiainen, M.K. Myosin-1C uses a novel phosphoinositide-dependent pathway for nuclear localization. *EMBO Rep.* **2018**, *19*, 290–304. [CrossRef] [PubMed]

193. De Lanerolle, P. Nuclear actin and myosins at a glance. *J. Cell Sci.* **2012**, *125*, 4945–4949. [CrossRef] [PubMed]

194. Ye, J.; Zhao, J.; Hoffmann-Rohrer, U.; Grummt, I. Nuclear myosin I acts in concert with polymeric actin to drive RNA polymerase I transcription. *Genes Dev.* **2008**, *22*, 322–330. [CrossRef] [PubMed]

195. Percipalle, P.; Fomproix, N.; Cavellan, E.; Voit, R.; Reimer, G.; Kruger, T.; Thyberg, J.; Scheer, U.; Grummt, I.; Farrants, A.K. The chromatin remodelling complex WSTF-SNF2h interacts with nuclear myosin 1 and has a role in RNA polymerase I transcription. *EMBO Rep.* **2006**, *7*, 525–530. [CrossRef] [PubMed]

196. Hofmann, W.A.; Vargas, G.M.; Ramchandran, R.; Stojiljkovic, L.; Goodrich, J.A.; de Lanerolle, P. Nuclear myosin I is necessary for the formation of the first phosphodiester bond during transcription initiation by RNA polymerase II. *J. Cell. Biochem.* **2006**, *99*, 1001–1009. [CrossRef] [PubMed]

197. Kysela, K.; Philimonenko, A.A.; Philimonenko, V.V.; Janacek, J.; Kahle, M.; Hozak, P. Nuclear distribution of actin and myosin I depends on transcriptional activity of the cell. *Histochem. Cell Biol.* **2005**, *124*, 347–358. [CrossRef] [PubMed]

198. Philimonenko, V.V.; Zhao, J.; Iben, S.; Dingova, H.; Kysela, K.; Kahle, M.; Zentgraf, H.; Hofmann, W.A.; de Lanerolle, P.; Hozak, P.; et al. Nuclear actin and myosin I are required for RNA polymerase I transcription. *Nat. Cell Biol.* **2004**, *6*, 1165–1172. [CrossRef] [PubMed]

199. Fomproix, N.; Percipalle, P. An actin-myosin complex on actively transcribing genes. *Exp. Cell Res.* **2004**, *294*, 140–148. [CrossRef] [PubMed]

200. Pestic-Dragovich, L.; Stojiljkovic, L.; Philimonenko, A.A.; Nowak, G.; Ke, Y.; Settlage, R.E.; Shabanowitz, J.; Hunt, D.F.; Hozak, P.; de Lanerolle, P. A myosin I isoform in the nucleus. *Science* **2000**, *290*, 337–341. [CrossRef] [PubMed]

201. Obrdlik, A.; Louvet, E.; Kukalev, A.; Naschekin, D.; Kiseleva, E.; Fahrenkrog, B.; Percipalle, P. Nuclear myosin 1 is in complex with mature rRNA transcripts and associates with the nuclear pore basket. *FASEB J.* **2010**, *24*, 146–157. [CrossRef] [PubMed]

202. Cisterna, B.; Necchi, D.; Prosperi, E.; Biggiogera, M. Small ribosomal subunits associate with nuclear myosin and actin in transit to the nuclear pores. *FASEB J.* **2006**, *20*, 1901–1903. [CrossRef] [PubMed]

203. Mehta, I.S.; Elcock, L.S.; Amira, M.; Kill, I.R.; Bridger, J.M. Nuclear motors and nuclear structures containing A-type lamins and emerin: Is there a functional link? *Biochem. Soc. Trans.* **2008**, *36*, 1384–1388. [CrossRef] [PubMed]

204. Chuang, C.H.; Carpenter, A.E.; Fuchsova, B.; Johnson, T.; de Lanerolle, P.; Belmont, A.S. Long-range directional movement of an interphase chromosome site. *Curr. Biol.* **2006**, *16*, 825–831. [CrossRef] [PubMed]

205. Venit, T.; Dzijak, R.; Kalendova, A.; Kahle, M.; Rohozkova, J.; Schmidt, V.; Rulicke, T.; Rathkolb, B.; Hans, W.; Bohla, A.; et al. Mouse nuclear myosin I knock-out shows interchangeability and redundancy of myosin isoforms in the cell nucleus. *PLoS ONE* **2013**, *8*, e61406. [CrossRef] [PubMed]

206. Lu, Q.; Li, J.; Ye, F.; Zhang, M. Structure of myosin-1c tail bound to calmodulin provides insights into calcium-mediated conformational coupling. *Nat. Struct. Mol. Biol.* **2015**, *22*, 81–88. [CrossRef] [PubMed]

207. Pruschy, M.; Ju, Y.; Spitz, L.; Carafoli, E.; Goldfarb, D.S. Facilitated nuclear transport of calmodulin in tissue culture cells. *J. Cell Biol.* **1994**, *127*, 1527–1536. [CrossRef] [PubMed]

208. Visuttijai, K.; Pettersson, J.; Mehrbani Azar, Y.; van den Bout, I.; Orndal, C.; Marcickiewicz, J.; Nilsson, S.; Hornquist, M.; Olsson, B.; Ejeskar, K.; et al. Lowered Expression of Tumor Suppressor Candidate MYO1C Stimulates Cell Proliferation, Suppresses Cell Adhesion and Activates AKT. *PLoS ONE* **2016**, *11*, e0164063. [CrossRef] [PubMed]

209. Letellier, E.; Schmitz, M.; Ginolhac, A.; Rodriguez, F.; Ullmann, P.; Qureshi-Baig, K.; Frasquilho, S.; Antunes, L.; Haan, S. Loss of Myosin Vb in colorectal cancer is a strong prognostic factor for disease recurrence. *Br. J. Cancer* **2017**, *117*, 1689–1701. [CrossRef] [PubMed]

210. Buschman, M.D.; Field, S.J. MYO18A: An unusual myosin. *Adv. Biol. Regul.* **2018**, *67*, 84–92. [CrossRef] [PubMed]

International Journal of
Molecular Sciences

MDPI

Review

Role of KCa3.1 Channels in Modulating Ca^{2+} Oscillations during Glioblastoma Cell Migration and Invasion

Luigi Catacuzzeno * and Fabio Franciolini *

Department of Chemistry, Biology and Biotechnology, University of Perugia, 06134 Perugia, Italy
* Correspondence: luigi.catacuzzeno@unipg.it (L.C.); fabio.franciolini@unipg.it (F.F)

Received: 30 August 2018; Accepted: 25 September 2018; Published: 29 September 2018

Abstract: Cell migration and invasion in glioblastoma (GBM), the most lethal form of primary brain tumors, are critically dependent on Ca^{2+} signaling. Increases of [Ca^{2+}]$_i$ in GBM cells often result from Ca^{2+} release from the endoplasmic reticulum (ER), promoted by a variety of agents present in the tumor microenvironment and able to activate the phospholipase C/inositol 1,4,5-trisphosphate PLC/IP$_3$ pathway. The Ca^{2+} signaling is further strengthened by the Ca^{2+} influx from the extracellular space through Ca^{2+} release-activated Ca^{2+} (CRAC) currents sustained by Orai/STIM channels, meant to replenish the partially depleted ER. Notably, the elevated cytosolic [Ca^{2+}]$_i$ activates the intermediate conductance Ca^{2+}-activated K (KCa3.1) channels highly expressed in the plasma membrane of GBM cells, and the resulting K$^+$ efflux hyperpolarizes the cell membrane. This translates to an enhancement of Ca^{2+} entry through Orai/STIM channels as a result of the increased electromotive (driving) force on Ca^{2+} influx, ending with the establishment of a recurrent cycle reinforcing the Ca^{2+} signal. Ca^{2+} signaling in migrating GBM cells often emerges in the form of intracellular Ca^{2+} oscillations, instrumental to promote key processes in the migratory cycle. This has suggested that KCa3.1 channels may promote GBM cell migration by inducing or modulating the shape of Ca^{2+} oscillations. In accordance, we recently built a theoretical model of Ca^{2+} oscillations incorporating the KCa3.1 channel-dependent dynamics of the membrane potential, and found that the KCa3.1 channel activity could significantly affect the IP$_3$ driven Ca^{2+} oscillations. Here we review our new theoretical model of Ca^{2+} oscillations in GBM, upgraded in the light of better knowledge of the KCa3.1 channel kinetics and Ca^{2+} sensitivity, the dynamics of the Orai/STIM channel modulation, the migration and invasion mechanisms of GBM cells, and their regulation by Ca^{2+} signals.

Keywords: KCa3.1 channels; glioblastoma; cell migration; calcium oscillations; mathematical model

---

1. The Glioblastoma

The large majority (more than 90%) of cancer deaths are due not to the primary tumor per se, but to relapses arising from new foci established in distant organs via metastasis [1]. Glioblastoma (GBM), the most common and aggressive form of primary brain tumors, is no exception, though it does not metastasize in the classical way (that is, by colonizing other tissues via the bloodstream), but invades brain parenchyma by detaching from the original tumor mass and infiltrating into the healthy tissue by degrading the extracellular matrix or squeezing through the brain interstitial spaces. The urgency of tackling the migration and invasion issues of GBM tumors is clear.

The 2016 World Health Organization classification of the various types of brain tumors regards the presence of isocitrate dehydrogenase gene (*IDH1/2*) mutations as one of the most critical biomarkers. Accordingly, *IDH1/2* wildtype gliomas are categorized as glioblastoma (formerly primary glioblastoma) and *IDH1/2*-mutated gliomas (including formerly classified secondary glioblastoma) as astrocytic glioma and oligodendroglioma. GBMs are further subdivided into four groups (Proneural,

Neural, Classical, and Mesenchymal), mainly based on the abnormally high levels of mutated genes (i.e., *EGFR* is highly upregulated in >98% of Classical GBM, whereas *TP53* (p53), which is most frequently mutated in Proneural GBM (50–60% of patients) is rarely mutated in Classical GBM). In spite of the intensive basic and clinical studies carried out over the past decades, and modern diagnostics and treatments, the average life expectancy for GBM patients is still only around 15 months. The major obstacle with GBM remains its high migratory and invasive potential into healthy brain parenchyma, which prevents complete surgical removal of tumor cells. Even with full clinical treatment (temozolomide-based chemotherapy and radiation therapy), tumors normally recur at some distance from the site of resection, establishing new tumor lesions that are by far the primary cause of mortality in GBM patients. Arguably, at the time of surgery, large numbers of cells have already detached from the original tumor mass and invaded normal brain tissue. Although GBM cell migration and invasion have been deeply investigated, many aspects of these processes are still poorly understood.

GBM cell migration is a highly regulated multistep process that initiates with GBM cells losing adhesion with surrounding elements, avoiding the cell death often associated with extracellular matrix (ECM) disconnection, and acquiring a highly migratory phenotype, which is a critical feature of the invasive process. The basic mechanisms underlying migration of GBM cells are common to most types of migratory cells. Migration is a property of many non-tumor cells, although it is often restricted to specific developmental stages or environmental conditions; the migration of tumor cells could be viewed as the result of mutation-induced dysregulation of specific biochemical pathways that in healthy tissue keep cell migration dormant.

2. Glioblastoma Cell Migration and Ca^{2+} Signaling

2.1. Cell Migration

The basic mechanisms of cell locomotion are now fairly well established. Locomotion can be described as the cyclical repeating of two main processes: (i) protrusion of the cell front due to local gain of cell volume mostly generated by active $Na^+/K^+/2Cl^-$ cotransport accompanied by isoosmotically obliged water, and actin polymerization, with formation of pseudopods; (ii) retraction of the rear cell body in the direction of motion, due to forces produced by actomyosin contraction, accompanied by loss of cell volume generated by passive ion (mainly K^+ and Cl^-) fluxes and osmotic water [2,3]. These two processes involve the coordinated and localized formation of integrin-dependent cell adhesions at the leading edge, and their disassembly at the cell rear [4,5].

Protrusion of the cell front is sustained by localized polymerization of submembrane actin-based cytoskeleton that generates the pushing force and forms flat lamellipodia or needle-like filipodia. A large variety of signaling molecules have been shown to play a leading role in these processes, including the Rho GTPases family (that act as molecular switches to control downstream transduction pathways), and their effector proteins CDC42, RAC1, and RhoA. PI3 kinases have also been deeply implicated in controlling actin polymerization and lamellipodium extension. Activation of PI3 kinase by the pro-invasive signal molecules present in the tumor microenvironment functions as the trigger of the process, in that its activation initiates actin polymerization and generates membrane protrusion [5].

The retraction of the cell rear depends on the contractile forces generated by the activation of the myosin motors along crosslinked actin filaments. The myosin molecule is formed by two heavy chains that make up the two heads and the coiled-coil tail, and four—two essential and two regulatory—myosin light chains (MLCs). The myosin motor is primarily activated by a Ca^{2+}-dependent cascade whereby a cytosolic Ca^{2+} increase activates a MLC kinase (MLCK) that leads to the regulatory MLC phosphorylation, which allows the myosin motor to crosslink with actin and produce tension to pull the rear end of the cell (however, some studies claim that the motor activity of myosin is not required for cell migration [6,7], playing instead a role in the establishment of cell polarity and in the coordination between different cell domains [7,8]). Clearly, to make the rear cell effectively move, the focal adhesions that anchor the actin cytoskeleton to the extracellular matrix (ECM) must be disassembled, normally through a proteolytic process that involves calpain, but also actin microtubules and focal adhesion kinase (FAK)-recruited dynamin that internalizes the integrins. Increasing actomyosin tension disrupts the possible residual resistances of focal adhesions at the cell rear [2]. Myosin can also be modulated by other pathways, including a Rho-dependent cascade (Rho → ROCK → MLC phosphorylation, or MLC phosphatase activation) that does not depend on cytosolic Ca^{2+}.

This locomotion cycle can be initiated by a variety of external stimuli or directional cues which are sensed and decoded by specialized receptors on the plasma membrane and transferred to the cell interior via distinct signaling pathways to actuate the mechanical response. Most common migration stimuli rely on cells sensing (by specific G protein-coupled receptors) local gradients in the concentration of chemical factors (chemoattractants) [9] present in the tumor microenvironment (for instance chemokines IL-8 or CXCL12), or certain growth factors, including the platelet-derived growth factor (PDGF) that plays an important role in tumor metastasis [10]. Cancer cells can also be guided by gradients of bound ligands. Generally attached to the ECM, these ligands are recognized by specific integrins that form oriented focal adhesion to direct the formation of invadopodia, thus the direction of movement. Common is also the observation of cancer cells being guided by mechanical stimuli (mechanotaxis), namely the stiffness of the extracellular matrix they try to penetrate [11]. It is important to note that in vivo cells can be simultaneously subjected to multiple types of cues that need be evaluated as a whole in order to provide appropriate responses.

2.2. Cell Volume Changes Associated with Migration

It is important to recognize that these two steps—cell front protrusion and cell rear retraction—normally occur in succession, implying that cells are subjected to major changes in cell volume, especially the cell rear retraction. This requires compensatory ion fluxes (and osmotic water flow) across the plasma membrane in order to maintain proper cytosol osmolarity. Along the line proposed by [12] with regard to cell migration in general, during the protrusion of the cell front the maintenance of a stable osmolarity is sustained by cotransporters and ion exchangers, for instance $Na^+/K^+/2Cl^-$ cotransport, followed by osmotically driven water. By contrast, the control of osmolarity during cell rear retraction critically depends on the activation of K channels and ensuing efflux of K^+ ions (accompanied by Cl^- fluxes, to which the membrane has become highly permeant, and osmotic water). Following the opening of K channels the membrane potential hyperpolarizes to a value between the equilibrium potentials of Cl^- and K^+ (E_{Cl} and E_K), a condition that allows a significant efflux of KCl with consequent loss of water and cell volume decrease that facilitates cell body retraction. This view can explain the slowing of cell locomotion following the inhibition of K and Cl channels [13]. A number of papers indicate that the channel types primarily involved in osmolarity control are the intermediate conductance Ca^{2+}-activated K channel (KCa3.1) and the Cl channel ClC3 [13,14].

2.3. Ca^{2+} Oscillations

Much work shows that cell migration is strictly regulated by Ca^{2+}, which is no surprise given that many proteins involved in migration such as myosin, myosin light chain kinase (MLCK), Ca^{2+}/calmodulin-dependent protein kinase II are Ca^{2+} sensitive [15]. In the resting cell Ca^{2+} concentration is kept very low (<100 nM) to prevent activation of Ca^{2+}-sensitive proteins that act as Ca^{2+} sensor and initiation of unwanted biological processes. Ca^{2+} concentration can easily increase more than ten-fold (well over 1 µM) following several types of cell stimulation, and this would bring many of the major cytoplasmic Ca^{2+}-sensitive proteins (calmodulin, PKA, Ca^{2+}-activated channels, etc.) to activate. This opens the question of how Ca^{2+} signaling, under these conditions, can selectively activate a specific Ca^{2+} sensor protein and the associated biochemical cascade. Several strategies have evolved in this respect, but specificity is mostly attained by confining the signal at sub-cellular level, or by regulating kinetics and magnitude of the Ca^{2+} signal. Since in the cytoplasm Ca^{2+} diffusion is extremely low and buffering high, the opening of Ca^{2+} permeant channels results in a very localized

Ca^{2+} increase (the Ca^{2+} microdomain), attaining spatial discrimination of the target proteins and linked downstream pathways. Another strategy relies on the temporal features of Ca^{2+} signals (rising rate, duration, repetition of spikes, etc.) that biological systems are capable to interpret.

In non-excitable cells, including GBM, typical Ca^{2+} signals induced experimentally by robust chemical (hormone) stimulation are slow, large and generally sustained. Hormones binding to their specific G-protein-coupled receptors (GPCRs) often results in the PLC-dependent synthesis of inositol 1,4,5-trisphosphate (IP$_3$), and consequent Ca^{2+} release from the endoplasmic reticulum (ER) via IP$_3$ receptors. Eventually, the sustained Ca^{2+} increase subsides as result of the activation of Ca^{2+} pumps placed on the sarco/endoplasmic reticulum (SERCA), or the plasma membrane (PMCA) that transfer Ca^{2+} ions from the cytosol to intracellular stores or the extracellular space.

A physiologically more meaningful type of Ca^{2+} signal, especially in relation to cell migration, where critical Ca^{2+}-dependent processes are cyclical, comes in the form of Ca^{2+} oscillations. It has recently become clear that this type of oscillatory Ca^{2+} increases is what normally occurs when using physiological agonist stimulations. The mechanisms underlying Ca^{2+} oscillations are not fully understood, and often depend on the cell model and agonist used. In some systems Ca^{2+} oscillations are secondary to oscillations of the cytoplasmic IP$_3$ level, which may be due to various types of feedback control that for brief intervals uncouple the GTP-coupled receptor from PLC. For instance, the reported negative feedback of the PLC products, IP$_3$ and diacylglycerol (DAG), on PLC itself, or upstream on the GPCR, would result in IP$_3$ oscillations [16], which would in turn generate cyclical release of Ca^{2+} from intracellular stores, and the Ca^{2+} oscillations [17].

More commonly, however, Ca^{2+} oscillations can be observed in the presence of a constant level of IP$_3$. The prevailing view is that, under these conditions, the repetitive Ca^{2+} oscillations secondary to moderate (physiological) hormone or agonist activation of membrane PLC and production of a constant amount of IP$_3$ arises from the biphasic effects of cytosolic Ca^{2+} on the IP$_3$ receptor gating, with the first phase being embodied by the establishment of the Ca^{2+}-induced Ca^{2+} release (CICR) mechanism, whereby a moderate increase of [Ca^{2+}] in the cytosol via IP$_3$ receptor causes a positive feedback activation of the IP$_3$ receptor that results in a higher Ca^{2+} released from internal stores ([18]; Figure 1A).

The second phase occurs when the cytosolic [Ca^{2+}] reaches significantly higher levels, shifting the previous positive feedback on IP$_3$ receptors into a delayed negative feedback that results in their closing and [Ca^{2+}]$_i$ returning to resting level by the action of both SERCA and PMCA Ca^{2+} pumps (Figure 1A). This view finds support from the observation that at constant levels of IP$_3$, the activity of the IP$_3$ receptor and the cytosolic Ca^{2+} concentration display a bell-shaped function, whereby low Ca^{2+} concentrations activate the receptor, whereas high Ca^{2+} concentrations inhibit it [19–22]. Ca^{2+} oscillations generated according to this scheme (i.e., Figure 1A) would however show a rapid decrease of spike amplitude with time, because part of the Ca^{2+} released from the ER during each spike will be pumped out of the cell by PMCA, and be no longer available for refilling the ER and contributing to the next Ca^{2+} spike. This Ca^{2+} oscillations time course is observed experimentally upon removal of external Ca^{2+}, or blockade of Ca^{2+} influx from the extracellular space. To have sustained (or slow decaying) Ca^{2+} oscillations, external Ca^{2+} pumped out of the cell by PMCAs must be allowed to re-enter the cell. This is most commonly accomplished through the mechanism of store-operated Ca^{2+} entry (SOCE).

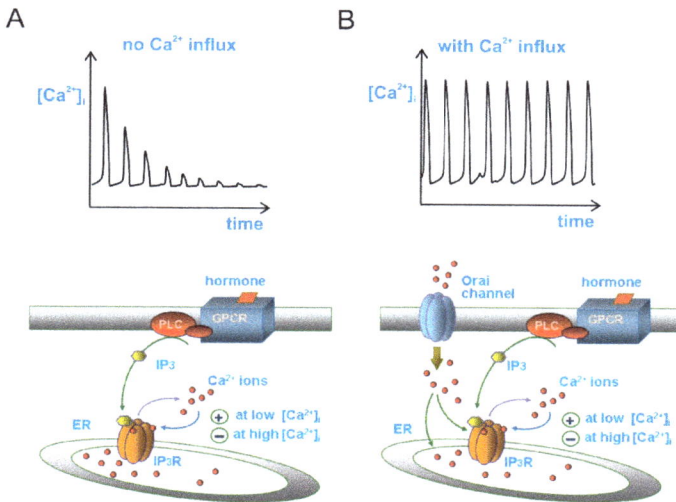

Figure 1. Ca^{2+} oscillations in response to inositol thriphosphate (IP_3) increase, with and without Ca^{2+} influx from extracellular space. (**A**) Bottom, drawing illustrating the hormone-based production of IP_3 that activates the IP_3 receptor to release Ca^{2+} from endoplasmic reticulum (ER). The biphasic effects of cytosolic Ca^{2+} on IP_3 receptor gating (the basic mechanism for Ca^{2+} oscillations), whereby Ca^{2+} modulates positively the receptor at low $[Ca^{2+}]$ but negatively at high $[Ca^{2+}]$, is also illustrated. Top, Ca^{2+} oscillations as produced from the schematics below. Note the decaying trend of Ca^{2+} spikes due to the absence of Ca^{2+} influx from extracellular space; (**B**) Here the drawing has been enriched with a Ca^{2+} influx apparatus from extracellular space through ER-depletion activated Orai channels on the plasma membrane (bottom), which generates sustained Ca^{2+} oscillations (top). For clarity, SERCA and PMCA Ca^{2+} pumps have not been sketched in the drawing, although their activity has always been taken into account. For the same reason, we omitted to draw STIM protein of the ER.

2.4. Store-Operated Ca²⁺ Entry (SOCE)

When Ca^{2+} is released from the ER, i.e., following hormone stimulation, the Ca^{2+} concentration within the organelle decreases. It is this decrease to trigger the process (known as SOCE) whereby Ca^{2+} channels in the plasma membrane are activated, allowing entry of Ca^{2+} into the cell. Although the process of SOCE was described some 30 years ago, its mechanisms and molecular counterparts—the STIM and Orai proteins—have been identified much more recently [23–25]. The stromal interaction molecules (STIMs) are a family of one-passage integral proteins of the ER that act as sensors of the Ca^{2+} concentration within the ER. Orai proteins form instead a group of highly selective Ca^{2+} channels placed in the plasma membrane. Orai channels are neither activated by voltage, nor by agonists (in the classical sense), and are normally silent when the cell is at rest and the ER filled with Ca^{2+} ions. Orai Ca^{2+} channels are opened by the interaction with activated STIM proteins, which occurs following depletion of Ca^{2+} in the ER, and their opening results in Ca^{2+} ions flowing inside the cell (the mechanism has been recently reviewed in depth on neurons by [26]). This Ca^{2+} entry into the cytoplasm enables the refilling of Ca^{2+} depleted ER through the SERCA pumps, and make the Ca^{2+} oscillations amplitude to be kept stable with time (Figure 1B). In the context of our review topic, this feedback mechanism of store-operated Ca^{2+} entry gains further relevance for two more reasons: it contributes to further modulating the IP_3 receptor (this modulation is also dependent on the cell type—for instance, in U87-MG human GBM cells Ca^{2+} oscillations are affected only marginally by removal of Ca^{2+} influx [27], whereas in rat C6 GBM cells this maneuver fully suppresses Ca^{2+}

oscillations [28]), and it represents the target on which the KCa3.1 channel exerts its modulation of Ca^{2+} oscillations, as we will see later.

It needs to be recalled that the store-operated Ca^{2+} entry is not the only suggested mechanism—although it is by far the most widely accepted—that links hormone stimulation to increased plasma membrane Ca^{2+} entry. For instance, heteromers of Orai1 and Orai3 can be activated by arachidonic acid [26], and the Ca^{2+}-permeant TRPC6 by receptor-induced PLC activation via direct action of DAG. Notably, neither pathway involves IP_3 receptor or Ca^{2+} depletion from the ER.

2.5. Ca^{2+} Oscillations and Cell Migration

Numerous studies have shown that migrating GBM cells display evident Ca^{2+} oscillations having a period of several tens of seconds, whose presence appears to be essential to the process of migration. For example, in U87-MG cells Ca^{2+} oscillations may be induced by the pro-migratory fetal calf serum (FCS), and trigger focal adhesion disassembly during cell rear retraction [29,30]. Similarly, in D54-MG cells prolonged exposure to bradykinin induces Ca^{2+} oscillations, which significantly enhance cell motility [31]. Mechanistically, the Ca^{2-} signal may translate into actomyosin contraction and rear cell retraction by first binding to calmodulin (CaM). The activated complex, Ca^{2+}/CaM, then activates MLCK, the kinase that phosphorylates the regulatory myosin light chains (MLC), thereby triggering cross-bridge movements and actomyosin contraction. It has been shown that MLCK can be modulated by several other protein kinases including PKA, PKC, and RhoK, indicating that this critical enzyme for cell migration is under the influence and control of many extracellular signals and cytoplasmic pathways. Although the consensus on this mechanism and the role of MLCK is widespread, a little caution is needed as most studies used a pharmacological approach, but really specific MLCK inhibitors were not available [32]. A recent report showed, for instance, that knockout cells for MLCK maintained unaltered (or increased) their migratory ability [33]. This shows that if the critical role of Ca^{2+} in cell motility is undisputed, uncertainties may remain as to how Ca^{2+} is linked to the cell migratory machinery. The role of the Ca^{2+}-activated K channels in cell volume changes and the dynamics of Ca^{2+} signals during migration may be a valid alternative.

3. The Intermediate Conductance Ca^{2+}-Activated K Channel

The intermediate conductance Ca^{2+}-activated K channel, KCa3.1, belongs to the Ca^{2+}-activated K channel (KCa) family, which comprises the large- (KCa1.1), intermediate (KCa3.1), and small conductance (KCa2.1-3) K channels, as originally classified according to their single-channel conductance. Genetic relationship and Ca^{2+} activation mechanisms have later shown that these channels form two well-defined and distantly related groups. One, including only the large-conductance KCa1.1 channel, is gated by the cooperative action of membrane depolarization and $[Ca^{2+}]_i$, while the other group includes both the intermediate conductance KCa3.1 and small-conductance KCa2.1-3 channels, gated solely by cytosolic Ca^{2+} increase.

3.1. Biophysics, Pharmacology, and Gating of the KCa3.1 Channel

The biophysical properties of the KCa3.1 channel have been investigated using various experimental approaches and cell models. Work on cultured human glioblastoma cells shows high K^+ selectivity, moderate inward rectification, and single-channel conductance of 20–60 pS in symmetrical 150 mM K^+. The KCa3.1 channel is voltage-insensitive, but highly sensitive to $[Ca^{2+}]_i$ showing an EC_{50} of <200 nM. Ca^{2+} ions activate the KCa3.1 channel via the Ca^{2+}-binding protein CaM, which is constitutively bound to the membrane-proximal region of the intracellular C terminus of the channel. Binding of Ca^{2+} to CaM results in conformational changes of the channel, and its opening. In addition to serving as a Ca^{2+} sensor for channel opening, CaM also regulates the assembly and trafficking of the channel protein to the plasma membrane [34,35].

The KCa3.1 channel can be blocked by peptidic toxins isolated from various scorpions, such as charybdotoxin (ChTx) and maurotoxin (MTx), which display high affinity (IC_{50} in the nM range).

Small synthetic molecules have also been developed, many derived from the clotrimazole template, the classical KCa3.1 channel blocker. The most widely used TRAM-34, developed by Wulff's group, inhibits KCa3.1 channels with an $IC_{50} < 20$ nM and displays high selectivity over the other KCa channels. As for KCa3.1 channel activators, we recall the benzimidazolone 1-EBIO that activates the channels with an $EC_{50} < 30$ μM and DC-EBIO that exhibits 10-fold higher potency. Two structurally similar and still more potent molecules are the oxime NS309 and the benzothiazole SKA-31, both showing an EC_{50} for KCa3.1 channels < 20 nM. We finally mention riluzole, another potent activator of KCa3.1 channels, which is the only FDA-approved drug for treatment of amyotrophic lateral sclerosis (ALS).

3.2. KCa3.1 Channel Expression and Impact on GBM Migration and Invasion

Initial biochemical and electrophysiological studies showed that KCa3.1 channels were diffusely expressed in virtually all cell types investigated. Notably, the KCa3.1 channel was not found in the brain (later the KCa3.1 channel was described in microglia [7], several types of brain neurons [8,36], and the nodes of Ranvier of cerebellar Purkinje neurons [37]), although it was highly expressed in established cell lines from brain tumors, as well as in brain tumors in situ. Moreover, KCa3.1 channel expression was reported in many other tumor types, and implicated in malignancy (increased cell growth, migration, invasion, apoptosis evasion). The KCa3.1 mRNA and protein expression were also found to be significantly enhanced in cancer stem cells derived from both the established cell line U87-MG and primary cell line FCN9 [38]. The REMBRANDT database shows that the gene KCNN4, which encodes the KCa3.1 channel, is overexpressed in more than 30% of gliomas, and its expression is associated with a poor prognosis.

To test the role of the KCa3.1 channel in GBM cells' infiltration into brain parenchyma, human GL-15 GBM cells were xenografted into the brains of SCID mice and later treated with the specific KCa3.1 blocker TRAM-34. Immunofluorescence analyses of cerebral slices after a five-week treatment revealed a significant reduction of tumor infiltration, compared with TRAM-34 untreated mice [39]. Reduction of tumor infiltration was also observed in the brain of mice transplanted with KCa3.1-silenced GL-15 cells, indicating a direct role of KCa3.1 channels in GBM tumor infiltration [40]. Similarly, KCa3.1 channel block with TRAM-34 or silencing by short hairpin RNA (shRNA) completely abolished CXCL12-induced GL-15 cell migration [40]. A strong correlation of KCa3.1 channel expression with migration was reported by Calogero's group in GBM-derived cancer stem cells (CSC) [38]. Blockage of the KCa3.1 channel with TRAM-34 was found to have a much greater impact on the motility of CSCs (75% reduction) that express a high level of KCa3.1 channel than on the FCN9 parental population (32% reduction), where the KCa3.1 channel is expressed at much lower levels. Similar results were also observed with the CSCs derived from U87-MG [38] and with mesenchymal glioblastoma stem cells [41]. On the same line stand the results later obtained by Sontheimer's group showing that pharmacological inhibition of the KCa3.1 channel (with TRAM-34) in U251 glioma cells, or silencing it with inducible siRNA, resulted in a significant reduction of tumor cells' migration in vitro and invasion into surrounding brain parenchyma of SCID mice [42]. On a retrospective study of a patient genomic database, they further showed that KCa3.1 channel expression was inversely correlated with patient survival.

3.3. Basic Functions of KCa3.1 Channel: Regulation of Cell Ca^{2+} Signaling

KCa3.1 channels control a number of basic cellular processes involved in the modulation of several higher-order biological functions critical to brain tumors' malignancy, including migration and invasion. The most relevant and common basic cellular processes controlled by KCa3.1 channels are the modulation of Ca^{2+} signaling and the control of cell volume. Here we will concentrate on the KCa3.1 channel modulation of Ca^{2+} signaling, the focus of this review.

In virtually all cells a robust stimulation of PLC-coupled membrane receptors triggers an initial IP_3-mediated release of Ca^{2+} from intracellular stores, followed by Ca^{2+} influx through store-operated Orai Ca^{2+} channels, which are activated in response to Ca^{2+} depletion of the ER.

One consequence of this Ca^{2+} influx, besides the activation of Ca^{2+}-dependent target proteins (ion channels included), is membrane depolarization, which, if left unchecked, would more and more strongly inhibit further Ca^{2+} influx due to the reduced electrochemical driving force for Ca^{2+} ions. It has been shown that the efflux of K$^+$ ions following KCa3.1 channels' activation by incoming Ca^{2+} hyperpolarizes the membrane towards E$_K$, increasing the electromotive force on incoming Ca^{2+} ions [43,44]. This represents an indirect modulatory mechanism of Ca^{2+} entry and, as a result, of cell Ca^{2+} signaling. The Ca^{2+}-dependent and voltage-independent gating of KCa3.1 channels appears to be particularly well suited for this function, since the coupling of Ca^{2+} influx and activation of K$^+$ channels will create a positive feedback loop whereby more Ca^{2+} influx will increase the K$^+$ efflux, hyperpolarize the plasma membrane, and further stimulate Ca^{2+} entry, thus amplifying the signal transduction.

This paradigm was first demonstrated in human macrophages, where KCa3.1 channels have been shown to hyperpolarize the membrane and increase the driving force for Ca^{2+} ions during store-operated Ca^{2+} entry [43], and in T cells, where KCa3.1 channels are rapidly upregulated following cell activation, and supposedly used to maximize Ca^{2+} influx during the reactivation of memory T cells [45,46]. Activated T cells isolated from KCa3.1$^{-/-}$ mice show a defective Ca^{2+} response to T cell receptor activation [47]. Also, mast cells appear to use KCa3.1 channels to hyperpolarize the membrane and increase the Ca^{2+} influx following antigen-mediated stimulation. In this case KCa3.1 channels have been found to physically interact with the Orai1 subunit, suggesting that they may be activated by the Ca^{2+} microdomain that forms close to the store-operated Ca^{2+} channel [28,48]. A similar conclusion has been reached for rat microglial cells, where the Orai1-KCa3.1 functional coupling could be interrupted by BAPTA but not by EGTA [49]. Since the two Ca^{2+} chelators have similar Ca^{2+} affinity but different Ca^{2+} binding rates (10 times higher for BAPTA), only a physical coupling between Orai1 and KCa3.1, with a separating distance of few nanometers, would explain these results.

Evidence of cross-talk between the KCa3.1 channel and store depletion-activated Ca^{2+} influx has also been reported for GBM cells. An increase of [Ca^{2+}]$_i$, consisting of a fast peak due to IP$_3$-dependent Ca^{2+} release from intracellular stores followed by a sustained phase of Ca^{2+} influx across the plasma membrane, supposedly activated by intracellular stores' depletion, was found upon prolonged application of histamine on GL-15 GBM cells [44]. The enhancing role of the KCa3.1 channel in sustaining the protracted influx of external Ca^{2+} was shown by the marked reduction of the sustained histamine-induced [Ca^{2+}]$_i$ increase following application of TRAM-34. This observation could be significant with regard to the KCa3.1 channels' contribution to GBM cell migration exerted through modulation of Ca^{2+} signals.

In another GBM cell line—U87-MG—we also observed that the promigratory FCS is able to promote Ca^{2+} oscillations, and these oscillations cyclically activate the KCa3.1 channels during cell migration [13]. Using a modeling approach, we also found that a channel activity with the properties of KCa3.1 channels could modulate these Ca^{2+} oscillations (it increased the amplitude, duration, and frequency of each Ca^{2+} spike [50]). The results we obtained, which are illustrated in the following section, were unexpectedly interesting, and also able to explain old experiments showing that the KCa3.1 channel inhibition by ChTx abolishes the bradykinin-induced Ca^{2+} oscillations in C6 glioma cells [51].

3.4. Modulation of Ca^{2+} Oscillations by KCa3.1 Channel Activity

We saw earlier that the mechanism underlying Ca^{2+} oscillations is essentially based on the biphasic effects of Ca^{2+} on IP$_3$ receptor gating—activatory at low and inhibitory at high concentrations. We also discussed the modulation that the KCa3.1 channel could exert on Ca^{2+} oscillations. This modulation relies on the high Ca^{2+} sensitivity and voltage independence of the KCa3.1 channel, and on the output of its activity, that is, the hyperpolarization of the cell membrane potential (V_m) as result of the K$^+$ efflux. As KCa3.1 channels are activated by Ca^{2+} concentrations well within the range observed during hormone-induced Ca^{2+} oscillations (150–600 nM; cf. above), they are expected to cyclically open (during the Ca^{2+} spikes) and cause cyclic membrane hyperpolarizations in phase with the Ca^{2+} oscillations.

We have observed experimentally such KCa3.1 channel oscillatory activity and associated membrane potential oscillations in U87-MG cells in response to FCS [13], an agent known to induce Ca^{2+} oscillations in these cells [29,30]. Notably, because of their voltage dependence and much lower Ca^{2+} sensitivity, the KCa1.1 channels, also highly expressed in GBM cells, could not be activated under these conditions, even at the peak Ca^{2+} concentration of the oscillations ([13], but see [52] for a case in which the KCa1.1 channel activity may control Ca^{2+} influx). In conclusion, since V_m controls the driving force for external Ca^{2+} entry, KCa3.1-dependent V_m oscillations are expected to cause oscillations in the amplitude of Ca^{2+} influx through the hormone-activated plasma membrane Ca^{2+} channels, which will in turn feedback onto, and modulate the Ca^{2+} oscillations themselves.

We recently implemented basic theoretical models of Ca^{2+} oscillations with the oscillating membrane V_m, as produced by the cyclic activation of KCa3.1 channels. The resulting model was used to predict how the hormone-induced Ca^{2+} oscillations would be influenced by the KCa3.1 channels-induced V_m oscillations in phase with Ca^{2+} oscillations. We found that the cyclic activation of KCa3.1 channels by Ca^{2+} oscillations induces V_m fluctuations that in turn determine oscillations in the Ca^{2+} influx from the extracellular medium, as a result of the oscillating V_m-dependent changes in the driving force for Ca^{2+} ions influx. Since Ca^{2+} influx peaks are in phase with the Ca^{2+} oscillations, the Ca^{2+} spikes will also be increased by the in-phase Ca^{2+} influx. Our model calculations in fact show that KCa3.1-induced V_m oscillations strengthen the hormone-induced $[Ca^{2+}]_i$ signals by increasing the amplitude, duration, and oscillatory frequency of Ca^{2+} spikes (Figure 2; see also [50]).

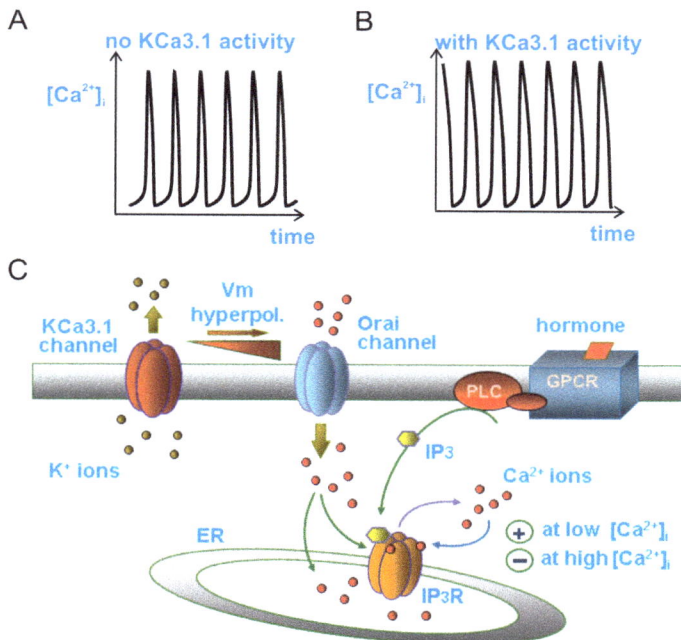

Figure 2. Modulation of Ca^{2+} oscillations by the KCa3.1 channel. Drawing (**C**) illustrating the main ion fluxes generating or modulating the Ca^{2+} oscillations in our model when the KCa3.1 channel is cut off (**A**) or introduced into the system (**B**). Please notice the much longer duration (width) and slightly higher amplitude of the Ca^{2+} oscillations in the presence of KCa3.1 channels.

3.5. KCa3.1 Channels Switch Ca²⁺ Oscillations On and Off

Arguably, the KCa3.1 channels can do more than just modulate the amplitude and frequency of Ca²⁺ oscillations: they can trigger or suppress them. This has been an unexpected prediction of our model that we later found to have been experimentally observed. As experimentally found, our model predicts that Ca²⁺ oscillations can be generated only within a specific range of IP₃ concentrations, while a stable, non-oscillating Ca²⁺ level is present for both lower and higher IP₃ concentrations (Figure 3A). Surprisingly, it was found that when the KCa3.1 conductance is changed in the system, the resulting alteration of the KCa3.1 channels-induced V_m oscillations significantly change the IP₃ concentrations needed to produce Ca²⁺ oscillations. In particular, increasing the activity of KCa3.1 channels shifts leftward the IP₃ range within which Ca²⁺ oscillations are produced. This implies that there are ranges of IP₃ levels where the presence or absence of Ca²⁺ oscillations are only dictated by the activity (conductance) of KCa3.1 channels (Figure 3B bottom, shaded areas).

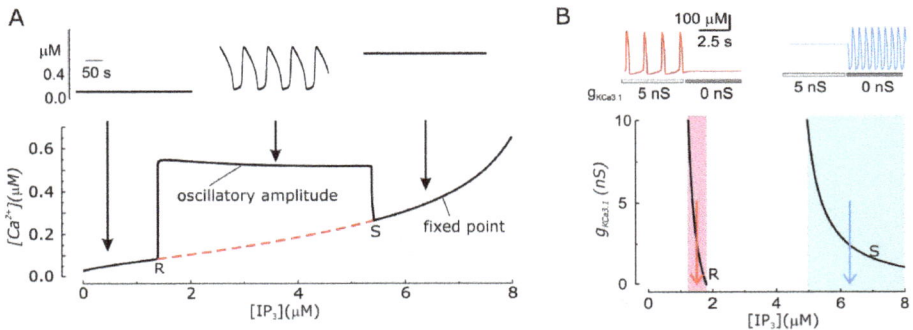

Figure 3. Ca²⁺ oscillation development depends on the IP₃ concentration and the activity of KCa3.1 channels. (**A**) Bottom. Bifurcation diagram showing that IP₃ concentration determines the establishment of the Ca²⁺ oscillations. The red dashed line represents the unstable equilibrium [Ca²⁺]ᵢ. The computation was performed with $g_{KCa3.1}$ = 5 nS. R and S indicate the Hopf bifurcations where Ca²⁺ oscillations appear and disappear, respectively, upon increasing the IP₃ levels. Top. The traces show the temporal changes of the [Ca²⁺]ᵢ for three different concentrations of IP₃ (indicated by arrows in the bottom part). (**B**) Bottom. Plot of the two Hopf bifurcations (R and S) as a function of the IP₃ concentration and $g_{KCa3.1}$. Three different ranges of IP₃ concentrations, where the KCa3.1-induced V_m oscillations appear to have different effects in the modulation of Ca²⁺ oscillations are evidenced (shaded areas; see text). More specifically in the pink region the presence of KCa3.1 channels is necessary for the existence of Ca²⁺ oscillations, in the light blue region KCa3.1 channels prevents Ca²⁺ oscillations, and finally in the white region in between KCa3.1 channels modulate the shape and duration of oscillation. Top. Simulated Ca²⁺ oscillations obtained with two IP₃ concentrations (1.5 left and 6.2 μM right, as indicated by the red and blue arrows below) within the regions where removing KCa3.1 channels make Ca²⁺ oscillations disappear or appear. (Modified from [48].)

These observations could explain a number of experimental results where KCa3.1 channel modulation was shown to have an effect on Ca²⁺ oscillations. In both activated T lymphocytes and C6 glioma cells, inhibition of V_m oscillations by blocking KCa3.1 channels is able to suppress Ca²⁺ oscillations [51,53]. Moreover *ras*-induced transformation of fibroblasts, which has been reported to upregulate KCa3.1 channels, has also been shown to trigger Ca²⁺ oscillations [54,55]. Finally, KCa3.1 inhibition leads to induction of Ca²⁺ oscillations in about half of the tested glioblastoma cells [52].

3.6. Glioblastoma KCa3.1 Channels Are Activated by Serum-Induced Ca²⁺ Oscillations and Participate to Cell Migration

Glioblastoma cells in vivo are exposed to a variety of tumor microenvironment components and soluble factors that can markedly affect their migratory ability. Among them there are unknown serum components that infiltrate into the tumor area of high-grade gliomas as result of the blood brain barrier breakdown [6,56]. As already shown, FCS enhances migration of U87-MG glioblastoma cells, and does so by inducing Ca^{2+} oscillations that are critically involved in the detachment of focal adhesions (by activating the focal adhesion kinase), and subsequent retraction of the cell rear [29,30]. On the same cell model we further reported that FCS, besides Ca^{2+} oscillations, induces an oscillatory activity of KCa3.1 channels, and this KCa3.1 channel activity is a necessary step to promote U87-MG cell migration by FCS [13]. In the same study, we also found a stable activation of Cl^- currents upon FCS stimulation. Altogether these observations sketch a coherent picture of the cell migratory process, in that, the retraction of the cell rear after the detachment of focal adhesions involves a major cell shape rearrangement and volume reduction. In these instances, the (oscillatory) KCa3.1 channel activity and the consequent cyclical K^+ efflux, in combination with Cl^- efflux (and essential osmotic water), form part of the whole machinery for achieving the cyclical volume reduction needed.

Cell locomotion is thought to be promoted by the cycling replication of four main steps: (i) protrusion of the cell front, associated to asymmetric polymerization of the actin-based cytoskeleton; (ii) adhesion of the protruded cell front to the substratum, mediated by the binding of focal adhesions to the extracellular polymers in contact with the cell; (iii) unbinding of the focal adhesions at the cell rear; (iv) retraction of the rear cell body in the direction of motion (cf. Figure 4). Two of these steps, namely cell front protrusion and cell rear retraction, involve major changes in cell volume and thus require compensatory ion fluxes (and osmotic water flow) across the membrane. Along the lines proposed by [12], during the protrusion of the cell front the maintenance of a stable osmolarity is sustained by cotransporters and ion exchangers—for instance, $Na^+/K^+/2Cl^-$ cotransport—followed by osmotically driven water. During this phase $[Ca^{2+}]_i$ is relatively low (red segment in the Ca^{2+} oscillation, Figure 4a), a condition in which KCa3.1 channels are closed and the resting membrane potential is near E_{Cl}. This condition prevents or strongly limits the loss of KCl via K and Cl channels that would nullify the osmotic work of the $Na^+/K^+/2Cl^-$ cotransport.

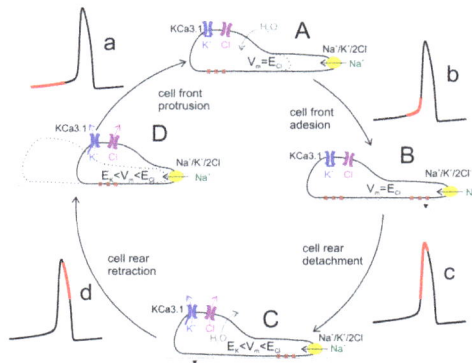

Figure 4. Schematic diagram of cell migratory cycle. The classic view of the cell migration process can be split down into four main cyclical steps. The cycle begins with the cell front protrusion, due to the activity of $Na^+/K^+/2Cl^-$ cotransport (yellow) (**A**) and establishment of adhesion structures (**B**). The elongated cell then removes/weakens the rear adhesions (**C**) so that ensuing contraction can pull the rear cell portion forward (**D**). The concomitant values of $[Ca^{2+}]_i$ is indicated by the red portion on the associated Ca^{2+} oscillation. V_m, ion and water fluxes are also illustrated (see text for details).

Int. J. Mol. Sci. **2018**, *19*, 2970

By contrast, the control of osmolarity during cell rear retraction critically depends on the activation of KCa3.1 channels and ensuing efflux of K$^+$ ions (accompanied by Cl$^-$ fluxes, to which membrane has become highly permeant, and osmotic water). Following the increase in [Ca^{2+}]$_i$ towards the peak of the Ca^{2+} oscillation (Figure 4c,d), KCa3.1 channels open and the membrane potential hyperpolarizes to a value between E$_{Cl}$ and E$_K$, a condition that would allow a significant efflux of KCl with consequent loss of water and cell volume decrease (Figure 4C,D). This would facilitate the process of cell body retraction. This view can explain the slowing of cell locomotion following the inhibition of K and Cl channels [13].

4. Conclusions

Substantial evidence suggests that the oscillatory activity of the KCa3.1 channel during Ca^{2+} oscillations may have both direct and indirect modulatory effects on GBM cell migration. As has long been recognized, the K$^+$ efflux through KCa3.1 channels directly participates in the cell volume changes during cell rear retraction. The same oscillatory K$^+$ efflux, however, will also determine the membrane potential oscillations that alter the driving force for Ca^{2+} influx and indirectly modulate the properties and even the presence of Ca^{2+} oscillations, and thus the timing of the migratory machinery. Both modulatory roles make KCa3.1 channels pivotal in GBM cell migration, and thus potential pharmacological targets for this deadly tumor. Although this review concentrates on the role of KCa3.1 and STIM/Orai channels in modulating Ca^{2+} oscillations and cell migration, it needs to be said that other channels including TRPM8 and KCa1.1, also abundantly expressed in glioblastoma cells, may also be important in these processes. Accordingly, TRPM8 [41] and KCa1.1 channel signaling [57] have been demonstrated to be required for basal and radiation-induced migration, respectively, and certainly contribute to Ca^{2+} oscillations.

Author Contributions: Both L.C. and F.F. performed literature search and wrote the paper.

Funding: This research received no external funding.

Conflicts of Interest: The authors declare no conflict of interest.

References

1. Hanahan, D.; Weinberg, R.A. Hallmarks of cancer: The next generation. *Cell* **2011**, *144*, 646–674. [CrossRef] [PubMed]
2. Ridley, A.J.; Schwartz, M.A.; Burridge, K.; Firtel, R.A.; Ginsberg, M.H.; Borisy, G.; Parsons, J.T.; Horwitz, A.R. Cell migration: Integrating signals from front to back. *Science* **2003**, *302*, 1704–1709. [CrossRef] [PubMed]
3. Ridley, A.J. Life at the leading edge. *Cell* **2011**, *145*, 1012–1022. [CrossRef] [PubMed]
4. Carragher, N.O.; Frame, M.C. Focal adhesion and actin dynamics: A place where kinases and proteases meet to promote invasion. *Trends Cell Biol.* **2004**, *14*, 241–249. [CrossRef] [PubMed]
5. Parsons, J.T.; Horwitz, A.R.; Schwartz, M.A. Cell adhesion: Integrating cytoskeletal dynamics and cellular tension. *Nat. Rev. Mol. Cell Biol.* **2010**, *11*, 633–643. [CrossRef] [PubMed]
6. Lund, C.V.; Nguyen, M.T.; Owens, G.C.; Pakchoian, A.J.; Shaterian, A.; Kruse, C.A.; Eliceiri, B.P. Reduced glioma infiltration in Src-deficient mice. *J. Neurooncol.* **2006**, *78*, 19–29. [CrossRef] [PubMed]
7. Wessels, D.; Soll, D.R.; Knecht, D.; Loomis, W.F.; De Lozanne, A.; Spudich, J. Cell motility and chemotaxis in Dictyostelium amebae lacking myosin heavy chain. *Dev. Biol.* **1988**, *128*, 164–177. [CrossRef]
8. Lombardi, M.L.; Knecht, D.A.; Dembo, M.; Lee, J. Traction force microscopy in Dictyostelium reveals distinct roles for myosin II motor and actin-crosslinking activity in polarized cell movement. *J. Cell Sci.* **2007**, *120 Pt 9*, 1624–1634. [CrossRef]
9. Roca-Cusachs, P.; Sunyer, R.; Trepat, X. Mechanical guidance of cell migration: Lessons from chemotaxis. *Curr. Opin. Cell Biol.* **2013**, *25*, 543–549. [CrossRef] [PubMed]
10. Ostman, A.; Heldin, C.H. PDGF receptors as targets in tumor treatment. *Adv. Cancer Res.* **2007**, *97*, 247–274. [PubMed]

11. Lo, C.M.; Wang, H.B.; Dembo, M.; Wang, Y.L. Cell movement is guided by the rigidity of the substrate. *Biophys. J.* **2000**, *79*, 144–152. [CrossRef]

12. Schwab, A.; Fabian, A.; Hanley, P.J.; Stock, C. Role of ion channels and transporters in cell migration. *Physiol. Rev.* **2012**, *92*, 1865–1913. [CrossRef] [PubMed]

13. Catacuzzeno, L.; Aiello, F.; Fioretti, B.; Sforna, L.; Castigli, E.; Ruggieri, P.; Tata, A.M.; Calogero, A.; Franciolini, F. Serum-activated K and Cl currents underlay U87-MG glioblastoma cell migration. *J. Cell Physiol.* **2011**, *226*, 1926–1933. [CrossRef] [PubMed]

14. Cuddapah, V.A.; Habela, C.W.; Watkins, S.; Moore, L.S.; Barclay, T.T.; Sontheimer, H. Kinase activation of ClC-3 accelerates cytoplasmic condensation during mitotic cell rounding. *Am. J. Physiol. Cell Physiol.* **2012**, *302*, C527–C538. [CrossRef] [PubMed]

15. Wei, C.; Wang, X.; Zheng, M.; Cheng, H. Calcium gradients underlying cell migration. *Curr. Opin. Cell Biol.* **2012**, *24*, 254–261. [CrossRef] [PubMed]

16. Cuthbertson, K.S.; Chay, T.R. Modelling receptor-controlled intracellular calcium oscillators. *Cell Calcium* **1991**, *12*, 97–109. [CrossRef]

17. Gaspers, L.D.; Bartlett, P.J.; Politi, A.; Burnett, P.; Metzger, W.; Johnston, J.; Joseph, S.K.; Höfer, T.; Thomas, A.P. Hormone-induced calcium oscillations depend on cross-coupling with inositol 1,4,5-trisphosphate oscillations. *Cell Rep.* **2014**, *9*, 1209–1218. [CrossRef] [PubMed]

18. Dupont, G.; Combettes, L.; Bird, G.S.; Putney, J.W. Calcium oscillations. *Cold Spring Harb. Perspect. Biol.* **2011**, *3*, a004226. [CrossRef] [PubMed]

19. Iino, M. Biphasic Ca^{2+} dependence of inositol 1,4,5-trisphosphate-induced Ca release in smooth muscle cells of the guinea pig taenia caeci. *J. Gen. Physiol.* **1990**, *95*, 1103–1122. [CrossRef] [PubMed]

20. Bezprozvanny, I.; Watras, J.; Ehrlich, B.E. Bell-shaped calcium-response curves of Ins(1,4,5)P3- and calcium-gated channels from endoplasmic reticulum of cerebellum. *Nature* **1991**, *351*, 751–754. [CrossRef] [PubMed]

21. Finch, E.A.; Turner, T.J.; Goldin, S.M. Calcium as a coagonist of inositol 1,4,5-trisphosphate-induced calcium release. *Science* **1991**, *252*, 443–446. [CrossRef] [PubMed]

22. Kaznacheyeva, E.; Lupu, V.D.; Bezprozvanny, I. Single-channel properties of inositol (1,4,5)-trisphosphate receptor heterologously expressed in HEK-293 cells. *J. Gen. Physiol.* **1998**, *111*, 847–856. [CrossRef] [PubMed]

23. Liou, J.; Kim, M.L.; Heo, W.D.; Jones, J.T.; Myers, J.W.; Ferrell, J.E., Jr.; Meyer, T. STIM is a Ca^{2+} sensor essential for Ca^{2+}-store-depletion-triggered Ca^{2+} influx. *Curr. Biol.* **2005**, *15*, 1235–1241. [CrossRef] [PubMed]

24. Roos, J.; DiGregorio, P.J.; Yeromin, A.V.; Ohlsen, K.; Lioudyno, M.; Zhang, S.; Safrina, O.; Kozak, J.A.; Wagner, S.L.; Cahalan, M.D.; et al. STIM1, an essential and conserved component of store-operated Ca^{2+} channel function. *J. Cell Biol.* **2005**, *169*, 435–445. [CrossRef] [PubMed]

25. Peinelt, C.; Vig, M.; Koomoa, D.L.; Beck, A.; Nadler, M.J.; Koblan-Huberson, M.; Lis, A.; Fleig, A.; Penner, R.; Kinet, J.P. Amplification of CRAC current by STIM1 and CRACM1 (Orai1). *Nat. Cell Biol.* **2006**, *8*, 771–773. [CrossRef] [PubMed]

26. Moccia, F.; Ruffinatti, F.A.; Zuccolo, E. Intracellular Ca^{2+} Signals to Reconstruct a Broken Heart: Still a Theoretical Approach? *Curr. Drug Targets* **2015**, *6*, 793–815. [CrossRef]

27. Dubois, C.; Vanden Abeele, F.; Lehen'kyi, V.; Gkika, D.; Guarmit, B.; Lepage, G.; Slomianny, C.; Borowiec, A.S.; Bidaux, G.; Benahmed, M.; et al. Remodeling of channel-forming ORAI proteins determines an oncogenic switch in prostate cancer. *Cancer Cell* **2014**, *26*, 19–32. [CrossRef] [PubMed]

28. Duffy, S.M.; Ashmole, I.; Smallwood, D.T.; Leyland, M.L.; Bradding, P. Orai/CRACM1 and KCa3.1 ion channels interact in the human lung mast cell plasma membrane. *Cell Commun. Signal.* **2015**, *13*, 32. [CrossRef] [PubMed]

29. Rondé, P.; Giannone, G.; Gerasymova, I.; Stoeckel, H.; Takeda, K.; Haiech, J. Mechanism of calcium oscillations in migrating human astrocytoma cells. *Biochim. Biophys. Acta* **2000**, *1498*, 273–280. [CrossRef]

30. Giannone, G.; Rondé, P.; Gaire, M.; Haiech, J.; Takeda, K. Calcium oscillations trigger focal adhesion disassembly in human U87 astrocytoma cells. *J. Biol. Chem.* **2002**, *277*, 26364–26371. [CrossRef] [PubMed]

31. Montana, V.; Sontheimer, H. Bradykinin promotes the chemotactic invasion of primary brain tumors. *J. Neurosci.* **2011**, *31*, 4858–4867. [CrossRef] [PubMed]

32. Totsukawa, G.; Wu, Y.; Sasaki, Y.; Hartshorne, D.J.; Yamakita, Y.; Yamashiro, S.; Matsumura, F. Distinct roles of MLCK and ROCK in the regulation of membrane protrusions and focal adhesion dynamics during cell migration of fibroblasts. *J. Cell Biol.* **2004**, *164*, 427–439. [CrossRef] [PubMed]

33. Chen, C.; Tao, T.; Wen, C.; He, W.Q.; Qiao, Y.N.; Gao, Y.Q.; Chen, X.; Wang, P.; Chen, C.P.; Zhao, W.; et al. Myosin light chain kinase (MLCK) regulates cell migration in a myosin regulatory light chain phosphorylation-independent mechanism. *J. Biol. Chem.* **2014**, *289*, 28478–28488. [CrossRef] [PubMed]

34. Sforna, L.; Megaro, A.; Pessia, M.; Franciolini, F.; Catacuzzeno, L. Structure, Gating and Basic Functions of the Ca^{2+}-activated K Channel of Intermediate Conductance. *Curr. Neuropharmacol.* **2018**, *16*, 608–617. [CrossRef] [PubMed]

35. Catacuzzeno, L.; Franciolini, F. Editorial: The Role of Ca^{2+}-activated K^{+} Channels of Intermediate Conductance in Glioblastoma Malignancy. *Curr. Neuropharmacol.* **2018**, *16*, 607. [CrossRef] [PubMed]

36. Vicente-Manzanares, M.; Koach, M.A.; Whitmore, L.; Lamers, M.L.; Horwitz, A.F. Segregation and activation of myosin IIB creates a rear in migrating cells. *J. Cell Biol.* **2008**, *183*, 543–554. [CrossRef] [PubMed]

37. Kaushal, V.; Koeberle, P.D.; Wang, Y.; Schlichter, L.C. The Ca^{2+}-activated K^{+} channel KCNN4/KCa3.1 contributes to microglia activation and nitric oxide-dependent neurodegeneration. *J. Neurosci.* **2007**, *27*, 234–244. [CrossRef] [PubMed]

38. Ruggieri, P.; Mangino, G.; Fioretti, B.; Catacuzzeno, L.; Puca, R.; Ponti, D.; Miscusi, M.; Franciolini, F.; Ragona, G.; Calogero, A. The inhibition of KCa3.1 channels activity reduces cell motility in glioblastoma derived cancer stem cells. *PLoS ONE* **2012**, *7*, e47825. [CrossRef] [PubMed]

39. D'Alessandro, G.; Catalano, M.; Sciaccaluga, M.; Chece, G.; Cipriani, R.; Rosito, M.; Grimaldi, A.; Lauro, C.; Cantore, G.; Santoro, A.; et al. KCa3.1 channels are involved in the infiltrative behavior of glioblastoma in vivo. *Cell Death Dis.* **2013**, *4*, e773. [CrossRef] [PubMed]

40. Sciaccaluga, M.; Fioretti, B.; Catacuzzeno, L.; Pagani, F.; Bertollini, C.; Rosito, M.; Catalano, M.; D'Alessandro, G.; Santoro, A.; Cantore, G.; et al. CXCL12-induced glioblastoma cell migration requires intermediate conductance Ca^{2+}-activated K^{+} channel activity. *Am. J. Physiol. Cell Physiol.* **2010**, *299*, C175–C184. [CrossRef] [PubMed]

41. Klumpp, L.; Sezgin, E.C.; Skardelly, M.; Eckert, F.; Huber, S.M. KCa3.1 Channels and Glioblastoma: In Vitro Studies. *Curr. Neuropharmacol.* **2018**, *16*, 627–635. [CrossRef] [PubMed]

42. Turner, K.L.; Honasoge, A.; Robert, S.M.; McFerrin, M.M.; Sontheimer, H. A proinvasive role for the Ca(2+)-activated K(+) channel KCa3.1 in malignant glioma. *Glia* **2014**, *62*, 971–981. [CrossRef] [PubMed]

43. Gao, Y.D.; Hanley, P.J.; Rinné, S.; Zuzarte, M.; Daut, J. Calcium-activated K^{+} channel (K$_{Ca}$3.1) activity during Ca^{2+} store depletion and store-operated Ca^{2+} entry in human macrophages. *Cell Calcium* **2010**, *48*, 19–27. [CrossRef] [PubMed]

44. Fioretti, B.; Catacuzzeno, L.; Sforna, L.; Aiello, F.; Pagani, F.; Ragozzino, D.; Castigli, E.; Franciolini, F. Histamine hyperpolarizes human glioblastoma cells by activating the intermediate-conductance Ca^{2+}-activated K^{+} channel. *Am. J. Physiol. Cell Physiol.* **2009**, *297*, C102–C110. [CrossRef] [PubMed]

45. Ghanshani, S.; Wulff, H.; Miller, M.J.; Rohm, H.; Neben, A.; Gutman, G.A.; Cahalan, M.D.; Chandy, K.G. Up-regulation of the IKCa1 potassium channel during T-cell activation. Molecular mechanism and functional consequences. *J. Biol. Chem.* **2000**, *275*, 37137–37149. [CrossRef] [PubMed]

46. Wulff, H.; Beeton, C.; Chandy, K.G. Potassium channels as therapeutic targets for autoimmune disorders. *Curr. Opin. Drug Discov. Dev.* **2003**, *6*, 640–647.

47. Di, L.; Srivastava, S.; Zhdanova, O.; Ding, Y.; Li, Z.; Wulff, H.; Lafaille, M.; Skolnik, E.Y. Inhibition of the K^{+} channel KCa3.1 ameliorates T cell-mediated colitis. *Proc. Natl. Acad. Sci. USA* **2010**, *107*, 1541–1546. [CrossRef] [PubMed]

48. Duffy, M.S.; Berger, P.; Cruse, G.; Yang, W.; Bolton, S.J.; Bradding, P. The K^{+} channel iKCA1 potentiates Ca^{2+} influx and degranulation in human lung mast cells. *J. Allergy Clin. Immunol.* **2004**, *114*, 66–72. [CrossRef] [PubMed]

49. Ferreira, R.; Schlichter, L.C. Selective activation of KCa3.1 and CRAC channels by P2Y2 receptors promotes Ca(2+) signaling, store refilling and migration of rat microglial cells. *PLoS ONE* **2013**, *8*, e62345. [CrossRef] [PubMed]

50. Catacuzzeno, L.; Fioretti, B.; Franciolini, F. A theoretical study on the role of Ca^{2+}-activated K^{+} channels in the regulation of hormone-induced Ca^{2+} oscillations and their synchronization in adjacent cells. *J. Theor. Biol.* **2012**, *309*, 103–112. [CrossRef] [PubMed]

51. Reetz, G.; Reiser, G. [Ca2+]i oscillations induced by bradykinin in rat glioma cells associated with Ca^{2+} store-dependent Ca^{2+} influx are controlled by cell volume and by membrane potential. *Cell Calcium* **1996**, *19*, 143–156. [CrossRef]

52. Stegen, B.; Klumpp, L.; Misovic, M.; Edalat, L.; Eckert, M.; Klumpp, D.; Ruth, P.; Huber, S.M. K$^+$ channel signaling in irradiated tumor cells. *Eur. Biophys. J.* **2016**, *45*, 585–598. [CrossRef] [PubMed]

53. Verheugen, J.A.; Vijverberg, H.P. Intracellular Ca^{2+} oscillations and membrane potential fluctuations in intact human T lymphocytes: Role of K$^+$ channels in Ca^{2+} signaling. *Cell Calcium* **1995**, *17*, 287–300. [CrossRef]

54. Hashii, M.; Nozawa, Y.; Higashida, H. Bradykinin-induced cytosolic Ca^{2+} oscillations and inositol tetrakisphosphate-induced Ca^{2+} influx in voltage-clamped ras-transformed NIH/3T3 fibroblasts. *J. Biol. Chem.* **1993**, *268*, 19403–19410. [PubMed]

55. Rane, S.G. A Ca2(+)-activated K$^+$ current in ras-transformed fibroblasts is absent from nontransformed cells. *Am. J. Physiol.* **1991**, *260*, C104–C112. [CrossRef] [PubMed]

56. Seitz, R.J.; Wechsler, W. Immunohistochemical demonstration of serum proteins in human cerebral gliomas. *Acta Neuropathol.* **1987**, *73*, 145–152. [CrossRef] [PubMed]

57. Steinle, M.; Palme, D.; Misovic, M.; Rudner, J.; Dittmann, K.; Lukowski, R.; Ruth, P.; Huber, S.M. Ionizing radiation induces migration of glioblastoma cells by activating BK K$^+$ channels. *Radiother. Oncol.* **2011**, *101*, 122–126. [CrossRef] [PubMed]

International Journal of
Molecular Sciences

MDPI

Review

Endothelial Ca^{2+} Signaling and the Resistance to Anticancer Treatments: Partners in Crime

Francesco Moccia

Laboratory of General Physiology, Department of Biology and Biotechnology "L. Spallanzani",
University of Pavia, I-27100 Pavia, Italy; francesco.moccia@unipv.it; Tel.: +39-382-39-0382-987527

Received: 6 December 2017; Accepted: 10 January 2018; Published: 11 January 2018

Abstract: Intracellular Ca^{2+} signaling drives angiogenesis and vasculogenesis by stimulating proliferation, migration, and tube formation in both vascular endothelial cells and endothelial colony forming cells (ECFCs), which represent the only endothelial precursor truly belonging to the endothelial phenotype. In addition, local Ca^{2+} signals at the endoplasmic reticulum (ER)–mitochondria interface regulate endothelial cell fate by stimulating survival or apoptosis depending on the extent of the mitochondrial Ca^{2+} increase. The present article aims at describing how remodeling of the endothelial Ca^{2+} toolkit contributes to establish intrinsic or acquired resistance to standard anti-cancer therapies. The endothelial Ca^{2+} toolkit undergoes a major alteration in tumor endothelial cells and tumor-associated ECFCs. These include changes in TRPV4 expression and increase in the expression of P2X7 receptors, Piezo2, Stim1, Orai1, TRPC1, TRPC5, Connexin 40 and dysregulation of the ER Ca^{2+} handling machinery. Additionally, remodeling of the endothelial Ca^{2+} toolkit could involve nicotinic acetylcholine receptors, gasotransmitters-gated channels, two-pore channels and Na$^+$/H$^+$ exchanger. Targeting the endothelial Ca^{2+} toolkit could represent an alternative adjuvant therapy to circumvent patients' resistance to current anti-cancer treatments.

Keywords: Ca^{2+} signaling; tumor; endothelial cells; endothelial progenitor cells; endothelial colony forming cells; anticancer therapies; VEGF; resistance to apoptosis

---

1. Introduction

An increase in intracellular Ca^{2+} concentration ([Ca^{2+}]$_i$) has long been known to play a crucial role in angiogenesis and arterial remodeling [1–5]. Accordingly, growth factors and cytokines, such as vascular endothelial growth factor (VEGF), epidermal growth factor (EGF), basic fibroblast growth factor (bFGF), insulin-like growth factor-1 (IGF-1), angiopoietin and stromal derived factor-1α (SDF-1α), trigger robust Ca^{2+} signals in vascular endothelial cells [6–12], which recruit a number of downstream Ca^{2+}-dependent pro-angiogenic decoders. These include, but are not limited to, the transcription factors, Nuclear factor of activated T-cells (NFAT), Nuclear factor-kappaB (NF-κB) and cAMP responsive element binding protein (CREB) [8,13,14], myosin light chain kinase (MLCK) and myosin 2 [8,15], endothelial nitric oxide synthase (eNOS) [16,17], extracellular signal–regulated kinases $\frac{1}{2}$ (ERK 1/2) [18,19] and Akt [19,20]. Not surprisingly, therefore, subsequent studies clearly revealed that endothelial Ca^{2+} signals may also drive tumor angiogenesis, growth and metastasis [3,21–24]. However, the process of tumor vascularization is far more complex than originally envisaged [25]. Accordingly, the angiogenic switch, which is the initial step in the multistep process that ensures cancer cells with an adequate supply of oxygen and nutrients and provides them with an escape route to enter peripheral circulation, is triggered by the recruitment of bone marrow-derived endothelial progenitor cells (EPCs), according to a process termed vasculogenesis [26–28]. Similar to mature endothelial cells, EPCs require an increase in [Ca^{2+}]$_i$ to proliferate, assembly into capillary-like tubular networks in vitro and form patent neovessels in vivo [29–31]. Of note, intracellular Ca^{2+} signals finely regulate proliferation and in vitro tubulogenesis also in tumor-derived EPCs (T-EPCs) [23,32,33]. An established tenet of neoplastic transformation is the remodeling of the Ca^{2+} machinery in malignant cells, which contributes to the distinct hallmarks of cancer described by Hanahan and Weinberg [34–36]. Tumor endothelial cells (T-ECs) and T-EPCs do not derive from the malignant clone, but they display a dramatic dysregulation of their Ca^{2+} signaling toolkit [29,32,37]. The present article surveys the most recent updates on the remodeling of endothelial Ca^{2+} signals during tumor vascularization. In particular, it has been outlined which Ca^{2+}-permeable channels and Ca^{2+}-transporting systems are up- or down-regulated in T-ECs and T-EPCs and how they impact on neovessel formation and/or apoptosis resistance in the presence of anti-cancer drugs. Finally, the hypothesis that the remodeling of endothelial Ca^{2+} signals may be deeply involved in tumor resistance to standard therapeutic treatments, including chemotherapy, radiotherapy and anti-angiogenic therapy is widely discussed.

2. Ca^{2+} Signaling in Normal Endothelial Cells: A Brief Introduction

The resting $[Ca^{2+}]_i$ in vascular endothelial cells is set at around 100–200 nM by the concerted interaction of three Ca^{2+}-transporting systems, which extrude Ca^{2+} across the plasma membrane, such as the Plasma-Membrane Ca^{2+}-ATPase and the Na^+/Ca^{2+} exchanger (NCX), or sequester cytosolic Ca^{2+} into the endoplasmic reticulum (ER), the largest intracellular Ca^{2+} reservoir [2,38–40], such as the SarcoEndoplasmic Reticulum Ca^{2+}-ATPase (SERCA). Endothelial cells lie at the interface between the vascular wall and the underlying tissue; therefore, they are continuously exposed to a myriad of low levels soluble factors, including growth factors, hormones and transmitters, which may induce highly localized events of inositol-1,4,5-trisphosphate (InsP$_3$)-dependent Ca^{2+} release from the ER even in the absence of global cytosolic elevations in $[Ca^{2+}]_i$ [41–45]. These spontaneous InsP$_3$-dependent Ca^{2+} microdomains are redirected towards the mitochondrial matrix through the direct physical association specific components of the outer mitochondrial membrane (OMM) with specialized ER regions, which are known as mitochondrial-associated membranes (MAMs) [46]. This constitutive ER-to-mitochondria Ca^{2+} shuttle drives cellular bioenergetics by activating intramitochondrial Ca^{2+}-dependent dehydrogenases, such as pyruvate dehydrogenase, NAD-isocitrate dehydrogenase and oxoglutarate dehydrogenase [47–49]. This pro-survival Ca^{2+} transfer may be switched into a pro-death Ca^{2+} signal by various apoptotic stimuli [46,47,50]. For instance, hydrogen peroxide (H_2O_2), menadione, resveratrol, ceramide, and etoposide boost the InsP$_3$-dependent ER-to-mitochondria Ca^{2+} communication, thereby causing a massive increase in mitochondrial Ca^{2+} concentration ($[Ca^{2+}]_{mit}$), which ultimately results in the opening of mitochondrial permeability transition pore and in the release of pro-apoptotic factors into the cytosol [46,51–53]. The hypoxic microenvironment of a growing tumor may then trigger an oxygen (O_2)-sensitive transcriptional program in tumor cells by activating two basic helix-loop-helix transcription factors, i.e., the hypoxia-inducible factors HIF-1 and HIF-2, which drive the expression of a myriad of growth factors and cytokines [54]. These include, but are not limited to, VEGF, EGF, bFGF, IGF-1, angiopoietin and SDF-1α [27,54], which are liberated into peripheral circulation according to a concentration gradient, which delivers a strong pro-angiogenic signal to vascular endothelial cells residing in close proximity to the primary tumor site [3,33]. Growth factors bind to their specific tyrosine kinase receptors (TKRs), such as VEGFR-2 (KDR/Flk-1), EGFR (ErbB-1), and IGF-1R, thereby stimulating phospholipase Cγ (PLCγ) to cleave phosphatidylinositol 4,5-bisphosphate (PIP$_2$) into the two intracellular second messengers, InsP$_3$ and diacylglycerol (DAG) [1,2,55]. The following increase in cytosolic InsP$_3$ levels further stimulates ER-dependent Ca^{2+} release through InsP$_3$ receptors (InsP$_3$Rs), which can be amplified by the recruitment of adjoining ryanodine receptors (RyRs) through the process of Ca^{2+}-induced Ca^{2+} release (CICR) [1,2]. The following drop in ER Ca^{2+} concentration ($[Ca^{2+}]_{ER}$) is detected by Stromal Interaction Molecule 1 (Stim1), a sensor of ER Ca^{2+} levels, which is prompted to aggregate into oligomers and relocate towards ER-plasma membrane junctions, known as puncta and positioned in close vicinity to the plasma membrane (10–20 nm). Herein, Stim1 interacts with and gates the Ca^{2+}-permeable channel, Orai1, thereby triggering the so-called store-operated Ca^{2+} entry (SOCE), the most important Ca^{2+} entry route in endothelial cells [42,56–59]. In addition, Stim1 may recruit additional Ca^{2+}-permeable channels, which belong to the Canonical Transient Receptor Potential (TRPC) sub-family [2,59]. Accordingly, the TRP superfamily of cation channels comprises 28 members, subdivided into six sub-families: TRPC, TRPV (Vanilloid), TRPM (Melastatin), TRPP (Polycystin), TRPML (Mucolipin) and TRPA (Ankyrin) based on the homology of their amino acid sequences [60]. More specifically, endothelial SOCE could involve TRPC1 and TRPC4, which are recruited by Stim1 into a supermolecular heteromeric complex [61], whose Ca^{2+} selectivity is determined by Orai1 [62,63]. Moreover, TRPC3 and TRPC6 may mediate DAG-induced Ca^{2+} entry in several types of endothelial cells [64,65]. This toolkit of Ca^{2+} release/entry channels may be differently exploited by growth factors to stimulate angiogenesis by eliciting diverse patterns of Ca^{2+} signals depending on the vascular bed. For instance, VEGF triggers a biphasic increase in $[Ca^{2+}]_i$ in human umbilical vein endothelial cells (HUVECs), which consists in an initial InsP$_3$-dependent Ca^{2+} peak followed by a plateau phase of intermediate amplitude due to SOCE activation [57,58]. Likewise, InsP$_3$ and SOCE shape VEGF- and EGF-induced intracellular Ca^{2+} oscillations in sheep uterine artery endothelial cells [66] and in rat microvascular endothelial cells (CMECs) [7], respectively. VEGF-induced Ca^{2+} influx in HUVECs may, however, be sustained by TRPC3, which causes Na$^+$ accumulation beneath the plasma membrane and stimulates the forward (i.e., Ca^{2+} entry) mode of NCX [18]. Moreover, the DAG-gated channel, TRPC6, underlies the monotonic increase in $[Ca^{2+}]_i$ induced by VEGF in human dermal microvascular endothelial cells (HDMECs) [67]. Finally, TRPC1 is engaged by bFGF to mediate Ca^{2+} entry in HDMECs [68]. These data have been recently confirmed by directly monitoring angiogenesis in developing zebrafish; this model showed that VEGF stimulated biphasic Ca^{2+} signals to drive migration in stalk cells and intracellular Ca^{2+} oscillations to promote proliferation in tip cells [8]. Besides growth factors-activated channels, vascular endothelial cells dispose of a larger toolkit of plasmalemmal Ca^{2+}-permeable channels that can be recruited by a multitude of chemical and physical stimuli [2]. For instance, endothelial Ca^{2+} entry may be mediated by additional intracellular second messengers, such as arachidonic acid (AA) and AA metabolites, i.e., epoxyeicosatrienoic acids (EETs) and 2-arachidonoylglycerol, which activate TRPV4 [69,70]; NO, which gates TRPC5 [71]; adenosine 5'-diphosphoribose (ADPR) and low micromolorar doses of H_2O_2, which converge on TRPM2 activation [72]; and cyclic nucleotides [73]. Moreover, vascular endothelial cells are endowed with several Ca^{2+}-permeable ionotropic receptors, including ATP-sensitive P$_{2X}$ receptors [74], acetylcholine-sensitive nicotinic receptors [75], and *N*-methyl-D-aspartate (NMDA) receptors [76]. Finally, mechanical stimuli (e.g., laminar shear stress, pulsatile stretch, and changes in the local osmotic pressure) elicit Ca^{2+} influx by recruiting a variety of mechano-sensitive channels, such as TRPP2 [77], heteromeric TRPC1-TRPP2 [78], TRPV4 [79], TRPC1-TRPP2-TRPV4 [80], and Piezo1 [81]. Recently, the Ca^{2+} toolkit has also been explored in human EPCs [29]; most of the work has been carried out in endothelial colony forming cells (ECFCs), which represent the only EPC subset truly belonging to the endothelial, rather than the myeloid, lineage [82]. VEGF triggers pro-angiogenic intracellular Ca^{2+} oscillations in ECFCs by triggering the interaction between InsP$_3$-dependent Ca^{2+} release and SOCE, which is mediated by Stim1, Orai1 and TRPC1 [83,84]. Conversely, RyRs and the DAG-sensitive channels, TRPC3 and TRPC6, are absent and do not contribute to Ca^{2+} signaling [84–86]. Of note, AA may promote proliferation by directly activating TRPV4 and inducing NO release in the presence of extracellular growth factors and cytokines [86]. Finally, the InsP$_3$-dependent ER-to-mitochondria Ca^{2+} shuttle is at work and finely regulates the sensitivity to apoptotic stimuli in ECFCs, too [87].

Herein, the mechanisms whereby the remodeling of the endothelial transportome, i.e., the specific arsenal of ion channels and transporters expressed by vascular endothelial cells and ECFCs, confers resistance to anti-cancer therapies have been subdivided into two main categories: (1) enhanced neovascularization, which attenuates the therapeutic outcome of anticancer treatments by nourishing cancer cells with O_2 and nutrients and removing their catabolic waste, and further provides them with a direct access to peripheral circulation, thereby favoring metastasis (Figure 1 and Table 1); and (2) resistance to apoptosis, which hampers the cellular stress induced by chemo- and radiotherapy on tumor endothelial cells and interferes with the dismantling of cancer vasculature (Figure 2 and Table 2).

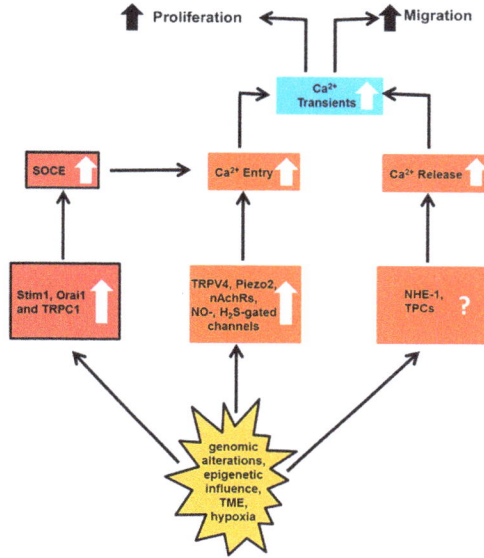

Figure 1. The endothelial Ca^{2+} transportome is remodeled to sustain tumor vascularization. The equence of events is illustrated by the black arrows. Upward arrows indicate the over-expression of a specific Ca^{2+}-permeable channel or transporter and the stimulation of a precise cellular process. See the text for further details.

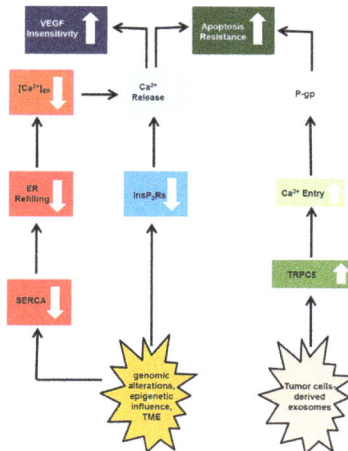

Figure 2. The endothelial Ca^{2+} transportome is remodeled to promote tumor endothelial cell resistance to apoptosis. The sequence of events is illustrated by the black arrows. Downward arrows indicate the down-regulation of a specific Ca^{2+}-permeable channel/transporter or of a precise cellular process. Upward arrows indicate the over-expression of a specific Ca^{2+}-permeable channel or the stimulation of a precise cellular process.

Table 1. Channels and transporters directly supporting tumor vascularization.

Channel/Transporter	Tumor and Cell Type (T-EC, T-ECFC T-EPC)	Expression Levels (Transcripts and/or Proteins)	Effect on Tumor Vascularization	Strategy to Target Tumor Vascularization	Ref.
TRPV4	Breast Cancer: T-ECs	↑	Stimulates B-TEC proliferation, migration and in vitro tubulogenesis	Channel blockade with shTRPV4 or with CAI (0.1–10 μM)	[88,89]
TRPV4	Lewis Lung Carcinoma: T-ECs (isolated from prostate adenocarcinoma)	↓	Inhibits T-EC mechanosensation, proliferation and migration in vitro and promotes the formation of a malfunctioning, leaky and exceedingly expanded vascular network in vivo	Injection of TRPV4 agonist GSK (10 μg/kg) to normalize tumor vasculature and favor cisplatin-induced tumor regression	[90–92]
Piezo2 proteins	Glioma: T-ECs	↑	Regulates tumor angiogenesis, vascular leakage and permeability	Blockade with siPiezo2	[93]
P2X7Rs	Breast cancer: T-ECs	↑	Inhibits B-TEC migration and normalizes B-TECs-derived vessels in vitro	Activated by BzATP (50 μM)	[94]
Stim1, Orai1, TRPC1	Renal cellular carcinoma: T-ECFCs	↑	Stimulate T-EPC proliferation and in vitro tubulogenesis	Blockade with siStim1 and siOrai1 and with YM-58483/BTP2 (20 μM), La^{3+} (10 μM), Gd^{3+} (10 μM), CAI (2–10 μM), 2-APB (50 μM), and genistein (50 μM)	[95]
Stiu1, Orai1, TRPC1	Breast cancer: T-ECFCs	=	Control T-ECFC proliferation and in vitro tubulogenesis	Blockade with YM-58483/BTP2 (20 μM), La^{3+} (10 μM), and CAI (10 μM)	[96]
Stim1, Orai1, TRPC1	Infantile hemangioma: T-ECFCs	↑	Control T-ECFCs proliferation in vitro	Blockade with with YM-58483/BTP2 (20 μM), La^{3+} (10 μM), and Pyr6 (10 μM)	[97]
α7-nAchRs	Lewis lung carcinoma: T-ECs and T-EPCs	Not determined	Controls tumor growth and angiogenesis in vivo	Blockade with mecamylamine (1.0 μg/kg) or hexamethonium (1.0 μg/kg)	[98,99]
			Stimulates EPC proliferation, migration and tubulogenesis in vitro and EPC recruitment in vivo	Blockade in vitro with mecamylamine (1 μM) and α-bungarotoxin (10 nM) and in vivo with mecamylamine (0.24 mg/kg per day)	[100,101]
Connexin40	Melanoma and urogenital cancers: T-EC	↑	Stimulates tumor angiogenesis and growth in vivo	Blockade in vivo with ^{40}Gap^{27} peptide (100 μg)	[102]
NHE-1	Breast cancer: TECs	Not determined	Stimulates B-TEC migration in vitro	Blocked with siNHE-1 and with cariporide (50 μM)	[103]

The generic term EPC, in this context, refers to circulating pro-angiogenic cells which cannot be grouped into the ECRC sub-family and are likely to belong to the myeloid lineage.

Table 2. Components of the endothelial Ca^{2+} toolkit that determine endothelial cell resistance to chemotherapeutic drugs.

Channel/Transporter	Tumor and Cell Type (T-EC and T-EPC)	Expression Levels	Effect on Tumor Vascularization	Strategy to Target Tumor Vascularization	Ref.
TRPC5	Breast Cancer: T-ECs	↑	Stimulates endothelial resistance to adriamycin	Channel blockade with the specific blocking antibody T5E3 (concentration not reported)	[104]
InsP$_3$Rs	RCC: T-ECFCs	↓	Favor T-ECFC resistance to rapamycin	Preventing InsP$_3$-dependent ER-mitochondria Ca^{2+} shuttle with selective InsP$_3$R inhibitors or cytosolic Ca^{2+} buffers (e.g., BAPTA)	[87]

3. Enhanced Neovascularization

3.1. Vanilloid Transient Receptor Potential 4 (TRPV4)

TRPV4 has been the first endothelial Ca^{2+}-permeable channel to be clearly involved in tumor vascularization [88]. TRPV4 is gated by an array of chemical and physical cues and represents, therefore, the archetypal of polymodal TRP channels [60]. For instance, TRPV4 may be activated by physiological stimuli, including AA and its cytochrome P450-derived metabolites mediators, i.e., EETs, acidic pH, hypotonic swelling, mechanical deformation, heat (>17–24 °C), and dimethylallyl pyrophosphate (DMAPP) [105,106]. Furthermore, TRPV4-mediated Ca^{2+} entry is elicited by manifold synthetic compounds, including the α-phorbol esters, phorbol 12-myristate 13-acetate (PMA) and 4α -phorbol 12,13-didecanoate (4α PDD), and the small molecule drugs, GSK1016790A (GSK) and JNc-440 [60,107]. TRPV4 has long been known to stimulate angiogenesis and arteriogenesis [4,5,108] by stimulating endothelial cell proliferation [5,109] and migration [110]. TRPV4-mediated Ca^{2+} entry is translated into a pro-angiogenic signal by several decoders, such as the Ca^{2+}-dependent transcription factors NFAT cytoplasmic 1 (NFATc1), myocyte enhancer factor 2C (MEF2C), and Kv channel interacting protein 3, calsenilin (KCNIP3/CSEN/DREAM), which drive endothelial cell proliferation, [4], β1-integrin and phosphatidylinositol 3-kinase (PI3-K), which promote endothelial cell motility [111]. The opening of only few TRPV4 channels, that tend to assemble into a four-channel cluster, results in spatially-restricted cytosolic Ca^{2+} microdomains, known as Ca^{2+} sparklets, which selectively recruit the downstream Ca^{2+}-dependent effectors [112,113]. A recent study revealed that TRPV4 was dramatically up-regulated in breast tumor-derived endothelial cells (B-TECs) and that TRPV4-mediated Ca^{2+} entry significantly increased the rate of cell migration as compared to control cells [88]. TRPV4 promoted B-TEC motility by eliciting local Ca^{2+} pulses at the leading edge of migrating cells [88], which were reminiscent of TRPV4-dependent Ca^{2+} sparklets [112]. TRPV4 was physiologically gated by AA [89], which is quite abundant in breast cancer microenvironment [114]. Likewise, cytosolic phospholipase A2 (PLA2), which cleaves AA from membrane phospholipids in response to physiological stimuli [115] is up-regulated and promotes cancer development by stimulating angiogenesis in several types of tumors, including breast cancer [116]. Therefore, TRPV4 might represent a novel and specific target to treat breast cancer as it is only barely expressed and does not drive migration in healthy endothelial cells [88].

Subsequently, the role of TRPV4 was investigated in prostate adenocarcinoma-derived endothelial cells (A-TECs). Unlike B-TECs, TRPV4 was down-regulated in A-TECs, which increased their sensitivity towards extracellular matrix stiffness, boosted their migration rate and favored the development of an aberrant (i.e., non-uniform, abnormally dilated and leaky) tumor vascular network [90]. This feature gains therapeutic relevance as the resultant hostile (i.e., low extracellular pH, hypoxia, and high interstitial pressure) microenvironment fuels tumor progression and hampers the efficacy of chemotherapy, radiation therapy, anti-angiogenic therapy immunotherapy [117,118]. Accordingly, overexpression or pharmacological activation of TRPV4 with GSK restored A-TEC mechanosensitivity and normalized their abnormal tube formation in vitro by inhibiting enhanced basal Rho activity [91]. Moreover, the daily intraperitoneal injection of GSK was able to normalize tumor vasculature in a xenograft mouse model of Lewis Lung Carcinoma (LLC), thereby improving cisplatin delivery and causing significant tumor shrinkage [91]. In addition, TRPV4-mediated Ca^{2+} entry reduced A-TEC proliferation in vitro by inhibiting the extracellular signal-regulated kinases 1/2 [92]. This mechanism further contributes to GSK-induced dismantling of LLC vasculature in vivo [92]. Therefore, remodeling of TRPV4-mediated Ca^{2+} entry may be used to effectively target tumor vascularization, although the most effective approach may depend on the tumor type. Accordingly, TRPV4 should be inhibited to halt tumor vascularization in breast cancer, while it must be stimulated to normalize tumor vasculature in LLC [23].

3.2. Piezo Proteins

Piezo1 and Piezo2 proteins are two recently identified non-selective cation channels that mediate mechanosensory transduction in mammalian cells [119,120]. Piezo proteins are gigantic homotetrameric complexes endowed with one or four ion-conducting pores: each subunit comprises over 2500 amino acids and presents 24–40 predicted transmembrane domains [119]. Piezo channels are Ca^{2+}-permeable and, therefore, lead to robust Ca^{2+} entry in response to mechanical deformation of the plasma membrane; unlike TRPV4 channels [105], Piezo-mediated Ca^{2+} entry is directly activated by tension within the lipid bilayer of the plasma membrane rather than by physical coupling to the sub-membranal cytoskeleton or intracellular second messengers [120]. A recent study demonstrated that the endothelial Piezo1 was activated by laminal shear stress to drive embryonic vascular development [81]. Piezo1 promoted vascular endothelial cell migration, alignment and re-alignment along the direction of blood flow by engaging the Ca^{2+}-dependent decoders, eNOS and calpain [81,121]. More recently, Piezo2 was found to be up-regulated in T-ECs from mouse xenografted with GL261 glioma cells [93]. Knocking down Piezo2 with a selective small interfering RNA (siRNA) reduced glioma angiogenesis and normalized tumor neovessels [93]. Moreover, suppressing Piezo2 expression decreased VEGF- and interleukin-1β-induced angiogenesis in the mouse corneal neovascularization model [93]. Finally, Piezo2-mediated Ca^{2+} entry elicited the Ca^{2+}-dependent transcription of Wnt11 and, consequently, the nuclear translocation of β-catenin in HUVECs, thereby promoting their angiogenic activity in vitro [93]. Although this mechanism remains to be confirmed in T-ECs, Piezo2 stands out as a crucial regulator of tumor angiogenesis and should be probed as a novel target for more effective anti-cancer treatments.

3.3. P2X7 Receptors

ATP and its metabolite, adenosine, are major constituents of tumor microenvironment and may differently affect tumor growth, immune cells and tumor-host interaction by activating a wealth of metabotropic (i.e., P2Y1, P2Y2, P2Y4, P2Y6, P2Y11, P2Y12, P2Y13, and P2Y14) and ionotropic (P2X1–2X8,) receptors [122]. Of note, ATP has long been known to stimulate angiogenesis though metabotropic P2y receptors [123]. Nevertheless, a recent investigation demonstrated that P2X7 stimulates tumor angiogenesis in vivo. Two different tumor cell lines, i.e., HEK293 and CT26 colon carcinoma cells, were transfected with P2X7 receptors and subsequently xenografted into immunodeficient or immunocompetent BALB/c mice, respectively. Tumor growth and angiogenesis were significantly enhanced by P_{2X7} expression; consequently, pharmacological inhibition (with AZ10606120) or genetic silencing of P_{2X7} decreased tumor growth and dramatically reduced vascular density [124]. This study further confirmed that P2X7 receptors were significantly up-regulated in several types of cancer cells, including those from breast cancer, and stimulated angiogenesis by promoting VEGF release [124]. More recently, it was found that P2X7 receptors were over-expressed also in B-TECs [94]. This study revealed that the activation of these purinergic receptors with high doses of ATP (>20 µM) and BzATP, a selective P_{2X7} agonist, inhibited B-TEC, but not HDMEC, migration in vitro. The anti-angiogenic effect of P_{2X7} was mediated by the Ca^{2+}-sensitive adenylate cyclase 10 (AC10), which increased cyclic adenosine monophosphate (cAMP) and recruited EPAC-1 to dampen cell migration by inducing cytoskeletal remodeling [94]. Moreover, P2X7 receptors-induced cAMP production stabilized bidimensional tumor vessels by favoring pericyte attraction towards B-TECs and reducing endothelial permeability [94]. Intriguingly, hypoxia prevented the anti-angiogenic ability of P_{2X7} receptors by likely reducing their expression [94,125]. These data, therefore, strongly suggest that stimulating P2X7 receptors could provide an efficient strategy to normalize tumor vasculature, thereby enhancing the delivery of cytotoxic drugs and of O_2 for radiotherapy. In this context, it should be pointed out that P2X7 receptors target hematopoietic EPCs to glioblastoma [126]. Although this investigation was conducted on healthy cells, and remains therefore to be validated in T-EPCs, it suggests that ATP may differentially affect tumor endothelial cells and T-EPCs. Alternatively, the effect exerted by P2X7 con T-EPC fate could be cancer-dependent and needs to be further investigated.

3.4. Stim1, Orai1 and Canonical Transient Receptor Potential 1 (TRPC1)

SOCE represents the most important Ca^{2+} entry pathway supporting the pro-angiogenic activity of human ECFCs [29,33,56]. Accordingly, TRPV4 boosted ECFC proliferation rate only when accompanied by the administration of a robust dose of growth factors [86,127], whereas TRPV1 stimulated ECFC proliferation and tubulogenesis by mediating the intracellular intake of anandamide in a Ca^{2+}-independent manner [128]. SOCE is activated by the pharmacological (by blocking SERCA-mediated Ca^{2+} sequestration) or physiological (by stimulating $InsP_3Rs$) depletion of the ER Ca^{2+} stores and is mediated by the dynamic interplay between Stim1, Orai1 and TRPC1 [29,56,58,84]. It is, however, still unknown whether Orai1 and TRPC1 form two independent Stim1-gated Ca^{2+}-permeable routes [129] or assemble into a unique heteromeric supermolecular complex in ECFCs [130]. A recent series of studies revealed that SOCE maintained VEGF-induced intracellular Ca^{2+} oscillations and promoted ECFC proliferation and in vitro tubulogenesis by recruiting the Ca^{2+}-dependent transcription factor, NF-κB [58,83]. Of note, SOCE was significantly enhanced in metastatic renal cellular carcinoma (RCC)-derived ECFCs (RCC-ECFCs) due to the up-regulation of Stim1, Orai1 and TRPC1 [95]. Similar to normal cells, the pharmacological blockade of SOCE with YM-58483/BTP2 or with low micromolar doses of lanthanides prevented proliferation and tube formation in RCC-ECFCs [95]. This finding strongly suggests that SOCE could provide an alternative target for the treatment of metastatic RCC [32,131], which develops either intrinsic or acquired refractoriness towards conventional treatments, such as anti-VEGF inhibitors and anti-mammalian target of rapamycin (mTOR) blockers [132]. As more extensively discussed below, the overall remodeling of the intracellular Ca^{2+} toolkit in T-ECFCs could indeed be responsible for the relative or complete failure of standard therapies in RCC patients. Conversely, SOCE was not significantly up-regulated in breast cancer-derived ECFCs (BC-ECFCs) [96]. Accordingly, Orai1 and TRPC1 expression were not significantly altered, while Stim1 was significantly more abundant as compared to control cells. Nevertheless, a tight stoichiometric ratio between Stim1, Orai1 and TRPC1 is required for SOCE to be activated [133]. If all Orai1 and TRPC1 channel proteins are gated by the physiological levels of Stim1, any increase in Stim1 expression will not be sufficient to enhance SOCE as there will be no further channels available on the plasma membrane. Similar to RCC-ECFCs, however, the pharmacological inhibition of SOCE abrogated BC-ECFC proliferation and tube formation, thereby confirming that Orai1 and TRPC1 could serve as reliable targets to interfere with tumor vascularization, although this hypothesis remains to be validated in vivo [134,135]. The strict requirement of Stim1 for tumor vascularization is further suggested by the recent finding that Stim1 transcription in hypoxic tumors is finely regulated by HIF-1 [136]. SOCE, in turn, was found to stimulate HIF-1 accumulation in hypoxic cancer cells by engaging Ca^{2+}/calmodulin-dependent protein kinase II and p300 [136]. Therefore, targeting SOCE could also affect the expression of the primary transcription factor responsible for RCC and breast cancer growth and metastasis [137,138]. Of note, HIF-1 has been shown to control also TRPC1 expression [139], although it is still unclear whether this regulation also occurs in tumor microenvironment and, if so, why TRPC1 is up-regulated in RCC-ECFCs, but not in RCC-ECFCs.

The role played by SOCE in tumor vascularization has, finally, been uncovered also in infantile hemangioma (IH), the most common childhood malignancy which may cause disfigurement, ulceration and obstruction and, if not treated, ultimately leads to patients' death [140]. IH is a vascular tumor that arises as a consequence of dysregulation of angiogenesis and vasculogenesis [140]. The clonal expansion of an endothelial progenitor/stem cell population, which is closely reminiscent of ECFCs, is deeply involved in IH vascularization [141,142]. A recent investigation provided the evidence that Stim1, Orai1 and TRPC1 drive the higher rate of IH-derived ECFC (IH-ECFC) growth as compared to control cells [97]. Stim1, Orai1 and TRPC1 were not up-regulated in IH-ECFCs; however, the ER Ca^{2+} store was depleted to such an extent that Stim1 was basally activated and gated the constitutive activation of Orai1 and TRPC1 [97]. Stim2 displays a lower Ca^{2+} affinity as respect to Stim1 and supports basal Ca^{2+} entry in HUVECs [143]. Nevertheless, the pharmacological abrogation of Stim2 silencing did not affect constitutive SOCE in IH-ECFCs [97]. Constitutive SOCE boosted IH-ECFC proliferation by enhancing NO release [97], thereby emerging as an alternative target to treat IH in propranolol-resistant patients [144].

3.5. Neuronal Nicotinic Receptors (nAchRs)

nAchRs belong to a super-family of Cys-loop ligand-gated non-selective cation channels that are physiologically activated by acetylcholine, mediate fast synaptic transmission in neurons and, by virtue of their resolvable Ca^{2+}-permeability, control a number of Ca^{2+}-dependent processes, including neurotransmitter release and synaptic plasticity [145,146]. However, nAchRs are also largely expressed in non-neuronal brain cells, such as astrocytes, in epithelial cells and in several types of vascular cells, including smooth muscle cells and endothelial cells [75,147–149]. It has been established that α7 homomeric nAchRs (α7-nAchRs) promote endothelial cell proliferation, migration and tube formation both in vitro and in vivo by recruiting an array of Ca^{2+}-dependent effectors [98,99]. These include eNOS, mitogen-activated protein kinase, phosphoinositide 3-kinase (PI3K), NF-κB, matrix metalloproteinase-2 and -9 [98,99,150]. In addition, α7-nAchRs were shown to induce the JAK2/STAT3 signaling cascade to promote endothelial cell survival [151]. Intriguingly, α7-nAchRs possess the highest Ca^{2+}-permeability among the known nAchR subtypes [152]. These pieces of evidence ignited the hypothesis that nicotine accelerated tumor growth by stimulating endothelial α7-nAchRs, thereby promoting angiogenesis and tumor vascularization [147,149]. In support of this model, nicotine induced tumor growth in a mouse model of LLC by stimulating endothelial cell proliferation and tube formation. Nicotine-induced tumor vascularization was significantly reduced by pharmacological blockade (with mecamylamine or hexamethonium) as well as by genetic silencing of α7-nAchRs. The signaling pathways recruited by α7-nAchRs to sustain tumor angiogenesis were not deeply investigated, but nicotine stimulated endothelial cells to release NO, prostacyclin and VEGF [98,99]. It should, however, be pointed out that the expression and role of α7-nAchRs in T-ECs has not been investigated, yet. Nevertheless, hypoxia has been shown to increase α7-nAchRs expression in a mouse model of hindlimb ischemia [98], whereas α7-nAchRs may stimulate HIF-1αtranscription [153]. These observations support the hypothesis that α7-nAchRs are actually over-expressed in T-ECs.

In addition to promoting angiogenesis, nicotine could recruit α7-nAchRs to boost vasculogenesis. A recent study revealed that nicotine induced proliferation, migration and tube formation also in ECFCs and that this effect was inhibited by mecamylamine or α-bungarotoxin [100]. Moreover, nicotine triggered EPC mobilization from bone marrow in a cohort of mice xenografted with colorectal cancer cells, thereby fostering tumor growth and vascularization [101]. Lastly, exposure to second hand smoke stimulated tumor angiogenesis and increased the number of circulating EPCs in a mouse model of LLC by enhancing VEGF release: mecamylamine, however, halted VEGF release, thereby reducing tumor size and capillary density. The pro-angiogenic effect of nicotine was, therefore, likely to be mediated by nAchRs [154]. We are yet to know whether and how α7-nAchRs are altered in T-ECs and T-EPCs. Nevertheless, these ionotropic receptors could be regarded as a promising target for alternative anti-angiogenic therapies.

3.6. Gasotransmitters-Activated Ca^{2+}-Permeable Channels

Gaseous mediators or gasotransmitters are endogenous signaling messengers that, although being toxic at high concentrations, regulate a multitude of physiological processes, ranging from the regulation of vascular tone to synaptic plasticity and mitochondrial bioenergetics [155–158]. The gasotransmitters NO and hydrogen sulphide (H_2S) have recently been shown to stimulate endothelial cells through an increase in $[Ca^{2+}]_i$ [6,156,159], while the role of CO in angiogenesis is less clear [160]. NO promotes angiogenesis and disease progression in several types of malignancies [161,162], including breast cancer [163]. The administration of two structurally unrelated NO donors, i.e., S-nitroso-N-acetylpenicillamine (SNAP) or sodium nitroprusside (SNP), was recently found to trigger Ca^{2+} influx and migration in B-TECs [164]. These effects were mimicked by elevating endogenous NO release with L-arginine [164], which is the physiological substrate for eNOS [156]. Of note, AA-induced TRPV4 activation in B-TECs was inhibited by preventing NO production with N^G-nitro-L-arginine methyl ester (L-NAME) [88,164]; moreover, AA- and NO-induced Ca^{2+} entry were both sensitive to protein kinase A (PKA) inhibition [164]. It is, therefore, likely that NO elicits Ca^{2+} entry in B-TECs by gating TRPV4. In agreement with this hypothesis, TRPV4 may be activated by NO through direct S-nitrosylation [71] and is phosphorylated by PKA upon AA stimulation in vascular endothelial cells [70]. Finally, NO-induced Ca^{2+} entry and migration were dramatically reduced in HDMECs [164], in which TRPV4 expression was significantly down-regulated [88]. Besides TRPV4, however, NO is able to recruit multiple TRP channels, such as TRPC1, TRPC4, TRPC5, TRPV1, and TRPV3 [71], some of which are up-regulated in T-ECFCs [95,165]. Unfortunately, it is still unclear whether NO elicits intracellular Ca^{2+} entry in these cells. Although future work is mandatory to understand whether NO stimulates TRP channels, as well as other Ca^{2+}-permeable channels, to promote tumor vascularization, endothelial Ca^{2+} signaling is emerging as an attractive target to prevent its pro-tumorigenic effect.

H_2S has also been shown to promote angiogenesis in a Ca^{2+}-dependent manner. For instance, H_2S mediated VEGF-induced Ea.hy926 cell proliferation and migration by inducing $InsP_3$-dependent ER Ca^{2+} release without the contribution of extracellular Ca^{2+} entry [6]. The components of the endothelial Ca^{2+} toolkit recruited by H_2S may, however, vary depending on the vascular bed [166]. H_2S induced ER-dependent Ca^{2+} release through $InsP_3Rs$ and RyRs followed by a sustained SOCE in primary cultures of human saphenous vein endothelial cells [167], whereas it recruited the reverse mode of NCX by gating a Na^+- and Ca^{2+}-permeable pathway in rat aortic endothelial cells [168] and HDMECs [21]. Conversely, NaHS did not elicit any resolvable elevation in $[Ca^{2+}]_i$ in ECFCs [6] and its role in neovasculogenesis in vivo operated by truly endothelial precursors remains to be elucidated [169]. H_2S-induced Ca^{2+} signals may be translated into a pro-angiogenic signal by multiple Ca^{2+}-dependent decoders, including the PI3K/Akt and the ERK/p38 signaling pathways [155,156]. Growing evidence demonstrated that H_2S drove disease progression and angiogenesis in several types of tumor, such as RCC and colorectal cancer [169]. Intriguingly, sodium hydrosulfide (NaHS), a widely employed H_2S donor, induced intracellular Ca^{2+} signals in both B-TECs and HDMECs; however, NaHS-elicited Ca^{2+} signals were enhanced and arose within a significantly lower range (nanomolar vs. micromolar) in B-TECs [21]. Consequently, NaHS promoted proliferation and migration in B-TECs, but not in control endothelial cells [21]. The Ca^{2+} response to H_2S was mediated by a Ca^{2+}-permeable non-selective cation channel and was sustained by membrane hyperpolarization through the activation of a K^+ conductance [21], likely an ATP-sensitive K^+ channel [157]. The molecular nature of this Ca^{2+}-permeable route is yet to be identified [166]. Nevertheless, H_2S is able to stimulate TRPV3 and TRPV6 in bone marrow-derived mesenchymal cells by direct sulfhydration of some of Cys residues within their protein structure [170]. Moreover, H_2S activated TRPA1 in RIN14B cells [171]. Deciphering the molecular target of H_2S in tumor endothelium is, therefore, mandatory to devise alternative anti-cancer treatments. In addition, both eNOS and cys-tathionine gamma lyase (CSE), the enzyme which catalyzes H_2S production in vascular endothelial cells, are Ca^{2+}-sensitive [16,42,172]. Therefore, targeting a Ca^{2+} entry/release pathway tightly coupled to either eNOS (i.e., Orai1, [173]) or CSE (yet to be identified) has the potential to interfere with multiple pro-angiogenic pathways and, therefore, exert a more profound anti-tumor effect.

3.7. Connexin 40 (Cx40)

Connexin (Cx) hemichannels, also termed connexons, have long been known to provide the building blocks of gap junctions, thereby enabling the transfer of small solutes, ions and signaling molecules, such as Ca^{2+} and $InsP_3$, between adjacent cells [174]. Three diverse Cx isoform exist in vascular endothelial cells, i.e., Cx37, Cx40, and Cx43, and synchronize robust NO release induced by extracellular autacoids by mediating intercellular Ca^{2+} communication [175,176]. In addition, unopposed Cx hemichannels were found to mediate extracellular Ca^{2+} entry and NO release in endothelial cells from different vascular beds [16,74,177–179]. Earlier work suggested that Cxs served as tumor suppressors and were down-regulated in cancer, thereby affecting vascular integrity and reducing vascular leakage [180–182]. However, a recent study challenged this dogma by showing that Cx40 was over-expressed in T-ECs and promoted disease progression and angiogenesis in melanoma and urogenital cancers [102]. Cx40 stimulated tumor growth by inducing eNOS recruitment, which strongly suggest that intracellular Ca^{2+} levels increased during the angiogenic process [102]. Intriguingly, targeting Cx40 function with ^{40}Gap27, a peptide that has long been use to inhibit Cx40-mediated intercellular communication and extracellular Ca^{2+} entry [1,16,178], normalized tumor vasculature and enhanced the efficacy of the chemotherapeutic drug, cyclophosphamide [102]. Therefore, although these findings remain to be confirmed in other tumor types, and the role served by Ca^{2+} is still unclear, Cx40 deserves careful consideration for the design of new anticancer drugs.

3.8. Na$^+$/H$^+$ Exchanger-1 (NHE-1)

The Na$^+$/H$^+$ exchanger NHE-1 is a reversible electroneutral antiporter that maintains cytosolic pH by expelling H$^+$ at expense of the inwardly directed Na$^+$ electrochemical gradient with a 1:1 stoichiometric ratio [183]. NHE-1 induces endothelial cell proliferation, migration and tube formation by means of several Ca^{2+}-dependent effectors, such as calpain [184], eNOS [185], and ERK 1/2 [186]. Accordingly, thrombin-induced NHE-1 activation was able to increase sub-membranal Na$^+$ levels, thereby switching NCX into the reverse mode and mediating extracellular Ca^{2+} entry in HUVECs [187,188]. Moreover, NHE-1-induced cytosolic alkalinization triggered ER-dependent Ca^{2+} release through InsP$_3$Rs in bovine aortic endothelial cells and human pulmonary artery endothelial cells [189,190]. NHE1 is constitutively activated in cancer cells to favor extracellular acidification and stimulate metastasis and invasion by facilitating protease-mediated degradation of the extracellular matrix [191]. In addition, NHE-1 is transcriptionally regulated by HIF-1 and is up-regulated in a multitude of carcinomas [191,192]. A recent series of studies demonstrated that NHE-1 was over-expressed in endothelial cells exposed to tumor microenvironment [193] and was able to boost vascularization, invasion and metastasis in several types of tumors, including breast cancer [191,194]. Accordingly, aldosterone-induced NHE-1 activation promoted B-TEC proliferation, migration and cytosolic alkalinization [103]. Further work is required to assess whether NHE-1 activation stimulates tumor vascularization through an increase in $[Ca^{2+}]_i$. However, NHE-1 blockers, including cariporide and the more specific 3-methyl-4-flouro analog of 5-aryl-4(4-(5-methyl-14-imidazol-4-yl) piperidin-1-yl)pyrimidine (Compound 9t), have been put forward as alternative anti-cancer drugs [195].

3.9. Two-Pore Channels (TPCs)

The ER is the largest endogenous Ca^{2+} store in vascular endothelial cells by accounting for ≈75% of the total Ca^{2+} storage capacity [2]. The remainder 25% of the total stored Ca^{2+} is located within the mitochondria and the acidic Ca^{2+} stores of the endolysosomal (EL) system [2]. As more widely illustrated in [196], the EL Ca^{2+} store releases Ca^{2+} through many Ca^{2+}-permeable channels, including Mucolipin TRP 1 (TRPML1), Melastatin TRPM 2 (TRPM2) and two-pore channels 1 and 2 (TPC1–2) [197,198]. The newly discovered second messenger, nicotinic acid adenine dinucleotide phosphate (NAADP), is the physiological stimulus that gates TPC1–2 in response to extracellular stimulation [196,199,200]. NAADP-induced EL Ca^{2+} release is, in turn, amplified by juxtaposed ER-embedded InsP$_3$Rs and RyRs through the CICR process, thereby initiating a regenerative Ca^{2+} wave [196,200,201]. NAADP-gated TPC2 channels are also expressed in vascular ECs [202,203], whereas *N*-ECFCs display larger amounts of TPC1 [86,204]. A recent study demonstrated that NAADP-induced Ca^{2+} signals promoted tumor vascularization and metastasis in murine models xenografted with B16 melanoma cells [205]. Of note, the pharmacological blockade of NAADP-induced Ca^{2+} release with Ned-19 dampened melanoma growth, vascularization and lung metastasis [205]. Future work will have to assess whether TPC2 channels are up-regulated in T-ECs and whether

NAADP-induced Ca^{2+} signaling also drive T-ECFC incorporation into tumor neovessels. However, TPCs stand out as promising targets to develop alternative anti-angiogenic treatments.

4. Resistance to Apoptosis

4.1. Canonical Transient Receptor Potential 5 (TRPC)

TRPC5 forms a homotetrameric Ca^{2+}-permeable channel that is gated upon PLCβ activation by Gq/11-coupled membrane receptors through a yet to be identified signaling cascade [2,206,207]. Accordingly, although some studies reported that TRPC5 is recruited in a store-dependent manner by Stim1 [208], it has been proposed that TRPC5 activation by PLCβ does not involve ER store depletion [209]. In addition, TRPC5-mediated Ca^{2+} entry is elicited by several physiological messengers, including reduced thioredoxin, protons, sphingosine-1-phosphate, lysophospholipids, NO and Ca^{2+} itself [207]. Finally, TRPC5 presents a spontaneous activity that is increased by lanthanides, cold temperatures (47 °C to 25 °C) and membrane stretch; consequently, TRPC5 serves as a cold sensor in the peripheral nervous system [210]. Of note, TRPC5 may establish physical associations with a multitude of molecular partners, including TRPC1, TRPC4 and TRPC6, which regulate its membrane localization and biophysical properties [207]. TRPC5 differently tunes angiogenesis depending on the vascular bed. For instance, TRPC5 promoted proliferation and tube formation by inducing intracellular Ca^{2+} oscillations in EA.hy926 cells [211]. Conversely, a TRPC6-TRPC5 channel interaction inhibited angiogenesis by decreasing the rate of migration in bovine aortic ECs (BOECs). In this context, TRPC6-mediated Ca^{2+} entry triggered an ERK-mediated phosphorylation cascade that leads to MLCK activation and TRPC5 externalization on the plasma membrane [212,213]. It has recently been shown that endothelial TRPC5 could underlie the development of chemoresistance to anticancer drugs in both breast cancer [104,214,215] and colorectal carcinoma [216]. P-glycoprotein (P-gp), also termed multidrug resistance protein 1 (MDR1), is a multidrug efflux transporter that expels xenobiotics out from the cytoplasm into the extracellular milieu [217]. P-gp overexpression, therefore, confers resistance to malignant cells, which become insensitive to a wide range of cancer chemotherapeutics, including adriamycin, vincristine, taxol, and anthracyclines [217]. Earlier evidence demonstrated that TRPC5 was up-regulated and induced P-gp overexpression by hyper-stimulating the Ca^{2+}-dependent transcription factor, NFATc3 in chemoresistant MCF-7 breast cancer cells [214]. In agreement with this observation, microRNA 320a (miR-320a), which is able to associate with and degrade TRPC5 and NFATc3 transcripts in normal cells, was down-regulated in chemoresistant breast cancer cells due to the hypermethylation of its promoter sequence [218]. TRPC5 up-regulation induced resistance to adriamycin, paclixatel, epirubicin, mitoxantrone and vincristine [214]. Additionally, TRPC5-mediated Ca^{2+} entry promoted transcription of HIF-1α gene, thereby boosting VEGF release and enhancing tumor angiogenesis [219]. Remarkably, TRPC5-based chemoresistance could be shuttled to tumor endothelial via intercellular communication. Adriamycin-resistant MCF-7 cells could pack the up-regulated TRPC5 channels into mobile extracellular vesicles (EVs), which are released in tumor microenvironment and transferred their signaling content to surrounding endothelial cells. This scenario is supported by the observations that HDMECs exposed to TRPC5-containing EVs, which were collected from adriamycin MCF-7 breast cancer cells, over-expressed the TRPC3-NFATc3-P-gp signaling pathway and developed resistance to adriamycin-induced apoptosis [104]. Moreover, TRPC5-containing vesicles were identified in peripheral blood of breast cancer patients receiving chemotherapy and of nude mice bearing adriamycin-resistant MCF-7 tumor xenografts [215]. Furthermore, P-gp production was enriched in tumor endothelium of adriamycin-resistant MCF-7 xenografts than in other sites and was sensitive to TRPC5 inhibition with a specific siRNA (siTRPC5) [104]. These data, therefore, suggest that TRPC5 provide a promising target to design alternative adjuvant anticancer treatments [220]. Accordingly, a blocking TRPC5 antibody reduced P-gp expression, retarded cancer growth and boosted paclitaxel-induced tumor regression in chemoresistant breast cancer in vivo [104,214,215]. The endothelial effects of TRPC5 in breast cancer are seemingly limited to T-ECs, as BC-ECFCs do not express this channel [96]. Future work will have to assess whether, besides conferring B-TECs with the resistance to chemotherapeutic drugs, TRPC5 up-regulation accelerates breast cancer angiogenesis.

4.2. Inositol-1,4,5-Trisphosphate (InsP₃) Receptors (InsP₃Rs)

InsP₃Rs are non-selective cation channels which mediate ER-dependent Ca^{2+} release, thereby controlling multiple endothelial cell functions, including bioenergetics, apoptosis, angiogenesis and vasculogenesis (see Paragraph 2. Ca^{2+} signaling in normal endothelial cells: a brief introduction). Three distinct InsP₃R isoforms exist in both vascular endothelial cells and ECFCs [2,29], i.e., InsP₃R1, InsP₃R2 and InsP₃R3, which may associate into homo- or hetero-tetrameric ER-embedded channels [2]. It has recently been shown that InsP₃Rs were dramatically down-regulated in RCC-ECFCs, thereby preventing the onset of VEGF-induced intracellular Ca^{2+} oscillations, proliferation and in vitro tubulogenesis [95] (Figure 3). More specifically, RCC-ECFCs only expressed InsP₃R1, while InsP₃R2 and InsP₃R3 were absent [95]. This result was surprising as InsP₃R1 was transcriptionally regulated by HIF-2 in human RCC cancer cell lines [221]. The failure of the pro-angiogenic Ca^{2+} response to VEGF also involves the chronic reduction in the ER Ca^{2+} concentration ($[Ca^{2+}]_{ER}$) in RCC-ECFCs, as monitored by using an ER-targeted aequorin Ca^{2+} indicator [87,222]. Therefore, in contrast with the widely accepted belief that VEGF sustains the angiogenic switch [223], VEGF does not stimulate ECFC-dependent neovessel formation in RCC patients [27,32]. This observation shed novel light on the refractoriness to anti-VEGF therapies in individuals suffering from RCC [224–226]. It has long been known that humanized monoclonal anti-VEGF antibodies, such as bevacizumab, or small molecule tyrosine kinase inhibitors, such as sorafenib and sunitinib, did not increase the overall survival of RCC patients, who ultimately developed secondary (acquired) resistance and succumbed because of tumor relapse and metastasis. In addition, targeting VEGF-dependent pathway proved to be ineffective in a large cohort of subjects, who displayed intrinsic refractoriness to these anti-VEGF drugs and did not show any improvement in their progression free survival [227,228]. ECFCs play a key role during the early phases of the angiogenic switch that supports tumor vascularization and metastasis [27,28,229]. If tumor vasculature is dismantled by anti-VEGF drugs, the following drop in P_{O2} will release in circulation a cytokine storm that attracts ECFCs from their bone marrow and/or vascular niches. ECFCs will home to the shrunk tumor, but, being insensitive to VEGF, will not be affected by the presence of anti-VEGF drugs. Consequently, they will proliferate in response to the mixture of growth factors liberated in tumor microenvironment and will restore blood supply to cancer cells [32]. Remodeling of the Ca^{2+} toolkit in RCC-ECFCs could, therefore, underlie the resistance to anti-angiogenic therapies in RCC patients. Similar data were obtained in BC-ECFCs, in which the significant reduction in $[Ca^{2-}]_{ER}$ prevented VEGF from triggering robust intracellular Ca^{2+} oscillations, proliferation and tube formation, although the pattern of InsP₃R expression remained unchanged [37,96]. Again, this result is consistent with notion that also breast cancer patients present intrinsic or secondary refractoriness to anti-VEGF therapies [54,230]. The reduction in $[Ca^{2+}]_{ER}$ observed in several types of tumor-associated ECFCs, including IH-ECFCs [95–97], was likely to reflect the down-regulation of SERCA2B activity [87,95]. Accordingly, ATP-induced InsP₃-dependent ER Ca^{2+} release in RCC-ECFCs decayed to resting Ca^{2+} levels with slower kinetics as compared to normal ECFCs [95]. Intriguingly, the gene expression profile of RCC- and BC-ECFCs resulted to be dramatically different with respect to normal cells [37]: BC-ECFCs and RCC-ECFCs presented, respectively, 382 and 71 differently expressed genes (DEGs) as compared to healthy cells, including TMTC1 [37]. TMTC1 is a tetratricopeptide repeat-containing adapter protein, which binds to and inhibits SERCA2B, thereby reducing ER Ca^{2+} levels and dampening agonist-induced intracellular Ca^{2+} release [231]. It is conceivable that TMTC1 up-regulation in T-ECFCs contributes to the chronic underfilling of their ER Ca^{2+} reservoir. In further agreement with this observation, electron microscopy revealed that both RCC- and BC-ECFCs presented dramatic ultrastructural differences as compared to control cells [87,96]. In particular, T-ECFCs presented a remarkable expansion of ER volume, whereas mitochondria were more abundant and very often elongated as compared to N-ECFCs [87,96]. This ultrastructural remodeling is consistent with the ER stress caused by the chronic reduction in $[Ca^{2+}]_{ER}$ [232,233].

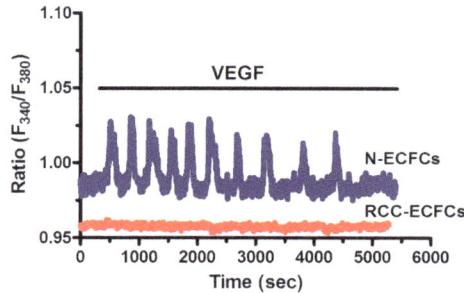

Figure 3. VEGF does not trigger pro-angiogenic Ca^{2+} oscillations in tumor-derived endothelial colony forming cells. VEGF (10 ng/mL) triggers intracellular Ca^{2+} oscillations in N-ECFCs, but not in RCC-ECFCs. Adapted from [95].

5. Targeting the Endothelial Ca^{2+} Toolkit to Circumvent the Resistance to Anticancer Treatments

Remodeling of the Ca^{2+} toolkit in tumor cells led many authors to search for alternative strategies to treat cancer. Intracellular Ca^{2+} signaling controls most, if not all, the so-called cancer hallmarks and could, therefore, be targeted to inhibit or, at least, retard tumor growth and metastasis [35,36,234–240]. As shown above, the Ca^{2+} transportome is also altered in stromal cancer cells [32,241,242], including endothelial cells and ECFCs. Remodeling of the endothelial Ca^{2+} toolkit could play a crucial role in the refractoriness to anticancer treatments, by supporting tumor vascularization and decreasing the susceptibility to pro-apoptotic stimuli. Therefore, the endothelial Ca^{2+} transportome might provide an efficient target for adjuvant therapies to conventional anti-cancer treatments. Three strategies could be pursued to improve the therapeutic outcome of standard therapies by interfering with the endothelial Ca^{2+} machinery: (1) blocking Ca^{2+} signaling to dampen angiogenesis and/or vasculogenesis; (2) stimulating Ca^{2+} entry to normalize tumor vessels, thereby improving the delivery and efficacy of chemo-, radio- and immunotherapy; and (3) manipulating Ca^{2+} signaling to endothelial cell apoptosis and dismantle tumor vasculature.

SOCE is, perhaps, the most suitable target to affect tumor vasculature by inhibiting both angiogenesis and vasculogenesis. Although there is no report of SOCE expression in T-ECs, Stim1 and Orai1 control proliferation and tube formation in normal endothelial cells, such as rat CMECs [7], bovine brain capillary endothelial cells [243], mouse lymphatic endothelial cells [244], and HUVEC [57,58]. Moreover, the pharmacological blockade of SOCE attenuates the rate of cell growth and abrogates in vitro tubulogenesis in RCC-, IH- and BC-derived ECFCs [95–97]. In addition, SOCE controls proliferation, migration and metastasis in a multitude of different cancer cell lines [134,135], which expands the cellular targets of SOCE inhibitors to the whole tumor microenvironment. We [131,135] and others [133,245] have recently described the Orai1 and TRPC1 inhibitors, some of which have been listed in Table 1, that could serve as a molecular template to design novel anticancer drugs. Unfortunately, none of these drugs have reached the milestone of being approved by US Food and Drug Administration (FDA) due to their scarce selectivity and high toxicity. For instance, carboxyamidotriazole (CAI), a non-selective blocker of Ca^{2+} signaling, was originally used to inhibit angiogenesis in vitro and tumor vascularization in vivo [131,246–248]. Depending on the cell type, CAI was able to block SOCE by occluding the mitochondrial Ca^{2+} uniporter [249,250] or reducing $InsP_3$ synthesis, which in turn prevents $InsP_3$-dependent Ca^{2+} release and Stim activation [56,251,252]. Intriguingly, CAI inhibited proliferation and tube formation also in RCC- and BC-ECFCs by preventing

InsP$_3$-dependent ER Ca^{2+} release [95,96]. Additionally, CAI was found to block growth and motility in several types of cancer cell lines [252,253]. Therefore, phase I-III clinical trials were launched to assess CAI toxicity and tolerability in patients suffering from several types of malignancies, including RCC, breast cancer, ovarian cancer, melanoma, non-small cell lung carcinoma, and gastrointestinal (stomach and pancreas) adenocarcinomas [56,248,254,255]. As discussed elsewhere [131], this drug caused disease stabilization when administrated alone or as adjuvant of chemo- or radio-therapy, and induced well tolerable side effects in most patients, such diarrhea, nausea and/or vomiting, fatigue and constipation. The therapy was discontinued only in RCC patients, who underwent disease progression and suffered from unacceptable toxicities, such as neuropsychiatric difficulties and asthenia [256]. As mentioned earlier, however, the effect of CAI is not directed towards the SOCE machinery, but is indirect. In addition to SOCE, CAI may also block TRPV4 and ER leakage channels [89,95,96,131]. A recent investigation, however, screened a library of >1800 FDA-approved drugs to search for specific SOCE blockers and identified five novel compounds, i.e., leflunomide, teriflunomide, lansoprazole, tolvaptan and roflumilast, that could be successfully used in therapy (leflunomide and teriflunomide) or provide the template to design more selective Orai1 inhibitors (i.e., lansoprazole, tolvaptan and roflumilast) [257].

An alternative strategy consists in stimulating endothelial Ca^{2+} signaling to induce tumor normalization by activating distinct Ca^{2+} entry routes depending on the tumor type. For instance, TRPV4-mediated Ca^{2+} entry drives tumor normalization in LLC [90,91], whereas P$_{2X7}$ receptors could be targeted to normalize tumor vasculature in breast cancer [94]. Tumor normalization, in turn, represents a promising adjuvant approach to facilitate cancer therapy by increasing the diffusion of chemotherapeutic drugs, improving radiotherapy efficiency and favoring the recruitment of tumor-killing immune cells [118,258]. Several synthetic agonists may selectively induce TRPV4 opening, such as 4αPDD derivatives, RN-1747, and JNc-440. Moreover, GlaxoSmithKline commercialized several patent applications of small molecule TRPV4 activators, the most famous of which is GSK [105,259]. Likewise, BzATP is regarded as the most potent P$_{2X7}$ receptor agonist, while 2-meSATP and ATPγS are only partial agonists and αβ-meATP and βγ-meATP exert a rather weaker on activation [260]. Clearly, further studies are required to uncover additional components of the endothelial Ca^{2+} toolkit potentially implicated in tumor normalization. Nevertheless, a recent investigation reported that angiopoietins, which induce vessel maturation by regulating the interaction between luminal endothelial cells and mural cells, such as vascular smooth muscle cells and pericytes, stimulate HUVEC migration by promoting ER-dependent Ca^{2+} release through InsP$_3$Rs and RyRs [261]. These findings lend further support to the hypothesis that targeting the endothelial Ca^{2+} signaling provides a suitable means to accelerate the dismantling of tumor vasculature by standard anticancer therapies.

Finally, the endothelial Ca^{2+} machinery could be properly manipulated to enhance the pro-apoptotic outcome of chemo- and radiation-therapy. For instance, TRPC5-mediated Ca^{2+} entry could be inhibited in B-TECs by taking advantage of a battery of novel small molecule inhibitors, such as Pico145 [262], 3,5,7-trihydroxy-2-(2-bromophenyl)-4H-chromen-4-one (AM12) [263], 2-aminobenzimidazole derivatives [264], ML204 [265], and neuroactive steroids [266]. Alternatively, the [Ca^{2+}]$_{ER}$ could be augmented to such an extent to induce the pro-apoptotic InsP$_3$-driven ER-to-mitochondria Ca^{2+} transfer. Pinton's group demonstrated that phototherapy induces a p53-dependent increase in [Ca^{2+}]$_{mit}$, which leads to tumor disruption in vivo [239,267]. Moreover, cytotoxic ER-dependent Ca^{2+} mobilization could be promoted by conjugating thapsigargin, a selective SERCA inhibitor, with a protease-specific peptide carrier, which is cleaved by the prostate-specific membrane antigen (PMSA) [268]. PMSA is widely expressed in the endothelium of many solid tumors [269,270], including RCC, thereby selectively favoring thapsigargin release in TME and inducing cancer and stromal cell apoptosis [268,271]. This prodrug has been termed mipsagargin or prodrug G202 and has recently been probed in a phase I clinical trials in patients suffering from refractory, advanced or metastatic solid tumors [272]. We do not know yet whether [Ca^{2+}]$_{ER}$ is also decreased in the endothelium of tumor neovessels, as ECFCs are likely to be replaced/diluted by local endothelial cells after the angiogenic switch [27]. Nevertheless, mipsagargin is likely to cause pro-apoptotic Ca^{2+} release in all stromal cells, including T-ECs.

As outlined elsewhere [23,32,135], caution is warranted when targeting a ubiquitous intracellular second messenger, such as Ca^{2+}. It should, however, be pointed out that several inhibitors of voltage-gated Ca^{2+} channels, such as verapamil, nifedipine and nitrendipin, are routinely employed in clinical practice to treat severe cardiovascular disorders, including hypertension, arrhythmia, acute myocardial infarction-induced heart failure and chronic stable angina [135,273]. In agreement with this observation, a phase I clinical trial is currently assessing the therapeutic outcome of Ca^{2+} electroporation on cutaneous metastases of solid tumors as compared to standard electrochemotherapy with bleomycin (https://clinicaltrials.gov/ct2/show/NCT01941901). Ca^{2+} electroporation is predicted to enhance the rate of cancer cell death by resulting in cytotoxic Ca^{2+} accumulation in the cytosol and in mitochondria [36]. Finally, the pharmacological inhibition of intracellular Ca^{2+} signaling did not elicit any intolerable side effects, such as immune depression, bleeding or neuropathic disorders, in least three distinct models of human cancer xenografts [24,205,274].

6. Conclusions

The present article discussed how remodeling of the endothelial Ca^{2+} toolkit (or transportome) could contribute to the resistance to anti-cancer treatments, which hampers from the very beginning their therapeutic outcome (intrinsic resistance) or leads to tumor relapse (acquired resistance) and patients' death. The intimate relationship between endothelial Ca^{2+} signaling and refractoriness to anti-cancer treatments cannot be fully appreciated by studying normal/healthy endothelial cells [32]. For instance, the role of VEGF in promoting tumor neovascularization has been extensively acknowledged based upon the observation that VEGF triggers pro-angiogenic Ca^{2+} signals in normal endothelial cells [6,58,251] and ECFCs. Nevertheless, VEGF does not stimulate proliferation and tube formation in T-ECFCs, which play a crucial role in sustaining the angiogenic switch and are likely to restore tumor vasculature prior to recurrence of disease progression. Therefore, to be effective in the patients, a strategy aiming at targeting the Ca^{2+} toolkit must be first probed on tumor-associated endothelial cells and ECFCs in vitro, as the their Ca^{2+} machinery could be different from that of naïve cells. The protocol to isolate ECFCs from peripheral blood does not require an unreasonable volume of blood ($\approx$40 mL), but it takes no less than three weeks [31] due to lack of ECFCs-specific membrane antigens. It will be imperative to speed up this procedure to accelerate the therapeutic translation of the findings generated by basic research. Isolating T-ECs represents a more technically demanding challenge, but several strategies were designed to collect and expand T-ECs from several types of solid cancers [275]. Further work on patients-derived T-ECs or T-ECFCs is mandatory to identify novel components of the endothelial Ca^{2+} toolkit involved in the refractoriness to anti-cancer therapies. Most of the attention is, of course, currently paid to Stim1 and Orai1 [33,131] and to the multiple TRP channel subfamilies that drive physiological angiogenesis [2,131,276,277]. Additional components of the endothelial Ca^{2+} transportome deserve careful investigation. For instance, Orai3 was found to up-regulated in several types of T-ECFCs [95,165], but its role in tumor vascularization is currently unknown. Of note, Orai3 may replace Orai1 as the pore-forming subunit of store-operated channels in cancer cells and could, therefore, emerge as a promising target for anti-cancer therapies [278]. NCX provides another unconventional Ca^{2+} entry that regulates proliferation and tube formation in healthy endothelial cells [18,279], but has been scarcely investigated in tumor neovessels. Finally, nAchRs are not the only ionotropic receptors expressed in vascular endothelial cells. *N*-methyl-D-aspartate receptors (NMDARs) are widely expressed in brain microvascular endothelial cells [280], in which they recruit eNOS and stimulate NO release in response to synaptic activity [281]. Aberrant glutamate signaling has been associated to glioma growth [282] and NMDARs-mediated Ca^{2+} entry could engage Ca^{2+}-dependent decoders other than eNOS in brain endothelium. Therefore, the expression and role of endothelial NMDARs in glioblastoma should be carefully evaluated.

Acknowledgments: The author gratefully acknowledges the contribution and support of all the colleagues and students who participated in the studies described in the present article. In particular, I would like to thank Vittorio Rosti and Germano Guerra. I am also grateful to Teresa Soda for her continuous support and comprehension.

Conflicts of Interest: The author declares no conflict of interest.

References

1. Moccia, F.; Tanzi, F.; Munaron, L. Endothelial remodelling and intracellular calcium machinery. *Curr. Mol. Med.* **2014**, *14*, 457–480. [CrossRef] [PubMed]
2. Moccia, F.; Berra-Romani, R.; Tanzi, F. Update on vascular endothelial Ca^{2+} signalling: A tale of ion channels, pumps and transporters. *World J. Biol. Chem.* **2012**, *3*, 127–158. [CrossRef] [PubMed]
3. Munaron, L.; Pla, A.F. Endothelial Calcium Machinery and Angiogenesis: Understanding Physiology to Interfere with Pathology. *Curr. Med. Chem.* **2009**, *16*, 4691–4703. [CrossRef] [PubMed]
4. Troidl, C.; Nef, H.; Voss, S.; Schilp, A.; Kostin, S.; Troidl, K.; Szardien, S.; Rolf, A.; Schmitz-Rixen, T.; Schaper, W.; et al. Calcium-dependent signalling is essential during collateral growth in the pig hind limb-ischemia model. *J. Mol. Cell. Cardiol.* **2010**, *49*, 142–151. [CrossRef] [PubMed]
5. Troidl, C.; Troidl, K.; Schierling, W.; Cai, W.J.; Nef, H.; Mollmann, H.; Kostin, S.; Schimanski, S.; Hammer, L.; Elsasser, A.; et al. Trpv4 induces collateral vessel growth during regeneration of the arterial circulation. *J. Cell. Mol. Med.* **2009**, *13*, 2613–2621. [CrossRef] [PubMed]
6. Potenza, D.M.; Guerra, G.; Avanzato, D.; Poletto, V.; Pareek, S.; Guido, D.; Gallanti, A.; Rosti, V.; Munaron, L.; Tanzi, F.; et al. Hydrogen sulphide triggers VEGF-induced intracellular Ca^{2+} signals in human endothelial cells but not in their immature progenitors. *Cell Calcium* **2014**, *56*, 225–236. [CrossRef] [PubMed]
7. Moccia, F.; Berra-Romani, R.; Tritto, S.; Signorelli, S.; Taglietti, V.; Tanzi, F. Epidermal growth factor induces intracellular Ca^{2+} oscillations in microvascular endothelial cells. *J. Cell. Physiol.* **2003**, *194*, 139–150. [CrossRef] [PubMed]
8. Noren, D.P.; Chou, W.H.; Lee, S.H.; Qutub, A.A.; Warmflash, A.; Wagner, D.S.; Popel, A.S.; Levchenko, A. Endothelial cells decode VEGF-mediated Ca^{2+} signaling patterns to produce distinct functional responses. *Sci. Signal.* **2016**, *9*, ra20. [CrossRef] [PubMed]
9. Munaron, L.; Fiorio Pla, A. Calcium influx induced by activation of tyrosine kinase receptors in cultured bovine aortic endothelial cells. *J. Cell. Physiol.* **2000**, *185*, 454–463. [CrossRef]
10. Gupta, S.K.; Lysko, P.G.; Pillarisetti, K.; Ohlstein, E.; Stadel, J.M. Chemokine receptors in human endothelial cells. Functional expression of CXCR4 and its transcriptional regulation by inflammatory cytokines. *J. Biol. Chem.* **1998**, *273*, 4282–4287. [CrossRef] [PubMed]
11. Moccia, F.; Bonetti, E.; Dragoni, S.; Fontana, J.; Lodola, F.; Romani, R.B.; Laforenza, U.; Rosti, V.; Tanzi, F. Hematopoietic progenitor and stem cells circulate by surfing on intracellular Ca^{2+} waves: A novel target for cell-based therapy and anti-cancer treatment? *Curr. Signal Trans. Ther.* **2012**, *7*, 161–176. [CrossRef]
12. Yang, C.; Ohk, J.; Lee, J.Y.; Kim, E.J.; Kim, J.; Han, S.; Park, D.; Jung, H.; Kim, C. Calmodulin Mediates Ca^{2+}-Dependent Inhibition of Tie2 Signaling and Acts as a Developmental Brake During Embryonic Angiogenesis. *Arterioscler. Thromb. Vasc. Biol.* **2016**, *36*, 1406–1416. [CrossRef] [PubMed]
13. Zhu, L.P.; Luo, Y.G.; Chen, T.X.; Chen, F.R.; Wang, T.; Hu, Q. Ca^{2+} oscillation frequency regulates agonist-stimulated gene expression in vascular endothelial cells. *J. Cell Sci.* **2008**, *121*, 2511–2518. [CrossRef] [PubMed]
14. Chen, F.; Zhu, L.; Cai, L.; Zhang, J.; Zeng, X.; Li, J.; Su, Y.; Hu, Q. A stromal interaction molecule 1 variant up-regulates matrix metalloproteinase-2 expression by strengthening nucleoplasmic Ca^{2+} signaling. *Biochim. Biophys. Acta* **2016**, *1863*, 617–629. [CrossRef] [PubMed]
15. Tsai, F.C.; Seki, A.; Yang, H.W.; Hayer, A.; Carrasco, S.; Malmersjo, S.; Meyer, T. A polarized Ca^{2+}, diacylglycerol and STIM1 signalling system regulates directed cell migration. *Nat. Cell Biol.* **2014**, *16*, 133–144. [CrossRef] [PubMed]
16. Berra-Romani, R.; Avelino-Cruz, J.E.; Raqeeb, A.; Della Corte, A.; Cinelli, M.; Montagnani, S.; Guerra, G.; Moccia, F.; Tanzi, F. Ca^{2+}-dependent nitric oxide release in the injured endothelium of excised rat aorta: A promising mechanism applying in vascular prosthetic devices in aging patients. *BMC Surg.* **2013**, *13* (Suppl. 2), S40. [CrossRef] [PubMed]

17. Charoensin, S.; Eroglu, E.; Opelt, M.; Bischof, H.; Madreiter-Sokolowski, C.T.; Kirsch, A.; Depaoli, M.R.; Frank, S.; Schrammel, A.; Mayer, B.; et al. Intact mitochondrial Ca^{2+} uniport is essential for agonist-induced activation of endothelial nitric oxide synthase (eNOS). *Free Radic. Biol. Med.* **2017**, *102*, 248–259. [CrossRef] [PubMed]

18. Andrikopoulos, P.; Eccles, S.A.; Yaqoob, M.M. Coupling between the TRPC3 ion channel and the NCX1 transporter contributed to VEGF-induced ERK1/2 activation and angiogenesis in human primary endothelial cells. *Cell. Signal.* **2017**, *37*, 12–30. [CrossRef] [PubMed]

19. Lyubchenko, T.; Woodward, H.; Veo, K.D.; Burns, N.; Nijmeh, H.; Liubchenko, G.A.; Stenmark, K.R.; Gerasimovskaya, E.V. P2Y1 and P2Y13 purinergic receptors mediate Ca^{2+} signaling and proliferative responses in pulmonary artery vasa vasorum endothelial cells. *Am. J. Physiol. Cell Physiol.* **2011**, *300*, C266–C275. [CrossRef] [PubMed]

20. Sameermahmood, Z.; Balasubramanyam, M.; Saravanan, T.; Rema, M. Curcumin modulates SDF-1alpha/CXCR4-induced migration of human retinal endothelial cells (HRECs). *Investig. Ophthalmol. Vis. Sci.* **2008**, *49*, 3305–3311. [CrossRef] [PubMed]

21. Pupo, E.; Pla, A.F.; Avanzato, D.; Moccia, F.; Cruz, J.E.; Tanzi, F.; Merlino, A.; Mancardi, D.; Munaron, L. Hydrogen sulfide promotes calcium signals and migration in tumor-derived endothelial cells. *Free Radic. Biol. Med.* **2011**, *51*, 1765–1773. [CrossRef] [PubMed]

22. Fiorio Pla, A.; Munaron, L. Functional properties of ion channels and transporters in tumour vascularization. *Philos. Trans. R. Soc. Lond. B Biol. Sci.* **2014**, *369*, 20130103. [CrossRef] [PubMed]

23. Moccia, F. Remodelling of the Ca^{2+} Toolkit in Tumor Endothelium as a Crucial Responsible for the Resistance to Anticancer Therapies. *Curr. Signal Trans. Ther.* **2017**, *12*. [CrossRef]

24. Chen, Y.F.; Chiu, W.T.; Chen, Y.T.; Lin, P.Y.; Huang, H.J.; Chou, C.Y.; Chang, H.C.; Tang, M.J.; Shen, M.R. Calcium store sensor stromal-interaction molecule 1-dependent signaling plays an important role in cervical cancer growth, migration, and angiogenesis. *Proc. Natl. Acad. Sci. USA* **2011**, *108*, 15225–15230. [CrossRef] [PubMed]

25. Jain, R.K.; Carmeliet, P. SnapShot: Tumor angiogenesis. *Cell* **2012**, *149*, 1408.e1. [CrossRef] [PubMed]

26. Gao, D.C.; Nolan, D.; McDonnell, K.; Vahdat, L.; Benezra, R.; Altorki, N.; Mittal, V. Bone marrow-derived endothelial progenitor cells contribute to the angiogenic switch in tumor growth and metastatic progression. *Biochim. Biophys. Acta* **2009**, *1796*, 33–40. [CrossRef] [PubMed]

27. Moccia, F.; Zuccolo, E.; Poletto, V.; Cinelli, M.; Bonetti, E.; Guerra, G.; Rosti, V. Endothelial progenitor cells support tumour growth and metastatisation: Implications for the resistance to anti-angiogenic therapy. *Tumour Biol.* **2015**, *36*, 6603–6614. [CrossRef] [PubMed]

28. Yoder, M.C.; Ingram, D.A. The definition of EPCs and other bone marrow cells contributing to neoangiogenesis and tumor growth: Is there common ground for understanding the roles of numerous marrow-derived cells in the neoangiogenic process? *Biochim. Biophys. Acta* **2009**, *1796*, 50–54. [CrossRef] [PubMed]

29. Moccia, F.; Guerra, G. Ca^{2+} Signalling in Endothelial Progenitor Cells: Friend or Foe? *J. Cell. Physiol.* **2016**, *231*, 314–327. [CrossRef] [PubMed]

30. Maeng, Y.S.; Choi, H.J.; Kwon, J.Y.; Park, Y.W.; Choi, K.S.; Min, J.K.; Kim, Y.H.; Suh, P.G.; Kang, K.S.; Won, M.H.; et al. Endothelial progenitor cell homing: Prominent role of the IGF2-IGF2R-PLCbeta2 axis. *Blood* **2009**, *113*, 233–243. [CrossRef] [PubMed]

31. Moccia, F.; Ruffinatti, F.A.; Zuccolo, E. Intracellular Ca^{2+} Signals to Reconstruct A Broken Heart: Still A Theoretical Approach? *Curr. Drug Targets* **2015**, *16*, 793–815. [CrossRef] [PubMed]

32. Moccia, F.; Poletto, V. May the remodeling of the Ca^{2+} toolkit in endothelial progenitor cells derived from cancer patients suggest alternative targets for anti-angiogenic treatment? *Biochim. Biophys. Acta* **2015**, *1853*, 1958–1973. [CrossRef] [PubMed]

33. Moccia, F.; Lodola, F.; Dragoni, S.; Bonetti, E.; Bottino, C.; Guerra, G.; Laforenza, U.; Rosti, V.; Tanzi, F. Ca^{2+} signalling in endothelial progenitor cells: A novel means to improve cell-based therapy and impair tumour vascularisation. *Curr. Vasc. Pharmacol.* **2014**, *12*, 87–105. [CrossRef] [PubMed]

34. Bergers, G.; Hanahan, D. Modes of resistance to anti-angiogenic therapy. *Nat. Rev. Cancer* **2008**, *8*, 592–603. [CrossRef] [PubMed]

35. Prevarskaya, N.; Ouadid-Ahidouch, H.; Skryma, R.; Shuba, Y. Remodelling of Ca^{2+} transport in cancer: How it contributes to cancer hallmarks? *Philos. Trans. R. Soc. Lond. B Biol. Sci.* **2014**, *369*, 20130097. [CrossRef] [PubMed]

36. Monteith, G.R.; Prevarskaya, N.; Roberts-Thomson, S.J. The calcium-cancer signalling nexus. *Nat. Rev. Cancer* **2017**, *17*, 367–380. [CrossRef] [PubMed]

37. Moccia, F.; Fotia, V.; Tancredi, R.; Della Porta, M.G.; Rosti, V.; Bonetti, E.; Poletto, V.; Marchini, S.; Beltrame, L.; Gallizzi, G.; et al. Breast and renal cancer-Derived endothelial colony forming cells share a common gene signature. *Eur. J. Cancer* **2017**, *77*, 155–164. [CrossRef] [PubMed]

38. Moccia, F.; Berra-Romani, R.; Baruffi, S.; Spaggiari, S.; Signorelli, S.; Castelli, L.; Magistretti, J.; Taglietti, V.; Tanzi, F. Ca^{2+} uptake by the endoplasmic reticulum Ca^{2+}-ATPase in rat microvascular endothelial cells. *Biochem. J.* **2002**, *364 Pt 1*, 235–244. [CrossRef] [PubMed]

39. Berra-Romani, R.; Raqeeb, A.; Guzman-Silva, A.; Torres-Jacome, J.; Tanzi, F.; Moccia, F. Na^+-Ca^{2+} exchanger contributes to Ca^{2+} extrusion in ATP-stimulated endothelium of intact rat aorta. *Biochem. Biophys. Res. Commun.* **2010**, *395*, 126–130. [CrossRef] [PubMed]

40. Paszty, K.; Caride, A.J.; Bajzer, Z.; Offord, C.P.; Padanyi, R.; Hegedus, L.; Varga, K.; Strehler, E.E.; Enyedi, A. Plasma membrane Ca^{2+}-ATPases can shape the pattern of Ca^{2+} transients induced by store-operated Ca^{2+} entry. *Sci. Signal.* **2015**, *8*, ra19. [CrossRef] [PubMed]

41. Cardenas, C.; Foskett, J.K. Mitochondrial Ca^{2+} signals in autophagy. *Cell Calcium* **2012**, *52*, 44–51. [CrossRef] [PubMed]

42. Zuccolo, E.; Lim, D.; Kheder, D.A.; Perna, A.; Catarsi, P.; Botta, L.; Rosti, V.; Riboni, L.; Sancini, G.; Tanzi, F.; et al. Acetylcholine induces intracellular Ca^{2+} oscillations and nitric oxide release in mouse brain endothelial cells. *Cell Calcium* **2017**, *66*, 33–47. [CrossRef] [PubMed]

43. Huang, T.Y.; Chu, T.F.; Chen, H.I.; Jen, C.J. Heterogeneity of $[Ca^{2+}](i)$ signaling in intact rat aortic endothelium. *FASEB J.* **2000**, *14*, 797–804. [PubMed]

44. Duza, T.; Sarelius, I.H. Localized transient increases in endothelial cell Ca^{2+} in arterioles in situ: Implications for coordination of vascular function. *Am. J. Physiol. Heart Circ. Physiol.* **2004**, *286*, H2322–H2331. [CrossRef] [PubMed]

45. Kansui, Y.; Garland, C.J.; Dora, K.A. Enhanced spontaneous Ca^{2+} events in endothelial cells reflect signalling through myoendothelial gap junctions in pressurized mesenteric arteries. *Cell Calcium* **2008**, *44*, 135–146. [CrossRef] [PubMed]

46. Pedriali, G.; Rimessi, A.; Sbano, L.; Giorgi, C.; Wieckowski, M.R.; Previati, M.; Pinton, P. Regulation of Endoplasmic Reticulum-Mitochondria Ca^{2+} Transfer and Its Importance for Anti-Cancer Therapies. *Front. Oncol.* **2017**, *7*, 180. [CrossRef] [PubMed]

47. De Stefani, D.; Rizzuto, R.; Pozzan, T. Enjoy the Trip: Calcium in Mitochondria Back and Forth. *Annu. Rev. Biochem.* **2016**, *85*, 161–192. [CrossRef] [PubMed]

48. Marcu, R.; Wiczer, B.M.; Neeley, C.K.; Hawkins, B.J. Mitochondrial matrix Ca^{2+} accumulation regulates cytosolic NAD(+)/NADH metabolism, protein acetylation, and sirtuin expression. *Mol. Cell. Biol.* **2014**, *34*, 2890–2902. [CrossRef] [PubMed]

49. Dong, Z.; Shanmughapriya, S.; Tomar, D.; Siddiqui, N.; Lynch, S.; Nemani, N.; Breves, S.L.; Zhang, X.; Tripathi, A.; Palaniappan, P.; et al. Mitochondrial Ca^{2+} Uniporter Is a Mitochondrial Luminal Redox Sensor that Augments MCU Channel Activity. *Mol. Cell* **2017**, *65*, 1014.e7–1028.e7. [CrossRef] [PubMed]

50. Bittremieux, M.; Parys, J.B.; Pinton, P.; Bultynck, G. ER functions of oncogenes and tumor suppressors: Modulators of intracellular Ca^{2+} signaling. *Biochim. Biophys. Acta* **2016**, *1863 Pt B*, 1364–1378. [CrossRef] [PubMed]

51. Pinton, P.; Giorgi, C.; Pandolfi, P.P. The role of PML in the control of apoptotic cell fate: A new key player at ER-mitochondria sites. *Cell Death Differ.* **2011**, *18*, 1450–1456. [CrossRef] [PubMed]

52. Zhu, H.; Jin, Q.; Li, Y.; Ma, Q.; Wang, J.; Li, D.; Zhou, H.; Chen, Y. Melatonin protected cardiac microvascular endothelial cells against oxidative stress injury via suppression of IP3R-$[Ca^{2+}]$c/VDAC-$[Ca^{2+}]$m axis by activation of MAPK/ERK signaling pathway. *Cell Stress Chaperones* **2018**, *23*, 101–113. [CrossRef] [PubMed]

53. Madreiter-Sokolowski, C.T.; Gottschalk, B.; Parichatikanond, W.; Eroglu, E.; Klec, C.; Waldeck-Weiermair, M.; Malli, R.; Graier, W.F. Resveratrol Specifically Kills Cancer Cells by a Devastating Increase in the Ca^{2+} Coupling Between the Greatly Tethered Endoplasmic Reticulum and Mitochondria. *Cell. Physiol. Biochem.* **2016**, *39*, 1404–1420. [CrossRef] [PubMed]

54. Carmeliet, P.; Jain, R.K. Molecular mechanisms and clinical applications of angiogenesis. *Nature* **2011**, *473*, 298–307. [CrossRef] [PubMed]

55. Berridge, M.J.; Bootman, M.D.; Roderick, H.L. Calcium signalling: Dynamics, homeostasis and remodelling. *Nat. Rev. Mol. Cell Biol.* **2003**, *4*, 517–529. [CrossRef] [PubMed]

56. Moccia, F.; Dragoni, S.; Lodola, F.; Bonetti, E.; Bottino, C.; Guerra, G.; Laforenza, U.; Rosti, V.; Tanzi, F. Store-dependent Ca^{2+} entry in endothelial progenitor cells as a perspective tool to enhance cell-based therapy and adverse tumour vascularization. *Curr. Med. Chem.* **2012**, *19*, 5802–5818. [CrossRef] [PubMed]

57. Abdullaev, I.F.; Bisaillon, J.M.; Potier, M.; Gonzalez, J.C.; Motiani, R.K.; Trebak, M. Stim1 and Orai1 mediate CRAC currents and store-operated calcium entry important for endothelial cell proliferation. *Circ. Res.* **2008**, *103*, 1289–1299. [CrossRef] [PubMed]

58. Li, J.; Cubbon, R.M.; Wilson, L.A.; Amer, M.S.; McKeown, L.; Hou, B.; Majeed, Y.; Tumova, S.; Seymour, V.A.L.; Taylor, H.; et al. Orai1 and CRAC channel dependence of VEGF-activated Ca^{2+} entry and endothelial tube formation. *Circ. Res.* **2011**, *108*, 1190–1198. [CrossRef] [PubMed]

59. Blatter, L.A. Tissue Specificity: SOCE: Implications for Ca^{2+} Handling in Endothelial Cells. *Adv. Exp. Med. Biol.* **2017**, *993*, 343–361. [PubMed]

60. Gees, M.; Colsoul, B.; Nilius, B. The role of transient receptor potential cation channels in Ca^{2+} signaling. *Cold Spring Harb. Perspect. Biol.* **2010**, *2*, a003962. [CrossRef] [PubMed]

61. Sundivakkam, P.C.; Freichel, M.; Singh, V.; Yuan, J.P.; Vogel, S.M.; Flockerzi, V.; Malik, A.B.; Tiruppathi, C. The Ca^{2+} sensor stromal interaction molecule 1 (STIM1) is necessary and sufficient for the store-operated Ca^{2+} entry function of transient receptor potential canonical (TRPC) 1 and 4 channels in endothelial cells. *Mol. Pharmacol.* **2012**, *81*, 510–526. [CrossRef] [PubMed]

62. Cioffi, D.L.; Wu, S.; Chen, H.; Alexeyev, M.; St Croix, C.M.; Pitt, B.R.; Uhlig, S.; Stevens, T. Orai1 determines calcium selectivity of an endogenous TRPC heterotetramer channel. *Circ. Res.* **2012**, *110*, 1435–1444. [CrossRef] [PubMed]

63. Xu, N.; Cioffi, D.L.; Alexeyev, M.; Rich, T.C.; Stevens, T. Sodium entry through endothelial store-operated calcium entry channels: Regulation by Orai1. *Am. J. Physiol. Cell Physiol.* **2015**, *308*, C277–C288. [CrossRef] [PubMed]

64. Antigny, F.; Jousset, H.; Konig, S.; Frieden, M. Thapsigargin activates Ca^{2+} entry both by store-dependent, STIM1/Orai1-mediated, and store-independent, TRPC3/PLC/PKC-mediated pathways in human endothelial cells. *Cell Calcium* **2011**, *49*, 115–127. [CrossRef] [PubMed]

65. Weissmann, N.; Sydykov, A.; Kalwa, H.; Storch, U.; Fuchs, B.; Mederos y Schnitzler, M.; Brandes, R.P.; Grimminger, F.; Meissner, M.; Freichel, M.; et al. Activation of TRPC6 channels is essential for lung ischaemia-reperfusion induced oedema in mice. *Nat. Commun.* **2012**, *3*, 649. [CrossRef] [PubMed]

66. Boeldt, D.S.; Grummer, M.A.; Magness, R.R.; Bird, I.M. Altered VEGF-stimulated Ca^{2+} signaling in part underlies pregnancy-adapted eNOS activity in UAEC. *J. Endocrinol.* **2014**, *223*, 1–11. [CrossRef] [PubMed]

67. Hamdollah Zadeh, M.A.; Glass, C.A.; Magnussen, A.; Hancox, J.C.; Bates, D.O. VEGF-mediated elevated intracellular calcium and angiogenesis in human microvascular endothelial cells in vitro are inhibited by dominant negative TRPC6. *Microcirculation* **2008**, *15*, 605–614. [CrossRef] [PubMed]

68. Antoniotti, S.; Lovisolo, D.; Fiorio Pla, A.; Munaron, L. Expression and functional role of bTRPC1 channels in native endothelial cells. *FEBS Lett.* **2002**, *510*, 189–195. [CrossRef]

69. Ho, W.S.; Zheng, X.; Zhang, D.X. Role of endothelial TRPV4 channels in vascular actions of the endocannabinoid, 2-arachidonoylglycerol. *Br. J. Pharmacol.* **2015**, *172*, 5251–5264. [CrossRef] [PubMed]

70. Zheng, X.; Zinkevich, N.S.; Gebremedhin, D.; Gauthier, K.M.; Nishijima, Y.; Fang, J.; Wilcox, D.A.; Campbell, W.B.; Gutterman, D.D.; Zhang, D.X. Arachidonic acid-induced dilation in human coronary arterioles: Convergence of signaling mechanisms on endothelial TRPV4-mediated Ca^{2+} entry. *J. Am. Heart Assoc.* **2013**, *2*, e000080. [CrossRef] [PubMed]

71. Yoshida, T.; Inoue, R.; Morii, T.; Takahashi, N.; Yamamoto, S.; Hara, Y.; Tominaga, M.; Shimizu, S.; Sato, Y.; Mori, Y. Nitric oxide activates TRP channels by cysteine S-nitrosylation. *Nat. Chem. Biol.* **2006**, *2*, 596–607. [CrossRef] [PubMed]

72. Dietrich, A.; Gudermann, T. Another TRP to endothelial dysfunction: TRPM2 and endothelial permeability. *Circ. Res.* **2008**, *102*, 275–277. [CrossRef] [PubMed]

73. Kwan, H.Y.; Cheng, K.T.; Ma, Y.; Huang, Y.; Tang, N.L.; Yu, S.; Yao, X. CNGA2 contributes to ATP-induced noncapacitative Ca^{2+} influx in vascular endothelial cells. *J. Vasc. Res.* **2010**, *47*, 148–156. [CrossRef] [PubMed]

74. Berra-Romani, R.; Raqeeb, A.; Avelino-Cruz, J.E.; Moccia, F.; Oldani, A.; Speroni, F.; Taglietti, V.; Tanzi, F. Ca^{2+} signaling in injured in situ endothelium of rat aorta. *Cell Calcium* **2008**, *44*, 298–309. [CrossRef] [PubMed]

75. Moccia, F.; Frost, C.; Berra-Romani, R.; Tanzi, F.; Adams, D.J. Expression and function of neuronal nicotinic ACh receptors in rat microvascular endothelial cells. *Am. J. Physiol. Heart Circ. Physiol.* **2004**, *286*, H486–H491. [CrossRef] [PubMed]

76. LeMaistre, J.L.; Sanders, S.A.; Stobart, M.J.; Lu, L.; Knox, J.D.; Anderson, H.D.; Anderson, C.M. Coactivation of NMDA receptors by glutamate and D-serine induces dilation of isolated middle cerebral arteries. *J. Cereb. Blood Flow Metab.* **2012**, *32*, 537–547. [CrossRef] [PubMed]

77. AbouAlaiwi, W.A.; Takahashi, M.; Mell, B.R.; Jones, T.J.; Ratnam, S.; Kolb, R.J.; Nauli, S.M. Ciliary polycystin-2 is a mechanosensitive calcium channel involved in nitric oxide signaling cascades. *Circ. Res.* **2009**, *104*, 860–869. [CrossRef] [PubMed]

78. Berrout, J.; Jin, M.; O'Neil, R.G. Critical role of TRPP2 and TRPC1 channels in stretch-induced injury of blood-brain barrier endothelial cells. *Brain Res.* **2012**, *1436*, 1–12. [CrossRef] [PubMed]

79. Filosa, J.A.; Yao, X.; Rath, G. TRPV4 and the regulation of vascular tone. *J. Cardiovasc. Pharmacol.* **2013**, *61*, 113–119. [CrossRef] [PubMed]

80. Du, J.; Ma, X.; Shen, B.; Huang, Y.; Birnbaumer, L.; Yao, X. TRPV4, TRPC1, and TRPP2 assemble to form a flow-sensitive heteromeric channel. *FASEB J.* **2014**, *28*, 4677–4685. [CrossRef] [PubMed]

81. Li, J.; Hou, B.; Tumova, S.; Muraki, K.; Bruns, A.; Ludlow, M.J.; Sedo, A.; Hyman, A.J.; McKeown, L.; Young, R.S.; et al. Piezo1 integration of vascular architecture with physiological force. *Nature* **2014**, *515*, 279–282. [CrossRef] [PubMed]

82. Medina, R.J.; Barber, C.L.; Sabatier, F.; Dignat-George, F.; Melero-Martin, J.M.; Khosrotehrani, K.; Ohneda, O.; Randi, A.M.; Chan, J.K.Y.; Yamaguchi, T.; et al. Endothelial Progenitors: A Consensus Statement on Nomenclature. *Stem Cells Transl. Med.* **2017**, *6*, 1316–1320. [CrossRef] [PubMed]

83. Dragoni, S.; Laforenza, U.; Bonetti, E.; Lodola, F.; Bottino, C.; Berra-Romani, R.; Carlo Bongio, G.; Cinelli, M.P.; Guerra, G.; Pedrazzoli, P.; et al. Vascular endothelial growth factor stimulates endothelial colony forming cells proliferation and tubulogenesis by inducing oscillations in intracellular Ca^{2+} concentration. *Stem Cells* **2011**, *29*, 1898–1907. [CrossRef] [PubMed]

84. Sanchez-Hernandez, Y.; Laforenza, U.; Bonetti, E.; Fontana, J.; Dragoni, S.; Russo, M.; Avelino-Cruz, J.E.; Schinelli, S.; Testa, D.; Guerra, G.; et al. Store-operated Ca^{2+} entry is expressed in human endothelial progenitor cells. *Stem Cells Dev.* **2010**, *19*, 1967–1981. [CrossRef] [PubMed]

85. Dragoni, S.; Laforenza, U.; Bonetti, E.; Lodola, F.; Bottino, C.; Guerra, G.; Borghesi, A.; Stronati, M.; Rosti, V.; Tanzi, F.; et al. Canonical transient receptor potential 3 channel triggers vascular endothelial growth factor-induced intracellular Ca^{2+} oscillations in endothelial progenitor cells isolated from umbilical cord blood. *Stem Cells Dev.* **2013**, *22*, 2561–2580. [CrossRef] [PubMed]

86. Zuccolo, E.; Dragoni, S.; Poletto, V.; Catarsi, P.; Guido, D.; Rappa, A.; Reforgiato, M.; Lodola, F.; Lim, D.; Rosti, V.; et al. Arachidonic acid-evoked Ca^{2+} signals promote nitric oxide release and proliferation in human endothelial colony forming cells. *Vascul. Pharmacol.* **2016**, *87*, 159–171. [CrossRef] [PubMed]

87. Poletto, V.; Dragoni, S.; Lim, D.; Biggiogera, M.; Aronica, A.; Cinelli, M.; De Luca, A.; Rosti, V.; Porta, C.; Guerra, G.; et al. Endoplasmic Reticulum Ca^{2+} Handling and Apoptotic Resistance in Tumor-Derived Endothelial Colony Forming Cells. *J. Cell. Biochem.* **2016**, *117*, 2260–2271. [CrossRef] [PubMed]

88. Pla, A.F.; Ong, H.L.; Cheng, K.T.; Brossa, A.; Bussolati, B.; Lockwich, T.; Paria, B.; Munaron, L.; Ambudkar, I.S. TRPV4 mediates tumor-derived endothelial cell migration via arachidonic acid-activated actin remodeling. *Oncogene* **2012**, *31*, 200–212. [CrossRef] [PubMed]

89. Fiorio Pla, A.; Grange, C.; Antoniotti, S.; Tomatis, C.; Merlino, A.; Bussolati, B.; Munaron, L. Arachidonic acid-induced Ca^{2+} entry is involved in early steps of tumor angiogenesis. *Mol. Cancer Res.* **2008**, *6*, 535–545. [PubMed]

90. Adapala, R.K.; Thoppil, R.J.; Ghosh, K.; Cappelli, H.C.; Dudley, A.C.; Paruchuri, S.; Keshamouni, V.; Klagsbrun, M.; Meszaros, J.G.; Chilian, W.M.; et al. Activation of mechanosensitive ion channel TRPV4 normalizes tumor vasculature and improves cancer therapy. *Oncogene* **2016**, *35*, 314–322. [CrossRef] [PubMed]

91. Thoppil, R.J.; Cappelli, H.C.; Adapala, R.K.; Kanugula, A.K.; Paruchuri, S.; Thodeti, C.K. TRPV4 channels regulate tumor angiogenesis via modulation of Rho/Rho kinase pathway. *Oncotarget* **2016**, *7*, 25849–25861. [CrossRef] [PubMed]

92. Thoppil, R.J.; Adapala, R.K.; Cappelli, H.C.; Kondeti, V.; Dudley, A.C.; Gary Meszaros, J.; Paruchuri, S.; Thodeti, C.K. TRPV4 channel activation selectively inhibits tumor endothelial cell proliferation. *Sci. Rep.* **2015**, *5*, 14257. [CrossRef] [PubMed]

93. Yang, H.; Liu, C.; Zhou, R.M.; Yao, J.; Li, X.M.; Shen, Y.; Cheng, H.; Yuan, J.; Yan, B.; Jiang, Q. Piezo2 protein: A novel regulator of tumor angiogenesis and hyperpermeability. *Oncotarget* **2016**, *7*, 44630–44643. [CrossRef] [PubMed]

94. Avanzato, D.; Genova, T.; Fiorio Pla, A.; Bernardini, M.; Bianco, S.; Bussolati, B.; Mancardi, D.; Giraudo, E.; Maione, F.; Cassoni, P.; et al. Activation of P2X7 and P2Y11 purinergic receptors inhibits migration and normalizes tumor-derived endothelial cells via cAMP signaling. *Sci. Rep.* **2016**, *6*, 32602. [CrossRef] [PubMed]

95. Lodola, F.; Laforenza, U.; Bonetti, E.; Lim, D.; Dragoni, S.; Bottino, C.; Ong, H.L.; Guerra, G.; Ganini, C.; Massa, M.; et al. Store-operated Ca^{2+} entry is remodelled and controls in vitro angiogenesis in endothelial progenitor cells isolated from tumoral patients. *PLoS ONE* **2012**, *7*, e42541. [CrossRef] [PubMed]

96. Lodola, F.; Laforenza, U.; Cattaneo, F.; Ruffinatti, F.A.; Poletto, V.; Massa, M.; Tancredi, R.; Zuccolo, E.; Khdar, A.D.; Riccardi, A.; et al. VEGF-induced intracellular Ca^{2+} oscillations are down-regulated and do not stimulate angiogenesis in breast cancer-derived endothelial colony forming cells. *Oncotarget* **2017**, *8*, 95223–95246. [PubMed]

97. Zuccolo, E.; Bottino, C.; Diofano, F.; Poletto, V.; Codazzi, A.C.; Mannarino, S.; Campanelli, R.; Fois, G.; Marseglia, G.L.; Guerra, G.; et al. Constitutive Store-Operated Ca^{2+} Entry Leads to Enhanced Nitric Oxide Production and Proliferation in Infantile Hemangioma-Derived Endothelial Colony-Forming Cells. *Stem Cells Dev.* **2016**, *25*, 301–319. [CrossRef] [PubMed]

98. Heeschen, C.; Weis, M.; Aicher, A.; Dimmeler, S.; Cooke, J.P. A novel angiogenic pathway mediated by non-neuronal nicotinic acetylcholine receptors. *J. Clin. Investig.* **2002**, *110*, 527–536. [CrossRef] [PubMed]

99. Heeschen, C.; Jang, J.J.; Weis, M.; Pathak, A.; Kaji, S.; Hu, R.S.; Tsao, P.S.; Johnson, F.L.; Cooke, J.P. Nicotine stimulates angiogenesis and promotes tumor growth and atherosclerosis. *Nat. Med.* **2001**, *7*, 833–839. [CrossRef] [PubMed]

100. Yu, M.; Liu, Q.; Sun, J.; Yi, K.; Wu, L.; Tan, X. Nicotine improves the functional activity of late endothelial progenitor cells via nicotinic acetylcholine receptors. *Biochem. Cell Biol.* **2011**, *89*, 405–410. [CrossRef] [PubMed]

101. Natori, T.; Sata, M.; Washida, M.; Hirata, Y.; Nagai, R.; Makuuchi, M. Nicotine enhances neovascularization and promotes tumor growth. *Mol. Cells* **2003**, *16*, 143–146. [PubMed]

102. Alonso, F.; Domingos-Pereira, S.; Le Gal, L.; Derre, L.; Meda, P.; Jichlinski, P.; Nardelli-Haefliger, D.; Haefliger, J.A. Targeting endothelial connexin40 inhibits tumor growth by reducing angiogenesis and improving vessel perfusion. *Oncotarget* **2016**, *7*, 14015–14028. [CrossRef] [PubMed]

103. Rigiracciolo, D.C.; Scarpelli, A.; Lappano, R.; Pisano, A.; Santolla, M.F.; Avino, S.; De Marco, P.; Bussolati, B.; Maggiolini, M.; De Francesco, E.M. GPER is involved in the stimulatory effects of aldosterone in breast cancer cells and breast tumor-derived endothelial cells. *Oncotarget* **2016**, *7*, 94–111. [CrossRef] [PubMed]

104. Dong, Y.; Pan, Q.; Jiang, L.; Chen, Z.; Zhang, F.; Liu, Y.; Xing, H.; Shi, M.; Li, J.; Li, X.; et al. Tumor endothelial expression of P-glycoprotein upon microvesicular transfer of TrpC5 derived from adriamycin-resistant breast cancer cells. *Biochem. Biophys. Res. Commun.* **2014**, *446*, 85–90. [CrossRef] [PubMed]

105. White, J.P.; Cibelli, M.; Urban, L.; Nilius, B.; McGeown, J.G.; Nagy, I. TRPV4: Molecular Conductor of a Diverse Orchestra. *Physiol. Rev.* **2016**, *96*, 911–973. [CrossRef] [PubMed]

106. Everaerts, W.; Nilius, B.; Owsianik, G. The vanilloid transient receptor potential channel TRPV4: From structure to disease. *Prog. Biophys. Mol. Biol.* **2010**, *103*, 2–17. [CrossRef] [PubMed]

107. He, D.; Pan, Q.; Chen, Z.; Sun, C.; Zhang, P.; Mao, A.; Zhu, Y.; Li, H.; Lu, C.; Xie, M.; et al. Treatment of hypertension by increasing impaired endothelial TRPV4-KCa2.3 interaction. *EMBO Mol. Med.* **2017**, *9*, 1491–1503. [CrossRef] [PubMed]

108. Chen, C.K.; Hsu, P.Y.; Wang, T.M.; Miao, Z.F.; Lin, R.T.; Juo, S.H. TRPV4 Activation Contributes Functional Recovery from Ischemic Stroke via Angiogenesis and Neurogenesis. *Mol. Neurobiol.* **2017**. [CrossRef] [PubMed]

109. Hatano, N.; Suzuki, H.; Itoh, Y.; Muraki, K. TRPV4 partially participates in proliferation of human brain capillary endothelial cells. *Life Sci.* **2013**, *92*, 317–324. [CrossRef] [PubMed]

110. Matthews, B.D.; Thodeti, C.K.; Tytell, J.D.; Mammoto, A.; Overby, D.R.; Ingber, D.E. Ultra-rapid activation of TRPV4 ion channels by mechanical forces applied to cell surface beta1 integrins. *Integr. Biol. (Camb.)* **2010**, *2*, 435–442. [CrossRef] [PubMed]

111. Thodeti, C.K.; Matthews, B.; Ravi, A.; Mammoto, A.; Ghosh, K.; Bracha, A.L.; Ingber, D.E. TRPV4 Channels Mediate Cyclic Strain-Induced Endothelial Cell Reorientation Through Integrin-to-Integrin Signaling. *Circ. Res.* **2009**, *104*, 1123–1130. [CrossRef] [PubMed]

112. Sonkusare, S.K.; Bonev, A.D.; Ledoux, J.; Liedtke, W.; Kotlikoff, M.I.; Heppner, T.J.; Hill-Eubanks, D.C.; Nelson, M.T. Elementary Ca^{2+} signals through endothelial TRPV4 channels regulate vascular function. *Science* **2012**, *336*, 597–601. [CrossRef] [PubMed]

113. Zhao, L.; Sullivan, M.N.; Chase, M.; Gonzales, A.L.; Earley, S. Calcineurin/nuclear factor of activated T cells-coupled vanilliod transient receptor potential channel 4 Ca^{2+} sparklets stimulate airway smooth muscle cell proliferation. *Am. J. Respir. Cell Mol. Biol.* **2014**, *50*, 1064–1075. [CrossRef] [PubMed]

114. Wen, Z.H.; Su, Y.C.; Lai, P.L.; Zhang, Y.; Xu, Y.F.; Zhao, A.; Yao, G.Y.; Jia, C.H.; Lin, J.; Xu, S.; et al. Critical role of arachidonic acid-activated mTOR signaling in breast carcinogenesis and angiogenesis. *Oncogene* **2013**, *32*, 160–170. [CrossRef] [PubMed]

115. Munaron, L. Shuffling the cards in signal transduction: Calcium, arachidonic acid and mechanosensitivity. *World J. Biol. Chem.* **2011**, *2*, 59–66. [CrossRef] [PubMed]

116. Kim, E.; Tunset, H.M.; Cebulla, J.; Vettukattil, R.; Helgesen, H.; Feuerherm, A.J.; Engebraten, O.; Maelandsmo, G.M.; Johansen, B.; Moestue, S.A. Anti-vascular effects of the cytosolic phospholipase A2 inhibitor AVX235 in a patient-derived basal-like breast cancer model. *BMC Cancer* **2016**, *16*, 191. [CrossRef] [PubMed]

117. Carmeliet, P.; Jain, R.K. Principles and mechanisms of vessel normalization for cancer and other angiogenic diseases. *Nat. Rev. Drug Discov.* **2011**, *10*, 417–427. [CrossRef] [PubMed]

118. Goel, S.; Duda, D.G.; Xu, L.; Munn, L.L.; Boucher, Y.; Fukumura, D.; Jain, R.K. Normalization of the vasculature for treatment of cancer and other diseases. *Physiol. Rev.* **2011**, *91*, 1071–1121. [CrossRef] [PubMed]

119. Bagriantsev, S.N.; Gracheva, E.O.; Gallagher, P.G. Piezo proteins: Regulators of mechanosensation and other cellular processes. *J. Biol. Chem.* **2014**, *289*, 31673–31681. [CrossRef] [PubMed]

120. Honore, E.; Martins, J.R.; Penton, D.; Patel, A.; Demolombe, S. The Piezo Mechanosensitive Ion Channels: May the Force Be with You! *Rev. Physiol. Biochem. Pharmacol.* **2015**, *169*, 25–41. [PubMed]

121. Ranade, S.S.; Qiu, Z.; Woo, S.H.; Hur, S.S.; Murthy, S.E.; Cahalan, S.M.; Xu, J.; Mathur, J.; Bandell, M.; Coste, B.; et al. Piezo1, a mechanically activated ion channel, is required for vascular development in mice. *Proc. Natl. Acad. Sci. USA* **2014**, *111*, 10347–10352. [CrossRef] [PubMed]

122. Di Virgilio, F.; Adinolfi, E. Extracellular purines, purinergic receptors and tumor growth. *Oncogene* **2017**, *36*, 293–303. [CrossRef] [PubMed]

123. Burnstock, G. Purinergic Signaling in the Cardiovascular System. *Circ. Res.* **2017**, *120*, 207–228. [CrossRef] [PubMed]

124. Adinolfi, E.; Raffaghello, L.; Giuliani, A.L.; Cavazzini, L.; Capece, M.; Chiozzi, P.; Bianchi, G.; Kroemer, G.; Pistoia, V.; Di Virgilio, F. Expression of P2X7 receptor increases in vivo tumor growth. *Cancer Res.* **2012**, *72*, 2957–2969. [CrossRef] [PubMed]

125. Azimi, I.; Beilby, H.; Davis, F.M.; Marcial, D.L.; Kenny, P.A.; Thompson, E.W.; Roberts-Thomson, S.J.; Monteith, G.R. Altered purinergic receptor-Ca^{2+} signaling associated with hypoxia-induced epithelial-mesenchymal transition in breast cancer cells. *Mol. Oncol.* **2016**, *10*, 166–178. [CrossRef] [PubMed]

126. Fang, J.; Chen, X.; Wang, S.; Xie, T.; Du, X.; Liu, H.; Li, X.; Chen, J.; Zhang, B.; Liang, H.; et al. The expression of P2X(7) receptors in EPCs and their potential role in the targeting of EPCs to brain gliomas. *Cancer Biol. Ther.* **2015**, *16*, 498–510. [CrossRef] [PubMed]

127. Dragoni, S.; Guerra, G.; Fiorio Pla, A.; Bertoni, G.; Rappa, A.; Poletto, V.; Bottino, C.; Aronica, A.; Lodola, F.; Cinelli, M.P.; et al. A functional Transient Receptor Potential Vanilloid 4 (TRPV4) channel is expressed in human endothelial progenitor cells. *J. Cell. Physiol.* **2015**, *230*, 95–104. [CrossRef] [PubMed]

128. Hofmann, N.A.; Barth, S.; Waldeck-Weiermair, M.; Klec, C.; Strunk, D.; Malli, R.; Graier, W.F. TRPV1 mediates cellular uptake of anandamide and thus promotes endothelial cell proliferation and network-formation. *Biol. Open* **2014**, *3*, 1164–1172. [CrossRef] [PubMed]

129. Ong, H.L.; Jang, S.I.; Ambudkar, I.S. Distinct contributions of Orai1 and TRPC1 to agonist-induced [Ca^{2+}](i) signals determine specificity of Ca^{2+}-dependent gene expression. *PLoS ONE* **2012**, *7*, e47146. [CrossRef] [PubMed]

130. Gueguinou, M.; Harnois, T.; Crottes, D.; Uguen, A.; Deliot, N.; Gambade, A.; Chantome, A.; Haelters, J.P.; Jaffres, P.A.; Jourdan, M.L.; et al. SK3/TRPC1/Orai1 complex regulates SOCE-dependent colon cancer cell migration: A novel opportunity to modulate anti-EGFR mAb action by the alkyl-lipid Ohmline. *Oncotarget* **2016**, *7*, 36168–36184. [CrossRef] [PubMed]

131. Moccia, F.; Dragoni, S.; Poletto, V.; Rosti, V.; Tanzi, F.; Ganini, C.; Porta, C. Orai1 and Transient Receptor Potential Channels as novel molecular targets to impair tumor neovascularisation in renal cell carcinoma and other malignancies. *Anticancer Agents Med. Chem.* **2014**, *14*, 296–312. [CrossRef] [PubMed]

132. Porta, C.; Giglione, P.; Paglino, C. Targeted therapy for renal cell carcinoma: Focus on 2nd and 3rd line. *Expert Opin. Pharmacother.* **2016**, *17*, 643–655. [CrossRef] [PubMed]

133. Prakriya, M.; Lewis, R.S. Store-Operated Calcium Channels. *Physiol. Rev.* **2015**, *95*, 1383–1436. [CrossRef] [PubMed]

134. Vashisht, A.; Trebak, M.; Motiani, R.K. STIM and Orai proteins as novel targets for cancer therapy. A Review in the Theme: Cell and Molecular Processes in Cancer Metastasis. *Am. J. Physiol. Cell Physiol.* **2015**, *309*, C457–C469. [CrossRef] [PubMed]

135. Moccia, F.; Zuccolo, E.; Poletto, V.; Turin, I.; Guerra, G.; Pedrazzoli, P.; Rosti, V.; Porta, C.; Montagna, D. Targeting Stim and Orai Proteins as an Alternative Approach in Anticancer Therapy. *Curr. Med. Chem.* **2016**, *23*, 3450–3480. [CrossRef] [PubMed]

136. Li, Y.; Guo, B.; Xie, Q.; Ye, D.; Zhang, D.; Zhu, Y.; Chen, H.; Zhu, B. STIM1 Mediates Hypoxia-Driven Hepatocarcinogenesis via Interaction with HIF-1. *Cell Rep.* **2015**, *12*, 388–395. [CrossRef] [PubMed]

137. Gudas, L.J.; Fu, L.; Minton, D.R.; Mongan, N.P.; Nanus, D.M. The role of HIF1alpha in renal cell carcinoma tumorigenesis. *J. Mol. Med. (Berl.)* **2014**, *92*, 825–836. [CrossRef] [PubMed]

138. Semenza, G.L. Regulation of the breast cancer stem cell phenotype by hypoxia-inducible factors. *Clin. Sci. (Lond.)* **2015**, *129*, 1037–1045. [CrossRef] [PubMed]

139. Wang, J.; Weigand, L.; Lu, W.; Sylvester, J.T.; Semenza, G.L.; Shimoda, L.A. Hypoxia inducible factor 1 mediates hypoxia-induced TRPC expression and elevated intracellular Ca^{2+} in pulmonary arterial smooth muscle cells. *Circ. Res.* **2006**, *98*, 1528–1537. [CrossRef] [PubMed]

140. Leaute-Labreze, C.; Harper, J.I.; Hoeger, P.H. Infantile haemangioma. *Lancet* **2017**, *390*, 85–94. [CrossRef]

141. Bischoff, J. Progenitor cells in infantile hemangioma. *J. Craniofac. Surg.* **2009**, *20* (Suppl. 1), 695–697. [CrossRef] [PubMed]

142. Khan, Z.A.; Melero-Martin, J.M.; Wu, X.; Paruchuri, S.; Boscolo, E.; Mulliken, J.B.; Bischoff, J. Endothelial progenitor cells from infantile hemangioma and umbilical cord blood display unique cellular responses to endostatin. *Blood* **2006**, *108*, 915–921. [CrossRef] [PubMed]

143. Brandman, O.; Liou, J.; Park, W.S.; Meyer, T. STIM2 is a feedback regulator that stabilizes basal cytosolic and endoplasmic reticulum Ca^{2+} levels. *Cell* **2007**, *131*, 1327–1339. [CrossRef] [PubMed]

144. Greenberger, S.; Bischoff, J. Infantile hemangioma-mechanism(s) of drug action on a vascular tumor. *Cold Spring Harb. Perspect. Med.* **2011**, *1*, a006460. [CrossRef] [PubMed]

145. Zoli, M.; Pucci, S.; Vilella, A.; Gotti, C. Neuronal and extraneuronal nicotinic acetylcholine receptors. *Curr. Neuropharmacol.* **2017**. [CrossRef]

146. Yakel, J.L. Nicotinic ACh receptors in the hippocampal circuit; functional expression and role in synaptic plasticity. *J. Physiol.* **2014**, *592*, 4147–4153. [CrossRef] [PubMed]

147. Egleton, R.D.; Brown, K.C.; Dasgupta, P. Nicotinic acetylcholine receptors in cancer: Multiple roles in proliferation and inhibition of apoptosis. *Trends Pharmacol. Sci.* **2008**, *29*, 151–158. [CrossRef] [PubMed]

148. Cooke, J.P.; Ghebremariam, Y.T. Endothelial nicotinic acetylcholine receptors and angiogenesis. *Trends Cardiovasc. Med.* **2008**, *18*, 247–253. [CrossRef] [PubMed]

149. Egleton, R.D.; Brown, K.C.; Dasgupta, P. Angiogenic activity of nicotinic acetylcholine receptors: Implications in tobacco-related vascular diseases. *Pharmacol. Ther.* **2009**, *121*, 205–223. [CrossRef] [PubMed]

150. Dom, A.M.; Buckley, A.W.; Brown, K.C.; Egleton, R.D.; Marcelo, A.J.; Proper, N.A.; Weller, D.E.; Shah, Y.H.; Lau, J.K.; Dasgupta, P. The alpha7-nicotinic acetylcholine receptor and MMP-2/-9 pathway mediate the proangiogenic effect of nicotine in human retinal endothelial cells. *Investig. Ophthalmol. Vis. Sci.* **2011**, *52*, 4428–4438. [CrossRef] [PubMed]

151. Smedlund, K.; Tano, J.Y.; Margiotta, J.; Vazquez, G. Evidence for operation of nicotinic and muscarinic acetylcholine receptor-dependent survival pathways in human coronary artery endothelial cells. *J. Cell Biochem.* **2011**, *112*, 1978–1984. [CrossRef] [PubMed]

152. Fucile, S. Ca^{2+} permeability of nicotinic acetylcholine receptors. *Cell Calcium* **2004**, *35*, 1–8. [CrossRef] [PubMed]

153. Shi, D.; Guo, W.; Chen, W.; Fu, L.; Wang, J.; Tian, Y.; Xiao, X.; Kang, T.; Huang, W.; Deng, W. Nicotine promotes proliferation of human nasopharyngeal carcinoma cells by regulating alpha7AChR, ERK, HIF-1alpha and VEGF/PEDF signaling. *PLoS ONE* **2012**, *7*, e43898.

154. Zhu, B.Q.; Heeschen, C.; Sievers, R.E.; Karliner, J.S.; Parmley, W.W.; Glantz, S.A.; Cooke, J.P. Second hand smoke stimulates tumor angiogenesis and growth. *Cancer Cell* **2003**, *4*, 191–196. [CrossRef]

155. Altaany, Z.; Moccia, F.; Munaron, L.; Mancardi, D.; Wang, R. Hydrogen sulfide and endothelial dysfunction: Relationship with nitric oxide. *Curr. Med. Chem.* **2014**, *21*, 3646–3661. [CrossRef] [PubMed]

156. Mancardi, D.; Pla, A.F.; Moccia, F.; Tanzi, F.; Munaron, L. Old and new gasotransmitters in the cardiovascular system: Focus on the role of nitric oxide and hydrogen sulfide in endothelial cells and cardiomyocytes. *Curr. Pharm. Biotechnol.* **2011**, *12*, 1406–1415. [CrossRef] [PubMed]

157. Wang, R. Physiological implications of hydrogen sulfide: A whiff exploration that blossomed. *Physiol. Rev.* **2012**, *92*, 791–896. [CrossRef] [PubMed]

158. Hartmann, C.; Nussbaum, B.; Calzia, E.; Radermacher, P.; Wepler, M. Gaseous Mediators and Mitochondrial Function: The Future of Pharmacologically Induced Suspended Animation? *Front. Physiol.* **2017**, *8*, 691. [CrossRef] [PubMed]

159. Munaron, L. Intracellular calcium, endothelial cells and angiogenesis. *Recent Pat. Anticancer Drug Discov.* **2006**, *1*, 105–119. [CrossRef] [PubMed]

160. Loboda, A.; Jozkowicz, A.; Dulak, J. Carbon monoxide: Pro- or anti-angiogenic agent? Comment on Ahmad et al. (Thromb Haemost 2015; 113: 329–337). *Thromb. Haemost.* **2015**, *114*, 432–433. [CrossRef] [PubMed]

161. Tran, A.N.; Boyd, N.H.; Walker, K.; Hjelmeland, A.B. NOS Expression and NO Function in Glioma and Implications for Patient Therapies. *Antioxid. Redox Signal.* **2017**, *26*, 986–999. [CrossRef] [PubMed]

162. Mocellin, S. Nitric oxide: Cancer target or anticancer agent? *Curr. Cancer Drug Targets* **2009**, *9*, 214–236. [CrossRef] [PubMed]

163. Basudhar, D.; Somasundaram, V.; de Oliveira, G.A.; Kesarwala, A.; Heinecke, J.L.; Cheng, R.Y.; Glynn, S.A.; Ambs, S.; Wink, D.A.; Ridnour, L.A. Nitric Oxide Synthase-2-Derived Nitric Oxide Drives Multiple Pathways of Breast Cancer Progression. *Antioxid. Redox Signal.* **2017**, *26*, 1044–1058. [CrossRef] [PubMed]

164. Fiorio Pla, A.; Genova, T.; Pupo, E.; Tomatis, C.; Genazzani, A.; Zaninetti, R.; Munaron, L. Multiple roles of protein kinase a in arachidonic acid-mediated Ca^{2+} entry and tumor-derived human endothelial cell migration. *Mol. Cancer Res.* **2010**, *8*, 1466–1476. [CrossRef] [PubMed]

165. Dragoni, S.; Laforenza, U.; Bonetti, E.; Reforgiato, M.; Poletto, V.; Lodola, F.; Bottino, C.; Guido, D.; Rappa, A.; Pareek, S.; et al. Enhanced expression of Stim, Orai, and TRPC transcripts and proteins in endothelial progenitor cells isolated from patients with primary myelofibrosis. *PLoS ONE* **2014**, *9*, e91099. [CrossRef] [PubMed]

166. Munaron, L.; Avanzato, D.; Moccia, F.; Mancardi, D. Hydrogen sulfide as a regulator of calcium channels. *Cell Calcium* **2013**, *53*, 77–84. [CrossRef] [PubMed]

167. Bauer, C.C.; Boyle, J.P.; Porter, K.E.; Peers, C. Modulation of Ca^{2+} signalling in human vascular endothelial cells by hydrogen sulfide. *Atherosclerosis* **2010**, *209*, 374–380. [CrossRef] [PubMed]

168. Moccia, F.; Bertoni, G.; Pla, A.F.; Dragoni, S.; Pupo, E.; Merlino, A.; Mancardi, D.; Munaron, L.; Tanzi, F. Hydrogen sulfide regulates intracellular Ca^{2+} concentration in endothelial cells from excised rat aorta. *Curr. Pharm. Biotechnol.* **2011**, *12*, 1416–1426. [CrossRef] [PubMed]

169. Katsouda, A.; Bibli, S.I.; Pyriochou, A.; Szabo, C.; Papapetropoulos, A. Regulation and role of endogenously produced hydrogen sulfide in angiogenesis. *Pharmacol. Res.* **2016**, *113 Pt A*, 175–185. [CrossRef] [PubMed]

170. Liu, Y.; Yang, R.; Liu, X.; Zhou, Y.; Qu, C.; Kikuiri, T.; Wang, S.; Zandi, E.; Du, J.; Ambudkar, I.S.; et al. Hydrogen Sulfide Maintains Mesenchymal Stem Cell Function and Bone Homeostasis via Regulation of Ca^{2+} Channel Sulfhydration. *Cell Stem Cell* **2014**, *15*, 66–78. [CrossRef] [PubMed]

171. Ujike, A.; Otsuguro, K.; Miyamoto, R.; Yamaguchi, S.; Ito, S. Bidirectional effects of hydrogen sulfide via ATP-sensitive K(+) channels and transient receptor potential A1 channels in RIN14B cells. *Eur. J. Pharmacol.* **2015**, *764*, 463–470. [CrossRef] [PubMed]

172. Yang, G.; Wu, L.; Jiang, B.; Yang, W.; Qi, J.; Cao, K.; Meng, Q.; Mustafa, A.K.; Mu, W.; Zhang, S.; et al. H2S as a physiologic vasorelaxant: Hypertension in mice with deletion of cystathionine gamma-lyase. *Science* **2008**, *322*, 587–590. [CrossRef] [PubMed]

173. Dedkova, E.N.; Blatter, L.A. Nitric oxide inhibits capacitative Ca^{2+} entry and enhances endoplasmic reticulum Ca^{2+} uptake in bovine vascular endothelial cells. *J. Physiol.* **2002**, *539 Pt 1*, 77–91. [CrossRef] [PubMed]

174. Saez, J.C.; Leybaert, L. Hunting for connexin hemichannels. *FEBS Lett.* **2014**, *588*, 1205–1211. [CrossRef] [PubMed]

175. Boeldt, D.S.; Krupp, J.; Yi, F.X.; Khurshid, N.; Shah, D.M.; Bird, I.M. Positive versus negative effects of VEGF165 on Ca^{2+} signaling and NO production in human endothelial cells. *Am. J. Physiol. Heart Circ. Physiol.* **2017**, *312*, H173–H181. [CrossRef] [PubMed]

176. Boittin, F.X.; Alonso, F.; Le Gal, L.; Allagnat, F.; Beny, J.L.; Haefliger, J.A. Connexins and M3 muscarinic receptors contribute to heterogeneous Ca^{2+} signaling in mouse aortic endothelium. *Cell. Physiol. Biochem.* **2013**, *31*, 166–178. [CrossRef] [PubMed]

177. Berra-Romani, R.; Raqeeb, A.; Torres-Jácome, J.; Guzman-Silva, A.; Guerra, G.; Tanzi, F.; Moccia, F. The mechanism of injury-induced intracellular calcium concentration oscillations in the endothelium of excised rat aorta. *J. Vasc. Res.* **2012**, *49*, 65–76. [CrossRef] [PubMed]

178. De Bock, M.; Culot, M.; Wang, N.; Bol, M.; Decrock, E.; De Vuyst, E.; da Costa, A.; Dauwe, I.; Vinken, M.; Simon, A.M.; et al. Connexin channels provide a target to manipulate brain endothelial calcium dynamics and blood-brain barrier permeability. *J. Cereb. Blood Flow Metab.* **2011**, *31*, 1942–1957. [CrossRef] [PubMed]

179. De Bock, M.; Wang, N.; Decrock, E.; Bol, M.; Gadicherla, A.K.; Culot, M.; Cecchelli, R.; Bultynck, G.; Leybaert, L. Endothelial calcium dynamics, connexin channels and blood-brain barrier function. *Prog. Neurobiol.* **2013**, *108*, 1–20. [CrossRef] [PubMed]

180. McLachlan, E.; Shao, Q.; Wang, H.L.; Langlois, S.; Laird, D.W. Connexins act as tumor suppressors in three-dimensional mammary cell organoids by regulating differentiation and angiogenesis. *Cancer Res.* **2006**, *66*, 9886–9894. [CrossRef] [PubMed]

181. Trosko, J.E.; Ruch, R.J. Gap junctions as targets for cancer chemoprevention and chemotherapy. *Curr. Drug Targets* **2002**, *3*, 465–482. [CrossRef] [PubMed]

182. Zhang, J.; O'Carroll, S.J.; Henare, K.; Ching, L.M.; Ormonde, S.; Nicholson, L.F.; Danesh-Meyer, H.V.; Green, C.R. Connexin hemichannel induced vascular leak suggests a new paradigm for cancer therapy. *FEBS Lett.* **2014**, *588*, 1365–1371. [CrossRef] [PubMed]

183. Counillon, L.; Bouret, Y.; Marchiq, I.; Pouyssegur, J. Na(+)/H(+) antiporter (NHE1) and lactate/H(+) symporters (MCTs) in pH homeostasis and cancer metabolism. *Biochim. Biophys. Acta* **2016**, *1863*, 2465–2480. [CrossRef] [PubMed]

184. Mo, X.G.; Chen, Q.W.; Li, X.S.; Zheng, M.M.; Ke, D.Z.; Deng, W.; Li, G.Q.; Jiang, J.; Wu, Z.Q.; Wang, L.; et al. Suppression of NHE1 by small interfering RNA inhibits HIF-1alpha-induced angiogenesis in vitro via modulation of calpain activity. *Microvasc. Res.* **2011**, *81*, 160–168. [CrossRef] [PubMed]

185. Ayajiki, K.; Kindermann, M.; Hecker, M.; Fleming, I.; Busse, R. Intracellular pH and tyrosine phosphorylation but not calcium determine shear stress-induced nitric oxide production in native endothelial cells. *Circ. Res.* **1996**, *78*, 750–758. [CrossRef] [PubMed]

186. Yuen, N.; Lam, T.I.; Wallace, B.K.; Klug, N.R.; Anderson, S.E.; O'Donnell, M.E. Ischemic factor-induced increases in cerebral microvascular endothelial cell Na/H exchange activity and abundance: Evidence for involvement of ERK1/2 MAP kinase. *Am. J. Physiol. Cell Physiol.* **2014**, *306*, C931–C942. [CrossRef] [PubMed]

187. Ghigo, D.; Bussolino, F.; Garbarino, G.; Heller, R.; Turrini, F.; Pescarmona, G.; Cragoe, E.J., Jr.; Pegoraro, L.; Bosia, A. Role of Na^+/H^+ exchange in thrombin-induced platelet-activating factor production by human endothelial cells. *J. Biol. Chem.* **1988**, *263*, 19437–19446. [PubMed]

188. Siffert, W.; Akkerman, J.W. Na^+/H^+ exchange and Ca^{2+} influx. *FEBS Lett.* **1989**, *259*, 1–4. [CrossRef]

189. Danthuluri, N.R.; Kim, D.; Brock, T.A. Intracellular alkalinization leads to Ca^{2+} mobilization from agonist-sensitive pools in bovine aortic endothelial cells. *J. Biol. Chem.* **1990**, *265*, 19071–19076. [PubMed]

190. Nishio, K.; Suzuki, Y.; Takeshita, K.; Aoki, T.; Kudo, H.; Sato, N.; Naoki, K.; Miyao, N.; Ishii, M.; Yamaguchi, K. Effects of hypercapnia and hypocapnia on $[Ca^{2+}]i$ mobilization in human pulmonary artery endothelial cells. *J. Appl. Physiol. (1985)* **2001**, *90*, 2094–2100. [CrossRef] [PubMed]

191. Amith, S.R.; Fliegel, L. Regulation of the Na^+/H^+ Exchanger (NHE1) in Breast Cancer Metastasis. *Cancer Res.* **2013**, *73*, 1259–1264. [CrossRef] [PubMed]

192. Reshkin, S.J.; Greco, M.R.; Cardone, R.A. Role of pHi, and proton transporters in oncogene-driven neoplastic transformation. *Philos. Trans. R. Soc. Lond. B Biol. Sci.* **2014**, *369*, 20130100. [CrossRef] [PubMed]

193. Pedersen, A.K.; Mendes Lopes de Melo, J.; Morup, N.; Tritsaris, K.; Pedersen, S.F. Tumor microenvironment conditions alter Akt and Na$^+$/H$^+$ exchanger NHE1 expression in endothelial cells more than hypoxia alone: Implications for endothelial cell function in cancer. *BMC Cancer* **2017**, *17*, 542. [CrossRef] [PubMed]

194. Orive, G.; Reshkin, S.J.; Harguindey, S.; Pedraz, J.L. Hydrogen ion dynamics and the Na$^+$/H$^+$ exchanger in cancer angiogenesis and antiangiogenesis. *Br. J. Cancer* **2003**, *89*, 1395–1399. [CrossRef] [PubMed]

195. Spugnini, E.P.; Sonveaux, P.; Stock, C.; Perez-Sayans, M.; De Milito, A.; Avnet, S.; Garcia, A.G.; Harguindey, S.; Fais, S. Proton channels and exchangers in cancer. *Biochim. Biophys. Acta* **2015**, *1848 Pt B*, 2715–2726. [CrossRef] [PubMed]

196. Morgan, A.J.; Platt, F.M.; Lloyd-Evans, E.; Galione, A. Molecular mechanisms of endolysosomal Ca^{2+} signalling in health and disease. *Biochem. J.* **2011**, *439*, 349–374. [CrossRef] [PubMed]

197. Jha, A.; Brailoiu, E.; Muallem, S. How does NAADP release lysosomal Ca^{2+}? *Channels* **2014**, *8*, 174–175. [CrossRef] [PubMed]

198. Patel, S.; Docampo, R. Acidic calcium stores open for business: Expanding the potential for intracellular Ca^{2+} signaling. *Trends Cell Biol.* **2010**, *20*, 277–286. [CrossRef] [PubMed]

199. Cosker, F.; Cheviron, N.; Yamasaki, M.; Menteyne, A.; Lund, F.E.; Moutin, M.J.; Galione, A.; Cancela, J.M. The ecto-enzyme CD38 is a nicotinic acid adenine dinucleotide phosphate (NAADP) synthase that couples receptor activation to Ca^{2+} mobilization from lysosomes in pancreatic acinar cells. *J. Biol. Chem.* **2010**, *285*, 38251–38259. [CrossRef] [PubMed]

200. Galione, A. A primer of NAADP-mediated Ca^{2+} signalling: From sea urchin eggs to mammalian cells. *Cell Calcium* **2015**, *58*, 27–47. [CrossRef] [PubMed]

201. Moccia, F.; Nusco, G.A.; Lim, D.; Kyozuka, K.; Santella, L. NAADP and InsP3 play distinct roles at fertilization in starfish oocytes. *Dev. Biol.* **2006**, *294*, 24–38. [CrossRef] [PubMed]

202. Favia, A.; Desideri, M.; Gambara, G.; D'Alessio, A.; Ruas, M.; Esposito, B.; Del Bufalo, D.; Parrington, J.; Ziparo, E.; Palombi, F.; et al. VEGF-induced neoangiogenesis is mediated by NAADP and two-pore channel-2-dependent Ca^{2+} signaling. *Proc. Natl. Acad. Sci. USA* **2014**, *111*, E4706–E4715. [CrossRef] [PubMed]

203. Brailoiu, G.C.; Gurzu, B.; Gao, X.; Parkesh, R.; Aley, P.K.; Trifa, D.I.; Galione, A.; Dun, N.J.; Madesh, M.; Patel, S.; et al. Acidic NAADP-sensitive calcium stores in the endothelium: Agonist-specific recruitment and role in regulating blood pressure. *J. Biol. Chem.* **2010**, *285*, 37133–37137. [CrossRef] [PubMed]

204. Di Nezza, F.; Zuccolo, E.; Poletto, V.; Rosti, V.; De Luca, A.; Moccia, F.; Guerra, G.; Ambrosone, L. Liposomes as a Putative Tool to Investigate NAADP Signaling in Vasculogenesis. *J. Cell. Biochem.* **2017**, *118*, 3722–3729. [CrossRef] [PubMed]

205. Favia, A.; Pafumi, I.; Desideri, M.; Padula, F.; Montesano, C.; Passeri, D.; Nicoletti, C.; Orlandi, A.; Del Bufalo, D.; Sergi, M.; et al. NAADP-Dependent Ca^{2+} Signaling Controls Melanoma Progression, Metastatic Dissemination and Neoangiogenesis. *Sci. Rep.* **2016**, *6*, 18925. [CrossRef] [PubMed]

206. Schaefer, M.; Plant, T.D.; Obukhov, A.G.; Hofmann, T.; Gudermann, T.; Schultz, G. Receptor-mediated regulation of the nonselective cation channels TRPC4 and TRPC5. *J. Biol. Chem.* **2000**, *275*, 17517–17526. [CrossRef] [PubMed]

207. Zholos, A.V. TRPC5. *Handb. Exp. Pharmacol.* **2014**, *222*, 129–156. [PubMed]

208. Yuan, J.P.; Zeng, W.; Huang, G.N.; Worley, P.F.; Muallem, S. STIM1 heteromultimerizes TRPC channels to determine their function as store-operated channels. *Nat. Cell Biol.* **2007**, *9*, 636–645. [CrossRef] [PubMed]

209. DeHaven, W.I.; Jones, B.F.; Petranka, J.G.; Smyth, J.T.; Tomita, T.; Bird, G.S.; Putney, J.W., Jr. TRPC channels function independently of STIM1 and Orai1. *J. Physiol.* **2009**, *587 Pt 10*, 2275–2298. [CrossRef] [PubMed]

210. Zimmermann, K.; Lennerz, J.K.; Hein, A.; Link, A.S.; Kaczmarek, J.S.; Delling, M.; Uysal, S.; Pfeifer, J.D.; Riccio, A.; Clapham, D.E. Transient receptor potential cation channel, subfamily C, member 5 (TRPC5) is a cold-transducer in the peripheral nervous system. *Proc. Natl. Acad. Sci. USA* **2011**, *108*, 18114–18119. [CrossRef] [PubMed]

211. Antigny, F.; Girardin, N.; Frieden, M. Transient Receptor Potential Canonical Channels Are Required for in Vitro Endothelial Tube Formation. *J. Biol. Chem.* **2012**, *287*, 5917–5927. [CrossRef] [PubMed]

212. Chaudhuri, P.; Colles, S.M.; Bhat, M.; Van Wagoner, D.R.; Birnbaumer, L.; Graham, L.M. Elucidation of a TRPC6-TRPC5 channel cascade that restricts endothelial cell movement. *Mol. Biol. Cell* **2008**, *19*, 3203–3211. [CrossRef] [PubMed]

213. Chaudhuri, P.; Rosenbaum, M.A.; Birnbaumer, L.; Graham, L.M. Integration of TRPC6 and NADPH oxidase activation in lysophosphatidylcholine-induced TRPC5 externalization. *Am. J. Physiol. Cell Physiol.* **2017**, *313*, C541–C555. [CrossRef] [PubMed]

214. Ma, X.; Cai, Y.; He, D.; Zou, C.; Zhang, P.; Lo, C.Y.; Xu, Z.; Chan, F.L.; Yu, S.; Chen, Y.; et al. Transient receptor potential channel TRPC5 is essential for P-glycoprotein induction in drug-resistant cancer cells. *Proc. Natl. Acad. Sci. USA* **2012**, *109*, 16282–16287. [CrossRef] [PubMed]

215. Ma, X.; Chen, Z.; Hua, D.; He, D.; Wang, L.; Zhang, P.; Wang, J.; Cai, Y.; Gao, C.; Zhang, X.; et al. Essential role for TrpC5-containing extracellular vesicles in breast cancer with chemotherapeutic resistance. *Proc. Natl. Acad. Sci. USA* **2014**, *111*, 6389–6394. [CrossRef] [PubMed]

216. Wang, T.; Chen, Z.; Zhu, Y.; Pan, Q.; Liu, Y.; Qi, X.; Jin, L.; Jin, J.; Ma, X.; Hua, D. Inhibition of transient receptor potential channel 5 reverses 5-Fluorouracil resistance in human colorectal cancer cells. *J. Biol. Chem.* **2015**, *290*, 448–456. [CrossRef] [PubMed]

217. Pokharel, D.; Roseblade, A.; Oenarto, V.; Lu, J.F.; Bebawy, M. Proteins regulating the intercellular transfer and function of P-glycoprotein in multidrug-resistant cancer. *Ecancermedicalscience* **2017**, *11*, 768. [CrossRef] [PubMed]

218. He, D.X.; Gu, X.T.; Jiang, L.; Jin, J.; Ma, X. A methylation-based regulatory network for microRNA 320a in chemoresistant breast cancer. *Mol. Pharmacol.* **2014**, *86*, 536–547. [CrossRef] [PubMed]

219. Zhu, Y.; Pan, Q.; Meng, H.; Jiang, Y.; Mao, A.; Wang, T.; Hua, D.; Yao, X.; Jin, J.; Ma, X. Enhancement of vascular endothelial growth factor release in long-term drug-treated breast cancer via transient receptor potential channel 5-Ca^{2+}-hypoxia-inducible factor 1alpha pathway. *Pharmacol. Res.* **2015**, *93*, 36–42. [CrossRef] [PubMed]

220. He, D.X.; Ma, X. Transient receptor potential channel C5 in cancer chemoresistance. *Acta Pharmacol. Sin.* **2016**, *37*, 19–24. [CrossRef] [PubMed]

221. Messai, Y.; Noman, M.Z.; Hasmim, M.; Janji, B.; Tittarelli, A.; Boutet, M.; Baud, V.; Viry, E.; Billot, K.; Nanbakhsh, A.; et al. ITPR1 protects renal cancer cells against natural killer cells by inducing autophagy. *Cancer Res.* **2014**, *74*, 6820–6832. [CrossRef] [PubMed]

222. Lim, D.; Bertoli, A.; Sorgato, M.C.; Moccia, F. Generation and usage of aequorin lentiviral vectors for Ca^{2+} measurement in sub-cellular compartments of hard-to-transfect cells. *Cell Calcium* **2016**, *59*, 228–239. [CrossRef] [PubMed]

223. De la Puente, P.; Muz, B.; Azab, F.; Azab, A.K. Cell trafficking of endothelial progenitor cells in tumor progression. *Clin. Cancer. Res.* **2013**, *19*, 3360–3368. [CrossRef] [PubMed]

224. Escudier, B.; Szczylik, C.; Porta, C.; Gore, M. Treatment selection in metastatic renal cell carcinoma: Expert consensus. *Nat. Rev. Clin. Oncol.* **2012**, *9*, 327–337. [CrossRef] [PubMed]

225. Porta, C.; Paglino, C.; Imarisio, I.; Canipari, C.; Chen, K.; Neary, M.; Duh, M.S. Safety and treatment patterns of multikinase inhibitors in patients with metastatic renal cell carcinoma at a tertiary oncology center in Italy. *BMC Cancer* **2011**, *11*. [CrossRef] [PubMed]

226. Porta, C.; Paglino, C.; Imarisio, I.; Ganini, C.; Sacchi, L.; Quaglini, S.; Giunta, V.; De Amici, M. Changes in circulating pro-angiogenic cytokines, other than VEGF, before progression to sunitinib therapy in advanced renal cell carcinoma patients. *Oncology* **2013**, *84*, 115–122. [CrossRef] [PubMed]

227. Ribatti, D. Tumor refractoriness to anti-VEGF therapy. *Oncotarget* **2016**, *7*, 46668–46677. [CrossRef] [PubMed]

228. Hiles, J.J.; Kolesar, J.M. Role of sunitinib and sorafenib in the treatment of metastatic renal cell carcinoma. *Am. J. Health Syst. Pharm.* **2008**, *65*, 123–131. [CrossRef] [PubMed]

229. Naito, H.; Wakabayashi, T.; Kidoya, H.; Muramatsu, F.; Takara, K.; Eino, D.; Yamane, K.; Iba, T.; Takakura, N. Endothelial Side Population Cells Contribute to Tumor Angiogenesis and Antiangiogenic Drug Resistance. *Cancer Res.* **2016**, *76*, 3200–3210. [CrossRef] [PubMed]

230. Loges, S.; Schmidt, T.; Carmeliet, P. Mechanisms of resistance to anti-angiogenic therapy and development of third-generation anti-angiogenic drug candidates. *Genes Cancer* **2010**, *1*, 12–25. [CrossRef] [PubMed]

231. Sunryd, J.C.; Cheon, B.; Graham, J.B.; Giorda, K.M.; Fissore, R.A.; Hebert, D.N. TMTC1 and TMTC2 are novel endoplasmic reticulum tetratricopeptide repeat-containing adapter proteins involved in calcium homeostasis. *J. Biol. Chem.* **2014**, *289*, 16085–16099. [CrossRef] [PubMed]

232. Sammels, E.; Parys, J.B.; Missiaen, L.; De Smedt, H.; Bultynck, G. Intracellular Ca²⁺ storage in health and disease: A dynamic equilibrium. *Cell Calcium* **2010**, *47*, 297–314. [CrossRef] [PubMed]

233. Schuck, S.; Prinz, W.A.; Thorn, K.S.; Voss, C.; Walter, P. Membrane expansion alleviates endoplasmic reticulum stress independently of the unfolded protein response. *J. Cell Biol.* **2009**, *187*, 525–536. [CrossRef] [PubMed]

234. Dubois, C.; Vanden Abeele, F.; Prevarskaya, N. Targeting apoptosis by the remodelling of calcium-transporting proteins in cancerogenesis. *FEBS J.* **2013**, *280*, 5500–5510. [CrossRef] [PubMed]

235. Prevarskaya, N.; Skryma, R.; Shuba, Y. Ion channels and the hallmarks of cancer. *Trends Mol. Med.* **2010**, *16*, 107–121. [CrossRef] [PubMed]

236. Prevarskaya, N.; Skryma, R.; Shuba, Y. Targeting Ca²⁺ transport in cancer: Close reality or long perspective? *Expert Opin. Ther. Targets* **2013**, *17*, 225–241. [CrossRef] [PubMed]

237. Vanoverberghe, K.; Vanden Abeele, F.; Mariot, P.; Lepage, G.; Roudbaraki, M.; Bonnal, J.L.; Mauroy, B.; Shuba, Y.; Skryma, R.; Prevarskaya, N. Ca²⁺ homeostasis and apoptotic resistance of neuroendocrine-differentiated prostate cancer cells. *Cell Death Differ.* **2004**, *11*, 321–330. [CrossRef] [PubMed]

238. Cui, C.; Merritt, R.; Fu, L.; Pan, Z. Targeting calcium signaling in cancer therapy. *Acta Pharm. Sin. B* **2017**, *7*, 3–17. [CrossRef] [PubMed]

239. Bonora, M.; Giorgi, C.; Pinton, P. Novel frontiers in calcium signaling: A possible target for chemotherapy. *Pharmacol. Res.* **2015**, *99*, 82–85. [CrossRef] [PubMed]

240. Leanza, L.; Manago, A.; Zoratti, M.; Gulbins, E.; Szabo, I. Pharmacological targeting of ion channels for cancer therapy: In vivo evidences. *Biochim. Biophys. Acta* **2016**, *1863 Pt B*, 1385–1397. [CrossRef] [PubMed]

241. Nielsen, N.; Lindemann, O.; Schwab, A. TRP channels and STIM/ORAI proteins: Sensors and effectors of cancer and stroma cell migration. *Br. J. Pharmacol.* **2014**, *171*, 5524–5540. [CrossRef] [PubMed]

242. Munaron, L. Systems biology of ion channels and transporters in tumor angiogenesis: An omics view. *Biochim. Biophys. Acta* **2015**, *1848 Pt B*, 2647–2656. [CrossRef] [PubMed]

243. Kito, H.; Yamamura, H.; Suzuki, Y.; Yamamura, H.; Ohya, S.; Asai, K.; Imaizumi, Y. Regulation of store-operated Ca²⁺ entry activity by cell cycle dependent up-regulation of Orai2 in brain capillary endothelial cells. *Biochem. Biophys. Res. Commun.* **2015**, *459*, 457–462. [CrossRef] [PubMed]

244. Choi, D.; Park, E.; Jung, E.; Seong, Y.J.; Hong, M.; Lee, S.; Burford, J.; Gyarmati, G.; Peti-Peterdi, J.; Srikanth, S.; et al. ORAI1 Activates Proliferation of Lymphatic Endothelial Cells in Response to Laminar Flow Through Kruppel-Like Factors 2 and 4. *Circ. Res.* **2017**, *120*, 1426–1439. [CrossRef] [PubMed]

245. Tian, C.; Du, L.; Zhou, Y.; Li, M. Store-operated CRAC channel inhibitors: Opportunities and challenges. *Future Med. Chem.* **2016**, *8*, 817–832. [CrossRef] [PubMed]

246. Luzzi, K.J.; Varghese, H.J.; MacDonald, I.C.; Schmidt, E.E.; Kohn, E.C.; Morris, V.L.; Marshall, K.E.; Chambers, A.F.; Groom, A.C. Inhibition of angiogenesis in liver metastases by carboxyamidotriazole (CAI). *Angiogenesis* **1998**, *2*, 373–379. [CrossRef] [PubMed]

247. Oliver, V.K.; Patton, A.M.; Desai, S.; Lorang, D.; Libutti, S.K.; Kohn, E.C. Regulation of the pro-angiogenic microenvironment by carboxyamido-triazole. *J. Cell. Physiol.* **2003**, *197*, 139–148. [CrossRef] [PubMed]

248. Patton, A.M.; Kassis, J.; Doong, H.; Kohn, E.C. Calcium as a molecular target in angiogenesis. *Curr. Pharm. Des.* **2003**, *9*, 543–551. [CrossRef] [PubMed]

249. Mignen, O.; Brink, C.; Enfissi, A.; Nadkarni, A.; Shuttleworth, T.J.; Giovannucci, D.R.; Capiod, T. Carboxyamidotriazole-induced inhibition of mitochondrial calcium import blocks capacitative calcium entry and cell proliferation in HEK-293 cells. *J. Cell Sci.* **2005**, *118 Pt 23*, 5615–5623. [CrossRef] [PubMed]

250. Enfissi, A.; Prigent, S.; Colosetti, P.; Capiod, T. The blocking of capacitative calcium entry by 2-aminoethyl diphenylborate (2-APB) and carboxyamidotriazole (CAI) inhibits proliferation in Hep G2 and Huh-7 human hepatoma cells. *Cell Calcium* **2004**, *36*, 459–467. [CrossRef] [PubMed]

251. Faehling, M.; Kroll, J.; Fohr, K.J.; Fellbrich, G.; Mayr, U.; Trischler, G.; Waltenberger, J. Essential role of calcium in vascular endothelial growth factor A-induced signaling: Mechanism of the antiangiogenic effect of carboxyamidotriazole. *FASEB J.* **2002**, *16*, 1805–1807. [CrossRef] [PubMed]

252. Wu, Y.; Palad, A.J.; Wasilenko, W.J.; Blackmore, P.F.; Pincus, W.A.; Schechter, G.L.; Spoonster, J.R.; Kohn, E.C.; Somers, K.D. Inhibition of head and neck squamous cell carcinoma growth and invasion by the calcium influx inhibitor carboxyamido-triazole. *Clin. Cancer Res.* **1997**, *3*, 1915–1921. [PubMed]

253. Moody, T.W.; Chiles, J.; Moody, E.; Sieczkiewicz, G.J.; Kohn, E.C. CAI inhibits the growth of small cell lung cancer cells. *Lung Cancer* **2003**, *39*, 279–288. [CrossRef]

254. Hussain, M.M.; Kotz, H.; Minasian, L.; Premkumar, A.; Sarosy, G.; Reed, E.; Zhai, S.; Steinberg, S.M.; Raggio, M.; Oliver, V.K.; et al. Phase II trial of carboxyamidotriazole in patients with relapsed epithelial ovarian cancer. *J. Clin. Oncol.* **2003**, *21*, 4356–4363. [CrossRef] [PubMed]

255. Griffioen, A.W.; Molema, G. Angiogenesis: Potentials for pharmacologic intervention in the treatment of cancer, cardiovascular diseases, and chronic inflammation. *Pharmacol. Rev.* **2000**, *52*, 237–268. [PubMed]

256. Stadler, W.M.; Rosner, G.; Small, E.; Hollis, D.; Rini, B.; Zaentz, S.D.; Mahoney, J.; Ratain, M.J. Successful implementation of the randomized discontinuation trial design: An application to the study of the putative antiangiogenic agent carboxyaminoimidazole in renal cell carcinoma—CALGB 69901. *J. Clin. Oncol.* **2005**, *23*, 3726–3732. [CrossRef] [PubMed]

257. Rahman, S.; Rahman, T. Unveiling some FDA-approved drugs as inhibitors of the store-operated Ca^{2+} entry pathway. *Sci. Rep.* **2017**, *7*, 12881. [CrossRef] [PubMed]

258. Viallard, C.; Larrivee, B. Tumor angiogenesis and vascular normalization: Alternative therapeutic targets. *Angiogenesis* **2017**, *20*, 409–426. [CrossRef] [PubMed]

259. Vincent, F.; Duncton, M.A. TRPV4 agonists and antagonists. *Curr. Top. Med. Chem.* **2011**, *11*, 2216–2226. [CrossRef] [PubMed]

260. Coddou, C.; Yan, Z.; Obsil, T.; Huidobro-Toro, J.P.; Stojilkovic, S.S. Activation and regulation of purinergic P2X receptor channels. *Pharmacol. Rev.* **2011**, *63*, 641–683. [CrossRef] [PubMed]

261. Pafumi, I.; Favia, A.; Gambara, G.; Papacci, F.; Ziparo, E.; Palombi, F.; Filippini, A. Regulation of Angiogenic Functions by Angiopoietins through Calcium-Dependent Signaling Pathways. *BioMed Res. Int.* **2015**, *2015*, 965271. [CrossRef] [PubMed]

262. Rubaiy, H.N.; Ludlow, M.J.; Bon, R.S.; Beech, D.J. Pico145—Powerful new tool for TRPC1/4/5 channels. *Channels* **2017**, *11*, 362–364. [CrossRef] [PubMed]

263. Naylor, J.; Minard, A.; Gaunt, H.J.; Amer, M.S.; Wilson, L.A.; Migliore, M.; Cheung, S.Y.; Rubaiy, H.N.; Blythe, N.M.; Musialowski, K.E.; et al. Natural and synthetic flavonoid modulation of TRPC5 channels. *Br. J. Pharmacol.* **2016**, *173*, 562–574. [CrossRef] [PubMed]

264. Zhu, Y.; Lu, Y.; Qu, C.; Miller, M.; Tian, J.; Thakur, D.P.; Zhu, J.; Deng, Z.; Hu, X.; Wu, M.; et al. Identification and optimization of 2-aminobenzimidazole derivatives as novel inhibitors of TRPC4 and TRPC5 channels. *Br. J. Pharmacol.* **2015**, *172*, 3495–3509. [CrossRef] [PubMed]

265. Miller, M.; Shi, J.; Zhu, Y.; Kustov, M.; Tian, J.B.; Stevens, A.; Wu, M.; Xu, J.; Long, S.; Yang, P.; et al. Identification of ML204, a novel potent antagonist that selectively modulates native TRPC4/C5 ion channels. *J. Biol. Chem.* **2011**, *286*, 33436–33446. [CrossRef] [PubMed]

266. Majeed, Y.; Amer, M.S.; Agarwal, A.K.; McKeown, L.; Porter, K.E.; O'Regan, D.J.; Naylor, J.; Fishwick, C.W.; Muraki, K.; Beech, D.J. Stereo-selective inhibition of transient receptor potential TRPC5 cation channels by neuroactive steroids. *Br. J. Pharmacol.* **2011**, *162*, 1509–1520. [CrossRef] [PubMed]

267. Giorgi, C.; Bonora, M.; Missiroli, S.; Poletti, F.; Ramirez, F.G.; Morciano, G.; Morganti, C.; Pandolfi, P.P.; Mammano, F.; Pinton, P. Intravital imaging reveals p53-dependent cancer cell death induced by phototherapy via calcium signaling. *Oncotarget* **2015**, *6*, 1435–1445. [CrossRef] [PubMed]

268. Doan, N.T.; Paulsen, E.S.; Sehgal, P.; Moller, J.V.; Nissen, P.; Denmeade, S.R.; Isaacs, J.T.; Dionne, C.A.; Christensen, S.B. Targeting thapsigargin towards tumors. *Steroids* **2015**, *97*, 2–7. [CrossRef] [PubMed]

269. Denmeade, S.R.; Mhaka, A.M.; Rosen, D.M.; Brennen, W.N.; Dalrymple, S.; Dach, I.; Olesen, C.; Gurel, B.; Demarzo, A.M.; Wilding, G.; et al. Engineering a prostate-specific membrane antigen-activated tumor endothelial cell prodrug for cancer therapy. *Sci. Transl. Med.* **2012**, *4*, 140ra86. [CrossRef] [PubMed]

270. Liu, H.; Moy, P.; Kim, S.; Xia, Y.; Rajasekaran, A.; Navarro, V.; Knudsen, B.; Bander, N.H. Monoclonal antibodies to the extracellular domain of prostate-specific membrane antigen also react with tumor vascular endothelium. *Cancer Res.* **1997**, *57*, 3629–3634. [PubMed]

271. Quynh Doan, N.T.; Christensen, S.B. Thapsigargin, Origin, Chemistry, Structure-Activity Relationships and Prodrug Development. *Curr. Pharm. Des.* **2015**, *21*, 5501–5517. [CrossRef] [PubMed]

272. Mahalingam, D.; Wilding, G.; Denmeade, S.; Sarantopoulas, J.; Cosgrove, D.; Cetnar, J.; Azad, N.; Bruce, J.; Kurman, M.; Allgood, V.E.; et al. Mipsagargin, a novel thapsigargin-based PSMA-activated prodrug: Results of a first-in-man phase I clinical trial in patients with refractory, advanced or metastatic solid tumours. *Br. J. Cancer* **2016**, *114*, 986–994. [CrossRef] [PubMed]

273. Monteith, G.R.; McAndrew, D.; Faddy, H.M.; Roberts-Thomson, S.J. Calcium and cancer: Targeting Ca^{2+} transport. *Nat. Rev. Cancer* **2007**, *7*, 519–530. [CrossRef] [PubMed]

274. Zhu, H.; Zhang, H.; Jin, F.; Fang, M.; Huang, M.; Yang, C.S.; Chen, T.; Fu, L.; Pan, Z. Elevated Orai1 expression mediates tumor-promoting intracellular Ca^{2+} oscillations in human esophageal squamous cell carcinoma. *Oncotarget* **2014**, *5*, 3455–3471. [CrossRef] [PubMed]

275. Dudley, A.C. Tumor endothelial cells. *Cold Spring Harb. Perspect. Med.* **2012**, *2*, a006536. [CrossRef] [PubMed]

276. Fiorio Pla, A.; Gkika, D. Emerging role of TRP channels in cell migration: From tumor vascularization to metastasis. *Front. Physiol.* **2013**, *4*, 311. [CrossRef] [PubMed]

277. Munaron, L.; Genova, T.; Avanzato, D.; Antoniotti, S.; Fiorio Pla, A. Targeting calcium channels to block tumor vascularization. *Recent Pat. Anticancer Drug Discov.* **2013**, *8*, 27–37. [CrossRef] [PubMed]

278. Motiani, R.K.; Abdullaev, I.F.; Trebak, M. A novel native store-operated calcium channel encoded by Orai3: Selective requirement of Orai3 versus Orai1 in estrogen receptor-positive versus estrogen receptor-negative breast cancer cells. *J. Biol. Chem.* **2010**, *285*, 19173–19183. [CrossRef] [PubMed]

279. Andrikopoulos, P.; Baba, A.; Matsuda, T.; Djamgoz, M.B.; Yaqoob, M.M.; Eccles, S.A. Ca^{2+} influx through reverse mode Na^+/Ca^{2+} exchange is critical for vascular endothelial growth factor-mediated extracellular signal-regulated kinase (ERK) 1/2 activation and angiogenic functions of human endothelial cells. *J. Biol. Chem.* **2011**, *286*, 37919–37931. [CrossRef] [PubMed]

280. Hogan-Cann, A.D.; Anderson, C.M. Physiological Roles of Non-Neuronal NMDA Receptors. *Trends Pharmacol. Sci.* **2016**, *37*, 750–767. [CrossRef] [PubMed]

281. Stobart, J.L.; Lu, L.; Anderson, H.D.; Mori, H.; Anderson, C.M. Astrocyte-induced cortical vasodilation is mediated by D-serine and endothelial nitric oxide synthase. *Proc. Natl. Acad. Sci. USA* **2013**, *110*, 3149–3154. [CrossRef] [PubMed]

282. Takano, T.; Lin, J.H.; Arcuino, G.; Gao, Q.; Yang, J.; Nedergaard, M. Glutamate release promotes growth of malignant gliomas. *Nat. Med.* **2001**, *7*, 1010–1015. [CrossRef] [PubMed]

International Journal of
Molecular Sciences

MDPI

Review

Calcium Ion Channels: Roles in Infection and Sepsis Mechanisms of Calcium Channel Blocker Benefits in Immunocompromised Patients at Risk for Infection

John A. D'Elia and Larry A. Weinrauch *

E P Joslin Research Laboratory, Kidney and Hypertension Section, Joslin Diabetes Center, Department of Medicine, Mount Auburn Hospital, Harvard Medical School, Boston and Cambridge, 521 Mount Auburn Street Watertown, MA 02472, USA; jd'elia@joslin.harvard.edu
* Correspondence: lweinrauch@hms.harvard.edu; Tel.: +1-617-923-0800; Fax: +1-617-926-5665

Received: 22 June 2018; Accepted: 14 August 2018; Published: 21 August 2018

Abstract: Immunosuppression may occur for a number of reasons related to an individual's frailty, debility, disease or from therapeutic iatrogenic intervention or misadventure. A large percentage of morbidity and mortality in immunodeficient populations is related to an inadequate response to infectious agents with slow response to antibiotics, enhancements of antibiotic resistance in populations, and markedly increased prevalence of acute inflammatory response, septic and infection related death. Given known relationships between intracellular calcium ion concentrations and cytotoxicity and cellular death, we looked at currently available data linking blockade of calcium ion channels and potential decrease in expression of sepsis among immunosuppressed patients. Notable are relationships between calcium, calcium channel, vitamin D mechanisms associated with sepsis and demonstration of antibiotic-resistant pathogens that may utilize channels sensitive to calcium channel blocker. We note that sepsis shock syndrome represents loss of regulation of inflammatory response to infection and that vitamin D, parathyroid hormone, fibroblast growth factor, and klotho interact with sepsis defense mechanisms in which movement of calcium and phosphorus are part of the process. Given these observations we consider that further investigation of the effect of relatively inexpensive calcium channel blockade agents of infections in immunosuppressed populations might be worthwhile.

Keywords: calcium ion channels; calcium channel blockade; sepsis; infection; immunosuppression

---

1. Introduction

Immunosuppression may occur for a number of reasons related to an individual's frailty, debility, disease or from therapeutic iatrogenic intervention or misadventure. A large percentage of morbidity and mortality in immunodeficient populations is related to an inadequate response to infectious agents with slow response to antibiotics, enhancements of antibiotic resistance and markedly increased prevalence of acute inflammatory response, septic and infection related death. Given the known relationships between intracellular calcium ion concentrations, cytotoxicity and cellular death, we review currently available data linking blockade of calcium ion channels and potential decrease in expression of sepsis among immunosuppressed patients. We consider possibilities for therapeutic interference with calcium ion channels that may alter immune responses to invading organisms in immunocompromised populations (Table 1).

Int. J. Mol. Sci. **2018**, *19*, 2465

Table 1. Pathophysiologic Interactions of calcium ion channels with the production of pathogen mediated sepsis syndrome: Potential for therapeutic interference with calcium ion channels.

Pathogenic Organism	Calcium Ion Channel Effect
Organ specific toxicity	Envelope protein increases toxic level of cytosolic calcium in host cell
Cell membrane or wall	Antibiotic efflux may be diminished by calcium channel blockade
Organism replication	Interference with calcium dependent RNA transcription
Cytoplasm (mitochondria)	Potential to interrupt intracellular calcium shifts and interrupt calcium efflux
Host Defense	**Calcium Ion Channel Effect**

Immunocompetence	Calcium activated potassium channels regulate
	    Lymphocyte activation
	    Mitogenesis
	    Cell volume

Cellular Immunity Permeability, necrosis, apoptosis	Release of intracellular calcium leads to mitochondrial permeability and influx of extracellular calcium and permeability, necrosis, apoptosis
Humoral Immunity	Calcium controls antibody formation
Inflammasome	Calcium has a role in the production of TNF alpha, IL-1 beta

2. Calcium, Calcium Channel, Vitamin D Mechanisms Associated with Sepsis

In a survey of the United Kingdom General Practice database (Table 2) over 500,000 patients were listed as hypertensive. In this survey, use of angiotensin converting enzyme inhibitors was associated with a significantly higher rate of hospitalization for sepsis as well as mortality at thirty days when compared to use of angiotensin receptor blockers or calcium channel blockers [1]. One retrospective cohort study revealed that when 387 patients hospitalized for pneumonia who had not been treated with calcium channel blockers, were compared to 387 similarly hospitalized patients with pneumonia who had received calcium channel blockers there was a significantly higher incidence of bacteremia, respiratory insufficiency, and transfer to intensive care unit compared among those not so treated [2]. One retrospective evaluation of immunosuppressed recipients of kidney allografts from the pre-angiotensin receptor blocker era noted significantly higher incidence of sepsis and shorter survivals of allograft function for 33 patients who had not received calcium channel blockers as opposed to 36 who did receive calcium channel blockers [3].

Table 2. Calcium Ion Channels and hypothetical role in expression of infections.

I. Relationship of clinical infections to calcium channels (retrospective studies)			
	A.	Pneumonia: acute infection without antibiotic resistance	
		1.	General population
		2.	Recipients of Kidney Transplant allografts
		a.	Positive outcome results with calcium channel blockers
		b.	Negative outcome results with calcium channel blockers
	B.	Sepsis with or without antibiotic-resistant pathogens	
		1.	Potential benefit of calcium channel blockers in combination with more expensive drugs in an attempt to reduce costs while minimizing side effects of higher doses
		a.	quinolone-resistant streptococcus
		b.	rifampicin-resistant mycobacterium
		c.	quinine-resistant plasmodium
		d.	praziquantel-resistant schistosome
		e.	amphotericin-resistant leishmania
		f.	eflornithine-resistant trypanosome
II. Role of calcium movement in white blood cell defense against pathogens			
	A.	Neutrophils and macrophages may benefit from calcium channel blockers	
		1.	Restoration of capacity to attack pathogen
		2.	Limitation of capacity of pathogen to extrude antibiotic

Table 2. *Cont.*

III. Sepsis with shock following trauma		
A.	Intracellular calcium movement from storage sites may be stabilized by calcium channel -blockers, limiting cell injury B. Capillary leak during sepsis1.	
	1.	Angiopoietin 2 as a factor in sepsis-related capillary leaking
	2.	Flunarizine, which blocks both calcium influx and calcium movement from intracellular stores, in prevention of angiopoietin 2—related capillary leaking
IV. Pathogen colony growth mechanisms may or may not involve calcium		
A.	Generation of anti-oxidants (catalase, dismutase)	
	1.	*Bortadella pertussis*
	2.	*Pseudomonas aeruginosa*
B.	Efflux of calcium from host cells	
	1.	*Bacillus anthracis*
	2.	*Clostridium perfringens*
	3.	*Streptococcus pneumoniae*
C.	Colony growth mechanisms inhibited by calcium channel blockers	
	1.	*Aspergillus fumigatus*
	2.	*Saccharomyces cerevisiae*
	3.	*Candida albicans*
	4.	*Cryptococcus neoformans*
V. Mechanisms of Calcium balance in kidney failure		
A.	Vitamin D (and cathelicidin) deficiency association with infection risk	
	1.	Promotes availability of calcium through intestinal absorption
	2.	Promotes bone calcification at the growing front of osteoid
B.	Parathyroid hormone excess promotes lysis of calcified bone with excess calcium/phosphorus that may deposit in soft tissue, including blood vessels with skin cellulitis (calciphylaxis).	
	1.	Inhibits resorption of phosphate in kidney proximal tubule.
	2.	Promotes activity of 1-alpha hydroxylase in renal proximal tubule, which converts 25(OH) Vitamin D to its most active form $1,25(OH)_2$ vitamin D.
C.	Fibroblast Growth Factor 23 minimizes the accumulation of phosphate which is associated with an increased mortality rate.	
	1.	Cooperates with parathyroid hormone in the inhibition of resorption of phosphate in kidney proximal tubule.
	2.	Competes with parathyroid hormone's action on 1-alpha hydroxylase in the proximal tubule to generate the most active form of Vitamin D.
D.	Klotho, the antiaging gene, cooperates with parathyroid hormone and fibroblast growth factor 23 by inhibiting phosphate resorption in the kidney proximal tubule.	

Since the patients reviewed in studies of angiotensin-active medications and calcium channel blockers would have had hypertension, a precise mechanism for protection from sepsis would have to include protection from injury to blood vessels supplying skin, bronchus, urinary bladder unless there had been a blood vessel indwelling catheter-line, suggesting the vascular protection hypothesis might be too narrow. Consequently, researchers have developed other approaches. An experimental model for testing the impact of calcium channel blockers on sepsis involving ligation of the cecum with puncture of the wall of the intestine was one such example. This septic shock model was used to [4] demonstrate a lesser accumulation of oxygen radicals [5] as well as longer survival if diltiazem were injected prior to the onset of septic shock.

2.1. Positive Results

Several studies have addressed the use of dihydropyridine and non-dihydropyridine calcium channel blockers in the peri-operative period and in longer-term follow up (Table 2), demonstrating improved allograft function [6,7] that did not consistently appear to be secondary to blood pressure control [8,9]. One study did demonstrate stable serum creatinine associated with a fall in renal vascular resistance calculated from mean arterial pressure + renal blood flow [10]. A unique study found the occurrence of acute graft dysfunction (acute tubular necrosis) by biopsy to be significantly lower with verapamil vs. a non-calcium channel medication [11].

Rapid calcium influx into cytoplasm is associated with cell death [12,13] Clinical use for medications affecting calcium channels may involve control of aberrant cardiac rhythm while moderating vascular spasm [14]. As multiple types of calcium ion channels (N-type, L type and T type voltage-dependent calcium channels) became recognizable and multiple pharmaceutical agents with differing action profiles became available, further studies to find a role for these agents were performed.

The fact that some of these agents could directly influence biosynthesis of aldosterone in human adrenocortical cells [15] led to an early interest in their use in hypertension. Animal studies indicated an inhibition of the mineralocorticoid receptor as an additional property of dihydropyridine calcium channel blockers with felodipine being twice as powerful as amlodipine [16]. The other classes of calcium channel blockers (diltiazem, verapamil) were not effective here. Medium-sized blood vessels experience pulsatile flow as pressure is conveyed down to points of resistance and reflected backwards. Arterioles experience non-pulsatile constant flow, which can be increased with vasoconstriction or decreased with vasodilatation. Nitric oxide regulates arteriole flow while calcium-activated potassium channels regulate conduit artery flow [17]. Pathological concentrations of the vasoconstrictor angiotensin can cause damage to the muscular layer of larger blood vessels and to the endothelium of smaller blood vessels [18].

Uncontrolled hypertension may result in loss of vascular integrity with impaired endothelial resistance to sepsis [19,20] through activation of inflammation cascades which inhibit expression of nitric oxide synthase as well as the loss of control of glucose disposal such that oxidative stress disrupts calcium/potassium vascular physiology. These same mechanisms are operative in chronic loss of kidney function through injury to blood vessels and interstitial matrix, which diminishes production of $1,25(OH)_2$ Vitamin D (calcitriol) in renal proximal tubules. Since calcitriol has been shown to be additive to the inotropic effects of norepinephrine and vasopressin [21] and since receptors for $1,25(OH)_2$ D are found in increased amounts in hypertrophied heart muscle [22], the fact that calcitriol synthesis in the proximal tubule decreases as kidney function is lost may be seen as protective from accelerated hypertension. Since vitamin D is useful in controlling excessive secretion of parathyroid hormone in kidney failure, replacement therapy is routine in hemodialysis units. But a further benefit may be protection from sepsis as found in experimental model studies [23–26].

Initial enthusiasm for calcium channel blockers in renal transplantation related to their role in control of hypertension as well as the possibility that calcium channel blockers might be organ protective from intracellular calcium infusion related cell death (as seen in the necrotic myocardium during acute coronary occlusion). Then chronic myocardial stress in a rat hypertension model demonstrated protection by a calcium channel blocker (mibefradil) from myocardial scarring with an increase in interstitial collagen thought to be the result of increased fibroblast collagen production [27]. But since outcome studies had not shown long-term benefits of vasodilation following myocardial infarction, there was concern for perioperative hypotension with acute kidney allograft injury, which eventually became an impediment to the study of calcium channel blockers in renal transplant centers, particularly as the potential treatment population has aged from less than 55 to greater than 75 years of age under certain circumstances.

Thus, observations connecting calcium channel blockers with prolonged survival of kidney transplant allografts may have been due to both preservation of circulation [8–10] to the allograft at risk for ischemia due to the vasoconstrictive effect of the immunosuppressive agent, cyclosporine, but perhaps also to inhibition of mechanisms in the immunological rejection process [6,7]. An unanswered question in multiple centers from several countries is the method of preservation of donor kidneys at a point in time after nephrectomy when they are most susceptible to acute injury. A powerful vasodilator, the calcium channel blocker, lidoflazine, was shown to protect Lewis rat donor kidneys preserved in University of Wisconsin perfusate [28]. Additive effects of cyclosporine and verapamil on suppression of cytotoxic T lymphocyte cell proliferation [29] are considered to be operating via the calcium/calmodulin/calcineurin pathway to nuclear direction of the expression of antibodies through nuclear factor kappa Beta. Additive effects of cyclosporine and verapamil may also be found in suppression of the generation of cytokines by lymphocytes through inhibition of the incorporation of thymidine into DNA or uridine into RNA or leucine into protein.

Eventually the possibility that calcium channel blockers have targets other than on slow calcium channels at the cell borders has been extended to include effects on calcium movement into and out of the cytosol from storage areas within the cell (endoplasmic reticulum, sarcoplasmic reticulum, mitochondria) or on the calcium extrusion technique of pathogen survival in a host whose levels of calcium could be fatal to the pathogen [30]. Our experience with infection in diabetic patients with nephropathy has been limited to end-stage renal disease. Use of peritoneal dialysis for urgent removal of fluid to prevent pulmonary edema [31] or for chronic removal of uremic toxins had been associated with frequent instances of fever, abdominal pain and cloudy peritoneal effluent. When this triad occurred together, the diagnosis of peritonitis with a positive culture of effluent fluid would be expected. However, if only one of the diagnostic pointers were present, there might be some time before a second or third would appear. Our prospective study to determine whether frequent culture of peritoneal dialysis effluent fluid might yield an early warning prior to the clinical syndrome did not produce positive culture results worthy of the extensive expense [32]. At the height of an incursion of illicit distribution of cocaine into the New England sector of the USA, we found the availability of clean needles did not protect diabetic dialysis patients from life-threatening cellulitis, abscess formation, and sepsis [33] All the instances of visceral infection occurred in the setting of cocaine intake. Bacterial infection was detected four times as often in this target population while evidence of spread of hepatitis virus was ten-fold greater amongst users of cocaine. In our studies of cardiovascular end-points in the setting of diabetes associated with cardio-renal complications, there were serendipitous findings in the field of kidney transplantation ten years after the introduction of cyclosporine, which had marked a time of significant improvement in prevention of allograft rejection.

But when our retrospective review of 70 consecutive diabetic recipients of kidney allografts looking for cardiovascular end-points did confirm outcome benefits, it was purely by serendipity that the benefits were mainly derived through protection from attacks of sepsis [3]. Of the 70 study subjects we reviewed from this center, data on 69 was available for analysis. There were 36 individuals who had received calcium channel blockers and 33 who had not received calcium channel blockers. Of the 36, 21 were treated with diltiazem, 10 with nifedipine, and 5 with verapamil. One study subject who initially received nifedipine was subsequently switched to diltiazem. There were no significant outcome differences in cardiovascular events which were found to be less common than sepsis-related events. So, sepsis became the focus of outcome analysis in our retrospective review. Of the 19 individuals with a prior history of cardiovascular events, six who received calcium channel blockers had two sepsis-related events while 13 who had not received calcium channel blockers had 22 sepsis-related events. Of the 50 individuals without a prior history of cardiovascular events, 30 who had received calcium channel blockers had 13 sepsis-related events while 20 individuals who had not received calcium channel blockers experienced 39 sepsis-related events. While these unadjusted differences are highly significant, a weakness of the retrospective analysis would be unequal follow-up times. On the other hand, Kaplan-Meier survival curves were able to demonstrate 1.6 ($p < 0.02$) and 1.5 ($p < 0.05$)-fold greater survivals for patients and allografts, respectively for calcium channel blocker-treated vs non-treated subgroups at 24 months. When both calcium channel blocker and beta blocker medications were used, diabetic non-living allograft kidney recipients having been rigorously evaluated by experienced cardiologists [34] were documented to have durations of survival equivalent to those of non-diabetic recipients of living donor allografts using data from the UNOS Registry [35] and the US Renal Data System [36].

2.2. Negative Results

More recently we had the opportunity to perform a post-hoc evaluation of kidney transplantation outcomes in the International Folic Acid for Vascular Outcomes Reduction in Transplant (FAVORIT) trial, finding the incidence of infection risks to be much higher in diabetic recipients of allografts from living or non-living donors [37]. Since follow-up in this study began on average three years post-transplantation surgery, years four to ten were the observation period when most of the acute rejection and cardiovascular issues would probably have been resolved. Thus, the FAVORIT study was useful in documenting the importance of infection as the most serious outcome measure under the unique conditions involved. Of the 4110 individuals recruited into the FAVORIT trial, there were 2447 non-diabetic study subjects of whom 199 (8.2%) died during the study; 166 Type 1 diabetic study subjects of whom 44 (26.5%) died during the study; and 1497 Type 2 diabetic study subjects of whom 250 (16.7%) died during the study. Thus, there were 493 deaths with 191(38.7%) from cardiovascular causes and 286 (58.0%) from non-cardiovascular sources. Of the 286 non-cardiovascular deaths, 113 (59.2%) were due to infection while 76 (39.8%) were due to malignancy associated with immune-suppression.

Our results confirmed those from the UK National Health Service which had reported a 70% greater infection-related mortality for diabetic compared to non-diabetic kidney transplant recipients [38]. Between 2002 and 2007 when FAVORIT trial recruited 4110 renal transplant recipients, 4009 were followed to completion of the portion reviewed herein. Of the 4009, 2323 had received a non-living donor allograft while 1868 had received a living donor organ. Of the 2323 non-living allograft recipients, 871 (37.5%) were taking a calcium channel blocker while 1452(62.4%) were not taking a calcium channel blocker. Of the 1686 living donor organ recipients, 531 (31.5%) were taking a calcium channel blocker while 1155 (68.5%) were not taking a calcium channel blocker. Kaplan-Meier survival curves demonstrated no benefit for patient or allograft survivals over six years of follow-up. This is approximately nine years post-transplant surgery, a point at which both patient and allograft survival rates were not less than 75% for several sub-groups [39]. We have not analyzed this data according to the impact of statin medication upon groups with or without use of calcium channel blockers. Statins appear to inhibit a process by which excess growth of vascular smooth muscle cells pathologically associated with a subendothelial collection of matrix metalloprotein and LDL cholesterol. Studies also indicate that statins activate an L-type calcium that can be antagonized by calcium channel blockers, thus amlodipine cooperates in this statin end-point [40].

Rescue of neutrophilic phagocytic function through use of nifedipine among hemodialysis patients has been reported [41]. When calcium channel blockers were tested for protection from infections in the intra-venous tubing used as temporary or permanent access for hemodialysis, there was no clinical benefit reported [42]. Another instance of failure to find benefit from calcium channel blockers in the infection risk situation, occurred with use of verapamil for the first ten post-operative days following kidney transplantation [43]. Among 152 renal transplant recipients 16 of 77 (21%) who received 240 mg of verapamil for the first ten post-operative days developed a significant infection over the next 2–6 months compared with 4 of 75 (5%) who did not. 20 patients were hospitalized for infection in this brief follow up, 4 died (20%), including 3 of the 16 (19%) who had received verapamil vs. 1 of 4 (25%) who did not receive verapamil. However, episodes of infection occurred 6–20 weeks after verapamil, leave some biologic doubt as to causality. Despite levels of cyclosporine being higher during verapamil treatment, the occurrence of biopsy -proven allograft rejection was not significantly different between the two groups: 19/77 (25%) for the verapamil group vs. 22/75 (29%) for the no-verapamil group. The incidence of acute graft dysfunction, however, from a non-rejection cause (acute tubular necrosis) was lower in the verapamil group: 9/77 (12%) vs. 31/75 (41%) in the no-verapamil group. Since episodes of acute renal failure occurred in the immediate post-operative period, a statistical relationship with plausible biologic validity to verapamil is noted. And post-hoc evaluation of infection risk studies in populations with multiple risk factors are difficult to control for analysis. In a post-hoc evaluation of the FAVORIT study, which had identified infection as the single greatest long-term risk factor to survival of allograft recipients [37], there was no evidence for protection from serious infection for allograft recipients who had received calcium channel blockers vs. those who had not received calcium channel blockers [39].

3. Antibiotic-Resistant Pathogens May Utilize Channels Sensitive to Calcium Channel Blockers

Specific mechanisms for protection from sepsis by calcium channel blockers may include control of concentration of cytosolic calcium of mono- and poly-morphonuclear cells for efficient chemotaxis, migration, adhesion, phagocytosis [44,45] while limiting excessive cytokine response with massive capillary leaking found in the adult respiratory distress syndrome [46,47].

3.1. Treatment Resistance through Antibiotic Extrusion by the Pathogen or Host Cell

Ultimately, an effect on invading pathogens to limit natural selection of antibiotic-resistant strains would be a cost-effect efficiency goal (Table 2). Calcium channel blockers have been studied in quinolone-resistant pneumococcal pneumonia [48], rifampicin-resistant tuberculosis [49], quinine-resistant *Plasmodium falciparum* malaria [50,51], and praziquantel-resistant *Schistosoma mansoni* infestation [52,53], amphotericin-resistant *Leishmania donovani* [54], and eflornithine-resistant *Trypanosoma cruzi* [55]. Benidipine may be useful for treatment of *Toxoplasma gondii* infestation for individuals allergic to sulfonamides like sulfadiazine or sulfamethoxazole [56].

Pathogens able to efficiently extrude antibiotics quickly become resistant since they are being treated with progressively higher doses. Calcium channel blockers have been useful through closing the channel through which antibiotics are extruded. *Mycobacterium tuberculosis* has the capacity to extrude antibiotics which have entered through its tough outer membrane. The efflux mechanism [57] can be demonstrated following engulfing of the pathogen by macrophage. Verapamil inhibits the efflux pumping of antibiotics by closing the critical channel [58–60], which is not a typical calcium channel as shown by comparison with isomers and metabolites that are very effective at the extrusion channel, but ineffective at the calcium channel.

Since treatment of resistant mycobacteria with new medications is costly, studies have utilized verapamil to achieve in vitro and in vivo eradication by bedaquiline at lower effective dose [61,62]. One of the dangerous side effects of bedaquiline, an ATP synthase inhibitor, is prolongation of QTc interval, increasing the risk for ventricular arrhythmias. Verapamil suppresses this side effect. The calcium channel blocker, nimodipine used in slow, continuous infusion has been found effective in normalizing the electrocardiographic changes found in experimental cerebral malaria [54]. Resistance to rifampicin and other antibiotics like ethambutol is a major impediment to recovery from pulmonary tuberculosis for individuals who are immune-suppressed by malnutrition, uncontrolled glycemia, uremia, cancer chemotherapy. In mice, the use of verapamil has been shown to reduce drug resistance by decreasing the extrusion of rifampicin. A similar mechanism for prevention of uncontrolled infestation with Plasmodium falciparum due to drug resistance has been described with effective use of a calcium channel blocker.

Two biologic defensive mechanisms are responsible for the antibiotic resistance of *Mycobacterium tuberculosis* and *M. smegmatis*. The first is the antibiotic efflux process which has now been demonstrated on the cell membrane where a protein is described that forms the structure of a channel with a preference for cations [63]. The second is the finding that the same protein channel which extrudes cations and antibiotics also produces a toxin which causes sufficient necrosis to allow the pathogen to escape from the macrophage which had recently engulfed it. This same channel also allows the pathogen to take up nutrients before it escapes [64]. Several gram negative and gram-positive bacteria have been found to have the potential for antibiotic efflux. Among the more well-known gram-negative bacteria are *Bacteroides, Brucella, Campylobacter, Enterobacter, haemophilus, Neisseria, Pseudomonas, Vibrio*. Among the more well-known gram-positive bacteria are *Bacillus, Clostridium, Listeria, Mycobacterium, Staphylococcus*. A rare *Mycobacterium, M. abscessus* is an important pathogen in cystic fibrosis, an inherited disorder with an increased risk of pneumonia associated with airway obstruction. Although this pathogen is fast-growing, it is frequently antibiotic-resistant [65]. Data on calcium channel blocker impact in this rare form of tuberculosis may be forth-coming from current studies, particularly in India. It is now known that antibiotic-resistant *M. tuberculosis* not only works by extrusion of antibiotics, but also by preventing their entry in the first place.

3.2. Other Mechanisms of Pathogen Resistance Involving Calcium Ion Channels

The normal concentration of calcium in the plasma is in milli-molar range (2 mMol/L) while that of the cytosol of both host cells [66] and invading pathogens [67] is in micromoles (0.1 micromol/L), a thousand-fold difference. In addition to this initial environmental shock, the pathogen may then be attacked by a sudden burst of intracellular calcium at an even higher concentration as the host cell attempts eradication (Table 2). Two protective responses have been demonstrated in pathogens. The first adaptive mechanism seeks to protect from an impending state of oxidative stress through generation of anti-oxidant proteins, such as catalase and superoxide dismutase. Catalase may be responsible for survival of antibiotic-resistant *P. aeruginosa* [68], but for *Bordetella pertussis*, dismutase is more important than catalase for survival within polymorphonuclear leukocytes [69]. The second adaptive mechanism involves influx of cytosolic calcium across the plasma membrane of the host [30] by means of energy generated by ATPase, which mechanism has been observed in *Bacillus anthracis*, *Clostridium perfringens*, and *Streptococcus pneumoniae* (Table 2). It would appear the first of these mechanisms would be of relatively short-term benefit to the pathogen. The second of these two would allow for a more interference with host cellular defenses. Thus, there are attempts to develop drugs that inhibit the calcium efflux transport function of the pathogen [57].

Calcium-related mechanisms have been identified in certain fungi which may be invasive to humans. *Aspergillus fumigatus*, an allergen and an invasive pulmonary pathogen, utilizes the calcium/calmodulin/calcineurin pathway [70] in colony growth through extension and branching of hyphae. Cyclosporine is a growth inhibitor. Calcium pathways are actively-employed by *Saccharomyces cerevisiae*, *Candida allicins* and *Cryptococcus neoformans*. Inhibition of growth of *C. neoformans* by the calcineurin inhibitor, tacrolimus, has been detected [71]. Inhibition of development of hyphae of *Candida albicans* (Table 2) has been reported for verapamil [72].

Measurement of cytosolic calcium of the Sprague-Dawley rat [73] or human [74] hepatocyte by the fluorescent indicator, Fura 4f, indicated a level of $\approx$1 micro-molar could be achieved by ATP stimulation. Thus, the normal concentration of 0.1 micro-molar would be elevated ten-fold during hepatitis virus cell invasion through a mobilization of calcium. But in a counter move, hepatitis virus pathogens may direct cytosolic calcium levels above 1 micro-molar into temporary stores in mitochondria. Since the leading intra-cellular store of calcium is endoplasmic reticulum, the first phase calcium release would occur from this store. High concentration of this cation might lead to host cell injury. An example of unregulated calcium entry into cytosol has been studied in retinal cells which are destroyed in rodents exposed to HIV-1 virus [75].

Therefore, in a second phase while a short-term bolus injection of calcium might eradicate an invading pathogen, to protect the host cell from prolonged excess calcium this cation must be removed from cytosol to either intra-cellular stores by sarco-endoplasmic reticulum ATPase pumping or to extra-cellular environment by means of plasma membrane ATPase pumping. As extra-cellular calcium intake proceeds to replenish endoplasmic reticulum stores, a temporary high-plateau of calcium concentration may be followed by a lower resting concentration ($\approx$0.1 micro-molar).

3.3. Store Operated Calcium Entry (SOCE)

With the advent of CCB it became apparent that cytotoxic levels of calcium entry across the plasma membrane could be blocked with beneficial results. Entry of calcium into the cytosol from intracellular storage zones in the endoplasmic or sarcoplasmic reticulum (Store operated calcium entry, SOCE) may have therapeutic implications [76]. The juxtaposition of the sarcoplasmic endoplasmic reticulum allows for calcium signaling to control entry for calcium into the cytosol through the plasma membrane. Studies of defense mechanisms in liver cells infected with hepatitis B virus identify SOCE as being therapeutic for a short period of time. SOCE mechanisms may be activated during times of cytogenetic, inflammatory and hemodynamic instability.

In the instance of the hepatitis B patient or the victim of an attack by poliovirus [77], calcium transfer most likely occurred in the setting of a single pathogen in an uncompromised host.

This pathogenesis is more clearly delineated than that found in a trauma victim with multi-organism sepsis and shock due to capillary leak, sometimes referred to as adult respiratory distress syndrome (ARDS) during which SOCE can be sudden and to a massive degree [46,47]. An instance of SOCE that can occur without invasion by a pathogen is that of cancer metastasis [78]. Inhibition of proliferation of human leiomyoma by means of calcium release from intracellular stores via voltage gated calcium channels has been shown to be enhanced by simvastatin [66].

3.4. Hyperglycemia and Calcium Ion Channels

A related series of observations involves activity of the cardiac ATPase enzyme that may demonstrate increased expression following exposure to the diabetes medication liraglutide, a member of the glucagon-like peptide class of anti-hyperglycemia agents [79]. Increased cardiomyocyte plasma membrane ATPase calcium pumping has been linked to initiation and acceleration of congestive heart failure symptoms thought to be due to elevated intra-cellular calcium concentration. On the other hand, several studies have shown improved heart function with use of the diabetes medications empagliflozin, canagliflozin, or dapagliflozin, members of the sodium/glucose transporter 2 (SGLT2) inhibition class of antihyperglycemic agents [80]. Reversal of elevated cardiomyocyte calcium concentration may explain the beneficial result in heart failure reported with these medications. This mechanism could apply as well to heart failure patients who do not have the problem of hyperglycemia at least on a short-term basis [81]. Infection outcome results with SGLT2 inhibition would certainly be of interest, particularly in the realm of antibiotic-resistant pathogens.

Restoration in the capacity for phagocytosis when hyperglycemia is reversed with anti-hyperglycemic agents is reproducible and confirmed. The mechanism of disarming of the neutrophil appears to have been an increase in calcium concentration in the cytosol. If the source of calcium were the external medium of the neutrophil, then the effect of nifedipine/amlodipine would have to be exerted at the plasma membrane. But if the source of cytosolic calcium were stores in the endoplasmic reticulum, then the mechanism would be the target of future research agents. Which-ever source of calcium were involved, it would appear the pathological increase in calcium concentration in the cytoplasm was brought under control by means of a calcium channel blocker. Dihydropyridine calcium channel blockers like amlodipine or nifedipine in hypertension or with sulfonylurea medications, like glyburide or glipizide, in hyperglycemia has been found in rescue of neutrophil phagocytic function [44,45]. While the untested hypothesis that use of statin for cholesterol control may lower cost of expensive antibiotics has some rational grounds due to calcium channel blocking activity it should be remembered that for the type 2 diabetes population it is this same mechanism which may be responsible for hyperglycemia due to inhibition of beta cell insulin secretion [82].

4. Loss of Regulation of Inflammatory Response to Infection: Mechanisms Associated with the Sepsis Shock Syndrome

An experimental model for testing the impact of calcium channel blockers on sepsis involves ligation of the cecum with puncture of the wall of the intestine. This model has demonstrated relatively longer survival if diltiazem were injected prior to the onset of septic shock [4] in association with decreased formation of oxygen radicals [5]. Specific mechanisms for protection from sepsis by calcium channel blockers include: a decrease in cytosolic calcium of inflammation mediating cells, thereby limiting excessive cytokine responses, such as occurs in the adult respiratory distress syndrome [46]; an improved capacity to combat pathogens (chemotaxis, movement, adhesion, phagocytosis) through an increase in cytosolic calcium of polymorphonuclear cells and macrophages by release from intracellular stores (endoplasmic reticulum, sarcoplasmic reticulum, mitochondria) through alternate channels while the slow L channels from the exterior are blocked [47]; and an effect on invading pathogens to limit their capacity to select strains capable of rapid development of resistance to antibiotics. More recent reports of protection from sepsis have emerged from studies of antibiotic-resistant *Myocobacteria tuberculosis*. Along these lines, one analog of a calcium channel blocker (verapamil) may be more effective than another in assisting rifampicin in its battle with *M. tuberculosis* for reasons not easily understood in terms of L-type calcium channels, but rather by efficiency of docking at a site critical to the rifampin efflux mechanism. It is entirely possible that no available calcium blocker can reverse capillary leak as a complication of infection. Newer analogs however may become available [83]. It is now known that antibiotic-resistant *M. tuberculosis* bacilli inhibit entry and accelerate extrusion of antibiotics. Control of cytosolic calcium entry or transport from internal sources has been found critical for macrophage and neutrophil defense functions. Channels from endo- and sarcoplasmic reticula may interact with trauma/infection stresses for release of calcium from relevant storage areas. The endoplasmic reticulum utilizes inositol-3-phosphate receptors while the sarcoplasmic reticulum utilizes ryanodine receptors.

Sepsis Leads to Capillary Leaking

Understanding of the of the sepsis syndrome has shifted focus from that of an overwhelming inflammation cascade to an emphasis upon leaking of capillary fluid, resulting in hemoconcentration with diminished blood flow to vital organs and eventually into a state of shock due to hypovolemic hypotension (Table 2). This capillary leak results from endothelial dysfunction secondary to changes in the angiopoietin system. Ordinarily, angiopoietin 1 maintains the vascular barrier, but the effect of bacterial invasion is to enhance the concentration of angiopoietin 2 through its release from storage sites in endothelial cells mediated by tumor necrosis factor alpha (TNF α). A higher concentration of angiopoietin 1 vs. angiopoietin 2 normally operates the vascular barrier through a growth factor receptor, Tie2, which is inhibited when angiopoietin 2 concentration increases. So, a marker for the severity of the state of sepsis would be the rising level of circulating angiopoietin 2, a for-warning of imminent hypovolemia due to interstitial space accumulation of electrolyte/protein-containing fluid with strong osmolar capacity. Although the local presence of angiopoietin 2 at the portal of entry may be beneficial as a way of washing out the invasive pathogen, the target of sepsis-related research would be counter-measures by which a prolonged elevated level of circulating angiopoietin 2 might be brought into balance [84,85]. Calcium entry into endothelial cells under stress appears to occur via T-type calcium channels rather than L-type. Thus, amlodipine might have no effect upon capillary leak while flunarizine might be highly effective. In addition, flunarizine has been shown to inhibit vasoconstriction by norepinephrine and to diminish the risk of small vessel thrombosis [86]. Angiotensin caused vasoconstriction can damage muscular layers of larger and endothelium of smaller blood vessels [18]. The result may be a loss of resistance to sepsis [19,20].

5. Vitamin D, Parathyroid Hormone, Fibroblast Growth Factor, and Klotho Interact with Sepsis Defense Mechanisms in Which Movement of Calcium and Phosphorus Are Part of the Process

Given various findings of calcium movement inside host defense cells (polymorphonuclears, monocytes, macrophages), the issue of interaction between Vitamin D treatment for protection is an obvious area of sepsis study since lack of vitamin D is associated with risk of pneumonia [87,88]. In this connection relevant roles of vitamin D consist of assisting parathyroid hormone in the absorption of calcium in the intestinal tract as well as an independent capacity in the absorption of the companion anion phosphate which does not appear to be a target for parathyroid hormone. This increased prevalence of upper respiratory tract infections has been noted in patients with low vitamin D levels without [87,88] or with [89,90] advanced kidney failure. Intensive Care Unit patients who are septic have been found to have significantly lower concentrations of Vitamin D binding protein than those who were not septic. And there was also a positive correlation between Vitamin D and the toll-like receptors needed for expression of cathelicidin LL-37 levels [25]. Cathelicidins and defensins are antimicrobial peptides which cooperate in protection from infection. Since these agents are found in bronchopulmonary sites under attack from pathogens, such as *Mycobacterium tuberculosis*, they can be expected to offer a measure of protection in the lung.

Genes are being identified which code for peptides from pathogens, setting in motion expression of receptors (called Toll-like) which assist in generation of cathelicidins [25]. Susceptibility to tuberculosis is associated with vitamin D deficiency as well as a lack of production of cathelicidins. Efficiently activated monocytes will be capable of producing cathelicidin protein plus advancing 25(OH) vitamin D to the intracellular active form 1,25(OH)$_2$ D [25]. Downstream from the Vitamin D receptor is the peptide, cathelicidin, with bactericidal capacity against *Mycobacterium tuberculosis* (Table 2). This suggests that hydroxylation of the cholecalciferol by the 1-alpha hydroxylase activates cathelicidin through toll like receptor activity. Raising the question as to whether correction of Vitamin D deficiency directly confers bactericidal potential. A question to be answered is the relationship between intracellular levels of calcium, vitamin D, and calcium channel blockers. The relationship of calcium channel blockers to expression of bactericidal proteins, cathelicidin and Beta-defensin in association with intracellular elaboration of 1,25(OH)$_2$ vitamin D3 remains to be fully elucidated.

Since the demonstration of lower levels of 25(OH) vitamin D, vitamin D binding protein, and cathelicidin (LL37) in the serum of critically patients in the ICU compared to healthy controls, details of these observations have become of interest in trauma centers (Table 2). Eleven healthy controls were compared to 25 non-septic ICU control subjects versus 24 septic ICU study subjects. The individuals with sepsis had significantly higher levels of BUN and creatinine with lower levels of serum albumin, consistent with acute kidney injury. There was a positive correlation between 25(OH) vitamin D and cathelicidin (LL-37) levels [89]. Filtered 25(OH) vitamin D in complex with vitamin D binding protein can be resorbed from the glomerular filtrate with the assistance of proximal tubular protein, megalin [91] while the binding protein is degraded, the 25(OH) D is converted into 1,25(OH)2 D, and both 25(OH) D + 1,25(OH)$_2$ D are resorbed into the circulation. Retention of phosphorus with lowering of calcium (despite reflex increases in parathyroid hormone) has been shown to be related to mortality [92]. Kidney failure with vitamin D deficiency can generate fibroblast growth factor to assist in excretion of retained phosphorus if vitamin D is replaced [93]. Klotho, the anti-aging gene seems to have a role in lowering the elevated level of phosphorus in kidney failure in cooperation with Fibroblast Growth Factor 23. Some insight into renal failure as a risk factor for infection might be gained (Table 2) by studying response to phosphorus retention in the setting of combined effects of parathyroid hormone, fibroblast growth factor 23, and Klotho [94] cooperating to exclude excess phosphorus at the sodium/phosphate co-transporter in the proximal tubule. But while PTH and FGF 23 are capable of cooperating at the proximal tubular site for phosphate transport to decrease phosphorus resorption, they are simultaneously capable of competing with each other at a different site where PTH assists in activation of 1-alpha hydroxylase (Table 2) for completion of synthesis if 1,25(OH)$_2$ cholecalciferol (calcitriol), while inhibition of 1-alpha hydroxylase by Fibroblast Growth Factor is occurring simultaneously by way of inhibiting excess intestinal absorption of calcium in the setting of risk for deposition of calcium phosphorus in vascular tissues (calciphylaxis). In terms of infection risk, if the function of white blood cells is inhibited at low levels of both phosphorus and vitamin D, then the suggestion might be that kidney dysfunction and infection are more closely associated with Fibroblast Growth Factor 23 than with parathyroid hormone.

6. Conclusions

It is doubtful that sponsors can be found to support placebo-controlled trials for long-term calcium channel blockade in immunosuppressed individuals (such as solid organ transplant recipients) for the purpose of determining whether there is a beneficial effect on non-cardiovascular health care outcomes. However, there are transplant registries and multicenter trial databases from which additional information might be uncovered. An effort should be made to elucidate whether benefits or risks associated with calcium channel blockers in transplant populations as an initial example of immunosuppressed study groups with equally well-recorded information. Since long-term risk of infection in recipients of kidney transplant allografts is more important than cardiovascular complications [37] and outcome results for protection from sepsis by calcium channel blockers are in conflict [3,39] further studies need to be explored in terms of the multiple mechanisms reviewed herein. With respect to other populations immunosuppressed either iatrogenically or by virtue of underlying disease, we consider that further observations suggesting that this inexpensive class of medications might enhance treatment of infectious diseases should direct future trials. Another untested possibility is the hypothesis that use of statins to lower doses and cost of expensive antibiotics [95] may be effective since these drugs also exhibit some calcium channel blocking activity [96].

Funding: This research received no external funding.

Conflicts of Interest: The authors declare no conflict of interest.

References

1. Dial, S.; Nessim, S.J.; Kezouh, A.; Benisty, J.; Suissa, S. Antihypertensive agents acting on the renin-angiotensin system, and the risk of sepsis. *Br. J. Clin. Pharmacol.* **2014**, *78*, 1151–1158. [CrossRef] [PubMed]
2. Zeng, L.; Hunter, K.; Gaughan, J.; Podder, S. Preadmission use of calcium channel blockers and outcomes after hospitalization with pneumonia: A retrospective propensity-matched cohort study. *Am. J. Ther.* **2017**, *24*, e30–e38. [CrossRef] [PubMed]
3. Weinrauch, L.A.; D'Elia, J.A.; Gleason, R.E.; Shaffer, D.; Monaco, A.P. Role of calcium channel blockers in diabetic renal transplant patients: Preliminary observations on protection from sepsis. *Clin. Nephrol.* **1995**, *44*, 185–192. [PubMed]
4. Meldrum, D.R.; Ayala, A.; Chaudry, I.H. Mechanism of diltiazem's immunomodulatory effects after hemorrhage and resuscitation. *Am. J. Physiol.* **1993**, *265*, C412–C421. [CrossRef] [PubMed]
5. Rose, S.; Baumann, H.; Tahques, G.F.; Sayeed, M.M. Diltiazem and superoxide dismutase modulate hepatic acute phase response in gram-negative sepsis. *Shock* **1994**, *1*, 87–93. [CrossRef] [PubMed]
6. Weir, M.R. Therapeutic benefits of calcium channel blockers in cyclosporine-treated organ transplant recipients: Blood pressure control and immunosuppression. *Am. J. Med.* **1991**, *90*, 32S–36S. [CrossRef]
7. Chrysostomou, A.; Walker, R.G.; Russ, G.R.; D'Apice, A.J.F.; Kincaid-Smith, I.; Mathew, T.H. Diltiazem in renal allograft recipients receiving cyclosporine. *Transplantation* **1993**, *55*, 300–304. [CrossRef] [PubMed]
8. Van Riemsdijk, I.C.; Mulder, P.; deFijter, J.W.; Bruijn, J.A.; van Hoof, J.P.; Hoitsma, A.J.; Tegzess, A.M.; Weimar, W. Addition of isradipine (Lomir) results in better renal function after renal transplantation: A double blind randomized placebo controlled multicenter study. *Transplantation* **2000**, *70*, 122–126. [PubMed]
9. Kuypers, D.R.; Neumayer, H.H.; Fritsche, K.; Rodicio, J.L.; Vanrenterghem, Y. Lacidipine Study Group. Calcium channel blockade and preservation of renal graft function in cyclosporine-treated recipients: A prospective randomized placebo-controlled 2-year study. *Transplantation* **2004**, *78*, 1204–1211. [CrossRef] [PubMed]
10. Ahmed, K.; Michael, B.; Burk, J.F. Effects of isradipine on renal hemodynamics in renal transplant patients treated with cyclosporine. *Clin. Nephrol.* **1997**, *48*, 307–310. [PubMed]
11. Neumayer, H.-H.; Kunzendorf, U.; Schreiber, M. Protective effects of calcium antagonists in human renal transplantation. *Kidney Int.* **1992**, *36*, S87–S93.
12. Varley, K.G.; Dhalla, N.S. Excitation-contraction coupling in heart. XII. Subcellular calcium transport in isoproterenol-induced myocardial necrosis. *Exp. Mol. Pathol.* **1973**, *19*, 94–105. [CrossRef]
13. Judah, J.D.; McLean, A.E.; McLean, E.K. Biochemical mechanisms of liver injury. *Am. J. Med.* **1970**, *49*, 609–616. [CrossRef]
14. Ellrodt, G.; Chew, C.Y.; Singh, B.N. Therapeutic implications of slow-channel blockade in cardio-circulatory disorders. *Circulation* **1980**, *62*, 669–679. [CrossRef] [PubMed]
15. Akizuki, O.; Inayoshi, A.; Kitayama, T.; Yao, K.; Shirakura, S.; Saski, K.; Kusaka, H.; Marshbara, M. Blockade of T-type voltage-dependent Ca^{2+} by benidipine, a dihydropyridine calcium channel blocker, inhibits aldosterone production in human adreno-cortical cell line NCI-H 295 R. *Eur. J. Pharmacol.* **2008**, *584*, 424–434. [CrossRef] [PubMed]
16. Dietz, J.D.; Du, S.; Bolten, C.W.; Payne, M.A.; Xia, C.; Blinn, J.R.; Funder, J.W.; Hu, X. A number of marketed dihydropyridine calcium channel blockers have mineralocorticoid receptor antagonist activity. *Hypertension* **2008**, *51*, 742–748. [CrossRef] [PubMed]

17. Bellien, J.; Joannides, R.; Iacob, M.; Anaud, P.; Thuillez, C. Calcium activated potassium channels and nitric oxide regulate human peripheral conduit artery mechanics. *Hypertension* **2005**, *46*, 210–216. [CrossRef] [PubMed]

18. Doerschug, K.C.; Delsing, A.S.; Schmidt, G.A.; Ashare, A. Renin-angiotensin system activation correlates with microvascular dysfunction in a prospective cohort study of clinical sepsis. *Crit. Care* **2010**, *14*, R24. [CrossRef] [PubMed]

19. Lund, D.D.; Brooks, R.M.; Faraci, F.M.; Heistad, D.D. Role of angiotensin II in endothelial dysfunction by lipopolysaccharide in mice. *Am. J. Physiol. Heart Circ. Physiol.* **2007**, *293*, H3726–H3731. [CrossRef] [PubMed]

20. Laesser, M.; Oi, Y.; Ewert, J.; Fandriks, L.; Aneman, A. The angiotensin II receptor blocker candesartan improves survival and mesenteric perfusion in an acute porcine endotoxic model. *Acta Anaethesiol. Scand.* **2004**, *48*, 198–204. [CrossRef]

21. Bukoski, R.D.; Xua, H. On the vascular inotropic action of 1,25(OH) vitamin D3. *Am. J. Hypertens.* **1993**, *6*, 388–396. [CrossRef] [PubMed]

22. Walthers, M.R.; Wicker, D.C.; Riggle, P.C. 1,25-dihydroxy D3 receptors identified in the rat heart. *J. Mol. Cell Cardiol.* **1986**, *18*, 67–72. [CrossRef]

23. Horiuchi, H.; Nagata, I.; Komorlya, K. Protective effect of vitamin D analogues on endotoxin shock in mice. *Agents Action* **1991**, *33*, 343–348. [CrossRef]

24. Asakura, H.; Aoshima, K.; Suga, Y.; Yamazaki, M.; Morishita, E.; Saito, M.; Miyamoto, K.-I.; Nakao, S. Beneficial effect of the active form of vitamin D3 against LPS-induced DIC, but not against tissue-factor-induced DIC in rats. *Thromb. Haemost.* **2001**, *85*, 287–290. [PubMed]

25. Liu, P.T.; Stenger, S.; Li, H.; Wenzel, L.; Tan, B.H.; Krutzik, S.R.; Ochoa, M.T.; Schauber, J.; Wu, K.; Meinken, C.; et al. Toll-like receptor triggering of a vitamin D-medicated human antimicrobial response. *Science* **2008**, *311*, 1770–1773. [CrossRef] [PubMed]

26. Moller, S.; Laigaard, F.; Olgaard, K.; Hemmingsen, C. Effect of 1,25-dihydroxy-vitaminD3 in experimental sepsis. *Int. J. Med. Sci.* **2007**, *4*, 190–195. [CrossRef] [PubMed]

27. Ramires, F.J.A.; Sun, Y.; Weber, K.T. Myocardial fibrosis associated with aldosterone or angiotensin ll administration: Attenuation by calcium channel blockade. *J. Mol. Cell. Cardiol.* **1998**, *30*, 475–483. [CrossRef] [PubMed]

28. Jacobsson, J.; Odlind, B.; Tufveson, G.; Wahlberg, J. Improvement of renal preservation by adding lidoflazine to University of Wisconsin solution. An experimental study in the rat. *Cryobiology* **1992**, *29*, 305–309. [CrossRef]

29. Grgic, H.; Wulff, I.; Eichler, C.; Flothmann, R.; Kohler, R.; Hoyer, J. Blockade of T-lymphocyte Kca 3.1 and Kv1.3 channels as novel immunosuppression strategy to prevent allograft rejection. *Transplant. Proc.* **2009**, *41*, 2601–2606. [CrossRef] [PubMed]

30. Roach, J.W.; Sublett, J.; Gao, G.; Wang, Y.-D.; Tuomanen, E.I. Calcium efflux is essential for bacterial survival in the eukaryotic host. *Mol. Microbiol.* **2008**, *70*, 435–444.

31. Weinrauch, L.A.; Kaldany, A.; Miller, D.G.; Yoburn, D.C.; Belok, S.; Healy, R.W.; Leland, O.S.; D'Elia, J.A. Cardio-renal failure: Treatment of refractory biventricular failure by peritoneal dialysis. *Uremia Investig.* **1984**, *8*, 1–8. [CrossRef]

32. Cooper, G.; White, J.; D'Elia, J.; DeGirolami, P.; Arkin, C.; Kaldany, A.; Platt, R. Lack of utility of routine screening tests for early detection of peritonitis in patients requiring intermittent peritoneal dialysis. *Infect. Control* **1984**, *5*, 321–325. [CrossRef] [PubMed]

33. D'Elia, J.A.; Weinrauch, L.A.; Paine, D.F.; Domey, P.E.; Smith-Ossman, S.; Williams, M.E.; Kaldany, A. Increased infection rate in diabetic dialysis patients exposed to cocaine. *Am. J. Kidney Dis.* **1991**, *17*, 349–352. [CrossRef]

34. Lindley, E.M.; Hall, A.K.; Hess, J.; Abraham, J.; Smith, B.; Hopkins, P.N.; Shihab, F.; Welt, F.; Owan, T.; Fang, J.C. Cardiovascular risk assessment and management in pre-renal transplant candidates. *Am. J. Cardiol.* **2016**, *117*, 146–150. [CrossRef] [PubMed]

35. Cecka, J.M.; Terasaki, P.I. The UNOS Scientific Renal Transplant Registry. *Clin. Transpl.* **1993**, *1*, 1–18.

36. Almond, P.S.; Matas, A.; Gillingham, K.; Dunn, D.L.; Payne, W.D.; Gores, P.; Gruessner, R.; Nagarian, J.S. Risk factors for chronic rejection in renal allograft recipients. *Transplantation* **1993**, *55*, 752–756. [CrossRef] [PubMed]

37. Weinrauch, L.A.; D'Elia, J.A.; Weir, M.R.; Bunnapradist, S.; Finn, P.; Liu, J.; Claggett, B.; Monaco, A.P. Infection and malignancy outweigh cardiovascular mortality in kidney transplant recipients: Post-hoc analysis of the FAVORIT trial. *Am. J. Med.* **2018**, *131*, 165–172. [CrossRef] [PubMed]

38. Hayer, M.K.; Ferrugia, D.; Begaj, I.; Ray, D.; Sharif, A. Infection-related mortality for kidney allograft recipients with pre-transplant diabetes mellitus. *Diabetologia* **2014**, *57*, 554–561. [CrossRef] [PubMed]

39. Weinrauch, L.A.; Liu, J.; Claggett, B.; Finn, P.V.; Weir, M.R.; D'Elia, J.A. Calcium channel blockade and survival in recipients of successful renal transplant: An analysis of the FAVORIT trial results. *Int. J. Nephrol. Renov. Dis.* **2018**, *11*, 1–7. [CrossRef] [PubMed]

40. Clunn, G.F.; Sever, P.S.; Hughes, A.D. Calcium channel regulation in vascular smooth muscle cells: Synergistic effects of statins and calcium channel blockers. *Int. J. Cardiol.* **2010**, *139*, 2–6. [CrossRef] [PubMed]

41. Alexiewicz, J.M.; Smogorzewski, M.; Klin, M.; Akmal, M.; Massry, S.G. Effective treatment of hemodialysis patients with nifedipine on metabolism and function of polymorphonuclear leukocytes. *Am. J. Kidney Dis.* **1995**, *25*, 440–444. [CrossRef]

42. Obialo, C.I.; Conner, A.C.; Lebon, L.F. Calcium blocking agents do not ameliorate hemodialysis catheter bacteremia. *Dial. Transplant.* **2002**, *31*, 848–854.

43. Nanni, G.; Pannochia, N.; Tacchino, R.; Foco, M.; Piccioni, E.; Castagneto, M. Increased incidence of Infection in verapamil-treated kidney transplant recipients. *Transpl. Proc.* **2000**, *32*, 551–553. [CrossRef]

44. Seyrek, N.; Markinkowski, W.; Smorgorzewski, M.; Demerdash, T.M.; Massry, S.G. Amlodipine prevents and reverses the elevation in [Ca2+] and the impaired phagocytosis of PMNL of diabetic rats. *Nephrol. Dial. Transpl.* **1997**, *12*, 265–272. [CrossRef]

45. Krol, E.; Agueel, R.; Smorgorzewski, M.; Kumar, D.; Massry, S.G. Amlodipine reverses the elevation in [Ca^{2+}] and the impairment in PMNLs of NIDDM patients. *Kidney Int.* **2003**, *64*, 2188–2195. [CrossRef] [PubMed]

46. Lee, C.; Xu, D.Z.; Feketova, E.; Nemeth, Z.; Kannan, K.B.; Hasko, G.; Deitch, E.A.; Hauser, C.J. Calcium entry inhibition during resuscitation from shock attenuates inflammatory lung injury. *Shock* **2008**, *30*, 29–35. [CrossRef] [PubMed]

47. Hotchkiss, R.S.; Karl, I.E. Ca2+, a regulator of the inflammatory response—The good, the bad, and the possibilities. *Shock* **1997**, *7*, 308–310. [CrossRef] [PubMed]

48. Pletz, M.W.; Mikhaylov, N.; Schumacher, U.; van der Liden, M.; Duesberg, C.B.; Fuehner, T.; Klugman, C.P.; Welte, T.; Makarewicz, O. Antihypertensives suppress the emergence of fluoroquinolone-resistant mutants in pneumococci: An in vitro study. *Int. J. Med. Microbiol.* **2013**, *303*, 176–181. [CrossRef] [PubMed]

49. Song, L.; Cui, R.; Yang, Y.; Wu, X. Role of calcium channels in cellular anti-tuberculosis effects: Potential of voltage-gated calcium-channel blockers in tuberculosis therapy. *J. Microbiol. Immunol. Infect.* **2015**, *48*, 471–476. [CrossRef] [PubMed]

50. Scheibel, L.W.; Colambani, P.M.; Hess, A.D.; Aikawa, M.; Atkinson, T.; Milhous, W.K. Calcium and calmodulin antagonists inhibit human malaria parasites (*Plasmodium falciparum*): Implications for drug design. *Proc. Natl. Acad. Sci. USA* **1987**, *84*, 7311–7314. [CrossRef]

51. Martins, Y.C.; Clemmer, L.; Orjuela-Sanchez, P.; Zanini, G.M.; Ong, P.K.; Frangos, J.A.; Carvalho, L.J. Slow and continuous delivery of a low dose of nimodipine improves survival and electrocardiogram parameters in rescue therapy of mice with experimental cerebral malaria. *Malar. J.* **2013**, *12*, 138–154. [CrossRef] [PubMed]

52. Gryseels, B.; Mbaye, A.; DeVias, S.J.; Stelma, F.F.; Guisse, F.; Van Lieshout, L.; Faye, D.; Diop, M.; Ly, A.; Tchuem-Tchuente, L.A.; et al. Are poor responses to praziquantel for treatment of Schistosoma mansoni infections in Senegal due to resistance? An overview of the evidence. *Trop. Med. Int. Health* **2001**, *6*, 864–873. [CrossRef] [PubMed]

53. Silva-Mores, V.; Couto, F.F.; Vasconcelos, M.M.; Araujo, N.; Coelho, P.M.; Katyz, N.; Grnfell, R.F. Anti-Schistosomal activity of a calcium channel antagonist on Schistosoma and adult Schistosoma Mansoni worms. *Memorias do Instituo Oswaldo Cruz* **2013**, *108*, 600–604. [CrossRef]

54. Kashif, M.; Manna, P.P.; Akhter, Y.; Alaidarous, M.; Rub, A. Screening of novel inhibitors against Leishmania donovani calcium channel ion channel to fight Leishmaniasis. *Infect. Disord. Drug Targets* **2017**, *17*, 120–129. [CrossRef] [PubMed]

55. Pollo, L.A.E.; deMoraes, M.H.; Cisilotto, J.; Creczynski-Pasa, T.B.; Biavatti, M.W.; Steindel, M.; Sandjo, L.P. Synthesis and in vitro evaluation of Ca^{2+} channel blockers 1,4-dihydropyridines analogues against Trypanosoma cruzi and Leishmania amazonensis: SAR analysis. *Parasitol. Int.* **2017**, *66*, 789–797. [CrossRef] [PubMed]

56. Kanatani, S.; Fuks, J.M.; Olafsson, E.B.; Westermark, L.; Chambers, B.; Varas-Godoy, M.; Ulen, P.; Barrigan, A. Voltage-dependent calcium channel signaling mediates GABA a receptor-induced migratory activation of dendritic cells infected by Toxoplasma gondii. *PLoS Pathog.* **2017**, *13*, e1006739. [CrossRef] [PubMed]

57. Liz, X.-Z.; Nikaido, H. Efflux-mediated drug resistance in bacteria: An update. *Drugs* **2009**, *69*, 1555–1623.

58. Adams, K.N.; Szumowski, J.D.; Ramakrishnan, L. Verapamil and its metabolite nor-verapamil inhibit macrophage-induced, bacterial efflux pump-mediated tolerance to multiple anti-tubercular drugs. *J. Infect. Dis.* **2014**, *210*, 456–466. [CrossRef] [PubMed]

59. Adams, K.N.; Takaki, K.; Connolly, L.E. Drug tolerance in replicating mycobacteria mediated by a macrophage-induced efflux mechanism. *Cell* **2011**, *145*, 39–53. [CrossRef] [PubMed]

60. Gupta, S.; Cohen, K.A.; Winglee, K.; Maiga, M.; Diarra BBishai, W.R. Efflux inhibition with verapamil potentiates bedaquiline in Mycobacterium tuberculosis. *Antimicrob. Agents Chemother.* **2014**, *58*, 574–576. [CrossRef] [PubMed]

61. Srikrishna, G.; Gupta, S.; Dooley, K.E.; Bishai, W.R. Can the addition of verapamil to bedaquiline-containing regimens improve tuberculosis treatment outcomes? A novel approach to optimizing TB treatment. *Future Microbiol.* **2015**, *10*, 1257–1260. [CrossRef] [PubMed]

62. Diacon, A.H.; Donald, P.R.; Pym, A.; Grobusch, M.; Patientia, R.F.; Mahanyele, R.; Bantubani, N.; Narasimooloo, R.; DeMarez, R.; van Heeswijk, R.; et al. Randomized pilot trial of eight weeks of bedaquiline (tme207) treatment for multidrug-resistant tuberculosis: Long-term outcome, tolerability, and effect on emergence of drug resistance. *Antimicrob. Agents Chemother.* **2012**, *56*, 3271–3276. [CrossRef] [PubMed]

63. Siroy, A.; Mailaender, C.; Harder, D.; Koerber, S.; Wolschendorph, F.; Danilchanka, O.; Wang, Y.; Heinz, C.; Niederweis, M. RV 1698 of microtubular proteins, a new class of channel-forming outer membrane proteins. *J. Biol. Chem.* **2008**, *283*, 17827–17837. [CrossRef] [PubMed]

64. Danichanka, O.; Sun, J.; Pavlemok, M.; Maueroder, C.; Speer, A.; Siroy, A.; Mayhew, N.; Doomlos, K.S.; Munez, L.E.; Herrmann, M.; et al. An outer membrane channel protein of Mycobacterium tuberculosis with exotoxin activity. *Proc. Natl. Acad. Sci. USA* **2014**, *111*, 6750–6755. [CrossRef] [PubMed]

65. Bryant, J.M.; Grogono, D.M.; Greaves, D.; Foweraker, J.; Roddick, I.; Inns, T.; Reacher, M.; Haworth, C.S.; Curran, M.D.; Harris, S.R.; et al. Whole-genome sequencing to identify transmission of Mycobacterium abscessus between patients with cystic fibrosis: A prospective cohort study. *Lancet* **2013**, *9877*, 1551–1560. [CrossRef]

66. Borahay, M.A.; Killic, G.S.; Yallampalli, C.; Snyder, R.R.; Hankins, G.D.V.; Al-Hendry, A.; Boehning, D. Simvastatin potentially induces calcium-dependent apoptosis of human leiomyoma cells. *J. Biol. Chem.* **2014**, *289*, 35075–35086. [CrossRef] [PubMed]

67. Gangola, P.; Rosen, B.P. Maintenance of intracellular calcium in *Escherichia coli*. *J. Biol. Chem.* **1987**, *262*, 12570–12574. [PubMed]

68. Orlandi, V.T.; Martegani, E.; Bolognese, F. Catalase A is involved in the response to photo-oxidative stress in *Pseudomonas aeruginosa*. *Photodiagn. Photodyn. Ther.* **2018**, *22*, 233–240. [CrossRef] [PubMed]

69. Khelef, N.; DeShager, D.; Friedman, R.L. In vitro and in vivo analogue of Bordetella pertussis catalase and Fe-superoxide dismutase mutants. *FEMS Lett.* **1996**, *142*, 231–235. [CrossRef]

70. DiMarco, T.M.; Freitas, F.Z.; Almeida, R.S.; Brown, N.A.; des Reis, T.F.; Zambelli-Ramalho, L.N.; Savoldi, M.; Goldman, M.H.S. Functional characterization of an Aspergillus fumigatus calcium transporter (PmcA) that is essential for fungal infection. *PLoS ONE* **2012**, *7*, e37591. [CrossRef]

71. Krauss, P.R.; Nichols, C.B.; Heitman, J. Calcium- and calcineurin-independent roles for calmodulin in Cryptococcus neoformans morphogenesis and high-temperature growth. *Eukaryot. Cell* **2005**, *4*, 1079–1087. [CrossRef] [PubMed]

72. Yu, Q.; Ding, X.; Bing, Z.; Xu, N.; Jia, C.; Mao, J.; Zhang, B.; Xing, L.; Li, M. Inhibitory effect of verapamil on Candida albicans hyphal development, adhesion, and gastrointestinal colonization. *FEMS Yeast Res.* **2014**, *14*, 633–641. [CrossRef] [PubMed]

73. Casciano, J.C.; Duchemin, N.J.; Lamontagne, J.; Steel, L.F.; Bouchard, M.J. Hepatitis B virus modulates store operated calcium entry to enhance viral replication in primary hepatocytes. *PLoS ONE* **2017**, *12*, e0168328. [CrossRef] [PubMed]

74. Casciano, J.C.; Bouchard, M.J. Hepatitis virus X protein modulates cytosolic Ca+ signaling in primary human hepatocytes. *Virus Res.* **2018**, *246*, 23–27. [CrossRef] [PubMed]

75. Dreyer, E.B.; Kaiser, P.K.; Offermann, J.T.; Lipton, S.A. HIV-1 coat protein neurotoxicity prevented by calcium channel antagonists. *Science* **1990**, *248*, 364–367. [CrossRef] [PubMed]

76. Reddish, F.N.; Miller, C.L.; Gorkhali, R.; Yang, J.J. Calcium dynamics mediated by the endoplasmic/ sarcoplasmic reticulum and related diseases. *Int. J. Mol. Sci.* **2017**, *18*, 1024. [CrossRef] [PubMed]

77. Liu, S.; Rodriguez, A.V.; Tosteson, M.T. Role of simvastatin and beta cyclodextrin on inhibition of poliovirus infection. *Biochem. Biophys. Res. Commun.* **2006**, *347*, 51–59. [CrossRef] [PubMed]

78. Xie, J.; Pan, H.; Yao, J.; Han, W. SOCE and cancer: Recent progress and new perspectives. *Int. J. Cancer* **2016**, *138*, 2067–2077. [CrossRef] [PubMed]

79. Packer, M. Should we be combining GLP-1 receptor agonists and SGLT2 inhibitors in treating diabetes? *Am. J. Med.* **2018**, *131*, 461–463. [CrossRef] [PubMed]

80. Packer, M.; Anker, S.D.; Butler, J.; Filippatos, G.; Zannad, F. Effects of sodium-glucose cotransporter 2 inhibitors for the treatment of patients with heart failure: Proposal of a novel mechanism of action. *J. Am. Med. Assoc.* **2017**, *2*, 1025–1029. [CrossRef] [PubMed]

81. Byrne, N.J.; Parajuli, N.; Levasseur, J.L.; Boisvenue, J.; Becker, D.I.; Masson, G.; Fedak, P.W.M.; Verma, S.; Dyck, J.R.B. Empagliflozin prevents worsening of cardiac function in an experimental model of pressure overload-induced heart failure. *J. Am. Coll. Cardiol. Basic Transl. Sci.* **2017**, *2*, 347–354. [CrossRef] [PubMed]

82. Brault, M.; Ray, J.; Gomez, Y.H.; Mantzoros, C.S.; Daskalopoulou, S.S. Statin treatment and new onset diabetes: A review of proposed mechanisms. *Metabolism* **2014**, *63*, 735–745. [CrossRef] [PubMed]

83. Singh, K.; Kumar, M.; Pavodai, E.; Naran, K.; Warner, D.F.; Rominski, P.G.; Chibale, K. Synthesis of new verapamil analogues and their evaluation in combination with rifampicin analogues against Mycobacterium tuberculosis and molecular docking studies in the binding of efflux protein Rv1258c. *Bioorgan. Med. Chem. Lect.* **2014**, *14*, 2985–2990. [CrossRef] [PubMed]

84. Parikh, S.M.; Mammoto, T.; Schultz, A.; Yuan, H.T.; Christiani, D.; Karamuchi, S.A.; Sukhatme, V.P. Excess Angiopoietin-2 may contribute to pulmonary vascular leak in sepsis in humans. *PLoS Med.* **2006**, *3*, e46. [CrossRef] [PubMed]

85. David, S.; Kumpers, P.; van Slyke, P.; Parikh, S.M. Mending leaky blood vessels: The angiopoietin-Tie2 pathway in sepsis. *J. Pharmacol. Exp. Ther.* **2013**, *345*, 2–6. [CrossRef] [PubMed]

86. Retzlaff, J.; Thamm, K.; Ghosh, C.C.; Ziegler, W.; Haller, H.; Parikh, S.M.; David, S. Flunarizine suppresses endothelial angiopoietin-2 in a calcium-dependent fashion. *Science* **2017**, *7*, 44113. [CrossRef] [PubMed]

87. Grinde, A.A.; Mansbach, J.M.; Carmargo, C.A., Jr. Association between serum 25-hydroxyvitamin D Level and upper respiratory infection in the Third National Health and Nutrition Examination Survey. *Arch. Intern. Med.* **2009**, *169*, 384–390. [CrossRef] [PubMed]

88. Laaksi, I.; Ruohola, J.P.; Tuohimaa, P.; Auviven, A.; Haataja, R.; Pihlajamaki, H.; Yikomi, T. An association of serum vitamin D concentrations <40 nmol/L, with acute respiratory tract infection in young Finnish men. *Am. J. Clin. Nutr.* **2007**, *86*, 714–717. [PubMed]

89. Jengh, L.; Yamshchikov, A.V.; Judd, S.E.; Blumberg, H.M.; Martin, G.S.; Ziegler, T.R.; Tangpricha, V. Alterations in vitamin D status and anti-microbial peptide levels in patients in the intensive care unit with sepsis. *J. Transl. Med.* **2009**, *7*, 28–36. [CrossRef] [PubMed]

90. Chonchol, M.; Greene, T.; Zang, Y.; Hoofnagle, A.N.; Cheung, A.K. Low vitamin D and high fibroblast growth factor 23 serum levels associate with infectious and cardiac deaths in the HEMO study. *J. Am. Soc. Nephrol.* **2016**, *27*, 227–237. [CrossRef] [PubMed]

91. Nykjaer, A.; Dragun, D.; Walther, D.; Vorum, H.; Jacobsen, C.; Herz, J.; Mersen, F.; Christensen, E.I.; Willnow, T.E. An endocytic pathway essential for renal uptake of the steroid 25-(OH) vitamin D. *Cell* **1999**, *96*, 507–515. [CrossRef]

92. Kestenbaum, B.; Sampson, J.N.; Rudser, K.D.; Patterson, D.J.; Seliger, S.L.; Young, B.; Sherrard, D.J.; Andress, D.L. Serum phosphate levels and mortality risk among people with chronic kidney disease. *J. Am. Soc. Nephrol.* **2005**, *16*, 520–528. [CrossRef] [PubMed]

93. Fliser, D.; Kolleritis, B.; Neyer, U.; Ankerst, D.P.; Lhotta, K.; Lingenhel, A.; Rita, E.; Kronenberg, F.; Kuen, E.; Konig, P.; et al. Fibroblast growth factor (FGF 23) predicts progression of chronic kidney disease: The mile to moderate kidney disease (MMKD) study. *J. Am. Soc. Nephrol.* **2007**, *18*, 2600–2608. [CrossRef] [PubMed]

94. Razzaque, M.S. The FGF 23-Klotho axis: Endocrine regulation of phosphate homeostasis. *Nat. Rev. Endocrinol.* **2009**, *5*, 611–619. [CrossRef] [PubMed]

95. Lee, M.-Y.; Lin, K.-D.; Hsu, W.-H.; Chang, H.-L.; Yang, Y.-H.; Hsiao, P.-J.; Shin, S.-J. Statin, calcium channel blocker and beta blocker therapy may decrease the incidence of tuberculosis infection in elderly Taiwanese patients with type 2 diabetes. *Int. J. Mol. Sci.* **2015**, *16*, 11369–11384. [CrossRef] [PubMed]
96. Ali, N.; Begum RFaisal, M.S.; Khan, A.; Mabi, M.; Shehzadi, G.; Ullah, S.; Ali, W. Current statin show calcium channel blocking activity through voltage gated channels. *BioMed. Cent. Pharmacol. Toxicol.* **2016**, *17*, 43–50. [CrossRef] [PubMed]

MDPI

St. Alban-Anlage 66

4052 Basel

Switzerland

Tel. +41 61 683 77 34

Fax +41 61 302 89 18

www.mdpi.com

International Journal of Molecular Sciences Editorial Office

E-mail: ijms@mdpi.com

www.mdpi.com/journal/ijms